College Physics

PhET Simulations

*Indicates an associated tutorial available in the MasteringPhysics Item Library.

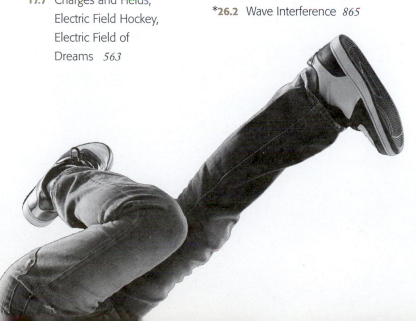

ActivPhysics OnLine™ Activities www.masteringphysics.com

About the Author   Hugh D. Young

Hugh D. Young is Emeritus Professor of Physics at Carnegie Mellon University. He earned both his undergraduate and graduate degrees from that university. He earned his Ph.D. in fundamental particle theory under the direction of the late Richard Cutkosky. He joined the faculty of Carnegie Mellon in 1956 and retired in 2004. He also had two visiting professorships at the University of California, Berkeley.

Dr. Young's career has centered entirely on undergraduate education. He has written several undergraduate-level textbooks, and in 1973 he became a coauthor with Francis Sears and Mark Zemansky for their well-known introductory texts. In addition to his authorship of Sears & Zemansky's *College Physics,* he is also coauthor, with Roger Freedman, of Sears & Zemansky's *University Physics*.

Dr. Young earned a bachelor's degree in organ performance from Carnegie Mellon in 1972 and spent several years as Associate Organist at St. Paul's Cathedral in Pittsburgh. He has played numerous organ recitals in the Pittsburgh area. Dr. Young and his wife, Alice, usually travel extensively in the summer, especially overseas and in the desert canyon country of southern Utah.

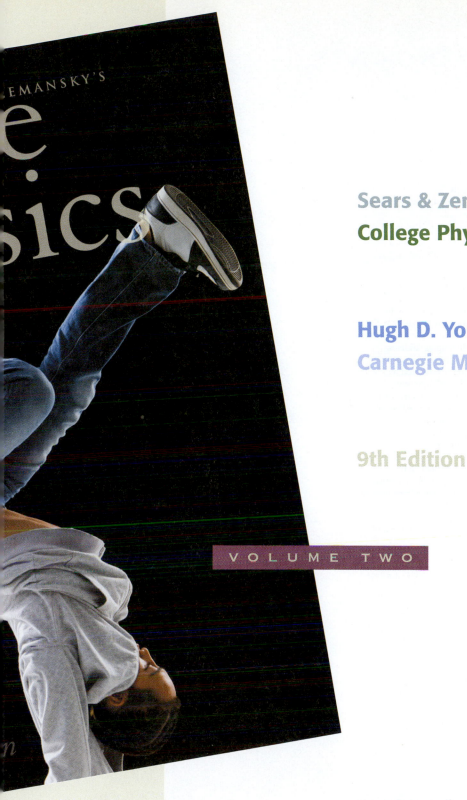

Sears & Zemansky's
College Physics

Hugh D. Young
Carnegie Mellon University

9th Edition

VOLUME TWO

Addison-Wesley

Boston Columbus Indianapolis
New York San Francisco Upper Saddle River
Amsterdam Cape Town Dubai London
Madrid Milan Munich Paris Montréal Toronto
Delhi Mexico City São Paulo Sydney
Hong Kong Seoul Singapore Taipei Tokyo

Publisher:	Jim Smith
Executive Editor:	Nancy Whilton
Editorial Manager:	Laura Kenney
Director of Development:	Michael Gillespie
Senior Development Editor:	Margot Otway
Editorial Assistant:	Steven Le
Associate Media Producer:	Kelly Reed
Managing Editor:	Corinne Benson
Production Project Manager:	Beth Collins
Production Management and Composition:	PreMediaGlobal
Proofreaders:	Elka Block and Frank Purcell
Interior Designers:	Gary Hespenheide Design, Derek Bacchus
Cover Designer:	Derek Bacchus
Illustrators:	Rolin Graphics
Senior Art Editor:	Donna Kalal
Photo Researcher:	Eric Shrader
Manufacturing Buyer:	Jeff Sargent
Senior Marketing Manager:	Kerry Chapman
Cover Photo Credit:	Mike Kemp/Rubberball/Corbis

Credits and acknowledgments borrowed from other sources and reproduced, with permission, in this textbook appear on the appropriate page within the text or on p. C-1.

Library of Congress Cataloging-in-Publication Data

Young, Hugh D.

  Sears & Zemansky's college physics. — 9th ed. / Hugh D. Young.

     p. cm.

  Includes bibliographical references and index.

  ISBN-13: 978-0-321-73317-7 (alk. paper)

  ISBN-10: 0-321-73317-7 (alk. paper)

  1. Physics—Textbooks. I. Sears, Francis Weston, 1898–1975. College physics. II. Title. III. Title: College physics. IV. Title: Sears and Zemansky's college physics.

  QC23.2.Y68 2012

  530—dc22

                        2010046658

Volume 2
ISBN 10: 0-321-76623-7; ISBN 13: 978-0-321-76623-6

1 2 3 4 5 6 7 8 9 10—WBC—14 13 12 11 10

Addison-Wesley
is an imprint of

PEARSON

www.pearsonhighered.com

Brief Contents

Build Skills

Learn basic and advanced skills that help solve a broad range of physics problems.

Problem-Solving Strategies coach students in how to approach specific types of problems. ▶

This text's uniquely extensive set of Examples enables students to explore problem-solving challenges in exceptional detail.

Consistent

The **Set Up / Solve / Reflect** format, used in all Examples, encourages students to tackle problems thoughtfully rather than skipping to the math.

Visual

Most Examples employ a diagram—often a **pencil sketch** that shows what a student should draw.

PROBLEM-SOLVING STRATEGY 24.1 **Image formation by mirrors**

SET UP

1. The principal-ray diagram is to geometric optics what the free-body diagram is to mechanics! When you attack a problem involving image formation by a mirror, *always* draw a principal-ray diagram first if you have enough information. (And apply the same advice to lenses in the sections that follow.) It's usually best to orient your diagrams consistently, with the incoming rays traveling from left to right. Don't draw a lot of other rays at random; stick with the principal rays—the ones that you know something about.

2. If your principal rays don't converge at a real image point, you may have to extend them straight backward to locate a virtual image point. We recommend drawing the extensions with broken lines. Another useful aid is to color-code your principal rays consistently; for example, referring to the preceding definitions of principal rays, Figure 24.19 uses purple for 1, green for 2, orange for 3, and pink for 4.

SOLVE

3. Identify the known and unknown quanties, such as s, s', R, and f. Make lists of the known and unknown quantities, and identify the relationships among them; then substitute the known values and solve for the ~~image distances~~, ~~and object and~~

~~work for all four~~ ~~plane and spheri-~~ ~~ities mentioned in~~ ~~the sign rules care-~~

EXAMPLE 24.1 **Image from a concave mirror**

A lamp is placed 10 cm in front of a concave spherical mirror that forms an image of the filament on a screen placed 3.0 m from the mirror. What is the radius of curvature of the mirror? If the lamp filament is 5.0 mm high, how tall is its image? What is the lateral magnification?

SOLUTION

SET UP Figure 24.14 shows our diagram.

SOLVE Both object distance and image distance are positive; we have $s = 10$ cm and $s' = 300$ cm. To find the radius of curvature, we use Equation 24.4:

$$\frac{1}{s} + \frac{1}{s'} = \frac{2}{R},$$
$$\frac{1}{10\text{ cm}} + \frac{1}{300\text{ cm}} = \frac{2}{R},$$

and $R = 19.4$ cm. To find the height of the image, we use Equation 24.7:

$$m = \frac{y'}{y} = -\frac{s'}{s},$$
$$\frac{y'}{5.0\text{ mm}} = -\frac{300\text{ cm}}{10\text{ cm}},$$

and $y' = -150$ mm. The lateral magnification m is

$$m = \frac{y'}{y} = \frac{-150\text{ mm}}{5\text{ mm}} = -30.$$

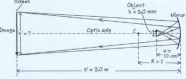

▲ **FIGURE 24.14** Our diagram for this problem (not to scale).

REFLECT The image is inverted (as we know because $m = -30$ is negative) and is 30 times taller than the object. Notice that the filament is *not* located at the mirror's focal point; the image is not formed by rays parallel to the optic axis. (The focal length of this mirror is $f = R/2 = 9.7$ cm.)

Practice Problem: A concave mirror has a radius of curvature $R = 25$ cm. An object of height 2 cm is placed 15 cm in front of the mirror. What is the image distance? What is the height of the image? *Answers:* $s' = 75$ cm, $y' = -10$ cm.

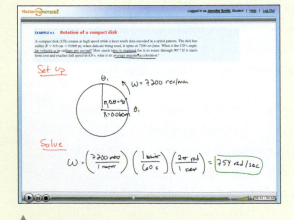

▲

NEW! Video Tutor Solution for Every Example
Each Example is explained and solved by an instructor in a Video Tutor solution provided in the Study Area of MasteringPhysics® and in the Pearson eText.

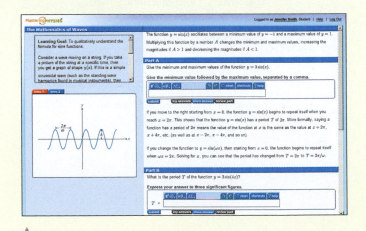

▲

NEW! Mathematics Review Tutorials
MasteringPhysics offers an extensive set of assignable mathematics review tutorials, covering all the areas in which students typically have trouble.

Build **Confidence**

Develop problem-solving confidence through a range of practice options—from guided to unguided.

NEW! **Passage Problems,** which use the same reading-passage format as most MCAT questions, develop students' ability to apply physics to a real-world situation (often biological or biomedical in nature).

About 20% of the **End-of-Chapter Problems** are new or revised. These revisions are driven by detailed student-performance data gathered nationally through MasteringPhysics.®

Problem difficulty is indicated by a three-dot ranking system based on data from MasteringPhysics.

22. •• A grasshopper leaps into the air from the edge of a vertical cliff, as shown in Figure 3.38. Use information from the figure to find (a) the initial speed of the grasshopper and (b) the height of the cliff.

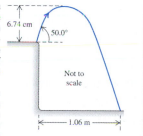

▲ **FIGURE 3.38** Problem 22.

23. •• Firemen are shooting a stream of water at a burning building. A high-pressure hose shoots out the water with a speed of 25.0 m/s as it leaves the hose nozzle. Once it leaves the hose, the water moves in projectile motion. The firemen adjust the angle of elevation of the hose until the water takes 3.00 s to reach a building 45.0 m away. You can ignore air resistance; assume that the end of the hose is at ground level. (a) Find the angle of elevation of the hose. (b) Find the speed and acceleration of the water at the highest point in its trajectory. (c) How high above the ground does the water strike the building, and how fast is it moving just before it hits the building?

24. •• Show that a projectile achieves its maximum range when it is fired at 45° above the horizontal if $y = y_0$.

Passage Problems

BIO Stimulating the brain.
Communication in the nervous system is based on propagating electrical signals called action potentials that travel along the extended nerve cell processes, the axons. Action potentials are generated when the electrical potential difference across the membrane changes so that the inside of the cell becomes more positive. Researchers in clinical medicine and neurobiology want to stimulate nerves noninvasively at specific locations in conscious subjects. But using electrodes to apply current on the skin is painful and requires large currents.

Anthony Barker and colleagues at the University of Sheffield in England developed a technique that is now widely used called transcranial magnetic stimulation (TMS). In the TMS technique, a coil positioned near the skull produces a time-varying magnetic field, which induces electric currents in the conductive brain tissue sufficient enough to cause action potentials in nerve cells. For example, if the coil is placed near the motor cortex, the region of the brain that controls voluntary movement, scientists can monitor the contraction of muscles and assess the state of the connections between the brain and the muscle.

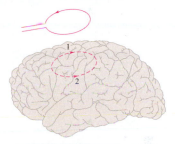

▲ **FIGURE 21.70** Problems 64–66.

64. In the diagram of TMS shown in Figure 21.70, a current pulse increases to a peak and then decreases to zero in the direction shown in the stimulating coil. What will be the direction (1 or 2) of the induced current (dotted line) in the brain tissue?
   A. 1
   B. 2
   C. 1 while the current increases in the stimulating coil and 2 while the current decreases
   D. 2 while the current increases in the stimulating coil, 1 while the current decreases

NEW! **Enhanced End-of-Chapter Problems in MasteringPhysics**
Select end-of-chapter problems will now offer additional support such as problem-solving strategy hints, relevant math review and practice, and links to the eText. These new enhanced problems bridge the gap between guided tutorials and traditional homework problems.

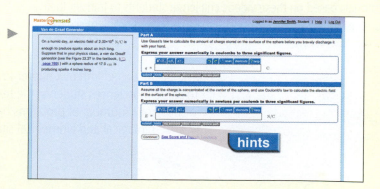

Bring Physics **to Life**

Deepen knowledge of physics by building connections to the real world.

ENHANCED! Applications of Physics ▶
Throughout the text, captioned photos apply physics to real situations, with particular emphasis on applications of biomedical and general interest.

▲ **BIO** Application **Real-time molecular biology.** The patch clamp technique is an ingenious way to investigate how cells work. To create a patch clamp, a polished microelectrode is carefully manipulated to the outer membrane of a cell to make a tight seal. Protein molecules called ion channel is isolated within the menter can then study the c properties of the single manipulate the voltage rane (and thus the chan- we know that an electrical ce across the membrane , and thus the movement of membrane. At the appro- channel flicks open for a ms) and then closes. The ing of voltage-gated chan- electrical signaling in most any aspects of cellular entors of the patch clamp warded the Nobel Prize in edicine in 1991.

▲ **BIO** Application **A long time coming.** Proportional reasoning may not make sense for two variables with a non-linear relationship. Albert Einstein's seminal work on Brownian motion showed that the distance that molecules diffuse is related to the time diffusion by a quadratic relation $x = kt^2$, where k incorporates information about the diffusing molecule and the medium through which the molecules are moving. Proteins and informational molecules can diffuse from the center of a typical spherical biological cell (diameter 20 μm) to the cell's periphery in about 1 s. In the case of a motor neuron, which the cell nucleus (in red) is in the spin cord, an extension of the cell called the axo is sent to the target muscle and may be 1 m long in humans. Proportional reasoning shows that the time required for molecules diffuse from the nucleus to the end of the ce where they are used would be 10^{10} s or abo 300 years. This is an impossibly long time, neurons have evolved sophisticated mecha- nisms for moving materials around much more rapidly.

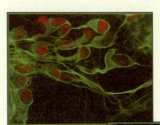

▲ **BIO** Application **Knowing up from down.** Plant roots have exquisitely developed mechanisms for sensing gravity and growing downward. If the direction of a root is changed, say by running into a rock in the soil, the root is forced to grow horizontally; however, as soon as it can, it again turns downward. This ability involves a number of cellular signals, but the primary sensor detects acceleration. The sensing cells contain statoliths, specialized starch-containing granules that are denser than the fluid in the cells and, in response to changes in the direction of gravity, fall (are accelerated) to a different location within the cells. This triggers an active localized response of the cells and causes the growth direction of the root to re-orient so that the root once again grows down.

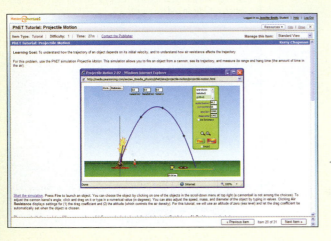

◀ NEW! PhET Simulations and Tutorials
76 PhET simulations are provided in the Study Area of the MasteringPhysics® website and in the Pearson eText. In addition, MasteringPhysics contains 16 new, assignable PhET-based tutorials.

NEW! Video Tutor Demonstrations and Tutorials
"Pause and predict" demonstration videos of key physics concepts engage students by asking them to submit a prediction before seeing the outcome. These videos are available through the Study Area of MasteringPhysics and in the Pearson eText. A set of assignable tutorials based on these videos challenge students to transfer their understanding of the demonstration to a related problem situation.

Biomedically Based End-of-Chapter Problems ▶
To serve biosciences students, the text offers a substantial number of problems based on biological and biomedical situations.

38. ● **Chin brace.** A person with an injured jaw has a brace
BIO below his chin. The brace is held in place by two cables directed at 65° above the horizontal. (See Figure 1.23.) The cables produce forces of equal magnitude having a vertical resultant of 2.25 N upward. (a) Make a scale drawing showing both the forces produced by the cables and the resultant force. Estimate the angle carefully or measure it with a protractor. (b) Use your scale drawing to estimate the magnitude of the force due to each cable.

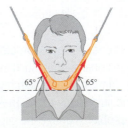

▲ **FIGURE 1.23** Problem 38.

Make a Difference with
MasteringPhysics®

www.masteringphysics.com

MasteringPhysics is the most effective and widely used online science tutorial, homework, and assessment system available.

NEW! Pre-Built Assignments

For every chapter in the book, MasteringPhysics now provides pre-built assignments that cover the material with a tested mix of tutorials and end-of-chapter problems of graded difficulty. Professors may use these assignments as-is or take them as a starting point for modification.

Gradebook

- Every assignment is graded automatically.
- Shades of red highlight vulnerable students and challenging assignments.

Class Performance on Assignment

Click on a problem to see which step your students struggled with most, and even their most common wrong answers. Compare results at every stage with the national average or with your previous class.

Gradebook Diagnostics

This screen provides your favorite weekly diagnostics. With a single click, charts summarize the most difficult problems, vulnerable students, grade distribution, and even improvement in scores over the course.

Real-World Applications

To the Student

How to Succeed in Physics

"Is physics hard? Is it too hard for me?" Many students are apprehensive about their physics course. However, while the course can be challenging, almost certainly it is *not* too hard for you. If you devote time to the course and use that time wisely, you can succeed.

Here's how to succeed in physics.

1. **Spend time studying.** The rule of thumb for college courses is that you should expect to study about 2 to 3 hours per week for each unit of credit, *in addition* to the time you spend in class. And budget your time: 3 hours every other day is much more effective than 33 hours right before the exam.

   The good news is that physics is consistent. Once you've learned how to tackle one topic, you'll use the same study techniques to tackle the rest of the course. So if you find you need to give the course extra time at first, do so and don't worry—it'll pay dividends as the course progresses.

2. **Don't miss class.** Yes, you could borrow a friend's notes, but listening and participating in class are far more effective. Of course, *participating* means paying active attention, and interacting when you have the chance!

3. **Make this book work for you.** This text is packed with decades of teaching experience—but to make it work for you, you must read and use it *actively*. *Think* about what the text is saying. *Use* the illustrations. Try to *solve* the Examples and the Quantitative and Conceptual Analysis problems on your own, before reading the solutions. If you *underline*, do so thoughtfully and not mechanically.

   A good practice is to skim the chapter before going to class to get a sense for the topic, and then read it carefully and work the examples after class.

4. **Approach physics problems systematically.** While it's important to attend class and use the book, your *real* learning will happen mostly as you work problems—*if* you approach them correctly. Physics problems aren't math problems. You need to approach them in a different way. (If you're "not good at math," this may be good news for you!) What you do before and after solving an equation is more important than the math itself. The worked examples in this book help you develop good habits by consistently following three steps—*Set Up*, *Solve*, and *Reflect*. (In fact, this global approach will help you with problem solving in all disciplines—chemistry, medicine, business, etc.)

5. **Use campus resources.** If you get stuck, get help. Your professor probably has office hours and email; use them. Use your TA or campus tutoring center if you have one. Partner with a friend to study together. But also try to get unstuck on your own *before* you go for help. That way, you'll benefit more from the help you get.

   So remember, you *can* succeed in physics. Just devote time to the job, work lots of problems, and get help when you need it. Your book is here to help. Have fun!

Preface

College Physics has been a success for over half a century, placing equal emphasis on quantitative, qualitative, and conceptual understanding.. Guided by tested and proven innovations in education research, we have revised and enhanced previous material and added new features focusing on more explicit problem-solving steps and techniques, conceptual understanding, and visualization and modeling skills. Our main objectives are to teach a solid understanding of the fundamentals, help students develop critical thinking and quantitative reasoning, teach sound problem-solving skills, and spark the students' interest in physics with interesting and relevant applications.

This text provides a comprehensive introduction to physics at the beginning college level. It is intended for students whose mathematics preparation includes high-school algebra and trigonometry but no calculus. The complete text may be taught in a two-semester or three-quarter course, and the book is also adaptable to a wide variety of shorter courses.

New to This Edition

- **New Chapter 0 (Mathematics Review)** covers math concepts that students will need to use throughout the course: Exponents; scientific notation and powers of 10; algebra; direct, inverse, and inverse-square relationships; logarithmic and exponential functions; areas and volumes; and plane geometry and trigonometry. This review chapter includes worked examples and end-of-chapter problems.
- **New margin applications** include over 40 new biosciences-related applications with photos added to the text, including those focused on cutting-edge technology. BIO icons signify the bio-related applications.
- **Changes to the end-of-chapter problems include the following:**
  - **15–20% of the problems are new.**
  - **Many additional biosciences-related problems.**
  - **One set of MCAT-style passage problems** added at the end of most chapters, many of them bio-related.

  The addition of new biological and biomedical real-world applications and problems gives this edition more coverage in the biosciences than nearly every other book on the market.

- **Over 70 PhET simulations** are linked to the Pearson eText and are provided in the study area of the MasteringPhysics website (with icons in the print text). These powerful simulations allow students to interact productively with the physics concepts they are learning. PhET clicker questions are also included on the Instructor Resource DVD.
- **Video Tutors bring key content to life throughout the text:**
  - **Dozens of Video Tutors feature "pause-and-predict" demonstrations of key physics concepts** and incorporate assessment as the student progresses, to actively engage the student in understanding the key conceptual ideas underlying the physics principles.
  - **Every Worked Example in the book is accompanied by a Video Tutor Solution** that walks students through the problem-solving process, providing a virtual teaching assistant on a round-the-clock basis.
  - **All of these Video Tutors play directly through links within the Pearson eText.** Many also appear in the Study area within MasteringPhysics.

Complete and Two-Volume Editions

With MasteringPhysics:
- **Complete Edition:** Chapters 1–30
  (ISBN 978-0-321-74980-2)

Without MasteringPhysics:
- **Complete Edition:** Chapters 1–30
  (ISBN 978-0-321-73317-7)
- **Volume 1:** Chapters 1–16
  (ISBN 978-0-321-76624-3)
- **Volume 2:** Chapters 17–30
  (ISBN 978-0-321-76623-6)

- **Assignable MasteringPhysics tutorials are based on the Video Tutor Demonstrations and PhET simulations.**

  - **Video Tutor Demonstrations will be expanded to tutorials in Mastering** by requiring the student to transfer their understanding to a new problem situation so that these will be gradable and distinct from the "pause and predict" demonstrations alone.

  - Sixteen new **PhET tutorials** enable students to not only explore the PhET simulations but also answer questions, helping them make connections between real-life phenomena and the underlying physics that explain such phenomena.

Key Features of *College Physics*

- **A systematic approach to problem solving.** To solve problems with confidence, students must learn to approach problems effectively at a global level, must understand the physics in question, and must acquire the specific skills needed for particular types of problems. The Ninth Edition provides research-proven tools for students to tackle each goal.

- The **worked examples** all follow a consistent and explicit **global problem-solving strategy** drawn from educational research. This three-step approach puts special emphasis on how to **set-up** the problem before trying to **solve** it, and the importance of how to **reflect** on whether the answer is sensible.

- Worked example solutions emphasize the steps and decisions students often skip. In particular, many worked examples include **pencil diagrams:** hand-drawn diagrams that show exactly what a student should draw in the **set-up** step of solving the problem.

- **Conceptual Analysis** and **Quantitative Analysis** problems help the students practice their qualitative and quantitative understanding of the physics. The Quantitative Analysis problems focus on skills of quantitative and proportional reasoning—skills that are key to success on the MCATs. The CAs and QAs use a multiple-choice format to elicit specific common misconceptions.

- **Problem-solving strategies** teach the students tactics for particular types of problems—such as problems requiring Newton's second law, energy conservation, etc.—and follow the same 3-step global approach (set-up, solve, and reflect).

- **Unique, highly effective figures incorporate the latest ideas from educational research.** Extraneous detail has been removed and color used only for strict pedagogical purposes—for instance, in mechanics, **color is used to identify the object of interest,** while all other objects are grayscale. **Illustrations include helpful blue annotated comments** to guide students in 'reading' graphs and physics figures. Throughout, **figures, models, and graphs are placed side by side** to help students 'translate' between multiple representations. **Pencil sketches** are used consistently in worked examples to emphasize what students should draw.

- **Visual chapter summaries** show each concept in words, math, and figures to reinforce how to 'translate' between different representations and address different student learning styles.

- **Rich and diverse end-of-chapter problem sets.** The renowned Sears & Zemansky problems, refined over five decades of use, have been revised, expanded and enhanced for today's courses, based on data from MasteringPhysics.

- Each chapter includes a set of **multiple-choice problems** that test the skills developed by the Qualitative Analysis and Quantitative Analysis problems in the chapter text. The multiple-choice format elicits specific common misconceptions, enabling students to pinpoint their misunderstandings.

- The General Problems contain many **context-rich problems** (also known as **real-world problems**), which require students to simplify and model more complex real-world situations. Many problems relate to the field of biology and medicine; these are all labeled BIO.
- **Connections of physics to the student's world.** In-margin photos with explanatory captions provide diverse, interesting, and self-contained examples of physics at work in the world. Many of these real-world "applications" are also related to the fields of biology and medicine and are labeled BIO.
- **Writing that is easy to follow and rigorous.** The writing is friendly yet focused; it conveys an exact, careful, straightforward understanding of the physics, with an emphasis on the connections between concepts.

Instructor Supplements

Note: For convenience, all of the following instructor supplements (except for the Instructor Resource DVD) can be downloaded from the Instructor Area, accessed via the left-hand navigation bar of MasteringPhysics (www.masteringphysics.com).

Instructor Solutions, prepared by A. Lewis Ford (Texas A&M University) and Forrest Newman (Sacramento City College) contain complete and detailed solutions to all end-of-chapter problems. All solutions follow consistently the same Set Up/Solve/Reflect problem-solving framework used in the textbook. Download only from the MasteringPhysics Instructor Area or from the Instructor Resource Center (www.pearsonhighered.com/irc).

The cross-platform **Instructor Resource DVD** (ISBN 978-0-321-76570-3) provides a comprehensive library of more than 420 applets from ActivPhysics OnLine as well as all line figures from the textbook in JPEG format. In addition, all the key equations, problem-solving strategies, tables, and chapter summaries are provided in editable Word format. Lecture outlines in PowerPoint are also included along with over 70 PhET simulations as well as Pause and Predict Video Tutors and Video Tutor Solutions.

MasteringPhysics® (www.masteringphysics.com) is the most advanced, educationally effective, and widely used physics homework and tutorial system in the world. Eight years in development, it provides instructors with a library of extensively pre-tested end-of-chapter problems and rich, multipart, multistep tutorials that incorporate a wide variety of answer types, wrong answer feedback, individualized help (comprising hints or simpler sub-problems upon request), all driven by the largest metadatabase of student problem-solving in the world. NSF-sponsored published research (and subsequent studies) show that Mastering-Physics has dramatic educational results. MasteringPhysics allows instructors to build wide-ranging homework assignments of just the right difficulty and length and provides them with efficient tools to analyze both class trends, and the work of any student in unprecedented detail.

MasteringPhysics routinely provides instant and individualized feedback and guidance to more than 100,000 students every day. A wide range of tools and support make MasteringPhysics fast and easy for instructors and students to learn to use. Extensive class tests show that by the end of their course, an unprecedented eight of nine students recommend MasteringPhysics as their preferred way to study physics and do homework.

MasteringPhysics enables instructors to:

- Quickly build homework assignments that combine regular end-of-chapter problems and tutoring (through additional multi-step tutorial problems that offer wrong-answer feedback and simpler problems upon request).
- Expand homework to include the widest range of automatically graded activities available–from numerical problems with randomized values, through algebraic answers, to free-hand drawing.

- Choose from a wide range of nationally pre-tested problems that provide accurate estimates of time to complete and difficulty.
- After an assignment is completed, quickly identify not only the problems that were the trickiest for students but the individual problem types where students had trouble.
- Compare class results against the system's worldwide average for each problem assigned, to identify issues to be addressed with just-in-time teaching.
- Check the work of an individual student in detail, including time spent on each problem, what wrong answers they submitted at each step, how much help they asked for, and how many practice problems they worked.

ActivPhysics OnLine™ (accessed through the Study Area within www. masteringphysics.com) provides a comprehensive library of more than 420 tried and tested ActivPhysics applets updated for web delivery using the latest online technologies. In addition, it provides a suite of highly regarded applet-based tutorials developed by education pioneers Alan Van Heuvelen and Paul D'Alessandris. Margin icons throughout the text direct students to specific exercises that complement the textbook discussion.

The online exercises are designed to encourage students to confront misconceptions, reason qualitatively about physical processes, experiment quantitatively, and learn to think critically. The highly acclaimed ActivPhysics OnLine companion workbooks help students work through complex concepts and understand them more clearly. More than 420 applets from the ActivPhysics OnLine library are also available on the Instructor Resource DVD for this text.

The **Test Bank** contains more than 2,000 high-quality problems, with a range of multiple-choice, true/false, short-answer, and regular homework-type questions. Test files are provided both in TestGen (an easy-to-use, fully networkable program for creating and editing quizzes and exams) and Word format. Download only from the MasteringPhysics Instructor Area or from the Instructor Resource Center (www.pearsonhighered.com/irc).

Five Easy Lessons: Strategies for Successful Physics Teaching (ISBN 978-0-8053-8702-5) by Randall D. Knight (California Polytechnic State University, San Luis Obispo) is packed with creative ideas on how to enhance any physics course. It is an invaluable companion for both novice and veteran physics instructors.

Student Supplements

The **Student Solutions Manual,** (ISBN 978-0-321-74769-3), written by Lewis Ford (Texas A&M University) and Forrest Newman (Sacramento City College), contains detailed, step-by-step solutions to more than half of the odd-numbered end-of-chapter problems from the textbook. All solutions consistently follow the same Set Up/Solve/Reflect problem-solving framework used in the textbook, reinforcing good problem-solving behavior.

MasteringPhysics® (www.masteringphysics.com) is a homework, tutorial, and assessment system based on years of research into how students work physics problems and precisely where they need help. Studies show that students who use MasteringPhysics significantly increase their scores compared to handwritten homework. MasteringPhysics achieves this improvement by providing students with instantaneous feedback specific to their wrong answers, simpler sub-problems upon request when they get stuck, and partial credit for their method(s). This individualized, 24/7 Socratic tutoring is recommended by 9 out of 10 students to their peers as the most effective and time-efficient way to study.

Pearson eText is available through MasteringPhysics, either automatically when MasteringPhysics is packaged with new books, or available as a purchased upgrade online. Allowing students access to the text wherever they have access to the Internet, Pearson eText comprises the full text, including figures that can be enlarged for better viewing. With eText, students are also able to pop up definitions and terms to help with vocabulary and the reading of the material. Students can also take notes in eText using the annotation feature at the top of each page.

Pearson Tutor Services (www.pearsontutorservices.com). Each student's subscription to MasteringPhysics also contains complimentary access to Pearson Tutor Services, powered by Smarthinking, Inc. By logging in with their MasteringPhysics ID and password, they will be connected to highly qualified e-instructors who provide additional interactive online tutoring on the major concepts of physics. Some restrictions apply; offer subject to change.

ActivPhysics OnLine (accessed through the Study Area within www. masteringphysics.com) provides students with a suite of highly regarded applet-based tutorials (see above). The following workbooks help students work through complex concepts and understand them more clearly.

ActivPhysics OnLine Workbook, Volume 1: Mechanics * Thermal Physics * Oscillations & Waves (978-0-8053-9060-5)

ActivPhysics OnLine Workbook, Volume 2: Electricity & Magnetism * Optics * Modern Physics (978-0-8053-9061-2)

Acknowledgments

I want to extend my heartfelt thanks to my colleagues at Carnegie Mellon for many stimulating discussions about physics pedagogy and for their support and encouragement during the writing of several successive editions of this book. I am equally indebted to the many generations of Carnegie Mellon students who have helped me learn what good teaching and good writing are, by showing me what works and what doesn't. I'm pleased to acknowledge also the contributors of problems, applications, and other essential elements for this new edition, including Ken Robinson, Charlie Hibbard, Forrest Newman, Larry Coleman, and Biman Das. Special thanks are due to the Addison-Wesley people, especially Laura Kenney and Margot Otway, who brought this all together, and to Nancy Whilton and Kerry Chapman. Thanks also to Jared Sterzer at PreMediaGlobal. Finally, and most importantly, it is always a joy and a privilege to express my gratitude to my wife Alice and our children Gretchen and Rebecca for their love, support, and emotional sustenance during the writing of several successive editions of this book. May all men and women be blessed with love such as theirs.

—H.D.Y.

Reviewers and Classroom Testers

Susmita Acharya, *Cardinal Stritch University*

Hamid Aidinejad, *Florida Community College, Jacksonville*

Alice Hawthorne Allen, *Virginia Tech*

Jim Andrews, *Youngstown State University*

Charles Bacon, *Ferris State University*

Jennifer Blue, *Miami University*

Richard Bone, *Florida International University*

Phillip Broussard, *Covenant College*

Young Choi, *University of Pittsburgh*

Orion Ciftja, *Prairie View A&M University*

Dennis Collins, *Grossmont College*

Lloyd Davis, *Montreat College*

Diana Driscoll, *Case Western Reserve University*

Laurencin Dunbar, *St. Louis Community College, Florissant Valley*

Alexander Dzyubenko, *California State University, Bakersfield*

Robert Ehrlich, *George Mason University*

Mark Fair, *Grove City College*

Shamanthi Fernando, *Northern Kentucky University*

Len Feuerhelm, *Oklahoma Christian University*

Carl Frederickson, *University of Central Arkansas*

Mikhail Goloubev, *Bowie State University*

Alan Grafe, *University of Michigan, Flint*

William Gregg, *Louisiana State University*

John Gruber, *San Jose State University*

Robert Hagood, *Washtenaw Community College*

Scott Hildreth, *Chabot College*

Andy Hollerman, *University of Louisiana, Lafayette*

John Hubisz, *North Carolina State University*

Manuel Huerta, *University of Miami*

Todd Hurt, *Chatham College*

Adam Johnston, *Weber State University*

Roman Kezerashvili, *New York City College of Technology*

Ju Kim, *University of North Dakota*

Jeremy King, *Clemson University*

David Klassen, *Rowan University*

Ichishiro Konno, *University of Texas, San Antonio*

Ikka Koskelo, *San Francisco State University*

Jon Levin, *University of Kentucky*

David Lind, *Florida State University*

Dean Livelybrooks, *University of Oregon, Eugene*

Estella Llinas, *University of Pittsburgh, Greensburg*

Craig Loony, *Merrimack College*

Rafael Lopez-Mobilia, *University of Texas, San Antonio*

Barbra Maher, *Red Rocks Community College*

Dan Mazilu, *Virginia Tech*

Randy McKee, *Tallahassee Community College*

Larry McRae, *Berry College*

William Mendoza, *Jacksonville University*

Anatoli Mirochnitchenko, *University of Toledo*

Charles Myles, *Texas Tech University*

Austin Napier, *Tufts University*

Erin O' Connor, *Allan Hancock College*

Christine O'Leary, *Wallace State College*

Jason Overby, *College of Charleston*

James Pazun, *Pfeiffer University*

Unil Perera, *Georgia State University*

David Potter, *Austin Community College*

Michael Pravica, *University of Nevada, Las Vegas*

Sal Rodano, *Harford Community College*

Rob Salgado, *Dillard University*

Surajit Sen, *SUNY Buffalo*

Bart Sheinberg, *Houston Community College*

Natalia Sidorovskaia, *University of Louisiana*

Chandralekha Singh, *University of Pittsburgh*

Marlina Slamet, *Sacred Heart University*

Daniel Smith, *South Carolina State University*

Gordon Smith, *Western Kentucky University*

Kenneth Smith, *Pennsylvania State University*

Zhiyan Song, *Savannah State University*

Sharon Stephenson, *Gettysburg College*

Chuck Stone, *North Carolina A&T State University*

George Strobel, *University of Georgia*

Chun Fu Su, *Mississippi State University*

Brenda Subramaniam, *Cypress College*

Mike Summers, *George Mason University*

Eric Swanson, *University of Pittsburgh*

Colin Terry, *Ventura College*

Vladimir Tsifrinovich, *Polytechnic University*

Gajendra Tulsian, *Daytona Beach Community College*

Paige Uozts, *Lander University*

James Vesenka, *University of New England*

Walter Wales, *University of Pennsylvania*

John Wernegreen, *Eastern Kentucky University*

Dan Whitmire, *University of Louisiana, Lafayette*

Sue Willis, *Northern Illinois University*

Jaehoon Yu, *University of Texas, Arlington*

Nouredine Zettili, *Jacksonville State University*

Bin Zhang, *Arkansas State University*

Detailed Contents

0 Mathematics Review

A study of physics at the level of this textbook requires some basic math skills. The relevant math topics are summarized in this chapter. We strongly recommend that you review this material, practice end-of-chapter problems, and become comfortable with these as quickly as possible, so that during your physics course, you can focus on the physics concepts and procedures that are being introduced, without being distracted by unfamiliarity with the math being used. Note that the beauty of physics cannot be enjoyed if you do not have adequate mastery of basic mathematical skills.

The arrangement of seeds in a sunflower is a classic example of how natural processes give rise to patterns that can be expressed by means of fairly simple mathematics. In this chapter, we will review most of the mathematics you will need for this course.

0.1 Exponents

Exponents are used frequently in physics. When we write 3^4, the superscript 4 is called an **exponent** and the **base number** 3 is said to be raised to the fourth power. The quantity 3^4 is equal to $3 \times 3 \times 3 \times 3 = 81$. Algebraic symbols can also be raised to a power—for example, x^4. There are special names for the operation when the exponent is 2 or 3. When the exponent is 2, we say that the quantity is **squared;** thus, x^2 means x is squared. When the exponent is 3, the quantity is **cubed;** hence, x^3 means x is cubed.

Note that $x^1 = x$, and the exponent 1 is typically not written. Any quantity raised to the zero power is defined to be unity (that is, 1). Negative exponents are used for reciprocals: $x^{-4} = 1/x^4$. An exponent can also be a fraction, as in $x^{1/4}$. The exponent $\frac{1}{2}$ is called a **square root,** and the exponent $\frac{1}{3}$ is called a **cube root.**

For example, $\sqrt{6}$ can also be written as $6^{1/2}$. Most calculators have special keys for calculating numbers raised to a power—for example, a key labeled y^x or one labeled x^2.

Exponents obey several simple rules, which follow directly from the meaning of raising a quantity to a power:

1. The product rule: $(x^m)(x^n) = x^{m+n}$.
   For example, $(3^3)(3^2) = 3^5 = 243$. To verify this result, note that $3^3 = 27$, $3^2 = 9$, and $(27)(9) = 243$.

2. The quotient rule: $\dfrac{x^m}{x^n} = x^{m-n}$.

   For example, $\dfrac{3^3}{3^2} = 3^{3-2} = 3^1 = 3$. To verify this result, note that $\dfrac{3^3}{3^2} = \dfrac{27}{9} = 3$.

   A special case of this rule is, $\dfrac{x^m}{x^m} = x^{m-m} = x^0 = 1$.

3. The first power rule: $(x^m)^n = x^{mn}$.
   For example, $(2^2)^3 = 2^6 = 64$. To verify this result, note that $2^2 = 4$, so $(2^2)^3 = (4)^3 = 64$.

4. Other power rules:

$$(xy)^m = (x^m)(y^m), \text{ and } \left(\frac{x}{y}\right)^m = \frac{x^m}{y^m}.$$

For example, $(3 \times 2)^4 = 6^4 = 1296$. To verify the first result, note that $3^4 = 81$, $2^4 = 16$, and $(81)(16) = 1296$.

If the base number is negative, it is helpful to know that $(-x)^n = (-1)^n x^n$, and $(-1)^n$ is $+1$ if n is even and -1 if n is odd. You can verify easily the other power rules for any x and y.

EXAMPLE 0.1   Simplifying an exponential expression

Simplify the expression $\dfrac{x^3 y^{-3} x y^{4/3}}{x^{-4} y^{1/3} (x^2)^3}$, and calculate its numerical value when $x = 6$ and $y = 3$.

SOLUTION

SET UP AND SOLVE We simplify the expression as follows:

$$\frac{x^3 x}{x^{-4}(x^2)^3} = x^3 x^1 x^4 x^{-6} = x^{3+1+4-6} = x^2;$$

$$\frac{y^{-3} y^{4/3}}{y^{1/3}} = y^{-3+\frac{4}{3}-\frac{1}{3}} = y^{-2}.$$

Therefore,

$$\frac{x^3 y^{-3} x y^{4/3}}{x^{-4} y^{1/3}(x^2)^3} = x^2 y^{-2} = x^2\left(\frac{1}{y}\right)^2 = \left(\frac{x}{y}\right)^2.$$

For $x = 6$ and $y = 3$, $\left(\dfrac{x}{y}\right)^2 = \left(\dfrac{6}{3}\right)^2 = 4.$

If we evaluate the original expression directly, we obtain

$$\frac{x^3 y^{-3} x y^{4/3}}{x^{-4} y^{1/3}(x^2)^3} = \frac{(6^3)(3^{-3})(6)(3^{4/3})}{(6^{-4})(3^{1/3})([6^2]^3)}$$

$$= \frac{(216)(1/27)(6)(4.33)}{(1/1296)(1.44)(46,656)} = 4.00,$$

which checks.

REFLECT This example demonstrates the usefulness of the rules for manipulating exponents.

EXAMPLE 0.2   Solving an exponential expression for the base number

If $x^4 = 81$, what is x?

SOLUTION

SET UP AND SOLVE We raise each side of the equation to the $\frac{1}{4}$ power: $(x^4)^{1/4} = (81)^{1/4}$. $(x^4)^{1/4} = x^1 = x$, so $x = (81)^{1/4}$ and $x = +3$ or $x = -3$. Either of these values of x gives $x^4 = 81$.

REFLECT Notice that we raised *both sides* of the equation to the $\frac{1}{4}$ power. As explained in Section 0.3, an operation performed on both sides of an equation does not affect the equation's validity.

0.2 Scientific Notation and Powers of 10

In physics, we frequently encounter very large and very small numbers, and it is important to use the proper number of significant figures when expressing a physical quantity. Both these issues are addressed by using **scientific notation,** in which a quantity is expressed as a decimal number with one digit to the left of the decimal point, multiplied by the appropriate power of 10. If the power of 10 is positive, it is the number of places the decimal point is moved to the right to obtain the fully written-out number. For example, $6.3 \times 10^4 = 63{,}000$. If the power of 10 is negative, it is the number of places the decimal point is moved to the left to obtain the fully written-out number. For example, $6.56 \times 10^{-3} = 0.00656$. In going from 6.56 to 0.00656, the decimal point is moved three places to the left, so 10^{-3} is the correct power of 10 to use when the number is written in scientific notation. Most calculators have keys for expressing a number in either decimal (floating-point) or scientific notation.

When two numbers written in scientific notation are multiplied (or divided), multiply (or divide) the decimal parts to get the decimal part of the result, and multiply (or divide) the powers of 10 to get the power-of-10 portion of the result. You may have to adjust the location of the decimal point in the answer to express it in scientific notation. For example,

$$
\begin{aligned}
(8.43 \times 10^8)(2.21 \times 10^{-5}) &= (8.43 \times 2.21)(10^8 \times 10^{-5}) \\
&= (18.6) \times (10^{8-5}) = 18.6 \times 10^3 \\
&= 1.86 \times 10^4.
\end{aligned}
$$

Similarly,

$$
\frac{5.6 \times 10^{-3}}{2.8 \times 10^{-6}} = \left(\frac{5.6}{2.8}\right) \times \left(\frac{10^{-3}}{10^{-6}}\right) = 2.0 \times 10^{-3-(-6)} = 2.0 \times 10^3.
$$

Your calculator can handle these operations for you automatically, but it is important for you to develop good "number sense" for scientific notation manipulations.

When adding, subtracting, multiplying, or dividing numbers, keeping the proper number of significant figures is important. See Section 1.5 to review how to keep the proper number of significant figures in these cases.

0.3 Algebra

Solving Equations

Equations written in terms of symbols that represent quantities are frequently used in physics. An **equation** consists of an equal sign and quantities to its left and to its right. Every equation tells us that the combination of quantities on the left of the equals sign has the same value as (that is, equals) the combination on the right of the equals sign. For example, the equation $y + 4 = x^2 + 8$ tells us that $y + 4$ has the same value as $x^2 + 8$. If $x = 3$, then the equation $y + 4 = x^2 + 8$ says that $y = 13$.

Often, one of the symbols in an equation is considered to be the *unknown,* and we wish to solve for the unknown in terms of the other quantities. For example, we might wish to solve the equation $2x^2 + 4 = 22$ for the value of x. Or we might wish to solve the equation $x = v_0 t + \frac{1}{2}at^2$ for the unknown a in terms of x, t, and v_0. Use the following rule to solve an equation:

An equation remains true if any valid operation performed on one side of the equation is also performed on the other side. The operations could be (a) adding or subtracting a number or symbol, (b) multiplying or dividing by a number or symbol, or (c) raising each side of the equation to the same power.

EXAMPLE 0.3 **Solving a numerical equation**

Solve the equation $2x^2 + 4 = 22$ for x.

SOLUTION

SET UP AND SOLVE First we subtract 4 from both sides. This gives $2x^2 = 18$. Then we divide both sides by 2 to get $x^2 = 9$. Finally, we raise both sides of the equation to the $\frac{1}{2}$ power. (In other words, we take the square root of both sides of the equation.) This gives $x = \pm\sqrt{9} = \pm 3$. That is, $x = +3$ or $x = -3$. We can verify our answers by substituting our result back into the original equation: $2x^2 + 4 = 2(\pm 3)^2 + 4 = 2(9) + 4 = 18 + 4 = 22$, so $x = \pm 3$ does satisfy the equation.

REFLECT Notice that a square root always has *two* possible values, one positive and one negative. For instance, $\sqrt{4} = \pm 2$, because $(2)(2) = 4$ and $(-2)(-2) = 4$. Your calculator will give you only a positive root; it's up to you to remember that there are actually two. Both roots are correct mathematically, but in a physics problem only one may represent the answer. For instance, if you can get dressed in $\sqrt{4}$ minutes, the only physically meaningful root is 2 minutes!

EXAMPLE 0.4 **Solving a symbolic equation**

Solve the equation $x = v_0 t + \frac{1}{2}at^2$ for a.

SOLUTION

SET UP AND SOLVE We subtract $v_0 t$ from both sides. This gives $x - v_0 t = \frac{1}{2}at^2$.

Now we multiply both sides by 2 and divide both sides by t^2, giving

$$a = \frac{2(x - v_0 t)}{t^2}.$$

REFLECT As we've indicated, it makes no difference whether the quantities in an equation are represented by variables (such as x, v, and t) or by numerical values.

The Quadratic Formula

Using the methods of the previous subsection, we can easily solve the equation $ax^2 + c = 0$ for x:

$$x = \pm\sqrt{\frac{-c}{a}}.$$

For example, if $a = 2$ and $c = -8$, the equation is $2x^2 - 8 = 0$ and the solution is

$$x = \pm\sqrt{\frac{-(-8)}{2}} = \pm\sqrt{4} = \pm 2.$$

The equation $ax^2 + bx = 0$ is also easily solved by factoring out an x on the left side of the equation, giving $x(ax + b) = 0$. (To *factor out* a quantity means to isolate it so that the rest of the expression is either multiplied or divided by that quantity.) The equation $x(ax + b) = 0$ is true (that is, the left side equals zero) if either $x = 0$ or $x = -\frac{b}{a}$. These are the two solutions of the equation. For example, if $a = 2$ and $b = 8$, the equation is $2x^2 + 8x = 0$ and the solutions are $x = 0$ and $x = -\frac{8}{2} = -4$.

But if the equation is in the form $ax^2 + bx + c = 0$, with a, b, and c all nonzero, we cannot use the previous simple methods to solve for x. Such an equation is called a **quadratic equation,** and its solutions are expressed by the **quadratic formula:**

Quadratic formula

For a quadratic equation in the form $ax^2 + bx + c = 0$, where a, b, and c are real numbers and $a \neq 0$, the solutions are given by the quadratic formula:

$$x = \frac{-b \pm \sqrt{b^2 - 4ac}}{2a}$$

In general, a quadratic equation has two roots (solutions), which may be real or complex numbers.

If $b^2 - 4ac = 0$, then the two roots are equal and real numbers.

If $b^2 > 4ac$, that is, $b^2 - 4ac$ is positive, then the two roots are unequal and real numbers.

By contrast, if $b^2 < 4ac$, that is, $b^2 - 4ac$ is negative, then the roots are unequal complex numbers and cannot represent physical quantities. In such a case, the quadratic equation has mathematical solutions, but no physical solutions.

EXAMPLE 0.5 **Solving a quadratic equation**

Find the values of x that satisfy the equation $2x^2 - 2x = 24$.

SOLUTION

SET UP AND SOLVE First we write the equation in the standard form $ax^2 + bx + c = 0$: $2x^2 - 2x - 24 = 0$. Then $a = 2$, $b = -2$, and $c = -24$. Next, the quadratic formula gives the two roots as

$$x = \frac{-(-2) \pm \sqrt{(-2)^2 - 4(2)(-24)}}{(2)(2)}$$

$$= \frac{+2 \pm \sqrt{4 + 192}}{4} = \frac{2 \pm 14}{4},$$

so $x = 4$ or $x = -3$. If x represents a physical quantity that takes only nonnegative values, then the negative root $x = -3$ is nonphysical and is discarded.

REFLECT As we've mentioned, when an equation has more than one mathematical solution or root, it's up to *you* to decide whether one or the other or both represent the true physical answer. (If neither solution seems physically plausible, you should review your work.)

Simultaneous Equations

If a problem has two unknowns—for example, x and y—then it takes two independent equations in x and y (that is, two equations for x and y, where one equation is not simply a multiple of the other) to determine their values uniquely. Such equations are called **simultaneous equations** because you solve them together. A typical procedure is to solve one equation for x in terms of y and then substitute the result into the second equation to obtain an equation in which y is the only unknown. You then solve this equation for y and use the value of y in either of the original equations in order to solve for x. A pair of equations in which all quantities are symbols can be combined to eliminate one of the common unknowns. In general, to solve for n unknowns, we must have n independent equations. Simultaneous equations can also be solved **graphically** by plotting both equations using the same scale on the same graph paper. The solutions are the coordinates of the points of intersection of the graphs.

Solving two equations in two unknowns

Solve the following pair of equations for x and y:

$$x + 4y = 14$$
$$3x - 5y = -9$$

SOLUTION

SET UP AND SOLVE The first equation gives $x = 14 - 4y$. Substituting this for x in the second equation yields, successively, $3(14 - 4y) - 5y = -9$, $42 - 12y - 5y = -9$, and $-17y = -51$. Thus, $y = \frac{-51}{-17} = 3$. Then $x = 14 - 4y = 14 - 12 = 2$. We can verify that $x = 2$, $y = 3$ satisfies both equations.

An alternative approach is to multiply the first equation by -3, yielding $-3x - 12y = -42$. Adding this to the second equation gives, successively, $3x - 5y + (-3x) + (-12y) =$ $-9 + (-42)$, $-17y = -51$, and $y = 3$, which agrees with our previous result.

REFLECT As shown by the alternative approach, simultaneous equations can be solved in more than one way. The basic methods we describe are easy to keep straight; other methods may be quicker, but may require more insight or forethought. Use the method you're comfortable with.

Solving two symbolic equations in two unknowns

Use the equations $v = v_0 + at$ and $x = v_0 t + \frac{1}{2}at^2$ to obtain an equation for x that does not contain a.

SOLUTION

SET UP AND SOLVE We solve the first equation for a:

$$a = \frac{v - v_0}{t}.$$

We substitute this expression into the second equation:

$$x = v_0 t + \frac{1}{2}\left(\frac{v - v_0}{t}\right)t^2 = v_0 t + \frac{1}{2}vt - \frac{1}{2}v_0 t$$

$$= \frac{1}{2}v_0 t + \frac{1}{2}vt = \left(\frac{v_0 + v}{2}\right)t.$$

REFLECT When you solve a physics problem, it's often best to work with symbols for all but the final step of the problem. Once you've arrived at the final equation, you can plug in numerical values and solve for an answer.

0.4 Direct, Inverse, and Inverse-Square Relationships

The essence of physics is to describe and verify the relationships among physical quantities. The relationships are often simple. For example, two quantities may be directly proportional to each other, inversely proportional to each other, or one quantity may be inversely proportional to the square of the other quantity.

Direct Relationship

Two quantities are said to be **directly proportional** to one another if an increase (or decrease) of the first quantity causes an increase (or decrease) of the second quantity by the same factor. If y is directly proportional to x, the direct proportionality is written as $y \propto x$. The ratio y/x is a constant, say, k. That is, $\frac{y_1}{x_1} = \frac{y_2}{x_2} = k$. For example, the ratio of the circumference C to the diameter d of a circle is always 3.14, known as π (pi). Therefore, the circumference of a circle is directly proportional to its diameter as $C = \pi d$, where π is the constant of proportionality. Another simple example of direct proportionality is the stretching or compression of an ordinary helical spring (discussed in Section 5.4). The spring has a certain length at rest, which increases when it is suspended vertically with weights attached to the bottom. If the amount of stretch is not too long, the amount of force F (measured in pounds or newtons) on the spring and the amount of stretch x are directly proportional to each other (Figure 0.1). Thus, $F = kx$ where k is the constant of proportionality.

In general, in a direct proportion, $\dfrac{a}{b} = \dfrac{c}{d}$. Multiplying both sides by bd, we find

$$\cancel{b}d \cdot \frac{a}{\cancel{b}} = b\cancel{d} \cdot \frac{c}{\cancel{d}} \quad \text{or} \quad a \cdot d = b \cdot c$$

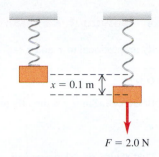

▲ FIGURE 0.1

Graph of Direct Proportionality Relationship

When y is directly proportional to x, $y = kx$, and the graph of y versus x is a straight line passing through the origin as shown in Figure 0.2. In the graph, the change of the quantity x is labeled as Δx (which is often called "run") and the corresponding change in y is labeled as Δy (which is often called "rise").

Here,

$$\Delta x = x_2 - x_1$$
$$\Delta y = y_2 - y_1,$$

where (x_1, y_1) and (x_2, y_2) are coordinates of the two points on the line. The constant of proportionality between Δy and Δx is also k. Thus,

$$\Delta y = k\, \Delta x$$

The steepness of the line is measured by the ratio $\Delta y/\Delta x$ and is called the *slope* of the line. Thus,

$$\text{Slope} = \frac{\Delta y}{\Delta x} = k$$

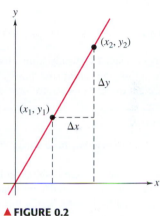

▲ FIGURE 0.2

The slope of a line can be *positive*, *negative*, *zero*, or *undefined*, as shown in Figure 0.3 (a–d).

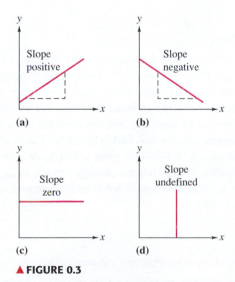

▲ FIGURE 0.3

Note that: slope *positive* means *y increases as x increases*; slope *negative* means *y decreases as x increases*; slope *zero* means *y does not change*—that is, the line is *parallel to the x-axis*; slope *undefined* means *x does not change*—that is, the line is *parallel to the y-axis*.

EXAMPLE 0.8   Solving for a quantity in direct proportion

If y is directly proportional to x, and $x = 2$ when $y = 8$, what is y when $x = 10$?

SOLUTION

SET UP AND SOLVE Since y is directly proportional to x,

$$\frac{y_1}{x_1} = \frac{y_2}{x_2} = k.$$

Substituting the values we get $\dfrac{y}{10} = \dfrac{8}{2}$.

Multiplying both sides by 2 and 10 to get rid of the fractions gives $2y = 10 \cdot 8 = 80$.

Divide by 2 to isolate y: $y = \dfrac{80}{2} = 40$.

REFLECT Although this is a simple problem, this gives you the strategy for how to solve problems in direct proportion. Note that x has increased by a factor of 5, so y must also increase by the same factor.

EXAMPLE 0.9   Solving for the stiffness constant of a spring

A spring is suspended vertically from a fixed support. When a weight of 2.0 newtons is attached to the bottom of the spring, the spring stretches by 0.1 m. (The newton, abbreviated N, is the SI unit for force; the hanging weight exerts a downward force on the spring.) Determine the spring constant of the spring.

SOLUTION

SET UP AND SOLVE We have force, $F = 2.0\,\text{N}$ and stretch, $x = 0.10\,\text{m}$.

The sketch of the problem is shown in Figure 0.1. The applied weight and the amount of stretch are related by a direct proportion, as expressed by $F = kx$. Using this equation we can solve for the stiffness constant:

$$k = \frac{F}{x} = \frac{2.0\,\text{N}}{0.10\,\text{m}} = 20\,\text{N/m}$$

REFLECT The stiffness constant in this equation is the constant of proportionality in the direct proportion between the force and the amount of stretch. Its unit is the ratio of the units of F and x.

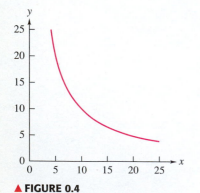

▲ **FIGURE 0.4**

Inverse Proportion

When one quantity increases and the other quantity decreases in such a way that their product stays the same, they are said to be in **inverse proportion**. In inverse proportion, when one quantity approaches zero, the other quantity becomes extremely large, so that the product remains the same. For example, the product of the pressure and volume of an ideal gas remains constant if the temperature of the gas is maintained constant, as you will find in Section 15.2. Mathematically, if y is in inverse proportion to x, $y \propto 1/x$. This gives $y = k/x$, or $xy = k$, where k is the constant of proportionality. That is, when x changes from x_1 to x_2, y changes from y_1 to y_2 so that $x_1 y_1 = x_2 y_2 = k$. This type of behavior is illustrated in Figure 0.4 (for an arbitrary choice of $k = 100$).

EXAMPLE 0.10   Solving for a quantity (volume of an ideal gas) in inverse proportion

According to Boyle's law, if the temperature of an ideal gas is kept constant, its pressure, P, is *inversely proportional* to its volume, V. A cylindrical flask is fitted with an airtight piston and contains an ideal gas. Initially, the pressure of the inside gas is 11×10^4 pascals (Pa) and the volume of the gas is $8.0 \times 10^{-3}\,\text{m}^3$. Assuming that the system is always at the temperature of 330 kelvins (K), determine the volume of the gas when its pressure increases to 24×10^4 Pa.

SOLUTION

SET UP AND SOLVE Since the pressure P is inversely proportional to the volume V according to Boyle's law, the product PV remains constant.

That is,

$$P_1 V_1 = P_2 V_2.$$

We divide by P_2 to solve for V_2. $V_2 = \dfrac{P_1 V_1}{P_2}.$

In this problem, $P_1 = 11 \times 10^4$ Pa, $V_1 = 8.0 \times 10^{-3}$ m³, and $P_2 = 24 \times 10^4$ Pa.

Substituting the values of P_1, V_1, and P_2, we solve for V_2.

$$V_2 = \frac{(11 \times 10^4\,\text{Pa}) \times (8.0 \times 10^{-3}\,\text{m}^3)}{24 \times 10^4\,\text{Pa}} = \frac{11 \times 8.0}{24} \times 10^{-3}\,\text{m}^3$$

$$= 3.7 \times 10^{-3}\,\text{m}^3$$

REFLECT Since the pressure increased, the final volume has decreased, as expected in an inverse proportion. Note that the pascal (Pa) and K/kelvin (K) are the SI units for pressure and temperature, respectively.

Inverse Square Proportion

Inverse square dependence is common in the laws of nature. For example, the force of gravity due to a body decreases as the inverse square of the distance from the body, as expressed by Newton's law of gravitation in Section 6.3. Similarly, the electrostatic force due to a point electric charge decreases as the square of the distance from the charge, as expressed by Coulomb's law in Section 17.4. The intensity of sound and of light also decreases as the inverse square of the distance from a point source, as you will find in Section 12.10. (Intensity in these cases is a measure of the power of the sound or light per unit area.) Mathematically, if y varies inversely with the square of x, then

$$y \propto \frac{1}{x^2} \quad \text{or} \quad y = \frac{k}{x^2} \quad \text{or} \quad x^2 y = k,$$

where k is the constant of proportionality. That is, when x changes from x_1 to x_2, y changes from y_1 to y_2 so that $x_1^2 y_1 = x_2^2 y_2 = k.$ Or

$$\frac{y_1}{y_2} = \frac{x_2^2}{x_1^2}$$

This relationship is illustrated in Figure 0.5 (for an arbitrary choice of $k = 100$).

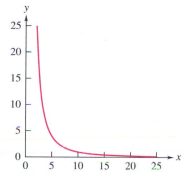

▲ **FIGURE 0.5**

EXAMPLE 0.11 **Solving for a quantity (sound intensity) that varies as the inverse square**

A small source of sound emits sound equally in all directions. The intensity of emitted sound at a point is given by the equation $I = k/r^2$, where k is a constant of proportionality and r is the distance of the point from the source. If the sound intensity is 0.05 watt/meter² (W/m²) at a distance 0.5 m from the source, find the sound intensity at a distance of 20 m from the source.

SOLUTION

SET UP AND SOLVE The intensity of sound is given by the equation $I = k/r^2$. We apply this equation to solve the problem. However, we do not need to know the value of the constant k. In this problem, we have an initial distance, $r_1 = 0.5$ m, sound intensity, $I_1 = 0.05$ W/m², and final distance, $r_2 = 20$ m, where intensity I_2 is to be determined.

From the inverse square relationship, we have (where the variables x and y have been replaced by r and I in the equation)

$$\frac{I_1}{I_2} = \frac{r_2^2}{r_1^2}$$

We take the reciprocal of both sides and multiply by I_1 to solve for I_2.

$$I_2 = I_1 \frac{r_1^2}{r_2^2} = (0.05\,\text{W/m}^2)\,\frac{(0.5\,\text{m})^2}{(20\,\text{m})^2}$$

$$= 3.1 \times 10^{-5}\,\text{W/m}^2$$

REFLECT As the distance increases, the intensity decreases. Note that because the intensity decreases as the square of the distance, the result for intensity is less than it would have been for the case of a simple inverse proportion. Thus, the result makes sense.

0.5 Logarithmic and Exponential Functions

The base-10 logarithm, or **common logarithm** (log), of a number y is the power to which 10 must be raised to obtain y: $y = 10^{\log y}$. For example, $1000 = 10^3$, so $\log(1000) = 3$; you must raise 10 to the power 3 to obtain 1000. Most calculators have a key for calculating the log of a number.

Sometimes we are given the log of a number and are asked to find the number. That is, if $\log y = x$ and x is given, what is y? To solve for y, write an equation in which 10 is raised to the power equal to either side of the original equation: $10^{\log y} = 10^x$. But $10^{\log y} = y$, so $y = 10^x$. In this case, y is called the **antilog** of x. For example, if $\log y = -2.0$, then $y = 10^{-2.0} = 1.0 \times 10^{-2} = 0.010$.

The log of a number is positive if the number is greater than 1. The log of a number is negative if the number is less than 1, but greater than zero. The log of zero or of a negative number is not defined, and $\log 1 = 0$.

Another base that occurs frequently in physics is the quantity $e = 2.718\ldots$. The **natural logarithm** (ln) of a number y is the power to which e must be raised to obtain y: $y = e^{\ln y}$. If $x = \ln y$, then $y = e^x$, which is called an **exponential function**, also written as $\exp(x)$. Most calculators have keys for $\ln x$ and for e^x. For example, $\ln 10.0 = 2.30$, and if $\ln x = 3.00$, then $x = e^{3.00} = 20.1$. Note that $\ln 1 = 0$. The plot of the function e^x is shown in Figure 0.6. The plot of e^{-x} will be a curve that is a reflection of Figure 0.6 about the y-axis.

The exponential function and natural logarithm occur in natural phenomena (such as, radioactive decay, Section 30.3) where the rate of increase or decrease of some quantity depends on the quantity. Exponential growth and decay of electric charges and electric current are also common in electric circuits as described in Sections 19.8 and 21.11.

Logarithms with any choice of base, including base 10 or base e, obey several simple and useful rules:

1. $\log(ab) = \log a + \log b$.

2. $\log\left(\dfrac{a}{b}\right) = \log a - \log b$.

3. $\log(a^n) = n\log a$.

A particular example of the second rule is

$$\log\left(\frac{1}{a}\right) = \log 1 - \log a = -\log a,$$

since $\log 1 = 0$.

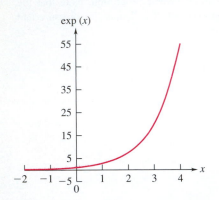

exp (x)

▲ **FIGURE 0.6**

EXAMPLE 0.12 **Solving a logarithmic equation**

If $\frac{1}{2} = e^{-\alpha T}$, solve for T in terms of α.

SOLUTION

SET UP AND SOLVE We take the natural logarithm of both sides of the equation: $\ln\left(\frac{1}{2}\right) = -\ln 2$ and $\ln\left(e^{-\alpha T}\right) = -\alpha T$. The equation thus becomes $-\alpha T = -\ln 2$, and it follows that $T = \frac{\ln 2}{\alpha}$.

REFLECT The equation $y = e^{\alpha x}$ expresses y in terms of the exponential function $e^{\alpha x}$. The general rules for exponents in Section 0.1 apply when the base is e, so $e^x e^y = e^{x+y}$, $e^x e^{-x} = e^{x+(-x)} = e^0 = 1$, and $(e^x)^2 = e^{2x}$.

0.6 Areas and Volumes

Figure 0.7 illustrates the formulas for the areas and volumes of common geometric shapes:

- A rectangle with length a and width b has area $A = ab$.
- A rectangular solid (a box) with length a, width b, and height c has volume $V = abc$.

- A circle with radius r has diameter $d = 2r$, circumference $C = 2\pi r = \pi d$, and area $A = \pi r^2 = \pi d^2 / 4$.
- A sphere with radius r has surface area $A = 4\pi r^2$ and volume $V = \frac{4}{3}\pi r^3$.
- A cylinder with radius r and height h has volume $V = \pi r^2 h$.

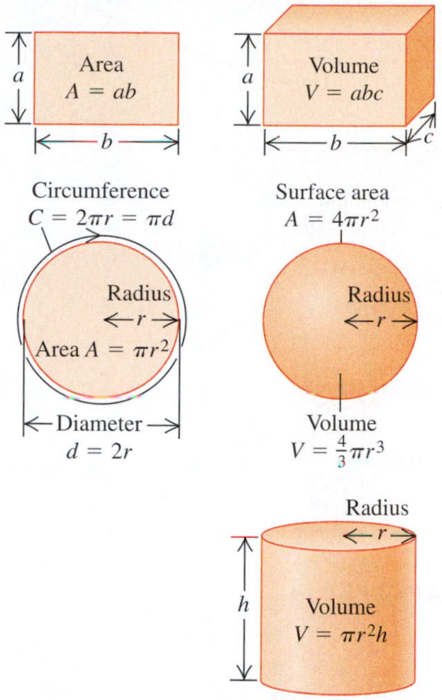

0.7 Plane Geometry and Trigonometry

Following are some useful results about angles:

1. Interior angles formed when two straight lines intersect are equal. For example, in Figure 0.8, the two angles θ and ϕ are equal.
2. When two parallel lines are intersected by a diagonal straight line, the alternate interior angles are equal. For example, in Figure 0.9, the two angles θ and ϕ are equal.
3. When the sides of one angle are each perpendicular to the corresponding sides of a second angle, then the two angles are equal. For example, in Figure 0.10, the two angles θ and ϕ are equal.
4. The sum of the angles on one side of a straight line is $180°$. In Figure 0.11, $\theta + \phi = 180°$.
5. The sum of the angles in any triangle is $180°$.

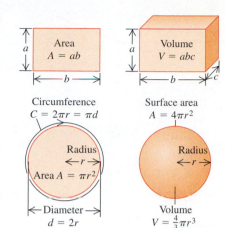

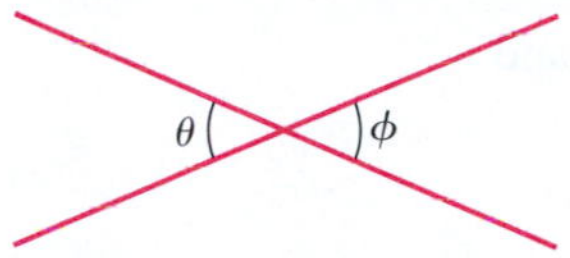

▲ **FIGURE 0.7**

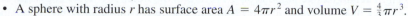

Interior angles formed when two straight lines intersect are equal:
$\theta = \phi$

▲ **FIGURE 0.8**

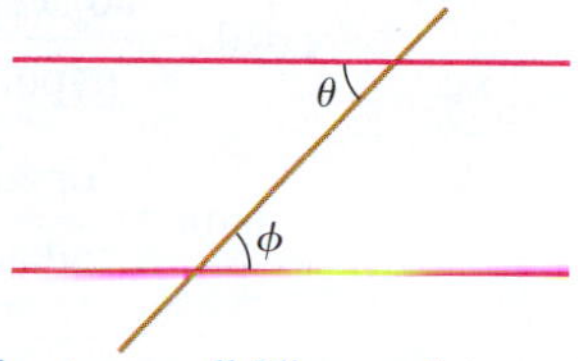

When two parallel lines are intersected by a diagonal straight line, the alternate interior angles are equal:
$\theta = \phi$

▲ **FIGURE 0.9**

Similar Triangles

Triangles are **similar** if they have the same shape, but different sizes or orientations. Similar triangles have equal corresponding angles and equal ratios of corresponding sides. If the two triangles in Figure 0.12 are similar, then $\theta_1 = \theta_2$, $\phi_1 = \phi_2$, $\gamma_1 = \gamma_2$, and $\dfrac{a_1}{a_2} = \dfrac{b_1}{b_2} = \dfrac{c_1}{c_2}$.

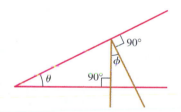

When the sides of one angle are each perpendicular to the corresponding sides of a second angle, then the two angles are equal:
$\theta = \phi$

▲ **FIGURE 0.10**

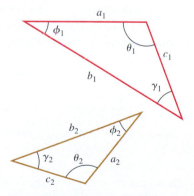

Two similar triangles: Same shape but not necessarily the same size.

▲ **FIGURE 0.12**

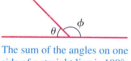

The sum of the angles on one side of a straight line is $180°$:
$\theta + \phi = 180°$

▲ **FIGURE 0.11**

If two similar triangles have the same size, they are said to be **congruent.** If triangles are congruent, one can be flipped and rotated so that it can be placed precisely on top of the other.

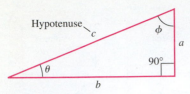

For a right triangle:
$\theta + \phi = 90°$
$c^2 = a^2 + b^2$ (Pythagorean theorem)

▲ **FIGURE 0.13**

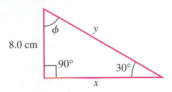

▲ **FIGURE 0.14**

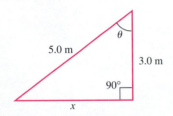

▲ **FIGURE 0.15**

Right Triangles and Trig Functions

In a **right triangle,** one angle is 90°. Therefore, the other two acute angles (*acute* means less than 90°) have a sum of 90°. In Figure 0.13, $\theta + \phi = 90°$. The side opposite the right angle is called the **hypotenuse** (side c in the figure). In a right triangle, the square of the length of the hypotenuse equals the sum of the squares of the lengths of the other two sides. For the triangle in Figure 0.13, $c^2 = a^2 + b^2$. This formula is called the **Pythagorean Theorem.**

If two right triangles have the same value for one acute angle, then the two triangles are similar and have the same ratio of corresponding sides. This true statement allows us to define the functions **sine, cosine,** and **tangent** that are ratios of a pair of sides. These functions, called **trigonometric** functions or **trig functions,** depend only on one of the angles in the right triangle. For an angle θ, these functions are written $\sin\theta$, $\cos\theta$, and $\tan\theta$.

In terms of the triangle in Figure 0.13, the sine, cosine, and tangent of the angle θ are as follows:

$$\sin\theta = \frac{\text{opposite side}}{\text{hypotenuse}} = \frac{a}{c},$$

$$\cos\theta = \frac{\text{adjacent side}}{\text{hypotenuse}} = \frac{b}{c}, \text{ and}$$

$$\tan\theta = \frac{\text{opposite side}}{\text{adjacent side}} = \frac{a}{b}.$$

Note that $\tan\theta = \dfrac{\sin\theta}{\cos\theta}$. For angle ϕ, $\sin\phi = \dfrac{b}{c}$, $\cos\phi = \dfrac{a}{c}$, and $\tan\phi = \dfrac{b}{a}$.

In physics, angles are expressed in either degrees or radians, where π radians $= 180°$. (For more on radians, see Section 9.1.) Most calculators have a key for switching between degrees and radians. Always be sure that your calculator is set to the appropriate angular measure.

Inverse trig functions, denoted, for example, by $\sin^{-1}x$ (or arcsin x) have a value equal to the angle that has the value x for the trig function. For example, $\sin 30° = 0.500$, so $\sin^{-1}(0.500) = \arcsin(0.500) = 30°$. Note that $\sin^{-1}x$ does *not* mean $\dfrac{1}{\sin x}$. Also, note that when you determine an angle using inverse trigonometric functions, the calculator will always give you the smallest correct angle, which may or may not be the right answer. Use the knowledge of which quadrant you are working in to determine the correct angle in the situation.

EXAMPLE 0.13 **Using trigonometry I**

A right triangle has one angle of 30° and one side with length 8.0 cm, as shown in Figure 0.14. What is the angle ϕ and what are the lengths x and y of the other two sides of the triangle?

SOLUTION

SET UP AND SOLVE $\phi + 30° = 90°$, so $\phi = 60°$.

$$\tan 30° = \frac{8.0 \text{ cm}}{x}, \text{ so } x = \frac{8.0 \text{ cm}}{\tan 30°} = 13.9 \text{ cm.}$$

To find y, we use the Pythagorean Theorem: $y^2 = (8.0 \text{ cm})^2 + (13.9 \text{ cm})^2$, so $y = 16.0$ cm.

Or we can say $\sin 30° = 8.0$ cm$/y$, so $y = 8.0$ cm$/\sin 30° = 16$ cm, which agrees with the previous result.

REFLECT Notice how we used the Pythagorean Theorem in combination with a trig function. You will use these tools constantly in physics, so make sure that you can employ them with confidence.

EXAMPLE 0.14 **Using trigonometry II**

A right triangle has two sides with lengths as specified in Figure 0.15. What is the length x of the third side of the triangle, and what is the angle θ, in degrees?

SOLUTION

SET UP AND SOLVE The Pythagorean Theorem applied to this right triangle gives $(3.0\,\text{m})^2 + x^2 = (5.0\,\text{m})^2$, so $x = \sqrt{(5.0\,\text{m})^2 - (3.0\,\text{m})^2} = 4.0\,\text{m}$. (Since x is a length, we take the positive root of the equation.) We also have

$$\cos\theta = \frac{3.0\,\text{m}}{5.0\,\text{m}} = 0.60,\ \text{so}\ \theta = \cos^{-1}(0.60) = 53°.$$

REFLECT In this case, we knew the lengths of two sides, but none of the acute angles, so we used the Pythagorean Theorem first and then an appropriate trig function.

In a right triangle, all angles are in the range from 0° to 90°, and the sine, cosine, and tangent of the angles are all positive. This must be the case, since the trig functions are ratios of lengths. But for other applications, such as finding the components of vectors, calculating the oscillatory motion of a mass on a spring, or describing wave motion, it is useful to define the sine, cosine, and tangent for angles outside that range. Graphs of $\sin\theta$ and $\cos\theta$ are given in Figure 0.16. The values of $\sin\theta$ and $\cos\theta$ vary between $+1$ and -1. Each function is periodic, with a period of 360°. Note the range of angles between 0° and 360° for which each function is positive and negative. The two functions $\sin\theta$ and $\cos\theta$ are 90° out of phase (that is, out of step). When one is zero, the other has its maximum magnitude (i.e., its maximum or minimum value).

For any triangle (see Figure 0.17)—in other words, not necessarily a right triangle—the following two relations apply:

1. $\dfrac{\sin\alpha}{a} = \dfrac{\sin\beta}{b} = \dfrac{\sin\gamma}{c}$ (law of sines).

2. $c^2 = a^2 + b^2 - 2ab\cos\gamma$ (law of cosines).

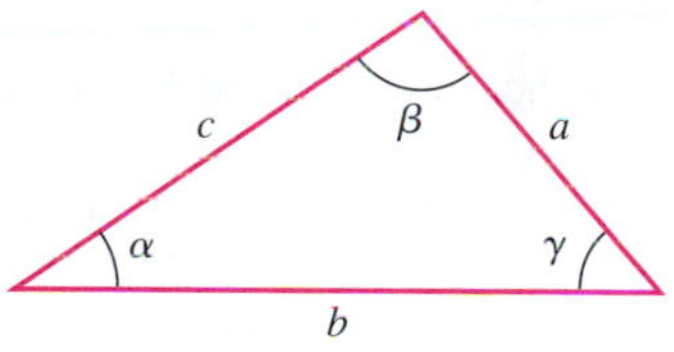

Some of the relations among trig functions are called trig identities. The following table lists only a few, those most useful in introductory physics:

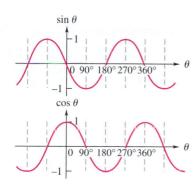

▲ **FIGURE 0.16**

▲ **FIGURE 0.17**

Useful trigonometric identities

$$\sin(-\theta) = -\sin(\theta)\quad (\sin\theta\ \text{is an odd function})$$

$$\cos(-\theta) = \cos(\theta)\quad (\cos\theta\ \text{is an even function})$$

$$\sin 2\theta = 2\sin\theta\cos\theta$$

$$\cos 2\theta = \cos^2\theta - \sin^2\theta = 2\cos^2\theta - 1 = 1 - 2\sin^2\theta$$

$$\sin(\theta \pm \phi) = \sin\theta\cos\phi \pm \cos\theta\sin\phi$$

$$\cos(\theta \pm \phi) = \cos\theta\cos\phi \mp \sin\theta\sin\phi$$

$$\sin(180° - \theta) = \sin\theta$$

$$\cos(180° - \theta) = -\cos\theta$$

$$\sin(90° - \theta) = \cos\theta$$

$$\cos(90° - \theta) = \sin\theta$$

For small angle θ (in radians),

$$\cos\theta \approx 1 - \frac{\theta^2}{2} \approx 1$$

$$\sin\theta \approx \theta$$

Problems

0.1 Exponents

Use the exponent rules to simplify the following expressions:

1. $(-3x^4y^2)^2$

2. $\dfrac{(2^3 4^4)^2}{(8)^4}$

3. $\left(\dfrac{4x^2}{2y^3}\right)^2$

4. $\left(-\dfrac{x^2y^4}{xy^{-2}}\right)^3$

0.2 Scientific Notation and Powers of 10

Express the following expressions in scientific notation:

5. 475000

6. 0.00000472

7. 123×10^{-6}

8. $\dfrac{8.3 \times 10^5}{7.8 \times 10^2}$

0.3 Algebra

Solve the following equations using any method:

9. $4x + 6 = 9x - 14$
10. $F = 9/5\,C + 32$ (solve for C)
11. $4x^2 + 6 = 3x^2 + 18$
12. $-196 = -9.8t^2$
13. $x^2 - 5x + 6 = 0$
14. $x^2 + x - 1 = 0$
15. $4.9t^2 + 2t - 20 = 0$
16. $5x - 4y = 1, 6y = 10x - 4$
17. $\dfrac{x}{2} + \dfrac{y}{3} = 2, 2x - y = 1$

0.4 Direct, Inverse, and Inverse-Square Relationships

18. If x is proportional to y and $x = 2$ when $y = 10$, what is the value of x when $y = 8$?
19. A hand exerciser uses a coiled spring and the force needed to compress the spring is *directly proportional* to the amount of compression. If a force of 80 N is needed to compress the spring by 0.02 m, determine the force required to compress the spring by 0.05 m.
20. According to the ideal gas equation (Chapter 15.2), the volume of an ideal gas is directly proportional to its temperature in kelvins (K) if the pressure of the gas is constant. An ideal gas occupies a volume of 4.0 liters at 100 K. Determine its volume when it is heated to 300 K while held at a constant pressure.
21. For a sound coming from a point source the amplitude of sound is inversely proportional to the distance. If the displacement amplitude of an air molecule in a sound wave is 4.8×10^{-6} m at a point

1.0 m from the source, what would be the displacement amplitude of the same sound when the distance increases to 4.0 m?
22. For a small lamp that emits light uniformly in all directions, the light intensity is given by the inverse square law, $I = k/r^2$. If the light intensity reaching an object at a distance of 0.40 m is 60.0 lux, determine the light intensity 1.80 m from the lamp.
23. The force of gravity on an object (which we experience as the object's weight) varies inversely as the square of the distance from the center of the earth. Determine the force of gravity on an astronaut when he is at a height of 6000 km from the surface of the earth if he weighs 700 newtons (N) when on the surface of the earth. The radius of the earth is 6.38×10^6 m. (If the astronaut is in orbit, he will float "weightlessly," but gravity still acts on him – he and his spaceship appear weightless because they are falling freely in their orbit around the earth.)

0.5 Logarithmic and Exponential Functions

24. Use the properties of logarithms and write each expression in terms of logarithms of x, y, and z.

    (a) $\log(x^4y^2z^8)$ (b) $\log\sqrt{x^3y^7}$ (c) $\log\sqrt[3]{\dfrac{x^2y^6}{z^3}}$

25. Simplify the expression.
    (a) $4\log x + \log y - 3\log(x + y)$
    (b) $\log(xy + x^2) - \log(xz + yz) + 2\log z$

26. $\beta = 10\log\left(\dfrac{I}{10^{-12}}\right)$, find β when $I = 10^{-4}$.

27. $\beta = 10\log\left(\dfrac{I}{10^{-12}}\right)$, find I when $\beta = 60$.

28. $N = N_0\,e^{-(0.210)t}$. If $N_0 = 2.00 \times 10^6$, solve for t when $N = 2.50 \times 10^4$.

0.6 Areas and Volumes

29. (a) Compute the circumference and area of a circle of radius 0.12 m. (b) Compute the surface area and volume of a sphere of radius 0.21 m. (c) Compute the total surface area and volume of a rectangular solid of length 0.18 m, width 0.15 m, and height 0.8 m. (d) Compute the total surface area and volume of a cylinder of radius 0.18 m and height 0.33 m.

0.7 Plane Geometry and Trigonometry

30. A right triangle has a hypotenuse of length 20 cm and another side of length 16 cm. Determine the third side of the triangle and the other two angles of the triangle.
31. In a stairway, each step is set back 30 cm from the next lower step. If the stairway rises at an angle of 36° with the horizontal, what is the height of each step?

17 Electric Charge and Electric Field

When you scuff your shoes across a carpet, you can get zapped by an annoying spark of static electricity. That same spark could totally destroy the function of a computer chip. Lightning, the same phenomenon on a vast scale, can destroy a lot more than chips. The clinging of newly laundered synthetic fabrics is related to such sparks. All these phenomena involve electric charge and electrical interactions, one of nature's fundamental classes of interactions.

In this chapter, we'll study the interactions among electric charges that are at rest in our frame of reference; we call these **electrostatic interactions.** They are governed by Coulomb's law and are most conveniently described using the concept of *electric field*. We'll find that charge is *quantized*; it can have only certain values. The total electric charge in a system must be an integer multiple of the charge of a single electron. Electric charge obeys a *conservation* law: The total electric charge in a closed system must be constant. Electrostatic interactions hold atoms, molecules, and our bodies together, but they also are constantly trying to tear apart the nuclei of atoms. We'll explore all these concepts in this chapter and the ones that follow.

In a thundercloud, collisions between ice and slush particles give the ice particles a slight positive charge, and powerful updrafts carry the lighter ice particles toward the cloud top. When the resulting charge separation becomes strong enough, a lightning bolt results.

17.1 Electric Charge

The ancient Greeks discovered as early as 600 B.C. that when they rubbed amber with wool, the amber could then attract other objects. Today we say that the amber has acquired a net **electric charge,** or has become *charged*. The word *electric* is derived from the Greek word *elektron*, meaning "amber." When you scuff your shoes across a nylon carpet, you become electrically charged, and you can charge a comb by passing it through dry hair.

545

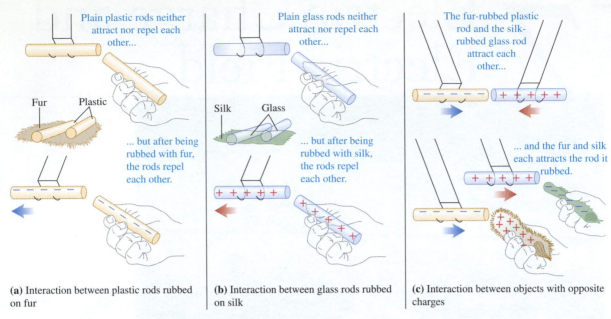

Plain plastic rods neither attract nor repel each other...

Fur   Plastic

... but after being rubbed with fur, the rods repel each other.

(a) Interaction between plastic rods rubbed on fur

Plain glass rods neither attract nor repel each other...

Silk   Glass

... but after being rubbed with silk, the rods repel each other.

(b) Interaction between glass rods rubbed on silk

The fur-rubbed plastic rod and the silk-rubbed glass rod attract each other...

... and the fur and silk each attracts the rod it rubbed.

(c) Interaction between objects with opposite charges

▲ **FIGURE 17.1** Experiments illustrating the nature of electric charge.

▲ **Application Run!** The person in this vacation snapshot, taken at a scenic overlook in Sequoia National Park, was amused to find her hair standing on end. Luckily, she and her companion left the overlook after taking the photo—and before it was hit by lightning. Just before lightning strikes, strong charges build up in the ground and in the clouds overhead. If you're standing on charged ground, the charge will spread onto your body. Because like charges repel, all your hairs tend to get as far from each other as they can. But the key thing is for *you* to get as far from that spot as *you* can!

Plastic rods and fur (real or fake) are particularly good for demonstrating electric-charge interactions. In Figure 17.1a, we charge two plastic rods by rubbing them on a piece of fur. We find that the rods repel each other. When we rub glass rods with silk (Figure 17.1b), the glass rods also become charged and repel each other. But a charged plastic rod *attracts* a charged glass rod (Figure 17.1c, top). Furthermore, the plastic rod and the fur attract each other, and the glass rod and the silk attract each other (Figure 17.1c, bottom).

These experiments and many others like them have shown that there are exactly two (no more) kinds of electric charge: the kind on the plastic rod rubbed with fur and the kind on the glass rod rubbed with silk. Benjamin Franklin (1706–1790) suggested calling these two kinds of charge *negative* and *positive*, respectively, and these names are still used.

Like and unlike charges

Two positive charges or two negative charges repel each other; a positive and a negative charge attract each other.

In the preceding discussion, the plastic rod and the silk have negative charge; the glass rod and the fur have positive charge.

When we rub a plastic rod with fur (or a glass rod with silk), *both* objects acquire net charges, and the net charges of the two objects are always equal in magnitude and opposite in sign. These experiments show that in the charging process we are not *creating* electric charge, but *transferring* it from one object to another. We now know that the plastic rod acquires extra electrons, which have negative charge. These electrons are taken from the fur, which is left with a deficiency of electrons (that is, fewer electrons than positively charged protons) and thus a net positive charge. The *total* electric charge on *both* objects does not change. This is an example of *conservation of charge;* we'll come back to this important principle later.

The sign of the charge

Three balls made of different materials are rubbed against different types of fabric—silk, polyester, etc. It is found that balls 1 and 2 repel each other and that balls 2 and 3 repel each other. From this result we can conclude that

A. Balls 1 and 3 carry charges of opposite sign.
B. Balls 1 and 3 carry charges of the same sign; ball 2 carries a charge of the opposite sign.
C. All three balls carry charges of the same sign.

SOLUTION Since balls 1 and 2 repel, they must be of the same sign, either both positive or both negative. Since 2 and 3 repel, they also must be of the same sign. This means that 1 and 3 both have the same sign as 2, so all three balls have the same sign.

The Physical Basis of Electric Charge

When all is said and done, we can't say what electric charge *is;* we can only describe its properties and its behavior. We *can* say with certainty that electric charge is one of the fundamental attributes of the particles of which matter is made. The interactions responsible for the structure and properties of atoms and molecules—and, indeed, of all ordinary matter—are primarily *electrical* interactions between electrically charged particles.

The structure of ordinary matter can be described in terms of three particles: the negatively charged **electron,** the positively charged **proton,** and the uncharged **neutron.** The protons and neutrons in an atom make up a small, very dense core called the **nucleus,** with dimensions on the order of 10^{-15} m (Figure 17.2). Surrounding the nucleus are the electrons, extending out to distances on the order of 10^{-10} m from the nucleus. If an atom were a few miles across, its nucleus would be the size of a tennis ball.

The negative charge of the electron has (within experimental error) *exactly* the same magnitude as the positive charge of the proton. In a neutral atom, the number of electrons equals the number of protons in the nucleus, and the net electrical charge (the algebraic sum of all the charges) is exactly zero (Figure 17.3a). The number of protons or electrons in neutral atoms of any element is called the **atomic number** of the element. If one or more electrons are removed, the remaining positively charged structure is called a **positive ion** (Figure 17.3b). A **negative ion** is an atom that has *gained* one or more electrons (Figure 17.3c). This gaining or losing of electrons is called **ionization.**

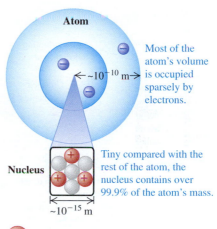

Atom

Most of the atom's volume is occupied sparsely by electrons.

← ~10^{-10} m →

Nucleus Tiny compared with the rest of the atom, the nucleus contains over 99.9% of the atom's mass.

~10^{-15} m

(+) **Proton:** Positive charge
Mass = 1.673×10^{-27} kg

○ **Neutron:** No charge
Mass = 1.675×10^{-27} kg

(−) **Electron:** Negative charge
Mass = 9.109×10^{-31} kg

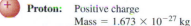
The charges of the electron and proton are equal in magnitude.

▲ **FIGURE 17.2** Schematic depiction of the structure and components of an atom.

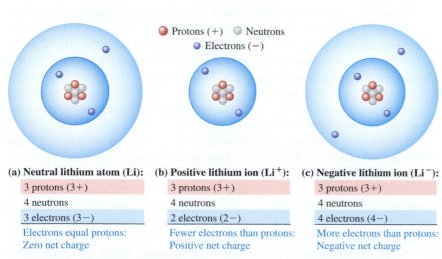

Protons (+)   Neutrons
Electrons (−)

(a) Neutral lithium atom (Li):	**(b) Positive lithium ion (Li⁺):**	**(c) Negative lithium ion (Li⁻):**
3 protons (3+)	3 protons (3+)	3 protons (3+)
4 neutrons	4 neutrons	4 neutrons
3 electrons (3−)	2 electrons (2−)	4 electrons (4−)
Electrons equal protons: Zero net charge	Fewer electrons than protons: Positive net charge	More electrons than protons: Negative net charge

▲ **FIGURE 17.3** The neutral lithium (Li) atom and positive and negative lithium ions.

The masses of the individual particles, to the precision that they are currently known, are as follows:

$$\text{Mass of electron} = m_e = 9.1093826(16) \times 10^{-31} \text{ kg};$$
$$\text{Mass of proton} = m_p = 1.67262171(29) \times 10^{-27} \text{ kg};$$
$$\text{Mass of neutron} = m_n = 1.67492728(29) \times 10^{-27} \text{ kg}.$$

The numbers in parentheses are the uncertainties in the last two digits. Note that the masses of the proton and neutron are nearly equal (within about 0.1%) and that the mass of the proton is roughly 2000 times that of the electron. Over 99.9% of the mass of any atom is concentrated in its nucleus.

When the number of protons in an object equals the number of electrons in the object, the total charge is zero, and the object as a whole is electrically neutral. To give a neutral object an excess negative charge, we may either *add negative* charges to it or *remove positive* charges from it. Similarly, we can give an excess positive charge to a neutral body by either *adding positive* charge or *removing negative* charge. When we speak of the charge on an object, we always mean its *net* charge.

An **ion** is an atom that has lost or gained one or more electrons. Ordinarily, when an ion is formed, the structure of the nucleus is unchanged. In a solid object such as a carpet or a copper wire, the nuclei of the atoms are not free to move about, so a net charge is due to an excess or deficit of electrons. However, in a liquid or a gas, net electrical charge may be due to movements of ions. Thus, a positively charged region in a fluid could represent an excess of positive ions, a deficit of negative ions, or both.

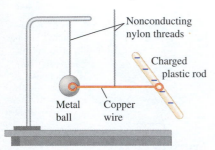

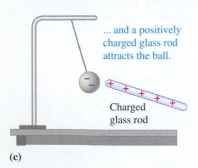

The wire conducts charge from the negatively charged plastic rod to the metal ball.

(a)

(b)

(c)

▲ **FIGURE 17.4** Charging by conduction. A copper wire is a good conductor. (a) The wire conducts charge between the plastic rod and the metal ball, giving the ball a negative charge. The charged ball is then (b) repelled by a like charge and (c) attracted by an unlike charge.

17.2 Conductors and Insulators

Some materials permit electric charge to move from one region of the material to another; others do not. For example, Figure 17.4 shows a copper wire supported by a nylon thread. Suppose you touch one end of the wire to a charged plastic rod and touch the other end to a metal ball that is initially uncharged. When you bring another charged object up close to the ball, the ball is attracted or repelled, showing that it has become electrically charged. Electric charge has been transferred through the copper wire between the ball and the surface of the plastic rod.

The wire is called a **conductor** of electricity. If you repeat the experiment, but this time using a rubber band or nylon thread in place of the wire, you find that *no* charge is transferred to the ball. These materials are called **insulators.** Conductors permit charge to move through them; insulators do not. Carpet fibers on a dry day are good insulators and allow charge to build up on us as we walk across the carpet. Coating the fibers with an antistatic layer that does not easily transfer electrons to or from our shoes is one solution to the charge-buildup problem; another is to wind some of the fibers around conducting cores.

Most of the materials we call *metals* are good conductors, and most *nonmetals* are insulators. Within a solid metal such as copper, one or more outer electrons in each atom become detached and can move freely throughout the material, just as the molecules of a gas can move through the spaces between the grains in a bucket of sand. The other electrons remain bound to the positively charged nuclei, which themselves are bound in fixed positions within the material. In an insulator, there are no, or at most very few, free electrons, and electric charge cannot move freely through the material. Some materials called *semiconductors* are intermediate in their properties between good conductors and good insulators.

We've noted that, in a liquid or gas, charge can move in the form of positive or negative ions. Ionic solutions are usually good conductors. For example, when ordinary table salt (NaCl) dissolves in water, each sodium (Na) atom loses an electron to become a positively charged sodium ion (Na^+), and each chlorine (Cl) atom gains an electron to become a negatively charged chloride ion (Cl^-). These charged particles can move freely in the solution and thus conduct charge from one region of the fluid to another, providing a mechanism for conductivity. Ionic solutions are the dominant conductivity mechanism in many biological processes.

Induction

When we charge a metal ball by touching it with an electrically charged plastic rod, some of the excess electrons on the rod move from it to the ball, leaving the rod with a smaller negative charge. There is also a technique in which the plastic rod can give another object a charge of *opposite* sign without losing any of its own charge. This process is called charging by **induction.**

Figure 17.5 shows an example of charging by induction. A metal sphere is supported on an insulating stand (step 1). When you bring a negatively charged rod near the sphere, without actually touching it (step 2), the free electrons on the surface of the sphere are repelled by the excess electrons on the rod, and they shift toward the right, away from the rod. They cannot escape from the sphere because the supporting stand and the surrounding air are insulators. So we get excess negative charge at the right side of the surface of the sphere and excess positive charge (or a deficiency of negative charge) at the left side. These excess charges are called **induced charges.**

Not all of the free electrons move to the right side of the surface of the sphere. As soon as any induced charge develops, it exerts forces toward the *left* on the other free electrons. These electrons are repelled from the negative induced charge on the right and attracted toward the positive induced charge on the left. The system reaches an equilibrium state in which the force toward the right on an electron, due to the charged rod, is just balanced by the force toward the left due to the induced charge. If we remove the charged rod, the free electrons shift back to the left, and the original neutral condition is restored.

What happens if, while the plastic rod is nearby, you touch one end of a conducting wire to the right surface of the sphere and the other end to the earth (step 3 in Figure 17.5)? The earth is a conductor, and it is so large that it can act as a practically infinite source of extra electrons or sink of unwanted electrons. Some of the negative

▲ **Application Good conductor, bad conductor.** Salt water is salty because it contains an abundance of dissolved ions. These ions are charged and can move freely, so salt water is an excellent conductor of electricity. Ordinary tap water contains enough ions to conduct electricity reasonably well—which is why you should never, ever, use an electrical device in a bath. However, absolutely pure distilled water is an insulator, because it consists only of neutral water molecules.

MasteringPHYSICS

PhET: Balloons and Static Electricity
PhET: John Travoltage

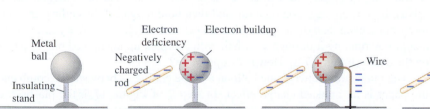

① Uncharged metal ball
② Negative charge on rod repels electrons, creating zones of negative and positive **induced charge.**
③ Wire lets electron buildup (induced negative charge) flow into ground.
④ Wire removed; ball now has only an electron-deficient region of positive charge.
⑤ Rod removed; positive charge spreads over ball.

▲ **FIGURE 17.5** Charging a metal ball by induction.

▲ **FIGURE 17.6** A charged plastic comb picks up *uncharged* bits of paper.

charge flows through the wire to the earth. Now suppose you disconnect the wire (step 4) and then remove the rod (step 5); a net positive charge is left on the sphere. The charge on the negatively charged rod has not changed during this process. The earth acquires a negative charge that is equal in magnitude to the induced positive charge remaining on the sphere.

Charging by induction would work just as well if the mobile charges in the sphere were positive charges instead of (negatively charged) electrons or even if both positive and negative mobile charges were present (as would be the case if we replaced the sphere with a flask of salt water). In this book, we'll talk mostly about metallic conductors, in which the mobile charges are negative electrons. However, even in a metal, we can describe conduction *as though* the moving charges were positive. In terms of transfer of charge in a conductor, a movement of electrons to the left is exactly equivalent to a movement of imaginary positive particles to the right. In fact, when we study electrical currents, we will find that, for historical reasons, currents in wires are described as though the moving charges were positive.

Conceptual Analysis 17.2

Charge and induction

If you charge a metal ball on an insulating stand *by induction,* using a charged glass or plastic rod, what happens to the charge on the rod?

A. It increases.
B. It decreases.
C. It stays the same.
D. The answer depends on whether the charge is positive or negative.

SOLUTION Look at Figure 17.5 to remind yourself how induction works. The rod *never touches the ball;* instead, its charge pushes or pulls the electrons in the ball so that they crowd (just slightly) to one or the other side of the ball, creating induced charges. A conducting wire lets the local excess of electrons drain into the ground and is then removed. No charges move from the rod to the ball (or to the ground), so the rod's charge is unchanged; thus, C is correct. It makes no difference whether the rod's charge is positive or negative.

When excess charge is placed on a solid conductor and is at rest (i.e., an electrostatic situation), the excess charge rests entirely on the surface of the conductor. If there were excess charge in the interior of the conductor, there would be electrical forces among the excess charges that would cause them to move, and the situation couldn't be electrostatic.

Polarization

A charged object can exert forces even on objects that are *not* charged themselves. If you rub a balloon on a rug and then hold it against the ceiling, the balloon sticks, even though the ceiling has no net electric charge. After you electrify a comb by running it through your hair, you can pick up uncharged bits of paper on the comb (Figure 17.6). How is this possible?

The interaction between the balloon and the ceiling or between the comb and the paper is an induced-charge effect. In step 2 of Figure 17.5, the plastic rod exerts a net attractive force on the sphere, even though the total charge on the sphere is zero, because the positive charges are closer to the rod than the negative charges are. Figure 17.7 shows this effect more clearly. The large ball *A* has a positive charge; the conducting metal ball *B* is uncharged. When we bring *B* close to *A*, the positive charge on *A* pulls on the electrons in *B*, setting up induced charges. Because the negative induced charge on the surface of *B* is closer to *A* than the positive induced charge is, *A* exerts a net attraction on *B*. (We'll study the dependence of electric forces on distance in Section 17.4.) Even in an insulator, the electric charges can shift back and forth a little when there is charge nearby. Figure 17.8 shows how a static charge enables a charged plastic comb to pick up

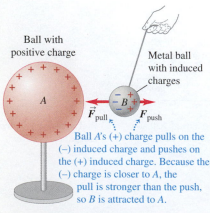

Ball with positive charge

Metal ball with induced charges

$\vec{F}_{pull}$ $\vec{F}_{push}$

Ball *A*'s (+) charge pulls on the (−) induced charge and pushes on the (+) induced charge. Because the (−) charge is closer to *A*, the pull is stronger than the push, so *B* is attracted to *A*.

▲ **FIGURE 17.7** The charge on ball *A* induces charges in ball *B*, resulting in a net attractive force between the balls.

uncharged bits of paper. Although the electrons in the paper are bound to their molecules and cannot move freely through the paper, they can still shift slightly to produce a net charge on one side and the opposite charge on the other. Thus, the comb causes each molecule in the paper to develop induced charges (an effect called **polarization).** The net result is that the scrap of paper shows a slight induced charge—enough to enable the comb to pick it up.

17.3 Conservation and Quantization of Charge

As we've discussed, an electrically neutral object is an object that has equal numbers of electrons and protons. The object can be given a charge by adding or removing either positive or negative charges. Implicit in this discussion are two very important principles. First is the principle of **conservation of charge:**

> **Conservation of charge**
> The algebraic sum of all the electric charges in any closed system is constant. Charge can be transferred from one object to another, and that is the only way in which an object can acquire a net charge.

Conservation of charge is believed to be a *universal* conservation law; there is no experimental evidence for any violation of this principle. Even in high-energy interactions in which charged particles are created and destroyed, the total charge of any closed system is exactly constant.

Second, the magnitude of the charge of the electron or proton is a natural unit of charge. Every amount of observable electric charge is always an integer multiple of this basic unit. Hence we say that charge is *quantized.* A more familiar example of quantization is money. When you pay cash for an item in a store, you have to do it in 1-cent increments. If grapefruits are selling three for a dollar, you can't buy one for $33\frac{1}{3}$ cents; you have to pay 34 cents. Cash can't be divided into smaller amounts than 1 cent, and electric charge can't be divided into smaller amounts than the charge of one electron or proton.

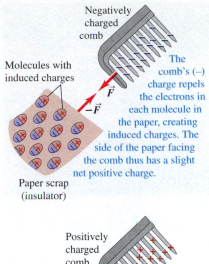

The comb's (−) charge repels the electrons in each molecule in the paper, creating induced charges. The side of the paper facing the comb thus has a slight net positive charge.

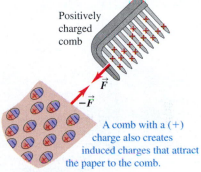

A comb with a (+) charge also creates induced charges that attract the paper to the comb.

▲ **FIGURE 17.8** A charged comb picks up uncharged paper by polarizing the paper's molecules.

Quantitative Analysis 17.3

Determine the charge

Three identical metal balls A, B, and C are mounted on insulating rods. Ball A has a charge $+q$, and balls B and C are initially uncharged (q is the usual symbol for electric charge). Ball A is touched first to ball B and then separately to ball C. At the end of this experiment, the charge on ball A is

A. $+q/2$.
B. $+q/3$.
C. $+q/4$.

SOLUTION When identical metal objects come in contact, any net charge they carry is shared equally between them. Thus, when A touches B, each ends up with a charge $+q/2$. When A then touches C, this charge is shared equally, leaving A and C each with a charge of $+q/4$.

Electric forces have many important practical applications. Electrostatic dust precipitators first create a charge on dust particles in the air and then use it to catch the dust. When a car is painted, electrostatic charges attract the droplets of sprayed paint to the body of the car. In a photocopy machine or a laser printer, charged areas on the paper attract the toner particles to the correct spots on the paper.

The forces that hold atoms and molecules together are fundamentally electrical in nature. The attraction between electrons and protons holds the electrons in atoms, holds atoms together to form polyatomic molecules, holds molecules together to form solids or liquids, and accounts for phenomena such as surface tension and the stickiness of glue. Within the atom, the electrons repel each other,

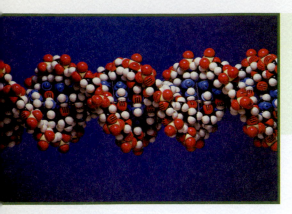

but they are held in the atom by the attractive force of the protons in the nucleus. But what keeps the positively charged protons together in the tiny nucleus despite their mutual repulsion? As we will learn in Chapter 30, they are held by another, even stronger interaction called the *nuclear force*.

17.4 Coulomb's Law

Charles Augustin de Coulomb (1736–1806) studied the interaction forces of charged particles in detail in 1784. He used a torsion balance (Figure 17.9a) similar to the one used 13 years later by Cavendish to study the (much weaker) gravitational interaction, as we discussed in Section 6.3. For *point charges* (charged bodies that are very small in comparison with the distance r between them), Coulomb found that the electric force is proportional to $1/r^2$. That is, when the distance doubles, the force decreases to one fourth of its initial value.

The force also depends on the quantity of charge on each object, which we'll denote by q or Q. To explore this dependence, Coulomb divided a charge into two equal parts by placing a small charged spherical conductor into contact with an identical but uncharged sphere; by symmetry, the charge is shared equally between the two spheres. (Note the essential role of the principle of conservation of charge in this procedure.) Thus, he could obtain one-half, one-quarter, and so on, of any initial charge. He found that the forces that two point charges q_1 and q_2 exert on each other are proportional to each charge and therefore are proportional to the *product* $q_1 q_2$ of the two charges.

The negatively charged ball attracts the positively charged one; the positive ball moves until the elastic forces in the torsion fiber balance the electrostatic attraction.

Torsion fiber

Charged pith balls

Scale

(a) A torsion balance of the type used by Coulomb to measure the electric force

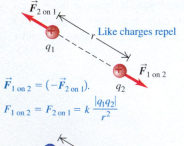

$\vec{F}_{2 \text{ on } 1}$

r Like charges repel

q_1

$\vec{F}_{1 \text{ on } 2}$

$\vec{F}_{1 \text{ on } 2} = (-\vec{F}_{2 \text{ on } 1}).$

q_2

$F_{1 \text{ on } 2} = F_{2 \text{ on } 1} = k \dfrac{|q_1 q_2|}{r^2}$

r Unlike charges attract

q_1 $\vec{F}_{2 \text{ on } 1}$

$\vec{F}_{1 \text{ on } 2}$

q_2

(b) Interaction of like and unlike charges

▲ **FIGURE 17.9** Schematic depiction of the apparatus Coulomb used to determine the forces between charged objects that can be treated as point charges.

> **Coulomb's law**
>
> The magnitude F of the force that each of two point charges q_1 and q_2 a distance r apart exerts on the other (Figure 17.9b) is directly proportional to the product of the charges $(q_1 q_2)$ and inversely proportional to the square of the distance between them (r^2). The relationship is expressed symbolically as
>
> $$F = k \frac{|q_1 q_2|}{r^2}. \tag{17.1}$$
>
> This relationship is called **Coulomb's law.**

The value of the proportionality constant k in Equation 17.1 depends on the system of units used. In the chapters on electricity and magnetism in this book, we'll use SI units exclusively. The SI electrical units include most of the familiar units, such as the volt, the ampere, the ohm, and the watt. (There is *no* British system of electrical units.) The SI unit of electric charge is called one **coulomb** (1 C). In this system, the constant k in Equation 17.1 is

$$k = 8.987551789 \times 10^9 \, \text{N} \cdot \text{m}^2/\text{C}^2.$$

In numerical calculations in problems, we'll often use the approximate value

$$k \approx 8.99 \times 10^9 \, \text{N} \cdot \text{m}^2/\text{C}^2,$$

which is in error by about 0.03%.

The forces that two charges exert on each other always act along the line joining the charges. The two forces are always equal in magnitude and opposite in direction, even when the charges are not equal. *The forces obey Newton's third law.*

As we've seen, q_1 and q_2 can be either positive or negative quantities. When the charges have the same sign (both positive or both negative), the forces are repulsive; when they are unlike, the forces are attractive. We need the absolute value bars in Equation 17.1 because F is the magnitude of a vector quantity. By definition, F is always positive, but the product $q_1 q_2$ is negative whenever the two charges have opposite signs.

The proportionality of the electrical force to $1/r^2$ has been verified with great precision. There is no experimental evidence that the exponent is anything different from precisely 2. The form of Equation 17.1 is the same as that of the law of gravitation, but electrical and gravitational interactions are two distinct classes of phenomena. The electrical interaction depends on electric charges and can be either attractive or repulsive; the gravitational interaction depends on mass and is always attractive (because there is no such thing as negative mass).

Strictly speaking, Coulomb's law, as we have stated it, should be used only for point charges *in vacuum*. If matter is present in the space between the charges, the net force acting on each charge is altered because charges are induced in the molecules of the intervening material. We'll describe this effect later. As a practical matter, though, we can use Coulomb's law unaltered for point charges in air; at normal atmospheric pressure, the presence of air changes the electrical force from its vacuum value by only about 1 part in 2000.

In SI units, the constant k in Equation 17.1 is often written as

$$k = \frac{1}{4\pi\epsilon_0},$$

where $\epsilon_0 = 8.854 \times 10^{-12} \, \text{C}^2/(\text{N} \cdot \text{m}^2)$ is another constant. This alternative form may appear to complicate matters, but it actually simplifies some of the formulas that we'll encounter later. When we study electromagnetic radiation in Chapter 23, we'll show that the numerical value of ϵ_0 is closely related to the speed of light.

The most fundamental unit of charge is the magnitude of the charge of an electron or a proton, denoted by e. The most precise value available, as of 2005, is

$$e = 1.60217653(14) \times 10^{-19} \, \text{C}.$$

The number (14) in parentheses represents the uncertainty in the last two digits.

One coulomb represents the total charge carried by about 6×10^{18} protons, or the negative of the total charge of about 6×10^{18} electrons. For comparison, the population of the earth is about 6×10^9 persons, and a cube of copper 1 cm on a side contains about 2.4×10^{24} electrons.

In electrostatics problems, charges as large as 1 coulomb are very unusual. Two charges with magnitude 1 C, at a distance 1 m apart, would exert forces of magnitude 9×10^9 N (about a million tons) on each other! A more typical range of magnitudes is 10^{-9} to 10^{-6} C. The *microcoulomb* $(1 \, \mu\text{C} = 10^{-6} \, \text{C})$ and the *nanocoulomb* $(1 \, \text{nC} = 10^{-9} \, \text{C})$ are often used as practical units of charge. The total charge of all the electrons in a penny is about 1.4×10^5 C. This number shows that we can't disturb electrical neutrality very much without using enormous forces.

▲ **Application Great balls of fire?** Before the invention of the cyclotron, which uses both electric and magnetic fields to accelerate subatomic particles, physicists used electric field generators in atom-smashing experiments. These generators, like the huge Van de Graaff generators shown here, can accumulate either positive or negative charges on the surface of a metal sphere, thus generating immense electric fields. Charged particles in such an electric field are acted upon by a large electrical force, which can be used to accelerate the particles to very high velocities. When all excess charge is located on the outer surface of a conductor in an electrostatic situation, the electric field inside is zero. Thus, scientists actually set up small laboratories *inside* each of the spheres of the generator to study subatomic particles subjected to millions of volts in a tube on the outside connecting the two spheres.

MasteringPHYSICS

ActivPhysics 11.1: Electric Force: Coulomb's Law
ActivPhysics 11.2: Electric Force: Superposition Principle
ActivPhysics 11.3: Electric Force: Superposition Principle (Quantitative)

Conceptual Analysis 17.4

Charged spheres in motion

Two small identical balls A and B are held a distance r apart on a frictionless surface; r is large compared with the size of the balls. Ball A has a net charge q; ball B has a net charge $4q$. The balls are released at the same instant and begin to move apart. The magnitudes of their accelerations are

A. constant.
B. equal and decreasing.
C. unequal and decreasing.

SOLUTION Coulomb's law states that the magnitude of the force between two charged objects that can be treated as particles is $F = (k|q_1 q_2|)/r^2$. Is this force somehow divided between the two objects? Does the object with the larger charge exert a stronger force? Should the force on each object be calculated separately? No; Newton's third law gives the answer. Whenever two objects interact, the forces that the two objects exert on each other are equal in magnitude (and opposite in direction). Since the balls experience the same magnitude of force and have the same mass, by Newton's second law they have the same magnitude of acceleration at any instant. As they move apart and r increases, the magnitude of acceleration decreases. The correct answer is B.

Superposition

When two charges exert forces simultaneously on a third charge, the total force acting on that charge is the *vector sum* of the forces that the two charges would exert individually. This important property, called the **principle of superposition,** holds for any number of charges. Coulomb's law, as we have stated it, describes only the interaction between two *point* charges, but by using the superposition principle, we can apply it to *any* collection of charges. Several of the examples that follow illustrate the superposition principle.

PROBLEM-SOLVING STRATEGY 17.1 **Coulomb's law**

SET UP

1. As always, consistent units are essential. With the value of k given earlier, distances *must* be in meters, charges in coulombs, and forces in newtons. If you are given distances in centimeters, inches, or furlongs, don't forget to convert! When a charge is given in microcoulombs, remember that $1\ \mu C = 10^{-6}\ C$.

SOLVE

2. When the forces acting on a charge are caused by two or more other charges, the total force on the charge is the *vector sum* of the individual forces. If you're not sure you remember how to do vector addition, you may want to review Sections 1.7 and 1.8. It's often useful to use components in an *x-y* coordinate system. As always, it's essential to distinguish between vectors, their magnitudes, and their components (using correct notation!) and to treat vectors properly as such.

3. Some situations involve a continuous distribution of charge along a line or over a surface. In this book, we'll consider only situations for which the vector sum described in Step 2 can be evaluated by using vector addition and symmetry considerations. In other cases, methods of integral calculus would be needed.

REFLECT

4. Try to think of particular cases where you can guess what the result should be, and compare your intuitive expectations with the results of your calculations.

EXAMPLE 17.1   Charge imbalance

(a) A large plastic block has a net charge of $-1.0 \ \mu\text{C} = -1.0 \times 10^{-6}$ C. How many more electrons than protons are in the block? **(b)** When rubbed with a silk cloth, a glass rod acquires a net positive charge of 1.0 nC. If the rod contains 1.0 mole of molecules, what fraction of the molecules have been stripped of an electron? Assume that at most one electron is removed from any molecule.

SOLUTION

SET UP AND SOLVE Part (a): We want to find the number of electrons N_e needed for a net charge of -1.0×10^{-6} C on the object. Each electron has charge $-e$. We divide the total charge by $-e$:

$$N_e = \frac{-1.0 \times 10^{-6} \text{ C}}{-1.60 \times 10^{-19} \text{ C}} = 6.2 \times 10^{12} \text{ electrons.}$$

Part (b): First we find the number N_{ion} of positive ions needed for a total charge of 1.0 nC if each ion has charge $+e$. The number of molecules in a mole is Avogadro's number, 6.02×10^{23}, so the rod contains 6.02×10^{23} molecules. As in part (a), we divide the total charge on the rod by the charge of one ion. Remember that 1 nC $= 10^{-9}$ C. Thus,

$$N_{\text{ion}} = \frac{1.0 \times 10^{-9} \text{ C}}{1.6 \times 10^{-19} \text{ C}} = 6.25 \times 10^{9}.$$

The fraction of all the molecules that are ionized is

$$\frac{N_{\text{ion}}}{6.02 \times 10^{23}} = \frac{6.25 \times 10^{9}}{6.02 \times 10^{23}} = 1.0 \times 10^{-14}.$$

REFLECT A charge imbalance of about 10^{-14} is typical for charged objects. Common objects contain a huge amount of charge, but they have very nearly equal amounts of positive and negative charge.

Practice Problem: A tiny object contains 5.26×10^{12} protons and 4.82×10^{12} electrons. What is the net charge on the object? *Answer:* 7.0×10^{-8} C.

EXAMPLE 17.2   Gravity in the hydrogen atom

A hydrogen atom consists of one electron and one proton. In an early, simple model of the hydrogen atom called the *Bohr model*, the electron is pictured as moving around the proton in a circular orbit with radius $r = 5.29 \times 10^{-11}$ m. (In Chapter 29, we'll study the Bohr model and also more sophisticated models of atomic structure.) What is the ratio of the magnitude of the electric force between the electron and proton to the magnitude of the gravitational attraction between them? The electron has mass $m_e = 9.11 \times 10^{-31}$ kg, and the proton has mass $m_p = 1.67 \times 10^{-27}$ kg.

SOLUTION

SET UP Figure 17.10 shows our sketch. The distance between the proton and electron is the radius r. Each particle has charge of magnitude e. The electric force is given by Coulomb's law and the gravitational force by Newton's law of gravitation.

SOLVE Coulomb's law gives the magnitude F_e of the electric force between the electron and proton as

$$F_e = k\frac{|q_1 q_2|}{r^2} = k\frac{e^2}{r^2},$$

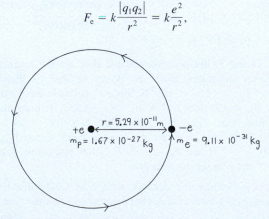

▲ **FIGURE 17.10** Our sketch for this problem.

where $k = 8.99 \times 10^9 \text{ N} \cdot \text{m}^2/\text{C}^2$. The gravitational force $\vec{F}_g$ has magnitude F_g:

$$F_g = G\frac{m_1 m_2}{r^2} = G\frac{m_e m_p}{r^2},$$

where $G = 6.67 \times 10^{-11} \text{ N} \cdot \text{m}^2/\text{kg}^2$. The ratio of the two forces is

$$\frac{F_e}{F_g} = \left(\frac{ke^2}{r^2}\right)\left(\frac{r^2}{Gm_e m_p}\right) = \frac{ke^2}{Gm_e m_p}$$

$$= \left(\frac{8.99 \times 10^9 \text{ N} \cdot \text{m}^2/\text{C}^2}{6.67 \times 10^{-11} \text{ N} \cdot \text{m}^2/\text{kg}^2}\right)$$

$$\times \frac{(1.60 \times 10^{-19} \text{ C})^2}{(9.11 \times 10^{-31} \text{ kg})(1.67 \times 10^{-27} \text{ kg})},$$

$$\frac{F_e}{F_g} = 2.27 \times 10^{39}.$$

REFLECT In our expression for the ratio, all the units cancel and the ratio is dimensionless. The astonishingly large value of F_e/F_g—about 10^{39}—shows that, in atomic structure, the gravitational force is completely negligible compared with the electrostatic force. The reason gravitational forces dominate in our daily experience

Continued

is that positive and negative electric charges are always nearly equal in number and thus cancel nearly completely. Since there is no "negative" gravitation, gravitational forces always add. Note also that because both F_e and F_g are proportional to $1/r^2$, the ratio F_e/F_g does not depend on the distance between the two particles.

Practice Problem: A hydrogen atom is at the earth's surface. The electron and proton in the atom are separated by a distance of 5.29×10^{-11} m. What is the ratio of the magnitude of the electrical force exerted by the proton on the electron to the weight of the electron? *Answer:* 9.2×10^{21}.

For all fundamental particles, the gravitational attraction is always much, much weaker than the electrical interaction. But suppose that the electric force were a million (10^6) times weaker than it really is. In that case, the ratio of electric to gravitational forces between an electron and a proton would be about $10^{39} \times 10^{-6} = 10^{33}$ and the universe would be a very different place. Materials would be a million times weaker than we are used to, because they are held together by electrostatic forces. Insects would need to have much thicker legs to support the same mass. In fact, animals couldn't get much larger than an insect unless they were made of steel, and even a hypothetical animal made of steel could be only a few centimeters in size before collapsing under its own weight. More significantly, if the electric force were a million times weaker, the lifetime of a typical star would decrease from 10 billion years down to 10 thousand years! This is hardly enough time for *any* living organisms—much less such complicated ones as insects or humans—to evolve.

EXAMPLE 17.3   Adding forces

Two point charges are located on the positive x axis of a coordinate system. Charge $q_1 = 3.0$ nC is 2.0 cm from the origin, and charge $q_2 = -7.0$ nC is 4.0 cm from the origin. What is the total force (magnitude and direction) exerted by these two charges on a third point charge $q_3 = 5.0$ nC located at the origin?

SOLUTION

SET UP We sketch the situation and draw a free-body diagram for charge q_3, using $\vec{F}_1$ to denote the force exerted by q_1 on q_3 and $\vec{F}_2$ for the force exerted by q_2 on q_3 (Figure 17.11). The directions of these forces are determined by the rule that like charges repel and unlike charges attract, so $\vec{F}_1$ points in the $-x$ direction and $\vec{F}_2$ points in the $+x$ direction. We don't yet know their relative magnitudes, so we draw them to arbitrary length.

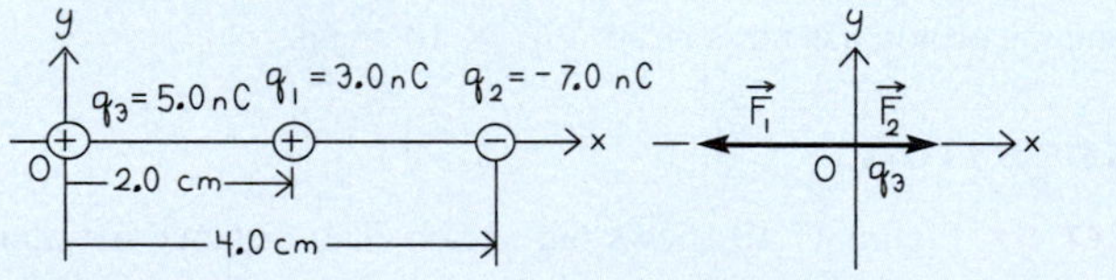

SOLVE We use Coulomb's law to find the magnitudes of the forces $\vec{F}_1$ and $\vec{F}_2$; then we add these two forces (as vectors) to find the resultant force on q_3:

$$\vec{F}_{total} = \vec{F}_1 + \vec{F}_2,$$
so $\quad F_{total,x} = F_{1x} + F_{2x}$
and $\quad F_{total,y} = F_{1y} + F_{2y}.$

$$F_1 = k\frac{|q_1 q_3|}{r_{12}^2}$$

$$= (8.99 \times 10^9 \text{ N} \cdot \text{m}^2/\text{C}^2)\frac{(3.0 \times 10^{-9} \text{ C})(5.0 \times 10^{-9} \text{ C})}{(0.020 \text{ m})^2}$$

$$= 3.37 \times 10^{-4} \text{ N},$$

(a) Our diagram of the situation

(b) Free-body diagram for q_3

▲ FIGURE 17.11 Our sketches for this problem.

$$F_2 = k\frac{|q_2 q_3|}{r_{23}^2}$$

$$= (8.99 \times 10^9 \text{ N} \cdot \text{m}^2/\text{C}^2)\frac{(7.0 \times 10^{-9} \text{ C})(5.0 \times 10^{-9} \text{ C})}{(0.040 \text{ m})^2}$$

$$= 1.97 \times 10^{-4} \text{ N}.$$

Both F_1 and F_2 are positive, because they are the magnitudes of vector quantities. Since $\vec{F}_1$ points in the $-x$ direction, $F_{1x} = -F_1 = -3.37 \times 10^{-4}$ N. Since $\vec{F}_2$ points in the $+x$ direction, $F_{2x} = +F_2 = +1.97 \times 10^{-4}$ N. Adding x components, we find that $F_{total,x} = -3.37 \times 10^{-4}$ N $+ 1.97 \times 10^{-4}$ N $= -1.40 \times 10^{-4}$ N. There are no y components. Since $F_{total,x} = -1.40 \times 10^{-4}$ N and $F_{total,y} = 0$, $\vec{F}_{total}$ has magnitude 1.40×10^{-4} N and is in the $-x$ direction.

REFLECT Because the distance term r in Coulomb's law is squared, F_1 is greater than F_2 even though $|q_2|$ is greater than $|q_1|$.

Practice Problem: In Example 17.3, what is the total force (magnitude and direction) exerted on q_1 by q_2 and q_3? *Answer:* 8.1×10^{-4} N, in the $+x$ direction.

EXAMPLE 17.4   Vector addition of forces

A point charge $q_1 = 2.0\ \mu\text{C}$ is located on the positive y axis at $y = 0.30$ m, and an identical charge q_2 is at the origin. Find the magnitude and direction of the total force that these two charges exert on a third charge $q_3 = 4.0\ \mu\text{C}$ that is on the positive x axis at $x = 0.40$ m.

SOLUTION

SET UP As in the previous example, we sketch the situation and draw a free-body diagram for q_3 (Figure 17.12), using $\vec{F}_1$ and $\vec{F}_2$ for the forces exerted on q_3 by q_1 and q_2, respectively. The directions of $\vec{F}_1$ and $\vec{F}_2$ are determined by the fact that like charges repel.

SOLVE As in Example 17.3, the net force acting on q_3 is the vector sum of $\vec{F}_1$ and $\vec{F}_2$. We use Coulomb's law to find the magnitudes F_1 and F_2 of the forces:

$$F_1 = k\frac{|q_1 q_3|}{r_{13}^2}$$

$$= (8.99 \times 10^9\ \text{N} \cdot \text{m}^2/\text{C}^2)\frac{(2.0 \times 10^{-6}\ \text{C})(4.0 \times 10^{-6}\ \text{C})}{(0.50\ \text{m})^2}$$

$$= 0.288\ \text{N},$$

$$F_2 = k\frac{|q_2 q_3|}{r_{23}^2}$$

$$= (8.99 \times 10^9\ \text{N} \cdot \text{m}^2/\text{C}^2)\frac{(2.0 \times 10^{-6}\ \text{C})(4.0 \times 10^{-6}\ \text{C})}{(0.40\ \text{m})^2}$$

$$= 0.450\ \text{N}.$$

We now calculate the x and y components of $\vec{F}_1$ and add them to the x and y components of $\vec{F}_2$, respectively, to obtain the components of the total force $\vec{F}_\text{total}$ on q_3. From Figure 17.12a, $\sin\theta = (0.30\ \text{m})/(0.50\ \text{m}) = 0.60$ and $\cos\theta = (0.40\ \text{m})/(0.50\ \text{m}) = 0.80$. Since the y component of $\vec{F}_2$ is zero and its x

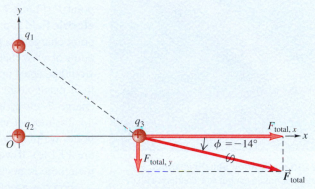

▲ **FIGURE 17.13** The total force on q_3.

component is positive, the x component of $\vec{F}_2$ is $F_{2x} = F_2 = 0.450$ N. The total x and y components are

$$F_{\text{total},x} = F_{1x} + F_{2x}$$
$$= (0.288\ \text{N})\cos\theta + 0.450\ \text{N}$$
$$= (0.288\ \text{N})(0.80) + 0.450\ \text{N} = 0.680\ \text{N},$$
$$F_{\text{total},y} = F_{1y} + F_{2y} = -(0.288\ \text{N})\sin\theta + 0$$
$$= -(0.288\ \text{N})(0.60) = -0.173\ \text{N}.$$

These components combine to form $\vec{F}_\text{total}$, as shown in Figure 17.13:

$$F_\text{total} = \sqrt{F_{\text{total},x}^2 + F_{\text{total},y}^2}$$
$$= \sqrt{(0.680\ \text{N})^2 + (-0.173\ \text{N})^2} = 0.70\ \text{N},$$
$$\tan\phi = \frac{F_{\text{total},y}}{F_{\text{total},x}} = \frac{-0.173\ \text{N}}{0.680\ N} \quad \text{and} \quad \phi = -14°.$$

The resultant force has magnitude 0.70 N and is directed at $14°$ below the $+x$ axis.

REFLECT The forces exerted by q_1 and q_2 both have components in the $+x$ direction, so these components add. The force exerted by q_1 also has a component in the $-y$ direction, so the net force is in the fourth quadrant. Even though q_1 and q_2 are identical, the force exerted by q_2 is larger than the force exerted by q_1 because q_3 is closer to q_2.

Practice Problem: In Example 17.4, what is the net force on q_3 if $q_1 = 2.0\ \mu\text{C}$, as in the example, but $q_2 = -2.0\ \mu\text{C}$? *Answer:* 0.28 N, $38°$ below the $-x$ axis.

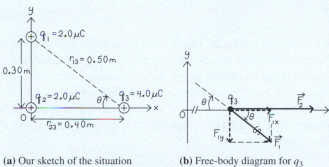

(a) Our sketch of the situation   **(b)** Free-body diagram for q_3

▲ **FIGURE 17.12** Our sketches for this problem.

17.5 Electric Field and Electric Forces

When two electrically charged particles in empty space interact, how does each one "know" that the other is there? What goes on in the space between them to transmit the effect of each one to the other? We can begin to answer these questions, and at the same time reformulate Coulomb's law in a very useful way, by using the concept of *electric field*. To introduce this concept, let's look at the mutual repulsion of two positively charged objects A and B (Figure 17.14a).

ActivPhysics 11.9: Motion of a Charge in an Electric Field: Introduction
ActivPhysics 11.10: Motion in an Electric Field: Problems

Suppose B is a point charge q', and let $\vec{F}'$ be the force on B, as shown in the figure. One way to think about this force is as an "action-at-a-distance" force—that is, as a force that acts across empty space without needing any matter (such as a pushrod or a rope) to transmit it through the intervening space.

Now think of object A as having the effect of somehow modifying the properties of the space around it. We remove object B and label its former position as point P (Figure 17.14b). We say that the charged object A produces or causes an **electric field** at point P (and at all other points in the neighborhood). Then, when point charge B is placed at point P and is acted upon by the force $\vec{F}'$, we take the point of view that the force is exerted on B *by the electric field at P*. Because B would be acted upon by a force at *any* point in the neighborhood of A, the electric field exists at all points in the region around A. (We could also say that point charge B sets up an electric field, which in turn exerts a force on object A.)

To find out experimentally whether there is an electric field at a particular point, we place a charged object, which we call a **test charge,** at the point (Figure 17.14c). If we find that the test charge experiences a non-zero electric force, then there is an electric field at that point.

Force is a vector quantity, so electric field is also a vector quantity. (Note the use of boldface letters with arrows on top of them in the discussion that follows.) To define the *electric field* $\vec{E}$ at any point, we place a test charge q' at the point and measure the electric force $\vec{F}'$ on it (Figure 17.14c). We define $\vec{E}$ at this point to be equal to $\vec{F}'$ divided by q':

Definition of electric field

When a charged particle with charge q' at a point P is acted upon by an electric force $\vec{F}'$, the electric field $\vec{E}$ at that point is defined as

$$\vec{E} = \frac{\vec{F}'}{q'}. \qquad (17.2)$$

The test charge q' can be either positive or negative. If it is positive, the directions of $\vec{E}$ and $\vec{F}'$ are the same; if it is *negative,* they are opposite (Figure 17.15).

Unit: In SI units, in which the unit of force is the newton and the unit of charge is the coulomb, the unit of electric-field magnitude is 1 newton per coulomb $(1\ \text{N/C})$.

The force acting on the test charge q' varies from point to point, so the electric field is also different at different points. Be sure that you understand that $\vec{E}$ is not a single vector quantity, but an infinite set of vector quantities, one associated with each point in space. We call this situation a **vector field**—a vector quantity associated with every point in a region of space, different at different points. In general, each component of $\vec{E}$ at any point depends on (i.e., is a function of) all the coordinates of the point.

NOTE ▶ There's a slight difficulty with our definition of electric field: In Figure 17.14, the force exerted by the test charge q' on the charge distribution A may cause the distribution to shift around, especially if object A is a conductor, in which charge is free to move. So the electric field around A when q' is present may not be the same as when q' is absent. But if q' is very small, the redistribution of charge on object A is also very small. So we refine our definition of electric field by taking the limit of Equation 17.2 as the test charge q' becomes very small and its disturbing effect on the charge distribution becomes negligible:

$$\vec{E} = \lim_{q' \to 0} \frac{\vec{F}'}{q'}. \qquad (17.3) \blacktriangleleft$$

A and B exert electric forces on each other.

(a) A

q' $\vec{F}'$ B

$-\vec{F}'$

B is removed; point P marks its position.

$\bullet$ P

(b) A

A test charge placed at P is acted upon by a force $\vec{F}'$ due to the electric field $\vec{E}$ of charge A. $\vec{E}$ is the force per unit charge exerted on the test charge.

$\vec{E} = \vec{F}'/q'$

Test charge q'

(c) A

▲ **FIGURE 17.14** A charged object creates an electric field in the space around it.

▶ **FIGURE 17.15** The direction of the electric force on a positive and negative test charge relative to the direction of the electric field.

The force on a positive test charge points in the direction of the electric field.

The force on a negative test charge points opposite to the electric field.

If an electric field exists within a *conductor*, the field exerts a force on every charge in the conductor, causing the free charges to move. By definition, an *electrostatic* situation is a situation in which the charges *do not* move. We conclude that **in an electrostatic situation, the electric field at every point within the material of a conductor must be zero.** (In Section 17.9, we'll consider the special case of a conductor that has a central cavity.)

In general, the magnitude and direction of an electric field can vary from point to point. If, in a particular situation, the magnitude and direction of the field are *constant* throughout a certain region, we say that the field is *uniform* in that region.

Conceptual Analysis 17.5

A moving electron

A vacuum chamber contains a uniform electric field directed downward. If an electron is shot horizontally into this region, its acceleration is:

A. downward and constant.
B. upward and constant.
C. upward and changing.
D. downward and changing.

SOLUTION The electron has a negative charge, so it is acted upon by a force directed *opposite* to the electric field—that is, an upward force giving an upward acceleration. We're told that the electric field is *uniform*, meaning that its magnitude and direction are constant. Therefore, the force exerted on the electron by the electric field is constant ($\vec{F}' = \vec{E}/q'$). Because the force is constant, so is the acceleration (by Newton's second law). The correct answer is B.

EXAMPLE 17.5 ### Accelerating an electron

When the terminals of a 100 V battery are connected to two large parallel horizontal plates 1.0 cm apart, the resulting charges on the plates produce an electric field $\vec{E}$ in the region between the plates that is very nearly uniform and has magnitude $E = 1.0 \times 10^4$ N/C. Suppose the lower plate has positive charge, so that the electric field is vertically upward, as shown in Figure 17.16. (The thin pink arrows represent the field.) If an electron is released from rest at the upper plate, what is its speed just before it reaches the lower plate? How much time is required for it to reach the lower plate? The mass of an electron is $m_e = 9.11 \times 10^{-31}$ kg.

The thin arrows represent the uniform electric field.

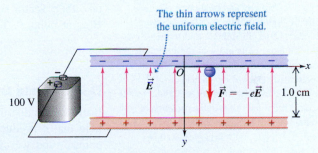

▲ **FIGURE 17.16**

SOLUTION

SET UP We place the origin of coordinates at the upper plate and take the $+y$ direction to be downward, toward the lower plate. The electron has negative charge, $q = -e$, so the direction of the force on the electron is downward, opposite to the electric field. The field is uniform, so the force on the electron is constant. Thus the electron has constant acceleration, and we can use the constant-acceleration equation $v_y^2 = v_{0y}^2 + 2a_y y$. The electron's initial velocity v_{0y} is zero, so

$$v_y^2 = 2a_y y.$$

SOLVE The force on the electron has only a y component, which is positive, and we can solve Equation 17.2 to find this component:

$$F_y = |q|E = (1.60 \times 10^{-19} \text{ C})(1.0 \times 10^4 \text{ N/C})$$
$$= 1.60 \times 10^{-15} \text{ N}.$$

Newton's second law then gives the electron's acceleration:

$$a_y = \frac{F_y}{m_e} = \frac{1.60 \times 10^{-15} \text{ N}}{9.11 \times 10^{-31} \text{ kg}} = +1.76 \times 10^{15} \text{ m/s}^2.$$

We want to find v_y when $y = 0.010$ m. The equation for v_y gives

$$v_y = \sqrt{2a_y y} = \sqrt{2(1.76 \times 10^{15} \text{ m/s}^2)(0.010 \text{ m})}$$
$$= 5.9 \times 10^6 \text{ m/s}.$$

Finally, $v_y = v_{0y} + a_y t$ gives the total travel time t:

$$t = \frac{v_y - v_{0y}}{a_y} = \frac{5.9 \times 10^6 \text{ m/s} - 0}{1.76 \times 10^{15} \text{ m/s}^2} = 3.4 \times 10^{-9} \text{ s}.$$

REFLECT The acceleration produced by the electric field is enormous; to give a 1000 kg car this acceleration, we would need a force of about 2×10^{18} N, or about 2×10^{14} tons. The effect of

Continued

gravity is completely negligible. The electron's final speed is only 2% of the speed of light, so we don't have to include relativistic effects. Note again that negative charges gain speed when they move in a direction opposite to the direction of the electric field.

Practice Problem: In this example, suppose a proton $(m_p = 1.67 \times 10^{-27} \text{ kg})$ is released from rest at the positive plate. What is its speed just before it reaches the negative plate? *Answer:* 1.38×10^5 m/s.

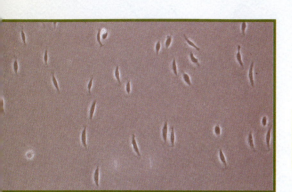

▲ **BIO** Application **They got their electrical marching orders.** Many cells, including nerve cells and skin cells, are remarkably sensitive to electrical fields. The photograph shows cultured skin cells of the zebrafish (an important experimental animal for biology and medicine). These cells are highly mobile in culture, moving at speeds of 10 μm/minute. Left to their own devices, these cells move at random, independently of each other; however, when exposed to a modest electrical field of 100 N/C, they align their long axes perpendicular to the field lines and move in the direction of the field. These cells respond to fields as small as 7 N/C, which is well within the range of electrical fields that have been measured near skin wounds in vertebrates. It may be that the wound healing response is controlled in part by natural electrical fields.

17.6 Calculating Electric Fields

In this section, we'll discuss several situations in which electric fields produced by specific charge distributions can be determined with fairly simple calculations. The key to these calculations is the principle of superposition, which we mentioned in Section 17.4 Restated in terms of electric fields, the principle is as follows:

Principle of superposition
The total electric field at any point due to two or more charges is the vector sum of the fields that would be produced at that point by the individual charges.

To find the field caused by several charges or an extended distribution of charge, we imagine the source to be made up of many point charges. We call the location of one of these points a **source point** (denoted by S, possibly with a subscript), and the point where we want to find the field is called the **field point** (denoted by P). We calculate the fields $\vec{E}_1$, $\vec{E}_2$, $\vec{E}_3$, $\cdots$ at point P caused by the individual point charges q_1, q_2, q_3, $\cdots$ located at points S_1, S_2, S_3, $\cdots$ and take their vector sum (using the superposition principle) to find the total field $\vec{E}_{\text{total}}$ at point P. That is,

$$\vec{E}_{\text{total}} = \vec{E}_1 + \vec{E}_2 + \vec{E}_3 + \cdots.$$

Electric Field Due to a Point Charge

If the source distribution is a single point charge q, it is easy to find the electric field that it produces. As before, we call the location of the charge the source point S, and we call the point P where we are determining the field the field point. If we place a small test charge q' at the field point P, at a distance r from the source point, the magnitude of the force $\vec{F}'$ is given by Coulomb's law, Equation 17.1:

$$F' = k\frac{|qq'|}{r^2}.$$

From Equation 17.3, we find the magnitude E of the electric field at P:

Electric field due to a point charge
The magnitude E of the electric field $\vec{E}$ at point P due to a point charge q at point S, a distance r from P, is given by

$$E = k\frac{|q|}{r^2}. \tag{17.4}$$

By definition, the electric field produced by a positive point charge always points *away from* it, but the electric field produced by a negative point charge points *toward* it.

Magnitude and direction

A small object S with a charge of magnitude q creates an electric field. At a point P located 0.36 m west of S, the field has a value of 40 N/C directed to the west. At a point 0.36 m east of S, the field is

A. 40 N/C, directed westward.
B. 40 N/C, directed eastward.
C. 80 N/C, directed westward.
D. 80 N/C, directed eastward.

SOLUTION As we just saw, the electric field of a positive point charge is directed radially away from the charge; that of a negative point charge is directed radially toward the charge. The fact that the field at P is directed to the west (away from S) means that S has a positive charge. Thus, at a point east of S, the field will point east. Equation 17.4, $E = k(|q|/r^2)$, tells us that the field has the same magnitude at all points that are the same radial distance r from S. Therefore, the correct answer is B.

Spherical Charge Distributions

In applications of electrostatics, we often encounter charge distributions that have spherical symmetry. Familiar examples include electric charge distributed uniformly over the surface of a conducting sphere and charge distributed uniformly throughout the volume of an insulating sphere. It turns out that the electric field produced by *any* spherically symmetric charge distribution, at all points outside this distribution, is the same as though all the charge were concentrated at a point at the center of the sphere. In field calculations, the field outside any spherical charge distribution can be obtained by replacing the distribution with a single point charge at the center of the sphere and equal to the total charge of the sphere.

EXAMPLE 17.6   **Electric field in a hydrogen atom**

(a) In the Bohr model of the hydrogen atom (described in Example 17.2), when the atom is in its lowest-energy state, the distance from the proton to the electron is 5.29×10^{-11} m. Find the electric field due to the proton at this distance. **(b)** A device called a Van de Graaff generator (a staple in science museums) can build up a large static charge on a metal sphere. Suppose the sphere of a Van de Graaff generator has a radius of 0.50 m and a net charge of 1.0 μC. What is the magnitude of the electric field 1.0 m from the center of the sphere? Compare this electric field with the field calculated in part (a).

SOLUTION

SET UP Figure 17.17 shows our diagrams for these cases.

SOLVE Part (a): We are asked to calculate the electric-field magnitude E at a distance of 5.29×10^{-11} m from a point charge (the proton). We use Equation 17.4; a proton has charge $q = +e = 1.60 \times 10^{-19}$ C, so

$$E = k\frac{|q|}{r^2}$$

$$= (8.99 \times 10^9 \text{ N} \cdot \text{m}^2/\text{C}^2)\frac{(1.60 \times 10^{-19} \text{ C})}{(5.29 \times 10^{-11} \text{ m})^2}$$

$$= 5.14 \times 10^{11} \text{ N/C}.$$

Part (b) To calculate the field of the van de Graaff sphere, we use the principle discussed above: a uniform spherical distribution of charge creates the same field as an equal point charge located at the center of the sphere. Thus, we can again use Equation 17.4:

$$E = k\frac{|q|}{r^2} = (8.99 \times 10^9 \text{ N} \cdot \text{m}^2/\text{C}^2)\frac{(1.0 \times 10^{-6} \text{ C})}{(1.0 \text{ m})^2}$$

$$= 9.0 \times 10^3 \text{ N/C}.$$

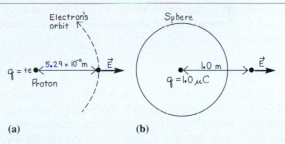

▲ FIGURE 17.17 Our sketches for this problem.

The electric field in part (a) is larger than that in part (b) by a factor of 5.7×10^7.

REFLECT The electric field in an atom is extremely large compared with the electric fields of macroscopic objects with easily obtainable electric charges.

Practice Problem: At what distance from a proton does the electric field of the proton have magnitude 9.0×10^3 N/C? How does this distance compare with the Bohr orbit radius ($r = 5.29 \times 10^{-11}$ m) of the electron in the lowest-energy state of the hydrogen atom? *Answers:* 4.0×10^{-7} m; 7.6×10^3 times larger.

EXAMPLE 17.7 Electric field of an electric dipole

Point charges q_1 and q_2 of $+12$ nC and -12 nC, respectively, are placed 10.0 cm apart (Figure 17.18). This combination of two charges with equal magnitude and opposite sign is called an **electric dipole.** Compute the resultant electric field (magnitude and direction) at **(a)** point a, midway between the charges, and **(b)** point b, 4.0 cm to the left of q_1. **(c)** What is the direction of the resultant electric field produced by these two charges at points along the perpendicular bisector of the line connecting the charges? Consider points both above and below the line connecting the charges.

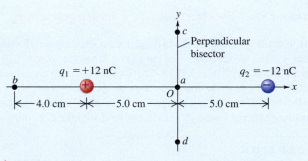

▲ **FIGURE 17.18**

SOLUTION

SET UP We use a coordinate system with the origin midway between the two charges and with the $+x$ axis toward q_2, as shown in Figure 17.18. The perpendicular bisector then lies along the y axis. We use $\vec{E}_1$ and $\vec{E}_2$ to denote the electric fields due to q_1 and q_2, respectively; the resultant electric field is the vector sum of these fields. The point charges are the source points, and points a, b, c, and d are the field points.

SOLVE For a point charge, the magnitude of the electric field is given by $E = k\dfrac{|q|}{r^2}$.

Part (a): The electric fields at point a are shown in Figure 17.19. $\vec{E}_1$ points away from q_1 (because q_1 is positive), and $\vec{E}_2$ points toward q_2 (because q_2 is negative). Thus,

$$E_1 = E_2 = k\frac{|q_1|}{r_1^2} = (8.99 \times 10^9 \,\text{N} \cdot \text{m}^2/\text{C}^2)\frac{(12 \times 10^{-9}\,\text{C})}{(0.050\,\text{m})^2}$$

$$= 4.32 \times 10^4\,\text{N/C}.$$

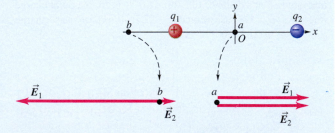

▲ **FIGURE 17.19** The electric fields due to the two charges at points a and b.

Since $\vec{E}_1$ and $\vec{E}_2$ point in the same direction, $E_{\text{total}} = E_1 + E_2 = 8.6 \times 10^4\,\text{N/C}$ and E_{total} points in the $+x$ direction, from the positive charge toward the negative charge.

Continued

Part (b): The electric fields at point b are shown in Figure 17.19. Again, $\vec{E}_1$ points away from q_1 and $\vec{E}_2$ points toward q_2. Hence,

$$E_1 = k\frac{|q_1|}{r_1^2} = (8.99 \times 10^9 \text{ N} \cdot \text{m}^2/\text{C}^2)\frac{(12 \times 10^{-9} \text{ C})}{(0.040 \text{ m})^2}$$
$$= 6.74 \times 10^4 \text{ N/C,}$$

$$E_2 = k\frac{|q_2|}{r_2^2} = (8.99 \times 10^9 \text{ N} \cdot \text{m}^2/\text{C}^2)\frac{(12 \times 10^{-9} \text{ C})}{(0.140 \text{ m})^2}$$
$$= 5.50 \times 10^3 \text{ N/C.}$$

E_1 is larger than E_2 because point b is closer to q_1 than to q_2.

Since $\vec{E}_1$ and $\vec{E}_2$ point in opposite directions, $E_{\text{total}} = E_1 - E_2 = 6.2 \times 10^4 \text{ N/C}$. $\vec{E}_{\text{total}}$ points to the left, in the direction of the stronger field.

Part (c): At point c in Figure 17.18, the two electric fields are directed as shown in Figure 17.20a. In Figure 17.20b, each electric field is replaced by its x and y components. Point c is equidistant from the two charges, and $|q_1| = |q_2|$, so $E_1 = E_2$. The y components of $\vec{E}_1$ and $\vec{E}_2$ are equal in magnitude and opposite in direction, and their sum is zero. The x components are equal in magnitude and are both in the $+x$ direction, so the resultant field is in the $+x$ direction.

At point d in Figure 17.18, the two electric fields are directed as shown in Figure 17.20c. The resultant field is again in the $+x$ direction. At all points along the perpendicular bisector of the line connecting the two charges, the resultant field is in the $+x$ direction, parallel to the direction from the positive charge toward the negative charge.

REFLECT Our general result in part (c) is consistent with the direction of the electric field calculated at point a. The resultant electric field has the same direction at every point along the perpendicular bisector, but the magnitude decreases at points farther from the charges.

Practice Problem: Repeat the calculations of this example, using the same value of q_1 as previously, but with $q_2 = +12$ nC (so that both charges are positive). *Answers:* (a) 0; (b) 7.3×10^4 N/C, in the $-x$ direction; (c) along the y axis and away from the charges.

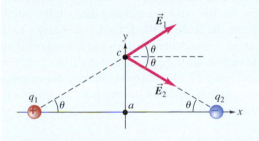

(a) The electric fields at point c

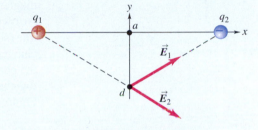

(b) The electric fields at point c and their components

(c) The electric fields at point d

▲ **FIGURE 17.20**

17.7 Electric Field Lines

The concept of an electric field may seem rather abstract; you can't see or feel one (although some animals can). It's often useful to draw a diagram that helps to visualize electric fields at various points in space. A central element in such a diagram is the concept of **electric field lines.** An electric field line is an imaginary line drawn through a region of space so that, at every point, it is tangent to the direction of the electric field vector at that point. The basic idea is shown in Figure 17.21. Michael Faraday (1791–1867) first introduced the concept of field lines. He called them "lines of force," but the term "field lines" is preferable.

Electric field lines show the direction of $\vec{E}$ at each point, and their spacing gives a general idea of the *magnitude* of $\vec{E}$ at each point. Where $\vec{E}$ is strong, we draw lines bunched closely together; where $\vec{E}$ is weaker, they are farther apart. At any particular point, the electric field has a unique direction, so only one field line can pass through each point of the field. In other words, *field lines never intersect.*

Electric field lines always have these characteristics:

1. At every point in space, the electric field vector $\vec{E}$ at that point is tangent to the electric field line through that point.
2. Electric field lines are close together in regions where the magnitude of $\vec{E}$ is large, farther apart where it is small.
3. Field lines point away from positive charges and toward negative charges.

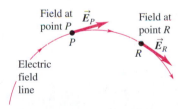

▲ **FIGURE 17.21** The direction of the electric field at any point is tangent to the field line through that point.

MasteringPHYSICS

PhET: Charges and Fields
PhET: Electric Field Hockey
PhET: Electric Field of Dreams
ActivPhysics 11.4: Electric Field: Point Charge
ActivPhysics 11.5: Electric Field Due to a Dipole
ActivPhysics 11.6: Electric Field: Problems

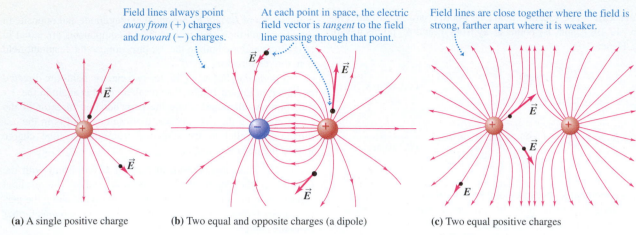

Field lines always point *away from* (+) charges and *toward* (−) charges.

At each point in space, the electric field vector is *tangent* to the field line passing through that point.

Field lines are close together where the field is strong, farther apart where it is weaker.

(a) A single positive charge   **(b)** Two equal and opposite charges (a dipole)   **(c)** Two equal positive charges

▲ **FIGURE 17.22** Electric field lines for several charge distributions.

Figure 17.22 shows some of the electric field lines in a plane containing (a) a single positive charge, (b) two equal charges, one positive and one negative (a dipole), and (c) two equal positive charges. These are cross sections of the actual three-dimensional patterns. The direction of the total electric field at every point in each diagram is along the tangent to the electric field line passing through the point. Arrowheads on the field lines indicate the sense of the $\vec{E}$ field vector along each line (showing that the field points away from positive charges and toward negative charges). In regions where the field magnitude is large, such as the space between the positive and negative charges in Figure 17.22b, the field lines are drawn close together. In regions where it is small, such as between the two positive charges in Figure 17.22c, the lines are widely separated.

NOTE ▶ There may be a temptation to think that when a charged particle moves in an electric field, its path always follows a field line. Resist that temptation; the thought is erroneous. The direction of a field line at a given point determines the direction of the particle's *acceleration,* not its velocity. We've seen several examples of motion in which the velocity and acceleration vectors have different directions. ◀

Parallel-Plate Capacitor

In a *uniform* electric field, the field lines are straight, parallel, and uniformly spaced, as in Figure 17.23. When two conducting sheets carry opposite charges and are close together compared with their size, the electric field in the region between them is approximately uniform. This arrangement is often used when a uniform electric field is needed, as in setups to deflect electron beams. A similar configuration of conductors, consisting of two sheets separated by a thin insulating layer, forms a device called a **parallel-plate capacitor,** which is widely used in electronic circuits and which we'll study in the next chapter.

Between the plates of the capacitor, the electric field is nearly uniform, pointing from the positive plate toward the negative one.

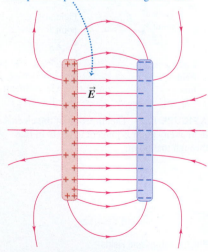

▲ **FIGURE 17.23** The electric field produced by a parallel-plate capacitor (seen in cross section). Between the plates, the field is nearly uniform.

17.8 Gauss's Law and Field Calculations

Gauss's law is an alternative formulation of the principles of electrostatics. It is logically equivalent to Coulomb's law, but for some problems it provides a useful alternative approach to calculating electric fields. Coulomb's law enables us to find the field at a *point P* caused by a single *point* charge q. To calculate fields produced by an *extended* charge distribution, we have to represent that distribution as an assembly of point charges and use the superposition

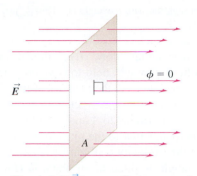

Electric field $\vec{E}$ is perpendicular to area A; the angle between $\vec{E}$ and a line perpendicular to the surface is zero. The flux is $\Phi_E = EA$.

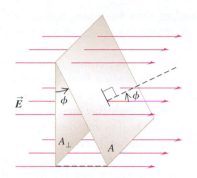

Area A is tilted at an angle ϕ from the perpendicular to $\vec{E}$. The flux is $\Phi_E = EA \cos \phi$.

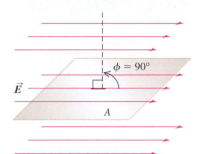

Area A is parallel to $\vec{E}$ (tilted at 90° from the perpendicular to $\vec{E}$). The flux is $\Phi_E = EA \cos 90° = 0$.

▲ **FIGURE 17.24** The electric flux through a flat surface at various orientations relative to a uniform electric field.

principle. Gauss's law takes a more global view. Given any general distribution of charge, we surround it with an imaginary closed surface (often called a *Gaussian surface*) that encloses the charge. Then we look at the electric field at various points on this imaginary surface. Gauss's law is a relation between the field at *all* the points on the surface and the total charge enclosed within the surface.

Gauss's law is part of the key to using symmetry considerations in electric-field calculations. Calculations with a system that has symmetry properties can nearly always be simplified if we can make use of the symmetry, and Gauss's law helps us do just that.

Electric Flux

In formulating Gauss's law, we'll use the concept of **electric flux,** also called *flux of the electric field.* We'll define this concept first, and then we'll discuss an analogy with fluid flow that will help you to develop intuition about it.

The definition of electric flux involves an area A and the electric field at various points in the area. The area needn't be the surface of a real object; in fact, it will usually be an imaginary area in space. Consider first a small, flat area A perpendicular to a uniform electric field $\vec{E}$ (Figure 17.24a). We denote electric flux by Φ_E; we define the electric flux Φ_E through the area A to be the product of the magnitude E of the electric field and the area A:

$$\Phi_E = EA.$$

Roughly speaking, we can picture Φ_E in terms of the number of field lines that pass through A. More area means more lines through the area, and a stronger field means more closely spaced lines and therefore more lines per unit area.

If the area element A isn't perpendicular to the field $\vec{E}$, then fewer field lines pass through it. In this case, what counts is the area of the silhouette of A that we see as we look along the direction of $\vec{E}$; this is the area $A_\perp$ in Figure 17.24b, the *projection* of the area A onto a surface perpendicular to $\vec{E}$. Two sides of the projected rectangle have the same length as the original one, but the other two are foreshortened by a factor $\cos \phi$; so the projected area $A_\perp$ is equal to $A \cos \phi$. We generalize our definition of electric flux for a uniform electric field to

$$\Phi_E = EA \cos \phi. \tag{17.5}$$

Thus, $E \cos \phi$ is the component of the vector $\vec{E}$ perpendicular to the area. Calling this component $E_\perp$, we can rewrite Equation 17.5 as

$$\Phi_E = E_\perp A. \tag{17.6}$$

▲ **BIO Application Feeling my way.** These African elephant-nose fish "feel" their way through their murky freshwater environment by producing an electric field and sensing how objects distort the field. The field is produced in pulses by an electric organ near each fish's tail; it is detected by receptors covering portions of the fish's skin. An object that conducts electricity better than fresh water, such as an animal or a plant, causes the nearby field lines to bunch together, creating a spot of stronger field on the fish's skin. An object that conducts less well than water, such as a rock, causes the field lines to spread apart, which the fish sense as a spot of weaker field. By integrating the information from their receptors, the fish can perceive their surroundings. Several groups of fish generate and use electric fields in this way.

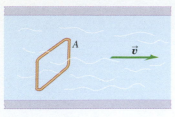

(a)

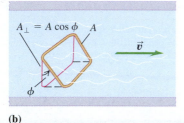

(b)

◀**FIGURE 17.25** The volume flow rate of water through the wire rectangle is $vA \cos\phi$, just as the electric flux through an area A is $EA \cos\phi$.

The word *flux* comes from a Latin word meaning "flow." Even though an electric field is *not* a flow, an analogy with fluid flow will help to develop your intuition about electric flux. Imagine that $\vec{E}$ is analogous to the velocity of flow $\vec{v}$ of water through the imaginary area bounded by the wire rectangle in Figure 17.25. The flow rate through the area A is proportional to A and v, and it also depends on the angle between $\vec{v}$ and a line perpendicular to the plane of the rectangle. When the area is perpendicular to the flow velocity $\vec{v}$ (Figure 17.25a), the volume flow rate is just vA. When the rectangle is tilted at an angle ϕ (Figure 17.25b), the area that counts is the silhouette area that we see when looking in the direction of $\vec{v}$. That area is $A \cos\phi$, as shown, and the volume flow rate through A is $vA \cos\phi$. This quantity is called the *flux* of $\vec{v}$ through the area A; *flux* is a natural term because it represents the volume rate of flow of fluid through the area. In the electric-field situation, *nothing is flowing,* but the analogy to fluid flow may help you to visualize the concept.

EXAMPLE 17.8 Electric flux through a disk

A disk with radius 0.10 m is oriented with its axis (the line through the center, perpendicular to the disk's surface) at an angle of 30° to a uniform electric field $\vec{E}$ with magnitude 2.0×10^3 N/C (Figure 17.26). **(a)** What is the total electric flux through the disk? **(b)** What is the total flux through the disk if it is turned so that its plane is parallel to $\vec{E}$? **(c)** What is the total flux through the disk if it is turned so that its axis (marked by the dashed line perpendicular to the disk in the figure) is parallel to $\vec{E}$?

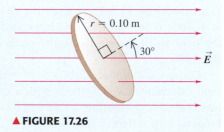

▲ **FIGURE 17.26**

SOLUTION

SET UP AND SOLVE Part (a): The area is $A = \pi(0.10\text{ m})^2 = 0.0314\text{ m}^2$. From Equation 17.5,

$$\Phi_E = EA \cos\phi = (2.0 \times 10^3\text{ N/C})(0.0314\text{ m}^2)(\cos 30°)$$
$$= 54\text{ N} \cdot \text{m}^2/\text{C}.$$

Part (b): The axis of the disk is now perpendicular to $\vec{E}$, so $\phi = 90°$, $\cos\phi = 0$, and $\Phi_E = 0$.

Part (c): The axis of the disk is parallel to $\vec{E}$, so $\phi = 0$, $\cos\phi = 1$, and, from Equation 17.5,

$$\Phi_E = EA \cos\phi = (2.0 \times 10^3\text{ N/C})(0.0314\text{ m}^2)(1)$$
$$= 63\text{ N} \cdot \text{m}^2/\text{C}.$$

REFLECT The flux through the disk is greatest when its axis is parallel to $\vec{E}$, and it is zero when $\vec{E}$ lies in the plane of the disk. That is, it is greatest when the most electric field lines pass through the disk, and it is zero when no lines pass through it.

Practice Problem: What is the flux through the disk if its axis makes an angle of 45° with $\vec{E}$? *Answer:* 44 N · m²/C.

EXAMPLE 17.9 Electric flux through a sphere

A positive point charge with magnitude 3.0 μC is placed at the center of a sphere with radius 0.20 m (Figure 17.27). Find the electric flux through the sphere due to this charge.

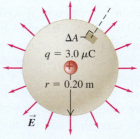

▶ **FIGURE 17.27**

Continued

SOLUTION

SET UP AND SOLVE At any point on the sphere, the magnitude of $\vec{E}$ is

$$E = \frac{kq}{r^2} = \frac{(8.99 \times 10^9\,\text{N} \cdot \text{m}^2/\text{C}^2)(3.0 \times 10^{-6}\,\text{C})}{(0.20\,\text{m})^2}$$
$$= 6.75 \times 10^5\,\text{N/C}.$$

From symmetry, the field is perpendicular to the spherical surface at every point (so that $E_\perp = E$), and it has the same magnitude at every point. The flux through any area element ΔA on the sphere is just $E\,\Delta A$, and the flux through the entire surface is E times the total surface area $A = 4\pi r^2$. Thus, the total flux coming out of the sphere is

$$\Phi_E = EA = (6.75 \times 10^5\,\text{N/C})(4\pi)(0.20\,\text{m})^2$$
$$= 3.4 \times 10^5\,\text{N} \cdot \text{m}^2/\text{C}.$$

REFLECT The symmetry of the sphere plays an essential role in this calculation. We made use of the facts that E has the same value at every point on the surface and that at every point $\vec{E}$ is perpendicular to the surface.

Practice Problem: Repeat this calculation for the same charge, but a radius of 0.10 m. You should find that the result is the same as the one you obtained previously. We would have obtained the same result with a sphere of radius 2.0 m or 200 m. There's a good physical reason for this, as we'll soon see. *Answer:* $3.4 \times 10^5\,\text{N} \cdot \text{m}^2/\text{C}$

Gauss's Law

Gauss's law is an alternative to Coulomb's law for expressing the relationship between electric charge and electric field. It was formulated by Karl Friedrich Gauss (1777–1855), one of the greatest mathematicians of all time. Many areas of mathematics, from number theory and geometry to the theory of differential equations, bear the mark of his influence, and he made equally significant contributions to theoretical physics.

> **Gauss's law**
>
> The total electric flux Φ_E coming out of any closed surface (that is, a surface enclosing a definite volume) is proportional to the total (net) electric charge Q_{encl} inside the surface, according to the relation
>
> $$\sum E_\perp \Delta A = 4\pi k Q_{\text{encl}}. \tag{17.7}$$
>
> The sum on the left side of this equation represents the operations of dividing the enclosing surface into small elements of area ΔA, computing $E_\perp \Delta A$ for each one, and adding all these products.

To develop Gauss's law, we'll start with the field due to a single positive point charge q. The field lines radiate out equally in all directions. We place this charge at the center of an imaginary spherical surface with radius R. The magnitude E of the electric field at every point on the surface is given by

$$E = k\frac{q}{R^2}.$$

At each point on the surface, $\vec{E}$ is perpendicular to the surface, and its magnitude is the same at every point, just as in Example 17.9. The total electric flux is just the product of the field magnitude E and the total area $A = 4\pi R^2$ of the sphere:

$$\Phi_E = EA = k\frac{q}{R^2}(4\pi R^2) = 4\pi k q. \qquad \text{(spherical surface)} \tag{17.8}$$

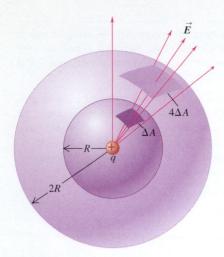

▲ FIGURE 17.28 Projection of an element of area ΔA, on a spherical surface of radius R, onto a sphere of radius $2R$. The projection multiplies each linear dimension by two, so the area element on the larger sphere is $4\Delta A$. The same number of field lines and the same flux pass through each area element.

We see that *the flux is independent of the radius R of the sphere.* It depends only on the charge q enclosed by the sphere.

We can also interpret this result in terms of field lines. We consider two spheres with radii R and $2R$, respectively (Figure 17.28). According to Coulomb's law, the field magnitude is one-fourth as great on the larger sphere as on the smaller, so the number of field lines per unit area should be one-fourth as great. But the area of the larger sphere is four times as great, so the *total* number of field lines passing through is the same for both spheres.

We've derived Equation 17.8 only for spherical surfaces, but we can generalize it to *any* closed surface surrounding an electric charge. We imagine the surface as being divided into small elements of area ΔA. If the electric field $\vec{E}$ is perpendicular to a particular element of area, then the number of field lines passing through that area is proportional to $E\,\Delta A$—that is, to the flux through A. If $\vec{E}$ is not perpendicular to the given element of area, we take the component of $\vec{E}$ perpendicular to ΔA; we call this component $E_\perp$, as before. Then the number of lines passing through ΔA is proportional to $E_\perp\,\Delta A$. (We don't consider the component of $\vec{E}$ *parallel* to the surface, because it doesn't correspond to any lines passing *through* the surface.)

To get the *total* number of field lines passing outward through the surface, we add up all the products $E_\perp\,\Delta A$ for all the surface elements that together make up the whole surface. This sum is the total flux through the entire surface. The total number of field lines passing through the surface is the same as that for the spherical surfaces we have discussed. Therefore, this sum is again equal to $4\pi kq$, just as in Equation 17.8, and our generalized relation is

$$\Phi_E = \Sigma E_\perp\,\Delta A = 4\pi kq. \qquad \text{(for any closed surface)} \qquad (17.9)$$

There is one further detail: We have to keep track of which lines point *into* the surface and which ones point *out*; we may have both types in some problems. Let's agree that $E_\perp$ and Φ_E are positive when the vector $\vec{E}$ has a component pointing *out of* the surface and negative when the component points *into* the surface.

Here's a further generalization: Suppose the surface encloses not just one point charge q, but several charges q_1, q_2, $q_3, \cdots$. Then the total (resultant) electric field $\vec{E}$ at any point is the vector sum of the $\vec{E}$ fields of the individual charges. Let $Q_{\text{encl}} = q_1 + q_2 + q_3 + \cdots$ be the *total* charge enclosed by the surface, and let $E_\perp$ be the component of the *total* field perpendicular to ΔA. Then the general statement of Gauss's law is

$$\Sigma E_\perp\,\Delta A = 4\pi k Q_{\text{encl}}. \qquad (17.10)$$

Gauss's law is usually written in terms of the constant ϵ_0 we introduced in Section 17.4, defined by the relation $k = 1/4\pi\epsilon_0$. In terms of ϵ_0,

$$\Sigma E_\perp\,\Delta A = \frac{Q_{\text{encl}}}{\epsilon_0}. \qquad \text{(for any closed surface)} \qquad (17.11)$$

In Equations 17.7, 17.10, and 17.11, Q_{encl} is always the algebraic sum of all the (positive and negative) charges enclosed by the surface, and $\vec{E}$ is the *total* field at each point on the surface. Also, note that this field is in general caused partly by charges inside the surface and partly by charges outside. The outside charges don't contribute to the total (net) flux through the surface, so Equation 17.11 is still correct even when there are additional charges outside the surface that contribute to the electric field at the surface. When $Q_{\text{encl}} = 0$, the total flux through the surface must be zero, even though some areas may have positive flux and others negative.

NOTE ▶ The surface used for applications of Gauss's law need not be a real physical surface; in fact, it is usually an imaginary surface, enclosing a definite volume and a definite quantity of electric charge. ◀

EXAMPLE 17.10 Field due to a spherical shell of charge

A positive charge q is spread uniformly over a thin spherical shell of radius R (Figure 17.29). Find the electric field at points inside and outside the shell.

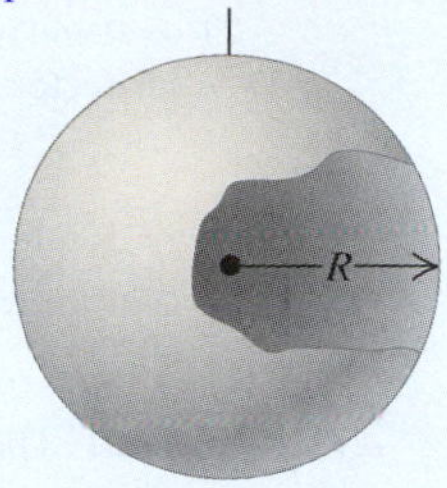

Thin spherical shell with total charge q

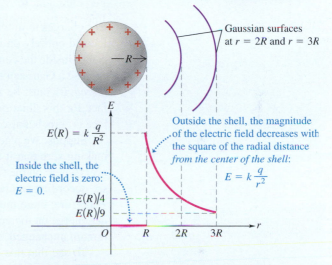

▶ **FIGURE 17.29**

SOLUTION

SET UP The system is spherically symmetric. This means that it is unchanged if we rotate it through any angle about an axis through its center. The field pattern of the rotated system must be identical to that of the original system. If the field had a component at some point that was perpendicular to the radial direction, that component would have to be different after at least some rotations. Thus, there can't be such a component, and the field must be radial.

We conclude that at every point outside the shell, the electric field due to the charge on the shell must be along a radial line—that is, along a line from the center of the shell to the field point. For the same reason, the magnitude E of the field depends only on the distance r from the center. Thus, the magnitude E is the same at all points on a spherical surface with radius r, concentric with the conductor.

SOLVE Because of the spherical symmetry, we take our Gaussian surface to be an imaginary sphere with radius r and concentric with the shell. We'll locate this surface first inside and then outside the shell of charge.

Inside the shell $(r < R)$: The Gaussian surface has area $4\pi r^2$. Since, by symmetry, the electric field is uniform over the Gaussian sphere and perpendicular to it at each point, the electric flux is $\Phi_E = EA = E(4\pi r^2)$. The Gaussian surface is inside the shell and encloses none of the charge on the shell, so $Q_{encl} = 0$.

Gauss's law $\Phi_E = Q_{encl}/\epsilon_0$ then says that $\Phi_E = E(4\pi r^2) = 0$, so $E = 0$. The electric field is zero at all points inside the shell.

Outside the shell $(r > R)$: Again, $\Phi_E = E(4\pi r^2)$. But now all of the shell is inside the Gaussian surface, so $Q_{encl} = q$. Gauss's law $\Phi_E = Q_{encl}/\epsilon_0$ then gives $E(4\pi r^2) = q/\epsilon_0$, and it follows that

$$E = \frac{q}{4\pi\epsilon_0 r^2} = k\frac{q}{r^2}.$$

▲ **FIGURE 17.30** The electric field of a charged spherical shell as a function of distance from the center. Outside the sphere, the field is the same as though the sphere's charge were all located at the center of the sphere.

REFLECT Figure 17.30 shows a graph of the field magnitude E as a function of r. The electric field is zero at all points inside the shell. At points outside the shell, the field drops off as $1/r^2$. Note that the magnitude of the electric field due to a point charge q is $E = kq/r^2$, so at points outside the shell the field is the same as if all the charge were concentrated at the center of the shell.

Practice Problem: What total charge q must be distributed uniformly over a spherical shell of radius $R = 0.50$ m to produce an electric field with magnitude 680 N/C at a point just outside the surface of the shell? *Answer:* ±19 nC.

17.9 Charges on Conductors

Early in our discussion of electric fields, we made the point that in an electrostatic situation (where there is no net motion of charge) the electric field at every point within a conductor is zero. (If it were not, the field would cause the conductor's

The charge q' is distributed over the surface of the conductor. The situation is electrostatic, so $\vec{E} = 0$ within the conductor.

q'

$\vec{E} = 0$ within conductor

(a) Solid conductor with charge q'

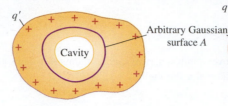

Because $\vec{E} = 0$ at all points within the conductor, the electric field at all points on the Gaussian surface must be zero.

q'

Cavity

Arbitrary Gaussian surface A

(b) The same conductor with an internal cavity

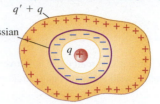

For $\vec{E}$ to be zero at all points on the Gaussian surface, the surface of the cavity must have a total charge $-q$.

$q' + q$

q

(c) An isolated charge q is placed in the cavity

▲ **FIGURE 17.31** The charge on a solid conductor, on a conductor with a cavity, and on a conductor with a cavity that contains a charge.

free charges to move.) We've also learned that the charge on a solid conductor is located entirely on its surface, as shown in Figure 17.31a. But what if there is a cavity inside the conductor (Figure 17.31b)? If there is no charge in the cavity, we can use a Gaussian surface such as A to show that the net charge on the surface *of the cavity* must be zero because $\vec{E} = 0$ everywhere on the Gaussian surface. In fact, for this situation, we can prove not only that the *total* charge on the cavity surface is zero, but also that there can't be any charge *anywhere* on the cavity surface.

Suppose we place a small object with a charge q inside a cavity in a conductor, making sure that it does not touch the conductor (Figure 17.31c). Again, $\vec{E} = 0$ everywhere on the Gaussian surface A (because the situation is still electrostatic), so, according to Gauss's law, the *total* charge inside this surface must be zero. Therefore, there must be a total charge $-q$ on the cavity surface. Of course, the *net* charge on the conductor (counting both the inner and the outer surface) must remain unchanged, so a charge $+q$ must appear on its outer surface.

To see that this charge must be on the outer surface and not in the material, imagine first shrinking surface A so that it's just barely bigger than the cavity. The field everywhere on A is still zero, so, according to Gauss's law, the total charge inside A is zero. Now let surface A expand until it is just inside the outer surface of the conductor. The field is still zero everywhere on surface A, so the total charge enclosed is still zero. We have not enclosed any additional charge by expanding surface A; therefore, there must be no charge in the interior of the material. We conclude that the charge $+q$ must appear on the outer surface. By the same reasoning, if the conductor originally had a charge q', then the total charge on the outer surface after the charge q is inserted into the cavity must be $q + q'$.

EXAMPLE 17.11 Location of net charge on conductors

A hollow conductor carries a net charge of $+7$ nC. In its cavity, insulated from the conductor, is a small, isolated object with a net charge of -5 nC. How much charge is on the outer surface of the hollow conductor? How much is on the wall of the cavity?

SOLUTION

SET UP Figure 17.32 shows our sketch. We know that in this electrostatic situation the electric field in the conducting material must be zero. We draw a Gaussian surface within the material of the conductor. and apply Gauss's law.

SOLVE We apply Gauss's law $\Phi_E = Q_{encl}/\epsilon_0$ to the Gaussian surface shown in Figure 17.32. The Gaussian surface lies within the conducting material, so $E = 0$ everywhere on that surface. By Gauss's law, $\Phi_E = Q_{encl}/\epsilon_0$. Thus, $\Phi_E = 0$, so $Q_{encl} = 0$. But then, in order to have $Q_{encl} = 0$, there must be a charge of $+5$ nC

Continued

on the inner surface of the cavity, to cancel the charge in the cavity. The conductor carries a total charge of $+7$ nC, and all of its net charge is on its surfaces. So, if there is $+5$ nC on the inner surface, the remaining $+2$ nC must be on the outer surface, as shown in our sketch.

REFLECT Field lines pass between the $+5$ nC on the inner surface of the cavity and the -5 nC on the object in the cavity. Each field line going to the -5 nC charge originated on the $+5$ nC charge; the field lines don't continue into the conducting material, since $E = 0$ there. There is an electric field outside the conductor, due to the $+2$ nC on its surface.

Practice Problem: Repeat this example for the case where the conductor has a net charge of $+3$ nC. *Answers:* inner surface: $+5$ nC; outer surface: -2 nC.

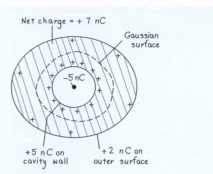

▲ **FIGURE 17.32** Our sketch for this problem.

Faraday Ice Pail

We can now consider a historic experiment, shown in Figure 17.33a. We mount a conducting container, such as a metal pail with a lid, on an insulating stand. The container is initially uncharged. Then we hang a charged metal ball from an insulating thread, lower it into the pail, and put the lid on (Figure 17.33b). Charges are induced on the walls of the container as shown. But now we let the ball *touch* the inner wall (Figure 17.33c). The surface of the ball becomes, in effect, part of the cavity surface. The situation is now the same as Figure 17.31b; if Gauss's law is correct, the net charge on this surface must be zero. Thus, the ball must lose all its charge. Finally, we pull the ball out, to find that it has indeed lost all its charge.

This experiment was performed by Michael Faraday, using a metal ice pail with a lid, and it is called **Faraday's ice-pail experiment.** (Similar experiments had been carried out earlier by Benjamin Franklin and Joseph Priestley, although with much less precision.) The result confirms the validity of Gauss's law and therefore of Coulomb's law. Faraday's result was significant because Coulomb's experimental method, using a torsion balance and dividing the charges, was not very precise. It is quite difficult to confirm the $1/r^2$ dependence of the electrostatic force with great precision by direct force measurements. Faraday's experiment tests the validity of Gauss's law, and therefore of Coulomb's law, with potentially much greater precision.

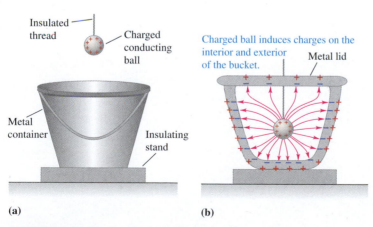

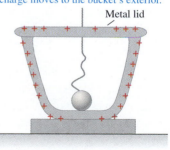

▲ **FIGURE 17.33** The Faraday ice-pail experiment.

The field induces charges on the left and right sides of the conducting box.

The total electric field inside the box is zero; the presence of the box distorts the field in adjacent regions.

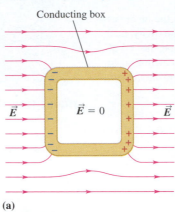

Conducting box

$\vec{E}$ $\vec{E} = 0$ $\vec{E}$

(a)

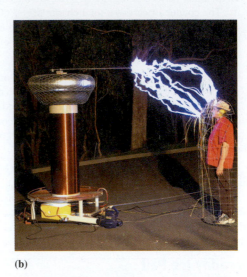

(b)

▲ **FIGURE 17.34** (a) The effect of putting a conducting box (an electrostatic shield) in a uniform electric field. (b) The conducting cage keeps the operator of this exhibit perfectly safe.

▲ **Application A Faraday cage when you need one.** If you find yourself in a thunderstorm while driving, *stay in your car*. If it gets hit by lightning, it will act as a Faraday cage and keep you safe.

Electrostatic Shielding

Suppose we have a highly sensitive electronic instrument that we want to protect from stray electric fields that might give erroneous measurements. We surround the instrument with a conducting box, or we line the walls, floor, and ceiling of the room with a conducting material such as sheet copper. The external electric field redistributes the free electrons in the conductor, leaving a net positive charge on the outer surface in some regions and a net negative charge in others (Figure 17.34). This charge distribution causes an additional electric field such that the *total* field at every point inside the box is zero, as Gauss's law says it must be. The charge distribution on the box also alters the shapes of the field lines near the box, as the figure shows. Such a setup is often called a *Faraday cage*.

SUMMARY

Electric Charge; Conductors and Insulators

(Sections 17.1–17.3) The fundamental entity in electrostatics is electric charge. There are two kinds of charge: positive and negative. Like charges repel each other; unlike charges attract. **Conductors** are materials that permit electric charge to move within them. **Insulators** permit charge to move much less readily. Most metals are good conductors; most nonmetals are insulators.

All ordinary matter is made of atoms consisting of protons, neutrons, and electrons. The protons and neutrons form the nucleus of the atom; the electrons surround the nucleus at distances much greater than its size. Electrical interactions are chiefly responsible for the structure of atoms, molecules, and solids.

Electric charge is conserved: It can be transferred between objects, but isolated charges cannot be created or destroyed. Electric charge is quantized: Every amount of observable charge is an integer multiple of the charge of an electron or proton.

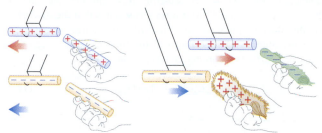

Like charges repel.   Unlike charges attract.

Coulomb's Law

(Section 17.4) **Coulomb's law** is the basic law of interaction for point electric charges. For point charges q_1 and q_2 separated by a distance r, the magnitude F of the force each charge exerts on the other is

$$F = k\frac{|q_1 q_2|}{r^2}. \qquad (17.1)$$

The force on each charge acts along the line joining the two charges. It is repulsive if q_1 and q_2 have the same sign, attractive if they have opposite signs. The forces form an action–reaction pair and obey Newton's third law.

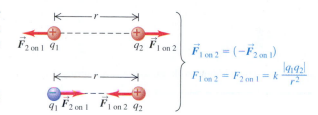

Electric Field and Electric Forces

(Sections 17.5 and 17.6) **Electric field,** a vector quantity, is the force per unit charge exerted on a test charge at any point, provided that the test charge is small enough that it does not disturb the charges that cause the field. The principle of superposition states that the electric field due to any combination of charges is the vector sum of the fields caused by the individual charges. From Coulomb's law, the magnitude of the electric field produced by a point charge is

$$E = k\frac{|q|}{r^2}. \qquad (17.4)$$

$\vec{E} = \vec{F}'/q'$

Test charge q'

Electric Field Lines

(Section 17.7) **Field lines** provide a graphical representation of electric fields. A field line at any point in space is tangent to the direction of $\vec{E}$ at that point, and the number of lines per unit area (perpendicular to their direction) is proportional to the magnitude of $\vec{E}$ at the point. Field lines point away from positive charges and toward negative charges.

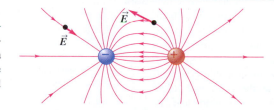

Continued

Gauss's Law

(Section 17.8) For a uniform electric field with component $E_\perp$ perpendicular to area A, the **electric flux** through the area is $\Phi_E = E_\perp A$ (Equation 17.6). **Gauss's law** states that the total electric flux Φ_E out of any closed surface (that is, a surface enclosing a definite volume) is proportional to the total electric charge Q_{encl} inside the surface, according to the relation

$$\sum E_\perp \, \Delta A = 4\pi k Q_{encl}. \tag{17.7}$$

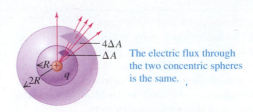

The electric flux through the two concentric spheres is the same.

Charges on Conductors

(Section 17.9) In a static configuration with no net motion of charge, the electric field is always zero within a conductor. The charge on a solid conductor is located entirely on its outer surface. If there is a cavity containing a charge $+q$ within the conductor, the surface of the cavity has a total induced charge $-q$.

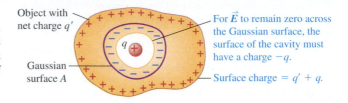

Object with net charge q'

For $\vec{E}$ to remain zero across the Gaussian surface, the surface of the cavity must have a charge $-q$.

Gaussian surface A

Surface charge $= q' + q$.

 For instructor-assigned homework, go to www.masteringphysics.com

Conceptual Questions

1. Bits of paper are attracted to an electrified comb or rod even though they have no net charge. How is this possible?

2. When you walk across a nylon rug and then touch a large metal object, you may get a spark and a shock. What causes this to happen?

3. What similarities does the electrical force have to the gravitational force? What are the most significant differences?

4. In a common physics demonstration, a rubber rod is first rubbed vigorously on silk or fur. It is then brought close to a small Styrofoam™ ball, which it attracts. If you then touch the ball with the rod, it suddenly repels the ball. Why does it first attract the ball, and why does it then repel the same ball?

5. How do we know that protons have positive charge and electrons have negative charge, rather than the reverse? Is there anything *inherently* positive about the proton's charge or inherently negative about the electron's charge?

6. Gasoline transport trucks sometimes have chains that hang down and drag on the road at the rear end of the truck. What are the chains for and how do they work?

7. A gold leaf electroscope, which is often used in physics demonstrations, consists of a metal tube with a metal ball at the top and a sheet of extremely thin gold leaf fastened at the other end. (See Fig. 17.35.) The gold leaf is attached in such a way that it can pivot about its upper edge. (a) If a charged rod is brought close to (but does not touch) the ball at the top, the gold leaf pivots outward, away from the

Metal tube and ball

Gold leaf

▲ **FIGURE 17.35** Question 7.

tube. Why? (b) What will the gold leaf do when the charged rod is removed? Why? (c) Suppose that the charged rod touches the metal ball for a second or so. What will the gold leaf do when the rod is removed in this case? Why?

8. Show how it is possible for *neutral* objects to attract each other electrically.

9. Suppose the disk in Example 17.8, instead of having its normal vector oriented at just two or three particular angles to the electric field, began to rotate continuously, so that its normal vector was first parallel to the field, then perpendicular to it, then opposite to it, and so on. Sketch a graph of the resulting electric flux versus time, for an entire rotation of 360°.

10. Atomic nuclei are made of protons and neutrons, a fact that, by itself, shows that there must be another kind of force in addition to the electrical and gravitational forces. Explain how we know this.

11. If an electric dipole is placed in a uniform electric field, what is the net force on it? Will the same thing necessarily be true if the field is *not* uniform?

12. *Why* do electric field lines point away from positive charges and toward negative charges?

13. A lightning rod is a pointed copper rod mounted on top of a building and welded to a heavy copper cable running down into the ground. Lightning rods are used in prairie country to protect houses and barns from lightning; the lightning current runs through the copper rather than through the building. Why does it do this?

14. A rubber balloon has a single point charge in its interior. Does the electric flux through the balloon depend on whether or not it is fully inflated? Explain your reasoning.

15. Explain how the electrical force plays an important role in understanding each of the following: (a) the friction force between two objects, (b) the hardness of steel, and (c) the bonding of amino acids to form proteins.

Multiple-Choice Problems

1. Just after two identical point charges are released when they are a distance D apart in outer space, they have an acceleration a. If you release them from a distance $D/2$ instead, their acceleration will be
   A. $a/4$.   B. $a/2$.   C. $2a$.   D. $4a$.

2. If the electric field is E at a distance d from a point charge, its magnitude will be $2E$ at a distance
   A. $d/4$.   B. $d/2$.   C. $d/\sqrt{2}$.
   D. $d\sqrt{2}$.   E. $2d$.

3. Two *unequal* point charges are separated as shown in Figure 17.36. The electric field due to this combination of charges can be zero
   A. only in region 1.
   B. only in region 2.
   C. only in region 3.
   D. in both regions 1 and 3.

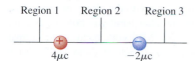

▲ **FIGURE 17.36** Multiple-choice problem 3.

4. Two protons close to each other are released from rest and are completely free to move. After being released (there may be more than one correct choice),
   A. their speeds gradually decrease to zero as they move apart.
   B. their speeds gradually increase as they move apart.
   C. their accelerations gradually decrease to zero as they move apart.
   D. their accelerations gradually increase as they move apart.

5. A spherical balloon contains a charge $+Q$ uniformly distributed over its surface. When it has a diameter D, the electric field at its surface has magnitude E. If the balloon is now blown up to twice this diameter without changing the charge, the electric field at its surface is
   A. $4E$.   B. $2E$.   C. E.
   D. $E/2$.   E. $E/4$.

6. Two microscopic bags each contain two protons. When they are separated by a distance d, the electrical force on each bag due to the other bag is F. You now transfer a proton from one bag to another without changing anything else. The electrical force on each bag is now
   A. F.   B. $\frac{3}{4}F$.   C. $\frac{1}{2}F$.   D. $\frac{1}{4}F$.

7. An electron is moving horizontally in a laboratory when a uniform electric field is suddenly turned on. This field points vertically downward. Which of the paths shown will the electron follow, assuming that gravity can be neglected?

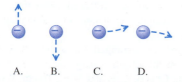

A.   B.   C.   D.

8. Point P in Figure 17.37 is equidistant from two point charges $\pm Q$ of equal magnitude. If a negative point charge is placed at P without moving the original charges, the net electrical force the charges $\pm Q$ will exert on it is
   A. directly upward.
   B. directly downward.
   C. zero.
   D. directly to the right.
   E. directly to the left.

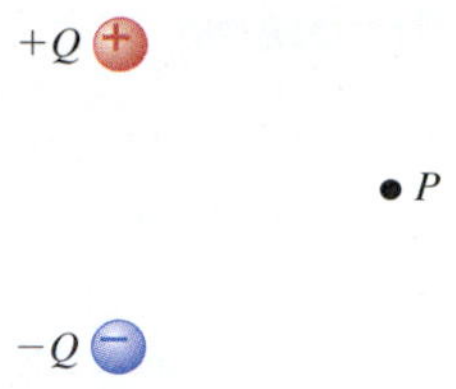

▲ **FIGURE 17.37** Multiple-choice problem 8.

9. A charge $+Q$ is suspended by a silk thread inside of a neutral metal box without touching the metal. What is true about the charge on the inner and outer surfaces of the box?
   A. The charge on both the inner and the outer surfaces is zero.
   B. The charge is $-Q$ on the inner surface and $+Q$ on the outer surface.
   C. The charge is $+Q$ on the inner surface and $-Q$ on the outer surface.
   D. The charge on both the inner and the outer surfaces is $+Q$.

10. A charge Q and a charge $3Q$ are released in a uniform electric field. If the force this field exerts on $3Q$ is F, the force it will exert on Q is
    A. F.   B. $F/3$.   C. $F/9$.

11. Three equal point charges are held in place as shown in Figure 17.38. If F_1 is the force on q due to Q_1 and F_2 is the force on q due to Q_2, how do F_1 and F_2 compare?
    A. $F_1 = 2F_2$.   B. $F_1 = 3F_2$.
    C. $F_1 = 4F_2$.   D. $F_1 = 9F_2$.

q Q_1 Q_2
⊕ ←d→ ⊕ ←——— $2d$ ———→ ⊕

▲ **FIGURE 17.38** Multiple-choice problem 11.

12. An electric field of magnitude E is measured at a distance R from a point charge Q. If the charge is doubled to $2Q$ and the electric field is now measured at a distance of $2R$ from the charge, the new measured value of the field will be:
    A. $2E$ B. E
    C. $E/2$ D. $E/4$

13. A very small ball containing a charge $-Q$ hangs from a light string between two vertical charged plates, as shown in Figure 17.39. When released from rest, the ball will
    A. swing to the right.
    B. swing to the left.
    C. remain hanging vertically.

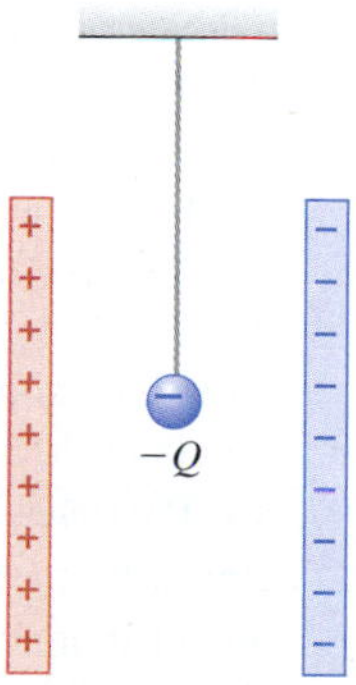

14. A point charge Q at the center of a sphere of radius R produces an electric flux of Φ_E coming out of the sphere. If the charge remains the same but the radius of the sphere is doubled, the electric flux coming out of it will be:
    A. $\Phi_E/2$ B. $\Phi_E/4$ C. $2\Phi_E$
    D. $4\Phi_E$ E. Φ_E

▲ **FIGURE 17.39** Multiple-choice problem 13.

15. Two charged small spheres are a distance R apart and exert an electrostatic force F on each other. If the distance is halved to $R/2$, the force exerted on each sphere will be
    A. $4F$.   B. $2F$.   C. $F/2$.   D. $F/4$.

Problems

17.1 Electric Charge
17.2 Conductors and Insulators

1. • A positively charged glass rod is brought close to a *neutral* sphere that is supported on a nonconducting plastic stand as shown in Figure 17.40. Sketch the distribution of charges on the sphere if it is made of (a) aluminum, (b) nonconducting plastic.

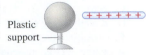 Plastic support

▲ **FIGURE 17.40** Problem 1.

2. • A positively charged rubber rod is moved close to a *neutral* copper ball that is resting on a nonconducting sheet of plastic. (a) Sketch the distribution of charges on the ball. (b) With the rod still close to the ball, a metal wire is briefly connected from the ball to the earth and then removed. After the rubber rod is also removed, sketch the distribution of charges (if any) on the copper ball.

3. • Two iron spheres contain excess charge, one positive and the other negative. (a) Show how the charges are arranged on these spheres if they are *very* far from each other. (b) If the spheres are now brought close to each other, but do not touch, sketch how the charges will be distributed on their surfaces. (c) In part (b), show how the charges would be distributed if both spheres were negative.

4. • **Electrical storms.** During an electrical storm, clouds can build up very large amounts of charge, and this charge can induce charges on the earth's surface. Sketch the distribution of charges at the earth's surface in the vicinity of a cloud if the cloud is positively charged and the earth behaves like a conductor.

17.3 Conservation and Quantization of Charge
17.4 Coulomb's Law

5. • In ordinary laboratory circuits, charges in the μC and nC range are common. How many excess electrons must you add to an object to give it a charge of (a) $-2.50\ \mu$C, (b) -2.50 nC?

6. • **Signal propagation in neurons.** *Neurons* are components **BIO** of the nervous system of the body that transmit signals as electrical impulses travel along their length. These impulses propagate when charge suddenly rushes into and then out of a part of the neutron called an *axon*. Measurements have shown that, during the inflow part of this cycle, approximately 5.6×10^{11} Na$^+$ (sodium ions) per meter, each with charge $+e$, enter the axon. How many coulombs of charge enter a 1.5 cm length of the axon during this process?

7. •• **Particles in a gold ring.** You have a pure (24-karat) gold ring with mass 17.7 g. Gold has an atomic mass of 197 g/mol and an atomic number of 79. (a) How many protons are in the ring, and what is their total positive charge? (b) If the ring carries no net charge, how many electrons are in it?

8. • Two equal point charges of $+3.00 \times 10^{-6}$ C are placed 0.200 m apart. What are the magnitude and direction of the force each charge exerts on the other?

9. • At what distance would the repulsive force between two electrons have a magnitude of 2.00 N? Between two protons?

10. • A negative charge of $-0.550\ \mu$C exerts an upward 0.200 N force on an unknown charge 0.300 m directly below it. (a) What is the unknown charge (magnitude and sign)? (b) What are the magnitude and direction of the force that the unknown charge exerts on the $-0.550\ \mu$C charge?

11. • **Forces in an atom.** The particles in the nucleus of an atom are approximately 10^{-15} m apart, while the electrons in an atom are about 10^{-10} m from the nucleus. (a) Calculate the electrical repulsion between two protons in a nucleus if they are 1.00×10^{-15} m apart. If you were holding these protons, do you think you could feel the effect of this force? How many pounds would the force be? (b) Calculate the electrical attraction that a proton in a nucleus exerts on an orbiting electron if the two particles are 1.00×10^{-10} m apart. If you were holding the electron, do you think you could feel the effect of this force?

12. •• (a) What is the total negative charge, in coulombs, of all the electrons in a small 1.00 g sphere of carbon? One mole of C is 12.0 g, and each atom contains 6 protons and 6 electrons. (b) Suppose you could take out all the electrons and hold them in one hand, while in the other hand you hold what is left of the original sphere. If you hold your hands 1.50 m apart at arms length, what force will each of them feel? Will it be attractive or repulsive?

13. • As you walk across a synthetic-fiber rug on a cold, dry winter day, you pick up an excess charge of $-55\ \mu$C. (a) How many excess electrons did you pick up? (b) What is the charge on the rug as a result of your walking across it?

14. •• Two small plastic spheres are given positive electrical charges. When they are 15.0 cm apart, the repulsive force between them has magnitude 0.220 N. What is the charge on each sphere (a) if the two charges are equal? (b) if one sphere has four times the charge of the other?

15. •• Two small aluminum spheres, each having mass 0.0250 kg, are separated by 80.0 cm. (a) How many electrons does each sphere contain? (The atomic mass of aluminum is 26.982 g/mol, and its atomic number is 13.) (b) How many electrons would have to be removed from one sphere and added to the other to cause an attractive force between the spheres of magnitude 1.00×10^4 N (roughly 1 ton)? Assume that the spheres may be treated as point charges. (c) What fraction of all the electrons in each sphere does this represent?

16. •• Two small spheres spaced 20.0 cm apart have equal charge. How many excess electrons must be present on each sphere if the magnitude of the force of repulsion between them is 4.57×10^{-21} N?

17. •• An average human weighs about 650 N. If two such generic humans each carried 1.0 coulomb of excess charge, one positive and one negative, how far apart would they have to be for the electric attraction between them to equal their 650-N weight?

18. •• If a proton and an electron are released when they are 2.0×10^{-10} m apart (typical atomic distances), find the initial acceleration of each of them.

19. •• Three point charges are arranged on a line. Charge $q_3 = +5.00$ nC and is at the origin. Charge $q_2 = -3.00$ nC and is at $x = +4.00$ cm. Charge q_1 is at $x = +2.00$ cm. What is q_1 (magnitude and sign) if the net force on q_3 is zero?

20. •• If two electrons are each 1.50×10^{-10} m from a proton, as shown in Figure 17.41, find the magnitude and direction of the net electrical force they will exert on the proton.

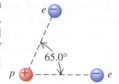

▲ **FIGURE 17.41** Problem 20.

21. •• Two point charges are located on the y axis as follows: charge $q_1 = -1.50$ nC at $y = -0.600$ m, and charge $q_2 = +3.20$ nC at the origin ($y = 0$). What is the net force (magnitude and direction) exerted by these two charges on a third charge $q_3 = +5.00$ nC located at $y = -0.400$ m?

22. •• Two point charges are placed on the x axis as follows: Charge $q_1 = +4.00$ nC is located at $x = 0.200$ m, and charge $q_2 = +5.00$ nC is at $x = -0.300$ m. What are the magnitude and direction of the net force exerted by these two charges on a negative point charge $q_3 = -0.600$ nC placed at the origin?

23. •• Three charges are at the corners of an isosceles triangle as shown in Figure 17.42. The $\pm 5.00\ \mu$C charges form a dipole. (a) Find the magnitude and direction of the net force that the $-10.0\ \mu$C charge exerts on the dipole. (b) For an axis perpendicular to the line connecting the two charges of the dipole at its midpoint and perpendicular to the plane of the paper, find the magnitude and direction of the torque exerted on the dipole by the $-10.0\ \mu$C charge.

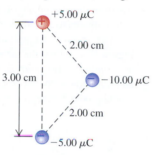

▲ **FIGURE 17.42** Problem 23.

24. •• **Base pairing in DNA, I.** The two sides of the DNA dou-
BIO ble helix are connected by pairs of bases (adenine, thymine, cytosine, and guanine). Because of the geometric shape of these molecules, adenine bonds with thymine and cytosine bonds with guanine. Figure 17.43 shows the thymine–adenine

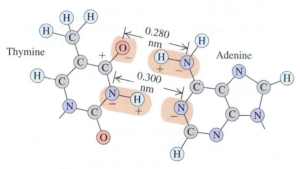

▲ **FIGURE 17.43** Problem 24.

bond. Each charge shown is $\pm e$, and the H — N distance is 0.110 nm. (a) Calculate the *net* force that thymine exerts on adenine. Is it attractive or repulsive? To keep the calculations fairly simple, yet reasonable, consider only the forces due to the O — H — N and the N — H — N combinations, assuming that these two combinations are parallel to each other. Remember, however, that in the O — H — N set, the O⁻ exerts a force on both the H⁺ and the N⁻, and likewise along

the N — H — N set. (b) Calculate the force on the electron in the hydrogen atom, which is 0.0529 nm from the proton. Then compare the strength of the bonding force of the electron in hydrogen with the bonding force of the adenine–thymine molecules.

25. •• **Base pairing in DNA, II.** Refer to the previous problem.
BIO Figure 17.44 shows the bonding of the cytosine and guanine molecules. The O — H and H — N distances are each 0.110 nm. In this case, assume that the bonding is due only to the forces along the O — H — O, N — H — N, and O — H — N combinations, and assume also that these three combinations are parallel to each other. Calculate the *net* force that cytosine exerts on guanine due to the preceding three combinations. Is this force attractive or repulsive?

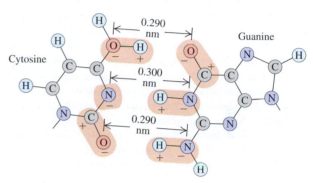

▲ **FIGURE 17.44** Problem 25.

26. •• **Surface tension.** Surface tension is the force that causes the surface of water (and other liquids) to form a "skin" that resists penetration. Because of this force, water forms into beads, and insects such as water spiders can walk on water. As we shall see, the force is electrical in nature. The surface of a polar liquid, such as water, can be viewed as a series of dipoles strung together in the stable arrangement in which the dipole moment vectors are parallel to the surface, all pointing in the same direction. Suppose now that something presses inward on the surface, distorting the dipoles as shown in Figure 17.45. Show that the two slanted dipoles exert a net upward force on the dipole between them and hence oppose the downward external force. Show also that the dipoles attract each other and thus resist being separated. Notice that the force between dipoles opposes penetration of the liquid's surface and is a simple model for surface tension.

▲ **FIGURE 17.45** Problem 26.

27. •• If the central charge shown in Figure 17.46 is displaced 0.350 nm to the right while the other charges are held in place, find the magnitude and direction of the net force that the other two charges exert on it.

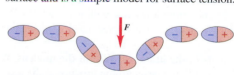

▲ **FIGURE 17.46** Problem 27.

28. •• Two unequal charges repel each other with a force F. If both charges are doubled in magnitude, what will be the new force in terms of F?

29. •• In an experiment in space, one proton is held fixed and another proton is released from rest a distance of 2.50 mm away. (a) What is the initial acceleration of the proton after it is released? (b) Sketch qualitative (no numbers!) acceleration–time and velocity–time graphs of the released proton's motion.

30. ••• A charge $+Q$ is located at the origin and a second charge, $+4Q$, is at distance d on the x-axis. Where should a third charge, q, be placed, and what should be its sign and magnitude, so that all three charges will be in equilibrium?

17.5 Electric Field and Electric Forces

31. • A small object carrying a charge of -8.00 nC is acted upon by a downward force of 20.0 nN when placed at a certain point in an electric field. (a) What are the magnitude and direction of the electric field at the point in question? (b) What would be the magnitude and direction of the force acting on a proton placed at this same point in the electric field?

32. • (a) What must the charge (sign and magnitude) of a 1.45 g particle be for it to remain balanced against gravity when placed in a downward-directed electric field of magnitude 650 N/C? (b) What is the magnitude of an electric field in which the electric force it exerts on a proton is equal in magnitude to the proton's weight?

33. •• A uniform electric field exists in the region between two oppositely charged plane parallel plates. An electron is released from rest at the surface of the negatively charged plate and strikes the surface of the opposite plate, 3.20 cm distant from the first, in a time interval of 1.5×10^{-8} s. (a) Find the magnitude of this electric field. (b) Find the speed of the electron when it strikes the second plate.

17.6 Calculating Electric Fields

34. • A particle has a charge of -3.00 nC. (a) Find the magnitude and direction of the electric field due to this particle at a point 0.250 m directly above it. (b) At what distance from the particle does its electric field have a magnitude of 12.0 N/C?

35. • The electric field caused by a certain point charge has a magnitude of 6.50×10^3 N/C at a distance of 0.100 m from the charge. What is the magnitude of the charge?

36. • At what distance from a particle with a charge of 5.00 nC does the electric field of that charge have a magnitude of 4.00 N/C?

37. • **Electric fields in the atom.** (a) **Within the nucleus.** What strength of electric field does a proton produce at the distance of another proton, about 5.0×10^{-15} m away? (b) **At the electrons.** What strength of electric field does this proton produce at the distance of the electrons, approximately 5.0×10^{-10} m away?

38. •• A proton is traveling horizontally to the right at 4.50×10^6 m/s. (a) Find the magnitude and direction of the weakest electric field that can bring the proton uniformly to rest over a distance of 3.20 cm. (b) How much time does it take the proton to stop after entering the field? (c) What minimum field (magnitude and direction) would be needed to stop an electron under the conditions of part (a)?

39. •• **Electric field of axons.** A nerve signal is transmitted **BIO** through a neuron when an excess of Na$^+$ ions suddenly enters the axon, a long cylindrical part of the neuron. Axons are approximately 10.0 μm in diameter, and measurements show that about 5.6×10^{11} Na$^+$ ions per meter (each of charge $+e$) enter during this process. Although the axon is a long cylinder, the charge does not all enter everywhere at the same time. A plausible model would be a series of nearly point charges moving along the axon. Let us look at a 0.10 mm length of the axon and model it as a point charge. (a) If the charge that enters each meter of the axon gets distributed uniformly along it, how many coulombs of charge enter a 0.10 mm length of the axon? (b) What electric field (magnitude and direction) does the sudden influx of charge produce at the surface of the body if the axon is 5.00 cm below the skin? (c) Certain sharks can respond to electric fields as weak as 1.0 μN/C. How far from this segment of axon could a shark be and still detect its electric field?

40. •• Two point charges are separated by 25.0 cm (see Figure 17.47). Find the net electric field these charges produce at (a) point A, (b) point B. (c) What would be the magnitude and direction of the electric force this combination of charges would produce on a proton at A?

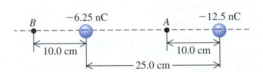

▲ **FIGURE 17.47** Problem 40.

41. •• A point charge of -4.00 nC is at the origin, and a second point charge of $+6.00$ nC is on the x axis at $x = 0.800$ m. Find the magnitude and direction of the electric field at each of the following points on the x axis: (a) $x = 20.0$ cm, (b) $x = 1.20$ m, (c) $x = -20.0$ cm.

42. •• In a rectangular coordinate system, a positive point charge $q = 6.00$ nC is placed at the point $x = +0.150$ m, $y = 0$, and an identical point charge is placed at $x = -0.150$ m, $y = 0$. Find the x and y components and the magnitude and direction of the electric field at the following points: (a) the origin; (b) $x = 0.300$ m, $y = 0$; (c) $x = 0.150$ m, $y = -0.400$ m, (d) $x = 0$, $y = 0.200$ m.

43. •• Two particles having charges of $+0.500$ nC and $+8.00$ nC are separated by a distance of 1.20 m. (a) At what point along the line connecting the two charges is the net electric field due to the two charges equal to zero? (b) Where would the net electric field be zero if one of the charges were negative?

44. •• Three negative point charges lie along a line as shown in Figure 17.48. Find the magnitude and direction of the electric field this combination of charges produces at point P, which lies 6.00 cm from the -2.00 μC charge measured perpendicular to the line connecting the three charges.

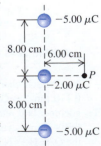

▲ **FIGURE 17.48** Problem 44.

45. •• **Torque and force on a dipole.** An electric dipole is in a uniform external electric field $\vec{E}$ as shown in Figure 17.49. (a) What is the net force this field exerts on the dipole? (b) Find the orientations of the dipole for which the torque on

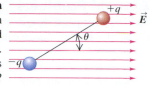

▲ **FIGURE 17.49** Problem 45.

it about an axis through its center perpendicular to the plane of the figure is zero. (c) Which of the orientations in part (b) is stable, and which is unstable? (*Hint:* Consider a small displacement away from the equilibrium position, and see what happens.) (d) Show that, for the stable orientation in part (c), the dipole's own electric field *opposes* the external field for points between the charges.

46. •• (a) An electron is moving east in a uniform electric field of 1.50 N/C directed to the west. At point A, the velocity of the electron is 4.50×10^5 m/s toward the east. What is the speed of the electron when it reaches point B, 0.375 m east of point A? (b) A proton is moving in the uniform electric field of part (a). At point A, the velocity of the proton is 1.90×10^4 m/s, east. What is the speed of the proton at point B?

47. •• The electric field due to a certain point charge has a magnitude E at a distance of 1.0 cm from the charge. (a) What will be the magnitude of this field (in terms of E) if we move 1.0 cm farther away from the charge? (b) What will be the magnitude of the field (in terms of E) if we move an *additional* 1.0 cm farther away than in part (a)?

48. ••• For the dipole shown in Figure 17.50, show that the electric field at points on the x axis points vertically downward and has magnitude $kq(2a)/(a^2 + x^2)^{3/2}$. What does this expression reduce to when the distance between the two charges is much less than x?

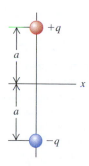

▲ **FIGURE 17.50** Problem 48.

17.7 Electric Field Lines

49. • Figure 17.51 shows some of the electric field lines due to three point charges arranged along the vertical axis. All three charges have the same magnitude. (a) What are the signs of the three charges? Explain your reasoning. (b) At what point(s) is the magnitude of the electric field the smallest? Explain your reasoning. Explain how the fields produced by each individual point charge combine to give a small net field at this point or points.

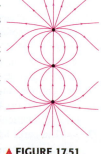

▲ **FIGURE 17.51** Problem 49.

50. • A proton and an electron are separated as shown in Figure 17.52. Points A, B, and C lie on the perpendicular bisector of the line connecting these two charges. Sketch the direction of the net electric field due to the two charges at (a) point A, (b) point B, and (c) point C.

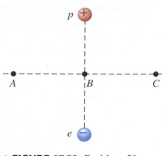

▲ **FIGURE 17.52** Problem 50.

51. •• Sketch electric field lines in the vicinity of two charges, Q and $-4Q$, located a small distance apart on the x-axis.

52. • Two point charges Q and $+q$ (where q is positive) produce the net electric field shown at point P in Figure 17.53. The field points parallel to the line connecting the two charges. (a) What can you conclude about the sign and magnitude of Q? Explain your reasoning. (b) If the lower charge were negative instead, would it be possible for the field to have the direction shown in the figure? Explain your reasoning.

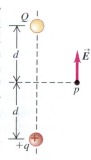

▲ **FIGURE 17.53** Problem 52.

53. •• Two very large parallel sheets of the same size carry equal magnitudes of charge spread uniformly over them, as shown in Figure 17.54. In each of the cases that follow, sketch the net pattern of electric-field lines in the region between the sheets, but far from their edges. (*Hint:* First sketch the field lines due to each sheet, and then add these fields to get the net field.) (a) The top sheet is positive and the bottom sheet is negative, as shown, (b) both sheets are positive, (c) both sheets are negative.

▲ **FIGURE 17.54** Problem 53.

17.8 Gauss's Law and Field Calculations

54. • (a) A closed surface encloses a net charge of 2.50 μC. What is the net electric flux through the surface? (b) If the electric flux through a closed surface is determined to be 1.40 N · m²/C, how much charge is enclosed by the surface?

55. • Figure 17.55 shows cross sections of five *closed* surfaces S_1, S_2, etc. Find the net electric flux passing through each of these surfaces.

56. •• A point charge 8.00 nC is at the center of a cube with sides of length 0.200 m. What is the electric flux through (a) the surface of the cube, (b) one of the six faces of the cube?

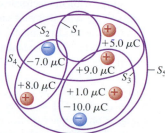

▲ **FIGURE 17.55** Problem 55.

57. • A charged paint is spread in a very thin uniform layer over the surface of a plastic sphere of diameter 12.0 cm, giving it a charge of -15.0 μC. Find the electric field (a) just inside the paint layer, (b) just outside the paint layer, and (c) 5.00 cm outside the surface of the paint layer.

58. •• On a humid day, an electric field of 2.00×10^4 N/C is enough to produce sparks about an inch long. Suppose that in your physics class, a van de Graaff generator (see Fig. 7.56) with a sphere radius of 15.0 cm is producing sparks 6 inches long. (a) Use Gauss's law to calculate the amount of charge stored on the surface of the sphere before you bravely discharge it with your hand. (b) Assume all the charge is concentrated at the center of the sphere, and use Coulomb's law to calculate the electric field at the surface of the sphere.

▲ **FIGURE 17.56**
Problem 58.

59. • (a) How many excess electrons must be distributed uniformly within the volume of an isolated plastic sphere 30.0 cm in diameter to produce an electric field of 1150 N/C just outside the surface of the sphere? (b) What is the electric field at a point 10.0 cm outside the surface of the sphere?

60. •• In a certain region of space, the electric field $\vec{E}$ is uniform; i.e., neither its direction nor its magnitude changes in the region. (a) Use Gauss's law to prove that this region of space must be electrically neutral; that is, there must be no charge in this region. (b) Is the converse true? That is, in a region of space where there is no charge, must $\vec{E}$ be uniform? Explain.

61. •• A total charge of magnitude Q is distributed uniformly within a *thick* spherical shell of inner radius a and outer radius b. (a) Use Gauss's law to find the electric field within the cavity $(r \le a)$. (b) Use Gauss's law to prove that the electric field outside the shell $(r \ge b)$ is exactly the same as if all the charge were concentrated as a point charge Q at the center of the sphere. (c) Explain why the result in part (a) for a *thick* shell is the same as that found in Example 17.10 for a *thin* shell. (*Hint:* A thick shell can be viewed as infinitely many thin shells.)

17.9 Charges on Conductors

62. • During a violent electrical storm, a car is struck by a falling high-voltage wire that puts an excess charge of $-850 \, \mu C$ on the metal car. (a) How much of this charge is on the inner surface of the car? (b) How much is on the outer surface?

63. • A neutral conductor completely encloses a hole inside of it. You observe that the outer surface of this conductor carries a charge of $-12 \, \mu C$. (a) Can you conclude that there is a charge inside the hole? If so, what is this charge? (b) How much charge is on the inner surface of the conductor?

64. •• An irregular neutral conductor has a hollow cavity inside of it and is insulated from its surroundings. An excess charge of $+16$ nC is sprayed onto this conductor. (a) Find the charge on the inner and outer surfaces of the conductor. (b) Without touching the conductor, a charge of -11 nC is inserted into the cavity through a small hole in the conductor. Find the charge on the inner and outer surfaces of the conductor in this case.

General Problems

65. •• Three point charges are arranged along the x axis. Charge $q_1 = -4.50$ nC is located at $x = 0.200$ m, and charge $q_2 = +2.50$ nC is at $x = -0.300$ m. A positive point charge q_3 is located at the origin. (a) What must the value of q_3 be for the net force on this point charge to have magnitude $4.00 \, \mu N$? (b) What is the direction of the net force on q_3? (c) Where along the x axis can q_3 be placed and the net force on it be zero, other than the trivial answers of $x = +\infty$ and $x = -\infty$?

66. •• An electron is released from rest in a uniform electric field. The electron accelerates vertically upward, traveling 4.50 m in the first 3.00 μs after it is released. (a) What are the magnitude and direction of the electric field? (b) Are we justified in ignoring the effects of gravity? Justify your answer quantitatively.

67. •• A charge $q_1 = +5.00$ nC is placed at the origin of an xy-coordinate system, and a charge $q_2 = -2.00$ nC is placed on the positive x axis at $x = 4.00$ cm. (a) If a third charge $q_3 = +6.00$ nC is now placed at the point $x = 4.00$ cm, $y = 3.00$ cm, find the x and y components of the total force exerted on this charge by the other two charges. (b) Find the magnitude and direction of this force.

68. •• A charge of -3.00 nC is placed at the origin of an xy-coordinate system, and a charge of 2.00 nC is placed on the y axis at $y = 4.00$ cm. (a) If a third charge, of 5.00 nC, is now placed at the point $x = 3.00$ cm, $y = 4.00$ cm, find the x and y components of the total force exerted on this charge by the other two charges. (b) Find the magnitude and direction of this force.

69. •• Point charges of 3.00 nC are situated at each of three corners of a square whose side is 0.200 m. What are the magnitude and direction of the resultant force on a point charge of $-1.00 \, \mu C$ if it is placed (a) at the center of the square, (b) at the vacant corner of the square?

70. •• An electron is projected with an initial speed $v_0 = 5.00 \times 10^6$ m/s into the uniform field between the parallel plates in Figure 17.57. The direction of the field is vertically downward, and the field is zero except in the space between the two plates. The electron enters the field at a point midway between the plates. If the electron just misses the upper plate as it emerges from the field, find the magnitude of the electric field.

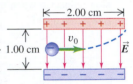

▲ **FIGURE 17.57**
Problem 70.

71. •• A small 12.3 g plastic ball is tied to a very light 28.6 cm string that is attached to the vertical wall of a room. (See Figure 17.58.) A uniform horizontal electric field exists in this room. When the ball has been given an excess charge of $-1.11 \, \mu C$, you observe that it remains suspended, with the string making an angle of 17.4° with the wall. Find the magnitude and direction of the electric field in the room.

▲ **FIGURE 17.58**
Problem 71.

72. •• A -5.00 nC point charge is on the x axis at $x = 1.20$ m. A second point charge Q is on the x axis at -0.600 m. What must be the sign and magnitude of Q for the resultant electric field at the origin to be (a) 45.0 N/C in the $+x$ direction, (b) 45.0 N/C in the $-x$ direction?

73. •• The earth has a downward-directed electric field near its surface of about 150 N/m. If a raindrop with a diameter of 0.020 mm is suspended, motionless, in this field, how many excess electrons must it have on its surface?

74. •• A 9.60-μC point charge is at the center of a cube with sides of length 0.500 m. (a) What is the electric flux through one of the six faces of the cube? (b) How would your answer to part (a) change if the sides were 0.250 m long? Explain.

75. •• Two point charges q_1 and q_2 are held 4.00 cm apart. An electron released at a point that is equidistant from both charges (see Figure 17.59) undergoes an initial acceleration of 8.25×10^{18} m/s^2 directly upward in the figure, parallel to the line connecting q_1 and q_2. Find the magnitude and sign of q_1 and q_2.

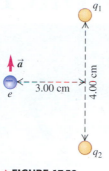

▲ **FIGURE 17.59** Problem 75.

76. •• **Electrophoresis.** Electrophoresis is a **BIO** process used by biologists to separate different biological molecules (such as proteins) from each other according to their ratio of charge to size. The materials to be separated are in a viscous solution that produces a drag force F_D proportional to the size and speed of the molecule. We can express this relationship as $F_D = KRv$, where R is the radius of the molecule (modeled as being spherical), v is its speed, and K is a constant that depends on the viscosity of the solution. The solution is placed in an external electric field E so that the electric force on a particle of charge q is $F = qE$. (a) Show that when the electric field is adjusted so that the two forces (electrical and viscous drag) just balance, the ratio of q to R is Kv/E. (b) Show that if we leave the electric field on for a time T, the distance x that the molecule moves during that time is $x = (ET/k)(q/R)$. (c) Suppose you have a sample containing three different biological molecules for which the molecular ratio q/R for material 2 is twice that of material 1 and the ratio for material 3 is three times that of material 1. Show that the distances migrated by these molecules after the same amount of time are $x_2 = 2x_1$ and $x_3 = 3x_1$. In other words, material 2 travels twice as far as material 1, and material 3 travels three times as far as material 1. Therefore, we have separated these molecules according to their ratio of charge to size. In practice, this process can be carried out in a special gel or paper, along which the biological molecules migrate. (See Figure 17.60.) The process can be rather slow, requiring several hours for separations of just a centimeter or so.

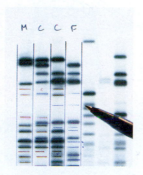

▲ **FIGURE 17.60** Problem 76.

77. •• An early model of the hydrogen atom viewed it as an electron orbiting a proton in a circular path with a radius of 5.29×10^{-11} m. What would be the speed of the electron in this model? You'll need some information from Appendix E, and may need to review Chapter 6 on circular motion.

Passage Problems

BIO How might cells respond to an electric field? Some cells can be observed to grow or migrate parallel to an applied electric field. (The field is applied by passing a current through the aqueous medium surrounding the cell. Because the cell membrane has a higher resistance than the medium, the current flows around the cell.) It has been hypothesized that this phenomenon may participate in the natural guidance of cells during embryonic development or wound repair. However, the mechanism by which the cells sense the electric field is not known.

In one proposed mechanism, the cell would use the distribution of cell-surface proteins to sense the electric field. The membrane surrounding the cell consists of a double layer of lipid molecules and has a viscosity similar to that of olive oil. The membrane is studded with protein molecules, which are free to move in the plane of the membrane. In the absence of a perturbing force, diffusion tends to distribute these molecules uniformly. However, for many membrane proteins, the portion of the protein that projects into the extracellular medium carries a net charge. An applied electric field will tend to move such charged proteins toward one or the other end of the cell. For a given type of charged protein, the resulting steady-state distribution depends on the applied field and on the concentration and net diffusion rate of the protein molecules. In theory, a cell could use an asymmetric protein distribution to sense and respond to an electric field. The graph in Figure 17.61 shows the steady-state distribution of a particular membrane protein in response to an electric field.

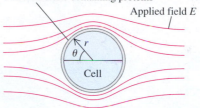

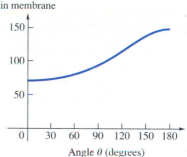

▲ **FIGURE 17.61** Problems 78–79.

78. What is the direction of the electrical field?
    A. It points from 0° to 180°.
    B. It points from 180° to 0°.
    C. Cannot tell without knowing the absolute value of the charge on the molecules.
    D. Cannot tell without knowing the sign of the charge on the molecules.

79. After the protein has reached the steady-state distribution shown by the graph, you turn the electric field off. Assuming diffusion acts unimpeded, the protein distribution will
    A. remain unchanged.
    B. become uniform, with the density at each location equal to that at 180° on the graph.
    C. become uniform with a density intermediate between those at 0° and 180° on the graph.
    D. become uniform, with the density at each location equal to that at 0° on the graph.

18 Electric Potential and Capacitance

This chapter is about energy associated with electrical interactions. Every time you turn on a light or an electric motor, you are making use of electrical energy, a familiar part of everyday life and an indispensable ingredient of our technological society. In Chapter 7, we introduced the concepts of *work* and *energy* in a mechanical context; now we combine these concepts with what we have learned about electric charge, Coulomb's law, and electric fields.

When a charged particle moves in an electric field created by charges at rest (i.e., an electrostatic field), the electric force does *work* on the particle. The force is *conservative;* the work can always be expressed in terms of a potential energy. This in turn is associated with a new concept called *electric potential,* or simply *potential.* In circuits, potential is often called *voltage.* The practical applications of this concept cover a wide range, including electric circuits, electron beams in TV picture tubes, high-energy particle accelerators, and many other devices and phenomena. The concept of potential is also essential for the analysis of a common circuit device called a *capacitor,* which we'll study later in the chapter.

The electrical potential between high-voltage wires and steel pylons is very high, so the wires are held away from the pylons by stacks of insulators.

18.1 Electric Potential Energy

The opening sections of this chapter are about work, potential energy, and conservation of energy. Let's begin by reviewing several essential points from Chapter 7. First, when a constant force $\vec{F}$ acts on a particle that moves in a straight line through a displacement $\vec{s}$ from point a to point b, the work $W_{a \to b}$ done by the force is

$$W_{a \to b} = Fs\cos\phi, \tag{18.1}$$

where ϕ is the angle between the force and displacement. We'll point the x axis in the direction of the particle's motion.

▶ **BIO Application Nanomachine.** This image shows the structure of a protein complex called a *voltage-gated potassium channel,* which participates in the functioning of nerve and muscle cells. The channel sits in the cell membrane and controls the flow of potassium ions out of the cell. Remarkably, the force that opens and closes the channel is provided by the electric field across the membrane. Each of the four subunits that make up the channel has an arm that carries positive charges. This arm is fairly mobile, so the electric field does work on it, pulling it toward one or the other side of the membrane, depending on the field's direction and strength. When the arm is pulled inward, the channel closes; when it is pulled outward, the channel opens.

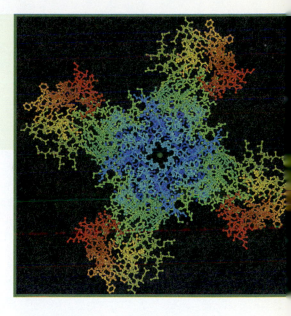

Second, because the force field is *conservative,* as we defined the term in Section 7.5, the work that is done can always be expressed in terms of a *potential energy U.* When the particle moves from a point where the potential energy is U_a to a point where it is U_b, the work $W_{a \to b}$ done by the force is

$$W_{a \to b} = U_a - U_b. \qquad (18.2)$$

When $W_{a \to b}$ is positive, U_a is greater than U_b, and the potential energy *decreases.* That's what happens when a baseball falls from a high point (a) to a lower point (b) under the action of the earth's gravity (Figure 18.1a). The force of gravity does positive work, and the gravitational potential energy decreases. When a ball is thrown upward, the gravitational force does negative work during the ascent, and the potential energy increases. Figure 18.1b shows the analogous situation for electric fields; we'll discuss it in detail later in this section.

Third, the work–energy theorem says that the change in kinetic energy $\Delta K = K_b - K_a$ during any displacement is equal to the total work done on the particle. So if Equation 18.2 gives the *total* work, then $K_b - K_a = U_a - U_b$, which we usually write as

$$K_a + U_a = K_b + U_b. \qquad (18.3)$$

Let's look at an electrical example of these basic concepts. In Figure 18.2, a pair of charged parallel metal plates sets up a uniform electric field with magnitude E. The field exerts a downward force with magnitude $F = q'E$ on a positive test charge q' as the charge moves a distance s from point a to point b. The force on the test charge is constant and independent of its location, so the work done by the electric field is

$$W_{a \to b} = Fs = q'Es. \qquad (18.4)$$

We can represent this work in terms of a potential energy U, just as we did for gravitational potential energy in Section 7.5. The y component of force, $F_y = -q'E$, is constant, and there is no x or z component, so the work is independent of the path the particle takes from a to b. Just as the potential energy for the gravitational force $F_y = -mg$ was $U = mgy$, the potential energy for the electric-field force $F_y = -q'E$ is

$$U = q'Ey. \qquad (18.5)$$

(We've chosen U to be zero at $y = 0$.) When the test charge moves from height y_a to height y_b, the work done on the charge by the field is given by

$$W_{a \to b} = U_a - U_b = q'Ey_a - q'Ey_b = q'E(y_a - y_b). \qquad (18.6)$$

When y_a is greater than y_b (Figure 18.3a), the particle moves in the same direction as the $\vec{E}$ field, U decreases, and the field does positive work. When y_a is less than y_b (Figure 18.3b), the particle moves in the opposite direction to $\vec{E}$, U

Object moving in a uniform gravitational field:

Charge moving in a uniform electric field:

$W = -\Delta U_{grav} = mgh$

$W = -\Delta U_E = qEs$

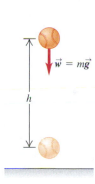

(a)   (b)

▲ **FIGURE 18.1** Because electric and gravitational forces are conservative, work done by either can be expressed in terms of a potential energy.

Work done on charge q' by the *constant* electric force between the plates: $W_{a \to b} = q'Es$

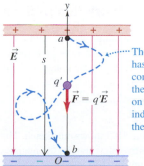

The electric force has only a y component, so the work it does on the charge is independent of the charge's path.

▲ **FIGURE 18.2** A test charge q' moves from point a to point b in a uniform electric field.

Positive charge moves in the direction of $\vec{E}$:
• Field does *positive* work on charge;
• *U decreases.*

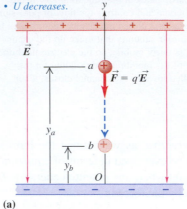

Positive charge moves opposite to $\vec{E}$:
• Field does *negative* work on charge;
• *U increases.*

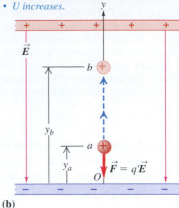

(a)   (b)

▲ **FIGURE 18.3** The work done by an electric field on a positive charge moving (a) in the direction of and (b) opposite to the electric field.

Negative charge moves in the direction of $\vec{E}$:
• Field does *negative* work on charge;
• *U increases.*

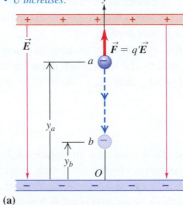

Negative charge moves opposite to $\vec{E}$:
• Field does *positive* work on charge;
• *U decreases.*

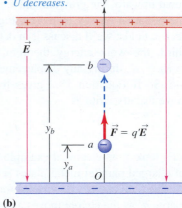

(a)   (b)

▲ **FIGURE 18.4** The work done by an electric field on a negative charge moving (a) in the direction of and (b) opposite to the electric field.

increases, and the field does negative work. In particular, if $y_a = s$ and $y_b = 0$, then Equation 18.6 gives $W_{a \to b} = q'Es$, in agreement with Equation 18.4.

If the test charge q' is negative, the potential energy increases when it moves with the field and decreases when it moves against the field (Figure 18.4).

EXAMPLE 18.1   **Work in a uniform electric field**

Two large conducting plates separated by 6.36 mm carry charges of equal magnitude and opposite sign, creating a uniform electric field with magnitude 2.80×10^3 N/C between the plates. An electron moves from the negatively charged plate to the positively charged plate. How much work does the electric field do on the electron?

SOLUTION

SET UP Figure 18.5 shows our sketch. The electric field is directed from the positive plate toward the negative plate. $\vec{F}_E = q\vec{E}$, so for an electron with negative charge $q = -e$, the electric force $\vec{F}_E$ points in the direction opposite to the electric field. Its magnitude is $F_E = eE$. The electric field is uniform, so the force it exerts on the electron is constant during the electron's motion.

Continued

SOLVE The force and displacement are parallel; the work W done by the electric-field force during a displacement of magnitude d is $W = F_e d \cos\phi$ with $\phi = 0$, so

$$W = F_e d = eEd$$
$$= (1.60 \times 10^{-19}\,\text{C})(2.80 \times 10^3\,\text{N/C})(6.36 \times 10^{-3}\,\text{m})$$
$$= 2.85 \times 10^{-18}\,\text{J}.$$

REFLECT The amount of work done is proportional to the electric field magnitude E and to the displacement magnitude d of the electron. The electric field does positive work on the electron. If there are no other forces, the electron's kinetic energy increases by the same amount as the work done on the electron by the electric field.

Practice Problem: In Example 18.1, how much work does the electric field do on the electron if the magnitude of the field is doubled, to $5.60 \times 10^3\,\text{N/C}$, and the separation between the plates is halved, to 3.18 mm? *Answer:* $2.85 \times 10^{-18}\,\text{J}$.

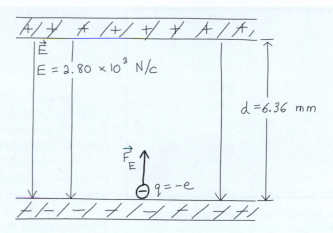

▲ **FIGURE 18.5** Our sketch for this problem.

Potential Energy of Point Charges

It's useful to calculate the work done on a test charge q' when it moves in the electric field caused by a single stationary point charge q. Suppose we place charge q at the origin of a coordinate system, and suppose the test charge q' moves along the x axis from point $x = a$ to point $x = b$ (Figure 18.6a). How much work does the force due to q do on q' during this displacement?

We can't simply multiply the force by the displacement, because the force isn't constant; it varies with the distance x according to the graph in Figure 18.6b. The work $W_{a \to b}$ done on the test charge is represented graphically by the area under the curve between $x = a$ and $x = b$. This area can be calculated with methods of integral calculus. The result is

$$W_{a \to b} = kqq'\left(\frac{1}{a} - \frac{1}{b}\right). \tag{18.7}$$

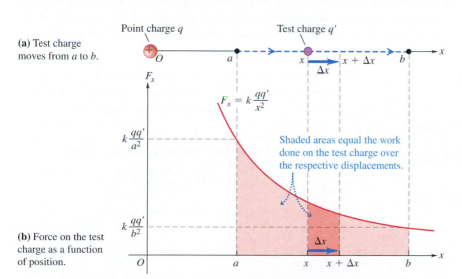

(a) Test charge moves from a to b.

(b) Force on the test charge as a function of position.

▲ **FIGURE 18.6** A test charge q' moves radially along a straight line extending from charge q. As it does so, the electric force on it decreases in magnitude.

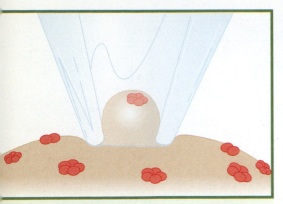

▲ BIO Application Real-time molecular biology. The patch clamp technique is an ingenious way to investigate how cells work. To create a patch clamp, a polished microelectrode is carefully manipulated to the outer membrane of a cell to make a tight seal. Protein molecules called ion channels float in the oily membrane, and often a single channel is isolated within the patch. An experimenter can then study the electrophysiologic properties of the single ion channel and manipulate the voltage across the membrane (and thus the channel). In this way, we know that an electrical potential difference across the membrane controls the pore, and thus the movement of ions through the membrane. At the appropriate voltage, a channel flicks open for a time (perhaps 10 ms) and then closes. The opening and closing of voltage-gated channels underlie all electrical signaling in most cells to control many aspects of cellular function. The inventors of the patch clamp technique were awarded the Nobel Prize in Physiology or Medicine in 1991.

Equation 18.7 also shows how we can define a potential energy for this interaction. We define

$$U_a = \frac{kqq'}{a} \quad \text{and} \quad U_b = \frac{kqq'}{b}.$$

This definition is consistent with the requirement that $W_{a \to b} = U_a - U_b$. If both charges are positive and b is greater than a, the electric-field force on q' does positive work as it moves away from q. Correspondingly, U_a is greater than U_b, so the potential energy decreases as the force does positive work, just as we expect.

This important result can be generalized to the case where q, point a, and point b don't all lie along the same line. It can be shown that the work $W_{a \to b}$ done on q' by the $\vec{E}$ field produced by q is the same for *all possible paths* from a to b, even if these points don't lie on the same radial line from charge q. The work depends only on the distances a and b, not on the details of the path. Also, if q' returns to its starting point a by a different path, the total work done in the round-trip displacement is zero. These are the needed characteristics for a *conservative* force field, as we defined it in Chapter 7. Thus, we've verified that the force that q exerts on q' is a *conservative* force, and we've obtained an expression for the potential energy U when the test charge q' is at *any* distance r from charge q:

> **Potential energy of point charges**
> The potential energy U of a system consisting of a point charge q' located in the field produced by a stationary point charge q, at a distance r from the charge, is
>
> $$U = k\frac{qq'}{r}. \tag{18.8}$$

We have *not* assumed anything about the signs of q and q'; Equation 18.8 is valid for any combination of signs.

Potential energy is always defined relative to some reference point at which $U = 0$. In Equation 18.8, U is zero when q and q' are infinitely far apart, or $r = \infty$. Therefore, U represents the work done on the test charge q' by the field of q when q' moves from an initial distance r to infinity. If q and q' have the same sign, the interaction is repulsive, the work is positive, and U is positive at any finite separation. If they have opposite signs, the interaction is attractive and U is negative.

Quantitative Analysis 18.1

Change in potential energy

Consider two positive point charges q_1 and q_2. Their potential energy is defined as zero when they are infinitely far apart, and it increases as they move closer. If q_2 starts at an initial distance r_i from q_1 and moves toward q_1 to a final distance $r_i - \Delta r$ (where Δr is positive), by how much does the system's potential energy change?

A. $kq_1q_2\left(\dfrac{1}{\Delta r} - \dfrac{1}{r_i - \Delta r}\right).$ B. $\Delta r\left(\dfrac{kq_1q_2}{r_i^2}\right).$

C. $kq_1q_2\left(\dfrac{1}{r_i - \Delta r} - \dfrac{1}{r_i}\right).$

SOLUTION The electric potential energy of the two charges depends on the distance r between them: $U = k(q_1q_2)/r$. Initially, the distance between them is r_i. After q_2 moves a distance Δr toward q_1, the distance is $r_i - \Delta r$. The change in potential energy depends on the reciprocal of these distances, so C must be the answer. More formally, the change in potential energy is

$$\Delta U = U_f - U_i = \frac{kq_1q_2}{r_i - \Delta r} - \frac{kq_1q_2}{r_i}.$$

Simplifying this equation yields expression C.

We can generalize Equation 18.8 to situations in which the $\vec{E}$ field is caused by *several* point charges q_1, q_2, q_3, ... at distances r_1, r_2, r_3, ..., respectively, from q'. The total electric field at each point is the *vector sum* of the fields due to the individual charges, and the total work done on q' during any displacement is the sum of the contributions from the individual charges. We conclude that the potential energy U associated with a test charge q' at point a in Figure 18.7, due to a collection of charges $q_1, q_2, q_3, \ldots$ at distances $r_1, r_2, r_3, \ldots$, respectively, from q' is the *algebraic* sum (*not* a vector sum)

$$U = kq'\left(\frac{q_1}{r_1} + \frac{q_2}{r_2} + \frac{q_3}{r_3} + \cdots\right).\qquad(18.9)$$

When q' is at a different point b, the potential energy is given by the same expression, but $r_1, r_2, \ldots$ are the distances from $q_1, q_2, \ldots$, respectively, to point b. The work $W_{a \to b}$ done on charge q' when it moves from a to b along *any* path is still equal to the difference $U_a - U_b$ between the potential energies when q' is at a and at b.

We can represent *any* charge distribution as a collection of point charges, so Equation 18.9 shows that we can always find a potential-energy function for *any* static electric field. It follows that **every electric field due to a static charge distribution is a conservative force field.**

Equations 18.8 and 18.9 define U to be zero when *all* the distances $r_1, r_2, \ldots$ are *infinite*—that is, when the test charge q' is very far away from all the charges that produce the field. As with any potential-energy function, the reference point is arbitrary: We can always add a constant to make U equal zero at any point we choose. Making $U = 0$ at infinity is a convenient reference level for electrostatic problems, but in circuit analysis other reference levels are often more convenient.

18.2 Potential

In Section 18.1, we looked at the potential energy U associated with a test charge q' in an electric field. Now we want to describe this potential energy on a "potential energy per unit charge" basis, just as the electric field describes the force on a charged particle in the field on a "force per unit charge" basis. Doing this leads us to the concept of **electric potential,** often called, simply, **potential.** The concept of electric potential is useful in calculations involving energies of charged particles. It also facilitates many electric-field calculations, because it is closely related to the concept of the $\vec{E}$ field. When we need to calculate an electric field, it is often easier to calculate the potential first and then find the field from it.

Definition of potential

The electric potential V at any point in an electric field is the electric potential energy U per unit charge associated with a test charge q' at that point:

$$V = \frac{U}{q'}, \qquad \text{or} \qquad U = q'V. \qquad (18.10)$$

Potential energy and charge are both scalars, so potential is a scalar quantity.

Unit: From Equation 18.10, the units of potential are energy divided by charge. The SI unit of potential, 1 J/C, is called one **volt** (1 V), in honor of the Italian scientist Alessandro Volta (1745–1827):

$$1\text{ V} = 1\text{ volt} = 1\text{ J/C} = 1\text{ joule/coulomb.}$$

In the context of electric circuits, potential is often called **voltage.** For instance, a 9 V battery has a difference in electric potential **(potential difference)** of 9 V between its two terminals. A 20,000 V power line has a potential difference of 20,000 V between itself and the ground.

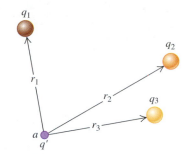

▲ **FIGURE 18.7** Potential energy associated with a charge q' at point a depends on charges $q_1, q_2,$ and q_3 and on their respective distances $r_1, r_2,$ and r_3 from point a.

MasteringPHYSICS

PhET: Charges and Fields
ActivPhysics 11.11: Electric Potential: Qualitative Introduction

▲ **Application** *Really* **high voltage.**
A lightning bolt occurs when the electric potential difference between cloud and ground becomes so great that the air between them ionizes and allows a current to flow. A typical bolt discharges about 10^9 J of energy across a potential difference of about 10^7 V. In a major electrical storm, the total potential energy accumulated and discharged is enormous.

To put Equation 18.2 on a "work per unit charge" basis, we divide both sides by q', obtaining

$$\frac{W_{a \to b}}{q'} = \frac{U_a}{q'} - \frac{U_b}{q'} = V_a - V_b, \tag{18.11}$$

where $V_a = U_a/q'$ is the potential energy per unit charge at point a and V_b is that at b. We call V_a and V_b the *potential at point a* and *potential at point b*, respectively. The potential difference $V_a - V_b$ is called *the potential of a with respect to b*.

EXAMPLE 18.2   Parallel plates and conservation of energy

A 9.0 V battery is connected across two large parallel plates that are separated by 4.5 mm of air, creating a potential difference of 9.0 V between the plates. **(a)** What is the electric field in the region between the plates? **(b)** An electron is released from rest at the negative plate. If the only force on the electron is the electric force exerted by the electric field of the plates, what is the speed of the electron as it reaches the positive plate? The mass of an electron is $m_e = 9.11 \times 10^{-31}$ kg.

SOLUTION

SET UP Figure 18.8 shows our sketch. We use a to designate the electron's starting position at the negative plate and b for its final position at the positive plate. Then $V_b - V_a = +9.0$ V. The electric field is directed from the positive plate b toward the negative plate a (i.e., from higher potential toward lower potential), and it is uniform between the plates.

SOLVE Part (a): The expression $V_b - V_a$ is the potential at point b with respect to point a. This quantity (work per unit charge) is related to the electric field E (force per unit charge) between the plates by $V_b - V_a = Ed$, where d is the separation between the plates and Ed is the work per unit charge on a positively charged particle that moves from b to a. Thus,

$$E = \frac{V_b - V_a}{d} = \frac{9.0 \text{ V}}{4.5 \times 10^{-3} \text{ m}} = 2.0 \times 10^3 \text{ V/m}.$$

Part (b): Conservation of energy applied to points a and b at the corresponding plates gives

$$K_a + U_a = K_b + U_b.$$

Also, $U = q'V$, where $q' = -e$, the charge of an electron. Using this expression to replace U in the conservation-of-energy equation gives

$$K_a + q'V_a = K_b + q'V_b.$$

The electron is released from rest from point a, so $K_a = 0$. We next solve for K_b:

$$K_b = q'(V_a - V_b) = -e(V_a - V_b) = +e(V_b - V_a)$$
$$= (1.60 \times 10^{-19} \text{ C})(9.0 \text{ V})$$
$$= 1.44 \times 10^{-18} \text{ J}.$$

Then $K_b = \frac{1}{2}m_e v_b^2$ gives

$$v_b = \sqrt{\frac{2K_b}{m_e}} = \sqrt{\frac{2(1.44 \times 10^{-18} \text{ J})}{9.11 \times 10^{-31} \text{ kg}}} = 1.8 \times 10^6 \text{ m/s}.$$

ALTERNATIVE SOLUTION Part (b) could also be done by calculating the acceleration of the electron. We use a y coordinate with the origin at point b and the $+y$ axis pointing toward a. Newton's second law then gives

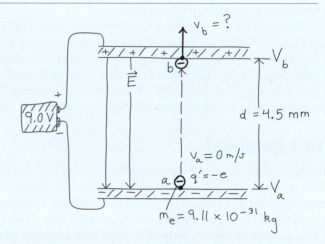

▲ **FIGURE 18.8** Our sketch for this problem.

$$a = \frac{F}{m} = \frac{eE}{m} = \frac{(1.60 \times 10^{-19} \text{ C})(2.0 \times 10^3 \text{ N/C})}{9.11 \times 10^{-31} \text{ kg}}$$
$$= 3.51 \times 10^{14} \text{ m/s}^2.$$

Also, $v_{0y} = 0$ and $y = 4.5 \times 10^{-3}$ m, so, from the relation $v^2 = v_{0y}^2 + 2a_y y$, we find

$$v_y = \sqrt{2a_y y} = \sqrt{2(3.51 \times 10^{14} \text{ m/s}^2)(4.5 \times 10^{-3} \text{ m})}$$
$$= 1.8 \times 10^6 \text{ m/s},$$

in agreement with the result from conservation of energy.

REFLECT Remember that electric fields point away from positive charges and toward negative charges and that they point in the direction of decreasing potential. Negative charges gain kinetic energy when they move to higher potential, because when V increases (becomes more positive), the electrical potential energy of a negative charge decreases (becomes less positive or more negative). When the electron moves from a to b, it loses potential energy and gains kinetic energy.

Practice Problem: Repeat the preceding problem, replacing the 9.0 V battery by an 18.0 V battery. *Answers:* (a) 4.0×10^3 V/m; (b) 2.5×10^6 m/s.

Potential of a Point Charge

The potential V (potential energy per unit charge) due to a point charge q, at any distance r from the charge, is obtained by dividing Equation 18.8 by the test charge q':

Potential of a point charge

When a test charge q' is a distance r from a point charge q, the potential V is

$$V = \frac{U}{q'} = k\frac{q}{r}, \qquad (18.12)$$

where k is the same constant as in Coulomb's law (Equation 17.1).

Similarly, to find the potential V at a point due to any collection of point charges q_1, q_2, q_3, ... at distances r_1, r_2, r_3, ..., respectively, from q', we divide Equation 18.9 by q':

$$V = \frac{U}{q'} = k\left(\frac{q_1}{r_1} + \frac{q_2}{r_2} + \frac{q_3}{r_3} + \cdots\right). \qquad (18.13)$$

In deriving Equations 18.8 and 18.9, we assumed that the potential energy of a point charge is zero at an infinite distance from the charge; thus, the V defined by Equation 18.13 is zero at points infinitely far away from *all* the charges. We could add any constant to Equation 18.13 without changing the meaning, because only *differences* between potentials at two points are physically significant.

As noted earlier, the difference $V_a - V_b$ is called the *potential of a with respect to b;* we sometimes abbreviate this difference as $V_{ab} = V_a - V_b$. This is sometimes called the *potential difference between a and b,* but that's ambiguous unless we specify which is the reference point (that is, which point is at higher potential). Note that potential, like electric field, is independent of the test charge q' that we use to define it. When a positive test charge moves from a point of higher to a point of lower potential (that is, $V_a > V_b$), the electric field does positive work on it. A positive charge tends to "fall" from a high-potential region to a region with lower potential. The opposite is true for a negative charge.

▲ **BIO Application Are you all right up there?** Have you ever wondered why birds can perch safely on power lines? The answer is that no current flows through the bird because the bird does not offer a path to a point at lower potential—provided that it touches only the wire. Large hawks and other raptors are big enough that an outstretched wing can touch the power pole, a transformer, or another wire. If the bird bridges two structures at sufficiently different potentials, it can be electrocuted.

Quantitative Analysis 18.2   **Force and potential**

A positively charged particle is placed on the x axis in a region where the electrical potential due to other charges increases in the $+x$ direction, but does not change in the y or z direction. The particle

A. is acted upon by a force in the $+x$ direction.

B. is acted upon by a force in the $-x$ direction.

C. is not acted upon by any force.

SOLUTION   Since potential is potential energy per unit charge, the direction of decreasing potential (the $-x$ direction) is the direction in which the particle's potential energy decreases. If the particle moves in this direction, the electric field does positive work on it, increasing its kinetic energy. Thus, that is the direction of the force exerted by the electric field on the particle, and B is correct. If this isn't clear, consider a gravitational analogy: When an object falls downward in a gravitational field, in the direction of the weight force, its potential energy U_{grav} decreases and its kinetic energy K increases.

EXAMPLE 18.3   **Potential of two point charges**

Two electrons are held in place 10.0 cm apart. Point a is midway between the two electrons, and point b is 12.0 cm directly above point a. **(a)** Calculate the electric potential at point a and at point b. **(b)** A third electron is released from rest at point b. What is the speed of this electron when it is far from the other two electrons? The mass of an electron is $m_e = 9.11 \times 10^{-31}$ kg.

Continued

SOLUTION

SET UP Figure 18.9 shows our sketch. Point b is a distance $r_b = \sqrt{(12.0\text{ cm}^2) + (5.0\text{ cm})^2} = 13.0$ cm from each electron.

SOLVE Part (a): The electric potential V at each point is the sum of the electric potentials of each electron: $V = V_1 + V_2 = k\dfrac{q_1}{r_1} + k\dfrac{q_2}{r_2}$, with $q_1 = q_2 = -e$. At point a, $r_1 = r_2 = r_a = 0.050$ m, so

$$V_a = -\frac{2ke}{r_a} = -\frac{2(8.99 \times 10^9\text{ N} \cdot \text{m}^2/\text{C}^2)(1.60 \times 10^{-19}\text{ C})}{0.050\text{ m}}$$
$$= -5.8 \times 10^{-8}\text{ V}.$$

At point b, $r_1 = r_2 = r_b = 0.130$ m, so

$$V_b = -\frac{2ke}{r_b} = -\frac{2(8.99 \times 10^9\text{ N} \cdot \text{m}^2/\text{C}^2)(1.60 \times 10^{-19}\text{ C})}{0.130\text{ m}}$$
$$= -2.2 \times 10^{-8}\text{ V}.$$

Part (b): Remember that our equation for potential assumes that U is zero when $r = \infty$. Thus, when the third electron is far from the other two (at a location we designate c), we can assume that $U = 0$. To find the electron's speed at point c, we use conservation of energy:

$$K_b + U_b = K_c + U_c.$$

We solve for K_c. First we use $U = q'V$ and $q' = -e$ to rewrite the preceding expression as

$$K_b - eV_b = K_c - eV_c.$$

We know that $V_c = 0$ because $V_c = \dfrac{kq}{r_c}$ and r_c is very large. Also, $K_b = 0$ because the electron is at rest before it is released. Then

$$K_c = -eV_b = -(1.60 \times 10^{-19}\text{ C})(-2.2 \times 10^{-8}\text{ V})$$
$$= +3.52 \times 10^{-27}\text{ J},$$
$$K_c = \tfrac{1}{2}m_e v_c^2,$$

so

$$v_c = \sqrt{\frac{2K_c}{m_e}} = \sqrt{\frac{2(3.52 \times 10^{-27}\text{ J})}{9.11 \times 10^{-31}\text{ kg}}} = 88\text{ m/s}.$$

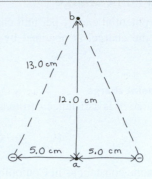

▲ **FIGURE 18.9** Our sketch for this problem.

REFLECT Remember that electric potential is a scalar quantity. We never talk about components of V; there is no such thing. When we add potentials caused by two or more point charges, the operation is simple scalar addition, not vector addition. But the sign of V, determined by the sign of the q that produces V, is important. Note that at point a the electric fields of the two electrons are equal in magnitude and opposite in direction and sum to zero. But the potentials for the electrons are both negative and *do not* add to zero. Make sure you understand this distinction.

When the third electron (the one that moves) is at point b, the electric potential energy is $U_b = -eV_b = +3.52 \times 10^{-27}$ J. At point c, the potential energy is zero. All of the initial electrical potential energy has been converted to kinetic energy because of the positive work done on it by the repulsive forces of the other two electrons. The negatively charged electron gains kinetic energy when it moves from a lower-potential point to a higher-potential point, in this case from $V_b = -2.2 \times 10^{-8}$ V to $V_c = 0$.

Note that the net force on the third electron decreases as that electron moves away from point b. Its acceleration is not constant, so constant-acceleration equations *cannot* be used to find its final speed. But conservation-of-energy principles are easy to apply.

Practice Problem: The electron at point b is replaced with a proton (mass $m_p = 1.67 \times 10^{-27}$ kg) that is released from rest and accelerates toward point a. What is the speed of the proton when it reaches point a? *Answer:* 2.6 m/s.

PROBLEM-SOLVING STRATEGY 18.1 **Calculation of potential**

SET UP AND SOLVE

1. Remember that potential is simply *potential energy per unit charge*. Understanding this simple statement can get you a long way.
2. To find the potential due to a collection of point charges, use Equation 18.13.
3. If you are given an electric field, or if you can find it by using any of the methods of Chapter 17, it may be easier to calculate the work done on a test charge during a displacement from point a to point b. When it's appropriate, make use of your freedom to define V to be zero at some convenient place. For point charges, this will usually be at infinity, but for other distributions of charge, it may be convenient or necessary to define V to be zero at some finite distance from the charge distribution—say, at point b. This is just like defining U to be zero at ground level in gravitational problems.

REFLECT

4. Remember that potential is a *scalar* quantity, not a *vector*. It doesn't have components. It would be seriously wrong to try to use components of potential.

EXAMPLE 18.4   **Parallel plates**

Find the potential at any height y between the two charged parallel plates discussed at the beginning of Section 18.1.

SOLUTION

SET UP Figure 18.10 shows our sketch. As before, we point the y axis upward. The electric field is uniform and directed vertically downward. We choose the potential V to be zero at $y = 0$ (point b in our sketch). The potential increases linearly as we move toward the upper plate.

SOLVE The potential energy U for a test charge q' at a distance y above the bottom plate is given by Equation 18.5, $U = q'Ey$. The potential V at point y is the potential energy per unit charge, $V = U/q'$, so

$$V = Ey.$$

Even if we had chosen a different reference level (at which $V = 0$), it would still be true that $V_y - V_b = Ey$. At point a, where $y = d$ and $V_y = V_a$, $V_a - V_b = Ed$ and

$$E = \frac{V_a - V_b}{d} = \frac{V_{ab}}{d}.$$

REFLECT The magnitude of the electric field equals the potential difference between the plates, divided by the distance between them. (*Caution!* This relation holds only for the parallel-plate

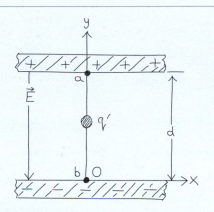

▲ FIGURE 18.10 Our sketch for this problem.

arrangement described here, in which the electric field is uniform.)

Practice Problem: Suppose that in this problem we had chosen the potential to be zero at the upper plate, where $y = d$. Derive an expression for the potential at any value of y. *Answer:* $V = E(y - d)$.

We have defined potential as potential energy per unit charge, so

$$1 \text{ V} = \frac{1 \text{ J}}{1 \text{ C}} = \frac{(1 \text{ N})(1 \text{ m})}{1 \text{ C}}, \quad \text{and} \quad 1 \text{ N/C} = 1 \text{ V/m}.$$

Thus, the unit of electric field can be expressed as 1 *volt per meter* (1 V/m), as well as 1 N/C:

$$1 \text{ V/m} = 1 \text{ N/C}.$$

That is, we can think of electric field either as force per unit charge or as potential difference per unit distance. In practice, the latter (the volt per meter) is the usual unit of E.

18.3 Equipotential Surfaces

Field lines (Section 17.7) help us visualize electric fields. In a similar way, the potential at various points in an electric field can be represented graphically by **equipotential surfaces.** An equipotential surface is defined as a surface on which the potential is the same at every point. In a region where an electric field is present, we can construct an equipotential surface through any point. In diagrams, we usually show only a few representative equipotentials, often with equal potential differences between adjacent surfaces. No point can be at two different potentials, so equipotential surfaces for different potentials can never touch or intersect. We can't draw three-dimensional surfaces on a two-dimensional diagram, so we draw lines representing the intersections of equipotential surfaces with the plane of the diagram.

The potential energy for a test charge is the same at every point on a given equipotential surface, so the $\vec{E}$ field does no work on a test charge when it moves from point to point on such a surface. It follows that the $\vec{E}$ field can never have a

 Electric field lines

—— Cross sections of equipotential
surfaces at 20 V intervals

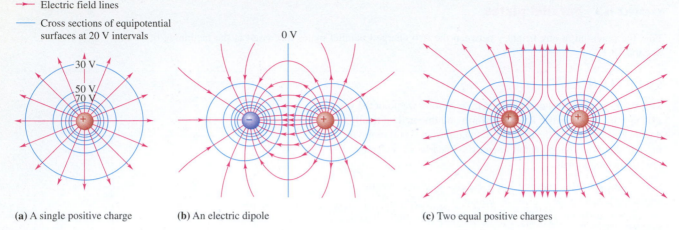

(a) A single positive charge   **(b)** An electric dipole   **(c)** Two equal positive charges

▲ **FIGURE 18.11** Equipotential surfaces and electric field lines for assemblies of point charges. How would the diagrams change if the charges were reversed?

component tangent to the surface; such a component would do work on a charge moving on the surface.

Therefore, $\vec{E}$ must be perpendicular to the surface at every point. **Field lines and equipotential surfaces are always mutually perpendicular.** In general, field lines are curves and equipotentials are curved surfaces. For the special case of a *uniform* field, in which the field lines are straight, parallel, and equally spaced, the equipotentials are parallel *planes* perpendicular to the field lines.

Figure 18.11 shows several arrangements of charges. The field lines in the plane of the charges are represented by red lines, and the intersections of the equipotential surfaces with that plane (that is, cross sections of those surfaces) are shown as blue lines. The actual field lines and equipotential surfaces are three dimensional. At each crossing of an equipotential and a field line, the two are perpendicular.

We can prove that when all charges are at rest, **the electric field just outside a conductor must be perpendicular to the surface at every point.** We know that $\vec{E} = 0$ at every point inside the conductor; otherwise, charges would move. In particular, the component of $\vec{E}$ tangent to the surface, just inside it at any point, is zero. It follows that the tangential component of $\vec{E}$ is also zero at every point just *outside* the surface. If it were not, a charge could move around a rectangular path partly inside and partly outside (Figure 18.12) and return to its starting point with a net amount of work having been done on it. This would violate the conservative nature of electrostatic fields. We conclude that the tangential component of $\vec{E}$ just outside the surface must be zero at every point on the surface. Thus, $\vec{E}$ is perpendicular to the surface at each point (Figure 18.13). It follows that, in an electrostatic situation, **a conducting surface is always an equipotential surface.**

Potential Gradient

We can draw equipotentials so that adjacent surfaces have equal potential differences. Then, in regions where the magnitude of $\vec{E}$ is large, the equipotential surfaces are close together because the field does a relatively large amount of work on a test charge in a relatively small displacement. Conversely, in regions where the field is weaker, the equipotential surfaces are farther apart.

To state this relationship more quantitatively, suppose we have two adjacent equipotential surfaces separated by a small distance Δs, with a potential difference ΔV between them. If Δs is very small, the electric field is approximately constant over that distance, so the work done by the electric field on a test charge q' that moves from one surface to the other in the direction of $\vec{E}$ is equal to $q'E\,\Delta s$. But from the definition of potential (potential energy per unit charge),

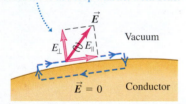

This doesn't happen!
If the electric field at the surface of a conductor had a tangential component $E_\parallel$, the electron could move in a loop with net work done.

▲ **FIGURE 18.12** At all points on the surface of a conductor, the electric field must be perpendicular to the surface. If it had a tangential component, conservation of energy could be violated, as shown here.

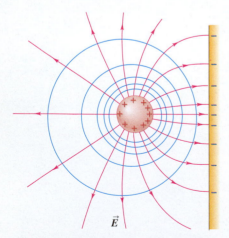

▲ **FIGURE 18.13** When charges are at rest, a conducting surface is always an equipotential surface. Field lines are perpendicular to a conducting surface.

this work is also equal to $-q'\,\Delta V$. Equating these two expressions, we find that $q'E\,\Delta s = -q'\,\Delta V$.

Electric field represented as potential gradient

The magnitude of the electric field at any point on an equipotential surface equals the rate of change of potential, ΔV, with distance Δs as the point moves perpendicularly from the surface to an adjacent one a distance Δs away:

$$E = -\frac{\Delta V}{\Delta s}. \qquad (18.14)$$

The negative sign shows that when a point moves in the direction of the electric field, the potential decreases. The quantity $\Delta V/\Delta s$, representing a rate of change of V with distance, is called the **potential gradient.** We see that this is an alternative name for electric field.

The relationship stated by Equation 18.14 can be expressed more precisely in terms of a coordinate system. If E_x represents the x component of electric field, then the correct relation is

$$E_x = -\frac{\Delta V}{\Delta x},$$

and similar equations hold for the y and z components of $\vec{E}$.

EXAMPLE 18.5 **Equipotential surfaces within a capacitor**

Suppose we have a parallel-plate capacitor like the one described at the beginning of this chapter. The plates are separated by 6.0 mm and carry charges of equal magnitude and opposite sign (Figure 18.14). The potential difference between the plates is 24.0 V. Let the potential of the negatively charged plate be zero; then the potential of the positive plate is $+24.0$ V. Draw an enlarged sketch of Figure 18.14; on it, sketch (1) the electric field lines between the plates and (2) the cross sections of the equipotential surfaces for which the potential is $+24.0$ V, $+18.0$ V, $+12.0$ V, $+6.0$ V, and 0.

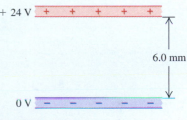

▲ **FIGURE 18.14**

SOLUTION

SET UP AND SOLVE Figure 18.15 shows our solution. Between the plates, the electric field is uniform and perpendicular to the plates, in the direction from the positive $(+)$ plate toward the negative $(-)$ plate. We draw the field lines evenly spaced to show that the field is uniform.

Conductors are equipotential surfaces, so the negative plate is an equipotential surface with $V = 0$, and the positive plate is an equipotential surface with $V = +24.0$ V. Also, $\Delta V = -E\,\Delta s$, so the equipotential surfaces between the plates are parallel to the plates. The potential increases linearly as we move from the negative to the positive plate; it changes by 24.0 V in the 6.0 mm between the plates. Thus, it changes by 4.0 V for every 1.0 mm, or by 6.0 V for every 1.5 mm. Thus, the $+6.0$ V, $+12.0$ V, and $+18.0$ V equipotential surfaces are located 1.5 mm, 3.0 mm, and 4.5 mm, respectively, from the negative plate.

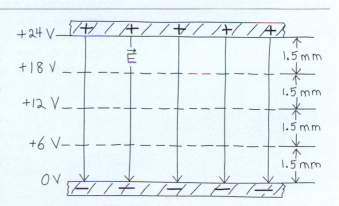

▲ **FIGURE 18.15** Our solution to this problem.

the equipotential surfaces would get closer together with increasing electric field magnitude.

REFLECT Our results demonstrate the general principle that electric field lines and equipotential surfaces are perpendicular to each other. Also, because the electric field is uniform between the plates, equipotential surfaces representing equal potential differences are equally spaced. If the electric field were not uniform,

Practice Problem: For the capacitor in this example, find (a) the magnitude of the electric field and (b) the potential (again, assuming that $V = 0$ at the negative plate) at a point 4.0 mm from the negative plate and 2.0 mm from the positive plate. *Answers:* 4.0×10^3 V/m; $+16.0$ V.

18.4 The Millikan Oil-Drop Experiment

In Section 17.3, we talked a little about the *quantization* of charge. Have you ever wondered how the charge of an individual electron can be measured? The first solution to this formidable experimental problem was the **Millikan oil-drop experiment,** a brilliant piece of work carried out at the University of Chicago during the years 1909–1913 by Robert Andrews Millikan. In 1923 Millikan was awarded the Nobel prize for this and other related fundamental research.

Millikan's apparatus is shown schematically in Figure 18.16a. Two parallel horizontal metal plates *a* and *b* are insulated from each other and separated by a few millimeters. Oil is sprayed in very fine drops (with a diameter of around 10^{-4} mm) from an atomizer above the upper plate. A few drops are allowed to fall through a small hole in this plate and are observed with a telescope. A scale in the telescope permits precise measurements of the vertical positions of the drops, so their speeds can also be measured. In the process of atomization, some of the oil drops acquire a small electric charge—usually negative, but occasionally positive.

Here's how Millikan measured the charge on a drop. Suppose a drop has a negative charge with absolute value *q* and the plates are maintained at a potential difference such that there is a downward electric field with magnitude *E* between them. The forces on the drop are then its weight *mg* and the upward force *qE*. By adjusting the field *E*, Millikan could make *qE* equal to *mg* (Figure 18.16b). The drop was then in static equilibrium, and $q = mg/E$. The electric-field magnitude *E* is the potential difference V_{ab} divided by the distance *d* between the plates, as we found in Example 18.2. The drops are pulled into spherical shapes by surface tension; the mass *m* of a drop can be determined if its radius *r* is known, because the mass equals the product of the density ρ and the volume $4\pi r^3/3$. So the expression for *q* can be rewritten as

$$q = \frac{4\pi}{3} \frac{\rho r^3 g d}{V_{ab}}.$$

Everything on the right side of this equation is easy to measure, except for the radius *r* of the drop, which is much too small to measure directly with any degree of precision. But here comes an example of Millikan's genius: He determined the drop's radius by cutting off the electric field and measuring the *terminal speed* v_t of the drop as it fell (Figure 18.16c). (You may want to review the concept of terminal speed in Example 5.13). At the terminal speed, the weight *mg* is just balanced by the drag force F_D due to air resistance. This force in turn depends on the radius of the drop, so measuring the terminal speed enabled Millikan to calculate

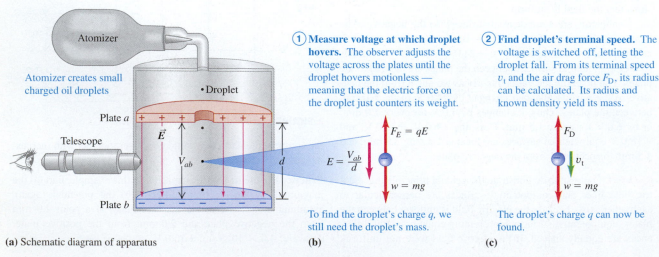

(a) Schematic diagram of apparatus

① **Measure voltage at which droplet hovers.** The observer adjusts the voltage across the plates until the droplet hovers motionless — meaning that the electric force on the droplet just counters its weight.

$$F_E = qE$$
$$E = \frac{V_{ab}}{d}$$
$$w = mg$$

To find the droplet's charge *q*, we still need the droplet's mass.

(b)

② **Find droplet's terminal speed.** The voltage is switched off, letting the droplet fall. From its terminal speed v_t and the air drag force F_D, its radius can be calculated. Its radius and known density yield its mass.

$$F_D$$
$$v_t$$
$$w = mg$$

The droplet's charge *q* can now be found.

(c)

▲ **FIGURE 18.16** The Millikan oil-drop experiment, which demonstrated that charge is quantized and provided the first determination of *e*.

the radius of a drop. Thus, he was able to determine the electric charge q of a drop in terms of its terminal speed!

Millikan and his coworkers measured the charges of thousands of drops. Within the limits of their experimental error, every drop had a charge equal to some small integer multiple of a basic charge e. That is, they found drops with charges of $\pm 2e$, $\pm 5e$, and so on, but never with a value such as $0.76e$ or $2.49e$. A drop with charge $-e$ has acquired one extra electron; if its charge is $-2e$, it has acquired two extra electrons, and so on.

As we stated in Section 17.4, the present best experimental value of the absolute value of the charge of the electron is

$$|e| = 1.60217653(14) \times 10^{-19} \text{ C},$$

where the (14) indicates the uncertainty in the last two digits, 53. This is far greater precision than Millikan was able to achieve.

The Electronvolt

The magnitude e of the charge of the electron can be used to define a unit of energy, the *electronvolt*, that is useful in many calculations with atomic and nuclear systems. When a particle with charge q moves from a point where the potential is V_a to a point where it is V_b, the change ΔU in the potential energy U is

$$\Delta U = q(V_b - V_a) = qV_{ba}.$$

If the charge q equals the magnitude e of the electron charge, namely, 1.602×10^{-19} C, and the potential difference is $V_{ba} = 1$ V, then the change in energy is

$$\Delta U = (1.602 \times 10^{-19} \text{ C})(1 \text{ V}) = 1.602 \times 10^{-19} \text{ J}.$$

This quantity of energy is defined to be 1 **electronvolt** (1 eV):

$$1 \text{ eV} = 1.602 \times 10^{-19} \text{ J}.$$

The multiples meV, keV, MeV, GeV, and TeV are often used.

When a particle with charge e moves through a potential difference of 1 V, its change in potential energy has magnitude 1 eV. If the charge is an integer multiple of e, such as Ne, the change in potential energy in electronvolts is N times the potential difference in volts. For example, when an alpha particle, which has charge $2e$, moves between two points with a potential difference of 1000 V, the change in its potential energy is $2(1000 \text{ eV}) = 2000$ eV.

Although we've defined the electronvolt in terms of *potential* energy, we can use it for *any* form of energy, such as the kinetic energy of a moving particle. When we speak of a "1-million-volt electron," we mean an electron with a kinetic energy of 1 million electronvolts (1 MeV), equal to $(10^6)(1.602 \times 10^{-19} \text{ J}) = 1.602 \times 10^{-13}$ J.

18.5 Capacitors

A capacitor is a device that stores electric potential energy and electric charge. For instance, a camera flash requires a brief burst of power much greater than a camera battery can deliver. The flash is powered by a capacitor, which stores up energy from the battery and delivers it in a pulse when needed. Capacitors are also used in energy-storage units for pulsed lasers, in computer chips that store information, in circuits that improve the efficiency of power transmission lines, and in thousands of other devices. The study of capacitors will help us to develop insight into the behavior of electric fields and their interactions with matter. Figure 18.17 shows several commercial capacitors.

In principle, a capacitor consists of any two conductors separated by vacuum or an insulating material (Figure 18.18). When charges with equal magnitude and

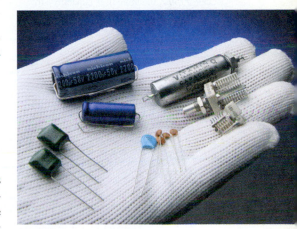

▲ **FIGURE 18.17** An assortment of practical capacitors.

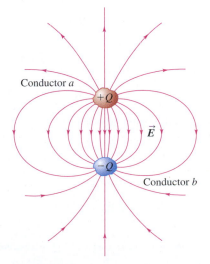

▲ **FIGURE 18.18** Any two conductors separated by vacuum or an insulating material form a capacitor.

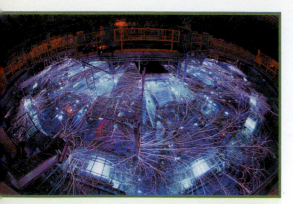

opposite sign are placed on the conductors, an electric field is established in the region between the conductors, and there is a potential difference between them. In most practical applications, the conductors have charges with equal magnitude and opposite sign, and the *net* charge on the capacitor is zero. We'll assume throughout this section that that is the case. When we say that a capacitor has charge Q, we mean that the conductor at higher potential has charge Q and the conductor at lower potential has charge $-Q$ (assuming that Q is positive). Keep this in mind in the discussion and examples that follow.

For a capacitor with a given charged surface area, the electric field at any point in the region between the conductors is proportional to the magnitude Q of the charge on each conductor. It follows that the *potential difference V_{ab}* between the conductors is also proportional to Q. If we double the charge Q on the capacitor, the electric field at each point and the potential difference between the conductors both double, but the *ratio* of charge Q to potential difference V_{ab} does not change.

We define the **capacitance C** of a capacitor as follows:

Definition of capacitance

The capacitance C of a capacitor is the ratio of the magnitude of the charge Q on *either* conductor to the magnitude of the potential difference V_{ab} between the conductors:

$$C = \frac{Q}{V_{ab}}. \tag{18.15}$$

Unit: The SI unit of capacitance is called 1 **farad** (1 F), in honor of Michael Faraday. From Equation 18.15, 1 farad is equal to 1 *coulomb per volt* (1 C/V):
$1 \text{ F} = 1 \text{ C/V}$.

In circuit diagrams, a capacitor is represented by either of these symbols:

Parallel-Plate Capacitors

The most common form of capacitor consists of two parallel conducting plates, each with area A, separated by a distance d that is small in comparison with their dimensions (Figure 18.19). Nearly all the field of such a capacitor is localized in the region between the plates, as shown. There is some "fringing" of the field at the edges, also shown in the figure, but if the distance between the plates is small in comparison to their size, we can neglect this effect. The field between the plates is then *uniform*, and the charges on the plates are uniformly distributed over their opposing surfaces. We call this arrangement a **parallel-plate capacitor.**

The electric-field magnitude in the region between the plates is directly proportional to the electric charge *per unit area* of the plate, since the charge on each plate distributes itself evenly over the surface that faces the opposite plate. This quantity is called the **surface charge density** and is denoted by the small Greek letter sigma (σ). For a pair of plates with area A and total charges Q and $-Q$, the surface charge densities on the plates are $\sigma = Q/A$ and $\sigma = -Q/A$. Gauss's law (Section 17.8) can be used to prove that, for the parallel-plate situation, the field magnitude E is related very simply to Q and σ:

$$E = \frac{\sigma}{\epsilon_0} = \frac{Q}{\epsilon_0 A}.$$

(We'll omit the details of this derivation.) We introduced the constant ϵ_0 in Section 17.4, where we said that the constant k in Coulomb's law can also be

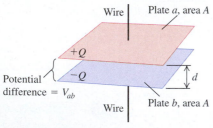

(a) A basic parallel-plate capacitor

(b) Electric field due to a parallel-plate capacitor

▲ **FIGURE 18.19** The elements of a parallel-plate capacitor.

written as $k = 1/4\pi\epsilon_0$, where $\epsilon_0 = 8.854 \times 10^{-12} \, \text{C}^2/\text{N}\cdot\text{m}^2$. The field is uniform; if the distance between the plates is d, then the potential difference (voltage) between them is

$$V_{ab} = Ed = \frac{1}{\epsilon_0}\frac{Qd}{A}.$$

From this relation, we can derive a simple expression for the capacitance of a parallel-plate capacitor:

Capacitance of a parallel-plate capacitor
The capacitance C of a parallel-plate capacitor in vacuum is directly proportional to the area A of each plate and inversely proportional to their separation d:

$$C = \frac{Q}{V_{ab}} = \epsilon_0\frac{A}{d}. \qquad (18.16)$$

The quantities ϵ_0, A, and d are constants for a given capacitor. The capacitance C is therefore a constant, independent of the charge on the capacitor.

In Equation 18.16, if A is in square meters and d in meters, C is in farads. The units of ϵ_0 are $\text{C}^2/(\text{N}\cdot\text{m}^2)$; it follows that

$$1 \, \text{F} = 1 \, \text{C}^2/(\text{N}\cdot\text{m}) = 1 \, \text{C}^2/\text{J}.$$

Because $1 \, \text{V} = 1 \, \text{J/C}$ (energy per unit charge), this relationship is consistent with our definition $1 \, \text{F} = 1 \, \text{C/V}$. Finally, the units of ϵ_0 can be expressed as

$$1 \, \text{C}^2/(\text{N}\cdot\text{m}^2) = 1 \, \text{F/m}.$$

This relation is useful in capacitance calculations, and it also helps us to verify that Equation 18.16 is dimensionally consistent.

The foregoing equations for parallel-plate capacitors are correct when there is only vacuum in the space between the plates. When matter is present, things are somewhat different. We'll return to this topic in Section 18.9.

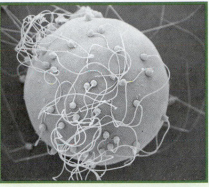

▲ **BIO Application Sea urchin contraception.** The sea urchin egg, when released into the ocean, is assaulted by thousands of sperm. If more than one sperm enters the egg, a condition known as polyspermy, the embryo will develop abnormally and die. In response to the entry of the first sperm, a protective extracellular envelope arises from the egg surface to prevent further sperm from entering. However, that process takes a minute or so, too slow to protect the egg from polyspermy. So a faster electrical response has evolved to block polyspermy. The cell membrane of the unfertilized egg has a potential difference of about 70 mV across it, with the inside of the egg negative with respect to the sea water. Since the membrane is quite thin, the electric field within the membrane is very high. When the first sperm enters the egg, ion channels in the egg membrane open to allow Na^+ to enter, rapidly reversing the potential inside the membrane to positive. A sperm can only fuse with the egg when the membrane's electric field is in the original orientation, so a second sperm trying to penetrate the egg is blocked. If an experimenter reverses the direction of the electrical field, no fertilization takes place.

Quantitative Analysis 18.3 **The effect of plate spacing**

Suppose two parallel-plate capacitors have the same area A and charge Q, but in capacitor 1 the spacing between the plates is d and in capacitor 2 it is $2d$. If the voltage between the plates in capacitor 1 is V, the voltage of capacitor 2 is

A. $\frac{1}{2}V$.
B. V.
C. $2V$.

SOLUTION There are two ways to approach this problem. First, since the surface charge density is the same for both capacitors, the electric field between their plates is the same. However, if you double the plate separation, a test charge gains or loses twice as much potential energy in moving from one plate to the other. Because voltage (potential) is potential energy divided by charge, doubling the plate separation doubles the voltage (answer C). Alternatively, we could use Equation 18.16, which tells us that $Q/V_{ab} = \epsilon_0 A/d$. Since Q and A are fixed (and ϵ_0 is a constant), doubling d also doubles V_{ab}.

EXAMPLE 18.6 **Size of a 1.0 F capacitor**

A parallel-plate capacitor has a capacitance of 1.0 F, and the plates are 1.0 mm apart. What is the area of the plates?

SOLUTION

SET UP AND SOLVE For a parallel-plate capacitor in air, $C = \epsilon_0 A/d$. Solving for A gives

$$A = \frac{Cd}{\epsilon_0} = \frac{(1.0 \, \text{F})(1.0 \times 10^{-3} \, \text{m})}{8.85 \times 10^{-12} \, \text{F/m}} = 1.1 \times 10^8 \, \text{m}^2.$$

Continued

REFLECT This area corresponds to a square about 10 km on a side, an area about a third larger than Manhattan Island! It used to be considered a good joke to send a newly graduated engineer to the stockroom for a 1.0 F capacitor. That is not as funny as it used to be; recently developed technology makes it possible to make 1.0 F capacitors that are a few *centimeters* on a side. One type uses activated carbon granules, of which 1 gram has a surface area of about 1000 m^2.

Practice Problem: What is the area of each plate of a parallel-plate air capacitor that has a capacitance of 1.0 pF = 1.0 $\times$ 10^{-12} F if the separation between the plates is 1.0 mm? If the plates are square, what is the length of each side? *Answers:* 1.1 $\times$ 10^{-4} m^2; 11 mm.

In many applications, the most convenient units of capacitance are the *microfarad* (1 μF = 10^{-6} F) and the *picofarad* (1 pF = 10^{-12} F).

EXAMPLE 18.7   Properties of a parallel-plate capacitor

The plates of a parallel-plate capacitor are 5.00 mm apart and 2.00 m^2 in area. A potential difference of 10.0 kV is applied across the capacitor. Compute **(a)** the capacitance, **(b)** the charge on each plate, and **(c)** the magnitude of the electric field in the region between the plates.

SOLUTION

SET UP Figure 18.20 shows our sketch.

SOLVE Part (a): To find the capacitance, we use Equation 18.16, which expresses capacitance in terms of the area and separation distance of the plates:

$$C = \frac{\epsilon_0 A}{d} = \frac{(8.85 \times 10^{-12}\ \text{F/m})(2.00\ \text{m}^2)}{5.00 \times 10^{-3}\ \text{m}}$$

$$= 3.54 \times 10^{-9}\ \text{F} = 0.00354\ \mu\text{F}.$$

Part (b): Now that we know the capacitance, we can use it and the voltage to find the charge on the plates. Remember that capacitance is charge divided by voltage: $C = Q/V_{ab}$. Solving for Q gives

$$Q = CV_{ab} = (3.54 \times 10^{-9}\ \text{F})(1.00 \times 10^4\ \text{V})$$

$$= 3.54 \times 10^{-5}\ \text{C} = 35.4\ \mu\text{C}.$$

Part (c): To find the electric field magnitude between the plates, we can use the relation $E = -\Delta V/\Delta s$; that is, the electric field magnitude equals the potential gradient. The minus sign in

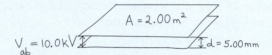

▲ **FIGURE 18.20** Our sketch for this problem.

this equation indicates that the electric field is in the direction of *decreasing* potential. Setting $\Delta V = V_{ab}$ and $\Delta s = d$, and considering only the magnitude of the electric field, we find that

$$E = \frac{V_{ab}}{d} = \frac{1.00 \times 10^4\ \text{V}}{5.00 \times 10^{-3}\ \text{m}} = 2.00 \times 10^6\ \text{V/m}.$$

REFLECT The physical dimensions of this capacitor are quite large. In practice, a capacitor with this C and large A can be constructed by rolling thin parallel conductors into a cylinder.

Practice Problem: Repeat the preceding problem for a capacitor with the same area and potential difference, but with a plate separation of 2.50 mm (half as large). Is the electric field magnitude between the plates larger or smaller than that for the original capacitor? *Answers:* (a) $C = 0.00708\ \mu$F; (b) 70.8 μC; (c) 4.00 $\times$ 10^6 V/m; larger.

18.6 Capacitors in Series and in Parallel

Capacitors are manufactured with certain standard capacitances and working voltages. However, these standard values may not be the ones you actually need in a particular circuit. You can obtain the values you need by combining capacitors; the simplest combinations are a series connection and a parallel connection.

Capacitors in Series

Figure 18.21a is a schematic diagram of a **series connection.** Two capacitors are connected in series (one after the other) between points a and b, and a constant potential difference V_{ab} is maintained. The capacitors are both initially uncharged. When a positive potential difference V_{ab} is applied between points a and b, the top plate of C_1 acquires a positive charge Q. The electric field of this positive charge pulls negative charge up to the bottom plate of C_1 until all of the field lines end on the bottom plate (Figure 18.21a) and the lower plate of C_1 has acquired charge $-Q$. These negative charges had to come from the top plate of C_2, which becomes positively charged with charge $+Q$. This positive charge then pulls negative charge $-Q$ from the connection at point b onto the bottom plate of C_2. The total charge on the lower plate of C_1 and the upper plate of C_2 must always be zero, because these plates aren't connected to anything except each other. **In a series connection, the magnitude of the charge on all of the plates is the same, because of conservation of charge. The total potential difference across all of the capacitors is the sum of the individual potential differences.**

Referring again to Figure 18.21a, we have

$$V_{ac} = V_1 = \frac{Q}{C_1}, \qquad V_{cb} = V_2 = \frac{Q}{C_2},$$

$$V_{ab} = V = V_1 + V_2 = Q\left(\frac{1}{C_1} + \frac{1}{C_2}\right),$$

$$\frac{V}{Q} = \frac{1}{C_1} + \frac{1}{C_2}.$$

The **equivalent capacitance** C_{eq} of the series combination is defined as the capacitance of a *single* capacitor for which the charge Q is the same as for the combination when the potential difference V is the same. For such a capacitor, shown in Figure 18.21b,

$$C_{eq} = \frac{Q}{V}, \qquad \text{or} \qquad \frac{1}{C_{eq}} = \frac{V}{Q}.$$

Combining the last two equations, we find that

$$\frac{1}{C_{eq}} = \frac{1}{C_1} + \frac{1}{C_2}.$$

We can extend this analysis to any number of capacitors in series:

Equivalent capacitance of capacitors in series
When capacitors are connected in series, **the reciprocal of the equivalent capacitance of a series combination equals the sum of the reciprocals of the individual capacitances:**

$$\frac{1}{C_{eq}} = \frac{1}{C_1} + \frac{1}{C_2} + \frac{1}{C_3} + \cdots. \qquad \text{(capacitors in series)} \qquad (18.17)$$

The magnitude of charge is the same on all of the plates of all of the capacitors, but the potential differences across individual capacitors are, in general, different.

Capacitors in Parallel

The arrangement shown in Figure 18.22a is called a **parallel connection.** Two capacitors are connected in parallel between points a and b. In this case, the upper plates of the two capacitors are connected together to form an equipotential surface, and the lower plates form another. **The potential difference is the same for both capacitors** and is equal to $V_{ab} = V$. The charges Q_1 and Q_2, which are not necessarily equal, are given by

$$Q_1 = C_1 V, \qquad Q_2 = C_2 V.$$

The *total* charge Q of the combination, and thus on the equivalent capacitor, is

$$Q = Q_1 + Q_2 = V(C_1 + C_2),$$

Capacitors in series:
• The capacitors have the same charge Q.
• Their potential differences add:
$V_{ac} + V_{cb} = V_{ab}$.

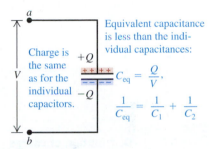

(a) Two capacitors in series

Equivalent capacitance is less than the individual capacitances:

Charge is the same as for the individual capacitors.

$$C_{eq} = \frac{Q}{V},$$

$$\frac{1}{C_{eq}} = \frac{1}{C_1} + \frac{1}{C_2}$$

(b) The equivalent single capacitor

▲ **FIGURE 18.21** The effect of connecting two capacitors in series.

Capacitors in parallel:
• The capacitors have the same potential V.
• The charge on each capacitor depends on its capacitance: $Q_1 = C_1V$, $Q_2 = C_2V$.

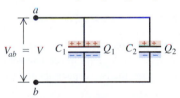

(a) Capacitors connected in parallel

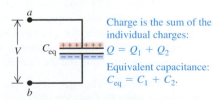

Charge is the sum of the individual charges:
$$Q = Q_1 + Q_2$$
Equivalent capacitance:
$$C_{eq} = C_1 + C_2.$$

(b) The equivalent single capacitor

▲ **FIGURE 18.22** The effect of connecting two capacitors in parallel.

so

$$\frac{Q}{V} = C_1 + C_2.$$

The *equivalent* capacitance C_{eq} of the parallel combination is defined as the capacitance of a single capacitor (Figure 18.22b) for which the total charge is the same as in Figure 18.22a. For this capacitor, $Q/V = C_{eq}$, so

$$C_{eq} = C_1 + C_2.$$

In the same way, we can derive an expression for the equivalent capacitance of any number of capacitors in parallel:

Equivalent capacitance of capacitors in parallel

When capacitors are connected in parallel, **the equivalent capacitance of the combination equals the *sum* of the individual capacitances:**

$$C_{eq} = C_1 + C_2 + C_3 + \cdots. \qquad \text{(capacitors in parallel)} \qquad (18.18)$$

In a parallel connection, the equivalent capacitance is always *greater than* any individual capacitance; in a series connection, it is always *less than* any individual capacitance.

Conceptual Analysis 18.4

Distribution of charge

The capacitors in Figure 18.23 have the same area and plate separation. The power source functions to separate charges, creating a buildup of positive charge on one side and an equal buildup of negative charge on the other. When the power source is switched on, which of the following choices describes the charge on the plates?

A. Plates A and D have equal and opposite charges; plates B and C are uncharged.
B. All plates have the same magnitude of charge; plates A and B have the same sign, plates C and D the opposite sign.
C. All plates have the same magnitude of charge; plates A and C have the same sign, plates B and D the opposite sign.

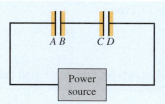

▲ **FIGURE 18.23**

SOLUTION The power source creates equal and opposite charges on plates A and D. The charge on plate A attracts an equal and opposite charge onto plate B, and the charge on plate D attracts an equal and opposite charge onto plate C. Thus, answer C is correct.

EXAMPLE 18.8 Capacitors in series and in parallel

Two capacitors, one with $C_1 = 6.0\ \mu F$ and the other with $C_2 = 3.0\ \mu F$, are connected to a potential difference of $V_{ab} = 18$ V. Find the equivalent capacitance, and find the charge and potential difference for each capacitor when the two capacitors are connected **(a)** in series and **(b)** in parallel.

SOLUTION

SET UP Figure 18.24 shows our sketches of the two situations. We remember that when capacitors are connected in series, the charges are the same on the two capacitors and the potential differences add. When they are connected in parallel, the potential differences are the same and the charges add.

SOLVE Part (a): The equivalent capacitance for the capacitors in series is given by Equation 18.17:

$$\frac{1}{C_{eq}} = \frac{1}{C_1} + \frac{1}{C_2} = \frac{C_1 + C_2}{C_1 C_2}.$$

Thus,

$$C_{eq} = \frac{C_1 C_2}{C_1 + C_2} = \frac{(6.0\ \mu F)(3.0\ \mu F)}{6.0\ \mu F + 3.0\ \mu F} = 2.0\ \mu F.$$

The charge is $Q = C_{eq} V = (2.0\ \mu F)(18\ V) = 36\ \mu C$, the same for both capacitors. The voltages are

$$V_1 = \frac{Q}{C_1} = \frac{36\ \mu C}{6.0\ \mu F} = 6.0\ V \quad \text{and} \quad V_2 = \frac{Q}{C_2} = \frac{36\ \mu C}{3.0\ \mu F} = 12.0\ V.$$

Note that $V_1 + V_2 = V_{ab}$ (i.e., 6.0 V + 12 V = 18 V).

Continued

Part (b): When capacitors are connected in parallel, the potential differences are the same and the charges add. The equivalent capacitance is given by Equation 18.18:

$$C_{eq} = C_1 + C_2 = 6.0 \ \mu\text{F} + 3.0 \ \mu\text{F} = 9.0 \ \mu\text{F}.$$

The potential difference for the equivalent capacitor is equal to the potential difference for each capacitor:

$$V_1 = V_2 = V_{ab} = 18 \text{ V}.$$

The charges of the capacitors are

$$Q_1 = C_1V = (6.0 \ \mu\text{F})(18 \text{ V}) = 108 \ \mu\text{C},$$
$$Q_2 = C_2V = (3.0 \ \mu\text{F})(18 \text{ V}) = 54 \ \mu\text{C}.$$

The total charge is $Q_1 + Q_2 = Q$, so the charge on the equivalent capacitor is $Q = C_{eq}V = (9.0 \ \mu\text{F})(18 \text{ V}) = 162 \ \mu\text{C}.$

REFLECT For capacitors in series, the *larger* potential difference appears across the capacitor with the *smaller* capacitance. For capacitors in parallel, the *greater* charge is on the capacitor with the *larger* capacitance. For capacitors in series, the equivalent capacitance is *smaller* than the capacitance of any of the individual capacitors. (Don't try to memorize rules like these; instead, focus on understanding how capacitors work so that the "rules" become self-evident.) Also, for capacitors in parallel, the equivalent capacitance is *larger* than the capacitance of any of the individual capacitors, so, for a given total potential difference, the parallel combination stores more charge than the series combination.

Practice Problem: Repeat this problem for $V_{ab} = 18$ V and $C_1 = C_2 = 9.0 \ \mu\text{F}$. Also, (c) find the ratio of total stored charge for the parallel combination to that for the series combination. *Answers:* (a) $C_{eq} = 4.5 \ \mu\text{F}$, $Q_1 = Q_2 = 81 \ \mu\text{C}$, $V_1 = V_2 = 9$ V. (b) $C_{eq} = 18 \ \mu\text{F}$, $V_1 = V_2 = 18$ V, $Q_1 = Q_2 = 162 \ \mu\text{C}$; (c) 4.0.

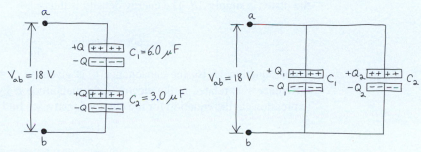

▲ **FIGURE 18.24** Our sketches for this problem.

18.7 Electric Field Energy

Many of the most important applications of capacitors depend on their ability to store energy. The capacitor plates, with opposite charges, separated and attracted toward each other, are analogous to a stretched spring or an object lifted in the earth's gravitational field. The potential energy corresponds to the energy input required to charge the capacitor and to the work done by the electrical forces when it discharges. This work is analogous to the work done by a spring or the earth's gravity when the system returns from its displaced position to the reference position.

One way to calculate the potential energy U of a charged capacitor is to calculate the work W required to charge it. The final charge Q and the final potential difference V are related by $Q = CV$. Let v and q be the varying potential difference and charge, respectively, during the charging process; then, at any instant, $v = q/C$. The work ΔW required to transfer an additional small element of charge Δq is

$$\Delta W = v \, \Delta q = \frac{q \, \Delta q}{C}.$$

Because the work needed to add Δq increases in direct proportion to the amount of charge q already present, and therefore to the potential difference v already created, the *total* work W needed to increase the charge q from zero to a final value Q is the *average* potential difference $V/2$ during the charging process, multiplied by the final charge Q:

$$U = W_{\text{total}} = \left(\frac{V}{2}\right)Q = \frac{Q^2}{2C} = \frac{1}{2}CV^2. \qquad (18.19)$$

If we define the potential energy of an *uncharged* capacitor to be zero, then W_{total} is equal to the potential energy U of the charged capacitor. When Q is in coulombs, C in farads (coulombs per volt), and V in volts (joules per coulomb), U is in joules.

A charged capacitor is the electrical analog of a stretched spring with elastic potential energy $U = \frac{1}{2}kx^2$. The charge Q is analogous to the elongation x, and the *reciprocal* of the capacitance, $1/C$, is analogous to the force constant k. The energy supplied to a capacitor in the charging process is analogous to the work we do on the spring when we stretch it.

Energy Density in an Electric Field

The energy stored in a capacitor is related directly to the electric field between the capacitor plates. In fact, we can think of the energy as stored *in the field* in the region between the plates. To develop this relation, let's find the energy *per unit volume* in the space between the plates of a parallel-plate capacitor with plate area A and separation d. We call this quantity the **energy density** and denote it by u. From Equation 18.19, the total energy is $U = \frac{1}{2}CV^2$, and the volume between the plates is simply Ad. The energy density is thus

$$u = \text{energy density} = \frac{\frac{1}{2}CV^2}{Ad}.$$

From Equation 18.16, the capacitance C is given by $C = \epsilon_0 A/d$. The potential difference V is related to the electric field magnitude E by $V = Ed$. Using these expressions in the equation for the energy density, we find that

$$u = \frac{1}{2}\epsilon_0 E^2. \tag{18.20}$$

We've derived Equation 18.20 only for one specific kind of capacitor, but it turns out to be valid for any capacitor in vacuum and, indeed, *for any electric field configuration in vacuum*. This result has an interesting implication: We think of vacuum as space with no matter in it, but vacuum can nevertheless have electric fields and therefore energy. In other words, it isn't necessarily just empty space. We'll use Equation 18.20 in Chapter 23 in connection with the energy transported through space by electromagnetic waves.

EXAMPLE 18.9 Stored energy

A capacitor with $C_1 = 8.0 \, \mu F$ is connected to a potential difference $V_0 = 120$ V, as shown in Figure 18.25a. **(a)** Find the magnitude of charge Q_0 and the total energy stored after the capacitor has become fully charged. **(b)** Without any charge being lost from the plates, the capacitor is disconnected from the source of potential difference and connected to a second capacitor $C_2 = 4.0 \, \mu F$ that is initially uncharged (Figure 18.25b). After the charge has finished redistributing between the two capacitors, find the charge and potential difference for each capacitor, and find the total stored energy.

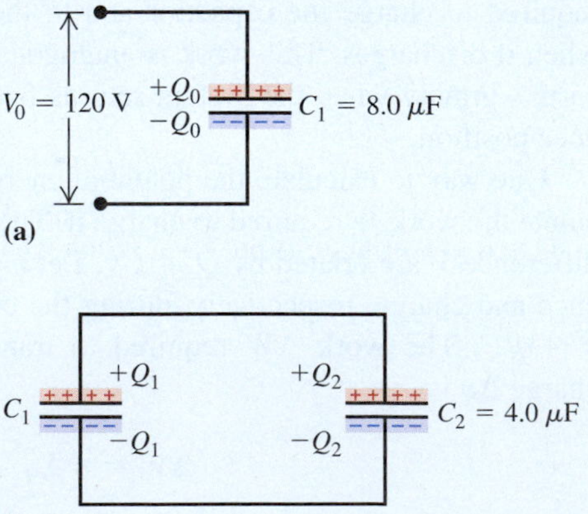

▶ **FIGURE 18.25** **(b)**

SOLUTION

SET UP After the two capacitors are connected (Figure 18.25b), the two upper plates of the capacitors are connected by a conducting wire. Therefore, they become a single conductor and form a single equipotential surface. Both lower plates are also at the same potential, different from that of the upper plates. The final potential difference between the plates, V, is thus the same for both capacitors. The final charges are given by $Q_1 = C_1 V$ and $Q_2 = C_2 V$.

Continued

The positive charge Q_0 originally on one plate of C_1 becomes distributed over the upper plates of both capacitors, and the negative charge $-Q_0$ originally on the other plate of C_1 becomes distributed over the lower plates of both.

SOLVE Part (a): For the original capacitor, we use the potential difference and the capacitance to find the charge: $Q_0 = C_1 V_0 = (8.0\ \mu\text{F})(120\ \text{V}) = 960\ \mu\text{C}$. To find the stored energy, we use Equation 18.19:

$$U = \frac{1}{2} Q_0 V_0 = \frac{1}{2}(960 \times 10^{-6}\ \text{C})(120\ \text{V}) = 0.058\ \text{J}.$$

Part (b): From conservation of charge, $Q_1 + Q_2 = Q_0$. Since V is the same for both capacitors, $Q_1 = C_1 V$ and $Q_2 = C_2 V$. When we substitute these equations into the conservation-of-charge equation, we find that $C_1 V + C_2 V = Q_0$ and

$$V = \frac{Q_0}{C_1 + C_2} = \frac{960\ \mu\text{C}}{12\ \mu\text{F}} = 80\ \text{V}.$$

Then $Q_1 = C_1 V = 640\ \mu\text{C}$ and $Q_2 = C_2 V = 320\ \mu\text{C}$.

The final total stored energy is the sum of the energies stored by each capacitor:

$$\frac{1}{2} Q_1 V + \frac{1}{2} Q_2 V = \frac{1}{2}(Q_1 + Q_2) V = \frac{1}{2} Q_0 V$$
$$= \frac{1}{2}(960 \times 10^{-6}\ \text{C})(80\ \text{V}) = 0.038\ \text{J}.$$

REFLECT As we would expect, the potential difference across C_1 decreases when the capacitors are connected, because the first capacitor gives some energy to the second one. Since the two capacitors have the same V, they are in parallel, but the concept of an equivalent capacitance is not needed here.

We see that the final stored energy is only 65% of the initial value. As charge moves to the new capacitor, some of the stored energy is converted to other forms: The conductors become a little warmer, and some energy is radiated as electromagnetic waves. We'll study these phenomena in later chapters.

Practice Problem: In this example, if $C_2 = 8.0\ \mu\text{F}$ (equal to C_1), what fraction of the original stored energy remains after C_1 is connected to C_2? *Answer: 50%.*

18.8 Dielectrics

Most capacitors have a nonconducting material, or **dielectric,** between their plates. A common type of capacitor uses two strips of metal foil for the plates and two strips of plastic sheet such as Mylar® for the dielectric. (Figure 18.26). The resulting sandwich is rolled up, forming a unit that can provide a capacitance of several microfarads in a compact package.

Placing a solid dielectric between the plates of a capacitor serves three functions. First, it solves the mechanical problem of maintaining two large metal sheets at a very small separation without actual contact.

Second, any dielectric material, when subjected to a sufficiently large electric field, undergoes *dielectric breakdown,* a partial ionization that permits conduction through a material that is supposed to be an insulator. Many insulating materials can tolerate stronger electric fields without breakdown than can air.

Third, the capacitance of a capacitor of given dimensions is *greater* when there is a dielectric material between the plates than when there is air or vacuum. We can demonstrate this effect with the aid of a sensitive electrometer, a device that measures the potential difference between two conductors without letting any appreciable charge flow from one to the other. Figure 18.27a shows an electrometer connected across a charged capacitor with charge of magnitude Q on each plate and potential difference V_0. When we insert a sheet of dielectric between the

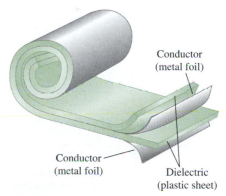

Conductor (metal foil)

Conductor (metal foil)

Dielectric (plastic sheet)

▲ **FIGURE 18.26** A common type of parallel-plate capacitor is made from a rolled-up sandwich of metal foil and plastic film.

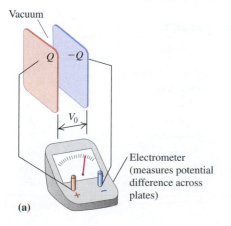

(a)

Vacuum

Q $-Q$

V_0

Electrometer (measures potential difference across plates)

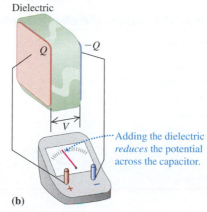

(b)

Dielectric

Q $-Q$

V

Adding the dielectric *reduces* the potential across the capacitor.

◀ **FIGURE 18.27** The effect of placing a dielectric between the plates of a parallel-plate capacitor.

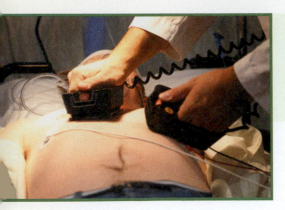

plates, the potential difference *decreases* to a smaller value V (Figure 18.27b). When we remove the dielectric, the potential difference returns to its original value V_0, showing that the original charges on the plates have not changed.

The original capacitance C_0 is given by $C_0 = Q/V_0$, and the capacitance C with the dielectric present is $C = Q/V$. The charge Q is the same in both cases, and V is less than V_0, so we conclude that the capacitance C with the dielectric present is *greater* than C_0. When the space between the plates is completely filled by the dielectric, the ratio of C to C_0 (equal to the ratio of V_0 to V) is called the **dielectric constant** of the material, K:

$$K = \frac{C}{C_0}. \qquad (18.21)$$

When the charge is constant, the potential difference is *reduced* by a factor K:

$$V = \frac{V_0}{K}.$$

The dielectric constant K is a pure number. Because C is always greater than C_0, K is always greater than 1. A few representative values of K are given in Table 18.1. For vacuum, $K = 1$ by definition. For air at ordinary temperatures and pressures, K is about 1.0006; this is so nearly equal to 1 that, for most purposes, a capacitor in air is equivalent to one in vacuum.

When a dielectric material is inserted between the plates of a capacitor while the charge is kept constant, the potential difference between the plates decreases by a factor K. Therefore, the electric field between the plates must decrease by the same factor. (Refer to Equation 18.14 if you don't see why.) If E_0 is the vacuum value and E the value with the dielectric, then

$$E = \frac{E_0}{K}. \qquad (18.22)$$

The fact that E is smaller when the dielectric is present means that the surface charge density is also smaller. The surface charge on the conducting plates does not change, but an *induced* charge of the opposite sign appears on each surface of the dielectric (Figure 18.28). This surface charge is a result of the redistribution of charge within the molecules of the dielectric material, a phenomenon called **polarization.** We'll discuss its molecular basis in Section 18.10.

For a given charge density σ_i, the induced charges on the dielectric's surfaces reduce the electric field between the plates.

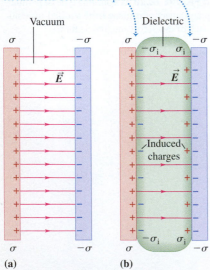

▲ **FIGURE 18.28** The effect of a dielectric on the electric field between the plates of a capacitor.

TABLE 18.1 Values of dielectric constant K at 20°C

Material	K	Material	K
Vacuum	1	Polyvinyl chloride	3.18
Air (1 atm)	1.00059	Plexiglas®	3.40
Air (100 atm)	1.0548	Glass	5–10
Teflon®	2.1	Neoprene	6.70
Polyethylene	2.25	Germanium	16
Benzene	2.28	Glycerin	42.5
Mica	3.–6	Water	80.4
Mylar®	3.1	Strontium titanate	310

EXAMPLE 18.10 **Effect of a dielectric**

The plates of an air-filled parallel-plate capacitor each have area 2.00×10^3 cm^2 and are 1.00 cm apart. The capacitor is connected to a power supply and charged to a potential difference of $V_0 = 3.00$ kV. The capacitor is then disconnected from the power supply without any charge being lost from its plates. After the capacitor has been disconnected, a sheet of insulating plastic material is inserted between the plates, completely filling the space between them. When this is done, the potential difference between the plates decreases to 1.00 kV, while the charge remains constant. **(a)** What is the capacitance of the capacitor before and after the dielectric is inserted? **(b)** What is the dielectric constant of the plastic? **(c)** What is the electric field between the plates before and after the dielectric is inserted?

SOLUTION

SET UP Figure 18.29 shows our before and after sketches.

SOLVE **Part (a):** The presence of the dielectric increases the capacitance. Without the dielectric, the capacitance is

$$C_0 = \frac{\epsilon_0 A}{d} = \frac{(8.85 \times 10^{-12} \text{ F/m})(0.200 \text{ m}^2)}{0.010 \text{ m}}$$
$$= 1.77 \times 10^{-10} \text{ F} = 177 \text{ pF}.$$

The original charge on the capacitor is

$$Q = C_0 V_0 = (1.77 \times 10^{-10} \text{ F})(3.00 \times 10^3 \text{ V})$$
$$= 5.31 \times 10^{-7} \text{ C} = 0.531 \text{ } \mu\text{C}.$$

After the dielectric is inserted, the charge is still $Q = 0.531$ μC, but now $V = 1.00$ kV, so

$$C = \frac{Q}{V} = \frac{0.531 \times 10^{-6} \text{ C}}{1.00 \times 10^3 \text{ V}} = 5.31 \times 10^{-10} \text{ F} = 531 \text{ pF}.$$

Part (b): By definition, the dielectric constant is

$$K = C/C_0 = (531 \text{ pF})/(177 \text{ pF}) = 3.00.$$

Note that this is also

$$K = V_0/V = (3.00 \text{ kV})/(1.00 \text{ kV}) = 3.00.$$

Part (c): Since the electric field is uniform between the plates, $V = Ed$. Hence,

$$E_0 = \frac{V_0}{d} = \frac{3.00 \text{ kV}}{0.0100 \text{ m}} = 3.00 \times 10^5 \text{ V/m}$$

before the dielectric is inserted, and

$$E = \frac{V}{d} = \frac{1.00 \text{ kV}}{0.0100 \text{ m}} = 1.00 \times 10^5 \text{ V/m}$$

after it is inserted.

REFLECT Both the potential difference and the electric field decrease by a factor of K. The charge Q is the same with and without the dielectric; the induced surface charge on the dielectric causes a decrease in the net field between the plates. The electric field and the potential difference are proportional, so when E decreases, V decreases.

Practice Problem: Suppose we repeat the preceding experiment, but this time keep the capacitor connected to the power supply while inserting the dielectric, so that the potential difference across the capacitor remains at 3.00 kV. Find (a) the charge on the plates and (b) the electric field between the plates before and after the dielectric is inserted? *Answers:* (a) 0.531 μC; 1.59 μC. (b) 3.00×10^5 V/m; 3.00×10^5 V/m.

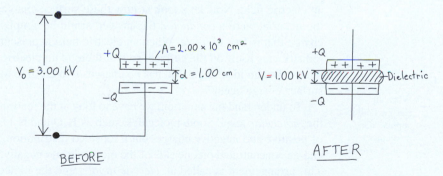

▲ **FIGURE 18.29** Our sketches for this problem.

Dielectric Breakdown

We mentioned earlier that when any dielectric material is subjected to a sufficiently strong electric field, it becomes a conductor. This phenomenon is called **dielectric breakdown.** Conduction occurs when the electric field is so strong that

▲ **FIGURE 18.30** Dielectric breakdown in the laboratory and in nature. The left-hand photo shows a block of Plexiglas® subjected to a very strong electric field; the pattern was etched by flowing charge.

electrons are ripped loose from their molecules and crash into other molecules, liberating even more electrons. This avalanche of moving charge, forming a spark or arc discharge, often starts quite suddenly. Figure 18.30a shows a beautiful laboratory example of dielectric breakdown. Lightning represents dielectric breakdown in air under the action of a sufficiently large potential difference between cloud and ground (Figure 18.30b).

Capacitors always have maximum voltage ratings. When a capacitor is subjected to excessive voltage, an arc may form through a layer of dielectric, burning or melting a hole in it. This hole then provides a conducting path between the conductors, creating a short circuit. If the path remains after the arc is extinguished, the device is rendered permanently useless as a capacitor. The maximum electric field a material can withstand without the occurrence of breakdown is called its **dielectric strength.** The dielectric strength of dry air is about 3×10^6 V/m. Values of dielectric strength for common insulating materials are typically in the range from 1 to 6×10^7 V/m, all substantially greater than that for air.

18.9 Molecular Model of Induced Charge

In the preceding section, we discussed induced surface charges on a dielectric in an electric field. Now let's look at how these surface charges can come about. If the material were a *conductor,* the answer would be simple: Conductors contain charge that is free to move. When an electric field is present, some of the charge redistributes itself on the surface, so that there is no electric field inside the conductor. But an ideal dielectric has *no* charges that are free to move, so how can a surface charge occur?

To understand the situation, we have to look at the rearrangement of charges at the *molecular* level. Some molecules, such as H_2O and N_2O, have equal amounts of positive and negative charge, but a lopsided distribution, with excess positive charge concentrated on one side of the molecule and negative charge on the other. This arrangement is called an *electric dipole,* and the molecule is called a *polar molecule.* When no electric field is present in a gas or liquid with polar molecules, the molecules are oriented randomly (Figure 18.31a). When they are placed in an electric field, however, they tend to orient themselves as in Figure 18.31b as a result of the electric-field forces.

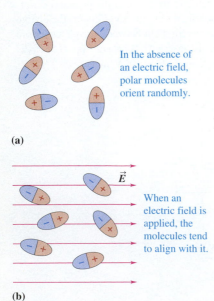

In the absence of an electric field, polar molecules orient randomly.

(a)

$\vec{E}$

When an electric field is applied, the molecules tend to align with it.

(b)

▲ **FIGURE 18.31** The effect of an electric field on a group of polar molecules.

Even a molecule that is *not* ordinarily polar *becomes* a dipole when it is placed in an electric field, because the field pushes the positive charges in the molecule in the direction of the field and pushes the negative charges in the opposite direction. This action causes a redistribution of charge within the molecules (Figure 18.32). Such dipoles are called *induced* dipoles.

With either polar or nonpolar molecules, the redistribution of charge caused by the field leads to the formation of a layer of charge on each surface of the dielectric material. The charges are not free to move indefinitely, as they would be in a conductor, because each charge is bound to a molecule. They are in fact called *bound charges,* to distinguish them from the *free charges* that are added to and removed from the conducting capacitor plates. In the interior of the material, the net charge per unit volume remains zero. This redistribution of charge is called **polarization,** and the material is said to be *polarized.*

The four parts of Figure 18.33 show the behavior of a slab of dielectric when it is inserted into the field between a pair of oppositely charged capacitor plates. Figure 18.33a shows the original field. Figure 18.33b is the situation after the dielectric has been inserted, but before any rearrangement of charges has occurred. The thinner arrows in Figure 18.33c show the additional field set up in the dielectric by the induced surface charges. This field is *opposite* to the original field, but it is not great enough to cancel it completely, because the charges in the dielectric are not free to move indefinitely. The field in the dielectric is therefore decreased in magnitude. The resultant field is shown in Figure 18.33d. In the field-line representation, some of the field lines leaving the positive plate go through the dielectric, and others terminate on the induced charges on the faces of the dielectric.

In the absence of an electric field, nonpolar molecules are not electric dipoles.

(a)

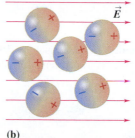

An electric field causes the molecules' positive and negative charges to separate slightly, making the molecule effectively polar.

(b)

▲ **FIGURE 18.32** The effect of an electric field on a group of nonpolar molecules.

Mastering**PHYSICS**

PhET: Molecular Motors
PhET: Optical Tweezers and Applications
PhET: Stretching DNA

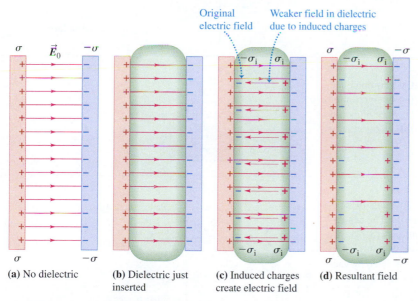

Original electric field

Weaker field in dielectric due to induced charges

(a) No dielectric

(b) Dielectric just inserted

(c) Induced charges create electric field

(d) Resultant field

▲ **FIGURE 18.33** How a dielectric reduces the electric field between capacitor plates.

SUMMARY

Electric Potential Energy

(Section 18.1) The work W done by the electric-field force on a charged particle moving in a field can be represented in terms of potential energy U: $W_{a \to b} = U_a - U_b$ (Equation 18.2). For a charge q' that undergoes a displacement $\vec{s}$ parallel to a uniform electric field, the change in potential energy is $U_a - U_b = q'Es$ (Equation 18.5). The potential energy for a point charge q' moving in the field produced by a point charge q at a distance r from q' is

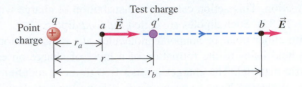

$$U = k\frac{qq'}{r}. \qquad (18.8)$$

Potential

(Section 18.2) **Potential,** a scalar quantity denoted by V, is potential energy per unit charge. The potential at any point due to a point charge is

$$V = \frac{U}{q'} = k\frac{q}{r}. \qquad (18.12)$$

A positive test charge tends to "fall" from a high-potential region to a low-potential region.

Equipotential Surfaces

(Section 18.3) An **equipotential surface** is a surface on which the potential has the same value at every point. At a point where a field line crosses an equipotential surface, the two are perpendicular. When all charges are at rest, the surface of a conductor is always an equipotential surface, and all points in the interior of a conductor are at the same potential.

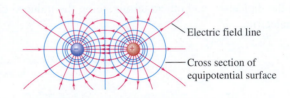

The Millikan Oil-Drop Experiment

(Section 18.4) The Millikan oil-drop experiment determined the electric charge of individual electrons by measuring the motion of electrically charged oil drops in an electric field. The size of a drop is determined by measuring its terminal speed of fall under gravity and the drag force of air.

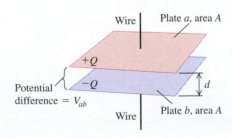

Capacitors

(Sections 18.5 and 18.6) A **capacitor** consists of any pair of conductors separated by vacuum or an insulating material. The **capacitance** C is defined as $C = Q/V_{ab}$ (Equation 18.14). A **parallel-plate capacitor** is made with two parallel plates, each with area A, separated by a distance d. If they are separated by vacuum, the capacitance is $C = \epsilon_0(A/d)$ (Equation 18.16).

When capacitors with capacitances $C_1, C_2, C_3, \ldots$ are connected in series, the equivalent capacitance C_{eq} is given by

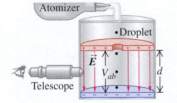

$$\frac{1}{C_{eq}} = \frac{1}{C_1} + \frac{1}{C_2} + \frac{1}{C_3} + \cdots. \qquad (18.17)$$

When they are connected in parallel, the equivalent capacitance is

$$C_{eq} = C_1 + C_2 + C_3 + \cdots. \qquad (18.18)$$

Continued

Electric Field Energy

(Section 18.7) The energy U required to charge a capacitor C to a potential difference V and a charge Q is equal to the energy stored in the capacitor and is given by

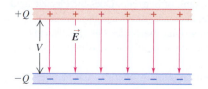

$$U = W_{\text{total}} = \left(\frac{V}{2}\right)Q = \frac{Q^2}{2C} = \frac{1}{2}CV^2. \qquad (18.19)$$

This energy can be thought of as residing in the electric field between the conductors; the energy density u (energy per unit volume) is $u = \frac{1}{2}\epsilon_0 E^2$ (Equation 18.20).

Dielectrics

(Section 18.8) When the space between the conductors is filled with a dielectric material, the capacitance *increases* by a factor K called the dielectric constant of the material. When the charges $\pm Q$ on the plates remain constant, charges induced on the surface of the dielectric *decrease* the electric field and potential difference between conductors by the same factor K. Under sufficiently strong fields, dielectrics become conductors, a phenomenon called dielectric breakdown. The maximum field that a material can withstand without breakdown is called its dielectric strength.

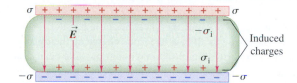

Molecular Model of Induced Charge

(Section 18.9) A *polar molecule* has equal amounts of positive and negative charge, but a lopsided distribution, with excess positive charge concentrated on one side of the molecule and negative charge on the other. When placed in an electric field, polar molecules tend to partially align with the field. For a material containing polar molecules, this microscopic alignment appears as an induced surface charge density. Even a molecule that is not ordinarily polar attains a lopsided charge distribution when it is placed in an electric field: The field pushes the positive charges in the molecule in the direction of the field and pushes the negative charges in the opposite direction.

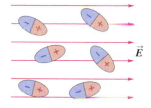

 For instructor-assigned homework, go to www.masteringphysics.com

Conceptual Questions

1. *Why* must electric-field lines be perpendicular to equipotential surfaces?

2. Which way do electric-field lines point, from high to low potential or from low to high? Explain your reasoning.

3. If the electric field is zero throughout a certain region of space, is the potential necessarily zero in that region? If not, what *can* be said about the potential?

4. The potential (relative to a point at infinity) midway between two charges of equal magnitude and opposite sign is zero. Can you think of a way to bring a test charge from infinity to this midpoint in such a way that no work is done in any part of the displacement?

5. A high-voltage dc power line falls on a car, putting the entire metal body of the car at a potential of 10,000 V with respect to the ground. What happens to the occupants (a) when they are sitting in the car and (b) if they step out of the car?

6. Since potential can have any value you want depending on the choice of reference level of zero potential, how does a voltmeter know what to read when you connect it between two points?

7. A capacitor is charged by being connected to a battery and is then disconnected from the battery. The plates are then pulled apart a little. How does each of the following quantities change as all this goes on? (a) the electric field between the plates, (b) the charge on the plates, (c) the potential difference across the plates, (d) the total energy stored in the capacitor.

8. A capacitor is charged by being connected to a battery of fixed potential and is kept connected to the battery. The plates are then pulled apart a little. How does each of the following quantities change as all this goes on? (a) the electric field between the plates, (b) the charge on the plates, (c) the potential difference across the plates, (d) the total energy stored in the capacitor?

9. Two parallel-plate capacitors, identical except that one has twice the plate separation of the other, are charged by the same voltage source. Which capacitor has a stronger electric field between the plates? Which capacitor has a greater charge? Which has greater energy density? Explain your reasoning.

10. The two plates of a capacitor are given charges $\pm Q$, and then they are immersed in a tank of benzene. Does the electric field between them increase, decrease, or remain the same?

11. Liquid dielectrics having polar molecules (such as water) have dielectric constants that *decrease* with increasing temperature. Why?

12. To store the maximum amount of energy in a parallel-plate capacitor with a given battery (voltage source), would it be better to have the plates far apart or closer together?

13. You have two capacitors and want to connect them across a voltage source (battery) to store the maximum amount of energy. Should they be connected in series or in parallel?

14. You have three capacitors, not necessarily equal, that you can connect across a battery of fixed potential. Show how you should connect these capacitors (in series or in parallel) so that (a) the capacitors will all have the same charge on their plates; (b) each capacitor will have the maximum possible charge on its plates; (c) you will store the most possible energy in the capacitor combination; (d) the capacitors will all have the same potential across them.

15. (a) If the potential (relative to infinity) is zero at a point, is the electric field necessarily zero at that point? (b) If the electric field is zero at a point, is the potential (relative to infinity) necessarily zero there? Prove your answers, using simple examples.

Multiple-Choice Problems

1. A surface will be an *equipotential* surface if (there may be more than one correct choice)
   A. the electric field is zero at all points on it.
   B. the electric field is tangent to the surface at all points.
   C. the electric field is perpendicular to the surface at all points.

2. In Figure 18.34, point P is equidistant from both point charges. At that point (there may be more than one correct choice),
   A. the electric field points directly to the right.
   B. the electric field is zero.
   C. the potential (relative to infinity) is zero.
   D. the potential (relative to infinity) points upward.

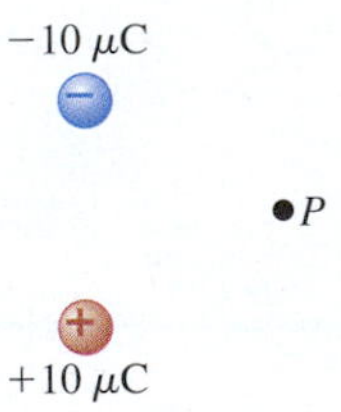

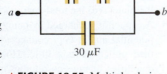

$-10\,\mu C$

$\bullet P$

$+10\,\mu C$

▲ **FIGURE 18.34** Multiple-choice problem 2.

3. For the capacitor network shown in Figure 18.35, a constant potential difference of 50 V is maintained across points a and b by a battery. Which of the following statements about this network is correct? (There may be more than one correct choice.)
   A. The $10\,\mu F$ and $20\,\mu F$ capacitors have equal charges.
   B. The charge on the $20\,\mu F$ capacitor is twice the charge on the $10\,\mu F$ capacitor.
   C. The potential difference across the $10\,\mu F$ capacitor is the same as the potential difference across the $20\,\mu F$ capacitor.
   D. The equivalent capacitance of the network is $60\,\mu F$.

$10\,\mu F$ $20\,\mu F$

a b

$30\,\mu F$

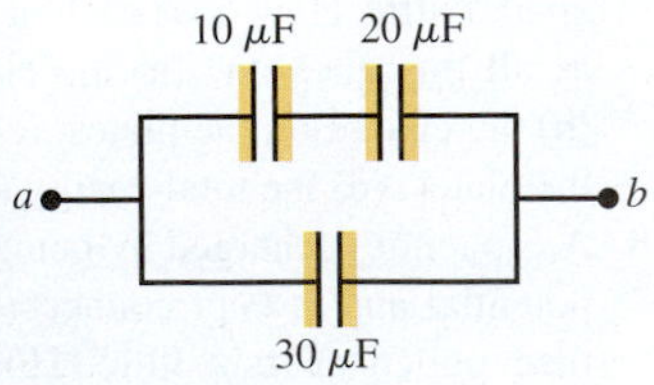

▲ **FIGURE 18.35** Multiple-choice problem 3.

4. A parallel-plate capacitor having circular plates of radius R and separation d is held at a fixed potential difference by a battery. If the plates are moved closer together while they are held at the same potential difference (there may be more than one correct choice),
   A. the amount of charge on each of them will increase.
   B. the amount of charge on each of them will decrease.
   C. the amount of charge on each of them will stay the same.
   D. the energy stored in the capacitor increases.

5. A parallel-plate capacitor having circular plates of radius R and separation d is charged to a potential difference by a battery. It is then *removed* from the battery. If the plates are moved closer together (there may be more than one correct choice),
   A. the amount of charge on each of them will increase.
   B. the amount of charge on each of them will decrease.
   C. the amount of charge on each of them will stay the same.
   D. the energy stored in the capacitor increases.

6. Two electrons close to each other are released from rest and are completely free to move. After being released (there may be more than one correct choice),
   A. their kinetic energies gradually decrease to zero as they move apart.
   B. their kinetic energies increase as they move apart.
   C. their electrical potential energy gradually decreases to zero as they move apart.
   D. their electrical potential energy increases as they move apart.
   E. their speeds gradually decrease to zero as they move apart.

7. The capacitor network shown in Figure 18.36 is connected across a fixed potential difference of 25 V. Which statements about this network must be true? (There may be more than one correct choice.)

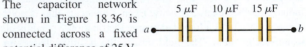

$5\,\mu F$ $10\,\mu F$ $15\,\mu F$

a b

▲ **FIGURE 18.36** Multiple-choice problem 7.

   A. The potential difference is the same across each capacitor.
   B. The charge is the same on each capacitor.
   C. The equivalent capacitance of the network is $30\,\mu F$.
   D. The equivalent capacitance of the network is less than $30\,\mu F$.

8. If the potential (relative to infinity) due to a point charge is V at a distance R from this charge, the distance at which the potential (relative to infinity) is $2V$ is
   A. $4R$.   B. $2R$.   C. $R/2$.   D. $R/4$.

9. If the electrical potential energy of two point charges is U when they are a distance d apart, their potential energy when they are twice as far apart will be
   A. $U/4$.   B. $U/2$.   C. $2U$.   D. $4U$.

10. An electron is released between the plates of a charged parallel-plate capacitor very close to the right-hand plate. Just as it reaches the left-hand plate, its speed is v. If the distance between the plates were *halved* without changing the electric potential difference between them, then the speed of the electron when it reached the left-hand plate would be
    A. $2v$.   B. $v\sqrt{2}$.   C. v.
    D. $v/\sqrt{2}$.   E. $v/2$.

11. The plates of a parallel-plate capacitor are connected across a battery of fixed potential difference and that produces a

uniform electric field E between the plates. If the plates are pulled twice as far apart, but are kept connected to the battery, the electric field between the plates will be

A. $4E$.   B. $2E$.   C. E.
D. $E/2$.   E. $E/4$.

12. At a point P a distance d from a point charge, the potential relative to infinity is V and the electric-field magnitude is E. If you now move to a point S at which the potential is $V/2$, the electric-field magnitude at S will be

A. $E/4$.   B. $E/2$.   C. $2E$.   D. $4E$.

13. When a certain capacitor carries charge of magnitude Q on each of its plates, it stores energy U. In order to store twice as much energy, how much charge should it have on its plates?

A. $\sqrt{2}Q$.   B. $2Q$.   C. $4Q$.   D. $8Q$.

14. Two large metal plates carry equal and opposite charges spread over their surfaces, as shown in Figure 18.37. Which statements about these plates are correct? (There may be more than one correct choice.)

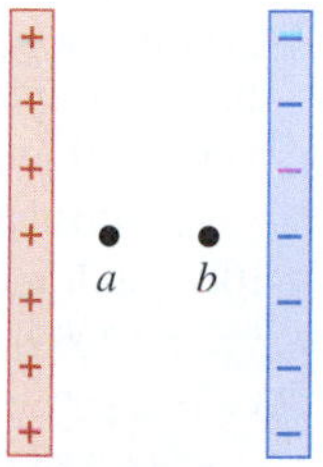

A. The electrical potential at point a is higher than the potential at point b.
B. The electrical potential at point a is equal to the potential at point b.
C. The electric-field strength at point a is equal to the field strength at point b.
D. If a positive point charge is released at point a, it will move with constant velocity toward point b.

▲ FIGURE 18.37 Multiple-choice problem 14.

15. The electric potential (relative to infinity) due to a single point charge Q is $+400$ V at a point that is 0.90 m to the right of Q. The electric potential (relative to infinity) at a point 0.90 m to the left of Q is

A. -400 V.   B. $+200$ V.   C. $+400$ V.

Problems

18.1 Electrical Potential Energy

1. • A charge of 28.0 nC is placed in a uniform electric field that is directed vertically upward and that has a magnitude of 4.00×10^4 N/C. What work is done by the electric force when the charge moves (a) 0.450 m to the right; (b) 0.670 m upward; (c) 2.60 m at an angle of 45.0° downward from the horizontal?

2. • Two very large charged parallel metal plates are 10.0 cm apart and produce a uniform electric field of 2.80×10^6 N/C between them. A proton is fired perpendicular to these plates with an initial speed of 5.20 km/s, starting at the middle of the negative plate and going toward the positive plate. How much work has the electric field done on this proton by the time it reaches the positive plate?

3. • How far from a $-7.20\ \mu$C point charge must a $+2.30\ \mu$C point charge be placed in order for the electric potential energy of the pair of charges to be -0.400 J? (Take the energy to be zero when the charges are infinitely far apart.)

4. •• A point charge $q_1 = +2.40\ \mu$C is held stationary at the origin. A second point charge $q_2 = -4.30\ \mu$C moves from the point $x = 0.150$ m, $y = 0$, to the point $x = 0.250$ m, $y = 0.250$ m. How much work is done by the electric force on q_2?

5. •• Two stationary point charges of $+3.00$ nC and $+2.00$ nC are separated by a distance of 50.0 cm. An electron is released from rest at a point midway between the charges and moves along the line connecting them. What is the electric potential energy for the electron when it is (a) at the midpoint and (b) 10.0 cm from the $+3.00$ nC charge?

6. •• **Energy of DNA base pairing, I.** (See Problem 24 in **BIO** Chapter 17; see also Figure 17.43.) (a) Calculate the electric potential energy of the adenine–thymine bond, using the same combinations of molecules $(O-H-N$ and $N-H-N)$ as in Problem 17.24. (b) Compare this energy with the potential energy of the proton–electron pair in the hydrogen atom.

7. •• **Energy of DNA base pairing, II.** (See Problem 25 in **BIO** Chapter 17; see also Figure 17.44.) Calculate the electric potential energy of the guanine–cytosine bond, using the same combinations of molecules $(O-H-O, N-H-N,$ and $O-H-N)$ as in Problem 17.25.

8. •• (a) A set of point charges is held in place at the vertices of an equilateral triangle of side 10.0 cm, as shown in Figure 18.38(a). Find the maximum amount of total kinetic energy that will be produced when the charges are released from rest in the frictionless void of outer space. (b) If the charges at the vertices of the right triangle in Figure 18.38(b) are released, how much total kinetic energy will they gain? When will this maximum kinetic energy be achieved, just following the release of the charges or after a very long time?

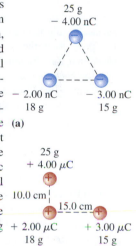

▲ FIGURE 18.38 Problem 8.

9. •• Three equal 1.20-μC point charges are placed at the corners of an equilateral triangle whose sides are 0.500 m long. What is the potential energy of the system? (Take as zero the potential energy of the three charges when they are infinitely far apart.)

10. •• When two point charges are a distance R apart, their potential energy is -2.0 J. How far (in terms of R) should they be from each other so that their potential energy is -6.0 J?

18.2 Potential

11. • Two large metal parallel plates carry opposite charges of equal magnitude. They are separated by 45.0 mm, and the potential difference between them is 360 V. (a) What is the magnitude of the electric field (assumed to be uniform) in the region between the plates? (b) What is the magnitude of the force this field exerts on a particle with charge $+2.40$ nC?

12. • A potential difference of 4.75 kV is established between parallel plates in air. If the air becomes ionized (and hence electrically conducting) when the electric field exceeds 3.00×10^6 V/m, what is the minimum separation the plates can have without ionizing the air?

13. • **Oscilloscope.** Oscilloscopes are found in most science laboratories. Inside, they contain deflecting plates consisting of more-or-less square parallel metal sheets, typically about 2.5 cm on each side and 2.0 mm apart. In many experiments,

the maximum potential across these plates is about 25 V. For this maximum potential, (a) what is the strength of the electric field between the plates, and (b) what magnitude of acceleration would this field produce on an electron midway between the plates?

14. • **Axons.** Neurons are the basic units of the nervous system. They contain long tubular structures called *axons* that propagate electrical signals away from the ends of the neurons. The axon contains a solution of potassium ions K$^+$ and large negative organic ions. The axon membrane prevents the large ions from leaking out, but the smaller K$^+$ ions are able to penetrate the membrane to some degree. (See Figure 18.39.) This leaves an excess negative charge on the inner surface of the axon membrane and an excess of positive charge on the outer surface, resulting in a potential difference across the membrane that prevents further K$^+$ ions from leaking out. Measurements show that this potential difference is typically about 70 mV. The thickness of the axon membrane itself varies from about 5 to 10 nm, so we'll use an average of 7.5 nm. We can model the membrane as a large sheet having equal and opposite charge densities on its faces. (a) Find the electric field inside the axon membrane, assuming (not too realistically) that it is filled with air. Which way does it point, into or out of the axon? (b) Which is at a higher potential, the inside surface or the outside surface of the axon membrane?

BIO

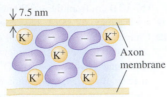
▲ FIGURE 18.39 Problem 14.

15. • **Electrical sensitivity of sharks.** Certain sharks can detect an electric field as weak as 1.0 μV/m. To grasp how weak this field is, if you wanted to produce it between two parallel metal plates by connecting an ordinary 1.5 V AA battery across these plates, how far apart would the plates have to be?

BIO

16. •• A particle with a charge of +4.20 nC is in a uniform electric field $\vec{E}$ directed to the left. It is released from rest and moves to the left; after it has moved 6.00 cm, its kinetic energy is found to be +1.50 × 10^{-6} J. (a) What work was done by the electric force? (b) What is the potential of the starting point with respect to the endpoint? (c) What is the magnitude of $\vec{E}$?

17. •• Two very large metal parallel plates are 20.0 cm apart and carry equal, but opposite, surface charge densities. Figure 18.40 shows a graph of the potential, relative to the negative plate, as a function of *x*. For this case, *x* is the distance from the inner surface of the negative plate, measured perpendicular to the

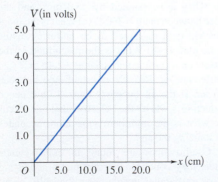
▲ FIGURE 18.40 Problem 17.

plates, and points from the negative plate toward the positive plate. Find the electric field between the plates.

18. •• A uniform electric field has magnitude E and is directed in the negative *x*-direction. The potential difference between point *a* (at *x* = 0.60 m) and point *b* (at *x* = 0.90 m) is 240 V. (a) Which point, *a* or *b*, is at the higher potential? (b) Calculate the value of E. (c) A negative point charge $q = -0.200$ μC is moved from *b* to *a*. Calculate the work done on the point charge by the electric field.

19. • A point charge has a charge of 2.50 × 10^{-11} C. At what distance from the point charge is the electric potential (a) 90.0 V? (b) 30.0 V? Take the potential to be zero at an infinite distance from the charge.

20. •• (a) An electron is to be accelerated from 3.00 × 10^6 m/s to 8.00 × 10^6 m/s. Through what potential difference must the electron pass to accomplish this? (b) Through what potential difference must the electron pass if it is to be slowed from 8.00 × 10^6 m/s to a halt?

21. •• A small particle has charge −5.00 μC and mass 2.00 × 10^{-4} kg. It moves from point *A*, where the electric potential is $V_A = +200$ V, to point *B*, where the electric potential is $V_B = +800$ V. The electric force is the only force acting on the particle. The particle has speed 5.00 m/s at point *A*. What is its speed at point *B*? Is it moving faster or slower at *B* than at *A*? Explain.

22. •• Two point charges $q_1 = +2.40$ nC and $q_2 = -6.50$ nC are 0.100 m apart. Point *A* is midway between them; point *B* is 0.080 m from q_1 and 0.060 m from q_2. (See Figure 18.41.) Take the electric potential to be zero at infinity. Find (a) the potential at point *A*; (b) the potential at point *B*; (c) the work done by the electric field on a charge of 2.50 nC that travels from point *B* to point *A*.

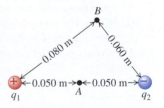
▲ FIGURE 18.41 Problem 22.

23. •• A point charge $Q = +4.60$ μC is held fixed at the origin. A second point charge $q = +1.20$ μC with mass of 2.80 × 10^{-4} kg is placed on the *x* axis, 0.250 m from the origin. (a) What is the electric potential energy U of the pair of charges? (Take U to be zero when the charges have infinite separation.) (b) The second point charge is released from rest. What is its speed when its distance from the origin is (i) 0.500 m; (ii) 5.00 m; (iii) 50.0 m?

24. •• Two protons are released from rest when they are 0.750 nm apart. (a) What is the maximum speed they will reach? When does this speed occur? (b) What is the maximum acceleration they will achieve? When does this acceleration occur?

25. •• **Cathode-ray tube.** A *cathode-ray tube* (CRT) is an evacuated glass tube. Electrons are produced at one end, usually by the heating of a metal. After being focused electromagnetically into a beam, they are accelerated through a potential difference, called the *accelerating potential*. The electrons then strike a coated screen, where they transfer their energy to the coating through collisions, causing it to glow. CRTs are found in oscilloscopes and computer monitors, as well as in earlier versions of television screens. (a) If an electron of mass m and charge $-e$ is accelerated from rest through an accelerating

potential V, show that the speed it gains is $v = \sqrt{2eV/m}$. (We are assuming that V is small enough that the final speed is much less than the speed of light.) (b) If the accelerating potential is 95 V, how fast will the electrons be moving when they hit the screen?

26. ●● **X-ray tube.** An X-ray tube is similar to a cathode-ray tube. (See previous problem.) Electrons are accelerated to high speeds at one end of the tube. If they are moving fast enough when they hit the target at the other end, they give up their energy as X-rays (a form of nonvisible light). (a) Through what potential difference should electrons be accelerated so that their speed is 1.0% of the speed of light when they hit the target? (b) What potential difference would be needed to give protons the same kinetic energy as the electrons? (c) What speed would this potential difference give to protons? Express your answer in m/s and as a percent of the speed of light.

27. ●● A gold nucleus has a radius of 7.3×10^{-15} m and a charge of $+79e$. Through what voltage must an α-particle, with its charge of $+2e$, be accelerated so that it has just enough energy to reach a distance of 2.0×10^{-14} m from the surface of a gold nucleus? (Assume the gold nucleus remains stationary and can be treated as a point charge.)

18.3 Equipotential Surfaces

28. ●● A parallel-plate capacitor having plates 6.0 cm apart is connected across the terminals of a 12 V battery. (a) Being as *quantitative* as you can, describe the location and shape of the equipotential surface that is at a potential of $+6.0$ V relative to the potential of the negative plate. Avoid the edges of the plates. (b) Do the same for the equipotential surface that is at $+2.0$ V relative to the negative plate. (c) What is the potential gradient between the plates?

29. ●● Two very large metal parallel plates that are 25 cm apart, oriented perpendicular to a sheet of paper, are connected across the terminals of a 50.0 V battery. (a) Draw *to scale* the lines where the equipotential surfaces due to these plates intersect the paper. Limit your drawing to the region between the plates, avoiding their edges, and draw the lines for surfaces that are 10.0 V apart, starting at the low-potential plate. (b) These surfaces are separated equally in potential. Are they also separated equally in distance? (c) *In words,* describe the shape and orientation of the surfaces you just found.

30. ●● (a) A $+5.00$ pC charge is located on a sheet of paper. (a) Draw *to scale* the curves where the equipotential surfaces due to these charges intersect the paper. Show only the surfaces that have a potential (relative to infinity) of 1.00 V, 2.00 V, 3.00 V, 4.00 V, and 5.00 V. (b) The surfaces are separated equally in potential. Are they also separated equally in distance? (c) *In words,* describe the shape and orientation of the surfaces you just found.

31. ●● A metal sphere carrying an evenly distributed charge will have spherical equipotential surfaces surrounding it. Suppose the sphere's radius is 50.0 cm and it carries a total charge of $+1.50\ \mu C$. (a) Calculate the potential of the sphere's surface. (b) You want to draw equipotential surfaces at intervals of 500 V outside the sphere's surface. Calculate the distance between the first and the second equipotential surfaces, and between the 20th and 21st equipotential surfaces. (c) What does the changing spacing of the surfaces tell you about the electric field?

32. ●● Figure 18.42 shows a set of electric-field lines for a particular distribution of charges. Use these lines to draw a series of equipotential surfaces for this system. Limit yourself to the plane of the paper.

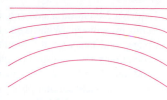

▲ **FIGURE 18.42** Problem 32.

33. ●● **Dipole.** A dipole is located on a sheet of paper. (a) In the plane of that paper, carefully sketch the electric field lines for this dipole. (b) Use your field lines in part (a) to sketch the equipotential curves where the equipotential surfaces intersect the paper.

18.4 The Millikan Oil-Drop Experiment

34. ●● In a particular Millikan oil-drop apparatus, the plates are 2.25 cm apart. The oil used has a density of 0.820 g/cm^3, and the atomizer that sprays the oil drops produces drops of diameter 1.00×10^{-3} mm. (a) What strength of electric field is needed to hold such a drop stationary against gravity if the drop contains five excess electrons? (b) What should be the potential difference across the plates to produce this electric field? (c) If another drop of the same oil requires a plate potential of 73.8 V to hold it stationary, how many excess electrons did it contain?

35. ● (a) If an electron and a proton each have a kinetic energy of 1.00 eV, how fast is each one moving? (b) What would be their speeds if each had a kinetic energy of 1.00 keV? (c) If they were each traveling at 1.00% the speed of light, what would be their kinetic energies in keV?

18.5 Capacitors

36. ● (a) You find that if you place charges of $\pm 1.25\ \mu C$ on two separated metal objects, the potential difference between them is 11.3 V. What is their capacitance? (b) A capacitor has a capacitance of $7.28\ \mu F$. What amount of excess charge must be placed on each of its plates to make the potential difference between the plates equal to 25.0 V?

37. ● The plates of a parallel-plate capacitor are 3.28 mm apart, and each has an area of 12.2 cm^2. Each plate carries a charge of magnitude 4.35×10^{-8} C. The plates are in vacuum. (a) What is the capacitance? (b) What is the potential difference between the plates? (c) What is the magnitude of the electric field between the plates?

38. ● The plates of a parallel-plate capacitor are 2.50 mm apart, and each carries a charge of magnitude 80.0 nC. The plates are in vacuum. The electric field between the plates has a magnitude of 4.00×10^6 V/m. (a) What is the potential difference between the plates? (b) What is the area of each plate? (c) What is the capacitance?

39. ● A parallel-plate air capacitor has a capacitance of 500.0 pF and a charge of magnitude $0.200\ \mu C$ on each plate. The plates are 0.600 mm apart. (a) What is the potential difference between the plates? (b) What is the area of each plate? (c) What is the electric-field magnitude between the plates? (d) What is the surface charge density on each plate?

40. ● **Capacitance of an oscilloscope.** Oscilloscopes have parallel metal plates inside them to deflect the electron beam. These plates are called the *deflecting plates.* Typically, they are

squares 3.0 cm on a side and separated by 5.0 mm, with vacuum in between. What is the capacitance of these deflecting plates and hence of the oscilloscope? (This capacitance can sometimes have an effect on the circuit you are trying to study and must be taken into consideration in your calculations.)

41. •• A 10.0 μF parallel-plate capacitor with circular plates is connected to a 12.0 V battery. (a) What is the charge on each plate? (b) How much charge would be on the plates if their separation were doubled while the capacitor remained connected to the battery? (c) How much charge would be on the plates if the capacitor were connected to the 12.0 V battery after the radius of each plate was doubled without changing their separation?

42. •• A 10.0 μF parallel-plate capacitor is connected to a 12.0 V battery. After the capacitor is fully charged, the battery is disconnected without loss of any of the charge on the plates. (a) A voltmeter is connected across the two plates without discharging them. What does it read? (b) What would the voltmeter read if (i) the plate separation were doubled; (ii) the radius of each plate was doubled, but the separation between the plates was unchanged?

43. •• You make a capacitor by cutting the 15.0-cm-diameter bottoms out of two aluminum pie plates, separating them by 3.50 mm, and connecting them across a 6.00-V battery. (a) What's the capacitance of your capacitor? (b) If you disconnect the battery and separate the plates to a distance of 3.50 cm without discharging them, what will be the potential difference between them?

44. •• A 5.00 pF parallel-plate air-filled capacitor with circular plates is to be used in a circuit in which it will be subjected to potentials of up to 1.00×10^2 V. The electric field between the plates is to be no greater than 1.00×10^4 N/C. As a budding electrical engineer for Live-Wire Electronics, your tasks are to (a) design the capacitor by finding what its physical dimensions and separation must be and (b) find the maximum charge these plates can hold.

45. •• How far apart would parallel pennies have to be to make a 1.00-pF capacitor? Does your answer suggest that you are justified in treating these pennies as infinite sheets? Explain.

46. •• A parallel-plate capacitor C is charged up to a potential V_0 with a charge of magnitude Q_0 on each plate. It is then disconnected from the battery, and the plates are pulled apart to twice their original separation. (a) What is the new capacitance in terms of C? (b) How much charge is now on the plates in terms of Q_0? (c) What is the potential difference across the plates in terms of V_0?

18.6 Capacitors in Series and in Parallel

47. • For the system of capacitors shown in Figure 18.43, find the equivalent capacitance (a) between b and c, (b) between a and c.

48. • **Electric eels.** Electric **BIO** eels and electric fish generate large potential differences that are used to stun enemies and prey. These potentials are produced by cells that each can generate 0.10 V. We can plausibly model such cells as charged capacitors. (a) How should

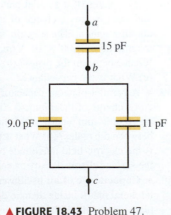

▲ **FIGURE 18.43** Problem 47.

these cells be connected (in series or in parallel) to produce a total potential of more than 0.10 V? (b) Using the connection in part (a), how many cells must be connected together to produce the 500 V surge of the electric eel?

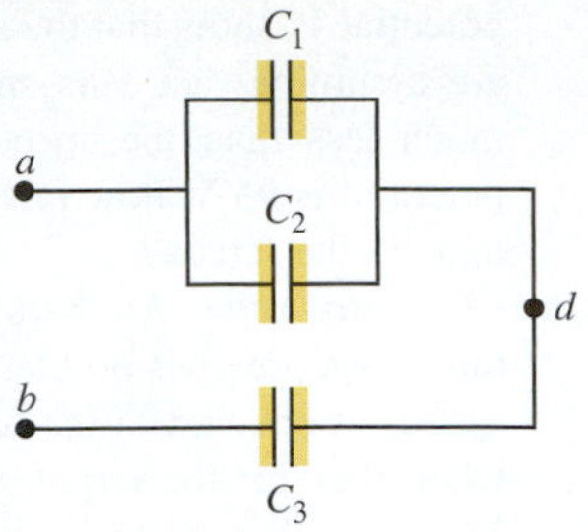

49. •• In Figure 18.44, $C_1 = 6.00 \mu$F, $C_2 = 3.00 \mu$F, and $C_3 = 5.00 \mu$F. The capacitor network is connected to an applied potential V_{ab}. After the charges on the capacitors have reached their final values, the charge on C_2 is 40.0 μC. (a) What are the charges on capacitors C_1 and C_3? (b) What is the applied voltage V_{ab}?

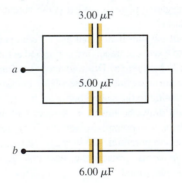

▲ **FIGURE 18.44** Problem 49.

50. •• You are working on an electronics project requiring a variety of capacitors, but have only a large supply of 100 nF capacitors available. Show how you can connect these capacitors to produce each of the following equivalent capacitances: (a) 50 nF, (b) 450 nF, (c) 25 nF, (d) 75 nF.

51. •• In Figure 18.44, $C_1 = 3.00 \mu$F and $V_{ab} = 120$ V. The charge on capacitor C_1 is 150 μC. Calculate the voltage across the other two capacitors.

52. •• A 4.00 μF and a 6.00 μF capacitor are connected in series, and this combination is connected across a 48.0 V potential difference. Calculate (a) the charge on each capacitor and (b) the potential difference across each of them.

53. •• In the circuit shown in Figure 18.45, the potential difference across ab is $+24.0$ V. Calculate (a) the charge on each capacitor and (b) the potential difference across each capacitor.

3.00 μF

a

5.00 μF

b

6.00 μF

▲ **FIGURE 18.45** Problem 53.

54. •• In Figure 18.46, each capacitor has $C = 4.00 \mu$F and $V_{ab} = +28.0$ V. Calculate (a) the charge on each capacitor and (b) the potential difference across each capacitor.

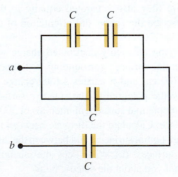

▲ **FIGURE 18.46** Problem 54.

55. •• Figure 18.47 shows a system of four capacitors, where the potential difference across *ab* is 50.0 V. (a) Find the equivalent capacitance of this system between *a* and *b*. (b) How much charge is stored by this combination of capacitors? (c) How much charge is stored in each of the 10.0 μF and the 9.0 μF capacitors?

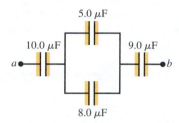

▲ **FIGURE 18.47** Problem 55.

56. •• For the system of capacitors shown in Figure 18.48, a potential difference of 25 V is maintained across *ab*. (a) What is the equivalent capacitance of this system between *a* and *b*? (b) How much charge is stored by this system? (c) How much charge does the 6.5 nF capacitor store? (d) What is the potential difference across the 7.5 nF capacitor?

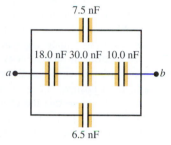

▲ **FIGURE 18.48** Problem 56.

18.7 Electric Field Energy

57. • How much charge does a 12 V battery have to supply to fully charge a 2.5 μF capacitor and a 5.0 μF capacitor when they're (a) in parallel, (b) in series? (c) How much energy does the battery have to supply in each case?

58. • A 5.80 μF parallel-plate air capacitor has a plate separation of 5.00 mm and is charged to a potential difference of 400 V. Calculate the energy density in the region between the plates, in units of J/m^3.

59. • (a) How much charge does a battery have to supply to a 5.0 μF capacitor to create a potential difference of 1.5 V across its plates? How much energy is stored in the capacitor in this case? (b) How much charge would the battery have to supply to store 1.0 J of energy in the capacitor? What would be the potential across the capacitor in that case?

60. • In the text, it was shown that the energy stored in a capacitor *C* charged to a potential *V* is $U = \frac{1}{2}QV$. Show that this energy can also be expressed as (a) $U = Q^2/2C$ and (b) $U = \frac{1}{2}CV^2$.

61. •• A parallel-plate vacuum capacitor has 8.38 J of energy stored in it. The separation between the plates is 2.30 mm. If the separation is decreased to 1.15 mm, what is the energy stored (a) if the capacitor is disconnected from the potential source so the charge on the plates remains constant, and (b) if the capacitor remains connected to the potential source so the potential difference between the plates remains constant?

62. •• (a) How many excess electrons must be added to one plate and removed from the other to give a 5.00 nF parallel-plate capacitor 25.0 μJ of stored energy? (b) How could you modify the geometry of this capacitor to get it to store 50.0 μJ of energy without changing the charge on its plates?

63. •• For the capacitor network shown in Figure 18.49, the potential difference across *ab* is 36 V. Find (a) the total charge stored in this network, (b) the charge on each capacitor, (c) the total energy stored in the network, (d) the energy stored in each capacitor, and (e) the potential difference across each capacitor.

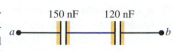

▲ **FIGURE 18.49** Problem 63.

64. •• For the capacitor network shown in Figure 18.50, the potential difference across *ab* is 220 V. Find (a) the total charge stored in this network, (b) the charge on each capacitor, (c) the total energy stored in the network, (d) the energy stored in each capacitor, and (e) the potential difference across each capacitor.

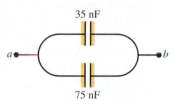

▲ **FIGURE 18.50** Problem 64.

65. •• A 20.0 μF capacitor is charged to a potential difference of 800 V. The terminals of the charged capacitor are then connected to those of an uncharged 10.0 μF capacitor. Compute (a) the original charge of the system, (b) the final potential difference across each capacitor, (c) the final energy of the system, and (d) the decrease in energy when the capacitors are connected.

66. •• For the capacitor network shown in Figure 18.51, the potential difference across *ab* is 12.0 V. Find (a) the total energy stored in this network and (b) the energy stored in the 4.80 μF capacitor.

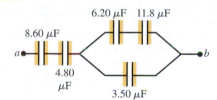

▲ **FIGURE 18.51** Problem 66.

67. •• A parallel-plate air capacitor has a capacitance of 920 pF. The charge on each plate is 2.55 μC. (a) What is the potential difference between the plates? (b) If the charge is kept constant, what will be the potential difference between the plates if the separation is doubled? (c) How much work is required to double the separation?

18.8 Dielectrics

68. • A parallel-plate capacitor has capacitance $C_0 = 5.00$ pF when there is air between the plates. The separation between the plates is 1.50 mm. (a) What is the maximum magnitude of charge Q that can be placed on each plate if the electric field in the region between the plates is not to exceed 3.00×10^4 V/m? (b) A dielectric with $K = 2.70$ is inserted between the plates of the capacitor, completely filling the volume between the plates. Now what is the maximum magnitude of charge on each plate if the electric field between the plates is not to exceed 3.00×10^4 V/m?

69. • **Cell membranes.** Cell mem-
BIO branes (the walled enclosure around a cell) are typically about 7.5 nm thick. They are partially permeable to allow charged material to pass in and out, as needed. Equal but oppo-site charge densities build up

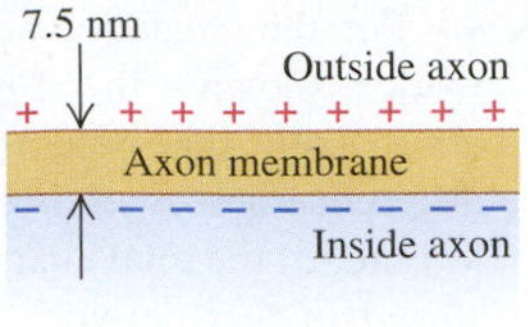

7.5 nm

Outside axon

Axon membrane

Inside axon

▲ **FIGURE 18.52** Problem 69.

on the inside and outside faces of such a membrane, and these charges prevent additional charges from passing through the cell wall. We can model a cell membrane as a parallel-plate capacitor, with the membrane itself containing proteins embedded in an organic material to give the membrane a dielectric constant of about 10. (See Figure 18.52.) (a) What is the capacitance per square centimeter of such a cell wall? (b) In its normal resting state, a cell has a potential difference of 85 mV across its membrane. What is the electric field inside this membrane?

70. •• A parallel-plate capacitor is to be constructed by using, as a dielectric, rubber with a dielectric constant of 3.20 and a dielectric strength of 20.0 MV/m. The capacitor is to have a capacitance of 1.50 nF and must be able to withstand a maxi-mum potential difference of 4.00 kV. What is the minimum area the plates of this capacitor can have?

71. •• A 12.5 μF capacitor is connected to a power supply that keeps a constant potential difference of 24.0 V across the plates. A piece of material having a dielectric constant of 3.75 is placed between the plates, completely filling the space between them. (a) How much energy is stored in the capacitor before and after the dielectric is inserted? (b) By how much did the energy change during the insertion? Did it increase or decrease?

72. •• The paper dielectric in a paper-and-foil capacitor is 0.0800 mm thick. Its dielectric constant is 2.50, and its dielectric strength is 50.0 MV/m. Assume that the geometry is that of a parallel-plate capacitor, with the metal foil serving as the plates. (a) What area of each plate is required for a 0.200 μF capacitor? (b) If the electric field in the paper is not to exceed one-half the dielectric strength, what is the maxi-mum potential difference that can be applied across the capacitor?

73. •• A constant potential difference of 12 V is maintained between the terminals of a 0.25-μF, parallel-plate, air capaci-tor. (a) A sheet of Mylar is inserted between the plates of the capacitor, completely filling the space between the plates. When this is done, how much additional charge flows onto the positive plate of the capacitor (see Table 18.1)? (b) What is the total induced charge on either face of the Mylar sheet? (c) What effect does the Mylar sheet have on the electric field between the plates? Explain how you can reconcile this with the increase in charge on the plates, which acts to *increase* the electric field.

General Problems

74. •• (a) If a spherical raindrop of radius 0.650 mm carries a charge of −1.20 pC uniformly distributed over its volume, what is the potential at its surface? (Take the potential to be zero at an infinite distance from the raindrop.) (b) Two identi-cal raindrops, each with radius and charge specified in part (a),

collide and merge into one larger raindrop. What is the radius of this larger drop, and what is the potential at its surface, if its charge is uniformly distributed over its volume?

75. •• At a certain distance from a point charge, the potential and electric-field magnitude due to that charge are 4.98 V and 12.0 V/m, respectively. (Take the potential to be zero at infin-ity.) (a) What is the distance to the point charge? (b) What is the magnitude of the charge? (c) Is the electric field directed toward or away from the point charge?

76. •• Two oppositely charged identical insulating spheres, each 50.0 cm in diameter and carrying a uniform charge of magnitude 175 μC, are placed 1.00 m apart center to center

▲ **FIGURE 18.53** Problem 76.

(Fig. 18.53). (a) If a voltmeter is connected between the near-est points (a and b) on their surfaces, what will it read? (b) Which point, a or b, is at the higher potential? How can you know this without any calculations?

77. •• **Potential in human cells.** Some cell walls in the human
BIO body have a layer of negative charge on the inside surface and a layer of positive charge of equal magnitude on the outside surface. Suppose that the charge density on either surface is $\pm 0.50 \times 10^{-3}$ C/m^2, the cell wall is 5.0 nm thick, and the cell-wall material is air. (a) Find the magnitude of $\vec{E}$ in the wall between the two layers of charge. (b) Find the potential difference between the inside and the outside of the cell. Which is at the higher potential? (c) A typical cell in the human body has a volume of 10^{-16} m^3. Estimate the total electric-field energy stored in the wall of a cell of this size. (*Hint:* Assume that the cell is spherical, and calculate the vol-ume of the cell wall.) (d) In reality, the cell wall is made up, not of air, but of tissue with a dielectric constant of 5.4. Repeat parts (a) and (b) in this case.

78. •• An alpha particle with a kinetic energy of 10.0 MeV makes a head-on collision with a gold nucleus at rest. What is the distance of closest approach of the two particles? (Assume that the gold nucleus remains stationary and that it may be treated as a point charge. The atomic number of gold is 79, and an alpha particle is a helium nucleus consisting of two protons and two neutrons.)

79. •• In the *Bohr model* of the hydrogen atom, a single electron revolves around a single proton in a circle of radius r. Assume that the proton remains at rest. (a) By equating the electric force to the electron mass times its acceleration, derive an expression for the electron's speed. (b) Obtain an expression for the electron's kinetic energy, and show that its magnitude is just half that of the electric potential energy. (c) Obtain an expression for the total energy, and evaluate it using $r = 5.29 \times 10^{-11}$ m. Give your numerical result in joules and in electron volts.

80. •• A proton and an alpha particle are released from rest when they are 0.225 nm apart. The alpha particle (a helium nucleus) has essentially four times the mass and two times the charge of a proton. Find the maximum *speed* and maximum *acceleration* of each of these particles. When do these maxima occur, just following the release of the particles or after a very long time?

81. •• A parallel-plate air capacitor is made from two plates 0.200 m square, spaced 0.800 cm apart. It is connected to a 120-V battery. (a) What is the capacitance? (b) What is the

charge on each plate? (c) What is the electric field between the plates? (d) What is the energy stored in the capacitor? (e) If the battery is disconnected and then the plates are pulled apart to a separation of 1.60 cm, what are the answers to parts (a), (b), (c), and (d)?

82. •• In the previous problem, suppose the battery remains connected while the plates are pulled apart. What are the answers then to parts (a), (b), (c), and (d) after the plates have been pulled apart?

83. •• A capacitor consists of two parallel plates, each with an area of 16.0 cm², separated by a distance of 0.200 cm. The material that fills the volume between the plates has a dielectric constant of 5.00. The plates of the capacitor are connected to a 300-V battery. (a) What is the capacitance of the capacitor? (b) What is the charge on either plate? (c) How much energy is stored in the charged capacitor?

84. •• Electronic flash units for cameras contain a capacitor for storing the energy used to produce the flash. In one such unit, the flash lasts for $\frac{1}{675}$ s with an average light power output of 2.70×10^5 W. (a) If the conversion of electrical energy to light is 95% efficient (the rest of the energy goes to thermal energy), how much energy must be stored in the capacitor for one flash? (b) The capacitor has a potential difference between its plates of 125 V when the stored energy equals the value calculated in part (a). What is the capacitance?

85. •• In Figure 18.54, each capacitance C_1 is 6.9 μF and each capacitance C_2 is 4.6 μF. (a) Compute the equivalent capacitance of the network between points a and b. (b) Compute the charge on each of the three capacitors nearest a and b when $V_{ab} = 420$ V.

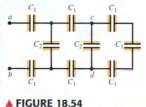

▲ **FIGURE 18.54** Problem 85.

86. •• A parallel-plate capacitor is made from two plates 12.0 cm on each side and 4.50 mm apart. Half of the space between these plates contains only air, but the other half is

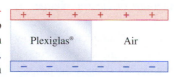

▲ **FIGURE 18.55** Problem 86.

filled with Plexiglas® of dielectric constant 3.40. (See Figure 18.55.) An 18.0 V battery is connected across the plates. (a) What is the capacitance of this combination? (*Hint:* Can you think of this capacitor as equivalent to two capacitors in parallel?) (b) How much energy is stored in the capacitor? (c) If we remove the Plexiglas®, but change nothing else, how much energy will be stored in the capacitor?

87. •• A parallel-plate capacitor with plate separation d has the space between the plates filled with two slabs of dielectric, one with constant K_1 and the other with constant K_2, and each having thickness $d/2$. (a) Show that the capacitance is given by

$$C = \frac{2\epsilon_o A}{d}\left(\frac{K_1 K_2}{K_1 + K_2}\right).$$ (*Hint:* Can you think of this combination as two capacitors in series?) (b) To see if your answer is reasonable, check it in the following cases: (i) There is only one dielectric, with constant K, and it completely fills the space between the plates. (ii) The plates have nothing but air, which we can treat as vacuum, between them.

Passage Problems

BIO The electric egg. The eggs of many species undergo a rapid change in the electrical potential difference across the outer membrane when they are fertilized. This change in potential difference affects the physiological development of the eggs. The potential difference across the membrane is called the membrane potential, V_m, defined as the inside potential minus the outside potential. The membrane potential V_m arises when protein enzymes use the energy available in ATP to actively expel sodium ions (Na⁺) and accumulate potassium ions (K⁺). Because the membrane of the unfertilized egg is selectively permeable to K⁺, the V_m of the resting sea urchin egg is about −70 mV; that is, the inside has a potential of 70 mV less than that of the outside. The egg membrane behaves as a capacitor with a specific capacitance of about 1 μF/cm². When a sea urchin egg is fertilized, Na⁺ channels in the membrane are opened, Na⁺ enters the egg, and V_m rapidly changes to +30 mV, where it remains for several minutes. The concentration of Na⁺ in the egg's interior is about 30 mmoles/liter (30 mM) and 450 mM in the surrounding sea water. The inside K⁺ concentration is about 200 mM and the outside K⁺ is 10 mM. A useful constant that connects electrical and chemical units is the Faraday number, which has a value of approximately 10^5 coulomb/mole. That is, an Avogadro number (a mole) of monovalent ions such as Na⁺ or K⁺ carries a charge of 10^5 C.

88. How many moles of Na⁺ must move per unit area of membrane to change V_m from −70 mV to +30 mV, making the assumption that the membrane behaves purely as a capacitor?
    A. 10^{-4} mole/cm²
    B. 10^{-9} mole/cm²
    C. 10^{-12} mole/cm²
    D. 10^{-14} mole/cm²

89. Suppose the egg has a diameter of 200 μm. What fractional change in internal Na⁺ concentration results from the fertilization-induced change in V_m? Assume that the Na⁺ ions are distributed throughout the cell volume.
    A. Increases by 1 part in 10^4
    B. Increases by 1 part in 10^5
    C. Increases by 1 part in 10^6
    D. Increases by 1 part in 10^7

90. Suppose the change in V_m was caused by the entry of Ca²⁺ instead of Na⁺. How many Ca²⁺ ions would have to enter the cell per unit membrane to produce the change?
    A. Half as many as for Na⁺
    B. The same as for Na⁺
    C. Twice as many as for Na⁺
    D. Cannot say without knowing the inside and outside concentrations of Ca²⁺

91. What is the minimum amount of work that must be done by the cell to restore V_m to its original value?
    A. 3 mJ
    B. 3 μJ
    C. 3 nJ
    D. 3 pJ

19 Current, Resistance, and Direct-Current Circuits

Electric circuits are at the heart of all radio and television transmitters and receivers, CD players, household and industrial power distribution systems, flashlights, computers, and the nervous systems of animals. In the past two chapters, we've studied the interactions of electric charges *at rest;* now we're ready to study charges *in motion.*

In this chapter, we'll study the basic properties of electrical conductors and how their behavior depends on temperature. We'll learn why a short, fat, cold copper wire is a better conductor than a long, skinny, hot steel wire. We'll study the properties of batteries and how they cause current and energy transfer in a circuit. In this analysis, we'll use the concepts of current, potential difference ("voltage"), resistance, and electromotive force.

This reliably sunny beachside roof is covered with electricity-generating photovoltaic panels.

19.1 Current

In the preceding two chapters, our primary emphasis was on electrostatics, the study of electrical interactions with charges at rest. We learned that, in electrostatic situations, there can be no electric field $\vec{E}$ in the interior of a conducting material; otherwise the mobile charges in the conductor would move, and the situation wouldn't be electrostatic.

In this chapter, we shift our emphasis to situations in which non-zero electric fields exist inside conductors, causing motion of the mobile charges within the conductors. A **current** (also called *electric current*) is any motion of charge from one region of a conductor to another. To maintain a steady flow of charge in a conductor, we have to maintain a steady force on the mobile charges, either with an electrostatic field or by other means that we'll consider later. For now, let's assume that there is an electric field $\vec{E}$ within the conductor, so a particle with charge q is acted upon by a force $\vec{F} = q\vec{E}$.

▶ **Application** **The secret of semiconductors.** Metals conduct electricity well because each atom gives up one or more outer electrons, which become free charge carriers. Insulators don't conduct electricity because all their electrons are firmly bound. *Semiconductors* such as silicon have loosely bound electrons that occasionally become free charge carriers. The positive "hole" left behind by a free electron can also migrate. A semiconductor can be modified so as to have mainly electrons (N-type) or mainly holes (P-type) as its free charge carriers. A junction between N and P semiconductors has the remarkable property of conducting electricity preferentially in one direction, from P to N. This property is central to the working of computer chips (like the one shown here) and many other electronic devices.

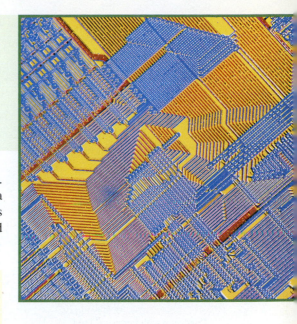

To help visualize current, let's think of conductors in the shape of wires. (Figure 19.1). Current is defined to be the amount of charge that moves through a given cross section of conductor per unit time. Thus, if a net charge ΔQ flows through a cross section during a time Δt, the current through that area, denoted by I, is defined as follows:

Definition of current

When a net charge ΔQ passes through a cross section of conductor during time Δt, the current is

$$I = \frac{\Delta Q}{\Delta t}. \tag{19.1}$$

Unit: 1 coulomb/second = 1 C/s = 1 ampere = 1 A.

Current is a *scalar* quantity. The SI unit of current is the **ampere;** one ampere is defined to be 1 *coulomb per second* $(1 \text{ A} = 1 \text{ C/s})$. This unit is named in honor of the French scientist André Marie Ampère (1775–1836). When an ordinary flashlight (D-cell size) is turned on, the current in the bulb is about 0.5 to 1 A; the current in the starter motor of a car engine is around 200 A. Currents in radio and television circuits are usually expressed in *milliamperes* $(1 \text{ mA} = 10^{-3} \text{ A})$ or *microamperes* $(1 \mu\text{A} = 10^{-6} \text{ A})$, and currents in computer circuits are expressed in *nanoamperes* $(1 \text{ nA} = 10^{-9} \text{ A})$ or *picoamperes* $(1 \text{ pA} = 10^{-12} \text{ A})$.

In Figure 19.2a, the moving charges are positive and the current flows in the same direction as the moving charges. In Figure 19.2b, they are negative, and the current is opposite to the motion of the moving charges. But the *current* is still from left to right, because in both cases the result is a net transfer of positive charge from left to right. Particles flowing out at an end of the cylindrical section are continuously replaced by particles flowing *in* at the opposite end.

When there is a steady current in a closed loop (a "complete circuit"), the total charge in every segment of the conductor is constant. From the principle of

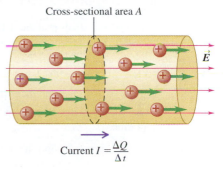

Current $I = \dfrac{\Delta Q}{\Delta t}$

▲ **FIGURE 19.1** The definition of current.

A **conventional current** is treated as a flow of positive charges, regardless of whether the free charges in the conductor are positive, negative, or both.

In a metallic conductor, the moving charges are electrons — but the *current* still points in the direction positive charges would flow.

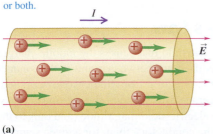

(a)

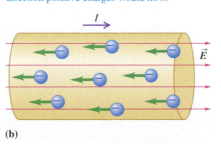

(b)

▲ **FIGURE 19.2** Conventional and electron currents.

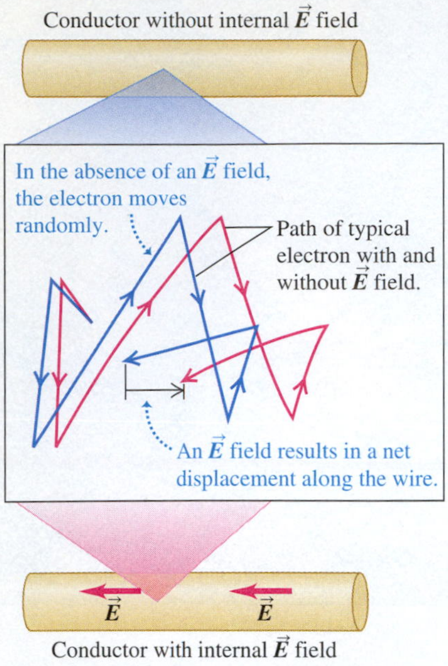

Conductor without internal $\vec{E}$ field

In the absence of an $\vec{E}$ field, the electron moves randomly.

Path of typical electron with and without $\vec{E}$ field.

An $\vec{E}$ field results in a net displacement along the wire.

$\vec{E}$ $\qquad$ $\vec{E}$

Conductor with internal $\vec{E}$ field

▲ **FIGURE 19.3** The presence of an electric field imposes a small drift (greatly exaggerated here) on an electron's random motion.

conservation of charge (Section 17.3), the rate of flow of charge *out* at one end of a segment at any instant equals the rate of flow of charge *in* at the other end of the segment at that instant, and *the current at any instant is the same at all cross sections.* Current is *not* something that squirts out of the positive terminal of a battery and is consumed or used up by the time it reaches the negative terminal.

In circuit analysis, we'll always describe currents *as though* they consisted entirely of positive charge flow, even in cases in which we know the actual current is due to (negatively charged) electrons. In metals, the moving charges are always (negative) electrons, but in an ionic solution, both positive and negative ions are moving. We speak of "conventional current" to describe a flow of positive charge that is equivalent to the actual flow of charge of either sign.

In a metal, the free electrons have a lot of random motion, somewhat like the molecules in a gas. When an electric field is applied to the metal, the forces that it exerts on the electrons lead to a small net motion, or *drift,* in the direction of the force, in addition to the random motion (Figure 19.3). Thus, the motion consists of random motion with very large average speeds (on the order of 10^6 m/s) and a much slower drift speed (often on the order of 10^{-4} m/s) in the direction of the electric-field force. But when you turn on a light switch, the light comes on almost instantaneously because the electric fields in the conductors travel with a speed approaching the speed of light. You don't have to wait for individual electrons to travel from the switch to the bulb!

EXAMPLE 19.1   How many electrons?

One of the circuits in a small portable CD player operates on a current of 2.5 mA. How many electrons enter and leave this part of the player in 1.0 s?

SOLUTION

SET UP Conservation of charge tells us that when a steady current flows, the same amount of current enters and leaves the player per unit time.

SOLVE We use the current to find the total charge that flows in 1.0 s. We have

$$I = \frac{\Delta Q}{\Delta t}, \quad \text{so}$$

$$\Delta Q = I\,\Delta t = (2.5 \times 10^{-3}\,\text{A})(1.0\,\text{s}) = 2.5 \times 10^{-3}\,\text{C}.$$

Each electron has charge of magnitude $e = 1.60 \times 10^{-19}$ C. The number N of electrons is the total charge ΔQ, divided by the magnitude of the charge e of one electron:

$$N = \frac{\Delta Q}{e} = \frac{2.5 \times 10^{-3}\,\text{C}}{1.60 \times 10^{-19}\,\text{C}} = 1.6 \times 10^{16}.$$

REFLECT It's important to realize that charge is not "used up" in a CD player (or any other electrical device); the same amount that flows in also flows out. However, as we will see, the charge loses *potential energy* as it flows through the player; this is how it powers the player's operation.

Practice Problem: The current in a wire is 2.00 A. How much time is needed for 1 mole of electrons (6.02×10^{23} electrons) to pass a point in the wire? *Answer:* 13.4 h.

Mastering PHYSICS

PhET: Conductivity
PhET: Ohm's Law
PhET: Resistance in a Wire

19.2 Resistance and Ohm's Law

In a conductor carrying a current, the electric field that causes the mobile charges to move also is associated with a potential difference between points in the conductor. The electric field $\vec{E}$ always points in the direction from higher to lower potential. The current I is proportional to the average drift speed of the moving charges. If this speed is in turn proportional to the electric-field magnitude (and thus to the potential difference between the ends of the conductor), then, for a given segment of conductor, the current I is approximately proportional to the potential difference

V between the ends. In this case, the ratio V/I is approximately constant. This ratio is called the **resistance** of the conductor; it is defined as follows:

Definition of resistance

When the potential difference V between the ends of a conductor is proportional to the current I in the conductor, the ratio V/I is called the resistance of the conductor:

$$R = \frac{V}{I}. \qquad (19.2)$$

Unit: The SI unit of resistance is the **ohm,** equal to 1 volt per ampere. The ohm is abbreviated with a capital Greek omega, Ω. Thus, $1\ \Omega = 1\ \text{V}/\text{A}$. The *kilohm* $(1\ \text{k}\Omega = 10^3\ \Omega)$ and the *megohm* $(1\ \text{M}\Omega = 10^6\ \Omega)$ are also in common use.

The observation that for many conducting materials, current is proportional to potential difference (voltage) is called **Ohm's law:**

Ohm's law

The potential difference V between the ends of a conductor is proportional to the current I through the conductor; the proportionality factor is the resistance R.

We see that when Ohm's law is obeyed, the resistance defined by Equation 19.2 is a constant, independent of V and I. Like the ideal-gas equation or Hooke's law, Ohm's law represents an *idealized model;* it describes the behavior of some materials quite well, but it isn't a general description of all materials. We'll return to this point later.

A 100 W, 120 V light bulb has a resistance of 140 Ω at the bulb's operating temperature. A 100 m length of 12 gauge copper wire (the size usually used in household wiring) has a resistance of about 0.5 Ω at room temperature. Resistors with designated resistances are used in a wide variety of electric circuits; we'll study some of their applications later in this chapter. Resistors in the range from 0.01 Ω to 10^7 Ω can be bought off the shelf. Figure 19.4 illustrates a color code that is used to label the resistance of resistors.

Resistivity

For a conductor in the form of a uniform cylinder (such as a wire with uniform cross section), the resistance R is found to be proportional to the length L of the conductor and inversely proportional to its cross-sectional area A. These relationships are reasonable: For a given current, if we double the length of conductor, the total potential difference between the ends must double; and if the cross-sectional area is doubled, a given potential difference causes twice as great a current flow. This relationship can be expressed as follows:

Definition of resistivity

The resistance R is proportional to the length L and inversely proportional to the cross-sectional area A, with a proportionality factor ρ called the **resistivity** of the material. That is,

$$R = \rho \frac{L}{A}, \qquad (19.3)$$

where ρ, in general different for different materials, characterizes the conduction properties of a material.

Unit: The SI unit of resistivity is 1 ohm $\cdot$ meter $= 1\ \Omega \cdot \text{m}$.

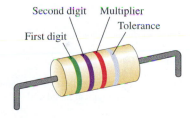

First digit Second digit Multiplier Tolerance

▲ **FIGURE 19.4** Commercial resistors use a code consisting of colored bands to indicate their resistance.

TABLE 19.1 **Resistivities at room temperature**

Substance	$\rho \ (\Omega \cdot m)$	Substance	$\rho \ (\Omega \cdot m)$
Conductors:		Mercury	95×10^{-8}
Silver	1.47×10^{-8}	Nichrome alloy	100×10^{-8}
Copper	1.72×10^{-8}	Insulators:	
Gold	2.44×10^{-8}	Glass	$10^{10} - 10^{14}$
Aluminum	2.63×10^{-8}	Lucite	$> 10^{13}$
Tungsten	5.51×10^{-8}	Quartz (fused)	75×10^{16}
Steel	20×10^{-8}	Teflon®	$> 10^{13}$
Lead	22×10^{-8}	Wood	$10^{8} - 10^{11}$

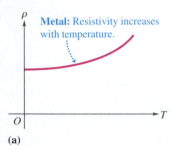

(a)

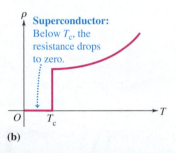

(b)

▲ **FIGURE 19.5** The temperature dependence of resistance in (a) a typical metal and (b) a superconductor.

▲ **FIGURE 19.6** A maglev train in Shanghai. Maglev ("magnetic-levitation") trains use superconducting electromagnets to create magnetic fields strong enough to levitate a train off the tracks.

Equation 19.3 shows that the resistance R of a wire or other conductor with uniform cross section is directly proportional to the length of the wire and inversely proportional to its cross-sectional area. It is also proportional to the resistivity of the material of which the conductor is made. This behavior is analogous to water flowing through a hose. A thin hose offers more resistance to flow than a fat one. We can increase the resistance to flow by stuffing the hose with cotton or sand; this corresponds to increasing the resistivity. The flow rate is approximately proportional to the pressure difference between the ends. Flow rate is analogous to current, pressure difference to potential difference (voltage). Let's not stretch this analogy too far, though; the flow rate of water in a pipe is usually *not* proportional to the pipe's cross-sectional area.

NOTE ▶ It's important to distinguish between resistivity and resistance. Resistivity is a property of a material, independent of the shape and size of the specimen, while resistance depends on the size and shape of the specimen or device, as well as on its resistivity. ◀

A few representative values of resistivity are given in Table 19.1. A perfect conductor would have zero resistivity, and a perfect insulator would have infinite resistivity. Metals and alloys have the smallest resistivities and are the best conductors. Note that the resistivities of insulators are greater than those of metals by an enormous factor, on the order of 10^{18} to 10^{22}.

Temperature Dependence of Resistance

The resistance of every conductor varies somewhat with temperature. The resistivity of a *metallic* conductor nearly always increases with increasing temperature (Figure 19.5a). Over a small temperature range (up to 100 C° or so), the change in resistivity of a metal is approximately proportional to the temperature change. If R_0 is the resistance at a reference temperature T_0 (often taken as 0°C or 20°C) and R_T is the resistance at temperature T, then the variation of R with temperature is described approximately by the equation

$$R_T = R_0[1 + \alpha(T - T_0)]. \tag{19.4}$$

The factor α is called the **temperature coefficient of resistivity.** For common metals, α typically has a value of 0.003 to 0.005 $(C°)^{-1}$. That is, an increase in temperature of 1 C° increases the resistance by 0.3% to 0.5%.

A small semiconductor crystal called a *thermistor* can be used to make a sensitive electronic thermometer. Its resistance is used as a thermometric property.

Superconductivity

Some materials, including several metallic alloys and oxides, show a phenomenon called *superconductivity*. In these materials, as the temperature decreases, the resistivity at first decreases smoothly, like that of any metal. But then, at a certain critical transition temperature T_c, a phase transition occurs, and the resistivity suddenly drops to zero, as shown in Figure 19.5b. Once a current has been established in a superconducting ring, it continues indefinitely without the presence of any driving field.

Superconductivity was discovered in 1911 in the laboratory of H. Kamerlingh-Onnes. He had just discovered how to liquefy helium, which has a boiling temperature of 4.2 K at atmospheric pressure. Measurements of the resistance of mercury at very low temperatures showed that below 4.2 K, its resistivity suddenly dropped to zero. In recent years, complex oxides of yttrium, copper, and barium have been found that have a much higher superconducting transition temperature. The current (2010) record for T_c is about 160 K, and materials that are superconductors at room temperature may well become a reality. The implications of these discoveries for power-distribution systems, computer design, and transportation are enormous. Meanwhile, superconducting electromagnets cooled by liquid helium are used in particle accelerators and some experimental magnetic-levitation (maglev) trains (Figure 19.6).

In 1913, Kamerlingh-Onnes was awarded a Nobel Prize for his research on properties of materials at very low temperatures. In 1987, J. Georg Bednorz and Karl Alexander Müller were awarded a Nobel Prize for their discovery of high-temperature superconductivity in ceramic materials.

Non-ohmic Conductors

If a conductor obeys Ohm's law, a graph of current versus voltage is a straight line with a slope of $1/R$ (Figure 19.7a). Such a material is said to be *ohmic*. By contrast, Figure 19.7b shows the graph for a semiconductor diode, a device that is decidedly *non*-ohmic. Notice that the resistance of a diode depends on the *direction* of the current. Diodes act like one-way valves for current; they are used to perform a wide variety of logic functions in computer circuitry.

Ohmic resistor (e.g., typical metal wire): At a given temperature, current is proportional to voltage.

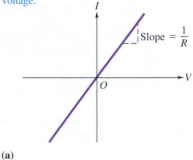

(a)

Semiconductor diode: a non-ohmic resistor

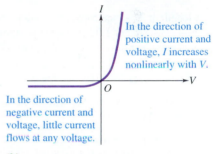

In the direction of positive current and voltage, I increases nonlinearly with V.

In the direction of negative current and voltage, little current flows at any voltage.

(b)

▲ **FIGURE 19.7** Graphs of current versus voltage for (a) a typical ohmic resistor and (b) a semiconductor diode.

Conceptual Analysis 19.1 **Resistance and resistivity**

Two wires made of pure copper have different resistances. These wires may differ in

A. length.
B. cross-sectional area.
C. resistivity.
D. temperature.

SOLUTION Since the two wires are made of the same material, their resistivities at a given temperature are the same. Any difference in resistance at a given temperature must be due to a difference in length or cross-sectional area (or both). If the wires are at significantly different temperatures however, their difference in resistance could be due to the resulting difference in resistivity, as well as to a difference in their length or cross-sectional area. Thus, *all* of the choices are correct.

EXAMPLE 19.2 **Resistance in your stereo system**

Suppose you're hooking up a pair of stereo speakers. **(a)** You happen to have on hand some 20-m-long pieces of 16 gauge copper wire (diameter 1.3 mm); you use them to connect the speakers to the amplifier. These wires are longer than needed, but you just coil up the excess length instead of cutting them. What is the resistance of one of these wires? **(b)** To improve the performance of the system, you purchase 3.0-m-long speaker cables that are made with 8 gauge copper wire (diameter 3.3 mm). What is the resistance of one of these cables?

Continued

SOLUTION

SET UP Figure 19.8 shows our sketch. The resistivity of copper at room temperature is $\rho = 1.72 \times 10^{-8}\ \Omega \cdot m$ (Table 19.1). The cross-sectional area A of a wire is related to its radius by $A = \pi r^2$.

SOLVE To find the resistances, we use Equation 19.3, $R = \rho L/A$.

Part (a): $R = \dfrac{(1.72 \times 10^{-8}\ \Omega \cdot m)(20\ m)}{\pi (6.5 \times 10^{-4}\ m)^2} = 0.26\ \Omega.$

Part (b): $R = \dfrac{(1.72 \times 10^{-8}\ \Omega \cdot m)(3.0\ m)}{\pi (1.65 \times 10^{-3}\ m)^2} = 6.0 \times 10^{-3}\ \Omega.$

REFLECT The shorter, fatter wires offer over forty times less resistance than the longer, skinnier ones.

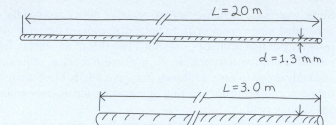

▲ **FIGURE 19.8** Our sketch for this problem.

Practice Problem: 14 gauge copper wire has a diameter of 1.6 mm. What length of this wire has a resistance of 1.0 Ω? *Answer:* 120 m.

EXAMPLE 19.3 Warm wires and cold wires

A length of 18 gauge copper wire with a diameter of 1.02 mm and a cross-sectional area of $8.20 \times 10^{-7}\ m^2$ has a resistance of 1.02 Ω at a temperature of 20°C. Find the resistance at 0°C and at 100°C. The temperature coefficient of resistivity of copper is $0.0039\ (C°)^{-1}$.

SOLUTION

SET UP AND SOLVE We use Equation 19.4, with $T_0 = 20°C$ and $R_0 = 1.02\ \Omega$. At $T = 0°C$,

$\begin{aligned} R &= R_0[1 + \alpha(T - T_0)] \\ &= (1.02\ \Omega)(1 + [0.0039(C°)^{-1}][0°C - 20°C]) \\ &= 0.94\ \Omega. \end{aligned}$

At $T = 100°C$,

$\begin{aligned} R &= (1.02\ \Omega)(1 + [0.0039(C°)^{-1}][100°C - 20°C]) \\ &= 1.34\ \Omega. \end{aligned}$

REFLECT Between the freezing and boiling temperatures of water, the resistance increases by about 40%.

Practice Problem: On a hot summer day in Death Valley, the resistance is 1.14 Ω. What is the temperature? *Answer:* 51°C.

MasteringPHYSICS®

PhET: Battery Voltage

19.3 Electromotive Force and Circuits

For a conductor to have a steady current, it must be part of a path that forms a closed loop, or **complete circuit.** But the path cannot consist entirely of resistance. In a resistor, charge always moves in the direction of decreasing potential energy. There must be some part of the circuit where the potential energy *increases.*

The situation is analogous to the fountain in Figure 19.9a. The water emerges at the top, cascades down to the basin, and then is pumped back to the top for another trip. The pump does work on the water, raising it to a position of higher gravitational potential energy; without the pump, the water would simply fall to the basin and stay there. Similarly, an electric circuit must contain a battery or other device that does work on electric charges to bring them to a position of higher electric potential energy so that they can flow through a circuit to a lower potential energy. The situation of Figure 19.9b doesn't occur in the real world!

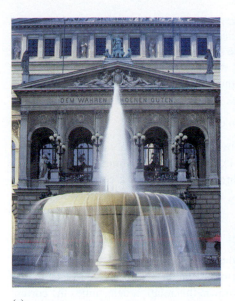

(a)   (b)

▲ **FIGURE 19.9** (a) A pump is needed to keep the water circulating in this fountain. (b) A current in an ordinary circuit with no source of emf would be analogous to this famous impossible scene by Escher.

Electromotive Force

The influence that moves charge from lower to higher potential (despite the electric-field forces in the opposite direction) is called **electromotive force** (abbreviated **emf** and pronounced "ee-em-eff"). Every complete circuit with a continuous current must include some device that provides emf. "Electromotive force" is a poor term because emf is *not* a force, but an "energy per unit charge" quantity, like potential. The SI unit of emf is the same as the unit for potential, the volt $(1\text{ V} = 1\text{ J/C})$. A battery with an emf of 1.5 V does 1.5 J of work on every coulomb of charge that passes through it. We'll use the symbol $\mathcal{E}$ for emf.

Batteries, electric generators, solar cells, thermocouples, and fuel cells are all sources of emf. Each such device converts energy of some form (mechanical, chemical, thermal, and so on) into electrical potential energy and transfers it into the circuit where the device is connected. An *ideal* source of emf maintains a constant potential difference between its terminals, independently of the current through it. We define emf quantitatively as the magnitude of this potential difference. As we will see, such an ideal source is a mythical beast, like the unicorn, the frictionless plane, and the free lunch. Nevertheless, it's a useful idealization.

Figure 19.10 is a schematic diagram of a source of emf that maintains a potential difference between points *a* and *b*, called the *terminals* of the device. Terminal *a*, marked +, is maintained at *higher* potential than terminal *b*, marked −. Associated with this potential difference is an electric field $\vec{E}$ in the region around the terminals, both inside and outside the source. The electric field inside the device is directed from *a* to *b* as shown. A charge *q* within the source is acted upon by an electric force $\vec{F}_E = q\vec{E}$. The source has to provide some additional influence, which we represent as a non-electrostatic force $\vec{F}_n$ that pushes charge from *b* to *a* inside the device (opposite to the direction of $\vec{F}_E$) and maintains the potential difference. The nature of this additional influence depends on the source. In a battery, it is due to chemical processes; in an electric generator, it results from magnetic forces. It could even be you, rubbing plastic on fur, as we saw in Chapter 17.

▲ **BIO Application Shocking!** As we saw in the last chapter, some fishes generate weak electric fields, which they use to probe their environment. A few fishes go farther: Their electric organs are capable of delivering shocks that can stun or kill prey and repel attackers. A large electric eel such as the one shown can generate voltages of up to 500 V. Their emf source—the electric organ—consists of massive stacks of highly modified muscle cells. Each cell contributes to the total emf by moving ions across membranes so as to create a tiny voltage in one direction. The cells are stacked in series to yield a large voltage, and there are many stacks in parallel, giving a large current.

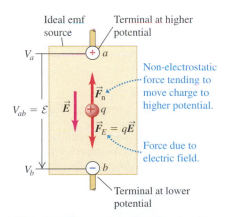

When the emf source is not part of a closed circuit, $F_n = F_E$ and there is no net motion of charge between the terminals.

▲ **FIGURE 19.10** Schematic diagram of an ideal emf source in an "open-circuit" condition.

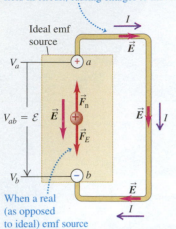

Potential across terminals creates electric field in circuit, causing charges to move.

When a real (as opposed to ideal) emf source is connected to a circuit, V_{ab} and thus F_E fall, so that $F_n > F_E$ and F_n does work on the charges.

▲ **FIGURE 19.11** Schematic diagram of an ideal emf source in a complete circuit.

The potential V_{ab} of point a with respect to point b is defined, as always, as the work per unit charge done by the electrostatic force $\vec{F}_E = q\vec{E}$ on a charge q that moves from a to b. The emf $\mathcal{E}$ of the source is the energy per unit charge supplied by the source during an "uphill" displacement from b to a. For the ideal source of emf that we have described, the potential difference V_{ab} is equal to the electromotive force $\mathcal{E}$:

$$V_{ab} = \mathcal{E}. \qquad \text{(no complete circuit)} \qquad (19.5)$$

Now let's make a complete circuit by connecting a wire with resistance R to the terminals of a source (Figure 19.11). The charged terminals a and b of the source set up an electric field in the wire, and this causes a current in the wire, directed from a toward b. From $V = IR$, the current I in the circuit is determined by

$$\mathcal{E} = V_{ab} = IR. \qquad \text{(ideal source of emf)} \qquad (19.6)$$

That is, when a charge q flows around the circuit, the potential rise $\mathcal{E}$ as it passes through the source is numerically equal to the potential drop $V_{ab} = IR$ as it passes through the resistor. Once $\mathcal{E}$ and R are known, this relation determines the current in the circuit. The current is the same at every point in the circuit.

Conceptual Analysis 19.2

Current in a circuit

The circuit in Figure 19.12 contains a battery and two identical lightbulbs, which are resistors. The battery maintains a constant potential difference between its terminals, independently of the current through it. The bulbs are equally bright when the switch is closed, meaning that they receive the same current. When the switch is opened, what happens to the brightness of bulb A?

A. It increases (because bulb A now receives the current that formerly went through bulb B).
B. It stays the same (meaning that the current through bulb A is unchanged).

SOLUTION The key to this problem is the fact that a battery or other emf source is a source of constant *voltage*, not constant *current*. A 9 V battery produces whatever current is determined by the resistance of the external circuit, but the emf is always 9 V. Be sure you understand this point; don't treat an emf source as though it produces constant current.

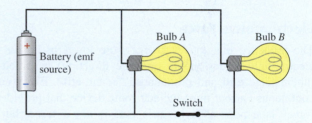

▲ **FIGURE 19.12**

The current drawn by each bulb depends on its resistance and the voltage across it. Since bulb B is connected in parallel with bulb A, both experience the full voltage produced by the battery. The presence or absence of B thus makes no difference to the current through bulb A, so opening the switch does not alter A's brightness. What *does* change is the amount of current supplied by the battery. When the switch is closed, the two bulbs receive equal current from the battery, draining it twice as fast as when only one bulb is lit.

EXAMPLE 19.4 **Electrical hazards in heart surgery**

A patient is undergoing open-heart surgery. A sustained current as small as 25 μA passing through the heart can be fatal. Assume that the heart has a constant resistance of 250 Ω; determine the minimum voltage that poses a danger to the patient.

SOLUTION

SET UP AND SOLVE Figure 19.13 shows our sketch. $1 \mu A = 1 \times 10^{-6}$ A. We use Ohm's law to relate voltage, resistance, and current:

$$V = IR = (25 \times 10^{-6}\,\text{A})(250\,\Omega) = 6.25 \times 10^{-3}\,\text{V}$$
$$= 6.25\,\text{mV}.$$

▶ **FIGURE 19.13** Our sketch for this problem.

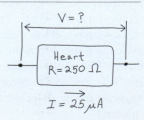

Continued

REFLECT Our result shows that even a small voltage can be dangerous if applied directly to the heart. As a safety measure, any electrical equipment near a patient during surgery must be "grounded." Grounding means that there is a low-resistance conducting path to the earth, so any undesirable current will pass into the ground instead of through the person. During some surgeries, even the patient is grounded to prevent electrical shock.

Practice Problem: A voltage of 12.0 V across the terminals of a device produces a current of 3.00 mA through the device. What is the resistance of the device? *Answer:* 4.00 kΩ.

Internal Resistance in a Source of emf

Real sources of emf don't behave exactly like the ideal sources we've described because charge that moves through the material of any real source encounters *resistance*. We call this the **internal resistance** of the source, denoted by r. If this resistance behaves according to Ohm's law, r is constant. The current through r has an associated drop in potential equal to Ir. The terminal potential difference V_{ab} is then

$$V_{ab} = \mathcal{E} - Ir. \quad \text{(source with internal resistance)} \quad (19.7)$$

The potential V_{ab}, called the **terminal voltage,** is less than the emf $\mathcal{E}$ because of the term Ir representing the potential drop across the internal resistance r.

The current in the external circuit is still determined by $V_{ab} = IR$. Combining this relationship with Equation 19.7, we find that $\mathcal{E} - Ir = IR$, and it follows that

$$I = \frac{\mathcal{E}}{R + r}. \quad \text{(source with internal resistance)} \quad (19.8)$$

That is, the current I equals the source emf $\mathcal{E}$, divided by the *total* circuit resistance $(R + r)$. Thus, we can describe the behavior of a source in terms of two properties: an emf $\mathcal{E}$, which supplies a constant potential difference that is independent of the current, and a series internal resistance r.

To summarize, a circuit is a closed conducting path containing resistors, sources of emf, and possibly other circuit elements. Equation 19.7 shows that the algebraic sum of the potential differences and emf's around the path is zero. Also, the current in a simple loop is the same at every point. Charge is conserved; if the current were different at different points, there would be a continuing accumulation of charge at some points, and the current couldn't be constant.

Table 19.2 shows the symbols usually used in schematic circuit diagrams. We'll use these symbols in most of the circuit analysis in the remainder of this chapter.

TABLE 19.2 Circuit symbols used in this chapter

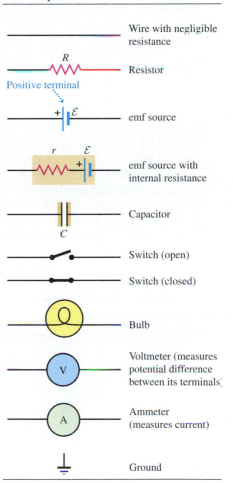

	Wire with negligible resistance
R	Resistor
Positive terminal $+\ \mathcal{E}$	emf source
$r\quad\mathcal{E}$	emf source with internal resistance
C	Capacitor
	Switch (open)
	Switch (closed)
	Bulb
V	Voltmeter (measures potential difference between its terminals)
A	Ammeter (measures current)
	Ground

EXAMPLE 19.5 A dim flashlight

As a flashlight battery ages, its emf stays approximately constant, but its internal resistance increases. A fresh battery has an emf of 1.5 V and negligible internal resistance. When the battery needs replacement, its emf is still 1.5 V, but its internal resistance has increased to 1000 Ω. If this old battery is supplying 1.0 mA to a lightbulb, what is its terminal voltage?

SOLUTION

SET UP AND SOLVE The terminal voltage of a new battery is 1.5 V. The terminal voltage of the old, worn-out battery is given by $V_{ab} = \mathcal{E} - Ir$, so

$$V_{ab} = 1.5\ \text{V} - (1.0 \times 10^{-3}\ \text{A})(1000\ \Omega) = 0.5\ \text{V}.$$

REFLECT The terminal voltage is less than the emf because of the potential drop across the internal resistance of the battery.

When testing a battery, it is better to measure its terminal voltage when it is supplying current than just to measure its emf.

Practice Problem: An old battery with an emf of 9.0 V has a terminal voltage of 8.2 V when it is supplying a current of 2.0 mA. What is the internal resistance of the battery? *Answer:* 400 Ω.

NOTE ▶ In the examples that follow, it's important to understand how the meters in the circuit work. The symbol V in a circle represents an ideal voltmeter. It measures the potential difference between the two points in the circuit where it is connected, but *no current flows through the voltmeter*. The symbol A in a circle represents an ideal ammeter. It measures the current that flows through it, but *there is no potential difference between its terminals*. Thus, the behavior of a circuit doesn't change when an ideal ammeter or voltmeter is connected to it. ◀

EXAMPLE 19.6   A source in an open circuit

Figure 19.14 shows a source with an emf $\mathcal{E}$ of 12 V and an internal resistance r of 2.0 Ω. (For comparison, the internal resistance of a commercial 12 V lead storage battery is only a few thousandths of an ohm.) Determine the readings of the ideal voltmeter V and ammeter A.

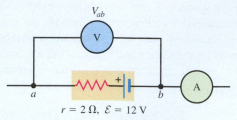

▶ **FIGURE 19.14**

SOLUTION

SET UP AND SOLVE First we see that the diagram does not show a complete circuit. (Remember that there is no current through an ideal voltmeter; thus, the loop containing the voltmeter does *not* represent a circuit.) Because there is no current through the battery, there is no potential difference across its internal resistance. The potential difference V_{ab} across its termi-

nals is equal to the emf $(V_{ab} = \mathcal{E} = 12 \text{ V})$, and the voltmeter reads $V = 12$ V. Because there is no complete circuit, there is no current, so the ammeter reads zero.

REFLECT As soon as this battery is put into a complete circuit, its internal resistance causes its terminal voltage to be less than its emf.

EXAMPLE 19.7   A source in a complete circuit

Using the battery in Example 19.6, we add a 4.0 Ω resistor to form the complete circuit shown in Figure 19.15. What are the voltmeter and ammeter readings now?

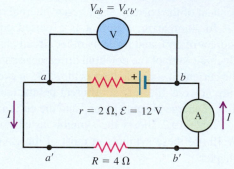

▶ **FIGURE 19.15**

SOLUTION

SET UP AND SOLVE We now have a complete circuit and a current I through the resistor R, determined by Equation 19.8:

$$I = \frac{\mathcal{E}}{R + r} = \frac{12 \text{ V}}{4.0 \ \Omega + 2.0 \ \Omega} = 2.0 \text{ A}.$$

The ammeter A reads $I = 2.0$ A.

Our idealized conducting wires have zero resistance, so there is no potential difference between points a and a' or between b and b'. That is, $V_{ab} = V_{a'b'}$, and the voltmeter reading is this potential difference. We can find V_{ab} by considering a and b either as the terminals of the resistor or as the terminals of the

source. Considering them as the terminals of the resistor, we use Ohm's law $(V = IR)$ to obtain

$$V_{a'b'} = IR = (2.0 \text{ A})(4.0 \ \Omega) = 8.0 \text{ V}.$$

Considering them as the terminals of the source, we have

$$V_{ab} = \mathcal{E} - Ir = 12 \text{ V} - (2.0 \text{ A})(2.0 \ \Omega) = 8.0 \text{ V}.$$

Either way, we conclude that the voltmeter reads $V_{ab} = 8.0$ V.

REFLECT The terminal voltage V_{ab} of the battery is less than the battery emf because of the potential drop across the battery's internal resistance r.

Practice Problem: If a different resistor R is used and the voltmeter reads 6.0 V, what is the ammeter reading? *Answer:* 3.0 A.

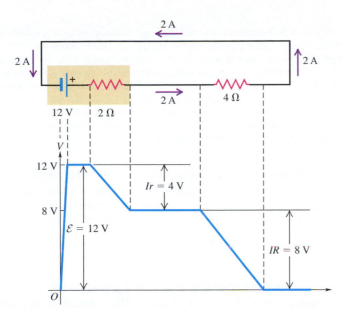

▲ FIGURE 19.16 Potential rises and drops in the circuit.

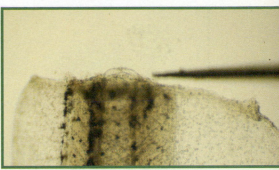

Figure 19.16 is a graph showing how the potential varies around the complete circuit of Figure 19.15. The horizontal axis doesn't necessarily represent actual distances; rather, it shows various points in the loop. If we take the potential to be zero at the negative terminal of the battery, then we have a rise $\mathcal{E}$ and a drop Ir in the battery and an additional drop IR in the external resistor. As we finish our trip around the loop, the potential is back where it started.

The difference between a fresh flashlight battery and an old one is not so much in the emf, which decreases only slightly with use, but mostly in the internal resistance, which may increase from a few ohms when the battery is fresh to as much as 1000 Ω or more after long use. Similarly, a car battery can deliver less current to the starter motor on a cold morning than when the battery is warm, not because the emf is appreciably less, but because the internal resistance increases with decreasing temperature. Residents of northern Iowa have been known to soak their car batteries in warm water to provide greater starting power on very cold mornings!

19.4 Energy and Power in Electric Circuits

Let's now look at some energy and power relations in electric circuits. The box in Figure 19.17 represents a circuit element with potential difference $V_a - V_b = V_{ab}$ between its terminals and current I passing through it in the direction from a toward b. This element might be a resistor, a battery, or something else; the details don't matter. As charge passes through the circuit element, the electric field does work on the charge.

When a charge q passes through the circuit element, the work done on the charge is equal to the product of q and the potential difference V_{ab} (work per unit charge). When V_{ab} is positive, a positive amount of work qV_{ab} is done on the charge as it "falls" from potential V_a to the lower potential V_b. If the current is I, then in a time interval Δt an amount of charge $\Delta Q = I\,\Delta t$ passes through. The work ΔW done on this amount of charge is

$$\Delta W = V_{ab}\,\Delta Q = V_{ab}I\,\Delta t.$$

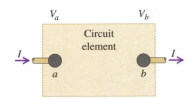

▲ FIGURE 19.17 The power input P to the portion of the circuit between a and b is $P = V_{ab}I$.

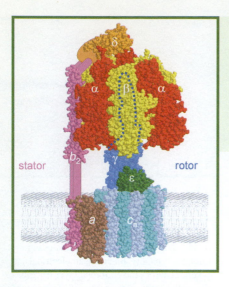

stator

rotor

This work represents electrical energy transferred *into* the circuit element. The time rate of energy transfer is *power,* denoted by P. Dividing the preceding equation by Δt, we obtain the *time rate* at which the rest of the circuit delivers electrical energy to the circuit element:

$$\frac{\Delta W}{\Delta t} = P = V_{ab}I. \tag{19.9}$$

If the potential at b is higher than at a, then V_{ab} is negative, and there is a net transfer of energy *out of* the circuit element. The element is then acting as a source, delivering electrical energy into the circuit to which it is connected. This is the usual situation for a battery, which converts chemical energy into electrical energy and delivers it to the external circuit.

The unit of V_{ab} is 1 volt, or 1 joule per coulomb, and the unit of I is 1 ampere, or 1 coulomb per second. We can now confirm that the SI unit of power is 1 watt:

$$(1\ \text{J/C})(1\ \text{C/s}) = 1\ \text{J/s} = 1\ \text{W}.$$

Conceptual Analysis 19.3

Buying electricity

What do you buy from the power company?

A. Only energy.
B. Electrons and energy.
C. Only electrons.

SOLUTION The wires to your house provide two things: a current and a potential difference (usually either 110 V or 220 V, depending on where you live). Remember that a current is a flow of charge. Because charge is a conserved quantity, the rate at which charge enters a circuit element at any instant must equal the rate at which it leaves the element. Therefore, the electrons are more like a conveyor belt than they are like the material on the belt. You don't buy electrons and use them up. However, you *do* use the energy represented by the flow of charge down a potential difference. Thus, the power company sells you energy only.

PhET: Battery-Resistor Circuit
PhET: Signal Circuit

Pure Resistance

When current flows through a resistor, electrical energy is transformed into thermal energy. An electric toaster is an obvious example. We calculate the power dissipated through a resistor as follows: The potential difference across the resistor is $V_{ab} = IR$. From Equation 19.9, the electric power delivered to the resistor by the circuit is

$$P = V_{ab}I = I^2R = \frac{V_{ab}^2}{R}. \tag{19.10}$$

What becomes of this energy? The moving charges collide with atoms in the resistor and transfer some of their energy to these atoms, increasing the internal energy of the material. Either the temperature of the resistor increases, or there is a flow of heat out of it, or both. We say that energy is *dissipated* in the resistor at a rate I^2R. Too high a temperature can change the resistance unpredictably; the resistor may melt or even explode. Of course, some devices, such as electric heaters, are designed to get hot and transfer heat to their surroundings. But every resistor has a *power rating:* the maximum power that the device can dissipate without becoming overheated and damaged. In practical applications, the power rating of a resistor is often just as important a characteristic as its resistance.

- The emf source converts non-electrical to electrical energy at a rate $\mathcal{E}I$.
- Its internal resistance *dissipates* energy at a rate I^2r.
- The difference $\mathcal{E}I - I^2r$ is its power output.

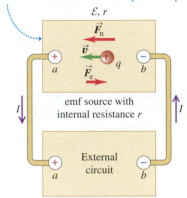

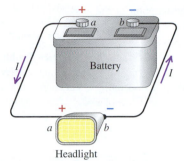

(a) Diagrammatic circuit

(b) A real circuit of the type shown in (a)

▲ **FIGURE 19.18** Power output of an emf source.

▲ **Application Cheap light** If you've had incandescent flashlights or bicycle lights and changed to lights that use light-emitting diodes (LEDs), you know the large difference in energy consumption. A halogen bicycle headlight might go through a set of batteries in 3 hours, but an even brighter LED headlight will last 30 hours. Why the difference? The answer is that any incandescent bulb (including a halogen bulb) works by using the dissipation of electrical energy to heat a filament white hot. Some of the energy is converted to visible light, but most is lost as heat. In an LED, electrical energy is used to move semiconductor electrons to a region where they emit light. Most of the electrical energy, then, emerges as light; little is lost as heat.

Power Output of a Source

The upper box in Figure 19.18a represents a source with emf $\mathcal{E}$ and internal resistance r, connected by ideal (resistanceless) conductors to an external circuit represented by the lower box. This arrangement could describe a car battery connected to the car's headlights (Figure 19.18b). Point a is at higher potential than point b; that is, $V_a > V_b$. But now the current I is *leaving* the device at the higher-potential terminal (rather than entering there). Energy is being delivered to the external circuit, and the rate of its delivery to the circuit is given by Equation 19.10: $P = V_{ab}I$. For a source that can be described by an emf $\mathcal{E}$ and an internal resistance r, we may use Equation 19.7: $V_{ab} = \mathcal{E} - Ir$.

Multiplying this equation by I, we find that

$$P = V_{ab}I = \mathcal{E}I - I^2r. \qquad (19.11)$$

The emf $\mathcal{E}$ is the work per unit charge performed on the charges as they are pushed "uphill" from b to a in the source. When a charge ΔQ flows through the source during a time interval Δt, the work done on it is $\mathcal{E}\Delta Q = \mathcal{E}I\Delta t$. Thus, $\mathcal{E}I$ is the *rate* at which work is done on the circulating charges. This term represents the rate of conversion of non-electrical energy to electrical energy within the source. The term I^2r is the rate at which electrical energy is *dissipated* in the internal resistance of the source. The difference, $\mathcal{E}I - I^2r$, is the *net* electrical power output of the source—that is, the rate at which the source delivers electrical energy to the remainder of the circuit.

Quantitative Analysis 19.4

Power and current

A 1200 W floor heater, a 360 W television, and a hand iron operating at 900 W are all plugged into the same 120 volt circuit in a house (that is, the same pair of wires that come from the basement fuse box). What is the total current flowing through this circuit?

A. 20.5 A.
B. 17.5 A.
C. 15 A.
D. 12.5 A.

SOLUTION The electric power input to an appliance is given by $P = VI$. Each device is attached to the same 120 volt source. That is, the devices are in parallel, and their currents add to give the total current in the pair of wires coming from the fuse box. The current I through each device is given by $I = P/120$ V. The currents through each, in the order listed, are 10 A, 3 A, and 7.5 A. The sum of these currents is 20.5 A. This would likely trip the circuit breaker for that circuit.

EXAMPLE 19.8 **Lightbulb resistance**

The bulb for an interior light in a car is rated at 15.0 W and operates from the car battery voltage of 12.6 V. What is the resistance of the bulb?

SOLUTION

SET UP AND SOLVE The electric power consumption of the bulb is 15.0 W. Thus,

$$P = \frac{V^2}{R}, \quad \text{so} \quad R = \frac{V^2}{P} = \frac{(12.6 \text{ V})^2}{15.0 \text{ W}} = 10.6 \text{ }\Omega.$$

REFLECT This is the resistance of the lightbulb at its operating temperature. When the light first comes on, the filament is cooler

and the resistance is lower. Because $P = V^2/R$, the bulb has a greater rate of electrical energy consumption when it first comes on. Also, because $I = V/R$, the current drawn by the bulb is greater when it is cool and R is less.

Practice Problem: What is the resistance of a heating element when the current through it is 1.20 A and it is consuming electrical energy at a rate of 220 W? *Answer:* 153 Ω.

19.5 Resistors in Series and in Parallel

Resistors turn up in all kinds of circuits, ranging from hair dryers and space heaters to circuits that limit or divide current or that reduce or divide a voltage. Suppose we have three resistors with resistances R_1, R_2, and R_3. Figure 19.19 shows four different ways they might be connected between points a and b. In Figure 19.19a, the resistors provide only a single path between these points. When several circuit elements, such as resistors, batteries, and motors, are connected in sequence as in Figure 19.19a, we say that they are connected in **series.** Since no charge accumulates anywhere between point a and point b, the *current* is the same in all of the resistors when they are connected in series.

The resistors in Figure 19.19b are said to be connected in **parallel** between points a and b. Each resistor provides an alternative path between the points. For circuit elements that are connected in parallel, the *potential difference* is the same across each element. In Figure 19.19c, resistors R_2 and R_3 are in parallel, and this combination is in series with R_1. In Figure 19.19d, R_2 and R_3 are in series, and this combination is in parallel with R_1.

For any combination of resistors that obey Ohm's law, we can always find a single resistor that could replace the combination and result in the same total current and potential difference. The resistance of this single resistor is called the **equivalent resistance** of the combination. If any one of the networks in Figure 19.19 were replaced by its equivalent resistance R_{eq}, we could write

$$V_{ab} = IR_{eq}, \quad \text{or} \quad R_{eq} = \frac{V_{ab}}{I},$$

where V_{ab} is the potential difference between terminals a and b of the network and I is the current at point a or point b. To compute an equivalent resistance, we assume a potential difference V_{ab} across the actual network, compute the corresponding current I, and take the ratio V_{ab}/I.

Resistors in Series

For series and parallel combinations, we can derive general equations for the equivalent resistance. If the resistors are in *series,* as in Figure 19.19a, the current I must be the same in all of them. (Otherwise, charge would be piling up at the points where the terminals of two resistors are connected.) Applying $V = IR$ to each resistor, we have

$$V_{ax} = IR_1, \quad V_{xy} = IR_2, \quad V_{yb} = IR_3.$$

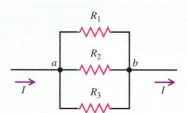

(a) Resistors in series

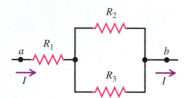

(b) Resistors in parallel

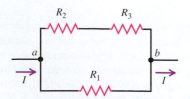

(c) R_1 in series with a parallel combination of R_2 and R_3

(d) R_1 in parallel with a series combination of R_2 and R_3

▲ **FIGURE 19.19** Four different ways of connecting three resistors.

The potential difference V_{ab} is the sum of these three quantities:

$$V_{ab} = V_{ax} + V_{xy} + V_{yb} = I(R_1 + R_2 + R_3),$$

or

$$\frac{V_{ab}}{I} = R_1 + R_2 + R_3.$$

But V_{ab}/I is, by definition, the equivalent resistance R_{eq}. Therefore,

$$R_{eq} = R_1 + R_2 + R_3.$$

It is easy to generalize this relationship to any number of resistors:

> **Equivalent resistance for resistors in series**
> The equivalent resistance of *any number* of resistors in series equals the sum of their individual resistances:
>
> $$R_{eq} = R_1 + R_2 + R_3 + \cdots. \qquad (19.12)$$
>
> The equivalent resistance is always *greater than* any individual resistance.

Resistors in Parallel

For resistors in parallel, as in Figure 19.19b, the potential difference between the terminals of each resistor must be the same and equal to V_{ab}. Let's call the currents in the three resistors I_1, I_2, and I_3, respectively. Then, from $I = V/R$,

$$I_1 = \frac{V_{ab}}{R_1}, \qquad I_2 = \frac{V_{ab}}{R_2}, \qquad I_3 = \frac{V_{ab}}{R_3}.$$

Charge is neither accumulating at, nor draining out of, point a; all charge that enters point a also leaves that point. Thus, the total current I must equal the sum of the three currents in the resistors:

$$I = I_1 + I_2 + I_3 = V_{ab}\left(\frac{1}{R_1} + \frac{1}{R_2} + \frac{1}{R_3}\right),$$

or

$$\frac{I}{V_{ab}} = \frac{1}{R_1} + \frac{1}{R_2} + \frac{1}{R_3}.$$

But by the definition of the equivalent resistance R_{eq}, $I/V_{ab} = 1/R_{eq}$, so

$$\frac{1}{R_{eq}} = \frac{1}{R_1} + \frac{1}{R_2} + \frac{1}{R_3}.$$

Again, it is easy to generalize this relationship to *any number* of resistors in parallel:

> **Equivalent resistance for resistors in parallel**
> For *any number* of resistors in parallel, the *reciprocal* of the equivalent resistance equals the *sum of the reciprocals* of their individual resistances:
>
> $$\frac{1}{R_{eq}} = \frac{1}{R_1} + \frac{1}{R_2} + \frac{1}{R_3} + \cdots. \qquad (19.13)$$
>
> The equivalent resistance is always *less than* any individual resistance.

Mastering PHYSICS

PhET: Circuit Construction Kit (DC Only)
ActivPhysics 12.1: DC Series Circuits (Qualitative)
ActivPhysics 12.2: DC Parallel Circuits
ActivPhysics 12.3: DC Circuit Puzzles

PROBLEM-SOLVING STRATEGY 19.1 **Resistors in series and in parallel**

SET UP

1. It helps to remember that when resistors are connected in series, the total potential difference across the combination is the sum of the individual potential differences. When resistors are connected in parallel, the potential difference is the same for every resistor and is equal to the potential difference across the parallel combination.

2. Also, keep in mind the analogous statements for current: When resistors are connected in series, the current is the same through every resistor and is equal to the current through the series combination. When resistors are connected in parallel, the total current through the combination is equal to the sum of currents through the individual resistors.

SOLVE

3. We can often consider networks such as those in Figures 19.19c and 19.19d as combinations of series and parallel arrangements. In Figure 19.19c, we first replace the parallel combination of R_2 and R_3 by its equivalent resistance; this then forms a series combination with R_1. In Figure 19.19d, the combination of R_2 and R_3 in series forms a parallel combination with R_1.

REFLECT

4. The rule for combining resistors in parallel follows directly from the principle of conservation of charge. The rule for combining resistors in series results from a fundamental principle about work: When a particle moves along a path, the total work done on it is the sum of the quantities of work done during the individual segments of the path.

EXAMPLE 19.9 A resistor network

Three identical resistors with resistances of 6.0 Ω are connected as shown in Figure 19.20 to a battery with an emf of 18.0 V and zero internal resistance. **(a)** Find the equivalent resistance of the resistor network. **(b)** Find the current in each resistor.

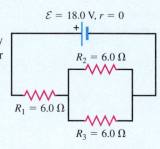

$\mathcal{E} = 18.0$ V, $r = 0$

$R_2 = 6.0\ \Omega$

$R_1 = 6.0\ \Omega$

$R_3 = 6.0\ \Omega$

▶ **FIGURE 19.20**

SOLUTION

SET UP AND SOLVE Part (a): To find the equivalent resistance, we identify series or parallel combinations of resistors and replace them by their equivalent resistors, continuing this process until the circuit has just a single resistor that is the equivalent resistor for the network. Figure 19.21 shows the procedure.

In this network, R_2 and R_3 are in parallel, so their equivalent resistance R_{23} is given by

$$\frac{1}{R_{23}} = \frac{1}{R_2} + \frac{1}{R_3} = \frac{1}{6.0\ \Omega} + \frac{1}{6.0\ \Omega} = \frac{2}{6.0\ \Omega} = \frac{1}{3.0\ \Omega}.$$

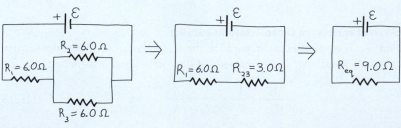

▶ **FIGURE 19.21** Our procedure for finding the equivalent resistance.

(a) Original circuit (b) Parallel resistors combined (c) Equivalent resistor

Continued

This gives $R_{23} = 3.0\ \Omega$. In Figure 19.21b, we've replaced the parallel combination of R_2 and R_3 with R_{23}. The circuit now has R_1 and R_{23} in series. Their equivalent resistance R_{eq} is given by

$$R_{eq} = R_1 + R_{23} = 6.0\ \Omega + 3.0\ \Omega = 9.0\ \Omega.$$

Thus, the equivalent resistance of the entire network is $9.0\ \Omega$ (Figure 19.21c).

Part (b): To find the currents and voltages, we work backward, as shown in Figure 19.22, starting with the single equivalent resistor. In Figure 19.22a, the voltage across R_{eq} is $\mathcal{E}$ (because the battery has no internal resistance). Also, $\mathcal{E} = IR_{eq}$, so $I = \mathcal{E}/R_{eq} = 18.0\ \text{V}/9.0\ \Omega = 2.0\ \text{A}$. This is the current through the battery and also the total current through the network.

In Figure 19.22b, R_1, R_{23}, and the battery are in series, so the same 2.0 A current passes through each. The voltage V_1 across R_1 is

$$V_1 = IR_1 = (2.0\ \text{A})(6.0\ \Omega) = 12.0\ \text{V}.$$

The voltage V_{23} across R_{23} is

$$V_{23} = IR_{23} = (2.0\ \text{A})(3.0\ \Omega) = 6.0\ \text{V}.$$

(Note that $V_1 + V_{23} = \mathcal{E}$.)

The voltage across the parallel combination of R_2 and R_3 is the potential difference between points c and d, which is $V_{23} = 6.0\ \text{V}$. Since the voltage across R_2 is 6.0 V, we can calculate the current I_2 through R_2: $I_2 = \dfrac{6.0\ \text{V}}{6.0\ \Omega} = 1.0\ \text{A}$. A similar calculation gives the current I_3 through R_3: $I_3 = \dfrac{6.0\ \text{V}}{6.0\ \Omega} = 1.0\ \text{A}$.

In summary, $I_1 = 2.0\ \text{A}$ and $I_2 = I_3 = 1.0\ \text{A}$, as shown in Figure 19.22c. (Note that $I_2 + I_3 = I_1$.)

REFLECT The two identical resistors R_2 and R_3 form a parallel combination; the 2.0 A current arriving at point c in the circuit divides equally: Half goes through R_2 and half through R_3. At point d, these two 1.0 A currents recombine into a 2.0 A current.

Practice Problem: What is the current through each resistor in the network in this example if the resistors aren't equal, but instead $R_1 = 4.0\ \Omega$, $R_2 = 6.0\ \Omega$, and $R_3 = 3.0\ \Omega$? *Answers:* $I_1 = 3.0\ \text{A}$, $I_2 = 1.0\ \text{A}$, $I_3 = 2.0\ \text{A}$.

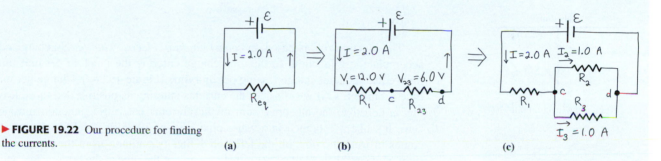

▶ **FIGURE 19.22** Our procedure for finding the currents.

(a)   (b)   (c)

19.6 Kirchhoff's Rules

Many practical networks cannot be reduced to simple series–parallel combinations. Figure 19.23 shows two examples. Figure 19.23a is a circuit with two emf sources and a resistor. (This circuit might represent a battery feeding current to a lightbulb while being charged by a battery charger.) We don't need any new *principles* to compute the currents in networks such as these, but there are several techniques that help us to handle them systematically. We'll describe one of these, first developed by Gustav Robert Kirchhoff (1824–1887).

MasteringPHYSICS

ActivPhysics 12.5: Using Kirchhoff's Laws

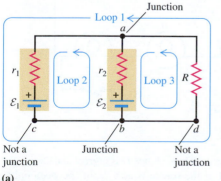

(a)   (b)

▲ **FIGURE 19.23** Two networks that cannot be reduced to simple series–parallel combinations of resistors.

First, here are some terms that we'll use often: A **junction** in a circuit is a point where three or more conductors meet. Junctions are also called *nodes* or *branch points*. The circuit in Figure 19.23a has two junctions: *a* and *b*. Points *c* and *d* are *not* junctions. A **loop** is any closed conducting path. Figure 19.23 shows the possible loops for the layout of each circuit.

Kirchhoff's rules consist of the following two statements:

> **Kirchhoff's junction (or point) rule:**
> **The algebraic sum of the currents into any junction is zero;** that is,
>
> $$\Sigma I = 0. \tag{19.14}$$

Currents *into* a junction are positive; currents *out of* a junction are negative.

> **Kirchhoff's loop rule:**
> **The algebraic sum of the potential differences in any loop,** including those associated with emf's and those of resistive elements, **must equal zero;** that is,
>
> $$\sum_{\text{around loop}} V = 0. \tag{19.15}$$

The junction rule is based on *conservation of electric charge*. No charge can accumulate at a junction, so the total charge entering the junction per unit time must equal the total charge leaving per unit time (Figure 19.24a). Charge per unit time is current, so if we consider the currents entering as positive and those leaving as negative, the algebraic sum of the currents entering a junction must be zero. It's like a T branch in a water pipe (Figure 19.24b); if 1 liter per minute comes in from the pipe on the left, and 1 liter per minute from the pipe on the right, you can't have 3 liters per minute going out the pipe at the bottom. We actually used the junction rule (without saying so) in Section 19.5, in the derivation of Equation 19.13 for resistors in parallel.

The loop rule is based on conservation of energy. The electrostatic field is a *conservative* force field. Suppose we go around a loop, measuring potential differences across successive circuit elements as we go. When we return to the starting point, we must find that the *algebraic sum* of these differences is zero. That is, when a charge travels in a loop and returns to its original location, the total change in the electric potential energy is zero; otherwise, the force wouldn't be conservative.

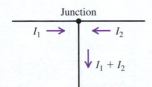

Junction

$I_1 \rightarrow \quad \leftarrow I_2$

$\downarrow I_1 + I_2$

(a) Kirchhoff's junction rule

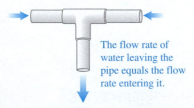

The flow rate of water leaving the pipe equals the flow rate entering it.

(b) Water-pipe analogy for Kirchhoff's junction rule

▲ **FIGURE 19.24** Kirchhoff's junction rule.

Quantitative Analysis 19.5

Throw the switch!

Bulbs *A*, *B*, and *C* in Figure 19.25 are identical. Closing the switch in the figure causes which of the following changes in the potential differences? More than one answer may be correct.

A. The potential differences across *A* and *B* are unchanged.
B. The potential difference across *C* drops by 50%.
C. The potential differences across *A* and *B* each increase by 50%.
D. The potential difference across *C* drops to zero.

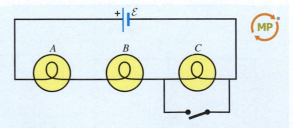

▲ **FIGURE 19.25**

SOLUTION When the switch is open, the same current passes through all the bulbs. Because the bulbs have the same resistance, the voltage $V = IR$ is also the same across all of them. When the switch is closed, the switch and bulb *C* form a loop. We assume that the switch is an ideal conductor, so the voltage drop across it is zero. Therefore, according to the loop rule, the voltage across

bulb *C* must also be zero. Thus, choice D is correct and choice B is incorrect. The emf of the source is now "split" between only two bulbs: *A* and *B*. Hence, the potential difference across each of these bulbs increases by 50% (so choice C is correct and choice A is incorrect).

Using Kirchhoff's laws to find the currents and potentials in a circuit can be tricky. We suggest that you study the following problem-solving strategy carefully. Often, the hardest part of the solution is not in understanding the basic principles, but in keeping track of algebraic signs!

PROBLEM-SOLVING STRATEGY 19.2 **Kirchhoff's rules**

SET UP

1. Draw a *large* circuit diagram so that you have plenty of room for labels. Label all quantities, known and unknown, including an assumed direction for each unknown current and emf. Often you won't know in advance the actual direction of an unknown current or emf, but this doesn't matter. Carry out your solution, using the assumed direction. If the actual direction of a particular quantity is opposite to your assumption, the result will come out with a negative sign. If you use Kirchhoff's rules correctly, they give you the directions as well as the magnitudes of unknown currents and emf's. We'll illustrate this point in the examples that follow.

2. Usually, when you label currents, it is best to use the junction rule immediately, to express the currents in terms of as few quantities as possible. For example, Figure 19.26a shows a correctly labeled circuit; Figure 19.26b shows the same circuit, relabeled by applying the junction rule to point a to eliminate I_3.

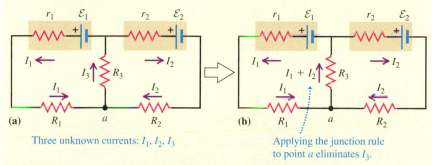

Three unknown currents: I_1, I_2, I_3 Applying the junction rule to point a eliminates I_3.

▲ **FIGURE 19.26** Using the junction rule to eliminate an unknown current.

SOLVE

3. Choose any closed loop in the network, and designate a direction (clockwise or counterclockwise) to go around the loop when applying the loop rule. The direction doesn't have to be the same as any assumed current's direction.

4. Go around the loop in the designated direction, adding potential differences as you cross them. An emf is counted as positive when you traverse it from − to + and negative when you traverse it from + to −. An IR product is negative if your path passes through the resistor in the *same* direction as the assumed current and positive if it passes through in the opposite direction. "Uphill" potential changes are always positive; "downhill" changes are always negative. Figure 19.27 summarizes these sign conventions. In each part of the figure, the direction of "travel" is the direction in which we are going around a loop, using Kirchhoff's loop rule, not necessarily the direction of the current.

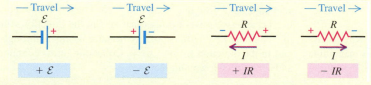

▲ **FIGURE 19.27** Sign conventions to use in traveling around a circuit loop when applying Kirchhoff's rules.

Continued

5. Apply Kirchhoff's loop rule to the potential differences obtained in Step 4: $\sum V = 0$.
6. If necessary, choose another loop to get a different relation among the unknowns, and continue until you have as many independent equations as unknowns or until every circuit element has been included in at least one of the chosen loops.
7. Finally, solve the equations, by substitution or some other means, to determine the unknowns. Be especially careful with algebraic manipulations; one sign error is fatal to the entire solution.
8. You can use this same bookkeeping system to find the potential V_{ab} of any point a with respect to any other point b. Start at b and add the potential changes that you encounter in going from b to a, using the same sign rules as in Step 4. The algebraic sum of these changes is $V_{ab} = V_a - V_b$.

REFLECT

9. Always remember that when you go around a loop, adding potential differences in accordance with Kirchhoff's loop rule, rises in potential are positive and drops in potential are negative. Follow carefully the procedure described; getting the signs right is absolutely essential.

EXAMPLE 19.10 Jump-start your car

The circuit shown in Figure 19.28 is used to start a car that has a dead (i.e., discharged) battery. It includes two batteries, each with an emf and an internal resistance, and two resistors. Find the current in the circuit and the potential difference V_{ab}.

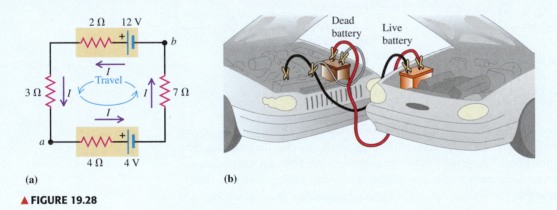

▲ FIGURE 19.28

SOLUTION

SET UP There is only one loop, so we don't need Kirchhoff's junction rule. To use the loop rule, we first assume a direction for the current. As shown in the figure, we assume that I is counterclockwise, coming from the + terminal of the battery with the larger emf. If this assumption is incorrect, then when we solve for I, we'll get a negative value. We choose to travel counterclockwise around the loop (but we could just as well have chosen to travel clockwise).

SOLVE To find the current in the circuit, we apply Kirchhoff's loop rule, starting at point a. The potential change has magnitude IR for a resistor and $\mathcal{E}$ for an emf. The sign of each potential change is determined as described in step 4 of Problem Solving Strategy 19.2. The loop rule gives

$$-I(4.0\ \Omega) - 4.0\ \text{V} - I(7.0\ \Omega) + 12\ \text{V} - I(2.0\ \Omega) - I(3.0\ \Omega) = 0.$$

Be absolutely sure that you understand the signs in this equation. Each IR term in the sum has a minus sign because we travel through each resistor in the assumed direction of the current, encountering a drop in potential in each. The 4.0 V term is negative because we travel through that emf in the direction from + to −, and the 12 V term is positive because we travel through that emf in the direction from − to +. Collecting terms containing I and solving for I, we find that

Continued

$$8.0 \text{ V} = I(16 \, \Omega) \quad \text{and} \quad I = 0.5 \text{ A}.$$

Our positive result for I tells us that the current actually is in the counterclockwise direction we assumed. If we had assumed that I is clockwise, the loop rule would have given us $I = -0.5$ A. The minus sign in the result would tell us that the actual current is opposite to this assumed direction.

To find V_{ab}, we first take the upper path from b to a, through the 12 V battery. We get

$$V_b + 12 \text{ V} - (0.50 \text{ A})(2.0 \, \Omega) - (0.50 \text{ A})(3.0 \, \Omega) = V_a,$$
$$V_{ab} = V_a - V_b = 12 \text{ V} - 1.0 \text{ V} - 1.5 \text{ V} = 9.5 \text{ V}.$$

Point a is at higher potential than point b. The potential rise through the 12 V emf is greater than the total drop through the resistors.

Taking the lower path, through the 4.0 V emf, gives

$$V_b + (0.50 \text{ A})(7.0 \, \Omega) + 4.0 \text{ V} + (0.50 \text{ A})(4.0 \, \Omega) = V_a,$$
$$V_{ab} = V_a - V_b = 3.5 \text{ V} + 4.0 \text{ V} + 2.0 \text{ V} = 9.5 \text{ V}.$$

The result for V_{ab} is the same as before. In this equation, the IR terms are positive because, as we travel along this path, we pass through each resistor in a direction opposite to the current direction.

REFLECT To jump-start a car, we create a circuit by connecting the + terminal of the live battery to the + terminal of the dead one and connecting the two − terminals together.

Practice Problem: In this example, if the emf of the 4 V battery is increased to 16 V and the rest of the circuit remains the same, what is the potential difference V_{ab}? *Answer:* 13.2 V.

EXAMPLE 19.11 Recharging a battery

In the circuit shown in Figure 19.29, a 12 V power supply with unknown resistance r (represented as a battery) is connected to a run-down rechargeable battery with unknown emf $\mathcal{E}$ and internal resistance 1 Ω and to a bulb with resistance 3 Ω. The current through the bulb is 2 A, and the current in the run-down battery is 1 A; the directions of both of these currents are shown in the figure. **(a)** Find the unknown current I in the 12 V battery, the internal resistance r of the power supply, and the emf $\mathcal{E}$ of the run-down battery. **(b)** Find the electrical power for each emf in the circuit.

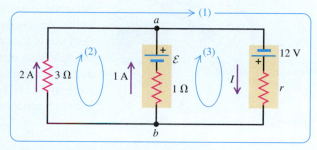

▲ **FIGURE 19.29**

SOLUTION

SET UP As shown in the figure, we assume that the current I is downward. We also identify the possible loops for applying the loop rule and choose a clockwise travel direction for each.

SOLVE Part (a): We first apply the junction rule to point a. When we sum the currents at a junction, currents into the junction are positive and currents out of the junction are negative. This gives

$$-I + 1 \text{ A} + 2 \text{ A} = 0 \quad \text{and} \quad I = 3 \text{ A}.$$

To find r, we use the loop rule for loop (1), since r is the only unknown quantity in that loop. Starting at point a and proceeding clockwise, we use the loop rule to obtain

$$+12 \text{ V} - (3 \text{ A})r - (2 \text{ A})(3 \, \Omega) = 0 \quad \text{and}$$
$$r = \frac{12 \text{ V} - 6 \text{ V}}{3 \text{ A}} = 2 \, \Omega.$$

The IR terms are negative because we pass through each resistance in the direction of the current. Note that the current may not be the same for all the resistors in a loop. To calculate $\mathcal{E}$, we use the loop rule with loop (2). Starting at point a and proceeding clockwise, we find that

$$-\mathcal{E} + (1 \text{ A})(1 \, \Omega) - (2 \text{ A})(3 \, \Omega) = 0 \quad \text{and}$$
$$\mathcal{E} = 1 \text{ V} - 6 \text{ V} = -5 \text{ V}.$$

The IR term for the 1 Ω resistor is positive because we travel through this resistor in the direction opposite to the current.

The fact that our result for $\mathcal{E}$ is negative shows that the actual polarity of this emf is opposite to what is shown in the figure.

As a check, we can use loop (3) in Figure 19.29. Starting at point a and again proceeding clockwise, we obtain the loop equation

$$+12 \text{ V} - Ir - (1 \text{ A})(1 \, \Omega) + \mathcal{E} = 0.$$

Since $Ir = (3 \text{ A})(2 \, \Omega) = 6 \text{ V}$, the equation becomes $+12 \text{ V} - 6 \text{ V} - 1 \text{ V} + \mathcal{E} = 0$. Solving gives $\mathcal{E} = -5 \text{ V}$, in agreement with the result from loop (2).

Part (b): The electrical power for an emf is $P = \mathcal{E}I$, where I is the current through the emf. When the current passes through the emf from − to +, positive charges *gain* electrical potential energy, so the emf adds energy to the circuit. If the current is in the opposite direction, the emf removes electrical energy from the circuit. In our circuit, the emf of the 12 V battery adds energy to the circuit at a rate

$$P_{\text{in}} = \mathcal{E}I = (12 \text{ V})(3 \text{ A}) = 36 \text{ W}.$$

The 5 V emf of the battery being recharged removes energy from the circuit at a rate

$$P_{\text{out}} = \mathcal{E}I = (5 \text{ V})(1 \text{ A}) = 5 \text{ W}.$$

When a battery is recharged, electrical energy is converted to chemical energy.

REFLECT The total rate at which the emfs supply energy to the circuit is $P_{\text{in}} - P_{\text{out}} = 36 \text{ W} - 5 \text{ W} = 31 \text{ W}.$ By energy

Continued

conservation, this rate must equal the total rate P_R at which electrical energy is dissipated in all the resistors. For a resistor, $P = IR^2$, so the total rate of electrical consumption in the resistors is

$$P_R = (2\text{ A})^2(3\ \Omega) + (1\text{ A})^2(1\ \Omega) + (3\text{ A})^2(2\ \Omega)$$
$$= 12\text{ W} + 1\text{ W} + 18\text{ W} = 31\text{ W},$$

which agrees with the power supplied by the emfs, confirming energy conservation.

Practice Problem: In this example, find I, r, and $\mathcal{E}$ if the current through the bulb is 3 A rather than 2 A. *Answers:* 4 A, 0.75 Ω, -8 V.

19.7 Electrical Measuring Instruments

The concepts of current, potential difference, and resistance have played a central role in our analysis of circuits in this chapter and the preceding one. But how do we *measure* these quantities? Many common devices, including car instrument panels, battery chargers, and inexpensive electrical instruments, measure potential difference (voltage), current, or resistance using a device containing a pivoted coil of wire placed in the magnetic field of a permanent magnet (Figure 19.30). Attached to the coil is a coiled spring; in the equilibrium position, with no current in the coil, the pointer is at zero. When there is a current in the coil, the magnetic field exerts a torque on the coil that is proportional to the current. Thus, the angular deflection of the coil and pointer is directly proportional to the coil current, and the device can be calibrated to measure current. The ideal behavior for a meter would be for it to measure the circuit quantities of interest without disturbing or changing those quantities by its presence.

Ammeters

An instrument that measures current is usually called an **ammeter.** The essential concept is that *an ammeter always measures the current passing through it.* An *ideal* ammeter would have *zero* resistance, so that including it in a branch of a circuit would not affect the current in that branch. All real ammeters have some finite resistance; it's always desirable for an ammeter to have as little resistance as possible, so that its presence disturbs the circuit behavior as little as possible.

Voltmeters

A **voltmeter** is a device that measures the potential difference (voltage) between two points. To make this measurement, a voltmeter must have its terminals connected to the two points in question. An ideal voltmeter would have *infinite* resistance, so that no current would flow through it. Then connecting it between two points in a circuit would not alter any of the circuit currents. Real voltmeters always have finite resistance, but a voltmeter should have a large enough resistance that connecting it in a circuit does not change the other currents appreciably.

For greater precision and mechanical ruggedness, pivoted-coil meters have been largely replaced by electronic instruments with direct digital readouts. These devices are more precise, stable, and mechanically rugged, and often more expensive, than the older devices. Digital voltmeters can be made with very high internal resistance, on the order of 100 MΩ. Figure 19.31 shows a *multimeter*—an instrument that can measure voltage, current, and resistance over a wide range.

19.8 Resistance–Capacitance Circuits

In our discussion of circuits, we've assumed that all the emf's and resistances are *constant.* As a result, all potentials and currents are constant and independent of time. Figure 19.32a shows a simple example of a circuit in which the current and voltages are *not* constant. We'll call this a resistance–capacitance (R–C) circuit. The capacitor is initially uncharged; at some initial time $t = 0$, we close the switch, completing the circuit and permitting current around the loop to begin

Scale　Spring

Pointer

0 ... 5 ... 10

Permanent magnet

Soft-iron core

Pivoted coil　Magnetic field

▲ **FIGURE 19.30** One type of galvanometer.

ActivPhysics 12.4: Using Ammeters and Voltmeters

▲ **FIGURE 19.31** A digital multimeter.

charging the capacitor (Figure 19.32b). The current begins at the same instant in every part of the circuit, and at each instant it is the same in every part.

We'll neglect the internal resistance of the battery, so its terminal voltage is constant and equal to the battery emf $\mathcal{E}$. The capacitor is initially uncharged; the potential difference across it is initially zero. At this time, from Kirchhoff's loop rule, the voltage across the resistor R is equal to the battery's terminal voltage $\mathcal{E}$. The initial current through the resistor, which we'll call I_0, is given by Ohm's law: $I_0 = \mathcal{E}/R$. The initial capacitor charge, which we'll call Q_0, is zero. We denote the time-varying current and charge as i and q, respectively.

As the capacitor charges, its voltage increases, so the potential difference across the resistor must decrease, corresponding to a decrease in current. From Kirchhoff's loop rule, the sum of these two is constant and equal to the battery emf $\mathcal{E}$:

$$\mathcal{E} = iR + \frac{q}{C}. \tag{19.16}$$

After a long time, the capacitor becomes fully charged,. The current decreases to zero, the potential difference across the resistor becomes zero, and the entire battery voltage $\mathcal{E}$ appears across the capacitor. Thus, the capacitor charge and current vary with time as shown by the graphs in Figure 19.33.

To obtain a detailed description of the time variation of charge and current in the circuit, we would need to solve Equation 19.16, which is a differential equation for the charge q and its rate of change i. The solutions contain the exponential function $e^{-t/RC}$, where $e = 2.718$ is the base of natural logarithms. We'll omit the details of these calculations; it turns out that if the switch is closed at time $t = 0$, the current i and charge q vary with time t according to the equations

$$i = I_0 e^{-t/RC}, \qquad q = Q_{\text{final}}\left(1 - e^{-t/RC}\right). \tag{19.17}$$

The graphs of Figure 19.33 are graphs of these equations. The graphs and the equations both show that as time goes on (and the exponential function $e^{-t/RC}$ approaches zero), the capacitor charge q approaches its final value, which we've called Q_{final}. The current decreases and eventually becomes zero. When $i = 0$, Equation 19.16 gives

$$\mathcal{E} = \frac{q}{C} \qquad \text{and} \qquad q = Q_{\text{final}} = C\mathcal{E}.$$

We note that the final charge Q_{final} doesn't depend on R, although the rate at which this final value is approached is slower when R is large than when it is small.

Returning to Figure 19.33, we note that at the instant the switch is closed $(t = 0)$ the current jumps from zero to its initial value $I_0 = \mathcal{E}/R$. After that, it decreases and gradually approaches zero. The capacitor charge starts at zero and gradually approaches the final value $Q_{\text{final}} = C\mathcal{E}$.

After a time $t = RC$, the exponential functions in Equations 19.17 have the value $e^{-1} = 1/e = 0.368$. At this time, the current has decreased from its initial value I_0 to $1/e$ of that value, and the capacitor charge has reached $(1 - 1/e) = 0.632$ of its final value $Q_{\text{final}} = C\mathcal{E}$. The product RC is therefore a measure of how quickly the capacitor charges; it is called the **time constant,** or the **relaxation time,** of the circuit, denoted by τ (the Greek letter tau):

$$\tau = RC. \tag{19.18}$$

When τ is small, the capacitor charges quickly; when τ is larger, the charging takes more time. The horizontal axis (where $i = 0$) is an *asymptote* for the curve in Figure 19.33a. Strictly speaking, i never becomes precisely zero, but the longer

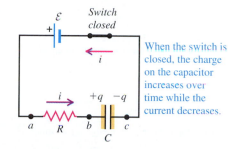

(a) Capacitor initially uncharged

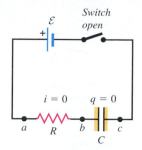

When the switch is closed, the charge on the capacitor increases over time while the current decreases.

(b) Charging the capacitor

▲ **FIGURE 19.32** Charging a capacitor in a resistance–capacitance circuit.

Mastering PHYSICS

ActivPhysics 12.6: Capacitance
ActivPhysics 12.7: Series and Parallel Capacitors
ActivPhysics 12.8: RC Circuit Time Constants

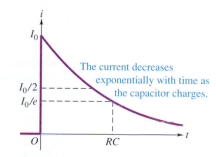

The current decreases exponentially with time as the capacitor charges.

(a) Graph of current versus time

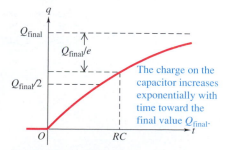

The charge on the capacitor increases exponentially with time toward the final value Q_{final}.

(b) Graph of capacitor charge versus time

▲ **FIGURE 19.33** Current i and capacitor charge q as functions of time for charging the capacitor in the circuit in Figure 19.32.

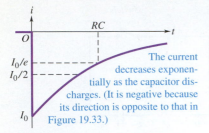

The current decreases exponentially as the capacitor discharges. (It is negative because its direction is opposite to that in Figure 19.33.)

(a) Graph of current versus time

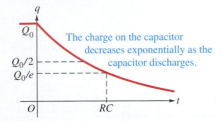

The charge on the capacitor decreases exponentially as the capacitor discharges.

(b) Graph of capacitor charge versus time

▲ **FIGURE 19.34** Current i and capacitor charge q as functions of time for discharging the capacitor in the circuit in Figure 19.32.

we wait, the closer it gets. For example, after a time equal to $10\tau = 10RC$, the current has decreased to about 0.00005 of its initial value. Similarly, the curve in Figure 19.33b approaches the horizontal broken line labeled Q_{final} as an asymptote. The charge q never attains precisely this value, but after a time equal to $10RC$, the difference between q and Q_{final} is about $0.00005Q_{final}$.

We could carry out a similar analysis for the *discharge* of a capacitor that is initially charged. We won't go into the details, but the results are similar to those obtained in the charging situation. The current i and the capacitor charge q vary with time according to the equations

$$i = I_0 e^{-t/RC} \qquad \text{and} \qquad q = Q_0 e^{-t/RC}, \qquad (19.19)$$

where I_0 and Q_0 are the initial values at time $t = 0$. Both I and q decrease to $1/e = 0.368$ of their initial values after a time equal to $\tau = RC$, as shown in Figure 19.34. After a time equal to $10\tau = 10RC$, they have both decreased to about 0.00005 of their initial values. So after a time equal to a few time constants, the capacitor is, for all practical purposes, completely discharged.

19.9 Physiological Effects of Currents

Electrical potential differences and currents play a vital role in the nervous systems of animals. Conduction of nerve impulses is basically an electrical process, although the mechanism of conduction is much more complex than in simple materials such as metals.

A nerve fiber, or *axon,* along which an electrical impulse can travel, includes a cylindrical membrane with one conducting fluid (electrolyte) inside and another outside (Figure 19.35a). Chemical systems similar to those in batteries maintain a potential difference on the order of 0.1 V between these fluids. When an electrical pulse is initiated, the nerve membrane temporarily becomes more permeable to the ions in the fluids, leading to a local drop in potential (Figure 19.35b). As the pulse propagates, the membrane recovers and overshoots briefly before returning the potential to its initial value (Figure 19.35c).

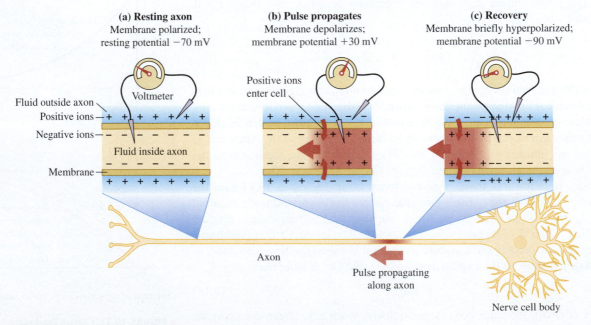

▲ **FIGURE 19.35** Propagation of a pulse along the axon of a nerve cell.

The electrical nature of nerve-impulse conduction is responsible for the great sensitivity of the body to externally supplied electric currents. (Impulses equivalent to those carried by nerve cells also occur in muscle cells, where they induce contraction. Thus, muscles as well as nerves are sensitive to currents.) Currents through the body as small as 0.1 A (much too small to produce significant heating) can be fatal because they interfere with nerve processes that are essential for vital functions, such as the heartbeat. The resistance of the human body is highly variable. Body fluids are usually quite good conductors because of their substantial ion concentrations, but the resistance of skin is relatively high, ranging from 500 kΩ for very dry skin to 1000 Ω or so for wet skin, depending also on the area of contact. If $R = 1000\ \Omega$, a current of 0.1 A requires a voltage of $V = IR = (0.1\ \text{A})(1000\ \Omega) = 100\ \text{V}$. This is within the range of household voltages, and it is one reason it is dangerous to receive a shock with wet skin.

Even smaller currents can be dangerous. A current of 0.01 A through an arm or leg causes strong, convulsive muscle action and considerable pain, and with a current of 0.02 A, a person who is holding the conductor that is inflicting the shock is typically unable to release it. Currents of this magnitude through the chest can cause ventricular fibrillation, a disorganized twitching of the heart muscles that pumps very little blood. Surprisingly, very large currents (over 0.1 A) are somewhat *less* likely to cause fatal fibrillation, because the heart muscle is "clamped" in one position. The heart actually stops beating and is more likely to resume normal beating when the current is removed. The electric defibrillators used for medical emergencies apply a large current pulse to stop the heart (and the fibrillation) and give it a chance to restore normal rhythm.

Thus, electric current poses three different kinds of hazards: interference with the nervous system, injury caused by convulsive muscle action, and burns from I^2R heating. The moral of this rather morbid story is that under certain conditions, voltages as small as 10 V can be dangerous. All electric circuits and equipment should always be approached with respect and caution.

On the positive side, rapidly alternating currents can have beneficial effects. Alternating currents with frequencies on the order of 10^6 Hz do not interfere appreciably with nerve processes and can be used for therapeutic heating for arthritic conditions, sinusitis, and a variety of other disorders. If one electrode is made very small, the resulting concentrated heating can be used for local destruction of tissue such as tumors or for cutting tissue in certain surgical procedures.

The study of particular nerve impulses is also an important *diagnostic* tool in medicine. The most familiar examples are electrocardiography (EKG) and electroencephalography (EEG). Electrocardiograms, obtained by attaching electrodes to the chest and back and recording the regularly varying potential differences, are used to study heart function. Similarly, electrodes attached to the scalp permit the study of potentials in the brain, and the resulting patterns can be helpful in diagnosing epilepsy, brain tumors, and other disorders.

19.10 Power Distribution Systems

We conclude this chapter with a brief discussion of practical household and automotive electric power distribution systems. Automobiles use direct-current (dc) systems, and nearly all household, commercial, and industrial systems use alternating current (ac) because of the ease of stepping voltage up and down with transformers. Most of the same basic wiring concepts apply to both systems. We'll talk about alternating-current circuits in greater detail in Chapter 22.

The various lamps, motors, and other appliances to be operated are always connected *in parallel* to the power source (the wires from the power company for houses or the battery and alternator for a car). The basic idea of house wiring is shown in Figure 19.36. One side of the "line," as the pair of conductors is called,

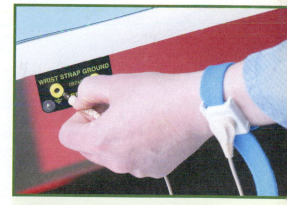

▲ **BIO Application Chained to your job.** This electronics worker is wearing a grounding strap around the wrist. People easily acquire a static charge by rubbing against objects, and because we are conductors, we can deliver this charge as a current to any conducting object we touch. Such a current won't harm a doorknob, but it's quite enough to fry sensitive electronics. The wrist strap prevents such mishaps. It makes direct electrical contact with the skin and is connected to a grounded conductor, so it keeps static charge from accumulating. If you've ever installed memory cards in your own computer, you may have received and used a disposable wrist strap.

▶ **FIGURE 19.36** Schematic diagram of part of a house wiring system. Only two branch circuits are shown; an actual system might have four to thirty branch circuits. Lamps and appliances may be plugged into the outlets. The grounding conductors, which normally carry no current, are not shown. Modern household systems usually have two "hot" lines with opposite polarity with respect to neutral, and a voltage of 240 V between them. (Actual wires use a different color-coding system.)

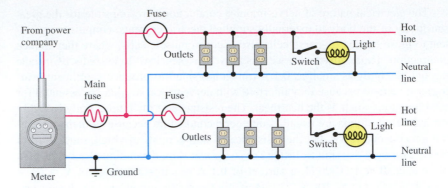

is called the *neutral* side; it is always connected to "ground" at the entrance panel. For houses, ground is an actual electrode driven into the earth (usually a good conductor) and also sometimes connected to the household cold-water pipes. Electricians speak of the "hot" side and the "neutral" side of the line. Most modern house wiring systems have *two* hot lines with opposite polarity with respect to the neutral; the voltage between them is twice the voltage between either hot line and the neutral line. (We'll return to this detail later.)

Household voltage is nominally 120 V in the United States and Canada, and often 240 V in Europe. (For alternating current, which varies sinusoidally with time, these numbers represent the *root-mean-square* voltage, which is $1/\sqrt{2}$ times the peak voltage. We'll discuss this further in Section 22.1.) The power input P to a device is given by Equation 19.9: $P = V_{ab}I$. For example, the current in a 100 W lightbulb is

$$I = \frac{P}{V} = \frac{100 \text{ W}}{120 \text{ V}} = 0.83 \text{ A}.$$

The resistance of this bulb at operating temperature is

$$R = \frac{V}{I} = \frac{120 \text{ V}}{0.83 \text{ A}} = 144 \ \Omega, \quad \text{or} \quad R = \frac{V^2}{P} = \frac{(120 \text{ V})^2}{100 \text{ W}} = 144 \ \Omega.$$

Similarly, a 1500 W waffle iron draws a current of $(1500 \text{ W})/(120 \text{ V}) = 12.5 \text{ A}$ and has a resistance of 9.6 Ω at its operating temperature. Because of the temperature dependence of resistivity, the resistances of these devices are considerably less when they are cold. If you measure the resistance of a lightbulb with an ohm-meter (whose small current causes very little temperature rise), you will probably get a value of about 10 Ω. When a light bulb is turned on, there is an initial surge of current as the filament heats up. That's why, when a light bulb gets ready to burn out, it nearly always happens when you turn it on.

The maximum current available from an individual circuit is limited by the resistance of the wires. The I^2R power loss in the wires causes them to become hot, and in extreme cases this can cause a fire or melt the wires. Ordinary lighting and outlet wiring in houses usually uses 12 gauge wire. This has a diameter of 2.05 mm and can carry a maximum current of 20 A safely (without overheating). Larger sizes, such as 8 gauge (3.26 mm) or 6 gauge (4.11 mm), are used for high-current appliances—for example, ranges and clothes dryers—and 2 gauge (6.54 mm) or larger is used for the main power lines entering a house.

Protection against overloading and overheating of circuits is provided by fuses or circuit breakers. A *fuse* contains a link of lead–tin alloy with a very low melting temperature; the link melts and breaks the circuit when its rated current is exceeded. A *circuit breaker* is an electromechanical device that performs the same function, using an electromagnet or a bimetallic strip to "trip" the breaker and interrupt the circuit when the current exceeds a specified value. Circuit breakers have the advantage that they can be reset after they are tripped. A blown fuse must be replaced, but fuses are somewhat more reliable in operation than circuit breakers are.

If your system has fuses and you plug too many high-current appliances into the same outlet, the fuse blows. *Do not* replace the fuse with one of larger rating; if you do, you risk overheating the wires and starting a fire. The only safe solution is to distribute the appliances among several circuits. Modern kitchens often have three or four separate 20 A circuits, each of which can carry a current of 20 A without overheating.

Contact between the hot and neutral sides of the line causes a *short circuit*. Such a situation, which can be caused by faulty insulation or by any of a variety of mechanical malfunctions, provides a very low resistance current path, permitting a large current that would quickly melt the wires and ignite their insulation if the current were not interrupted by a fuse or circuit breaker. An equally dangerous situation is a broken wire that interrupts the current path, creating an *open circuit*. This is hazardous because of the sparking that can occur at the point of intermittent contact.

In approved wiring practice, a fuse or breaker is placed *only* on the hot side of the line, never on the neutral side. Otherwise, if a short circuit should develop because of faulty insulation or other malfunction, the ground-side fuse could blow. The hot side would still be live and would pose a shock hazard if you touched the live conductor and a grounded object such as a water pipe. For similar reasons, the wall switch for a light fixture should always be on the hot side of the line, never the neutral side.

Further protection against shock hazard is provided by a third conductor called the *grounding wire*, included in all present-day wiring. This conductor corresponds to the long round or U-shaped prong of the three-prong connector plug on an appliance or power tool. It is connected to the neutral side of the line at the entrance panel, where the meter is. It normally carries no current, but it connects the metal case or frame of the device to ground. If a conductor on the hot side of the line accidentally contacts the frame or case, the grounding conductor provides a current path and the fuse blows. Without the ground wire, the frame could become "live"; that is, it could reach a potential 120 V above ground. Then, if you touched it and a water pipe (or even a damp basement floor) at the same time, you could get a dangerous shock (Figure 19.37). In some situations, especially for outlets located outdoors or near a sink or water pipes, a special kind of circuit breaker called a *ground-fault interrupter* (GFI or GFCI) is used. This device senses the difference in current between the hot and neutral conductors (which is normally zero) and trips when it exceeds some very small value, typically 5 mA.

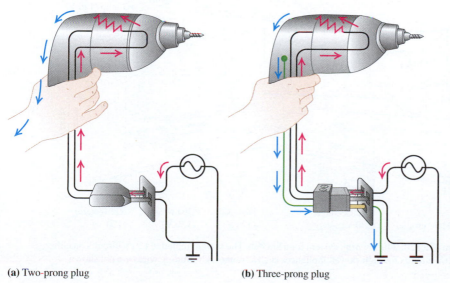

(a) Two-prong plug   **(b)** Three-prong plug

◀ **FIGURE 19.37** (a) If a malfunctioning electric drill is connected to a wall socket via a two-prong plug, a person may receive a shock. (b) When the drill malfunctions when connected via a three-prong plug, a person touching it receives no shock, since electric charge flows through the third prong and into the ground, rather than through the person's body. If the ground current is appreciable, the fuse blows.

Most household wiring systems actually use a slight elaboration of the system just described. Your power company provides *three* conductors (Figure 19.38). One is neutral; the other two are both at 120 V with respect to the neutral, but with opposite polarity, giving a voltage of 240 V between them. The power company calls this a *three-wire line,* in contrast to the 120 V two-wire (plus ground-wire) line described. With a three-wire line, 120 V lamps and appliances can be connected between the neutral conductor and either hot conductor, and high-power devices requiring 240 V are connected between the two hot lines. Ranges and dryers are usually designed for 240-V power input.

To help prevent wiring errors, household wiring uses a standardized color code in which the hot side of a line has black insulation (black and red for the two sides of a 240 V line), the neutral side has white insulation, and the grounding conductor is bare or has green insulation. In electronic devices and equipment, by contrast, the ground or neutral side of the line is usually black, so beware! (Our illustrations do not follow standard code, but use red for the hot line and blue for neutral.)

The preceding discussion can be applied directly to automobile wiring. The voltage is about 13 V (direct current); the power is supplied by the battery and the alternator, which charges the battery when the engine is running. The neutral side of the circuits is connected to the body and frame of the vehicle. For this low voltage, safety does not require a separate grounding conductor. The fuse or circuit breaker arrangement is the same, in principle, as in household wiring. Because of the lower voltage, more current is required for the same power; a 100 W headlight bulb requires a current of $(100 \text{ W})/(13 \text{ V}) = 7.7$ A.

Although we have spoken mostly of *power* in this section, what households really buy from their power company is *energy.* Power is energy transferred per unit time, so energy is average power multiplied by time. The usual unit of energy sold by the power company is the kilowatt-hour (1 kWh):

$$1 \text{ kWh} = (10^3 \text{ W})(3600 \text{ s}) = 3.6 \times 10^6 \text{ W} \cdot \text{s} = 3.6 \times 10^6 \text{ J}.$$

One kilowatt-hour typically costs 2 to 10 cents, depending on one's location and the quantity of energy purchased. To operate a 1500 W (1.5 kW) waffle iron continuously for 1 hour requires 1.5 kWh of energy; at 10 cents per kWh, the cost is 15 cents (not including the cost of the flour and eggs in the waffle batter). The cost of operating any lamp or appliance for a specified time can be calculated in the same way if the power rating is known. However, many electric cooking utensils (including waffle irons) cycle on and off to maintain a constant temperature, and the *average* power may be less than the power rating marked on the device.

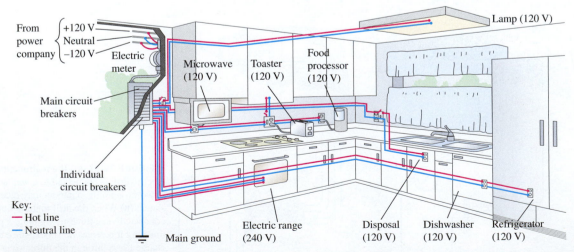

▲ **FIGURE 19.38** Schematic diagram of a 120–240 V wiring system for a kitchen. The system includes 120 V circuits on either side of the neutral line, as well as 240 V circuits for high-power appliances such as ranges. Grounding wires are not shown.

SUMMARY

Current

(Section 19.1) **Current** is the amount of charge flowing through a conductor per unit time. The SI unit of current is the ampere, equal to 1 coulomb per second ($1\text{ A} = 1\text{ C/s}$). If a net charge ΔQ flows through a wire in time Δt, the current through the wire is $I = \Delta Q / \Delta t$ (Equation 19.1).

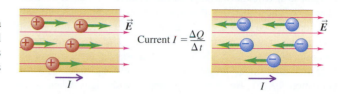

$$\text{Current } I = \frac{\Delta Q}{\Delta t}$$

Resistance and Ohm's Law

(Section 19.2) In a conductor, the **resistance** R is the ratio of voltage to current: $R = V/I$ (Equation 19.2). The SI unit of resistance is the **ohm** (Ω), equal to 1 volt per ampere. In materials that obey **Ohm's law,** the potential difference V between the ends of a conductor is proportional to the current I through the conductor; the proportionality factor is the resistance R.

For a given conducting material, resistance R is proportional to length and inversely proportional to cross-sectional area. For a specific material, this relationship can be expressed as $R = \rho(L/A)$ (Equation 19.3), where ρ is the **resistivity** of that material.

Resistance and resistivity vary with temperature; for metals, they usually increase with increasing temperature.

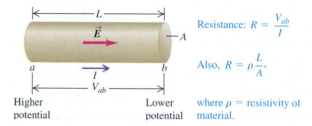

Higher potential

Lower potential

Resistance: $R = \dfrac{V_{ab}}{I}$

Also, $R = \rho\dfrac{L}{A}$,

where ρ = resistivity of material.

Electromotive Force and Circuits

(Section 19.3) A **complete circuit** is a conductor in the form of a loop providing a continuous current-carrying path. A complete circuit carrying a steady current must contain a source of electromotive force (emf), symbolized by $\mathcal{E}$. An ideal source of emf maintains a constant potential difference $V_{ab} = \mathcal{E}$ (Equation 19.5), but every real source of emf has some internal resistance r. The terminal potential difference V_{ab} then depends on current: $V_{ab} = \mathcal{E} - Ir$ (Equation 19.7).

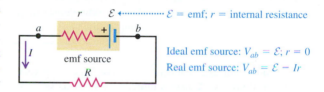

$\mathcal{E} =$ emf; $r =$ internal resistance

Ideal emf source: $V_{ab} = \mathcal{E}$; $r = 0$

Real emf source: $V_{ab} = \mathcal{E} - Ir$

Energy and Power in Electric Circuits

(Section 19.4) A circuit element with a potential difference V and a current I puts energy into a circuit if the current direction is from lower to higher potential in the device and takes energy out of the circuit if the current is opposite. The power P (rate of energy transfer) is $P = VI$ (Equation 19.9). A resistor R always takes energy out of a circuit, converting it to thermal energy at a rate given by $P = V_{ab}I = I^2R = V_{ab}^2/R$ (Equation 19.10).

Resistors in Series and in Parallel

(Section 19.5) When several resistors $R_1, R_2, R_3, \cdots$ are connected in series, the **equivalent resistance** R_{eq} is the sum of the individual resistances: $R_{eq} = R_1 + R_2 + R_3 + \cdots$. (Equation 19.12). When several resistors are connected in parallel, the equivalent resistance R_{eq} is given by

$$\frac{1}{R_{eq}} = \frac{1}{R_1} + \frac{1}{R_2} + \frac{1}{R_3} + \cdots. \qquad (19.13)$$

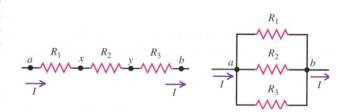

Continued

Kirchhoff's Rules

(Section 19.6) Kirchhoff's junction rule is based on conservation of charge. It states that the algebraic sum of the currents into any junction must be zero: $\sum I = 0$ (Equation 19.14). Kirchhoff's loop rule is based on conservation of energy and the conservative nature of electrostatic fields. It states that the algebraic sum of the potential differences around any loop must be zero: $\sum V = 0$ (Equation 19.15). Be especially careful with signs when using Kirchhoff's rules.

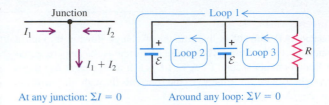

At any junction: $\sum I = 0$ Around any loop: $\sum V = 0$

Electrical Measuring Instruments

(Section 19.7) The ideal behavior for a meter is for it to measure the circuit quantities of interest without changing or disturbing them. An ammeter always measures the current passing through it. An *ideal* ammeter would have *zero* resistance, so that including it in a branch of a circuit would not affect the current in that branch. A voltmeter always measures the potential difference between two points. An ideal voltmeter would have *infinite* resistance, so that no current would flow through it.

Resistance–Capacitance Circuits

(Section 19.8) When a capacitor is charged by a battery in series with a resistor, the current and capacitor charge are not constant. The charge varies with time as $q = Q_{final}(1 - e^{-t/RC})$ (Equation 19.17). In a time $\tau = RC$, there is a significant change in the charge on the capacitor. This time is called the **time constant,** or **relaxation time,** and is the same for charging or discharging.

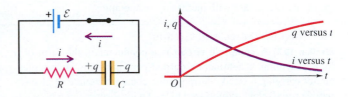

Applications of Currents

(Sections 19.9 and 19.10) The conduction of nerve impulses is basically an electrical process. Currents through the body as small as 0.1 A can be fatal because they interfere with this process (which also occurs in heart and other muscle cells).

In house wiring, one line entering the house is always *neutral,* or at the same voltage as the ground (to which it is connected). The other one or two wires are *hot.* The maximum current available from an individual circuit is limited by the resistance of the wires; if they carry too much current, I^2R power loss causes them to overheat. Protection against overloading of circuits is provided by fuses or circuit breakers.

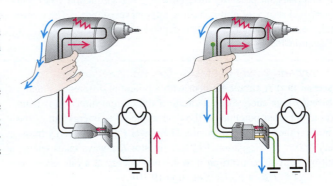

 For instructor-assigned homework, go to www.masteringphysics.com

Conceptual Questions

1. A rule of thumb used to determine the internal resistance of a source is that it is the open-circuit voltage divided by the short-circuit current. Is this rule correct?
2. The energy that can be extracted from a storage battery is always less than the energy that goes into it while it is being charged. Why?
3. A cylindrical rod has resistivity ρ. If we triple its length and diameter, what is its resistivity, in terms of ρ?
4. A fuse is a device designed to break a circuit, usually by melting, when the current exceeds a certain value. Fuses are widely used in electronic equipment, but have been replaced by circuit breakers in household wiring. In the "old days," people would sometimes replace a blown fuse with a penny, which happened to be the same size as a fuse. Was this a safe practice? Why?
5. True or false? (a) Adding more resistance to a circuit increases its resistance. (b) Removing resistance from a circuit decreases its resistance. Justify your answers with simple examples.
6. Why does the resistance of an object increase with temperature? Explain by looking at the movement of charges at the atomic level.

7. *How* does a capacitor store energy? Can a resistor store energy?
8. High-voltage power supplies are sometimes designed to have a rather large internal resistance as a safety precaution. Why is such a power supply with a large internal resistance safer than one with the same voltage, but lower internal resistance?
9. If a 1.5 V AA battery produces *less than* 1.5 V across its terminals while operating in a circuit, does this *necessarily* mean that the battery is running down and near the end of its life? Why?
10. Can all combinations of resistors be reduced to series and parallel combinations? Illustrate your answer with some examples. (*Hint:* Check out some of the circuits in this chapter.)
11. In a two-cell flashlight, the batteries are normally connected in series. Why not connect them in parallel?
12. Old-time Christmas tree lights had the property that, when one bulb burned out, all the lights went out. How were these lights connected, in series or in parallel? How could you rewire them to prevent all the lights from going out when one of them burned out?
13. You connect a number of identical light bulbs to a flashlight battery. (a) What happens to the brightness of each bulb as more and more bulbs are added to the circuit if you connect them (i) in series and (ii) in parallel? Will the battery last longer if the bulbs are in series or in parallel? Explain your reasoning.
14. For very large resistances, it is easy to construct resistance–capacitance circuits having time constants of several seconds or minutes. How might this fact be used to measure such very large resistances, too large to measure by more conventional means?
15. When you scuff your shoes across a nylon carpet, you can easily produce a potential of several thousand volts between your body and the carpet. Yet contact with a power line of comparable voltage would probably be fatal. Why the difference?

Multiple-Choice Problems

1. A cylindrical metal rod has a resistance R. If both its length and its diameter are tripled, its new resistance will be:
   A. R B. $9R$ C. $R/3$ D. $3R$
2. A resistor R and another resistor $2R$ are connected in series across a battery. If heat is produced at a rate of 10 W in R, then in $2R$ it is produced at a rate of
   A. 40 W.   B. 20 W.   C. 10 W.   D. 5 W.
3. A resistor R and another resistor $2R$ are connected in parallel across a battery. If heat is produced at a rate of 10 W in R, in $2R$ it is produced at a rate of
   A. 40 W.   B. 20 W.   C. 10 W.   D. 5 W.
4. Which statements about the circuit shown in Figure 19.39 are correct? All meters are considered to be ideal, the connecting leads have no resistance, and the battery has no internal resistance. (There may be more than one correct choice.)
   A. The reading in ammeter A_1 is greater than the reading in A_2 because current is lost in the resistor.
   B. The two ammeters have exactly the same readings.

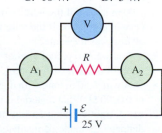

▲ **FIGURE 19.39** Multiple-choice problem 4.

C. The voltmeter reads less than 25 V because some voltage is lost in the resistor.
   D. The voltmeter reads exactly 25 V.
5. When the switch S in Figure 19.40 is closed, the reading of the voltmeter V will
   A. increase.   B. decrease.
   C. stay the same.

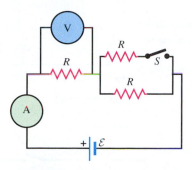

▲ **FIGURE 19.40** Multiple-choice problem 5.

6. When the switch S in the circuit in the previous question is closed, the reading of the ammeter A will
   A. increase.   B. decrease.
   C. stay the same.
7. Three identical lightbulbs are connected in the circuit shown in Figure 19.41. After the switch S is closed, what will be true about the brightness of these bulbs?
   A. B_1 will be brightest and B_3 will be dimmest.
   B. B_3 will be brightest and B_1 will be dimmest.
   C. All three bulbs will have the same brightness.

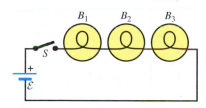

▲ **FIGURE 19.41** Multiple-choice problem 7.

8. A cylindrical metal rod of length L and diameter D is connected across a battery having no internal resistance. An ammeter in the circuit measures the current to be I. If we now double the diameter of the rod, but change nothing else, the ammeter will read
   A. $4I$.   B. $2I$.   C. $I/2$.   D. $I/4$.
9. Two identical metal rods are welded together end to end. If each rod has a length L and resistivity ρ, the resistivity of the combination will be
   A. 4ρ.   B. 2ρ.
   C. ρ.   D. $\rho/2$.
10. In the circuit shown in Figure 19.42, resistor A has three times the resistance of resistor B. Therefore,
   A. the current through A is three times the current through B.
   B. the current through B is three times the current through A.

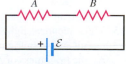

▲ **FIGURE 19.42** Multiple-choice problem 10.

C. the potential difference across A is three times the potential difference across B.

D. the potential difference is the same across both resistors.

11. In which of the two circuits shown in Figure 19.43 will the capacitors charge more rapidly when the switch is closed?

A. Circuit (a)
B. Circuit (b)
C. The capacitors will charge at the same rate in the two circuits.

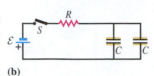

(a)

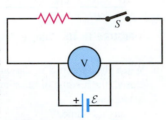

(b)

▲ **FIGURE 19.43** Multiple-choice problem 11.

12. The battery shown in the circuit in Figure 19.44 has some internal resistance. When we close S, the reading of the voltmeter V will

A. increase.
B. stay the same.
C. decrease.

13. A battery with no internal resistance is connected across identical lightbulbs as shown in Figure 19.45. When you close the switch S, bulbs B_1 and B_2 will be

A. brighter than before.
B. dimmer than before.
C. just as bright as before.

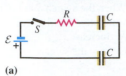

▲ **FIGURE 19.44** Multiple-choice problem 12.

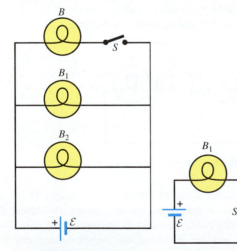

▲ **FIGURE 19.45** Multiple-choice problem 13.

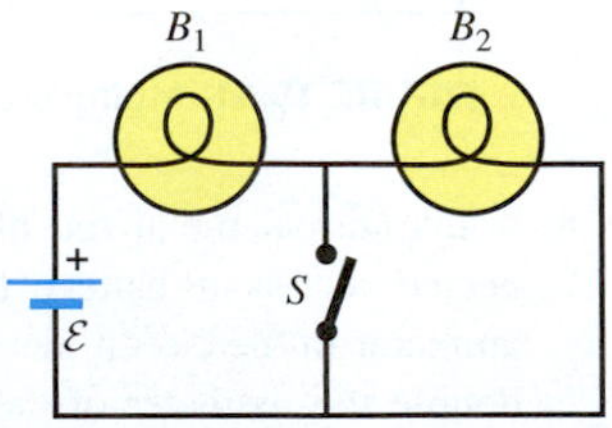

▲ **FIGURE 19.46** Multiple-choice problem 14.

14. The battery shown in the circuit in Figure 19.46 has no internal resistance. After you close the switch S, the brightness of bulb B_1 will

A. increase.   B. decrease.
C. remain the same.

15. Three identical light bulbs, A, B, and C, are connected in the circuit shown in Figure 19.47. When the switch S is closed,

A. the brightness of A and B remains the same as it was, but C goes out.
B. the brightness of A and B remains the same as it was, but C will be about half as bright as it was.

C. the brightness of A and B decreases, and C goes out.
D. the brightness of A and B increases, and C will be about half as bright as it was.
E. the brightness of A and B increases, but C goes out.

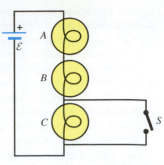

▲ **FIGURE 19.47** Multiple-choice problem 15.

Problems

19.1 Current

1. • Typical household currents are on the order of a few amperes. If a 1.50 A current flows through the leads of an electrical appliance, (a) how many electrons per second pass through it, (b) how many coulombs pass through it in 5.0 min, and (c) how long does it take for 7.50 C of charge to pass through?

2. • **Lightning strikes.** During lightning strikes from a cloud to the ground, currents as high as 25,000 A can occur and last for about 40 μs. How much charge is transferred from the cloud to the earth during such a strike?

3. • **Transmission of nerve impulses.** Nerve cells transmit electric signals through their long tubular axons. These signals **BIO** propagate due to a sudden rush of Na^+ ions, each with charge $+e$, into the axon. Measurements have revealed that typically about 5.6×10^{11} Na^+ ions enter each meter of the axon during a time of 10 ms. What is the current during this inflow of charge in a meter of axon?

4. •• In an ionic solution, a current consists of Ca^{2+} ions (of charge $+2e$) and Cl^- ions (of charge $-e$) traveling in opposite directions. If 5.11×10^{18} Cl^- ions go from A to B every 0.50 min, while 3.24×10^{18} Ca^{2+} ions move from B to A, what is the current (in mA) through this solution, and in which direction (from A to B or from B to A) is it going?

5. •• Copper has 8.5×10^{28} electrons per cubic meter. (a) How many electrons are there in a 25.0 cm length of 12-gauge copper wire (diameter 2.05 mm)? (b) If a current of 1.55 A is flowing in the wire, what is the average drift speed of the electrons along the wire? (There are 6.24×10^{18} electrons in 1 coulomb of charge.)

19.2 Resistance and Ohm's Law

6. • A 14 gauge copper wire of diameter 1.628 mm carries a current of 12.5 mA. (a) What is the potential difference across a 2.00 m length of the wire? (b) What would the potential difference in part (a) be if the wire were silver instead of copper, but all else was the same?

7. • You want to precut a set of 1.00 Ω strips of 14 gauge copper wire (of diameter 1.628 mm). How long should each strip be?

8. • A wire 6.50 m long with diameter of 2.05 mm has a resistance of 0.0290 Ω. What material is the wire most likely made of?

9. •• A tightly coiled spring having 75 coils, each 3.50 cm in diameter, is made of insulated metal wire 3.25 mm in diameter. An ohmmeter connected across opposite ends of the spring reads 1.74 Ω. What is the resistivity of the metal?

10. •• What diameter must a copper wire have if its resistance is to be the same as that of an equal length of aluminum wire with diameter 3.26 mm?

11. •• An aluminum bar 3.80 m long has a rectangular cross section 1.00 cm by 5.00 cm. (a) What is its resistance? (b) What is the length of a copper wire 1.50 mm in diameter having the same resistance?

12. •• If you triple the length of a cable and at the same time double its diameter, what will be its resistance if its original resistance was R?

13. •• A ductile metal wire has resistance R. What will be the resistance of this wire in terms of R if it is stretched to three times its original length, assuming that the density and resistivity of the material do not change when the wire is stretched. (*Hint:* The amount of metal does not change, so stretching out the wire will affect its cross-sectional area.)

14. • What is the resistance of a Nichrome™ wire at 0.0°C if its resistance is 100.00 Ω at 11.5°C? The temperature coefficient of resistivity for Nichrome™ is 0.00040 $(C°)^{-1}$.

15. •• A 1.50-m cylindrical rod of diameter 0.500 cm is connected to a power supply that maintains a constant potential difference of 15.0 V across its ends, while an ammeter measures the current through it. You observe that at room temperature (20.0°C) the ammeter reads 18.5 A, while at 92.0°C it reads 17.2 A. You can ignore any thermal expansion of the rod. Find (a) the resistivity and (b) the temperature coefficient of resistivity at 20°C for the material of the rod.

16. • A carbon resistor having a temperature coefficient of resistivity of $-0.00050 \, (C°)^{-1}$, is to be used as a thermometer. On a winter day when the temperature is 4.0°C, the resistance of the carbon resistor is 217.3 Ω. What is the temperature on a spring day when the resistance is 215.8 Ω? (Take the reference temperature T_0 to be 4.0°C.)

17. •• In a laboratory experiment, you vary the current through an object and measure the resulting potential difference across it in each case. Figure 19.48 shows a graph of this potential V as a function of the current I. (a) Does Ohm's law apply to this object? Why do you say so? (b) How is the resistance of the object related to the *slope* of the graph? Show why. (c) Use the slope of the graph to find the resistance of the object.

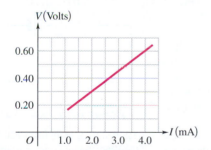

V(Volts)

0.60

0.40

0.20

O 1.0   2.0   3.0   4.0   → I(mA)

▲ **FIGURE 19.48** Problem 17.

18. •• The following measurements of current and potential difference were made on a resistor constructed of Nichrome™ wire, where V_{ab} is the potential difference across the wire and I is the current through it:

$I(A)$	0.50	1.00	2.00	4.00
$V_{ab}(V)$	1.94	3.88	7.76	15.52

(a) Graph V_{ab} as a function of I. (b) Does Ohm's law apply to Nichrome™? How can you tell? (c) What is the resistance of the resistor in ohms?

19. •• A battery-powered light bulb has a tungsten filament. When the switch connecting the bulb to the battery is first turned on and the temperature of the bulb is 20°C, the current in the bulb is 0.860 A. After the bulb has been on for 30 s, the current is 0.220 A. What is then the temperature of the filament?

19.3 Electromotive Force and Circuits

20. • When you connect an unknown resistor across the terminals of a 1.50 V AAA battery having negligible internal resistance, you measure a current of 18.0 mA flowing through it. (a) What is the resistance of this resistor? (b) If you now place the resistor across the terminals of a 12.6 V car battery having no internal resistance, how much current will flow? (c) You now put the resistor across the terminals of an unknown battery of negligible internal resistance and measure a current of 0.453 A flowing through it. What is the potential difference across the terminals of the battery?

21. • **Current in the body.** The resistance of the body varies
BIO from approximately 500 kΩ (when it is very dry) to about 1 kΩ (when it is wet). The maximum safe current is about 5.0 mA. At 10 mA or above, muscle contractions can occur that may be fatal. What is the largest potential difference that a person can safely touch if his body is wet? Is this result within the range of common household voltages?

22. • **"Current Baba."** According to a July 20, 2004, newspa-
BIO per article, a Hindu holy man known as "Current Baba" touches an electric wire three times daily to become "intoxicated." According to a doctor quoted in the article, "The human body can absorb currents up to 12 volts. In this case, however, repeated exposure to electricity seems to have built up ["Current Baba's"] body's tolerance levels to as much as 16 volts." (a) What is wrong with the doctor's statement? What do you think he really meant to say? (b) Since "Current Baba" was after the maximum "intoxication," he should have his body wet. In that case, how much current would he get with each jolt? Is this enough to be dangerous? (See the previous problem.)

23. • A copper transmission cable 100 km long and 10.0 cm in diameter carries a current of 125 A. What is the potential drop across the cable?

24. • A gold wire 6.40 m long and of diameter 0.840 mm carries a current of 1.15 A. Find (a) the resistance of this wire and (b) the potential difference between its ends.

25. •• When a solid cylindrical rod is connected across a fixed potential difference, a current I flows through the rod. What would be the current (in terms of I) if (a) the length were doubled, (b) the diameter were doubled, (c) both the length and the diameter were doubled?

26. • A 6.00 V lantern battery is connected to a 10.5 Ω lightbulb, and the resulting current in the circuit is 0.350 A. What is the internal resistance of the battery?

27. •• When switch S in Figure 19.49 is open, the voltmeter V across the battery reads 3.08 V. When the switch is closed, the voltmeter reading drops to 2.97 V and the ammeter A reads 1.65 A. Find the emf, the internal

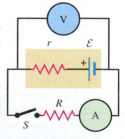

▲ **FIGURE 19.49** Problem 27.

resistance of the battery, and the circuit resistance R. Assume that the two meters are ideal, so that they don't affect the circuit.

28. •• A complete series circuit consists of a 12.0 V battery, a 4.70 Ω resistor, and a switch. The internal resistance of the battery is 0.30 Ω. The switch is open. What does an ideal voltmeter read when placed (a) across the terminals of the battery, (b) across the resistor, (c) across the switch? (d) Repeat parts (a), (b), and (c) for the case when the switch is closed.

29. •• With a 1500 MΩ resistor across its terminals, the terminal voltage of a certain battery is 2.50 V. With only a 5.00 Ω resistor across its terminals, the terminal voltage is 1.75 V. (a) Find the internal emf and the internal resistance of this battery. (b) What would be the terminal voltage if the 5.00 Ω resistor were replaced by a 7.00 Ω resistor?

30. • An automobile starter motor is connected to a 12.0 V battery. When the starter is activated it draws 150 A of current, and the battery voltage drops to 7.0 V. What is the battery's internal resistance?

31. •• Consider the circuit shown in Figure 19.50. The terminal voltage of the 24.0 V battery is 21.2 V. What is (a) the internal resistance r of the battery; (b) the resistance R of the circuit resistor?

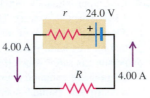

▲ FIGURE 19.50 Problem 31.

32. •• When switch S in Fig. 19.51 is open, the voltmeter V of the battery reads 3.08 V. When the switch is closed, the voltmeter reading drops to 2.97 V, and the ammeter A reads 1.65 A. Find the emf, the internal resistance of the battery, and the circuit resistance R. Assume that the two meters are ideal, so they don't affect the circuit.

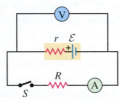

▲ FIGURE 19.51 Problem 32.

19.4 Energy and Power in Electric Circuits

33. • A resistor with a 15.0 V potential difference across its ends develops thermal energy at a rate of 327 W. (a) What is the current in the resistor? (b) What is its resistance?

34. • **Power rating of a resistor.** The *power rating* of a resistor is the maximum power it can safely dissipate without being damaged by overheating. (a) If the power rating of a certain 15 kΩ resistor is 5.0 W, what is the maximum current it can carry without damage? What is the greatest allowable potential difference across the terminals of this resistor? (b) If a 9.0 kΩ resistor is to be connected across a 120 V potential difference, what power rating is required for that resistor?

35. • An idealized voltmeter is connected across the terminals of a 15.0 V battery, and a 75.0 Ω appliance is also connected across its terminals. If the voltmeter reads 11.3 V: (a) how much power is being dissipated by the appliance, and (b) what is the internal resistance of the battery?

36. • **Treatment of heart failure.** A heart defibrillator is used to enable the heart to start beating if it has stopped. This is done by passing a large current of 12 A through the body at 25 V for a very short time, usually about 3.0 ms. (a) What power does the defibrillator deliver to the body, and (b) how much energy is transferred?

37. • **Lightbulbs.** The wattage rating of a lightbulb is the power it consumes *when it is connected across a 120 V potential difference.* For example, a 60 W lightbulb consumes 60.0 W of electrical power only when it is connected across a 120 V potential difference. (a) What is the resistance of a 60 W lightbulb? (b) Without doing any calculations, would you expect a 100 W bulb to have more or less resistance than a 60 W bulb? Calculate and find out.

38. • **Electrical safety.** This procedure is *not recommended!* You'll see why after you work the problem. You are on an aluminum ladder that is standing on the ground, trying to fix an electrical connection with a metal screwdriver having a metal handle. Your body is wet because you are sweating from the exertion; therefore, it has a resistance of 1.0 kΩ. (a) If you accidentally touch the "hot" wire connected to the 120 V line, how much current will pass through your body? Is this amount enough to be dangerous? (The maximum safe current is about 5 mA.) (b) How much electrical power is delivered to your body?

39. • **Electric eels.** Electric eels generate electric pulses along their skin that can be used to stun an enemy when they come into contact with it. Tests have shown that these pulses can be up to 500 V and produce currents of 80 mA (or even larger). A typical pulse lasts for 10 ms. What power and how much energy are delivered to the unfortunate enemy with a single pulse, assuming a steady current?

40. •• **Electric space heater.** A "540 W" electric heater is designed to operate from 120 V lines. (a) What is its resistance, and (b) what current does it draw? (c) At 7.4¢ per kWh, how much does it cost to operate this heater for an hour? (d) If the line voltage drops to 110 V, what power does the heater take, in watts? (Assume that the resistance is constant, although it actually will change because of the change in temperature.)

41. •• The battery for a certain cell phone is rated at 3.70 V. According to the manufacturer it can produce 3.15×10^4 J of electrical energy, enough for 5.25 h of operation, before needing to be recharged. Find the average current that this cell phone draws when turned on.

42. •• For the circuit in Fig. 19.52, find (a) the rate of conversion of internal (chemical) energy to electrical energy within the battery, (b) the rate of dissipation of electrical energy in the battery, (c) the rate of dissipation of electrical energy in the external resistor.

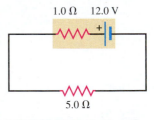

▲ FIGURE 19.52 Problem 42.

43. •• A 540-W electric heater is designed to operate from 120 V lines. (a) What is its resistance? (b) What current does it draw? (c) If the line voltage drops to 110 V, what power does the heater take? (Assume that the resistance is constant. Actually, it will change because of the change in temperature.) (d) The heater coils are metallic, so that the resistance of the heater decreases with decreasing temperature. If the change of resistance with temperature is taken into account, will the electrical power consumed by the heater be larger or smaller than what you calculated in part (c)? Explain.

44. •• **Electricity through the body, I.** A person with a body resistance of 10 kΩ between his hands accidentally grasps the terminals of a 14 kV power supply. (a) If the internal resistance of the power supply is 2000 Ω, what is the current through the person's body? (b) What is the power dissipated in

his body? (c) If the power supply is to be made safe by increasing its internal resistance, what should the internal resistance be for the maximum current in the situation just described to be 1.00 mA or less?

45. •• **Electricity through the body, II.** The average bulk resistivity of the human body (apart from surface resistance of the skin) is about $5.0\ \Omega \cdot m$. The conducting path between the hands can be represented approximately as a cylinder 1.6 m long and 0.10 m in diameter. The skin resistance can be made negligible by soaking the hands in salt water. (a) What is the resistance between the hands if the skin resistance is negligible? (b) What potential difference between the hands is needed for a lethal shock current of 100 mA? (Note that your result shows that small potential differences produce dangerous currents when the skin is damp.) (c) With the current in part (b), what power is dissipated in the body?

BIO

19.5 Resistors in Series and in Parallel

46. • Find the equivalent resistance of each combination shown in Figure 19.53.

47. • Calculate the (a) maximum and (b) minimum values of resistance that can be obtained by combining resistors of 36 Ω, 47 Ω, and 51 Ω.

48. •• Each of two identical uniform metal bars has a resistance R. If they are welded together along one-third of their lengths (see Figure 19.54), what is the resistance of this combination in terms of R?

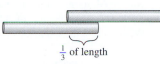

$\frac{1}{3}$ of length

▲ **FIGURE 19.54** Problem 48.

49. • A 40.0 Ω resistor and a 90.0 Ω resistor are connected in parallel, and the combination is connected across a 120-V dc line. (a) What is the resistance of the parallel combination? (b) What is the total current through the parallel combination? (c) What is the current through each resistor?

50. • Three resistors having resistances of 1.60 Ω, 2.40 Ω, and 4.80 Ω, respectively, are connected in parallel to a 28.0 V battery that has negligible internal resistance. Find (a) the equivalent resistance of the combination, (b) the current in each resistor, (c) the total current through the battery, (d) the voltage across each resistor, and (e) the power dissipated in each resistor. (f) Which resistor dissipates the most power, the one with the greatest resistance or the one with the least resistance? Explain why this should be.

51. • Now the three resistors of the previous problem are connected in series to the same battery. Answer the same questions for this situation.

52. •• Compute the equivalent resistance of the network in Figure 19.55, and find the current in each resistor. The battery has negligible internal resistance.

$\mathcal{E} = 60.0\ V,\ r = 0$

3.00 Ω 12.0 Ω

6.00 Ω 4.00 Ω

▲ **FIGURE 19.55** Problem 52.

53. •• Compute the equivalent resistance of the network in Figure 19.56, and find the current in each resistor. The battery has negligible internal resistance.

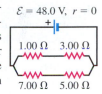

$\mathcal{E} = 48.0\ V,\ r = 0$

1.00 Ω 3.00 Ω

7.00 Ω 5.00 Ω

▲ **FIGURE 19.56** Problem 53.

54. •• **Lightbulbs in series, I.** The power rating of a lightbulb is the power it consumes when connected across a 120 V outlet. (a) If you put two 100 W bulbs in series across a 120 V outlet, how much power would each consume if its resistance were constant? (b) How much power does each one consume if you connect them in parallel across a 120 V outlet?

55. •• You absentmindedly solder a 69.8 kΩ resistor into a circuit where a 36.5 kΩ should be. How can you get the proper resistance without replacing the bigger resistor or removing anything from the circuit?

56. •• You need to connect a 68 kΩ resistor and one other resistor to a 110 V power line. If you want the two resistors to use 4 times as much power when connected in parallel as they use when connected in series, what should be the value of the unknown resistor?

19.6 Kirchhoff's Rules

57. •• The batteries shown in the circuit in Figure 19.57 have negligibly small internal resistances. Find the current through (a) the 30.0 Ω resistor, (b) the 20.0 Ω resistor, and (c) the 10.0 V battery.

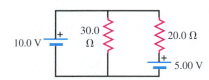

10.0 V   30.0 Ω 20.0 Ω 5.00 V

▲ **FIGURE 19.57** Problem 57.

58. •• Find the emf's $\mathcal{E}_1$ and $\mathcal{E}_2$ in the circuit shown in Figure 19.58.

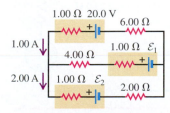

1.00 Ω 20.0 V   6.00 Ω

1.00 A   1.00 Ω $\mathcal{E}_1$

4.00 Ω

2.00 A   1.00 Ω $\mathcal{E}_2$ 2.00 Ω

▲ **FIGURE 19.58** Problem 58.

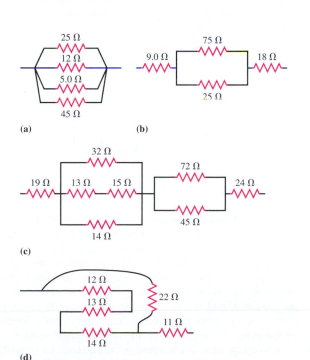

25 Ω
12 Ω
5.0 Ω
45 Ω

(a)

75 Ω
9.0 Ω 18 Ω
25 Ω

(b)

32 Ω
19 Ω 13 Ω 15 Ω 72 Ω 24 Ω
45 Ω
14 Ω

(c)

12 Ω
13 Ω 22 Ω
11 Ω
14 Ω

(d)

▲ **FIGURE 19.53** Problem 46.

59. •• In the circuit shown in Figure 19.59, ammeter A_1 reads 10.0 A and the batteries have no appreciable internal resistance. (a) What is the resistance of R? (b) Find the readings in the other ammeters.

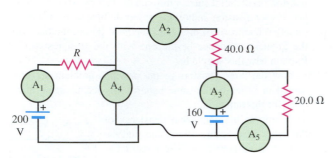

▲ **FIGURE 19.59** Problem 59.

60. •• In the circuit shown in Figure 19.60, find (a) the current in resistor R, (b) the value of the resistance R, and (c) the unknown emf $\mathcal{E}$.

61. •• In the circuit shown in Figure 19.61, current flows through the 5.00 Ω resistor in the direction shown, and this resistor is measured to be consuming energy at a rate of 20.0 W. The batteries have negligibly small internal resistance. What current does the ammeter A read?

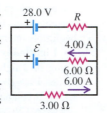

▲ **FIGURE 19.60** Problem 60.

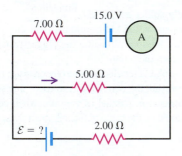

▲ **FIGURE 19.61** Problem 61.

62. •• In the circuit shown in Fig. 19.62, the 6.0 Ω resistor is consuming energy at a rate of 24 J/s when the current through it flows as shown. (a) Find the current through the ammeter A. (b) What are the polarity and emf of the battery $\mathcal{E}$, assuming it has negligible internal resistance?

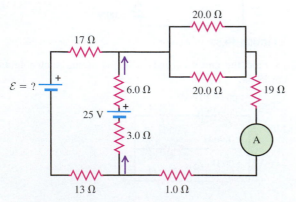

▲ **FIGURE 19.62** Problem 62.

19.8 Resistance-Capacitance Circuits

63. • A 500.0 Ω resistor is connected in series with a capacitor. What must be the capacitance of the capacitor to produce a time constant of 2.00 s?

64. • A fully charged 6.0 μF capacitor is connected in series with a 1.5×10^5 Ω resistor. What percentage of the original charge is left on the capacitor after 1.8 s of discharging?

65. •• A 12.4 μF capacitor is connected through a 0.895 MΩ resistor to a constant potential difference of 60.0 V. (a) Compute the charge on the capacitor at the following times after the connections are made: 0, 5.0 s, 10.0 s, 20.0 s, and 100.0 s. (b) Compute the charging currents at the same instants. (c) Graph the results of parts (a) and (b) for t between 0 and 20 s.

66. •• A 6.00 μF capacitor that is initially uncharged is connected in series with a 4500 Ω resistor and a 500 V emf source with negligible internal resistance. Just after the circuit is completed, what are (a) the voltage drop across the capacitor, (b) the voltage drop across the resistor, (c) the charge on the capacitor, and (d) the current through the resistor? (e) A long time after the circuit is completed (after many time constants), what are the values of the preceding four quantities?

67. •• A capacitor is charged to a potential of 12.0 V and is then connected to a voltmeter having an internal resistance of 3.40 MΩ. After a time of 4.00 s the voltmeter reads 3.0 V. What are (a) the capacitance and (b) the time constant of the circuit?

68. •• When a capacitor is being charged up, (a) how many time constants are required for it to receive 95% of its maximum charge, and (b) what is the current in the circuit at that time? (*Note:* You will have to solve an exponential equation.)

69. •• In the circuit shown in Figure 19.63, the capacitors are all initially uncharged and the battery has no appreciable internal resistance. After the switch S is closed, find (a) the maximum charge on each capacitor, (b) the maximum potential difference across each capacitor, (c) the maximum reading of the ammeter A, and (d) the time constant for the circuit.

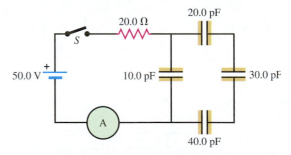

▲ **FIGURE 19.63** Problem 69.

70. •• **Charging and discharging a capacitor.** A 1.50 μF capacitor is charged through a 125 Ω resistor and then discharged through the same resistor by short-circuiting the battery. While the capacitor is being charged, find (a) the time for the charge on its plates to reach $1 - 1/e$ of its maximum value and (b) the current in the circuit at that time. (c) During the discharge of the capacitor, find the time for the charge on its plates to decrease to $1/e$ of its initial value. Also, find the time for the current in the circuit to decrease to $1/e$ of its initial value.

71. •• **Charging and discharging a capacitor.** An initially uncharged capacitor C charges through a resistor R for many time constants and then discharges through the same resistor.

Call Q_{max} the maximum charge on its plates and I_{max} the maximum current in the circuit. (a) Sketch clear graphs of the charge on the plates and the current in the circuit as functions of time for the charging process. (b) During the discharging process, the charge on the capacitor and the current both decrease exponentially from their maximum values. Use this fact to sketch graphs of the current in the circuit and the charge on the capacitor as functions of time.

General Problems

72. •• The circuit shown in Figure 19.64 contains two batteries, each with an emf and an internal resistance, and two resistors. Find (a) the current in the circuit (magnitude *and* direction) and (b) the terminal voltage V_{ab} of the 16.0 V battery.

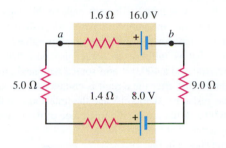

▲ **FIGURE 19.64** Problem 72.

73. •• If an ohmmeter is connected between points a and b in each of the circuits shown in Fig. 19.65, what will it read?

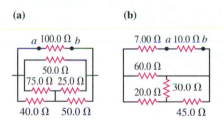

▲ **FIGURE 19.65** Problem 73.

74. •• A refrigerator draws 3.5 A of current while operating on a 120 V power line. If the refrigerator runs 50% of the time and electric power costs $0.12 per kWh, how much does it cost to run this refrigerator for a 30-day month?

75. •• A toaster using a Nichrome™ heating element operates on 120 V. When it is switched on at 20°C, the heating element carries an initial current of 1.35 A. A few seconds later, the current reaches the steady value of 1.23 A. (a) What is the final temperature of the element? The average value of the temperature coefficient of resistivity for Nichrome™ over the temperature range from 20°C to the final temperature of the element is $4.5 \times 10^{-4} \, (\text{C}°)^{-1}$. (b) What is the power dissipated in the heating element (i) initially; (ii) when the current reaches a steady value?

76. •• A piece of wire has a resistance R. It is cut into three pieces of equal length, and the pieces are twisted together parallel to each other. What is the resistance of the resulting wire in terms of R?

77. •• **Flashlight batteries.** A typical small flashlight contains two batteries, each having an emf of 1.5 V, connected in series with a bulb having resistance 17 Ω. (a) If the internal resistance of the batteries is negligible, what power is delivered to the bulb? (b) If the batteries last for 5.0 h, what is the total energy delivered to the bulb? (c) The resistance of real batteries increases as they run down. If the initial internal resistance is negligible, what is the combined internal resistance of both batteries when the power to the bulb has decreased to half its initial value? (Assume that the resistance of the bulb is constant. Actually, it will change somewhat when the current through the filament changes, because this changes the temperature of the filament and hence the resistivity of the filament wire.)

78. •• In the circuit of Figure 19.66, find (a) the current through the 8.0 Ω resistor and (b) the total rate of dissipation of electrical energy in the 8.0 Ω resistor and in the internal resistance of the batteries. (c) In one of the batteries, chemical energy is being converted into electrical energy. In which one is this happening and at what rate?

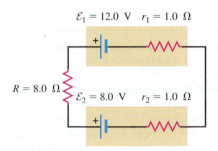

▲ **FIGURE 19.66** Problem 78.

79. •• **Struck by lightning.** Lightning strikes can involve currents as high as 25,000 A that last for about 40 μs. If a person is struck by a bolt of lightning with these properties, the current will pass through his body. We shall assume that his mass is 75 kg, that he is wet (after all, he is in a rainstorm) and therefore has a resistance of 1.0 kΩ, and that his body is all water (which is reasonable for a rough, but plausible, approximation). (a) By how many degrees Celsius would this lightning bolt increase the temperature of 75 kg of water? (b) Given that the internal body temperature is about 37°C, would the person's temperature actually increase that much? Why not? What would happen first?

80. •• **Navigation of electric fish.** Certain fish, such as the Nile fish (*Gnathonemus*), concentrate charges in their head and tail, thereby producing an electric field in the water around them. (See Figure 19.67.) This field creates a potential difference of a few volts between the head and tail, which in turn causes current to flow in the conducting seawater. As the fish swims, it passes near objects that have resistivities different from that of seawater, which in turn causes the current to vary. Cells in the skin of the fish are sensitive to this current and can detect changes in it. The changes in the current allow the fish to navigate. (In the next chapter, we shall investigate *how* the fish might detect this current.) Since the electric field is weak far from the fish, we shall consider only the field running directly from the head to the tail. We can model the seawater through which that field passes as a conducting tube of area

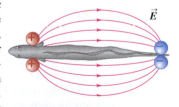

▲ **FIGURE 19.67** Problem 80.

1.0 cm^2 and having a potential difference of 3.0 V across its ends. The length of a Nile fish is about 20 cm, and the resistivity of seawater is 0.13 $\Omega \cdot$ m. (a) How large is the current through the tube of seawater? (b) Suppose the fish swims next to an object that is 10 cm long and 1.0 cm^2 in cross-sectional area and has half the resistivity of seawater. This object replaces the seawater for half the length of the tube. What is the current through the tube now? How large is the *change* in the current that the fish must detect? (*Hint:* How are this object and the remaining water in the tube connected, in series or in parallel?)

81. •• Each of three resistors in Figure 19.68 has a resistance of 2.00 Ω and can dissipate a maximum of 32.0 W without becoming excessively heated. What is the maximum power the circuit can dissipate?

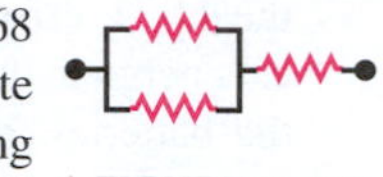

▲ **FIGURE 19.68** Problem 81.

82. •• **Leakage in a dielectric.** Two parallel plates of a capacitor have equal and opposite charges Q. The dielectric has a dielectric constant K and a resistivity ρ. Show that the "leakage" current I carried by the dielectric is given by $I = Q/K\epsilon_0\rho$. (*Note:* See Section 18.8 for a review of capacitors with dielectrics.)

83. •• **Energy use of home appliances.** An 1800 W toaster, a 1400 W electric frying pan, and a 75 W lamp are plugged into the same electrical outlet in a 20 A, 120 V circuit. (*Note:* When plugged into the same outlet, the three devices are in parallel with each other across the 120 V outlet.) (a) What current is drawn by each device? (b) Will this combination blow the circuit breaker?

84. •• Two identical 1.00 Ω wires are laid side by side and soldered together so that they touch each other for half of their lengths. (See Figure 19.69.) What is the equivalent resistance of this combination?

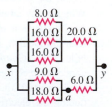

▲ **FIGURE 19.69** Problem 84.

85. •• Three identical resistors are connected in series. When a certain potential difference is applied across the combination, the total power dissipated is 27 W. What power would be dissipated if the three resistors were connected in parallel across the same potential difference?

86. •• (a) Calculate the equivalent resistance of the circuit of Figure 19.70 between x and y. (b) If a voltmeter is connected between points a and x when the current in the 8.0 Ω resistor is 2.4 A in the direction from left to right in the figure, what will it read?

▲ **FIGURE 19.70** Problem 86.

87. •• A power plant transmits 150 kW of power to a nearby town, through wires that have total resistance of 0.25 Ω. What percentage of the power is dissipated as heat in the wire if the power is transmitted at (a) 220 V and (b) 22 kV?

88. •• What must the emf $\mathcal{E}$ in Figure 19.71 be in order for the current through the 7.00 Ω resistor to be 1.80 A? Each emf source has negligible internal resistance.

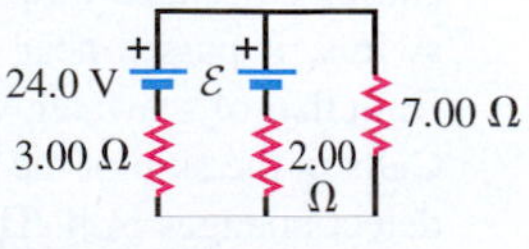

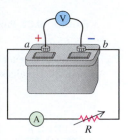

▲ **FIGURE 19.71** Problem 88.

89. •• For the circuit shown in Figure 19.72, if a voltmeter is connected across points a and b, (a) what will it read, and (b) which point is at a higher potential, a or b?

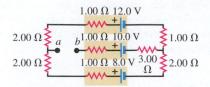

▲ **FIGURE 19.72** Problem 89.

90. •• A 4600 Ω resistor is connected across a charged 0.800 nF capacitor. The initial current through the resistor, just after the connection is made, is measured to be 0.250 A. (a) What magnitude of charge was initially on each plate of this capacitor? (b) How long after the connection is made will it take before the charge is reduced to $1/e$ of its maximum value?

91. •• A capacitor that is initially uncharged is connected in series with a resistor and a 400.0 V emf source with negligible internal resistance. Just after the circuit is completed, the current through the resistor is 0.800 mA and the time constant for the circuit is 6.00 s. What are (a) the resistance of the resistor and (b) the capacitance of the capacitor?

92. •• In the circuit shown in Fig. 19.73, R is a variable resistor whose value can range from 0 to ∞, and a and b are the terminals of a battery having an emf $\mathcal{E} = 15.0$ V and an internal resistance of 4.00 Ω. The ammeter and voltmeter are both idealized meters. As R varies over its full range of values, what will be the largest and smallest readings of (a) the voltmeter and (b) the ammeter? (c) Sketch qualitative graphs of the readings of both meters as functions of R, as R ranges from 0 to ∞.

▲ **FIGURE 19.73** Problem 92.

Passage Problems

BIO The nerve membrane as an R–C circuit. We now know that the electrical properties of nerve cells are governed by ion channels, which are protein molecules that span the limiting membrane of the axon. Each ion channel has a water-filled pore that electrically connects the interior of the axon to the outside bathing medium. The lipid-rich membrane in which the channels reside has very little electrical conductivity. Long before the existence of ion channels was known, scientists inferred many of their properties from electrical measurements on the squid giant axon, which could be easily impaled by electrodes and its electrical properties manipulated.

93. If we model the pore of a hypothetical ion channel spanning the membrane of a nerve cell as a cylinder 0.3 nm long with a radius of 0.3 nm filled with a fluid of resistivity 100 Ω cm, what is the conductance of the channel? (Be careful with units).
    A. 1 GΩ B. 1 nS   C. 100 GΩ D. 100 nS

94. If the actual conductance of an axon's ion channel is 10 pS and the peak current during a squid axon's action potential (the electrical disturbance that propagates down the axon) is 5 mA/cm^2, what is the density of channels in the membrane? Assume that the voltage across the membrane is 50 mV.
    A. $1/\text{m}^2$ B. $100/\text{cm}^2$ C. $1/\text{cm}^2$ D. $100/\mu\text{m}^2$

95. Cell membranes across a wide variety of organisms have a specific capacitance of $1 \text{ }\mu\text{F/cm}^2$. In order for the electrical signal of the nerve to propagate down the axon, the charge on the membrane capacitor must be changed. What is the characteristic time (time constant) required to do this?
    A. $1 \text{ }\mu\text{s}$ B. $10 \text{ }\mu\text{s}$ C. $100 \text{ }\mu\text{s}$ D. 1 ms

20 Magnetic Field and Magnetic Forces

In industrial settings, electromagnets are often used to pick up and move iron-containing material, such as this shredded scrap. How can electric currents cause magnetic forces? We'll learn in this chapter.

Everybody uses magnetic forces. Without them, there would be no electric motors or generators, no microwave ovens, and no computer printers or disk drives. The most familiar aspects of magnetism are those associated with permanent magnets, which attract unmagnetized iron objects and can also attract or repel other magnets. A compass needle interacting with the earth's magnetism is an example of this interaction. But the fundamental nature of magnetism is that it is an interaction associated with moving *electric* charges.

A magnetic field is established by a permanent magnet, by an electric current in a conductor, or by other moving charges. This magnetic field, in turn, exerts forces on moving charges and current-carrying conductors. Magnetic forces are an essential aspect of the interactions among electrically charged particles. In the first several sections of this chapter, we study magnetic forces and torques; then we examine the ways in which magnetic fields are *produced* by moving charges and currents.

20.1 Magnetism

Magnetic phenomena were first observed at least 2500 years ago in fragments of magnetized iron ore found near the ancient city of Magnesia (now Manisa, in western Turkey). It was discovered that when an iron rod is brought into contact with a natural magnet, the rod also becomes magnetized. When such a rod is suspended by a string from its center, it tends to line itself up in a north–south direction, like a compass needle. Magnets have been used for navigation at least since the 11th century. Magnets of this sort are called **permanent magnets.** Figure 20.1 shows a more contemporary example of the magnetic forces associated with a permanent magnet.

▲ **FIGURE 20.1** This bar magnet picks up steel filings—but not the copper filings in the pile. Later in this chapter, we'll learn why some metals are strongly magnetic and others are not.

Before the relation of magnetic interactions to moving charges was well understood, the interactions of bar magnets and of compass needles were described in terms of *magnetic poles.* The end of a bar magnet that points toward the earth's *geographic* north pole (which marks our planet's axis of rotation) is called a **north pole,** or **N pole,** and the other end is a **south pole,** or **S pole.** Two opposite poles attract each other, and two like poles repel each other (Figure 20.2). However, cutting a bar magnet in two does not give you two isolated poles. Instead, each half has N and S poles, as shown in Figure 20.3.

There is no evidence that a single isolated magnetic pole exists; poles always appear in pairs. The concept of magnetic poles is of limited usefulness and can be somewhat misleading. The existence of an isolated magnetic pole, or *magnetic monopole,* would have sweeping implications for theoretical physics. Extensive searches for magnetic monopoles have been carried out, so far without success.

A compass needle points north because the earth itself is a magnet; its geographical north pole is close to a magnetic *south* pole, as shown in Figure 20.4. The magnetic behavior of the earth is similar to that of a bar magnet with its axis not quite parallel to the geographic axis (the axis of rotation). Thus, a compass reading deviates somewhat from geographic north; this deviation, which varies with location, is called *magnetic declination.* Also, the earth's magnetic field is not horizontal at most points on the surface of our planet; its inclination up or down is described by the *angle of dip.*

Although permanent magnets are familiar to us, most of the magnets in our lives (and in the universe) are **electromagnets,** in which magnetic effects are produced by an electric current. Every appliance you own that has an electric motor contains an electromagnet. The phenomenon of electromagnetism was discovered

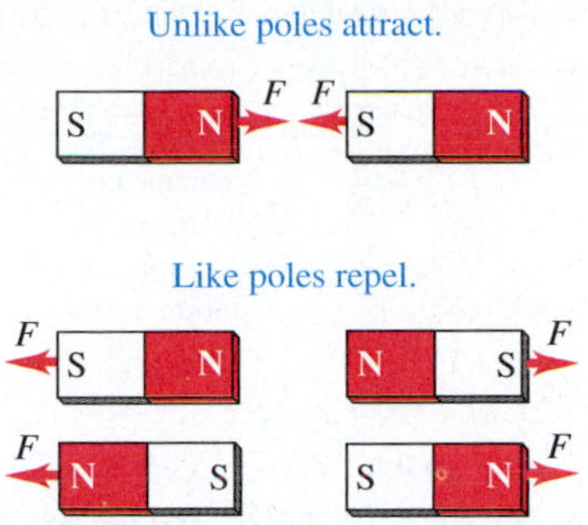

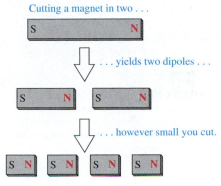

▲ **FIGURE 20.2** Unlike magnetic poles attract each other; like magnetic poles repel each other.

Differing from electric charges, magnetic poles always come paired and can't be isolated.

Cutting a magnet in two . . .

. . . yields two dipoles . . .

. . . however small you cut.

▲ **FIGURE 20.3** Magnets always have paired N and S poles.

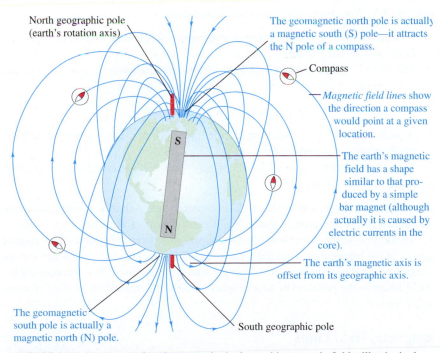

North geographic pole (earth's rotation axis)

The geomagnetic north pole is actually a magnetic south (S) pole—it attracts the N pole of a compass.

Compass

Magnetic field lines show the direction a compass would point at a given location.

The earth's magnetic field has a shape similar to that produced by a simple bar magnet (although actually it is caused by electric currents in the core).

The earth's magnetic axis is offset from its geographic axis.

The geomagnetic south pole is actually a magnetic north (N) pole.

South geographic pole

▲ **FIGURE 20.4** A compass placed at any point in the earth's magnetic field will point in the direction of the field line at that point. Representing the earth's field as that of a tilted bar magnet is only a crude approximation of its fairly complex configuration.

When the wire carries no current, the compass needle points north.

$I = 0$

When the wire carries a current, the compass needle deflects. The direction of deflection depends on the direction of the current.

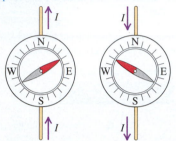

▲ **FIGURE 20.5** The behavior of a compass placed directly over a wire (seen from above). If the compass were placed directly *under* the wire, the deflection of the needle would be reversed.

Mastering PHYSICS

PhET: Magnet and Compass
PhET: Magnets and Electromagnets

At each point, the field line is tangent to the magnetic field vector $\vec{B}$.

The more densely the field lines are packed, the stronger the field is at that point.

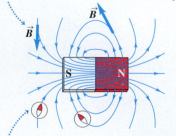

At each point, the field lines point in the same direction a compass would . . .

. . . therefore, magnetic field lines point *away from* N poles and *toward* S poles.

▲ **FIGURE 20.6** Magnetic field lines in a plane through the center of a permanent magnet.

in 1819 by the Danish scientist Hans Christian Oersted, who observed that a compass needle was deflected by a current-carrying wire, as shown in Figure 20.5. A few years later, it was found that moving a magnet near a conducting loop can cause a current in the loop and that a changing current in one conducting loop can cause a current in a separate loop.

We now know that electric and magnetic interactions are intimately intertwined. For example, the earth's magnetism results from electric currents circulating in its molten core. A compass needle points north because the earth itself is an electromagnet. The earth's outer core is made of conductive molten metal, which carries large electric currents while also circulating slowly. The currents in the core produce the earth's magnetic field.

This field changes through time, reflecting the complex dynamics of the circulating molten core. Geologic evidence shows that the magnetic poles wander relative to the geographic poles. Indeed, the earth's field periodically reverses, so that the former north magnetic pole becomes a south magnetic pole, and vice versa. Historically, reversals have occurred, on the average, a few times per million years. Currently, the earth's field is weakening; we may (or may not) be approaching a reversal.

Humans are hardly the first creatures to use earth's magnetic field for navigation. Many organisms, ranging from some bacteria to pigeons and perhaps even whales, can sense magnetic fields and use them for navigation or orientation. In many of these organisms, this sense appears to depend on tiny, intracellular crystals of an iron-containing mineral called magnetite—the same mineral that led to the human discovery of magnetism. These crystals presumably act as tiny bar magnets, orienting themselves in response to the local magnetic field.

20.2 Magnetic Field and Magnetic Force

To introduce the concept of a magnetic field, let's review our formulation of *electrical* interactions in Chapter 17, where we introduced the concept of an *electric* field. We represented electrical interactions in two steps:

1. A distribution of electric charge at rest creates an electric field $\vec{E}$ at all points in the surrounding space.
2. The electric field exerts a force $\vec{F} = q\vec{E}$ on any other charge q that is present in the field.

We can describe magnetic interactions in the same way:

1. A permanent magnet, a moving charge, or a current creates a **magnetic field** at all points in the surrounding space.
2. The magnetic field exerts a force $\vec{F}$ on any other moving charge or current that is present in the field.

Like an electric field, a magnetic field is a *vector field*—that is, a vector quantity associated with each point in space. We'll use the symbol $\vec{B}$ for magnetic field.

In the first several sections of this chapter, we'll concentrate on the *second* aspect of the interaction: Given the presence of a magnetic field, what force does it exert on a moving charge or a current? Then, in Sections 20.7 through 20.10, we'll return to the problem of how magnetic fields are *created* by moving charges and currents.

Magnetic Field Lines

We can represent any magnetic field by **magnetic field lines,** just as we did for the earth's magnetic field in Figure 20.4. As shown in Figure 20.6, we draw the lines so that the line through any point is tangent to the magnetic-field vector $\vec{B}$ at that point. Also, we draw the *number* of lines per unit area (perpendicular to the lines at a given point) to be proportional to the magnitude ("strength") of the field at that point. For a bar magnet, the field lines outside the magnet point from the N pole toward the S pole. As Figure 20.6 shows, a compass placed in a magnetic field always tends to orient its needle parallel to the magnetic field at each point; the *N* pole of the compass needle always tends to point in the direction of $\vec{B}$, and the *S* pole tends to point opposite to $\vec{B}$.

Magnetic field lines are sometimes called magnetic lines of force, but that's not a good name for them because, unlike electric field lines, they *do not* point in the direction of the force on a charge. At each point, the magnetic field line through the point *does* lie in the direction a compass needle would point when placed at that location; this may help you to visualize these lines. Just as with electric field lines, we draw only a few representative lines; otherwise, the lines would fill up all of space. Also, because the direction of $\vec{B}$ at each point is unique, field lines never intersect.

Figure 20.7 shows the magnetic field lines produced by several additional shapes of magnet. Notice especially the field of the C-shaped magnet in Figure 20.7a. In the space between its poles, the field lines are approximately straight, parallel, and equally spaced, showing that the magnetic field in this region is approximately *uniform* (constant in magnitude and direction). In Figure 20.7b, notice the graphical convention that we use to represent a magnetic field going into or out of the paper. Think of the dots as representing the points of vectors and the ×'s as the tail feathers of vectors. We'll use this scheme for diagrams later in the chapter. (The same convention can be used for electric fields.) Figure 20.7c shows that the magnetic field of a current-carrying loop resembles that of a bar magnet. Figure 20.4 shows that the magnetic field of the earth also resembles that of a bar magnet, with the field lines pointing from the N pole (near the geographic south pole) toward the S pole. Although the electric currents in the earth's core don't form simple loops or coils, the same principle is at work.

▲ **BIO Application Homeward bound.**
Homing pigeons are famous for being able to find their home roosts from thousands of kilometers away; they have been used since ancient times for carrying messages. But how do they do it? It appears that they use the earth's magnetic field, among other cues. Evidence indicates that they sense this field at least partly by means of small magnetite crystals located in their beaks. When the area of the beak containing the crystals is anaesthetized or the nerves to it are cut, the birds lose their ability to sense the magnetic field. The pigeon shown in this photo has a small magnet attached to its beak; this also interferes with the bird's ability to sense the earth's magnetic fields.

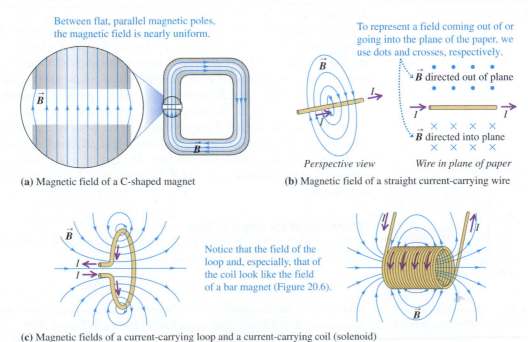

(a) Magnetic field of a C-shaped magnet

(b) Magnetic field of a straight current-carrying wire

(c) Magnetic fields of a current-carrying loop and a current-carrying coil (solenoid)

▲ **FIGURE 20.7** Some examples of magnetic fields.

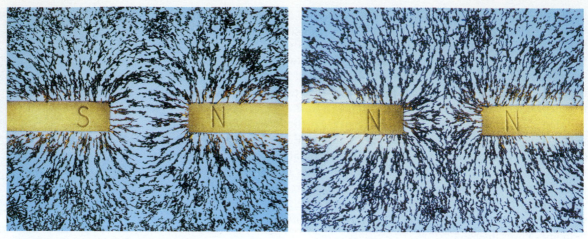

▲ **FIGURE 20.8** Magnetic field lines made visible by iron filings, which line up tangent to the field lines like little compass needles.

You may have had a chance to play with iron filings and a magnet. If you put the filings on a sheet of paper or plastic and hold the magnet beneath it, the filings line up tangent to magnetic field lines, as shown in Figure 20.8.

Magnetic Force

The most fundamental manifestation of a magnetic field is the force that it exerts on a moving charged particle. What are the characteristics of this force? First, its magnitude is proportional to the charge. If two particles, with charges q and $2q$, move in a given magnetic field with the same velocity, the force on the particle with charge $2q$ is twice as great in magnitude as that on charge q. The magnitude of the force is also proportional to the magnitude ("strength") of the field: If we double the magnitude of the field without changing the charge or its velocity, the magnitude of the force doubles.

The magnetic force is also proportional to the particle's speed. This is quite different from the electric-field force, which is the same whether the charge is moving or not. A charged particle at rest experiences no magnetic force. Furthermore, the magnetic force $\vec{F}$ on a moving charge does not have the same direction as the magnetic field $\vec{B}$; instead, it is always perpendicular to both $\vec{B}$ and the particle's velocity $\vec{v}$. The magnitude F of the force is found to be proportional to the component of $\vec{v}$ perpendicular to the field. When that component is zero (that is, when $\vec{v}$ and $\vec{B}$ are parallel or antiparallel), the force is zero.

Figure 20.9 shows these relationships. The direction of $\vec{F}$ is always perpendicular to the plane containing $\vec{v}$ and $\vec{B}$. Its magnitude F is given by the following expression:

A charge moving **parallel** to a magnetic field experiences **zero magnetic force.**

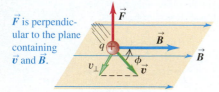

A charge moving at an angle ϕ to a magnetic field experiences a magnetic force with magnitude $F = |q|v_{\perp}B = |q|vB \sin\phi$.

$\vec{F}$ is perpendicular to the plane containing $\vec{v}$ and $\vec{B}$.

A charge moving **perpendicular** to a magnetic field experiences a maximal magnetic force with magnitude $F_{\text{max}} = qvB$.

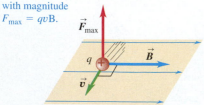

▲ **FIGURE 20.9** The magnetic force on a positive charge moving relative to a uniform magnetic field.

Magnitude of the magnetic force
When a charged particle moves with velocity $\vec{v}$ in a magnetic field $\vec{B}$, the magnitude F of the force exerted on it is

$$F = |q|v_{\perp}B = |q|vB\sin\phi, \qquad (20.1)$$

where $|q|$ is the magnitude of the charge and ϕ is the angle measured from the direction of $\vec{v}$ to the direction of $\vec{B}$, as shown in Figure 20.9.

Right-hand rule for the direction of magnetic force on a **positive** charge moving in a magnetic field:

① Place the $\vec{v}$ and $\vec{B}$ vectors tail to tail.

② Imagine turning $\vec{v}$ toward $\vec{B}$ in the $\vec{v}$-$\vec{B}$ plane (through the smaller angle).

③ The force acts along a line perpendicular to the $\vec{v}$-$\vec{B}$ plane. Curl the fingers of your *right hand* around this line in the same direction you rotated $\vec{v}$. Your thumb now points in the direction the force acts.

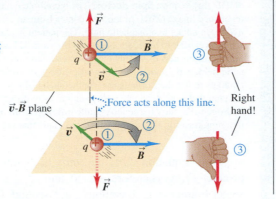

If the charge is negative, the direction of the force is *opposite* to that given by the right-hand rule.

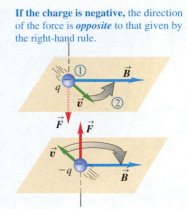

▲ **FIGURE 20.10** The right-hand rule for the direction of the magnetic force on a charge moving in a magnetic field.

Equation 20.1 can be interpreted in a different but equivalent way. Because ϕ is the angle between the directions of vectors $\vec{v}$ and $\vec{B}$, we may interpret $B\sin\phi$ as the component of $\vec{B}$ perpendicular to $\vec{v}$—that is, $B_\perp$. With this notation, the force expression (Equation 20.1) becomes

$$F = |q|vB_\perp. \tag{20.2}$$

This form is equivalent to Equation 20.1, but it's sometimes more convenient to use, especially in problems involving *currents* rather than individual particles. Later in the chapter, we'll discuss forces on conductors carrying currents.

This description of the magnetic-field force doesn't specify the direction of $\vec{F}$ completely; there are always two directions, opposite to each other and both perpendicular to the plane containing $\vec{v}$ and $\vec{B}$. To complete the description, we use a right-hand rule similar to the rule we used in connection with rotational motion in Chapter 10:

Right-hand rule for magnetic force

As shown in Figure 20.10, the right-hand rule for finding the direction of magnetic force on a *positive* charge is as follows:

1. Draw the vectors $\vec{v}$ and $\vec{B}$ with their tails together.
2. Imagine turning $\vec{v}$ in the plane containing $\vec{v}$ and $\vec{B}$ until it points in the direction of $\vec{B}$. Turn it through the smaller of the two possible angles.
3. The force then acts along a line perpendicular to the plane containing $\vec{v}$ and $\vec{B}$. Using your *right hand,* curl your fingers around this line in the same direction (clockwise or counterclockwise) that you turned $\vec{v}$. Your thumb now points in the direction of the force $\vec{F}$ on a *positive* charge.

If the charge is *negative,* the force on it points in the direction opposite to that given by the right-hand rule. In other words, for a negative charge, apply the right-hand rule and then reverse the direction of the force.

Figure 20.11 reinforces the fact that if two charges of opposite sign move with the same velocity in the same magnetic field, they experience magnetic forces in opposite directions. Figures 20.9 through 20.11 show several examples of the relationships of the directions of $\vec{F}$, $\vec{v}$, and $\vec{B}$ for both positive and negative charges. Be sure that you understand these relationships and can verify these figures for yourself.

Positive and negative charges moving in the same direction through a magnetic field experience magnetic forces in *opposite* directions.

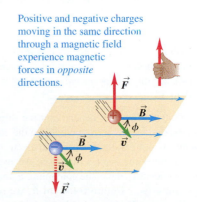

▲ **FIGURE 20.11** The effect of the sign of a moving charge on the magnetic force exerted on it.

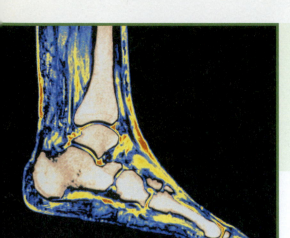

◄ **BIO Application Spin doctor?** The incredible detail shown in this false-color magnetic resonance image (MRI) of the foot comes from an analysis of the behavior of spinning hydrogen nuclei in a magnetic field. The patient is placed in a strong magnetic field of about 1.5 T, over 10,000 times stronger than the earth's. Each spinning hydrogen nucleus in the imaged tissue acts like a tiny electromagnet, aligning itself either with or against the magnetic field. A pulse of electromagnetic energy of about 50 MHz causes these tiny spinning magnets to flip their orientation. As the nucleii flip back following the pulse, they produce a signal that is proportional to the amount of hydrogen in any imaged tissue. Therefore, hydrogen-rich fatty tissue looks quite different from hydrogen-deficient bone, making MRI imaging ideal for analyzing soft-tissue details that are invisible in x-ray analysis.

From Equation 20.1, the *units* of B must be the same as the units of the product F/qv. Therefore, the SI unit of B is equivalent to $1 \, \text{N} \cdot \text{s}/(\text{C} \cdot \text{m})$, or, since 1 ampere is 1 coulomb per second $(1 \, \text{A} = 1 \, \text{C/s})$, it is equivalent to $1 \, \text{N}/(\text{A} \cdot \text{m})$. This unit is called the **tesla** (abbreviated T), in honor of Nikola Tesla (1857–1943), the prominent Serbian-American scientist and inventor.

Definition of the tesla

$$1 \text{ tesla} = 1 \text{ T} = 1 \, \text{N}/(\text{A} \cdot \text{m}).$$

The cgs unit of B, the **gauss** $(1 \, \text{G} = 10^{-4} \, \text{T})$, is also in common use. Instruments for measuring magnetic field are sometimes called gaussmeters or teslameters.

The magnetic field of the earth is about $0.5 \times 10^{-4} \, \text{T}$ (0.5 G). Magnetic fields on the order of 10 T occur in the interior of atoms, as shown by analysis of atomic spectra. The electromagnets in MRI machines used in medical imaging generate fields from less than a tesla to a few teslas. The largest values of steady magnetic field that have been achieved in the laboratory are about 45 T. The magnetic field at the surface of a neutron star is believed to be on the order of $10^8 \, \text{T}$.

Conceptual Analysis 20.1 **Direction of magnetic force**

Which of the three paths, 1, 2, or 3, does the electron in Figure 20.12 follow? (Remember that the blue ×'s represent a magnetic field pointing *into* the page, as explained in Figure 20.7b.)

A. Path 1.
B. Path 2, because the force on it is zero.
C. Path 2, because the force on it is perpendicular to the page. (We see the path projected onto the plane of the paper.)
D. Path 3.

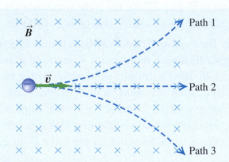

▲ **FIGURE 20.12**

SOLUTION The path depends on the direction of the force (if any) exerted by the magnetic field on the electron. The electron's velocity is not parallel to $\vec{B}$, so the electron experiences a force. To determine the force's direction, we use the right-hand rule. First, we identify the plane containing $\vec{v}$ and $\vec{B}$. (It is perpendicular to the page.) To turn $\vec{v}$ toward $\vec{B}$, we rotate it away from us into the page. Next, we hold our right hand so that the fingers can wrap around a line perpendicular to the plane of $\vec{v}$ and $\vec{B}$.

(This line is in the plane of the paper, parallel to the side of the page.) When we curl our fingers in the direction we turned $\vec{v}$, our thumb points toward the top margin of the page. But that is the direction of the force the magnetic field would exert on a *positive* charge. Since the electron is negative, the force exerted on it is in the plane of the paper, directed toward the bottom of the page. Thus, the electron follows path 3.

PROBLEM-SOLVING STRATEGY 20.1   **Magnetic forces**

SET UP

1. The biggest difficulty in determining magnetic forces is relating the directions of the vector quantities. In determining the direction of the magnetic-field force, draw the two vectors $\vec{v}$ and $\vec{B}$ with their tails together so that you can visualize and draw the plane in which they lie. This also helps you to identify the angle ϕ (always less than 180°) between the two vectors and to avoid getting its complement or some other erroneous angle. Then remember that $\vec{F}$ is always perpendicular to this plane, in a direction determined by the right-hand rule. Keep referring to Figures 20.9 and 20.10 until you're sure you understand this. If q is negative, the force is *opposite* to the direction given by the right-hand rule.

SOLVE AND REFLECT

2. Whenever you can, do the problem in two ways: using Equation 20.1 and then using Equation 20.2. Check that the results agree.

EXAMPLE 20.1   **A proton beam**

In Figure 20.13, a beam of protons moves through a uniform magnetic field with magnitude 2.0 T, directed along the positive z axis. The protons have a velocity of magnitude 3.0×10^5 m/s in the *x-z* plane at an angle of 30° to the positive z axis. Find the force on a proton. The charge of the proton is $q = +1.6 \times 10^{-19}$ C.

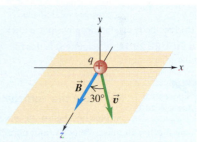

▶ **FIGURE 20.13**

SOLUTION

SET UP We use the right-hand rule to find the direction of the force. The force acts along the y axis, so we curl the fingers of our right hand around this axis in the direction from $\vec{v}$ toward $\vec{B}$. We find that the force acts in the $-y$ direction.

SOLVE To find the magnitude of the force, we use Equation 20.1:

$$F = qvB\sin\phi$$
$$= (1.6 \times 10^{-19}\,\text{C})(3.0 \times 10^5\,\text{m/s})(2.0\,\text{T})(\sin 30°)$$
$$= 4.8 \times 10^{-14}\,\text{N}.$$

REFLECT We could also obtain this result by finding $B_\perp$ and applying Equation 20.2: $F = |q|vB_\perp$. However, since we were given the angle ϕ, Equation 20.1 is more convenient. To check for consistency of units, we recall that $1\,\text{T} = 1\text{N}/(\text{A}\cdot\text{m})$.

Practice Problem: An electron beam moves through a uniform magnetic field with magnitude 3.8 T, directed in the $-z$ direction. The electrons have a velocity of 2.4×10^4 m/s in the *y-z* plane at an angle of 40° from the $-z$ axis toward the $+y$ axis. Find the force on an electron. *Answer:* $\vec{F} = 9.4 \times 10^{-15}$ N in the $+x$ direction.

Velocity Selector

To explore the principles we've learned, let's look at a device called a **velocity selector.** Many applications of present-day technology use a beam of charged particles that all have the same speed. Common sources of particle beams usually produce particles with a *range* of speeds. The velocity selector uses an arrangement of electric and magnetic fields that lets us select only particles with the desired speed.

ActivPhysics 13.8: Velocity Selector

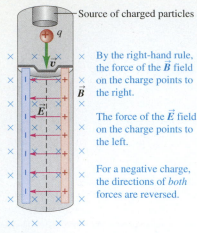

Source of charged particles

By the right-hand rule, the force of the $\vec{B}$ field on the charge points to the right.

The force of the $\vec{E}$ field on the charge points to the left.

For a negative charge, the directions of *both* forces are reversed.

(a) Schematic diagram of velocity selector

$F_E = qE$ $F_B = qvB$ Only if a charged particle has $v = E/B$ do the electric and magnetic forces cancel. All other particles are deflected.

(b) Free-body diagram for a positive particle

▲ **FIGURE 20.14** A velocity selector for charged particles.

The basic principle is shown in Figure 20.14. A charged particle with mass m, charge q, and speed v enters a region of space where the electric and magnetic fields are perpendicular to the particle's velocity and to each other, as shown in the figure. The electric field $\vec{E}$ points to the left, and the magnetic field $\vec{B}$ is into the plane of the page. If q is positive, the electric force is to the left, with magnitude qE, and the magnetic force is to the right, with magnitude qvB. By adjusting the magnitudes E and B, we can make these forces equal in magnitude. The *total* force is then zero, and the particle travels in a straight line with constant velocity. For zero total force, we need

$$\Sigma \vec{F} = 0, \qquad -qE + qvB = 0, \qquad \text{and} \qquad v = \frac{E}{B}. \qquad (20.3)$$

Only particles with speeds equal to E/B can pass through without being deflected by the fields. By adjusting E and B appropriately, we can select particles having a particular speed. Because q divides out, Equation 20.3 also works for electrons or other negatively charged particles. Do you understand why the electric and magnetic forces both have directions opposite to the preceding ones if q is negative?

Conceptual Analysis 20.2

Selecting the velocity of positive ions

A parallel beam of positive ions, differing from each other in mass, velocity, and charge, enters the magnetic and electric fields of the velocity selector shown in Figure 20.15. For the magnetic and electric forces on the particles to be in opposite directions, which of the charged plates has positive charge?

A. The upper plate.
B. The lower plate.

SOLUTION The magnetic field is into the page, the velocity of the charges points to the right, and the charges are positive. Therefore, by the right-hand rule, the force on the charges points toward the

Charged plates

▲ **FIGURE 20.15**

top of the page. We want the electric force to point toward the bottom of the page, however; thus, since the charges are positive, the electric field must also point toward the bottom, so the top plate must be positive and the bottom plate negative.

Thomson's e/m Experiment

In one of the landmark experiments in physics at the turn of the 20th century, Sir J. J. Thomson used the idea just described to measure the ratio of charge to mass for the electron. The speeds of electrons in a beam were determined from the potential difference used to accelerate them, and measurements of the electric and magnetic fields enabled Thomson to determine the ratio of electric charge magnitude (e) to mass (m) of the electrons. (The electron charge and mass were not measured separately until 15 years later.)

The most significant aspect of Thomson's e/m measurements was that he found that all particles in the beam had the *same value* for this quantity and that the value was independent of the materials used for the experiment. This independence showed that the particles in the beam, which we now call electrons, are a common constituent of all matter. Thus, Thomson is credited with the first discovery of a subatomic particle, the electron.

Fifteen years after Thomson's experiments, Millikan succeeded in measuring the charge of the electron. This result, together with the value of e/m, enabled

▶ Application **Radarange®** This photo shows the first commercial microwave oven, called a Radarange®, produced in 1947. It was the size of a modern refrigerator and cost nearly $3000. Microwave ovens were developed as an accidental offshoot of radar research. During a radar testing session, one scientist reached into his pocket to find his candy bar melted! This occurred because radar uses a device called a magnetron to generate electromagnetic radiation of a defined frequency, and certain frequencies can cook food. A filament in the center of the magnetron emits electrons, which then travel toward a surrounding cylindrical conductor. Magnets cause the electrons to curve in their path to the cylinder, where they encounter specially engineered resonant cavities that force the electrons into defined current oscillations. This arrangement generates just the proper frequency of electromagnetic radiation to pop your popcorn.

him to determine, for the first time, the *mass* of the electron. The most precise value available at present is

$$m_e = 9.1093826(16) \times 10^{-31} \text{ kg}.$$

20.3 Motion of Charged Particles in a Magnetic Field

When a charged particle moves in a magnetic field, the motion is determined by Newton's laws, with the magnetic force given by Equation 20.2. Figure 20.16 shows a simple example: A particle with positive charge q is at point O, moving with velocity $\vec{v}$ in a uniform magnetic field $\vec{B}$ directed into the plane of the figure. The vectors $\vec{v}$ and $\vec{B}$ are perpendicular, so the magnetic force $\vec{F}$ has magnitude $F = qvB$, and its direction is as shown in the figure. The force is *always* perpendicular to $\vec{v}$ so it cannot change the *magnitude* of the velocity, only its direction. (To put it differently, the force can't do work on the particle, so the force can't change the particle's kinetic energy.) Thus, the magnitudes of both $\vec{F}$ and $\vec{v}$ are constant. At points such as P and S in Figure 20.16a, the *directions* of force and velocity have changed, but not their magnitudes.

The particle therefore moves in a plane under the influence of a constant-magnitude force that is always at right angles to the velocity of the particle. Comparing these conditions with the discussion of circular motion in Chapter 6, we see that the particle's path is a *circle,* traced out with constant speed v (Figure 20.16b). The radial acceleration is v^2/R, and, from Newton's second law,

$$F = |q|vB = m\frac{v^2}{R},$$

where m is the mass of the particle. The radius R of the circular path is

$$R = \frac{mv}{|q|B}. \tag{20.4}$$

This result agrees with our intuition that it's more difficult to bend the paths of fast and heavy particles into a circle, so the radius is larger for fast, massive particles than for slower, less massive particles. Likewise, for a given charge, a larger magnetic field increases the force and pulls the particle into a smaller radius. If the charge q is negative, the particle moves *clockwise* around the orbit in Figure 20.16a.

The angular velocity ω of the particle is given by Equation 9.13: $\omega = v/R$. Combining this relationship with Equation 20.4, we get

$$\omega = \frac{v}{R} = v\frac{|q|B}{mv} = \frac{|q|B}{m}. \tag{20.5}$$

A charge moving at right angles to a uniform $\vec{B}$ field moves in a circle at constant speed because $\vec{F}$ and $\vec{v}$ are always perpendicular to each other.

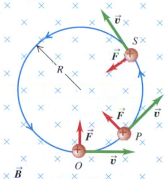

(a) The orbit of a charged particle in a uniform magnetic field

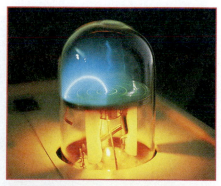

(b) An electron beam curving in a magnetic field

▲ **FIGURE 20.16** The circular orbit of a charged particle whose initial velocity is perpendicular to a magnetic field.

The number of revolutions per unit time (the frequency of revolution) is $\omega/2\pi$. This frequency is independent of the radius R of the path. It is called the **cyclotron frequency;** in a particle accelerator called a *cyclotron*, particles moving in nearly circular paths are given a boost twice each revolution (by an electric field), increasing their energy and their orbital radii, but not their angular velocity. Similarly, a *magnetron*—a common source of microwave radiation for microwave ovens and radar systems—emits radiation with a frequency proportional to the frequency of the circular motion of electrons in a vacuum chamber between the poles of a magnet.

Quantitative Analysis 20.3

Period of cyclotron motion

A charged particle enters a uniform magnetic field with a velocity $\vec{v}$ at right angles to the field. It moves in a circle with period T. If an identical particle enters the field with a velocity $2\vec{v}$, its period is

A. $4T$.
B. $2T$.
C. T.
D. $T/4$.

SOLUTION As we just found in deriving Equation 20.4, the radius of the particle's path is directly proportional to the particle's speed: $R = mv/|q|B$. Doubling the speed doubles the radius R and hence the circumference $2\pi R$ of the circle. Thus, to complete a circle, the particle that is moving twice as fast must go twice as far. But because it's moving twice as fast, it covers the distance in the same amount of time as the slower particle. Since the period is the amount of time it takes to complete one circle, the correct answer is C. The period is independent of the particle's speed.

This particle's motion has components both parallel ($v_{\parallel}$) and perpendicular ($v_{\perp}$) to the magnetic field, so it moves in a spiral.

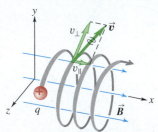

▲ **FIGURE 20.17** The spiral path of a charged particle whose initial velocity has components parallel and perpendicular to the magnetic field.

Helical Motion

If a particle moves in a uniform magnetic field and if the direction of its initial velocity is *not* perpendicular to the field, there is a component of velocity *parallel* to the field. This velocity component is constant, because there is no component of force parallel to the field. In this case, the particle moves in a helix (Figure 20.17). The radius R of the helix is given by Equation 20.4, where v is now the component of velocity perpendicular to the $\vec{B}$ field.

The motion we've just described is responsible for the earth's auroras (Figure 20.18). The sun emits a fast-moving "wind" of charged particles. When these particles encounter the earth's magnetic field far out in space, some are trapped and begin spiraling along magnetic field lines, which guide them toward the earth's poles. (See Figure 20.4.) In the upper atmosphere, these energetic charged particles collide with air molecules, causing the molecules to emit the light we call the auroras. The planets Jupiter and Saturn, which have strong magnetic fields, also have auroras.

▲ **FIGURE 20.18** Auroras on (a) earth and (b) Saturn.

Conceptual Analysis 20.4

The effects of magnetic force

A charged particle enters a uniform magnetic field. The field can change the particle's

A. velocity.
B. speed.
C. kinetic energy.

If the particle moves parallel to the magnetic field, the field has no effect on it. If the particle has a component of velocity $v_\perp$ perpendicular to the field, the field exerts a force on the particle. Because this force acts perpendicular to $v_\perp$, it changes the particle's direction but not its speed. Thus, a uniform magnetic field can change the velocity, but not the speed, of a charged particle. Because the kinetic energy of a given particle depends on speed $\left(K = \frac{1}{2}mv^2\right)$, the field also does not change the particle's kinetic energy. The correct answer is A.

EXAMPLE 20.2   Electron motion in a microwave oven

A magnetron in a microwave oven emits microwaves with frequency $f = 2450$ MHz. What magnetic field strength would be required for electrons to move in circular paths with this frequency?

SOLUTION

SET UP Figure 20.19 shows our diagram. Because the electron is negatively charged, the right-hand rule tells us that it circles clockwise. The frequency is $f = 2450$ MHz $= 2.45 \times 10^9$ s^{-1}. The corresponding angular velocity is $\omega = 2\pi f = (2\pi)(2.45 \times 10^9$ s$^{-1}) = 1.54 \times 10^{10}$ rad/s.

SOLVE From Equation 20.5,

$$B = \frac{m\omega}{|q|} = \frac{(9.11 \times 10^{-31}\text{ kg})(1.54 \times 10^{10}\text{ rad/s})}{1.60 \times 10^{-19}\text{ C}} = 0.0877\text{ T}.$$

REFLECT This is a moderate field strength, easily produced with a permanent magnet. Electromagnetic waves with this frequency can penetrate several centimeters into food with high water content.

Practice Problem: If the magnetron emits microwaves with frequency 2300 MHz, what magnetic field strength would be required

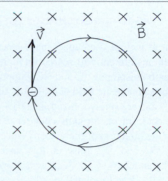

▲ **FIGURE 20.19** Our sketch for this problem.

for electrons to move in circular paths with that frequency? *Answer:* $B = 0.0823$ T.

20.4 Mass Spectrometers

Magnetic forces on charged particles play an important role in a family of instruments called **mass spectrometers.** These instruments are used to measure masses of positive ions and thus measure atomic and molecular masses. (Recall that a positive ion is an atom or molecule that has lost one or more electrons and hence has acquired a net positive charge.)

One type of mass spectrometer is shown in Figure 20.20. Ions from a source pass through the slits S_1 and S_2, forming a narrow beam. Then the ions pass through a *velocity selector* with crossed $\vec{E}$ and $\vec{B}$ fields, as we've described in Section 20.2, to block all ions except those with speeds v equal to E/B. Finally, the ions having this speed pass into a region with a magnetic field $\vec{B}'$ perpendicular to the figure, where they move in circular arcs with radius R determined by Equation 20.4, $R = mv/qB'$. Particles with different masses strike the detector at different points; a particle detector makes a scan to determine the number of particles collected at a given radius R. We assume that each ion has lost one electron, so the net charge of each ion is just e. With everything known in this equation except m, we can compute the mass m of the ion.

One of the earliest applications of the mass spectrometer was the discovery that the element neon has two species of atoms, with atomic masses 20 and 22 g/mol. We now call these species **isotopes** of the element. Later experiments

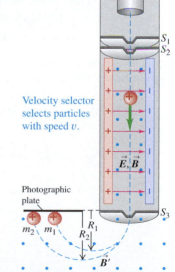

Velocity selector selects particles with speed v.

Photographic plate

Magnetic field separates particles by mass; the greater a particle's mass, the greater is the radius of its path.

▲ **FIGURE 20.20** One type of mass spectrometer.

have shown that many elements have several isotopes, with atoms that are very similar in chemical behavior, but different in mass. Each element has a definite number of electrons in its neutral, un-ionized atoms, and an equal number of protons in each nucleus. Mass differences between isotopes result from different numbers of *neutrons* in the nuclei.

For example, a neutral carbon atom always has six electrons, and there are always six protons in each carbon nucleus. But the nuclei can have a range of number of neutrons, most commonly six, but sometimes seven or eight. Using a prefixed superscript to indicate the total number of protons and neutrons (called the *mass number*), we denote these carbon isotopes as ^{12}C, ^{13}C, and ^{14}C.

An interesting application of radioactivity is the dating of archeological and geological specimens by measuring the concentration of radioactive isotopes of an element such as carbon (for once-living organisms) and potassium (for rock formations). We'll return to this topic in Chapter 30.

EXAMPLE 20.3   Mass spectrometer for a Mars rover

Scientists want to include a compact mass spectrometer on a future Mars rover. Among other things, the instrument will search for signs of life by measuring the relative abundances of the carbon isotopes ^{12}C and ^{13}C. Suppose the instrument is designed as shown in Figure 20.21, has a magnetic field of 0.0100 T, and selects carbon ions that have a speed of 5.00×10^3 m/s and are singly ionized (i.e., have a charge of $+e$). Each ion emerging from the velocity selector travels through a quarter circle before striking the detector plate (oriented at 45° to the direction of the velocity selector). **(a)** What are the radii R_{12} and R_{13} of the orbits of ^{12}C and ^{13}C ions in this spectrometer? These ions have masses of 1.99×10^{-26} kg and 2.16×10^{-26} kg, respectively. **(b)** How far apart are the spots these ions produce on the detector plate?

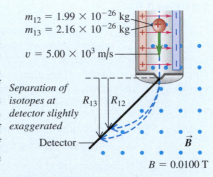

Separation of isotopes at detector slightly exaggerated

▲ **FIGURE 20.21**

SOLUTION

SET UP AND SOLVE Part (a): To find the radii of the orbits, we use Equation 20.4. The charge $+e$ is 1.60×10^{-19} C. We have

$$R_{12} = \frac{m_{12}v}{eB} = \frac{(1.99 \times 10^{-26} \text{ kg})(5.00 \times 10^3 \text{ m/s})}{(1.60 \times 10^{-19} \text{ C})(0.0100 \text{ T})} = 0.0622 \text{ m};$$

$$R_{13} = \frac{m_{13}v}{eB} = \frac{(2.16 \times 10^{-26} \text{ kg})(5.00 \times 10^3 \text{ m/s})}{(1.60 \times 10^{-19} \text{ C})(0.0100 \text{ T})} = 0.0675 \text{ m}.$$

Part (b): As the geometry of Figure 20.21 shows, the ions strike the detector plate at points separated by a distance of

$$\sqrt{2}(R_{13} - R_{12}) = \sqrt{2}(0.0675 \text{ m} - 0.0622 \text{ m})$$
$$= 7.50 \times 10^{-3} \text{ m} = 7.50 \text{ mm}.$$

REFLECT Notice that the radius is quite small, about 6 cm. Even though the complete mass spectrometer will be larger than this, it will still fit easily on a robotic rover. The quarter-circle design also helps reduce size and weight.

Practice Problem: If the isotope ^{14}C is also present in the sample, what will its radius be at the detection plate? The mass of ^{14}C is 2.33×10^{-26} kg. *Answer: R = 7.28 cm.*

20.5 Magnetic Force on a Current-Carrying Conductor

What makes an electric motor work? The forces that make it turn are forces that a magnetic field exerts on a conductor carrying a current. The magnetic forces on the moving charges within the conductor are transmitted to the material of the conductor, and the resulting force is distributed along the conductor's length. The moving-coil galvanometer that we described in Section 19.7 also uses magnetic forces on conductors.

We can compute the force on a current-carrying conductor, starting with the magnetic-field force on a single moving charge. The magnitude F of that force is given by Equation 20.1: $F = |q|vB\sin\phi$. Figure 20.22 shows a straight segment of a conducting wire, with length l and cross-sectional area A; the direction of the current is from bottom to top. The wire is in a uniform magnetic field $\vec{B}$; the field is perpendicular to the plane of the diagram and directed *into* the plane. Let's assume first that the moving charges are positive. Later we'll see what happens when they are negative.

The drift velocity $\vec{v}_d$ (the average velocity of the moving charges, discussed in Section 19.1) is upward, perpendicular to $\vec{B}$. According to the right-hand rule, the force $\vec{F}$ on each charge is directed to the left, as shown in the figure. In this case, $\vec{v}_d$ and $\vec{B}$ are perpendicular, and the magnitude F_{av} of the average force on each charge is $F_{av} = qv_d B$.

We can derive an expression for the *total* force on all the moving charges in a segment of conductor with length l in terms of the *current* in the conductor. Let Q be the magnitude of the *total* moving charge in this segment of conductor. The time Δt needed for a charge to move from one end of the segment to the other is $\Delta t = l/v_d$, and the total amount of charge Q that flows through the wire during this time interval is related to the current I by $Q = I \Delta t$.

From Equation 20.1 ($F = |q|v_\perp B$), the total force on all the moving charges Q in this segment has magnitude $F = Qv_d B$. Using $Q = I \Delta t$ and $\Delta t = l/v_d$, we can rewrite Equation 20.1 as

$$F = Qv_d B = (I\Delta t)(l/\Delta t)B,$$
$$F = IlB. \tag{20.6}$$

If the $\vec{B}$ field is not perpendicular to the wire, but makes an angle ϕ with it, we handle the situation the same way we did in Section 20.2 for a single charge. The component of $\vec{B}$ parallel to the wire (and to the drift velocities of the charges) exerts no force; the component perpendicular to the wire is $B_\perp = B\sin\phi$. Here is the general relation:

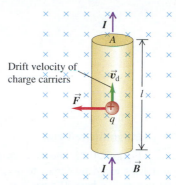

▲ **FIGURE 20.22** Magnetic force on a representative moving positive charge in a current-carrying conductor.

Force on a straight wire carrying a positive current and oriented at an angle ϕ to a magnetic field $\vec{B}$:

Magnitude is $F = I\,lB_\perp = I\,l\,B\sin\phi$

Direction is given by the right-hand rule.

> **Magnetic force on a current-carrying conductor**
>
> The magnetic-field force on a segment of conductor with length l, carrying a current I in a uniform magnetic field $\vec{B}$, is
>
> $$F = IlB_\perp = IlB\sin\phi. \tag{20.7}$$
>
> The force is always perpendicular to both the conductor and the field, with the direction determined by the same right-hand rule that we used for a moving positive charge (Figure 20.23).

Figure 20.24 shows the relations among the various directions for several cases.

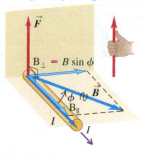

▲ **FIGURE 20.23** The magnetic force on a segment of current-carrying wire in a magnetic field.

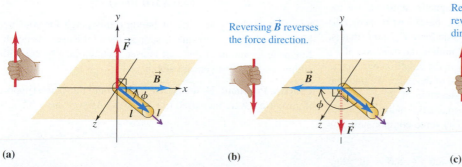

(a) (b)

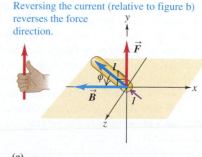

Reversing the current (relative to figure b) reverses the force direction.

(c)

▲ **FIGURE 20.24** The relation of the direction of the magnetic force on a current-carrying conductor to the directions of the current and the magnetic field.

Finally, what happens when the moving charges are negative, such as electrons in a metal? Then, in Figure 20.22, an upward current corresponds to a *downward* drift velocity. But because q is negative in this case, the direction of the force $\vec{F}$ is the same as before. Thus, Equations 20.6 and 20.7 are valid for both positive and negative charges and even when *both* signs of charge are present at once (as sometimes occurs in semiconductor materials and in ionic solutions).

Conceptual Analysis 20.5

Magnetic rail gun

Figure 20.25 shows a magnetic rail gun—a device that uses magnetic forces to accelerate a conducting object (in this case, a bar). The bar is placed across a pair of stationary conducting rails that are in a plane perpendicular to a magnetic field. Closing the switch sends a current through the circuit formed by the rails and the bar, causing the magnetic-field force to accelerate the bar along the rails. In which orientation, A or B, must the battery be connected if the bar is to accelerate to the right?

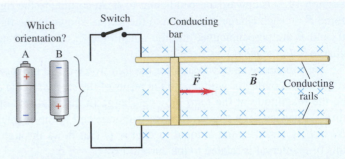

▲ **FIGURE 20.25**

SOLUTION To find the correct battery orientation, we use the right-hand rule in reverse. The magnetic force on the bar must point to the right, so we point the thumb of our right hand in that direction. The direction that our fingers curl is the direction in which the current direction in the bar would have to rotate to become parallel to $\vec{B}$. Thus, the current in the bar must point toward the bottom of the page. Since current flows from the positive to the negative terminal of a battery, we must connect the battery in orientation A.

EXAMPLE 20.4 **Magnetic bird perch**

The world's lightest bird is the bee hummingbird; its mass is 1.6 g (less than that of a penny). As shown in Figure 20.26, we design a perch for this bird that is supported by a magnetic force. The perch is 10 cm long, has a mass of 7.0 g, and contains a wire carrying a current $I = 4.0$ A. The perch is oriented at right angles to a uniform magnetic field that is just strong enough to support the perch and a bee hummingbird. What are the direction and magnitude of the magnetic field?

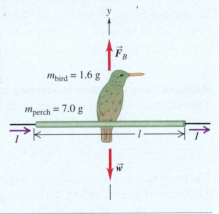

▶ **FIGURE 20.26**

SOLUTION

SET UP The magnetic field must generate an upward force on the perch equal to the combined weight of the perch and bird.

SOLVE To find the direction of the magnetic field, we use the right-hand rule in reverse. The magnetic field must be perpendicular to the plane containing the magnetic force $\vec{F}_B$ and I (that is, the plane of the page). We point our thumb in the direction of $\vec{F}_B$ and imagine rotating the current arrow in the direction our fingers curl until it is perpendicular to the plane of the page. We find that it points *into* the paper, so that is the direction of $\vec{B}$.

To find the magnitude of the magnetic field, we apply Newton's second law to the vertical (y) force components. The total mass is

$$M = m_{bird} + m_{perch} = 1.6\ g + 7.0\ g = 8.6\ g = 8.6 \times 10^{-3}\ kg.$$

The equilibrium condition $\Sigma F_y = 0$ gives

$$F_B - Mg = 0, \qquad IlB = Mg,$$
$$B = \frac{Mg}{Il} = \frac{(8.6 \times 10^{-3}\ kg)(9.8\ m/s^2)}{(4.0\ A)(0.10\ m)} = 0.21\ T.$$

REFLECT Before you think of building this perch for your pet bird, consider what happens as soon as the bird leaves the perch. Magnetic levitation of this sort is used technologically, but with additional systems to ensure stability.

Practice Problem: How massive a bird could this perch support if the current were increased to $I = 5.0$ A? *Answer: m = 3.7 g.*

20.6 Force and Torque on a Current Loop

Current-carrying conductors usually form closed loops, so it's worthwhile to use the results of Section 20.4 to find the *total* magnetic force and torque on a conductor in the form of a loop. As an example, let's look at a rectangular current loop in a uniform magnetic field. We can represent the loop as a series of straight line segments. We'll find that the total *force* on the loop is zero, but that there is a net *torque* acting on the loop, with some interesting properties.

Figure 20.27 shows a rectangular loop of wire with side lengths a and b. A line perpendicular to the plane of the loop (that is, a *normal* to the plane) makes an angle ϕ with the direction of the magnetic field $\vec{B}$, and the loop carries a current I. The wires leading the current into and out of the loop and the source of emf are omitted to keep the diagram simple.

The force $\vec{F}$ on the right side of the loop (length a) is in the direction of the x axis, toward the right, as shown. On this side, $\vec{B}$ is perpendicular to the direction of the current, and the force on this side has magnitude $F = IaB$. A force $-\vec{F}$ with the same magnitude, but opposite direction, acts on the opposite side, as shown in the figure.

The sides with length b make an angle equal to $90° - \phi$ with the direction of $\vec{B}$. The forces on these sides are the vectors $\vec{F}'$ and $-\vec{F}'$; their magnitude F' is given by

$$F' = IbB\sin(90° - \phi) = IbB\cos\phi.$$

The lines of action of both forces lie along the y axis.

The *total* force on the loop is zero because the forces on opposite sides cancel out in pairs. The two forces $\vec{F}'$ and $-\vec{F}'$ lie along the same line, so they have no net torque with respect to any axis. The two forces $\vec{F}$ and $-\vec{F}$, equal in magnitude and opposite in direction, but not acting along the same line, form what is called a *couple*. It can be shown that the torque of a couple, with respect to any axis, is the magnitude of either force, multiplied by the perpendicular distance between

▲ **Application Hello?** No matter what their size, dc motors operate on the same fundamental principle to convert electrical energy to mechanical energy. A current is sent through a movable wire coil in a magnetic field. The magnetic force produced acts as a torque at right angles to both the wire and the magnetic field, causing the coil to rotate. To keep the torque from reversing in direction as the coil rotates, a device called a commutator changes the direction of the current through the coil each half revolution. This mechanism provides a convenient, and nowadays indispensable, means of converting electrical energy into mechanical energy. The tiny motor shown here has an off-center weight attached to the shaft, causing it to move back and forth when the shaft rotates. You may own such a motor; it produces the familiar buzz of a cell phone in vibrate mode.

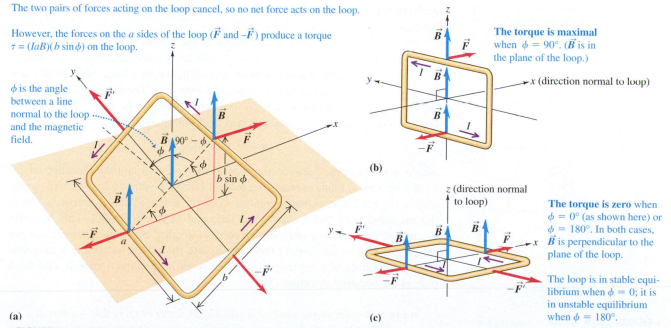

The two pairs of forces acting on the loop cancel, so no net force acts on the loop.

However, the forces on the *a* sides of the loop ($\vec{F}$ and $-\vec{F}$) produce a torque $\tau = (IaB)(b\sin\phi)$ on the loop.

ϕ is the angle between a line normal to the loop and the magnetic field.

(a)

The torque is maximal when $\phi = 90°$. ($\vec{B}$ is in the plane of the loop.)

x (direction normal to loop)

(b)

z (direction normal to loop)

The torque is zero when $\phi = 0°$ (as shown here) or $\phi = 180°$. In both cases, $\vec{B}$ is perpendicular to the plane of the loop.

The loop is in stable equilibrium when $\phi = 0$; it is in unstable equilibrium when $\phi = 180°$.

(c)

▲ **FIGURE 20.27** (a) Forces on the sides of a current-carrying loop rotating in a magnetic field. (b), (c) orientations at which the torque on the loop is maximal and zero, respectively.

ActivPhysics 13.6: Magnetic Torque on a Loop

the lines of action of the two forces. From Figure 20.27a, this distance is $b\sin\phi$, so the torque τ is

$$\tau = (IaB)(b\sin\phi). \tag{20.8}$$

The torque is greatest when $\phi = 90°$, B is in the plane of the loop, and the normal to this plane is perpendicular to $\vec{B}$ (Figure 20.27b). The torque is zero when ϕ is 0° or 180° ($\sin\phi = 0$) and the normal to the loop is parallel or antiparallel to the field (Figure 20.27c). The value $\phi = 0$ is a stable equilibrium position because the torque is zero there, and when the coil is rotated slightly from this position, the resulting torque tends to rotate it back toward $\phi = 0$. The position $\phi = 180°$ is an *unstable* equilibrium position.

The area A of the loop is equal to ab, so we can rewrite Equation 20.8 as follows:

Torque on a current-carrying loop

When a conducting loop with area A carries a current I in a uniform magnetic field of magnitude B, the torque exerted on the loop by the field is

$$\tau = IAB\sin\phi, \tag{20.8}$$

where ϕ is the angle between the normal to the loop and $\vec{B}$.

The torque τ tends to rotate the loop in the direction of *decreasing* ϕ—that is, toward its stable equilibrium position, in which $\phi = 0$ and the loop lies in the x-y plane, perpendicular to the direction of the field $\vec{B}$ (Figure 20.27c). The product IA is called the **magnetic moment** of the loop, denoted by μ:

$$\mu = IA. \tag{20.9}$$

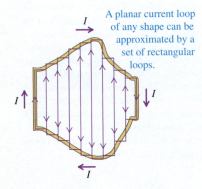

A planar current loop of any shape can be approximated by a set of rectangular loops.

▲ **FIGURE 20.28** An arbitrary planar shape can be approximated to any desired accuracy by a collection of rectangles. The narrower and more numerous the rectangles, the more closely they approximate the shape.

We've derived Equations 20.8 and 20.9 for a rectangular current loop, but it can be shown that these relations are valid for a plane loop of any shape. The proof rests on the fact that any planar loop may be approximated as closely as we wish by a large number of rectangular loops, as shown in Figure 20.28. Also, if we have N such loops wrapped close together, then, in place of Equation 20.9, we have $\mu = NIA$.

An arrangement of particular interest is the **solenoid,** a helical winding of wire, such as a coil wound on a circular cylinder (Figure 20.29). If the windings are closely spaced, the solenoid can be approximated by a large number of circular loops lying in planes at right angles to its long axis. The total torque on a solenoid in a magnetic field is simply the sum of the torques on the individual turns. For a solenoid with N turns in a uniform field with magnitude B,

$$\tau = NIAB\sin\phi,$$

where ϕ is the angle between the axis of the solenoid and the direction of the field. This torque tends to rotate the solenoid into a position where its axis is parallel to the field. The torque is greatest when the solenoid axis is perpendicular to the magnetic field and is zero when axis and field are parallel. This behavior resembles that of a bar magnet or compass needle: Both the solenoid and the magnet, if free to turn, orient themselves with their axes parallel to a magnetic field. The N pole of a bar magnet tends to point in the direction of $\vec{B}$, and the S pole in the opposite direction

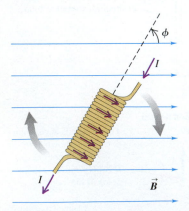

The torque tends to make the solenoid rotate clockwise in the plane of the page.

▲ **FIGURE 20.29** A current-carrying solenoid in a uniform magnetic field experiences a torque.

NOTE ▶ Beware of the symbol μ (the Greek letter mu). In this section, it's the symbol for magnetic moment, but later μ_0 will represent a constant called the *permeability of vacuum*. Be careful! ◀

Conceptual Analysis 20.6

Stability of a current loop

The loop in Figure 20.30 is oriented in the vertical plane at right angles to the magnetic field. Is this an equilibrium orientation? (That is, if you place the loop in exactly this orientation, will it stay that way?) If so, is the equilibrium stable? (That is, if you rotate the loop slightly away from this orientation, will it tend to rotate back?)

A. This is a stable equilibrium orientation.
B. This is an equilibrium orientation, but not a stable one.
C. This is not an equilibrium orientation.

SOLUTION The magnetic field points from the N to the S pole, to the right in the figure. The right-hand rule tells us that the force on the top part of the loop points toward the bottom of the page, and that the force on the bottom of the loop points toward the top

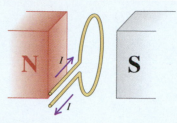

▲ **FIGURE 20.30**

of the page. These forces create no torque, so the loop is in equilibrium. However, if you displace the loop slightly, the forces produce a torque that tends to rotate the loop farther from its original position—not back toward it. Thus, this orientation is not stable; B is correct. If you rotate the loop 180°, it is then in a stable equilibrium position.

EXAMPLE 20.5 **Torque on a circular coil**

A circular coil of wire with average radius 0.0500 m and 30 turns lies in a horizontal plane. It carries a current of 5.00 A in a counterclockwise sense when viewed from above. The coil is in a uniform magnetic field directed toward the right, with magnitude 1.20 T. Find the magnetic moment and the torque on the coil. Which way does the coil tend to rotate?

SOLUTION

SET UP Figure 20.31 shows our diagram. The area of the coil is $A = \pi r^2 = \pi(0.0500 \text{ m})^2 = 7.85 \times 10^{-3} \text{ m}^2$; the angle ϕ between the direction of $\vec{B}$ and the *axis* of the coil (perpendicular to its plane) is 90°.

SOLVE The magnetic moment for one turn of the coil is $\mu = IA$ (Equation 20.9). Therefore, the total magnetic moment for all 30 turns is

$$\mu_{\text{total}} = 30\mu = 30IA = 30(5.00 \text{ A})(7.85 \times 10^{-3} \text{ m}^2)$$
$$= 1.18 \text{ A} \cdot \text{m}^2.$$

From Equation 20.8, the torque on each turn of the coil is

$$\tau = IAB\sin\phi = (5.00 \text{ A})(7.85 \times 10^{-3} \text{ m}^2)(1.20 \text{ T})(\sin 90°)$$
$$= 0.0471 \text{ N} \cdot \text{m},$$

and the total torque on the coil of 30 turns is

$$\tau = (30)(0.0471 \text{ N} \cdot \text{m}) = 1.41 \text{ N} \cdot \text{m}.$$

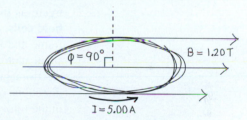

▲ **FIGURE 20.31** Our sketch for this problem.

Using the right-hand rule on the two sides of the coil, we find that the torque tends to rotate the right side down and the left side up.

REFLECT The torque tends to rotate the coil toward the stable equilibrium orientation, in which the normal to the plane is parallel to $\vec{B}$.

Practice Problem: Calculate the torque on the coil when it is placed in a magnetic field along the direction of the axis of the coil. *Answer:* $\tau = 0 \text{ N} \cdot \text{m}$.

The Direct-Current Motor

No one needs to be reminded of the importance of electric motors in contemporary society. Their operation depends on magnetic forces on current-carrying conductors. As an example, let's look at a simple type of direct-current motor, shown in Figure 20.32. The center part is the *armature,* or *rotor;* it is a cylinder of soft steel that rotates about its axis.

Embedded in slots in the rotor surface (parallel to its axis) are insulated copper conductors. Current is led into and out of these conductors through stationary graphite blocks called *brushes* that make sliding contact with a segmented

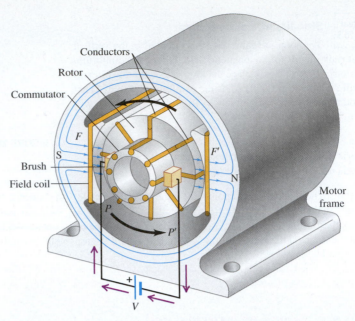

▲ **FIGURE 20.32** Schematic diagram of a dc motor. The rotor rotates on a shaft through its center, perpendicular to the plane of the figure.

cylinder, called the **commutator,** that turns with the rotor. The setup is shown in principle in Figure 20.33. The commutator is an automatic switching arrangement that maintains the currents in the conductors in the directions shown in the figure as the rotor turns. The coils F and F' shown in Figure 20.32 are called **field coils.** The steady current in these coils sets up a magnetic field in the motor frame and in the gap between the pole pieces P and P' and the rotor. (In some small motors, this magnetic field is supplied by permanent magnets instead of electromagnets.) Some of the magnetic field lines are shown in blue. With the directions of field and rotor currents as shown, the side thrust on each conductor in the rotor is such as to produce a *counterclockwise* torque on the rotor.

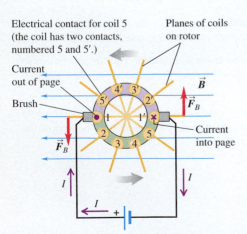

The brushes transmit current through contacts 1 and 1' to coil 1, which is oriented to receive maximal torque from the magnetic field.

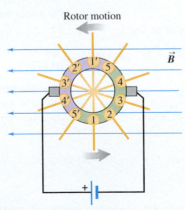

When the brushes are between contacts, inertia keeps the rotor turning.

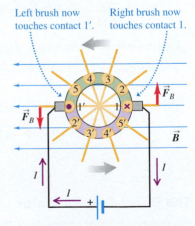

The brushes again contact coil 1, but with opposite polarity, so the torque on the coil is still counterclockwise.

▲ **FIGURE 20.33** By reversing the direction of the current in each coil once per half cycle, the commutator ensures that the torque on the coils always points in the same direction.

▶ **Application A clever solution.** As you know, a current traveling through a wire generates a magnetic field surrounding the wire. In a normal household extension cord, the two wires run side by side, and their fields add to produce a small net magnetic field. Because many types of sensitive electronic equipment cannot tolerate even these slight magnetic fields, the *coaxial* cable was developed. In a coaxial cable, one of the conductors has the form of a hollow tube, and the other runs through its center. (The cable is called "coaxial" because the conductors have the same axis.) As long as the currents in the two conductors are equal, the two magnetic fields cancel, so the cable produces no net magnetic field.

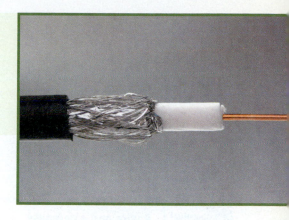

20.7 Magnetic Field of a Long, Straight Conductor

Thus far in this chapter, we've studied one aspect of the magnetic interaction of moving charges: the *forces* that a magnetic field exerts on moving charges and on currents in conductors. We didn't worry about what caused the magnetic field; we simply took its existence as a given fact. Now we're ready to return to the question of how magnetic fields are *produced* by moving charges and by currents. We begin by studying the magnetic fields produced by a few simple configurations, such as a long, straight wire and a circular loop. Then we look briefly at two more general methods for calculating the magnetic fields produced by more general configurations of currents in conductors.

The simplest current configuration to describe is a long, straight conductor carrying a steady current I. Experimentally, we find that the magnetic field produced by such a conductor has the general shape shown in Figure 20.34. The magnetic field lines are all *circles;* at each point, $\vec{B}$ is tangent to a circle centered on the conductor and lying in a plane perpendicular to it. Because of the axial symmetry, we know that $\vec{B}$ has the same *magnitude* at all points on a particular field line. The *magnitude* B at a distance r from the axis of the conductor is inversely proportional to r and is directly proportional to the current I. This I/r dependence can be derived from more general considerations, which we'll describe briefly in Section 20.10.

Mastering PHYSICS

ActivPhysics 13.1: Magnetic Field of a Wire

Right-hand rule for the magnetic field around a current-carrying wire: Point the thumb of your right hand in the direction of the current. Your fingers now curl around the wire in the direction of the magnetic field lines.

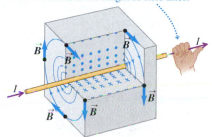

▲ **FIGURE 20.34** Right-hand rule for the direction of the magnetic field around a long, straight conductor carrying a current.

Magnetic field of a long, straight wire

The magnetic field $\vec{B}$ produced by a long, straight conductor carrying a current I, at a distance r from the axis of the conductor, has magnitude B given by

$$B = \frac{\mu_0 I}{2\pi r}. \tag{20.10}$$

In this equation, μ_0 is a constant called the *permeability of vacuum*. Its numerical value depends on the system of units we use. In SI units, the units of μ_0 are $(\text{T} \cdot \text{m/A})$. Its numerical value, which is related to the definition of the unit of current, is defined to be *exactly* $4\pi \times 10^{-7}$:

$$\mu_0 = 4\pi \times 10^{-7} \, \text{T} \cdot \text{m/A}.$$

A useful relationship for checks of unit consistency is $1 \, \text{T} \cdot \text{m/A} = 1 \, \text{N/A}^2$. We invite you to verify this equivalence.

NOTE ▶ The μ_0 just presented is a different quantity from the symbol μ for magnetic moment that we introduced in Section 20.6. This duplication in notation is unfortunately found everywhere in the literature. Beware! ◀

The shape of the magnetic field lines is completely different from that of the electric field lines in the analogous electrical situation. Electric field lines radiate outward from the charges that are their sources (or inward for negative charges).

By contrast, magnetic field lines *encircle* the current that acts as their source. Electric field lines begin and end at charges, while experiments have shown that magnetic field lines *never* have endpoints, no matter what shape the conductor is. (If lines *did* begin or end at a point, this point would correspond to a "magnetic charge," or a single magnetic pole. As we mentioned in Section 20.1, there is no experimental evidence that such a pole exists.)

As Figure 20.34 shows, the direction of the $\vec{B}$ lines around a straight conductor is given by a new right-hand rule: Grasp the conductor with your right hand, with your thumb extended in the direction of the current. Your fingers then curl around the conductor in the direction of the $\vec{B}$ lines.

Conceptual Analysis 20.7   A wire and a compass

Suppose you lay a magnetic compass flat on a table and stretch a wire horizontally under it, parallel to and a short distance below the compass needle (Figure 20.35). The wire is connected to a battery; when you close the switch, the N end of the compass needle deflects to the left. In which orientation (A or B) is the battery connected to the circuit?

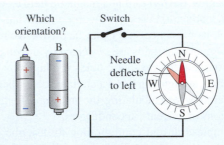

▲ **FIGURE 20.35**

SOLUTION Recall that a compass needle or bar magnet in a magnetic field tends to align so that a line from its S pole to its N pole points in the same direction as the magnetic field lines. Since the N pole of the compass needle deflects to the left (and the S pole to the right), we know that the magnetic field above the wire points from right to left. Next, we use our new right-hand rule (Figure 20.34) to find the direction of the current in the wire.

We hold our right hand so that the fingers curl in the direction of the magnetic field around the wire and find that our thumb points toward the bottom of the page. Thus the current flows clockwise in the loop, and the battery is connected in orientation A.

EXAMPLE 20.6   Magnetic field from power lines

A long, straight dc power line carries a current of 100 A. A swarm of bees builds a hive next to it. It is hypothesized that bees use the earth's magnetic field as a reference direction when orienting their honey-combs. At what distance from the power line is the magnitude of the magnetic field from the current equal to the magnitude of Earth's magnetic field, about 5.0×10^{-5} T?

SOLUTION

SET UP AND SOLVE We need to find the distance r at which the magnitude of the field B_{power} from the current in the power line equals the magnitude of field B_{earth} from the earth. Setting $B_{\text{power}} = B_{\text{earth}}$ and using Equation 20.10 for B_{power}, we find that

$$\frac{\mu_0 I}{2\pi r} = B_{\text{earth}}.$$

We solve this equation for the distance r from the power line:

$$r = \frac{\mu_0 I}{2\pi B_{\text{earth}}} = \frac{(4\pi \times 10^{-7}\,\text{T} \cdot \text{m/A})(100\,\text{A})}{2\pi(5.0 \times 10^{-5}\,\text{T})} = 0.40\,\text{m}.$$

REFLECT The magnetic field of the power line at a distance of half a meter is comparable in magnitude to the earth's field. Depending on the orientation of the power line relative to the earth's field, the current could cause a significant disruption in the bees' perception of direction.

Practice Problem: At what distance from a power line carrying a current of 110 A would it create a magnetic field with the same magnitude as the earth's? *Answer: r = 0.44 m.*

20.8 Force between Parallel Conductors

Let's look next at the interaction force between two long, straight, parallel current-carrying conductors. This problem comes up in a variety of practical situations, and it also has fundamental significance in connection with the definition

The magnetic field of the lower wire exerts an attractive force on the upper wire. By the same token, the upper wire attracts the lower one.

If the wires had currents in *opposite* directions, they would *repel* each other.

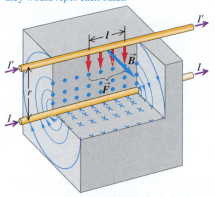

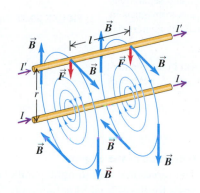

◀ **FIGURE 20.36** Attraction between parallel conductors carrying currents in the same direction.

of the ampere. Figure 20.36 shows segments of two long, straight, parallel conductors separated by a distance r and carrying currents I and I' in the same direction. Each conductor lies in the magnetic field set up by the other, so each is acted upon by a force. The diagram shows some of the field lines set up by the current in the *lower* conductor.

From Equation. 20.10, the magnitude of the $\vec{B}$ vector at points on the upper conductor is

$$B = \frac{\mu_0 I}{2\pi r}.$$

From Equation 20.6, the force on a length l of the upper conductor is

$$F = I'Bl = I'\left(\frac{\mu_0 I}{2\pi r}\right)l = \frac{\mu_0 l I I'}{2\pi r},$$

and the force *per unit length, F/l,* is

$$\frac{F}{l} = \frac{\mu_0 I I'}{2\pi r}. \tag{20.11}$$

The right-hand rule and Figure 20.36 show that the direction of the force on the upper conductor is *downward*. An equal (in magnitude) and opposite (in direction) upward force per unit length acts on the lower conductor; you can see that by looking at the field set up by the upper conductor. Therefore, the conductors *attract* each other. If the direction of either current is reversed, the forces reverse also. Parallel conductors carrying currents in *opposite* directions *repel* each other.

Mastering PHYSICS

ActivPhysics 13.5: Magnetic Force on a Wire

EXAMPLE 20.7 A superconducting cable

Two straight, parallel superconducting cables 4.5 mm apart (between centers) carry equal currents of 15,000 A in opposite directions. Find the magnitude and direction of the force per unit length exerted by one conductor on the other. Should we be concerned about the mechanical strength of these wires?

SOLUTION

SET UP Figure 20.37 shows our sketch. The two currents have opposite directions, so, from the preceding discussion, the forces are repulsive.

SOLVE To find the force per unit length, we use Equation 20.11:

$$\frac{F}{l} = \frac{\mu_0 I I'}{2\pi r} = \frac{(4\pi \times 10^{-7}\,\text{N/A}^2)(15{,}000\,\text{A})^2}{2\pi(4.5 \times 10^{-3}\,\text{m})}$$
$$= 1.0 \times 10^4\,\text{N/m}.$$

Continued

Therefore, the force exerted on a 1.0 m length of the conductor is 1.0×10^4 N.

REFLECT This is a large force, something over 1 ton per meter, so the mechanical strength of the conductors and insulating materials is certainly a significant consideration. Currents and separations of this magnitude are used in superconducting electromagnets in particle accelerators, and mechanical stress analysis is a crucial part of the design process.

Practice Problem: What is the maximum current in the conductors if the force per unit length is not to exceed 0.25×10^4 N/m? *Answer:* 7,500 A.

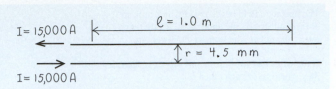

▲ **FIGURE 20.37** Our diagram for this problem.

20.9 Current Loops and Solenoids

In many practical devices, such as transformers and electromagnets, in which a current is used to establish a magnetic field, the wire carrying the current is wound into a *coil* consisting of many circular loops. So an expression for the magnetic field produced by a single circular conducting loop carrying a current is very useful. Figure 20.38 shows a circular conductor with radius R, carrying a current I. The current is led into and out of the loop through two long, straight wires side by side; the currents in these straight wires are in opposite directions, so their magnetic fields cancel each other.

   Experimentally, we find that the magnetic field at the center of a circular loop has the direction shown in the figure and that its magnitude is directly proportional to the current I and inversely proportional to the radius R of the loop. Specifically, we have the following relationship:

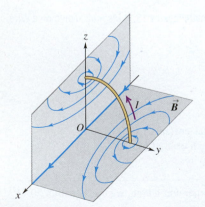

▲ **FIGURE 20.38** Magnetic field of a circular loop.

Magnetic field at center of circular loop

$$B = \frac{\mu_0 I}{2R} \qquad \text{(center of circular loop).} \qquad (20.12)$$

If we have a coil of N loops instead of a single loop, and if the loops are closely spaced and all have the same radius, then each loop contributes equally to the field, and the field at the center is just N times Equation 20.12:

$$B = \frac{\mu_0 N I}{2R} \qquad \text{(center of } N \text{ circular loops).} \qquad (20.13)$$

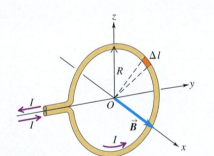

▲ **FIGURE 20.39** Magnetic field lines induced by the current in a circular loop. At points on the axis, the $\vec{B}$ field has the same direction as the magnetic moment of the loop.

Some of the magnetic field lines surrounding a circular loop and lying in planes through the axis are shown in Figure 20.39. The field lines encircle the conductor, and their directions are given by the same right-hand rule as that for a long, straight conductor. The magnetic field lines are *not* circles, but they are closed curves. At points along the axis of the loop, the $\vec{B}$ field is parallel to the axis; its magnitude is greatest at the center of the loop and decreases on both sides.

It's interesting to note the similarity of Equation 20.12 for a circular loop to Equation 20.10 for a long, straight conductor. The two expressions differ only by a factor of π: The field at the center of a circular loop with radius R is π times as great as the field at a distance R from a long, straight wire carrying the same current. Indeed, both equations can be derived from the same principles, which we'll mention briefly in Section 20.10.

Mastering PHYSICS

PhET: Faraday's Electromagnetic Lab
ActivPhysics 13.2: Magnetic Field of a Loop
ActivPhysics 13.3: Magnetic Field of a Solenoid

EXAMPLE 20.8   A current loop for an electron beam experiment

A coil used to produce a magnetic field for an electron beam experiment has 200 turns and a radius of 12 cm. **(a)** What current is needed to produce a magnetic field with a magnitude of 5.0×10^{-3} T at the center of the coil? **(b)** Figure 20.40 shows an electron being deflected as it moves through the coil. What is the direction of current in the coil?

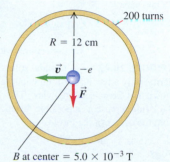

▶ **FIGURE 20.40** B at center $= 5.0 \times 10^{-3}$ T

SOLUTION

SET UP AND SOLVE To find the needed current, solve Equation 20.13 for I:

$$I = \frac{2RB}{\mu_0 N} = \frac{2(0.12 \text{ m})(5.0 \times 10^{-3} \text{ T})}{(4\pi \times 10^{-7} \text{ T} \cdot \text{m/A})(200)} = 4.8 \text{ A}.$$

From the right-hand rule, with the velocity vector of the electron pointing to the left and the force vector toward the bottom of the page, the direction of the magnetic field must be out of the page. (Remember that we're dealing with a *negative* charge.) The direction of the current must be counterclockwise.

REFLECT The current required is directly proportional to the radius of the coil; the greater the distance from the center to the conductor, the greater the current required. And the current required varies inversely with the number of turns; the more turns, the smaller required current.

Practice Problem: A proton moving through the coil to the right is deflected toward the bottom of the page. What is the direction of the current in the coil? *Answer:* Counterclockwise.

Magnetic Field of a Solenoid

A **solenoid** is a helical winding of wire, usually wound around the surface of a cylindrical form. Ordinarily, the turns are so closely spaced that each one is very nearly a circular loop. There may be several layers of windings. The solenoid in Figure 20.41 is drawn with only a few turns so that the field lines can be shown. All turns carry the same current I, and the total $\vec{B}$ field at every point is the vector sum of the fields caused by the individual turns. The figure shows field lines in the x-y and x-z planes. The field is found to be most intense in the center, less intense near the ends.

If the length L of the solenoid is large in comparison with its cross-sectional radius R, the $\vec{B}$ field inside the solenoid near its center is very nearly uniform and parallel to the axis, and the field outside, adjacent to the center, is very small. The magnetic-field magnitude B at the center depends on the number of turns *per unit length* of the solenoid; we'll call that quantity n. If there are N turns distributed uniformly over a total length L, then $n = N/L$. The field magnitude at the center of the solenoid is given by

$$B = \mu_0 n I \qquad \text{(center of long solenoid)}. \qquad (20.14)$$

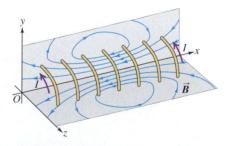

▲ **FIGURE 20.41** Magnetic field lines produced by the current in a solenoid. For clarity, only a few turns are shown.

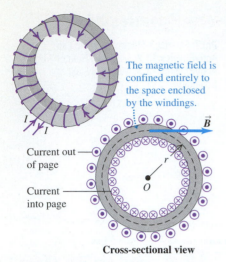

The magnetic field is confined entirely to the space enclosed by the windings.

Current out of page

Current into page

$\vec{B}$

O

r

Cross-sectional view

▲ **FIGURE 20.42** **(a)** A toroidal solenoid. For clarity, only a few turns of the winding are shown. **(b)** Cross-sectional view. The dashed black line represents a possible distance r from the center of the toroid.

A variation is the **toroidal** (doughnut-shaped) **solenoid,** more commonly called a *toroid,* shown in Figure 20.42. This shape has the interesting property that when there are many very closely spaced windings, the magnetic field is confined almost entirely to the space enclosed by the windings; there is almost no field at all outside this region. If there are N turns in all, then the field magnitude B at a distance r from the center of the toroid (*not* from the center of its cross section) is given by

$$B = \frac{\mu_0 N I}{2\pi r} \quad \text{(toroidal solenoid).} \tag{20.15}$$

The magnetic field is *not* uniform over a cross section of the core, but is inversely proportional to r. But if the radial thickness of the core is small in comparison with the overall radius of the toroid, the field is *nearly* uniform over a cross section. In that case, since $2\pi r$ is the circumference of the toroid and $N/2\pi r$ is the number of turns per unit length n, we can rewrite Equation 20.15 as

$$B = \mu_0 n I,$$

just as at the center of a long, *straight* solenoid.

Conceptual Analysis 20.8

Aligning a bar magnet

A bar magnet is suspended freely in line with the long axis of a solenoid, as shown in Figure 20.43. If the magnet aligns so that its N pole faces the solenoid, in which orientation (A or B) is the battery connected?

SOLUTION Because the magnet aligns with its N pole pointing left, the magnetic field along the axis of the solenoid also points left. (Remember that a magnet aligns so that a line from its S pole to its N pole points in the same direction as the local field.) To create this field, the current must enter the solenoid at its right end and exit at the left end. Thus, on the near side of each loop, the current is toward the top of the page. Therefore, the battery is connected in orientation A.

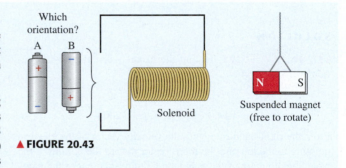

Which orientation?

A   B

Solenoid

N   S

Suspended magnet (free to rotate)

▲ **FIGURE 20.43**

20.10 Magnetic Field Calculations

In the previous three sections, we've stated without derivation several equations for the magnetic fields caused by currents in conductors. Deriving these relations can be quite complicated, so in this section we'll just sketch the principles on which the derivations are built. The basic relationships can be stated in two rather different forms. In the first, we begin with the $\vec{B}$ field produced by a current I in a short segment of conductor with length Δl. Then we add the contributions to $\vec{B}$ from all the segments of the conductor to find the *total* $\vec{B}$. This principle is called the *law of Biot and Savart.* In the second form, we consider the magnetic field at *all* points on a magnetic field line that encircles current-carrying conductors; this principle is called *Ampère's law.*

Law of Biot and Savart

The law of Biot and Savart gives the magnetic field $\Delta\vec{B}$ produced by a current I in a short segment of conductor with length Δl. The vector $\Delta\vec{l}$ is in the direction of the current. We call the location of the segment the **source point,** and the point P where we want to find the field is called the **field point.** The distance between the two points is r. The law is as follows:

Law of Biot and Savart

The magnitude ΔB of the magnetic field $\Delta \vec{B}$ due to a segment of conductor with length Δl, carrying a current I, is given by

$$\Delta B = \frac{\mu_0}{4\pi} \frac{I \, \Delta l \sin\theta}{r^2}. \tag{20.16}$$

This I/r^2 relationship reminds us of Coulomb's law. But the *direction* of $\Delta \vec{B}$ is *not* along the line from the source point to the field point. Instead, at each point it is perpendicular to the plane containing this line and the direction of the segment Δl, as shown in Figure 20.44. Furthermore, the field magnitude ΔB is proportional to the sine of the angle θ between these two directions.

Figure 20.44 shows the magnetic field $\vec{B}$ at several points in the vicinity of the segment Δl. At all points along the direction of Δl, the field is zero because, in Equation 20.16, $\theta = 0$ or $180°$ and $\sin\theta = 0$ at all such points. At any distance r from the segment, $\Delta \vec{B}$ has its greatest magnitude at points lying in the plane perpendicular to the segment; at all points in that plane, $\theta = 90°$ and $\sin\theta = 1$.

The magnetic field lines for the field $\Delta \vec{B}$ due to the current I in segment Δl are similar in character to those for a long, straight conductor; they are *circles* with centers along the line of Δl, lying in planes perpendicular to this line. We can use the same right-hand rule as for the long, straight conductor: Grasp the segment with your right hand so that your right thumb points in the direction of current flow; your fingers then curl around the segment in the same sense as the magnetic field lines.

To find the total magnetic field $\vec{B}$ at any point in space due to the current in a complete circuit, we represent the circuit as a large number of conducting segments Δl in series. We then use the *superposition principle,* which we've already encountered for *electric* fields, and find the *vector sum* of all the $\Delta \vec{B}$'s due to all the segments. This can become a sticky mathematical problem, but there are a few cases in which it is fairly simple.

For example, we can use the law of Biot and Savart to derive the expression for the magnetic field at the center of a circular conducting loop with radius R and current I (Equation 20.12). We represent the loop as a large number of segments with lengths Δl_1, Δl_2, and so on. A typical segment with length Δl is shown

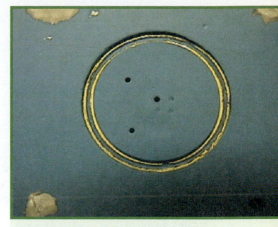

▲ Application **Micropump.** You may have heard of "lab on a chip" chemistry, in which complex procedures are performed on tiny samples as they move through microscale channels on a chip. For instance, the loop channel in this photo could be used for thermal cycling—you maintain different parts of the loop at different temperatures while pumping your sample around it. One challenge of microscale chemistry, however, is how to move fluids in tiny channels. *Magnetohydro-dynamics* provides an elegant answer. To create a magnetohydrodynamic (MHD) pump, you flank a channel with electrodes and magnets so that $\vec{E}$ and $\vec{B}$ fields cross the channel at right angles to each other. (In this photo, the gold rings are electrodes; the magnet poles are above and below the chip plane.) If the channel contains a conducting fluid, the effect of the two fields is to move the fluid *along* the conduit. There are no moving parts, and the pump is controlled by varying the voltage across the electrodes.

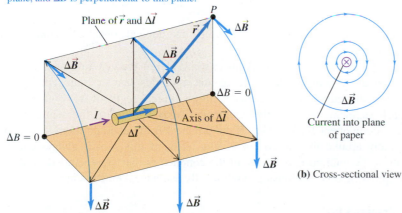

For these field points, $\vec{r}$ and $\Delta \vec{l}$ both lie in the tan-colored plane, and $\Delta \vec{B}$ is perpendicular to this plane.

Plane of $\vec{r}$ and $\Delta \vec{l}$

$\Delta \vec{B}$

θ

$\Delta B = 0$

Axis of $\Delta \vec{l}$

I

$\Delta \vec{l}$

$\Delta B = 0$

$\Delta \vec{B}$

For these field points, $\vec{r}$ and $\Delta \vec{l}$ both lie in the orange-colored plane, and $\Delta \vec{B}$ is perpendicular to this plane.

(a) Perspective view

Current into plane of paper

(b) Cross-sectional view

◄ **FIGURE 20.44** **(a)** Magnetic field vectors due to a short segment of current-carrying conductor, $\Delta \vec{l}$. At each point, $\Delta \vec{B}$ is perpendicular to the plane of $\vec{r}$ and $\Delta \vec{l}$ and its magnitude is proportional to the sine of the angle between them. **(b)** Magnetic field lines in a plane perpendicular to a short segment of current-carrying conductor with a current directed into the plane of the paper.

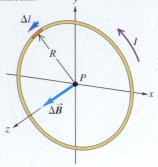

The segment Δl, the radial line R, and the magnetic field $\Delta \vec{B}$ are all mutually perpendicular.

▲ **FIGURE 20.45** Magnetic field $\Delta \vec{B}$ caused by a segment Δl of a circular conducting loop.

in Figure 20.45. All the segments are at the same distance R from the point P at the center, and each makes a right angle with the line joining it to P. The vectors $\Delta \vec{B}_1$, $\Delta \vec{B}_2$, and so on, due to the various segments, are all in the same direction, perpendicular to the plane of the loop, as shown. In Equation 20.16, $\theta = 90°$, $\sin \theta = 1$, and $r = R$, for every segment. The magnitude of the total $\vec{B}$ field is given by

$$B = \frac{\mu_0 I}{4\pi R^2}(\Delta l_1 + \Delta l_2 + \cdots).$$

But $\Delta l_1 + \Delta l_2 + \cdots$ is the total distance around the loop—that is, the *circumference* of the loop, $2\pi R$—so

$$B = \frac{\mu_0 I}{4\pi R^2}(2\pi R) = \frac{\mu_0 I}{2R},$$

in agreement with Equation 20.12.

This formulation is strictly valid only when the conductors are surrounded with vacuum. When air or any nonmagnetic material is present, the formulation is in error by only about 0.1% or less. In Section 20.11, we'll show how to modify the formulation to take account of the material around the conductors.

Ampère's Law

Ampère's law provides an alternative formulation of the relationship between a magnetic field and its sources. Ampère's law is analogous to Gauss's law, which offers an alternative to Coulomb's law for electric-field calculations. Like Gauss's law, Ampère's law is particularly useful when the problem at hand has symmetry properties that can be exploited.

To introduce the basic idea, let's consider the magnetic field caused by a long, straight conductor. We stated in Section 20.7, Equation 20.10, that the field at a distance r from a long, straight conductor carrying a current I has magnitude

$$B = \frac{\mu_0 I}{2\pi r}$$

and that the magnetic field lines are circles centered on the conductor (Figure 20.34). The circumference of a circular field line with radius r is $2\pi r$. Now we note that, for any circular field line with radius r, centered on the conductor, the product of B and the circumference is

$$B(2\pi r) = \frac{\mu_0 I}{2\pi r}(2\pi r) = \mu_0 I.$$

That is, this product is independent of r and depends only on the current I in the conductor.

This is a special case of Ampère's law. Here's the general statement: We construct an imaginary closed curve that encircles one or more conductors. We divide this curve into segments, calling a typical segment Δs (Figure 20.46). At each segment, we take the component of $\vec{B}$ parallel to the segment; we call this component $B_\parallel$. We take all the products $B_\parallel \Delta s$ and add them as we go completely around the closed curve. The result of this sum is always equal to μ_0 times the total current I_{encl} in all the conductors that are encircled by the curve. Expressing this relationship symbolically, we have Ampère's law:

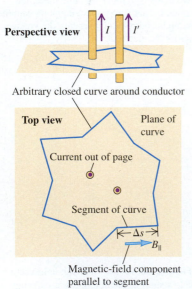

Ampère's law: If we take the products $B_\parallel \Delta s$ for all segments around the curve, their sum equals μ_0 times the total enclosed current:

$$\sum B_\parallel \Delta s = \mu_0 I_{encl}.$$

▲ **FIGURE 20.46** Ampère's law for an arbitrary closed curve of straight segments around a pair of conductors.

> **Ampère's law**
>
> When a path is made up of a series of segments Δs, and when that path links conductors carrying total current I_{encl},
>
> $$\sum B_\parallel \Delta s = \mu_0 I_{encl}. \qquad (20.17)$$

Our initial example, the field of a long, straight conductor, used a circular path for which $B_\parallel$ was equal to B and was the same at each point of the path, so $\sum B_\parallel \Delta s$ was just equal to B multiplied by the circumference of the path. As with Gauss's law, the path that we use doesn't have to be the outline of any actual physical object; usually, it is a purely geometric curve that we construct to apply Ampère's law to a specific situation.

If several conductors pass through the surface bounded by the path, the total magnetic field at any point on the path is the vector sum of the fields produced by the individual conductors. Then we evaluate Equation 20.17, using the *total* $\vec{B}$ field at each point and the total current I_{encl} enclosed by the path. The result equals μ_0 times the *algebraic sum* of the currents. We need a sign rule for the currents; here it is: For the surface bounded by our Ampère's-law path, take a line perpendicular to the surface and wrap the fingers of your right hand around this line so that your fingers curl around in the same direction you plan to go around the path when you evaluate the $B_\parallel \Delta s$ sum. Then your thumb indicates the positive current direction. Currents that pass through the surface in that direction are positive; those in the opposite direction are negative.

20.11 Magnetic Materials

In all of the preceding discussion of magnetic fields caused by currents, we've assumed that the space surrounding the conductors contains only vacuum. If matter is present in the surrounding space, the magnetic field is changed. The atoms that make up all matter contain electrons in motion, and these electrons form microscopic current loops that produce magnetic fields of their own. In many materials, these currents are randomly oriented, causing no net magnetic field. But in some materials, the presence of an externally caused field can cause the loops to become oriented preferentially with the field so that their magnetic fields *add* to the external field. We then say that the material is *magnetized*.

Paramagnetism

A material showing the behavior we've just described is said to be **paramagnetic.** The magnetic field at any point in such a material is greater by a numerical factor K_m than it would be in vacuum. The value of K_m is different for different materials; it is called the **relative permeability** of a material. For a given material, K_m depends on temperature; values of K_m for common paramagnetic materials at room temperature are typically 1.000002 to 1.0004.

All the equations in this chapter that relate magnetic fields to their sources can be adapted to the situation in which the conductor is embedded in a paramagnetic material by replacing μ_0 everywhere with $K_m\mu_0$. This product is usually denoted as μ; it is called the **permeability** of the material:

$$\mu = K_m\mu_0. \tag{20.18}$$

NOTE ▶ Remember that in this context μ is magnetic permeability, *not* the magnetic moment we defined in Section 20.6. Be careful! ◄

Diamagnetism

In some materials, the total field due to the electrons in each atom sums to zero when there is no external field; such materials have *no* net atomic current loops. But even in these materials, magnetic effects are present, because an external field causes slight distortion of the electron current loops. The additional field caused by this distortion is always *opposite* in direction to that of the external field. Such materials are said to be **diamagnetic;** they always have relative permeabilities very slightly less than unity, typically on the order of 0.9998 to 0.99999. Thus, the **susceptibility,** $K_m - 1$, of a diamagnetic material is very small and negative.

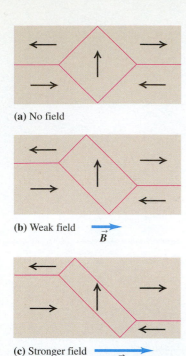

(a) No field

(b) Weak field $\vec{B}$

(c) Stronger field $\vec{B}$

▲ **FIGURE 20.47** In this drawing, adapted from a magnified photo, the arrows show the directions of magnetization in the domains of a single crystal of nickel. Domains magnetized in the direction of an applied magnetic field grow larger.

Ferromagnetism

In a third class of materials, called **ferromagnetic materials,** strong interactions between microscopic current loops cause the loops to line up parallel to each other in regions called **magnetic domains,** even when no external field is present. Figure 20.47 shows an example of magnetic domain structure. Within each domain, nearly all of the atomic current loops are parallel. When there is no externally applied field, the orientations of the domain magnetizations are random, and the net magnetization is zero. But when an externally applied field is present, it exerts torques on the atomic current loops, and they tend to orient themselves parallel to the field, like microscopic compass needles. The domain boundaries also shift; the domains magnetized in the direction of the field grow, and those magnetized in other directions shrink. This "cooperative" phenomenon leads to a relative permeability that is *much larger* than unity, typically on the order of 1000 to 10,000. Iron, cobalt, nickel, and many compounds and alloys containing these elements are ferromagnetic.

In ferromagnetic materials, as the external field increases, a point is reached at which nearly all the microscopic current loops have their axes parallel to that field. This condition is called *saturation magnetization;* after it is reached, a further increase in the external field causes no increase in magnetization. Some ferromagnetic materials retain their magnetization even after the external magnetic field is removed. These materials can thus become *permanent magnets.* Many kinds of steel and many alloys, such as alnico, are commonly used for permanent magnets. When such a material is magnetized to near saturation, the magnetic field in the material is typically on the order of 1 T. Magnetic tapes, strips, and computer disks use this same retention of magnetization.

Ferromagnetic materials are widely used in electromagnets, transformer cores, and motors and generators, where it is usually desirable to have as large a magnetic field as possible for a given current. In these applications, it is usually desirable for the material *not* to have permanent magnetization. Soft iron is often used because it has high permeability without appreciable permanent magnetization.

SUMMARY

Magnetism; Fields and Forces

(Sections 20.1 and 20.2) A bar magnet has a **north (N) pole** and a **south (S) pole.** Two opposite poles attract each other, and two like poles repel each other. A moving charge creates a **magnetic field** in the surrounding space. A moving charge, or current, experiences a force in the presence of a magnetic field. The direction of the force is given by the right-hand rule for magnetic forces, and the magnitude is given by Equation 20.1.

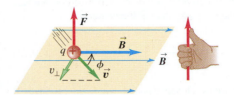

Motion of Charged Particles in Magnetic Fields

(Sections 20.3 and 20.4) The magnetic force is always perpendicular to $\vec{v}$; a particle moving under the action of a magnetic field alone moves with constant speed. In a uniform field, a particle with initial velocity perpendicular to the field moves in a circle with radius R given by $R = mv/|q|B$ (Equation 20.4). Mass spectrometers use this relationship to determine atomic masses. When a positive ion of known speed undergoes circular motion in a magnetic field of known strength, the mass can be determined by measuring the radius.

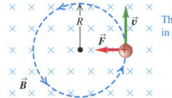

The orbit of a positive charge in a uniform magnetic field.

Magnetic Force on a Current-Carrying Conductor

(Section 20.5) When a current-carrying conductor is in the presence of a magnetic field, the field exerts a force on the conductor because each individual charge in the current is acted upon by a force given by $F = |q|vB\sin\phi$ (Equation 20.1). The direction of the force is determined by using the same right-hand rule that we used for a moving positive charge. The magnitude of the force is given by $F = IlB_\perp = IlB\sin\phi$ (Equation 20.7).

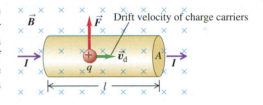

Force and Torque on a Current Loop; Direct-Current Motors

(Section 20.6) A current loop can be represented as connected segments of current-carrying conductors. In the presence of a magnetic field, each segment is acted upon by a force due to the field. In a uniform magnetic field, the total force on a current loop is zero, regardless of its shape, but the magnetic forces create a torque τ given by $\tau = IAB\sin\phi$ (Equation 20.8). A direct-current motor is driven by torques on current-carrying conductors. The key component is the **commutator,** which is used to reverse the direction of the current in order to maintain the torque.

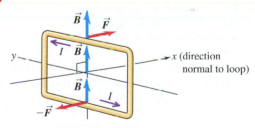

Magnetic Field of a Long, Straight Conductor

(Sections 20.7 and 20.8) Moving charges, and therefore currents, produce magnetic fields. When a current passes through a long, straight wire, the magnetic field lines are circles centered on the wire. Due to this symmetry of the field pattern, the magnitude of the magnetic field is the same at all points on a field line at radial distance r:

$$B = \frac{\mu_0 I}{2\pi r}. \tag{20.10}$$

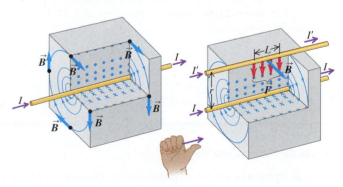

Current-carrying conductors can exert magnetic forces on each other. Thus, two parallel wires can attract or repel each other, depending on the direction of the currents they carry. For two long, straight parallel wires, the force per unit length is

$$\frac{F}{l} = \frac{\mu_0 II'}{2\pi r}. \tag{20.11}$$

Continued

Current Loops and Solenoids

(Section 20.9) Many practical devices depend on the magnetic field produced at the center of a circular coil of wire. If a coil of radius R consists of N loops, the magnetic field at the center is

$$B = \frac{\mu_0 NI}{2R}. \tag{20.13}$$

A long **solenoid** of many closely spaced windings produces a nearly uniform field in its interior, midway between its ends, having magnitude $B = \mu_0 nI$ (Equation 20.14).

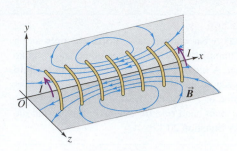

Magnetic Field Calculations

(Section 20.10) The magnetic field magnitude ΔB created by a short current-carrying segment Δl is given by the law of Biot and Savart:

$$\Delta B = \frac{\mu_0}{4\pi} \frac{I\,\Delta l\,\sin\theta}{r^2}. \tag{20.16}$$

To determine the magnetic field from an extended current-carrying wire, first consider the extended wire as being made of many smaller segments. Then, using the principle of superposition at a particular point in space, compute the vector sum of the magnetic field due to each small segment of current.

When a current I is enclosed by a path made of many small segments Δs, and when at each point the magnetic field has a component $B_\parallel$ parallel to the segments, **Ampère's law** states that

$$\sum B_\parallel\,\Delta s = \mu_0 I_{\text{encl}}. \tag{20.17}$$

If the products of the magnetic field component $B_\parallel$ and each segment Δs (that is, the components $B_\parallel\,\Delta s$) are summed around *any* path enclosing a total current I_{encl}, the result is $\mu_0 I_{\text{encl}}$.

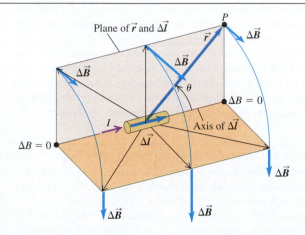

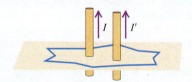

Magnetic Materials

(Section 20.11) For magnetic materials, the magnetization of the material causes an additional contribution to $\vec{B}$. For paramagnetic and diamagnetic materials, μ_0 is replaced in magnetic-field expressions by $\mu = K_m\mu_0$, where μ is the permeability of the material and K_m is its relative permeability. For **ferromagnetic** materials, K_m is much larger than unity. Some ferromagnetic materials are permanent magnets, retaining their magnetization even after the external magnetic field is removed.

 For instructor-assigned homework, go to www.masteringphysics.com

Conceptual Questions

1. If an electron beam in a cathode-ray tube travels in a straight line, can you be sure that no magnetic field is present?
2. Why is it *not* a good idea to call magnetic-field lines "magnetic lines of force"?
3. If the magnetic force does no work on a charged particle, how can it have any effect on the particle's motion? Can you think of other situations in which a force does no work, but has a significant effect on an object's motion?

4. A permanent magnet can be used to pick up a string of nails, tacks, or paper clips, even though these objects are not themselves magnets. How can this be?
5. Streams of charged particles emitted from the sun during unusual sunspot activity create a disturbance in the earth's magnetic field (called a *magnetic storm*). How can they cause such a disturbance?
6. A student once proposed to obtain an isolated magnetic pole by taking a bar magnet (N pole at one end, S at the other) and breaking it in half. Would this work? Why?

7. The magnetic force on a moving charged particle is always perpendicular to the magnetic field $\vec{B}$. Is the trajectory of a moving particle always perpendicular to the magnetic field lines? Explain your reasoning.

8. The text discusses the magnetic field of an infinitely long, straight conductor carrying a current. Of course, there is no such thing as an infinitely long *anything*. How would you decide whether a particular wire is long enough to be considered infinite?

9. Two parallel conductors carrying current in the same direction attract each other. If they are permitted to move toward each other, the forces of attraction do work. Where does the energy come from? Does this phenomenon contradict the assertion in this chapter that magnetic forces do no work on moving charges?

10. Household wires (such as lamp cords) often carry currents of several amps. Why do no magnetic effects show up near such wires?

11. You have a large bar magnet and want to identify its north and south poles. It is too heavy to show any effects from the Earth's weak magnetic field. You have an ordinary compass available, but alas, its poles are not marked. In spite of its lack of pole markings, how could you use this compass to determine the polarity of your bar magnet? (*Hint:* Can you first determine the north and south poles of the compass?)

12. When you bring a very strong permanent magnet close to an ordinary bar magnet, you often find that the north (or south) pole of the permanent magnet attracts *both* poles of the ordinary magnet. Why does this happen?

13. Students sometimes suggest that the reversals of the earth's magnetic field were caused when our planet's spin reversed its direction. Is this a plausible explanation? What are some good criticisms of it? (For example, would the energy required to cause this reversal be likely to leave some trace in the geological record?)

14. Can a charged particle move through a magnetic field that exerts no force on it? How? Could it move through an electric field that exerts no force on it?

15. If the magnitude of the magnetic field a distance R from a very long, straight, current-carrying wire is B, at what distance from the wire will the field have magnitude $3B$?

Multiple-Choice Problems

1. An electron traveling at a high speed enters a uniform magnetic field directed perpendicular to its path. Which of the following quantities will change while the electron travels through the field? (There may be more than one correct answer.)
   A. Its speed.
   B. Its velocity.
   C. Its acceleration.
   D. Its kinetic energy.
   E. Its potential energy.

2. A negatively charged particle shoots into a uniform magnetic field directed out of the paper, as shown in Figure 20.48. A possible path of this particle is
   A. 1.   B. 2.
   C. 3.   D. 4.

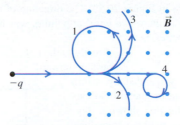

▲ **FIGURE 20.48** Multiple-choice problem 2.

3. A beam of protons is directed horizontally into the region between two bar magnets, as shown in Figure 20.49. The magnetic field in this region is horizontal. What is the effect of the magnetic field on the protons? (*Hint:* First refer to Figure 20.6 to find the direction of the magnetic field.)

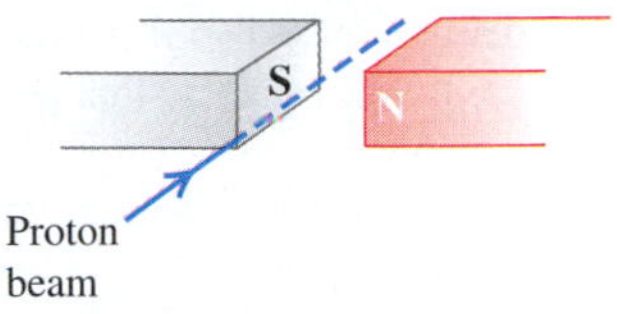

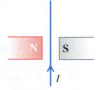

▲ **FIGURE 20.49** Multiple-choice problem 3.

   A. The protons are accelerated to the left, toward the S magnetic pole.
   B. The protons are accelerated to the right, toward the N magnetic pole.
   C. The protons are accelerated upward.
   D. The protons are accelerated downward.
   E. The protons are not accelerated, since the magnetic field does not change their speed.

4. A wire carrying a current in the direction shown in Fig. 20.50 passes between the poles of two bar magnets. What is the direction of the magnetic force on this wire due to the magnet? (*Hint:* Recall that magnetic-field lines point out of a north magnetic pole and into a south magnetic pole.)
   A. out of the paper.
   B. into the paper.
   C. toward the N pole of the magnet.
   D. toward the S pole of the magnet.

▲ **FIGURE 20.50** Multiple-choice problem 4.

5. A solenoid is connected to a battery as shown in Fig. 20.51, and a bar magnet is placed nearby. What is the direction of the magnetic force that this solenoid exerts on the bar magnet? (*Hint:* Think of the solenoid as a bar magnet, and identify what would be its north and south poles.)
   A. upward.
   B. downward.
   C. to the right, away from the solenoid.
   D. to the left, toward the solenoid

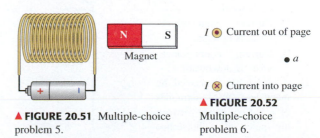

▲ **FIGURE 20.51** Multiple-choice problem 5.

▲ **FIGURE 20.52** Multiple-choice problem 6.

6. Two very long, straight parallel wires carry currents of equal magnitude, but opposite direction, perpendicular to the paper

in the directions shown in Figure 20.52. The direction of the net magnetic field due to these two wires at point *a*, which is equidistant from both wires, is

A. directly to the right.
B. directly to the left.
C. straight downward.
D. straight upward.

7. A light circular wire suspended by a thin silk thread in a uniform magnetic field carries a current in the direction shown in Figure 20.53. The magnetic field is perpendicular to the plane of the paper, and the wire is held at rest in that plane. If the wire is suddenly released so that it is free to rotate,
A. It will rotate so that point *a* goes into the paper.
B. It will rotate so that point *a* goes out of the paper.
C. It will not rotate.

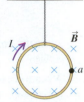

▲ **FIGURE 20.53** Multiple-choice problem 7.

8. An electron is moving directly toward you in a horizontal path when it suddenly enters a uniform magnetic field that is either vertical or horizontal. If the electron begins to curve upward in its motion just after it enters the field, you can conclude that the direction of the magnetic field is

A. upward.
B. downward.
C. to your left.
D. to your right.

9. The two coils shown in Figure 20.54 are parallel to each other and are connected to batteries. Coil *A* is held in place, but coil *C* is free to move. After the switch *S* is closed, coil *C* will initially move
A. toward coil *A*.
B. away from coil *A*.
C. upward.
D. downward.

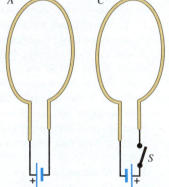

▲ **FIGURE 20.54** Multiple-choice problem 9.

10. A loose, floppy coil of wire is carrying current *I*. The loop of wire is placed on a horizontal table in a uniform magnetic field $\vec{B}$ perpendicular to the plane of the table. This causes the loop to expand into a circular shape while still lying on the table. What orientation of the current and magnetic field could cause this to happen? (There may be more than one correct answer.)
A. Current clockwise, $\vec{B}$ upward.
B. Current clockwise, $\vec{B}$ downward.
C. Current counterclockwise, $\vec{B}$ upward.
D. Current counterclockwise, $\vec{B}$ downward.

11. A metal bar connected by metal leads to the terminals of a battery hangs between the poles of a horseshoe magnet, as shown in Figure 20.55. Just after the switch *S* is closed, what will happen

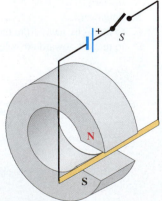

▲ **FIGURE 20.55** Multiple-choice problem 11.

to the bar? (*Hint:* Recall that magnetic-field lines point out of a north magnetic pole and into a south magnetic pole.)
A. It will be pushed upward, decreasing the tension in the leads.
B. It will be pushed downward, increasing the tension in the leads.
C. It will swing outward, away from the magnet.
D. It will swing inward, into the magnet.

12. A certain current produces a magnetic field *B* near the center of a solenoid. If the current is doubled, the field near the center will be
A. 4*B*.   B. 2*B*.   C. $B\sqrt{2}$.   D. *B*.

13. A coil is connected to a battery as shown in Figure 20.56. A bar magnet is suspended with its N pole just above the center of the coil. What will happen to the bar magnet just after the switch *S* is closed? (*Hint:* Think of the coil as a bar magnet, and identify what would be its north and south poles.)
A. It will be pulled toward the coil.
B. It will be pushed away from the coil.
C. It will be pushed out of the paper.
D. It will be pushed into the paper.

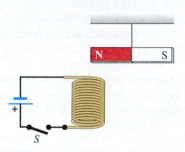

▲ **FIGURE 20.56** Multiple-choice problem 13.

14. The force exerted by a constant uniform magnetic field on a current-carrying wire of length *L* produces a force per unit length of F_L on the wire. If a wire twice as long and carrying the same current is placed in the same field with the same orientation, the force per unit length would be
A. $2F_L$.   B. F_L.   C. $F_L/2$.

15. A particle enters a uniform magnetic field initially traveling perpendicular to the field lines and is bent in a circular arc of radius *R*. If this particle were traveling twice as fast, the radius of its circular arc would be
A. 2*R*.   B. $R\sqrt{2}$.   C. $R/\sqrt{2}$.   D. $R/2$.

Problems

20.2 Magnetic Field and Magnetic Force

1. • In a 1.25 T magnetic field directed vertically upward, a particle having a charge of magnitude 8.50 μC and initially moving northward at 4.75 km/s is deflected toward the east. (a) What is the sign of the charge of this particle? Make a sketch to illustrate how you found your answer. (b) Find the magnetic force on the particle.

2. • An ion having charge +6*e* is traveling horizontally to the left at 8.50 km/s when it enters a magnetic field that is perpendicular to its velocity and deflects it downward with an initial magnetic force of 6.94×10^{-15} N. What are the direction and magnitude of this field? Illustrate your method of solving this problem with a diagram.

3. • A proton traveling at 3.60 km/s suddenly enters a uniform magnetic field of 0.750 T, traveling at an angle of 55.0° with the field lines (Figure 20.57). (a) Find the magnitude and direction of the force this magnetic field exerts on the proton. (b) If you can vary the direction of the proton's velocity, find the magnitude of the maximum and minimum forces you could achieve, and show how the velocity should be oriented to achieve these forces. (c) What would the answers to part (a) be if the proton were replaced by an electron traveling in the same way as the proton?

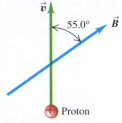

▲ **FIGURE 20.57** Problem 3.

4. • A particle having a mass of 0.195 g carries a charge of -2.50×10^{-8} C. The particle is given an initial horizontal northward velocity of 4.00×10^4 m/s. What are the magnitude and direction of the minimum magnetic field that will balance the earth's gravitational pull on the particle?

5. • At a given instant, a particle with a mass of 5.00×10^{-3} kg and a charge of 3.50×10^{-8} C has a velocity with a magnitude of 2.00×10^5 m/s in the $+y$ direction. It is moving in a uniform magnetic field that has magnitude 0.8 T and is in the $-x$ direction. What are (a) the magnitude and direction of the magnetic force on the particle and (b) its resulting acceleration?

6. •• If the magnitude of the magnetic force on a proton is F when it is moving at 15.0° with respect to the field, what is the magnitude of the force (in terms of F) when this charge is moving at 30.0° with respect to the field?

7. •• A ^{9}Be nucleus containing four protons and five neutrons has a mass of 1.50×10^{-26} kg and is traveling vertically upward at 1.35 km/s. If this particle suddenly enters a horizontal magnetic field of 1.12 T pointing from west to east, find the magnitude and direction of its acceleration vector the instant after it enters the field.

8. • A particle with a charge of -2.50×10^{-8} C is moving with an instantaneous velocity of magnitude 40.0 km/s in the xy-plane at an angle of 50° counterclockwise from the $+x$ axis. What are the magnitude and direction of the force exerted on this particle by a magnetic field with magnitude 2.00 T in the (a) $-x$ direction, and (b) $+z$ direction?

9. •• A particle with mass 1.81×10^{-3} kg and charge of $+1.22 \times 10^{-8}$ C has, at a given instant, a velocity of 3.00×10^4 m/s along the $+y$-axis, as shown in Figure 20.58. What are the magnitude and direction of the particle's acceleration produced by a magnetic field of magnitude 1.25 T in the xy-plane, directed at an angle of 45.0° counterclockwise from the $+x$-axis?

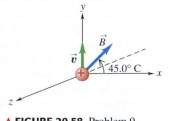

▲ **FIGURE 20.58** Problem 9.

10. •• A 150 V battery is connected across two parallel metal plates of area 28.5 cm^2 and separation 8.20 mm. A beam of alpha particles (charge $+2e$, mass 6.64×10^{-27} kg) is accelerated from rest through a potential difference of 1.75 kV and enters the region between the plates perpendicular to the electric field. What magnitude and direction of magnetic field are needed so that the alpha particles emerge undeflected from between the plates?

11. •• A velocity selector having uniform perpendicular electric and magnetic fields is shown in Figure 20.59. The electric field is provided by a 150 V DC battery connected across two large parallel metal plates that are 4.50 cm apart. (a) What must be the magnitude of the magnetic field so that charges having a velocity of 3.25 km/s perpendicular to the fields will pass through undeflected? (b) Show how the magnetic field should point in the region between the plates.

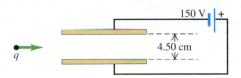

▲ **FIGURE 20.59** Problem 11.

12. •• An electron moves at 2.50×10^6 m/s through a region in which there is a magnetic field of unspecified direction and magnitude 7.40×10^{-2} T. (a) What are the largest and smallest possible magnitudes of the acceleration of the electron due to the magnetic field? (b) If the actual acceleration of the electron is one-fourth of the largest magnitude in part (a), what is the angle between the electron velocity and the magnetic field?

20.3 Motion of Charged Particles in a Magnetic Field

13. • In a cloud chamber experiment, a proton enters a uniform 0.250 T magnetic field directed perpendicular to its motion. You measure the proton's path on a photograph and find that it follows a circular arc of radius 6.13 cm. How fast was the proton moving?

14. • An alpha particle (a He nucleus, containing two protons and two neutrons and having a mass of 6.64×10^{-27} kg) traveling horizontally at 35.6 km/s enters a uniform, vertical, 1.10 T magnetic field. (a) What is the diameter of the path followed by this alpha particle? (b) What effect does the magnetic field have on the speed of the particle? (c) What are the magnitude and direction of the acceleration of the alpha particle while it is in the magnetic field? (d) Explain why the speed of the particle does not change even though an unbalanced external force acts on it.

15. • A deuteron particle (the nucleus of an isotope of hydrogen consisting of one proton and one neutron and having a mass of 3.34×10^{-27} kg) moving horizontally enters a uniform, vertical, 0.500 T magnetic field and follows a circular arc of radius 55.6 cm. (a) How fast was this deuteron moving just before it entered the magnetic field and just after it came out of the field? (b) What would be the radius of the arc followed by a proton that entered the field with the same velocity as the deuteron?

16. •• A beam of protons traveling at 1.20 km/s enters a uniform magnetic field, traveling perpendicular to the field. The beam exits the magnetic field in a direction perpendicular to its original direction (Fig. 20.60). The beam travels a distance of 1.18 cm *while in the field*. What is the magnitude of the magnetic field?

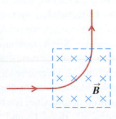

▲ **FIGURE 20.60** Problem 16.

17. •• A uniform magnetic field bends an electron in a circular arc of radius R. What will be the radius of the arc (in terms of R) if the field is tripled?

18. •• An electron at point A in Figure 20.61 has a speed v_0 of 1.41×10^6 m/s. Find (a) the magnitude and direction of the magnetic field that will cause the electron to follow the semi-circular path from A to B and (b) the time required for the electron to move from A to B. (c) What magnetic field would be needed if the particle were a proton instead of an electron?

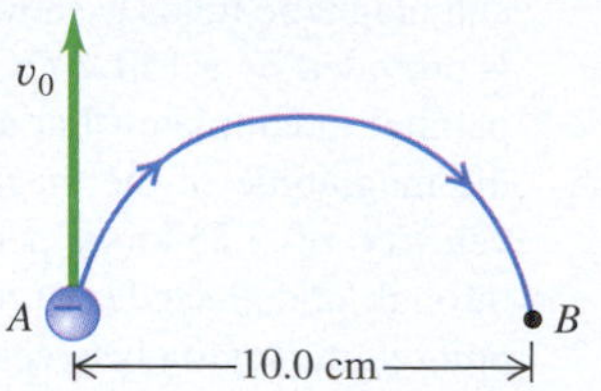

▲ **FIGURE 20.61** Problem 18.

19. •• A beam of protons is accelerated through a potential difference of 0.745 kV and then enters a uniform magnetic field traveling perpendicular to the field. (a) What magnitude of field is needed to bend these protons in a circular arc of diameter 1.75 m? (b) What magnetic field would be needed to produce a path with the same diameter if the particles were electrons having the same speed as the protons?

20. •• A 3.25 g bullet picks up an electric charge of 1.65 μC as it travels down the barrel of a rifle. It leaves the barrel at a speed of 425 m/s, traveling perpendicular to the earth's magnetic field, which has a magnitude of 5.50×10^{-4} T. Calculate (a) the magnitude of the magnetic force on the bullet and (b) the magnitude of the bullet's acceleration due to the magnetic force at the instant it leaves the rifle barrel.

21. •• An electron in the beam of a TV picture tube is accelerated through a potential difference of 2.00 kV. It then passes into a magnetic field perpendicular to its path, where it moves in a circular arc of diameter 0.360 m. What is the magnitude of this field?

20.4 Mass Spectrometers

22. •• (a) What is the speed of a beam of electrons when the simultaneous influence of an electric field of 1.56×10^4 V/m and a magnetic field of 4.62×10^{-3} T, with both fields normal to the beam and to each other, produces no deflection of the electrons? (b) In a diagram, show the relative orientation of the vectors $\vec{v}$, $\vec{E}$ and $\vec{B}$. (c) When the electric field is removed, what is the radius of the electron orbit? What is the period of the orbit?

23. • Singly ionized (one electron removed) atoms are accelerated and then passed through a velocity selector consisting of perpendicular electric and magnetic fields. The electric field is 155 V/m and the magnetic field is 0.0315 T. The ions next enter a uniform magnetic field of magnitude 0.0175 T that is oriented perpendicular to their velocity. (a) How fast are the ions moving when they emerge from the velocity selector? (b) If the radius of the path of the ions in the second magnetic field is 17.5 cm, what is their mass?

24. •• **Determining diet.** One method for determining the amount
BIO of corn in early Native American diets is the *stable isotope ratio analysis* (SIRA) technique. As corn photosynthesizes, it concentrates the isotope carbon-13, whereas most other plants concentrate carbon-12. Overreliance on corn consumption can then be correlated with certain diseases, because corn lacks the essential amino acid lysine. Archaeologists use a mass spectrometer to separate the ^{12}C and ^{13}C isotopes in samples of human remains. Suppose you use a velocity selector to obtain singly ionized (missing one electron) atoms of speed 8.50 km/s and want to bend them within a uniform magnetic field in a semicircle of diameter 25.0 cm for the ^{12}C. The measured masses of these isotopes are 1.99×10^{-26} kg (^{12}C) and 2.16×10^{-26} kg (^{13}C). (a) What strength of magnetic field is required? (b) What is the diameter of the ^{13}C semicircle? (c) What is the separation of the ^{12}C and ^{13}C ions at the detector at the end of the semicircle? Is this distance large enough to be easily observed?

25. •• **Ancient meat eating.** The amount of meat in prehistoric
BIO diets can be determined by measuring the ratio of the isotopes nitrogen-15 to nitrogen-14 in bone from human remains. Carnivores concentrate ^{15}N, so this ratio tells archaeologists how much meat was consumed by ancient people. Use the spectrometer of the previous problem to find the separation of the ^{14}N and ^{15}N isotopes at the detector. The measured masses of these isotopes are 2.32×10^{-26} kg (^{14}N) and 2.49×10^{-26} kg (^{15}N).

20.5 Magnetic Force on a Current-Carrying Conductor

26. • A straight vertical wire carries a current of 1.20 A downward in a region between the poles of a large electromagnet where the field strength is 0.588 T and is horizontal. What are the magnitude and direction of the magnetic force on a 1.00 cm section of this wire if the magnetic-field direction is (a) toward the east, (b) toward the south, (c) 30.0° south of west?

27. • **Magnetic force on a lightning bolt.** Currents during lightning strikes can be up to 50,000 A (or more!). We can model such a strike as a 50,000 A vertical current perpendicular to the earth's magnetic field, which is about $\frac{1}{2}$ gauss. What is the force on each meter of this current due to the earth's magnetic field?

28. • A horizontal rod 0.200 m long carries a current through a uniform horizontal magnetic field of magnitude 0.067 T that points perpendicular to the rod. If the magnetic force on this rod is measured to be 0.13 N, what is the current flowing through the rod?

29. • A straight 2.5 m wire carries a typical household current of 1.5 A (in one direction) at a location where the earth's magnetic field is 0.55 gauss from south to north. Find the magnitude and direction of the force that our planet's magnetic field exerts on this wire if is oriented so that the current in it is running (a) from west to east, (b) vertically upward, (c) from north to south. (d) Is the magnetic force ever large enough to cause significant effects under normal household conditions?

30. •• Between the poles of a powerful magnet is a cylindrical uniform magnetic field with a diameter of 3.50 cm and a strength of 1.40 T. A wire carries a current through the center of the field at an angle of 65.0° to the magnetic field lines. If the wire experiences a magnetic force of 0.0514 N, what is the current flowing in it?

31. • A rectangular 10.0 cm by 20.0 cm circuit carrying an 8.00 A current is oriented with its plane parallel to a uniform 0.750 T magnetic field (Figure 20.62). (a) Find the magnitude and direction of the magnetic force on each segment

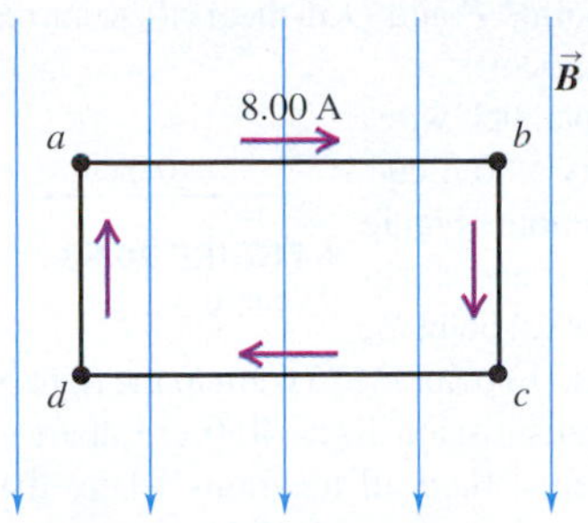

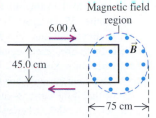

▲ **FIGURE 20.62** Problem 31.

(*ab*, *bc*, etc.) of this circuit. Illustrate your answers with clear diagrams. (b) Find the magnitude of the net force on the entire circuit.

32. •• A long wire carrying a 6.00 A current reverses direction by means of two right-angle bends, as shown in Figure 20.63. The part of the wire where the bend occurs is in a magnetic field of 0.666 T confined to the circular region of diameter 75 cm, as shown. Find the magnitude and direction of the net force that the magnetic field exerts on this wire.

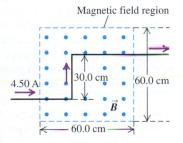

▲ **FIGURE 20.63** Problem 32.

33. •• A long wire carrying 4.50 A of current makes two 90° bends, as shown in Figure 20.64. The bent part of the wire passes through a uniform 0.240 T magnetic field directed as shown in the figure and confined to a limited region of space. Find the magnitude and direction of the force that the magnetic field exerts on the wire.

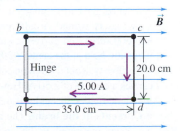

▲ **FIGURE 20.64** Problem 33.

20.6 Force and Torque on a Current Loop

34. • The 20.0 cm by 35.0 cm rectangular circuit shown in Figure 20.65 is hinged along side *ab*. It carries a clockwise 5.00 A current and is located in a uniform 1.20 T magnetic field oriented perpendicular to two of its sides, as shown. (a) Make a clear diagram showing the direction of the force that the magnetic field exerts on each segment of the circuit (*ab*, *bc*, etc.). (b) Of the four forces you drew in part (a), decide which ones exert a torque about the hinge *ab*. Then calculate only those forces that exert this torque. (c) Use your results from part (b) to calculate the torque that the magnetic field exerts on the circuit about the hinge axis *ab*.

▲ **FIGURE 20.65** Problem 34.

35. • The plane of a 5.0 cm by 8.0 cm rectangular loop of wire is parallel to a 0.19 T magnetic field, and the loop carries a current of 6.2 A. (a) What torque acts on the loop? (b) What is the magnetic moment of the loop?

36. •• A circular coil of wire 8.6 cm in diameter has 15 turns and carries a current of 2.7 A. The coil is in a region where the magnetic field is 0.56 T. (a) What orientation of the coil gives the maximum torque on the coil, and what is this maximum torque? (b) For what orientation of the coil is the magnitude of the torque 71% of the maximum found in part (a)?

37. •• A rectangular coil of wire 22.0 cm by 35.0 cm and carrying a current of 1.40 A is oriented with the plane of its loop perpendicular to a uniform 1.50 T magnetic field, as shown in Figure 20.66. (a) Calculate the net force and torque that the magnetic field exerts on this coil. (b) The coil is now rotated through a 30.0° angle about the axis shown, the left side coming out of the plane and the right side going into the plane. Calculate the net force and torque that the magnetic field exerts on the coil. (*Hint:* In order to help visualize this three-dimensional problem, make a careful drawing of the coil as viewed along its axis of rotation.)

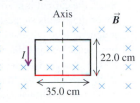

▲ **FIGURE 20.66** Problem 37.

38. • A solenoid having 165 turns and a cross-sectional area of 6.75 cm^2 carries a current of 1.20 A. If it is placed in a uniform 1.12 T magnetic field, find the torque this field exerts on the solenoid if its axis is oriented (a) perpendicular to the field, (b) parallel to the field, (c) at 35.0° with the field.

39. •• A circular coil of 50 loops and diameter 20.0 cm is lying flat on a tabletop, and carries a clockwise current of 2.50 A. A magnetic field of 0.450 T, directed to the north and at an angle of 45.0° from the vertical down through the coil and into the tabletop, is turned on. (a) What is the torque on the coil, and (b) which side of the coil (north or south) will tend to rise from the tabletop?

20.7 Magnetic Field of a Long, Straight Conductor

40. • You want to produce a magnetic field of magnitude 5.50 × 10^{-4} T at a distance of 0.040 m from a long, straight wire's center. (a) What current is required to produce this field? (b) With the current found in part (a), how strong is the magnetic field 8.00 cm from the wire's center?

41. • **Household magnetic fields.** Home circuit breakers typically have current capacities of around 10 A. How large a magnetic field would such a current produce 5.0 cm from a long wire's center? How does this field compare with the strength of the earth's magnetic field?

42. • (a) How large a current would a very long, straight wire have to carry so that the magnetic field 2.00 cm from the wire is equal to 1.00 G (comparable to the earth's northward-pointing magnetic field)? (b) If the wire is horizontal with the current running from east to west, at what locations would the magnetic field of the wire point in the same direction as the horizontal component of the earth's magnetic field? (c) Repeat part (b) except with the wire vertical and the current going upward.

43. • **Currents in the heart.** The body contains many small currents caused by the motion of ions in the organs and cells. Measurements of the magnetic field around the chest due to currents in the heart give values of about 1.0 μG. Although the

actual currents are rather complicated, we can gain a rough understanding of their magnitude if we model them as a long, straight wire. If the surface of the chest is 5.0 cm from this current, how large is the current in the heart?

44. • **Magnetic sensitivity of electric fish.** In a problem dealing
BIO with electric fish in Chapter 19, we saw that these fish navigate by responding to changes in the current in seawater. This current is due to a potential difference of around 3.0 V generated by the fish and is about 12 mA within a centimeter or so from the fish. Receptor cells in the fish are sensitive to the current. Since the current is at some distance from the fish, the sensitivity of these cells suggests that they might be responding to the magnetic field created by the current. To get some estimate of how sensitive the cells are, we can model the current as that of a long, straight wire with the receptor cells 2.0 cm away. What is the strength of the magnetic field at the receptor cells?

45. • In a conventional cheap flashlight, a straight copper strip runs along the tube of the flashlight to connect the bulb to the negative terminal of the battery at the bottom of the tube. If this strip carries a current of 0.65 A while you're holding the flashlight, what is the magnitude of the magnetic field at the surface of your hand, 0.30 cm from the strip? (You can treat the strip as a long, thin, straight wire.) How does your answer compare to the earth's magnetic field?

46. •• If the magnetic field due to a long, straight current-carrying wire has a magnitude B at a distance R from the wire's center, how far away must you be (in terms of R) for the magnetic field to decrease to $B/3$?

47. •• A current in a long, straight wire produces a magnetic field of 8.0 μT at 2.0 cm from the wire's center. Answer the following questions *without* finding the current: (a) What is the magnetic field strength 4.0 cm from the wire's center? (b) How far from the wire's center will the field be 1.0 μT? (c) If the current were doubled, what would the field be 2.0 cm from the wire's center?

48. •• **EMF.** Currents in DC transmission lines can be 100 A or
BIO more. Some people have expressed concern that the electromagnetic fields (EMFs) from such lines near their homes could cause health dangers. Using your own observations, estimate how high such lines are above the ground. Then use your estimate to calculate the strength of the magnetic field these lines produce at ground level. Express your answer in teslas and as a percent of the earth's magnetic field (which is 0.50 gauss). Does it seem that there is cause for worry?

49. • A long, straight telephone cable contains six wires, each carrying a current of 0.300 A. The distances between wires can be neglected. (a) If the currents in all six wires are in the same direction, what is the magnitude of the magnetic field 2.50 m from the cable? (b) If four wires carry currents in one direction and the other two carry currents in the opposite direction, what is the magnitude of the field 2.50 m from the cable?

50. •• Two insulated wires perpendicular to each other in the same plane carry currents as shown in Figure 20.67. Find the magnitude of the *net* magnetic field

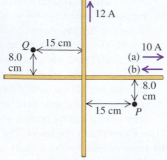

▲ **FIGURE 20.67** Problem 50.

these wires produce at points P and Q if the 10.0 A current is (a) to the right or (b) to the left.

51. •• Two long, straight parallel wires are 10.0 cm apart and carry 4.00 A currents in the same direction (Figure 20.68). Find the magnitude and direction of the magnetic field at (a) point P_1, midway between the wires, (b) point P_2, 25.0 cm to the right of P_1.

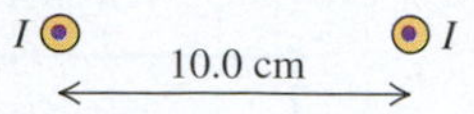

$I \odot \xleftarrow{\text{10.0 cm}} \odot I$

▲ **FIGURE 20.68** Problem 51.

52. •• Two long parallel transmission lines 40.0 cm apart carry 25.0 A and 75.0 A currents. Find all locations where the net magnetic field of the two wires is zero if these currents are in (a) the same direction, (b) opposite directions.

20.8 Force between Parallel Conductors

53. • Two high-current transmission lines carry currents of 25 A and 75 A in the same direction and are suspended parallel to each other 35 cm apart. If the vertical posts supporting these wires divide the lines into straight 15 m segments, what magnetic force does each segment exert on the other? Is this force attractive or repulsive?

54. •• Two long current-carrying wires run parallel to each other. Show that if the currents run in the same direction, these wires attract each other, whereas if they run in opposite directions, the wires repel.

55. •• A 2.0 m ordinary lamp extension cord carries a 5.0 A current. Such a cord typically consists of two parallel wires carrying equal currents in opposite directions. Find the magnitude and direction (attractive or repulsive) that the two segments of this cord exert on each other. (You will need to inspect an actual lamp cord at home and measure or reasonably estimate the quantities needed to do this calculation.)

56. •• An electric bus operates by drawing current from two parallel overhead cables, at a potential difference of 600 V, and spaced 55 cm apart. When the power input to the bus's motor is at its maximum power of 65 hp, (a) what current does it draw and (b) what is the attractive force per unit length between the cables?

20.9 Current Loops and Solenoids

57. • A circular metal loop is 22 cm in diameter. (a) How large a current must flow through this metal so that the magnetic field at its center is equal to the earth's magnetic field of 0.50 × 10^{-4} T? (b) Show how the loop should be oriented so that it can cancel the earth's magnetic field at its center.

58. • A closely wound circular coil with a diameter of 4.00 cm has 600 turns and carries a current of 0.500 A. What is the magnetic field at the center of the coil?

59. • A closely wound circular coil has a radius of 6.00 cm and carries a current of 2.50 A. How many turns must it have if the magnetic field at its center is 6.39 × 10^{-4} T?

60. • **Currents in the brain.** The magnetic field around the head
BIO has been measured to be approximately 3.0 × 10^{-8} gauss. Although the currents that cause this field are quite complicated, we can get a rough estimate of their size by modeling them as a single circular current loop 16 cm (the width of a typical head) in diameter. What is the current needed to produce such a field at the center of the loop?

61. • A closely wound, circular coil with radius 2.40 cm has 800 turns. What must the current in the coil be if the magnetic field at the center of the coil is 0.0580 T?

62. •• Two circular concentric loops of wire lie on a tabletop, one inside the other. The inner loop has a diameter of 20.0 cm and carries a clockwise current of 12.0 A, as viewed from above, and the outer wire has a diameter of 30.0 cm. What must be the magnitude and direction (as viewed from above) of the current in the outer loop so that the net magnetic field due to this combination of loops is zero at the common center of the loops?

63. •• Calculate the magnitude and direction of the magnetic field at point P due to the current in the semicircular section of wire shown in Figure 20.69. (*Hint:* The current in the long, straight section of wire produces no field at P. Can you relate the semicircle to a current loop?)

▲ **FIGURE 20.69** Problem 63.

64. • A solenoid contains 750 coils of very thin wire evenly wrapped over a length of 15.0 cm. Each coil is 0.800 cm in diameter. If this solenoid carries a current of 7.00 A, what is the magnetic field at its center?

65. • As a new electrical technician, you are designing a large solenoid to produce a uniform 0.150 T magnetic field near its center. You have enough wire for 4000 circular turns, and the solenoid must be 1.40 m long and 2.00 cm in diameter. What current will you need to produce the necessary field?

66. • A solenoid is designed to produce a 0.0279 T magnetic field near its center. It has a radius of 1.40 cm and a length of 40.0 cm, and the wire carries a current of 12.0 A. (a) How many turns must the solenoid have? (b) What total length of wire is required to make this solenoid?

67. •• A single circular current loop 10.0 cm in diameter carries a 2.00 A current. (a) What is the magnetic field at the center of this loop? (b) Suppose that we now connect 1000 of these loops in series within a 500 cm length to make a solenoid 500 cm long. What is the magnetic field at the center of this solenoid? Is it 1000 times the field at the center of the loop in part (a)? Why or why not?

68. •• A solenoid that is 35 cm long and contains 450 circular coils 2.0 cm in diameter carries a 1.75 A current. (a) What is the magnetic field at the center of the solenoid, 1.0 cm from the coils? (b) Suppose we now stretch out the coils to make a very long wire carrying the same current as before. What is the magnetic field 1.0 cm from the wire's center? Is it the same as you found in part (a)? Why or why not?

69. •• You have 25 m of wire, which you want to use to construct a 44 cm diameter coil whose magnetic field at its center will exactly cancel the earth's field of 0.55 gauss. What current will your coil require?

70. • A toroidal solenoid (see Figure 20.42) has inner radius $r_1 = 15.0$ cm and outer radius $r_2 = 18.0$ cm. The solenoid has 250 turns and carries a current of 8.50 A. What is the magnitude of the magnetic field at the following distances from the center of the torus: (a) 12.0 cm; (b) 16.0 cm; (c) 20.0 cm?

20.10 Magnetic Field Calculations

71. • A long, straight wire carries a current of 10.0 A, as shown in Figure 20.70. Use the law of Biot and Savart to find the magnitude and direction of the magnetic field at point P due to each of the following 2.00 mm segments of this wire: (a) segment A and (b) segment C.

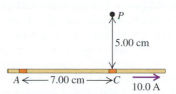

▲ **FIGURE 20.70** Problem 71.

72. • A long wire carrying a 5.00 A current makes an abrupt right-angle bend as shown in Figure 20.71. Use the law of Biot and Savart to determine the magnitude and direction of the magnetic field at point P due to the 1.50 cm bent segment if P is 15.0 cm from the midpoint of that segment.

▲ **FIGURE 20.71** Problem 72.

73. • Three long, straight electrical cables, running north and south, are tightly enclosed in an insulating sheath. One of the cables carries a 23.0 A current southward; the other two carry currents of 17.5 A and 11.3 A northward. Use Ampere's law to calculate the magnitude of the magnetic field at a distance of 10.0 m from the cables.

74. •• A long, straight, cylindrical wire of radius R carries a current uniformly distributed over its cross section. At what location is the magnetic field produced by this current equal to half of its largest value? Use Ampere's law and consider points inside and outside the wire.

20.11 Magnetic Materials

75. • Platinum is a paramagnetic metal having a relative permeability of 1.00026. (a) What is the magnetic permeability of platinum? (b) If a thin rod of platinum is placed in an external magnetic field of 1.3500 T, with its axis parallel to that field, what will be the magnetic field inside the rod?

76. • When a certain paramagnetic material is placed in an external magnetic field of 1.5000 T, the field inside the material is measured to be 1.5023 T. Find (a) the relative permeability and (b) the magnetic permeability of this material.

General Problems

77. •• A 150 g ball containing 4.00×10^8 excess electrons is dropped into a 125 m vertical shaft. At the bottom of the shaft, the ball suddenly enters a uniform horizontal 0.250 T magnetic field directed from east to west. If air resistance is negligibly small, find the magnitude and direction of the force that this magnetic field exerts on the ball just as it enters the field.

78. •• **Magnetic balance.** The circuit shown in Figure 20.72 is used to make a magnetic balance to weigh objects. The mass m to be measured is hung from the center of the bar, which is in a uniform magnetic field of 1.50 T directed into the plane of the figure. The battery voltage can be adjusted to vary the current in the circuit. The

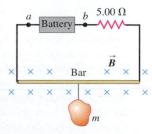

▲ **FIGURE 20.72** Problem 78.

horizontal bar is 60.0 cm long and is made of extremely light-weight material, so its mass can be neglected. It is connected to the battery by thin vertical wires that can support no appreciable tension; all the weight of the mass m is supported by the magnetic force on the bar. A 5.00 Ω resistor is in series with the bar, and the resistance of the rest of the circuit is negligibly small. (a) Which point, a or b, should be the positive terminal of the battery? (b) If the maximum terminal voltage of the battery is 175 V, what is the greatest mass m that this instrument can measure?

79. •• A thin 50.0-cm-long metal bar with mass 750 g rests on, but is not attached to, two metal supports in a uniform 0.450 T magnetic field, as shown in Figure 20.73. A battery and a 25.0 Ω resistor in series are connected to the supports. What is the largest terminal voltage the battery can have without breaking the circuit at the supports?

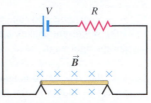

▲ **FIGURE 20.73** Problem 79.

80. •• A long, straight wire containing a semicircular region of radius 0.95 m is placed in a uniform magnetic field of magnitude 2.20 T as shown in Figure 20.74. What is the net magnetic force acting on the wire when it carries a current of 3.40 A? (*Hint:* In Figure 20.74, what does symmetry tell you about the forces on the upper and lower halves of the semicircular region?)

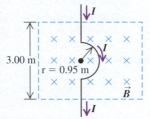

▲ **FIGURE 20.74** Problem 80.

81. •• A singly charged ion of ^{7}Li (an isotope of lithium containing three protons and four neutrons) has a mass of 1.16×10^{-26} kg. It is accelerated through a potential difference of 220 V and then enters a 0.723 T magnetic field perpendicular to the ion's path. What is the radius of the path of this ion in the magnetic field?

82. •• An insulated circular ring of diameter 6.50 cm carries a 12.0 A current and is tangent to a very long, straight insulated wire carrying 10.0 A of current, as shown in Figure 20.75. Find the magnitude and direction of the magnetic field at the center of the ring due to this combination of wires.

83. •• **The effect of transmission lines.** Two hikers are reading a compass under an overhead transmission line that is 5.50 m above the ground and carries a current of 0.800 kA in a horizontal direction from north to south. (a) Find the magnitude and direction of the magnetic field at a point on the ground directly under the transmission line. (b) One hiker suggests that they walk 50 m away from the lines to avoid inaccurate compass readings due to the current. Considering that the earth's magnetic field is on the order of 0.5×10^{-4} T, is the current really a problem?

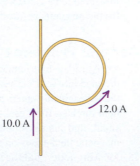

▲ **FIGURE 20.75** Problem 82.

84. •• A long, straight horizontal wire carries a current of 2.50 A directed toward the right. An electron is traveling in the vicinity of this wire. (a) At the instant the electron is 4.50 cm above the wire's center and moving with a speed of 6.00×10^4 m/s directly toward it, what are the magnitude and direction of the force that the magnetic field of the current exerts on the electron? (b) What would be the magnitude and direction of the magnetic force if the electron were instead moving parallel to the wire in the same direction as the current?

85. •• Two very long, straight wires carry currents as shown in Figure 20.76. For each case shown, find all locations where the net magnetic field due to these wires is zero.

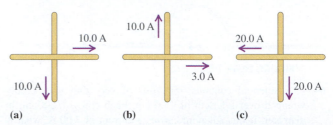

(a)   (b)   (c)

▲ **FIGURE 20.76** Problem 85.

86. •• **Bubble chamber, I.** Certain types of bubble chambers are filled with liquid hydrogen. When a particle (such as an electron or a proton) passes through the liquid, it leaves a track of bubbles, which can be photographed to show the path of the particle. The apparatus is immersed in a known magnetic field, which causes the particle to curve. Figure 20.77 is a trace of a bubble chamber image showing the path of an electron. (a) How could you determine the *sign* of the charge of a particle from a photograph of its path? (b) How can physicists determine the *momentum* and the *speed* of this electron by using measurements made on the photograph, given that the magnetic field is known and is perpendicular to the plane of the figure? (c) The electron is obviously spiraling into smaller and smaller circles. What properties of the electron must be changing to cause this behavior? Why does this happen? (d) What would be the path of a neutron in a bubble chamber? Why?

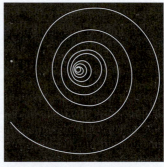

▲ **FIGURE 20.77** Problem 86.

87. •• A 3.00 N metal bar, 1.50 m long and having a resistance of 10.0 Ω, rests horizontally on conducting wires connecting it to the circuit shown in Figure 20.78. The bar is in a uniform, horizontal, 1.60 T magnetic field and is not attached to the wires in the circuit. What is the acceleration of the bar just after the switch S is closed?

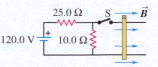

▲ **FIGURE 20.78** Problem 87.

88. •• A pair of long, rigid metal rods, each of length L, lie parallel to each other on a perfectly smooth table. Their ends are connected by identical, very light conducting springs of force constant k (Figure 20.79) and negligible unstretched length. If a

current I runs through this circuit, the springs will stretch. At what separation will the rods remain at rest? Assume that k is large enough so that the separation of the rods will be much less than L.

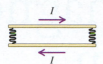
▲ **FIGURE 20.79** Problem 88.

89. •• **Atom smashers!** A *cyclotron particle accelerator* (sometimes called an "atom smasher" in the popular press) is a device for accelerating charged particles, such as electrons and protons, to speeds close to the speed of light. The basic design is quite simple. The particle is bent in a circular path by a uniform magnetic field. An electric field is pulsed periodically to increase the speed of the particle. The charged particle (or ion) of mass m and charge q is introduced into the cyclotron so that it is moving perpendicular to a uniform magnetic field $\vec{B}$. (a) Starting with the radius of the circular path of a charge moving in a uniform magnetic field, show that the time T for this particle to make one complete circle is $T = \dfrac{2\pi m}{|q|B}$. (*Hint:* You can express the speed v in terms of R and T because the particle travels through one circumference of the circle in time T.) (b) Which would take longer to complete one circle, an ion moving in a large circle or one moving in a small circle? Explain.

90. •• **Medical uses of cyclotrons.** The largest cyclotron (see
BIO previous problem) in the United States is the *Tevatron* at Fermilab, near Chicago, Illinois. It is called a Tevatron because it can accelerate particles to energies in the TeV range $(1 \text{ tera-eV} = 10^{12} \text{ eV})$. Its circumference is 6.4 km, and it currently can produce a maximum energy of 2.0 TeV. In a certain medical experiment, protons will be accelerated to energies of 1.25 MeV and aimed at a tumor to destroy its cells. (a) How fast are these protons moving when they hit the tumor? (b) How strong must the magnetic field be to bend the protons in the circle indicated?

91. •• A plastic circular loop has radius R, and a positive charge q is distributed uniformly around the circumference of the loop. The loop is now rotated around its central axis, perpendicular to the plane of the loop, with angular speed ω. If the loop is in a region where there is a uniform magnetic field $\vec{B}$ directed parallel to the plane of the loop, calculate the magnitude of the magnetic torque on the loop.

92. •• A long wire carrying 6.50 A of current makes two bends, as shown in Figure 20.80. The bent part of the wire passes through a uniform 0.280 T magnetic field directed as shown in the figure and confined to a limited region of space. Find the magnitude and direction of the force that the magnetic field exerts on the wire.

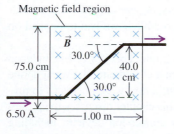
▲ **FIGURE 20.80** Problem 92.

Passage Problems

BIO **Magnetic fields and MRI.** Magnetic resonance imaging (MRI) is a powerful imaging method that, unlike x-ray imaging, allows sharp images of soft tissue to be made without exposure to potentially damaging radiation. While a full explanation of MRI is beyond the scope of an introductory physics textbook, some understanding can be achieved by the relatively simple application of the classical (that is, non-quantum) physics of magnetism. The starting point for MRI is nuclear magnetic resonance (NMR), a phenomenon that depends on the fact that protons in the atomic nucleus have a magnetic field, $\vec{B}$. The origin of the proton's magnetic field is the spin of the proton. Being charged, the spinning proton constitutes an electrical current analogous to a wire loop through which current flows. Like the wire loop, if the proton is subjected to an external magnetic field, $\vec{B}_0$, it experiences a torque and a magnetic moment, μ_p. The magnitude of the magnetic moment is about $1.4 \times 10^{-26} \text{J/T}$. The proton can be thought of as being in one of two states, with the magnetic moment oriented parallel or antiparallel to the applied magnetic field, and work must be done to flip the proton from the low energy state to the high energy state.

An important consideration is that the net magnetic field of any given nucleus, except for hydrogen, consists of protons and neutrons. The hydrogen atom, of course, has only a proton. If a nucleus has an even number of protons and neutrons, they will pair in a way such that half of the protons have spins in one orientation and half have spins in the other orientation, so net magnetic moment for the nucleus is zero. Only nuclei with a net magnetic moment are candidates for MRI. Hydrogen is the atom most commonly imaged.

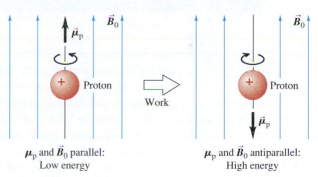

μ_p and $\vec{B}_0$ parallel: Low energy

μ_p and $\vec{B}_0$ antiparallel: High energy

▲ **FIGURE 20.81** Problems 93–95.

93. If a proton is exposed to an external magnetic field of 2T that has a direction perpendicular to the axis of the spin of the proton, what will be the torque on the proton?
  A. 0
  B. $1.4 \times 10^{-26} \text{N} \cdot \text{m}$
  C. $2.8 \times 10^{-26} \text{N} \cdot \text{m}$
  D. $0.7 \times 10^{-26} \text{N} \cdot \text{m}$

94. Which of following elements is a candidate for MRI?
  A. $^{12}C_6$ B. $^{16}O_8$
  C. $^{40}Ca_{20}$ D. $^{31}P_{15}$

21 Electromagnetic Induction

Have you ever wondered how a card reader reads the magnetically encoded data on a credit card? The information is stored in tiny magnetized regions on the card (on the order of 10^{-8} m). In a card reader, the motion of the credit card moves these tiny magnets past the read/write head, inducing currents in the head that convey the data to a computer.

What goes on in the read/write head is an example of electromagnetic induction: A changing magnetic field causes a current in a circuit, even though there's no battery or other obvious source of electromotive force (emf).

In this chapter, we discuss the electromotive force (emf) that results from *magnetic* interactions. Many of the components of present-day electric power systems, including generators, transformers, and motors, depend directly on magnetically induced emfs. These systems would not be possible if we had to depend on chemical sources of emf, such as batteries.

The central principle in this chapter is *Faraday's law.* This law relates induced emf to changing *magnetic flux* through a loop, often a closed circuit. (We'll define magnetic flux in Section 21.2.) We also discuss *Lenz's law,* which helps us to predict the directions of induced emf's and currents. This chapter discusses the principles that we need to understand electrical energy conversion devices, including motors, generators, and transformers. It also paves the way for the analysis of electromagnetic waves in Chapter 23.

This engineer examines the turbine rotor of a power-plant generator. The rotor converts the kinetic energy of high-pressure steam to rotational energy, which in turn is converted to electrical energy.

21.1 Induction Experiments

We begin our study of magnetically induced electromotive force with a look at several pioneering experiments carried out in the 1830s by Michael Faraday in England and Joseph Henry (the first director of the Smithsonian Institution) in the

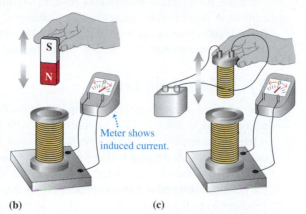
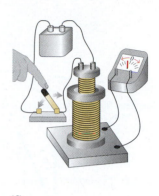

(a)

A stationary magnet does NOT induce a current in a coil.

(b)

Move the magnet toward or away from coil.

(c)

Move a second, current-carrying coil toward or away from the coil.

(d)

Vary the current in the second coil (by closing or opening a switch).

All these actions DO induce a current in the coil. What do they have in common? (They cause the magnetic field through the coil to *change*.)

▲ **FIGURE 21.1** Examples of experiments in which a magnetic field does or does not induce a current in a coil.

United States. Figure 21.1 shows several examples. In Figure 21.1a, a coil of wire is connected to a current-measuring device such as the moving-coil galvanometer we described in Section 19.7. When the nearby magnet is stationary, the meter shows no current. This isn't surprising; there's no source of emf in the circuit. But when we *move* the magnet either toward or away from the coil (Figure 21.1b), the meter shows current in the circuit, but *only* while the magnet is moving. If we hold the magnet stationary and move the coil, we again detect a current during the motion. So something about the *changing* magnetic field through the coil is causing a current in the circuit. We call this an **induced current,** and the corresponding emf that has to be present to cause such a current is called an **induced emf.**

In Figure 21.1c, we replace the magnet with a second coil connected to a battery. We find that when the second coil is stationary, there is no current in the first coil. But when we move the second coil toward or away from the first, or the first coil toward or away from the second, there is a current in the first coil, but again only while one coil is moving relative to the other.

Finally, using the two-coil setup in Figure 21.1d, we keep both coils stationary and vary the current in the second coil, either by opening and closing the switch or by changing the resistance in its circuit (with the switch closed). We find that as we open or close the switch, there is a momentary current pulse in the first circuit. When we vary the current in the second coil, there is an induced current in the first circuit, but only while the current in the second circuit is changing. The setup shown in Figure 21.2 can be used for more detailed experiments.

▲ **FIGURE 21.2** An apparatus for investigating induced currents. A change in the $\vec{B}$ field or the shape or location of the coil induces a current.

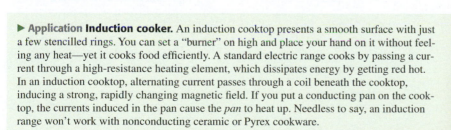

▶ **Application Induction cooker.** An induction cooktop presents a smooth surface with just a few stencilled rings. You can set a "burner" on high and place your hand on it without feeling any heat—yet it cooks food efficiently. A standard electric range cooks by passing a current through a high-resistance heating element, which dissipates energy by getting red hot. In an induction cooktop, alternating current passes through a coil beneath the cooktop, inducing a strong, rapidly changing magnetic field. If you put a conducting pan on the cooktop, the currents induced in the pan cause the *pan* to heat up. Needless to say, an induction range won't work with nonconducting ceramic or Pyrex cookware.

What do all these phenomena have in common? The answer, which we'll explore in detail in this chapter, is that a *changing magnetic flux* through the coil, from whatever cause, induces a current in the circuit. This statement forms the basis of Faraday's law of induction, the main subject of this chapter.

21.2 Magnetic Flux

The concept of *magnetic flux* is analogous to that of electric flux, which we encountered in Section 17.8 in connection with Gauss's law. It is also closely related to magnetic field lines, which we studied in Section 20.2. Magnetic flux, denoted by Φ_B, is defined with reference to an area and a magnetic field. We can divide any surface into elements of area ΔA, as shown in Figure 21.3. For each element, we determine the component $B_\perp$ of the magnetic field $\vec{B}$ normal (perpendicular) to the surface at the position of that element, as shown in the figure. From the figure, $B_\perp = B\cos\phi$, where ϕ is the angle between the direction of $\vec{B}$ and a line perpendicular to the surface. In general, this component is different for different points on the surface.

We define the magnetic flux $\Delta\Phi_B$ through the element of area ΔA as

$$\Delta\Phi_B = B_\perp \Delta A = B\cos\phi\,\Delta A. \tag{21.1}$$

The *total* magnetic flux through the surface is the sum of the contributions from the individual area elements. In the special case in which $\vec{B}$ is uniform over a plane surface with total area A, we can simplify the preceding expression as follows:

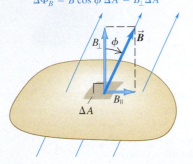

Magnetic flux through element of area ΔA:
$\Delta\Phi_B = B\cos\phi\,\Delta A = B_\perp\Delta A$

▲ **FIGURE 21.3** The magnetic flux through an area element ΔA.

Magnetic flux through a plane surface in a constant magnetic field

For a plane surface with area A in a uniform magnetic field $\vec{B}$, both $B_\perp$ and ϕ are the same at all points on the surface. The magnetic flux Φ_B through the surface is

$$\Phi_B = B_\perp A = BA\cos\phi. \tag{21.2}$$

If $\vec{B}$ happens to be perpendicular to the surface, then $\cos\phi = 1$ and Equation 21.2 reduces to $\Phi_B = BA$. Figure 21.4 shows three cases for a plane surface in a uniform magnetic field.

Unit: The SI unit of magnetic flux is the unit of magnetic field (1 T) times the unit of area (1 m^2):

$$(1\text{ T})(1\text{ m}^2) = [1\text{ N}/(\text{A}\cdot\text{m})](1\text{ m}^2) = 1\text{ N}\cdot\text{m}/\text{A}.$$

This unit is called 1 **weber** (1 Wb), in honor of Wilhelm Weber (1804–1891):

$$1\text{ weber} = 1\text{ Wb} = 1\text{ T}\cdot\text{m}^2 = 1\text{ N}\cdot\text{m}/\text{A}.$$

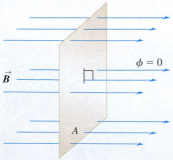

$\vec{B}$ perpendicular to A ($\phi = 0$): magnetic flux $\Phi_B = BA$.

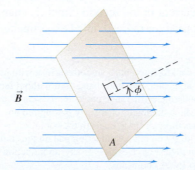

$\vec{B}$ at an angle ϕ to the perpendicular to A: magnetic flux $\Phi_B = BA\cos\phi$.

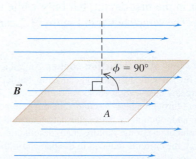

$\vec{B}$ parallel to A ($\phi = 90$): magnetic flux $\Phi_B = 0$.

▲ **FIGURE 21.4** The magnetic flux through a flat surface at various orientations relative to a uniform magnetic field.

If the element of area ΔA in Equation 21.1 is at right angles to the field lines, then $B_\perp = B$, and

$$B = \frac{\Delta \Phi_B}{\Delta A}. \qquad (21.3)$$

That is, magnetic field is *magnetic flux per unit area* across an area at right angles to the magnetic field. For this reason, magnetic field $\vec{B}$ is sometimes called **magnetic flux density.** The unit of magnetic flux is 1 weber, so the unit of magnetic field, 1 tesla, is also equal to 1 *weber per square meter:*

$$1\ \text{T} = 1\ \text{Wb/m}^2.$$

We can picture the total flux through a surface as proportional to the number of field lines passing through the surface and the field (the flux density) as proportional to the number of lines *per unit area.*

Conceptual Analysis 21.1

Flux through a flexible coil

Suppose you hold a circular wire coil perpendicular to a constant magnetic field and then squeeze it into a narrow oval, as shown in Figure 21.5. How does the flux through the squeezed coil compare with that through the circular coil?

A. The flux is less, because the area of the coil is smaller.
B. The flux is the same, because the length of wire is the same.
C. The flux is the same, because the angle of the coil relative to the magnetic field stays the same.

SOLUTION As Equation 21.2 shows, the magnetic flux through a coil depends on (1) the magnetic field magnitude B, (2) the area A of the coil, and (3) the angle ϕ of the coil's axis with respect to the magnetic field: $\Phi_B = BA\cos\phi$. A change in any of these variables changes the flux through the coil. Squeezing the coil reduces its area, so A is correct. (Neither B nor ϕ changes.) The magnetic flux through the coil does not depend on the length of wire in the coil, so B isn't correct. And the change in A results in a change in Φ_B, even though ϕ doesn't change; so C is also incorrect.

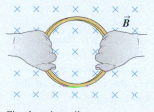

Circular wire coil   Coil squeezed into oval

▲ **FIGURE 21.5**

EXAMPLE 21.1   Magnetic flux

A plane surface with area 3.0 cm² is placed in a uniform magnetic field that is oriented at an angle of 30° to the surface. **(a)** What is the angle ϕ? **(b)** If the magnetic flux through this area is 0.90 mWb, what is the magnitude of the magnetic field?

SOLUTION

SET UP AND SOLVE **Part (a):** Figure 21.6 shows our sketch. As defined earlier, ϕ is the angle between the direction of $\vec{B}$ and a line *normal* to the surface, so $\phi = 60°$ (not 30°).

Part (b): Because B and ϕ are the same at all points on the surface, we can use Equation 21.2: $\Phi_B = BA\cos\phi$. We solve for B, remembering to convert the area to square meters:

$$B = \frac{\Phi_B}{A\cos\phi} = \frac{0.90 \times 10^{-3}\ \text{Wb}}{(3.0 \times 10^{-4}\ \text{m}^2)(\cos 60°)} = 6.0\ \text{T}.$$

REFLECT The flux through the surface depends on its angle relative to $\vec{B}$. Knowing this angle and the flux per unit area, we can find the magnitude of the magnetic field.

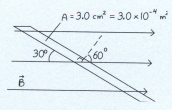

▲ **FIGURE 21.6** Our sketch for this problem.

Practice Problem: For the same B and A, find the angle ϕ at which the flux would have half the value given above (i.e., the angle ϕ at which $\Phi_B = 0.45 \times 10^{-3}$ Wb). *Answer:* $\phi = 76°$.

21.3 Faraday's Law

The common element in all induction effects is changing magnetic flux (defined in Section 21.2) through a circuit. *Faraday's law* states that the induced emf in a circuit is *directly proportional* to the time rate of change of the magnetic flux Φ_B through the circuit.

> **Faraday's law of induction:**
>
> The magnitude of the induced emf in a circuit equals the absolute value of the time rate of change of the magnetic flux through the circuit.
>
> In symbols, Faraday's law is
>
> $$\mathcal{E} = \left| \frac{\Delta \Phi_B}{\Delta t} \right|. \tag{21.4}$$
>
> In this definition, $\mathcal{E}$ is the magnitude of the emf and is always positive.

Here's a simple example of this law in action:

EXAMPLE 21.2 **Current induced in a single loop**

In Figure 21.7, the magnetic field in the region between the poles of the electromagnet is uniform at any time, but is increasing at the rate of 0.020 T/s. The area of the conducting loop in the field is 120 cm², and the total circuit resistance, including the meter, is 5.0 Ω. Find the magnitudes of the induced emf and the induced current in the circuit.

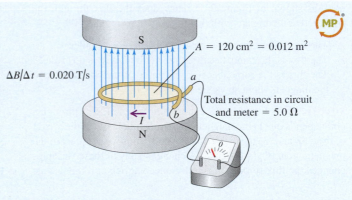

▶ **FIGURE 21.7**

SOLUTION

SET UP The area $A = 0.012$ m² is constant, so the absolute value of the rate of change of magnetic flux is A multiplied by the rate of change of the magnetic field magnitude B.

SOLVE The magnitude of the induced emf, $\mathcal{E}$, is equal to the absolute value of the rate of change of the magnetic flux, $|\Delta \Phi_B / \Delta t|$:

$$\mathcal{E} = \left| \frac{\Delta \Phi_B}{\Delta t} \right| = \frac{\Delta B}{\Delta t} A = (0.020 \text{ T/s})(0.012 \text{ m}^2) = 2.4 \times 10^{-4} \text{ V}$$

$$= 0.24 \text{ mV}.$$

The induced current in the circuit is

$$I = \frac{\mathcal{E}}{R} = \frac{2.4 \times 10^{-4} \text{ V}}{5.0 \ \Omega} = 4.8 \times 10^{-5} \text{ A} = 0.048 \text{ mA}.$$

REFLECT It's worthwhile to verify unit consistency in this calculation. There are many ways to do this; one is to note that we can rewrite the magnetic force relation $(F = qvB)$ as $B = F/qv$, so

$$1 \text{ T} = (1 \text{ N})/(1 \text{ C} \cdot \text{m/s}).$$

The units of magnetic flux can thus be expressed as

$$(1 \text{ T})(1 \text{ m}^2) = 1 \text{ N} \cdot \text{s} \cdot \text{m/C}$$

and the rate of change of magnetic flux as

$$1 \text{ N} \cdot \text{m/C} = 1 \text{ J/C} = 1 \text{ V}.$$

So the unit of $\Delta \Phi_B / \Delta t$ is the volt, as is required by Equation 21.4. Also, recall that the unit of magnetic flux is $1 \text{ T} \cdot \text{m}^2 = 1 \text{ Wb}$, so $1 \text{ V} = 1 \text{ Wb/s}$.

Practice Problem: Suppose we change the apparatus so that the magnetic field increases at a rate of 0.15 T/s, the area of the conducting loop in the field is 0.020 m², and the total circuit resistance is 7.5 Ω. Find the magnitude of the induced emf and the induced current in the circuit. *Answers:* $\mathcal{E} = 3.0$ mV; $I = 0.40$ mA.

If we have a coil with N identical turns, and the magnetic flux varies at the same rate through each turn, the induced emf's in the turns are all equal, are in *series,* and must be added. The total emf magnitude $\mathcal{E}$ is then

$$\mathcal{E} = N \left| \frac{\Delta \Phi_B}{\Delta t} \right|. \qquad (21.5)$$

Conceptual Analysis 21.2 A magnet and a coil

When a magnet is plunged into a coil with two turns (Figure 21.8a), a voltage is induced in the coil and a current flows in the circuit. If the magnet is plunged with the same speed into a coil with *four* turns (Figure 21.8b), the induced voltage is

A. twice as great.
B. four times as great.
C. half as great.

SOLUTION As we just saw, the emf is the same for each turn of the coil, and the emf's for the turns add (because the turns are in series). Thus, doubling the number of turns doubles the total induced voltage. Answer A is correct.

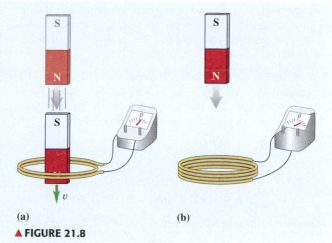

(a) (b)

▲ **FIGURE 21.8**

PROBLEM-SOLVING STRATEGY 21.1 **Faraday's law** (MP)

SET UP

1. To calculate the rate of change of magnetic flux, you first have to understand what is making the flux change. Is the conductor moving? Is it changing orientation? Is the magnetic field changing? Remember that it's not the flux itself, but its *rate of change,* that counts.

SOLVE

2. If your conductor has N turns in a coil, don't forget to multiply by N.
3. In the next section, we'll learn a method for finding the directions of induced currents and emf's.

REFLECT

4. The shape of the coil doesn't matter; it can be circular, rectangular, or some other shape. Only the total rate of change of flux through the coil, and its number of turns, are significant.

NOTE ▶ In the statement of Faraday's law given by Equations 21.4 and 21.5, we've assumed that $\mathcal{E}$ and Φ_B are positive quantities. However, Faraday's law can also be stated in a more general way as $\mathcal{E} = -\Delta \Phi_B / \Delta t$, in which $\mathcal{E}$ and Φ_B can be either positive or negative. This form, combined with a fairly elaborate set of sign rules, gives the direction as well as the magnitude of an induced emf or current. For most of our applications, we won't need this generalization; instead, we'll rely on a principle called *Lenz's law,* to be discussed in Section 21.4, to determine the directions of the various quantities. ◀

EXAMPLE 21.3 **Induced emf in a coil of wire**

A circular coil of wire with 500 turns and an average radius of 4.00 cm is placed at a 60° angle to the uniform magnetic field between the poles of a large electromagnet. The field changes at a rate of −0.200 T/s. What is the magnitude of the resulting induced emf?

SOLUTION

SET UP Figure 21.9 shows our sketch. We remember that the angle ϕ is the angle between the magnetic field and the *normal* to the coil and thus is 30°, not 60°. The area of the coil is $A = \pi(0.0400 \text{ m})^2 = 0.00503 \text{ m}^2$.

SOLVE To find the magnitude of the emf, we use Equation 21.5: $\mathcal{E} = N|\Delta\Phi_B/\Delta t|$. We first need to find the rate of change of magnetic flux, $\Delta\Phi_B/\Delta t$. The flux at any time is given by $\Phi_B = BA\cos\phi$, so the magnitude of the rate of change of flux is

$$\left|\frac{\Delta\Phi_B}{\Delta t}\right| = \left|\frac{\Delta B}{\Delta t}\right| A\cos\phi$$

$$= (0.200 \text{ T/s})(0.00503 \text{ m}^2)(\cos 30°)$$

$$= 0.000871 \text{ T} \cdot \text{m}^2/\text{s} = 0.000871 \text{ Wb/s}.$$

Now we calculate the induced emf:

$$\mathcal{E} = N\left|\frac{\Delta\Phi_B}{\Delta t}\right| = (500)|(-0.000871 \text{ Wb/s})| = 0.436 \text{ V}.$$

REFLECT We need the absolute value signs on the right side of Faraday's law because we have defined $\mathcal{E}$ to be the *magnitude* of

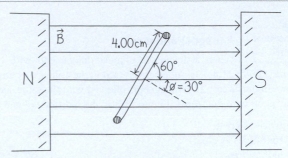

▲ FIGURE 21.9 Our sketch for this problem.

the emf, always a positive quantity. As we've noted, Faraday's law can be stated so that the emf is positive or negative, depending on its direction. We avoid that complexity by stating the law in terms of the *magnitude* of the emf. In the next section, we'll learn how to determine the directions of induced currents.

Practice Problem: A 300-turn coil with an average radius of 0.03 m is placed in a uniform magnetic field so that $\phi = 40°$. The field increases at a rate of 0.250 T/s. What is the magnitude of the resulting emf? *Answer:* $\mathcal{E} = 0.162 \text{ V}$.

EXAMPLE 21.4 **A slide-wire generator**

The U-shaped conductor in Figure 21.10 lies perpendicular to a uniform magnetic field $\vec{B}$ with magnitude $B = 0.60$ T, directed into the page. We lay a metal rod with length $L = 0.10$ m across the two arms of the conductor, forming a conducting loop, and move the rod to the right with constant speed $v = 2.5$ m/s. What is the magnitude of the resulting emf?

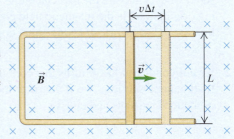

▶ FIGURE 21.10

SOLUTION

SET UP AND SOLVE The magnetic flux through the loop is changing because the area of the loop is increasing. During a time interval Δt, the sliding rod moves a distance $v\,\Delta t$ and the area increases by $\Delta A = Lv\,\Delta t$. Therefore, in time Δt, the magnetic flux through the loop increases by an amount $\Delta\Phi_B = B\,\Delta A = BLv\,\Delta t$. The magnitude of the induced emf is $\mathcal{E} = |\Delta\Phi_B/\Delta t| = BLv$. With the numerical values given, we find

$$\mathcal{E} = (0.60 \text{ T})(0.10 \text{ m})(2.5 \text{ m/s}) = 0.15 \text{ V}.$$

REFLECT As in Example 21.3, we've found the *magnitude* of the emf, but not the direction of the resulting induced current. Does the current flow clockwise or counterclockwise? We address this question in the next section.

Practice Problem: With the given magnetic field and rod length, what must the speed be if the induced emf has magnitude 0.75 V? *Answer:* 12.5 m/s.

Generators

A **generator** is a device that converts mechanical energy to electrical energy. A common design uses a conducting coil that rotates in a magnetic field, producing

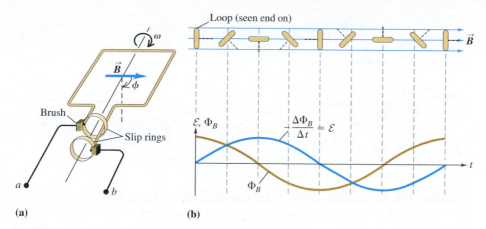

Loop (seen end on)

Brush

Slip rings

(a)   (b)

▲ **FIGURE 21.11** (a) Schematic diagram of a simple alternator, using a conducting loop rotating in a magnetic field. The loop connects to the external circuit via the slip rings. (b) Graphs of the flux through the loop (with loop position shown schematically) and of the resulting emf across terminals *ab*.

an induced emf according to Faraday's law. Figure 21.11a shows a simplified model of such a device. A rectangular loop with area A rotates with constant angular velocity ω about the axis shown. As the loop rotates, the magnetic flux through it changes sinusoidally with time, inducing a sinusoidally varying emf. If the shaft of the coil is connected, for instance, to a windmill, it converts wind power into electric power. The magnetic field $\vec{B}$ is uniform and constant. At time $t = 0$, $\phi = 0$; the figure shows the loop at the position $\phi = 90°$.

To analyze the behavior of this device, we need to use the generalized form of Faraday's law, mentioned above; we assume here that $\mathcal{E}$, Φ_B, and $\Delta\Phi_B/\Delta t$ can be either positive or negative. As the coil rotates, the signs of both $\mathcal{E}$ and Φ_B reverse twice during each revolution. The flux Φ_B through the loop equals its area A, multiplied by the component of B perpendicular to the plane of the loop—that is, $B\cos\phi$. Thus $\Phi_B = AB\cos\phi = AB\cos\omega t$, where now $\cos\phi$ and Φ_B can be either positive or negative.

Because Φ_B is a sinusoidal function of time, $\Delta\Phi_B/\Delta t$ isn't constant. To derive an expression for $\Delta\Phi_B/\Delta t$ would require calculus, but we can make an educated guess even without calculus. Figure 21.11b includes a graph of Φ_B as a function of time. At each point on the curve, the rate of change of Φ_B is equal to the *slope* of the curve. Where the curve is horizontal, Φ_B momentarily is not changing; where the slope is steepest, Φ_B is changing most rapidly with time. So, from the shape of the graph of Φ_B versus t, we can sketch the graph of $\Delta\Phi_B/\Delta t$ as a function of time. That sketch is shown in Figure 21.11b; it looks like a *sine* curve, and indeed it is. The rate of change of Φ_B turns out to equal $-\omega BA\sin\omega t$, so, from Faraday's law, the emf $\mathcal{E}$ is given by

$$\mathcal{E} = -\frac{\Delta\Phi_B}{\Delta t} = \omega AB\sin\omega t.$$

This equation shows that the induced emf $\mathcal{E}$ varies sinusoidally with time, as shown in Figure 21.11b. When the plane of the loop is perpendicular to $\vec{B}$ ($\phi = 0$ or 180°), Φ_B reaches its maximum (AB) and minimum ($-AB$) values. At these times, its instantaneous rate of change is zero and $\mathcal{E}$ is zero. Also, $|\mathcal{E}|$ is greatest in magnitude when the plane of the loop is parallel to $\vec{B}$ ($\phi = 90°$ or 270°) and Φ_B is changing most rapidly. Note that the induced emf depends, not on the *shape* of the loop, but only on its area. Finally, a real windmill generator of this type would use a coil with many turns, thus greatly increasing the induced emf.

The rotating loop is the prototype of a common kind of alternating-current (ac) generator, or *alternator;* it develops a sinusoidally varying emf. We can use

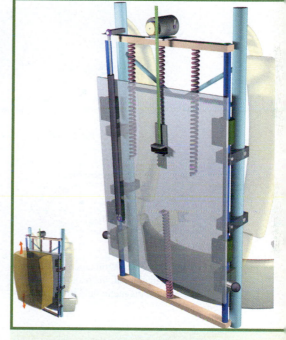

▲ **Application Geek alert!** Sure, you'd like to take your cell phone, GPS unit, PDA, and night-vision goggles on your two-month outback trek, but what about power? Who wants to crank a manual generator? The answer is this pack. Invented by Professor Larry Rome of the University of Pennsylvania, the device harnesses energy from the striding motion of walking that would otherwise be wasted. Instead of being fixed to the pack supports, the pack load is free to ride up and down as you stride. As it does so, a pinion uses the motion to turn a small generator. At a pace of 5.5 kilometers per hour, a 29 kilogram load can generate about 4 watts, plenty to power a number of small devices. Geek appeal aside, this device could be useful to field scientists, disaster-relief workers, and others operating in remote areas.

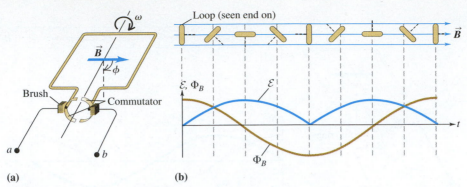

▲ **FIGURE 21.12** (a) Schematic diagram of a dc generator using a split-ring commutator. (b) Graph of the flux through the loop (with loop position shown schematically) and of the resulting emf across terminals ab.

it as a source of emf in an external circuit by adding two *slip rings,* which rotate with the loop, as shown in Figure 21.11a. Stationary contacts called *brushes* slide on the rings and are connected to the output terminals a and b.

We can use a similar scheme to obtain an emf that always has the same sign. The arrangement shown in Figure 21.12a is called a *commutator;* it reverses the connections to the external circuit at those angular positions where the emf reverses. The resulting emf is shown in Figure 21.12b. This device is the prototype of a direct-current (dc) generator. Commercial dc generators have a large number of coils and commutator segments; this arrangement smooths out the bumps in the emf, so the terminal voltage is not only unidirectional, but also practically constant. This brush-and-commutator arrangement is also used in the direct-current motor that we discussed in Section 20.6.

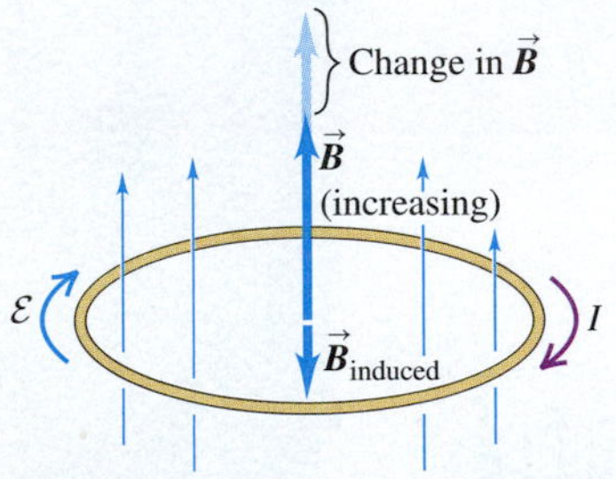

21.4 Lenz's Law

So far we've discussed the *magnitude,* but not the *direction,* of an induced emf and the associated current. To determine the direction of an induced current or emf, we use *Lenz's law.* This law is not an independent principle; it can be derived from Faraday's law with an appropriate set of sign conventions that we haven't discussed in detail. Lenz's law also helps us to gain an intuitive understanding of various induction effects and of the role of energy conservation. H. F. E. Lenz (1804–1865) was a German scientist who duplicated independently many of the discoveries of Faraday and Henry. **Lenz's law** is as follows:

Lenz's law
The direction of any magnetically induced current or emf is such as to oppose the direction of the phenomenon causing it.

The cause of a magnetically induced current or emf may be a changing flux through a stationary circuit, or the motion of a conductor in a magnetic field, or any combination of the two, as shown in Figure 21.1. A simple example of the first case is shown in Figure 21.13: A conducting loop is placed in an increasing magnetic field. According to Faraday's law, a current is induced in the loop; this current produces an additional magnetic field, as we discussed in Section 20.9 and as is shown in Figure 21.14. Lenz's law states that this additional field must

▲ **FIGURE 21.13** Relation among a changing external magnetic field, the resulting induced current, and the magnetic field induced by the current.

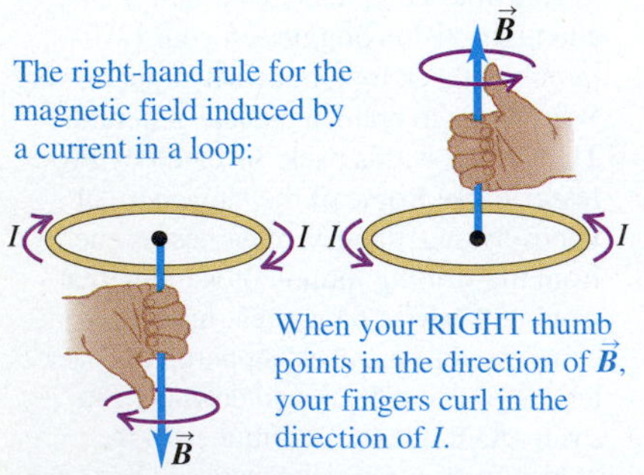

The right-hand rule for the magnetic field induced by a current in a loop:

When your RIGHT thumb points in the direction of $\vec{B}$, your fingers curl in the direction of I.

▲ **FIGURE 21.14** A review of the right-hand rule for the direction of the magnetic field induced by a current in a conducting loop.

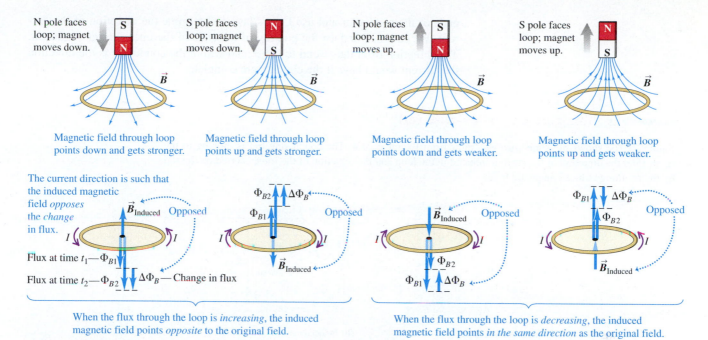

When the flux through the loop is *increasing*, the induced
magnetic field points *opposite* to the original field.

When the flux through the loop is *decreasing*, the induced
magnetic field points *in the same direction* as the original field.

▲ **FIGURE 21.15** Directions of induced currents as a bar magnet moves along the axis of a conducting loop.

oppose the direction in which the original field is increasing, so it must be downward in Figure 21.13. This in turn requires the induced current to be clockwise as
seen from above.

Figure 21.15 shows how the foregoing analysis works out for four situations.
Within the loop, the induced magnetic field points *opposite* to the original field if
the original field or flux is *increasing*. Conversely, if the original field is
decreasing, the additional field caused by the induced current points in the *same*
direction as the original field. In each case, as Lenz's law requires, the induced
magnetic field opposes the *change in flux* through the circuit (*not* the flux itself).

A similar analysis can be made for a conducting loop that moves in a magnetic field. If the magnetic flux through the loop changes, an induced current is
produced. As in the preceding case, the current generates its own induced magnetic field. According to Lenz's law, this induced magnetic field *opposes* the
change in flux that induced the current in the first place. Thus, if the flux is
increasing, the induced magnetic field points opposite to the original magnetic
field (opposing the increase in flux), and if the flux is decreasing, the induced
magnetic field points in the same direction as the original magnetic field (opposing the decrease in flux).

There is another, equivalent, way to look at this second case. Remember from
Chapter 20 that a current-carrying conductor is acted upon by a force when
placed in a magnetic field. (See Figures 20.10 and 20.24 to review the relevant
right-hand rule.) According to Lenz's law, this force is directed so as to *oppose*
the *motion* that induces the current. The induced current adopts the direction that
creates such a force.

NOTE ▸ In all these cases, the induced current tends to preserve the *status
quo* by opposing the motion or the change of flux that originally induced it. ◂

To have an induced current, we need a complete circuit. If a conductor does *not*
form a complete circuit, then we can mentally complete the circuit between the

ends of the conductor and use Lenz's law to determine the direction of the current. We can then deduce the polarity of the ends of the open-circuit conductor. The direction from the − end to the + end within the conductor is the direction the current would have if the circuit were complete.

EXAMPLE 21.5 **The force on a slide-wire rod**

Figure 21.16a shows the slide-wire generator from Example 21.4. The conducting rod is moving to the right. Find the direction of the current induced in the loop and the direction of the force exerted on the rod by the resulting induced magnetic field.

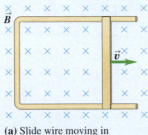

(a) Slide wire moving in magnetic field

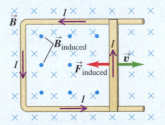

(b) Induced current, magnetic field, and magnetic force on slide wire

▲ **FIGURE 21.16**

SOLUTION

SET UP AND SOLVE The rod's motion increases the flux through the loop, so, by Lenz's law, the direction of the induced current must be such as to oppose this change—that is, to produce an induced magnetic field pointing out of the page (Figure 21.16b). By the right-hand rule for the magnetic field due to a current in a conductor (see Figure 21.14), the induced current is counterclockwise.

Lenz's law also tells us that the induced current creates a left-directed force on the moving bar, opposing the motion that causes the increase in flux. Does our counterclockwise current produce such a force? We apply the right-hand rule for the force on a current-carrying conductor in a magnetic field. (See Figure 20.24.) As predicted, the force points to the left. (Note that we use the direction of the *original* magnetic field in apply-

ing this right-hand rule, not the direction of the induced field. The original field is always stronger than the induced field and hence dominates.)

REFLECT We can apply Lenz's law to a moving conductor by looking at either the induced magnetic field (which opposes the change in magnetic flux) or the force on the conductor (which opposes the motion that causes the change in flux). The two perspectives give the same result and can be used to check each other.

The behavior described in this example is also directly related to energy conservation. If the induced current were in the opposite direction, the resulting magnetic force on the rod would accelerate it to an ever-increasing speed with no external energy source, despite the fact that electrical energy is being dissipated in the loop. This effect would be a clear violation of energy conservation.

EXAMPLE 21.6 **Induced current in a resistance–capacitance circuit**

Loop *A* in Figure 21.17 is part of a resistance–capacitance circuit. When the switch is closed, the capacitor discharges. While the capacitor is discharging, what are the directions of the induced emf, the induced current, and the induced magnetic field in loop *B*?

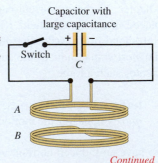

▶ **FIGURE 21.17**

Continued

21.5 Motional Electromotive Force

When a conductor moves in a magnetic field, we can gain added insight into the resulting induced emf by considering the magnetic forces on charges in the conductor. Figure 21.18a shows the same moving rod that we discussed in Example 21.5, separated for the moment from the U-shaped conductor. The magnetic field $\vec{B}$ is uniform and directed into the page, and we give the rod a constant velocity $\vec{v}$ to the right. A charged particle q in the rod then experiences a magnetic force with magnitude $F_B = |q|vB$. In the following discussion, we'll assume that q is positive; in that case, the direction of this force is *upward* along the rod, from b toward a. (The emf and the current are the same whether q is positive or negative.)

This magnetic force causes the free charges in the rod to move, creating an excess of positive charge at the upper end a and negative charge at the lower end b. This, in turn, creates an electric field $\vec{E}$ in the direction from a to b, which exerts a downward force $\vec{F}_E$ on the charges. The accumulation of charge at the ends of the rod continues until $\vec{E}$ is large enough for the downward electric force (with magnitude qE) to cancel exactly the upward magnetic force (magnitude qvB). Then $qE = qvB$ and the charges are in equilibrium, with point a at higher potential than point b.

What is the magnitude of the potential difference V_{ab}? It is equal to the electric field magnitude E, multiplied by the length L of the rod. From the preceding discussion, $E = vB$, so

$$V_{ab} = EL = vBL, \tag{21.6}$$

with point a at higher potential than point b.

Now suppose the moving rod slides along a stationary U-shaped conductor, forming a complete circuit (Figure 21.18b). No *magnetic* force acts on the charges in the stationary conductor, but there is an *electric* field caused by the accumulations of charge at a and b. Under the action of this field, a current is

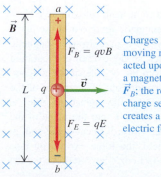

Charges in the moving rod are acted upon by a magnetic force $\vec{F}_B$; the resulting charge separation creates a canceling electric force $\vec{F}_E$.

(a) Isolated moving rod

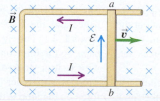

The motional emf $\mathcal{E}$ in the moving rod creates an electric field in the stationary conductor.

(b) Rod connected to stationary conductor

▲ **FIGURE 21.18** (a) A conducting rod moving in a uniform magnetic field. The rod, its velocity, and the magnetic field are mutually perpendicular. (b) The direction of the induced current when the rod is connected to a circuit.

◄ **Application Tether power** What to do about space debris? Every year, we put more satellites into low earth orbit. When their useful life is over, we should remove them to prevent collisions. One method would be to have the satellite use its thruster rocket to lower its orbit until it enters the atmosphere and burns up. But that requires lifting extra fuel into orbit and works only if the satellite is functional. The Terminator Tether offers a low-weight, reliable alternative. On a command from the ground, the tether module reels out a kilometers-long, very thin conducting cable. As the satellite orbits, the earth's magnetic field creates an upward emf in the cable. The cable can exchange electrons with the earth's ionosphere, so the emf drives an upward current in the cable. The resulting interaction with the earth's magnetic field produces a drag force tending to slow the satellite. As the satellite slows, its orbit decays until it enters the atmosphere and burns up.

established in the counterclockwise sense around the complete circuit. The moving rod has become a source of emf; within it, charge moves from lower to higher potential, and in the remainder of the circuit, charge moves from higher to lower potential. We call this emf a **motional emf;** and as with other emf's, we denote it as $\mathcal{E}$. When the velocity, magnetic field, and length are mutually perpendicular,

$$\mathcal{E} = vBL. \tag{21.7}$$

This is the same result that we obtained with Faraday's law in Section 21.3.

The emf associated with the moving rod in Figure 21.18 is analogous to that of a battery with its positive terminal at a and its negative terminal at b, although the origins of the two emf's are completely different. In each case, a non-electrostatic force acts on the charges in the device in the direction from b to a, and the emf is the work per unit charge done by this force when a charge moves from b to a in the device. When the device is connected to an external circuit, the direction of current is from b to a in the device and from a to b in the external circuit (in this case, the U-shaped conductor to the left of the rod).

If we express v in meters per second, B in teslas, and L in meters, $\mathcal{E}$ is in joules per coulomb, or volts. We invite you to verify this statement.

Conceptual Analysis 21.3

Forces on a square loop

The square loop in Figure 21.19 is pulled upward from between the poles of the magnet. How do the magnetic forces on the top and bottom wires compare?

A. The forces are directed downward on both wires and are equal in magnitude.
B. The forces are directed downward on both wires, but the force on the bottom wire is stronger.
C. The forces are directed upward on the top wire and downward on the bottom wire and are equal in magnitude.
D. The forces are directed upward on the top wire and downward on the bottom wire; the force on the bottom wire is stronger.

SOLUTION As the loop moves up from between the poles of the magnet, the flux through it decreases, inducing a current. Lenz's law tells us that a net downward force acts on the loop, opposing its upward motion. However, Lenz's law doesn't tell us about the relative magnitudes of the forces on the top and bottom wires. For that information, we must consider the magnetic forces acting on these two current-carrying wires.

First we determine the direction of the current. We could use Lenz's law, but instead let's use the right-hand rule for the force on a moving charge. The wires move upward and the magnetic

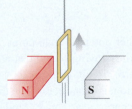

▲ **FIGURE 21.19**

field points to the right, so, by the right-hand rule, the free electrons in both wires experience a magnetic force out of the plane of the page. Because the bottom wire is in a stronger magnetic field, its electron forces are greater, so the overall conventional current is clockwise as viewed from the N pole of the magnet. The current on the top wire is therefore out of the plane of the page, and the force on this wire (by the right-hand rule) is upward. The current in the bottom wire is into the plane of the page, so the force on that wire is downward. Because the bottom wire is in a stronger magnetic field, the force exerted on it is stronger, so answer D is correct. As Lenz's law predicts, the net force on the loop is downward.

EXAMPLE 21.7   **Motional electromotive force**

Suppose the length L of the slide-wire rod in Figure 21.18b is 0.10 m, its speed v is 2.5 m/s, the total resistance of the loop is 0.030 Ω, and B is 0.60 T. Find the emf $\mathcal{E}$, the induced current, the force acting on the rod, and the mechanical power needed to keep the rod moving at constant speed.

SOLUTION

SET UP AND SOLVE To find the emf, we use Equation 21.7:

$$\mathcal{E} = vBL = (2.5 \text{ m/s})(0.60 \text{ T})(0.10 \text{ m}) = 0.15 \text{ V}.$$

From Ohm's law, the current in the loop is

$$I = \mathcal{E}/R = (0.15 \text{ V})/(0.030 \ \Omega) = 5.0 \text{ A}.$$

Lenz's law tells us that an induced force acts on the rod, opposite to the rod's direction of motion. The magnitude of this force is

$$F_{\text{induced}} = ILB = (5.0 \text{ A})(0.10 \text{ m})(0.60 \text{ T}) = 0.30 \text{ N}.$$

To keep the rod moving at constant speed, a force equal in magnitude and opposite in direction to $\vec{F}_{\text{induced}}$ must act on the rod. Therefore, the mechanical power P needed to keep the rod moving is

$$P = F_{\text{induced}}v = (0.30 \text{ N})(2.5 \text{ m/s}) = 0.75 \text{ W}.$$

REFLECT The expression for the emf is the same result we found in Example 21.4 from Faraday's law. The rate at which the induced emf delivers electrical energy to the circuit is

$$P = \mathcal{E}I = (0.15 \text{ V})(5.0 \text{ A}) = 0.75 \text{ W}.$$

This is equal to the mechanical power input, $F_{\text{induced}}v$, as we should expect. The system is converting mechanical energy (work) into electrical energy. Finally, the rate of *dissipation* of electrical energy in the circuit resistance is

$$P = I^2R = (5.0 \text{ A})^2(0.030 \ \Omega) = 0.75 \text{ W},$$

which is also to be expected.

Practice Problem: If $L = 0.15$ m, $v = 3.0$ m/s, the total resistance of the loop is 0.020 Ω, and $B = 0.5$ T, find the rate at which the induced emf delivers electrical energy to the circuit. *Answer:* $P = 2.5$ W.

21.6 Eddy Currents

In the examples of induction effects that we've studied, the induced currents have been confined to well-defined paths in conductors and other components, forming a *circuit*. However, many pieces of electrical equipment contain masses of metal moving in magnetic fields or located in changing magnetic fields. In situations like these, we can have induced currents that circulate throughout the volume of a conducting material. Because their flow patterns resemble swirling eddies in a river, we call these **eddy currents.**

As an example, consider a metallic disk rotating in a magnetic field perpendicular to the plane of the disk, but confined to a limited portion of the disk's area, as shown in Figure 21.20a. Sector *Ob* is moving across the field and has an emf induced in it. Sectors *Oa* and *Oc* are not in the field, but they provide conducting paths for charges displaced along *Ob* to return from *b* to *O*. The result is a circulation of eddy currents in the disk, somewhat as sketched in Figure 21.20b. The downward current in the neighborhood of sector *Ob* experiences a sideways magnetic-field force toward the right that *opposes* the rotation of the disk, as Lenz's law predicts. The return currents lie outside the field, so no such forces are exerted on them. The interaction between the eddy currents and the field causes a braking action on the disk and a conversion of electrical energy to heat because of the resistance of the material.

The shiny metal disk in the electric meter outside a house rotates as a result of eddy currents that are induced in it by magnetic fields caused by sinusoidally varying currents in a coil. Similar effects can be used to stop the rotation of a circular saw quickly when the power is turned off. Eddy current braking is also used on some electrically powered rapid-transit vehicles. Electromagnets mounted in the cars induce eddy currents in the rails; the resulting magnetic fields cause braking forces on the electromagnets and thus on the cars. Finally, the familiar metal detectors seen at security checkpoints in airports (Figure 21.21a) operate by detecting eddy currents induced in metallic objects. Similar devices (Figure 21.21b) are used to locate buried treasure such as bottle caps and lost pennies.

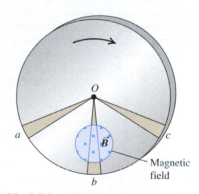

(a) Metal disk rotating through a magnetic field

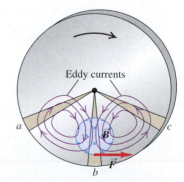

(b) Resulting eddy currents and braking force

▲ **FIGURE 21.20** The origin of eddy currents.

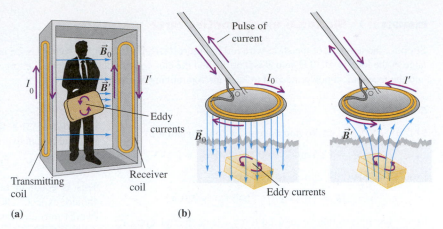

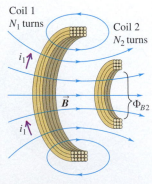

▲ **FIGURE 21.22** In mutual inductance, a portion of the magnetic flux set up by a current in coil 1 links with coil 2.

▲ **Application Plug-free power.** New technology may allow you to recharge one or more electronic devices simultaneously, while eliminating the need for specific plugs or holders for each device. A recharging pad plugs into a regular household outlet and generates a changing magnetic field just above its surface. When electronic devices with the appropriate circuitry are placed on this surface, the pad acts much like a transformer, inducing an emf in the devices. The resulting alternating current can be converted to direct current, which recharges the batteries of objects on the pad. A similar principle has been used for years in electric toothbrush rechargers, which operate without direct electrical contact between the toothbrush and the recharging stand, avoiding problems with moisture and possible short-circuiting of the charger.

▲ **FIGURE 21.21** (a) A metal detector at an airport security checkpoint detects eddy currents induced in conducting objects by an alternating magnetic field. (b) Portable metal detectors work on the same principle.

21.7 Mutual Inductance

How can a 12 volt car battery provide the thousands of volts needed to produce sparks across the gaps of the spark plugs in the engine? Electrical transmission lines often operate at 500,000 volts or more. Applied directly to your household wiring, this voltage would incinerate everything in sight. How can it be reduced to the relatively tame 120 or 240 volts required by familiar electrical appliances?

The solutions to both of these problems, and to many others concerned with varying currents in circuits, involve the *induction* effects we've studied in this chapter. A changing current in a coil induces an emf in an adjacent coil. The coupling between the coils is described by their *mutual inductance*. This interaction is the operating principle of a *transformer*, which we'll study in Section 21.9.

Figure 21.22 is a cross-sectional view of two coils of wire. A current i_1 in coil 1 sets up a magnetic field, as indicated by the blue lines, and some of these field lines pass through coil 2. We denote the magnetic flux through each turn of coil 2, caused by the current i_1 in coil 1, as Φ_{B2}. (We're using the lowercase letter i, with a subscript, for time-varying currents.) The magnetic field is proportional to i_1, so Φ_{B2} is also proportional to i_1. When i_1 changes, Φ_{B2} changes, inducing an emf $\mathcal{E}_2$ in coil 2 with magnitude

$$\mathcal{E}_2 = N_2 \left| \frac{\Delta \Phi_{B2}}{\Delta t} \right|. \tag{21.8}$$

We could represent the proportionality of Φ_{B2} and i_1 in the form $\Phi_{B2} = (\text{constant}) \, i_1$, but instead it is more convenient to include the number of turns, N_2, in the relation. Introducing a proportionality constant M_{21}, we write

$$N_2 |\Phi_{B2}| = M_{21} |i_1|. \tag{21.9}$$

From this relation, it follows that

$$N_2 \left| \frac{\Delta \Phi_{B2}}{\Delta t} \right| = M_{21} \left| \frac{\Delta i_1}{\Delta t} \right|,$$

and we can rewrite Equation 21.8 as

$$\mathcal{E}_2 = M_{21} \left| \frac{\Delta i_1}{\Delta t} \right|. \tag{21.10}$$

We can repeat this discussion for the opposite case, in which a changing current i_2 in coil 2 causes a changing flux Φ_{B1}, and an emf $\mathcal{E}_1$, in coil 1. The results, corresponding to Equations 21.8 and 21.10, are

$$\mathcal{E}_1 = N_1 \left| \frac{\Delta \Phi_{B1}}{\Delta t} \right| \quad \text{and} \quad \mathcal{E}_1 = M_{12} \left| \frac{\Delta i_2}{\Delta t} \right|.$$

We might expect that the corresponding constant M_{12} would be different from M_{21}, because, in general, the two coils are not identical and the flux through them is not the same. It turns out, however, that M_{12} is *always* equal to M_{21}, even when the two coils are not symmetric. We call this common value the **mutual inductance** M; it characterizes completely the induced-emf interaction of two coils.

Mutual inductance

The mutual inductance M of two coils is given by

$$M = M_{21} = M_{12} = \left| \frac{N_2 \Phi_{B2}}{i_1} \right| = \left| \frac{N_1 \Phi_{B1}}{i_2} \right|. \qquad (21.11)$$

From the preceding analysis, we can also write

$$\mathcal{E}_2 = M \left| \frac{\Delta i_1}{\Delta t} \right| \quad \text{and} \quad \mathcal{E}_1 = M \left| \frac{\Delta i_2}{\Delta t} \right|. \qquad (21.12)$$

Unit: The SI unit of mutual inductance is called the **henry** (1 H), in honor of Joseph Henry (1797–1878), one of the discoverers of electromagnetic induction. From Equation 21.11, one henry is equal to *one weber per ampere*. Other equivalent units, obtained by reference to Equation 21.10, are *one volt-second per ampere* and *one ohm-second*:

$$1 \text{ H} = 1 \text{ Wb/A} = 1 \text{ V} \cdot \text{s/A} = 1 \, \Omega \cdot \text{s}.$$

If the coils are in vacuum, M depends only on their geometry. If a magnetic material is present, M also depends on the magnetic properties of the material.

EXAMPLE 21.8 The Tesla coil

A Tesla coil is a type of high-voltage generator; you may have seen one in a science museum. In one form of Tesla coil, shown in Figure 21.23, a long solenoid with length l and cross-sectional area A is closely wound with N_1 turns of wire. A coil with N_2 turns surrounds it at its center. Find the mutual inductance of the coils.

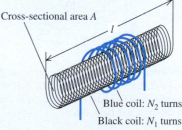

▶ **FIGURE 21.23** One form of Tesla coil, a long solenoid with cross-section area A and N_1 turns, surrounded at its center by a small coil with N_2 turns.

Cross-sectional area A

Blue coil: N_2 turns
Black coil: N_1 turns

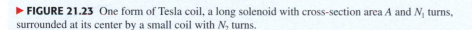

SOLUTION

SET UP A current i_1 in the solenoid sets up a magnetic field $\vec{B}_1$ at its center. According to Equation 20.14 (Section 20.9), the magnitude B of the magnetic field at the center of a solenoid is given by $B = \mu_0 nI$, where $n = N/l$ is the number of turns per unit length. Using this equation, we find that

$$B_1 = \mu_0 ni = \frac{\mu_0 N_1 i_1}{l}.$$

The flux through each turn at the center of coil 1, Φ_{B1}, equals $B_1 A$. All of this flux also passes through coil 2, so $\Phi_{B2} = \Phi_{B1} = B_1 A$.

SOLVE From Equation 21.11, the mutual inductance M is

$$M = \left| \frac{N_2 \Phi_{B2}}{i_1} \right| = \left| \frac{N_2}{i_1} B_1 A \right| = \left| \frac{N_2}{i_1} \frac{\mu_0 N_1 i_1}{l} A \right| = \frac{\mu_0 A N_1 N_2}{l}.$$

REFLECT Here's a numerical example to give you an idea of magnitudes. Suppose $l = 0.50$ m, $A = 10 \text{ cm}^2 = 1.0 \times 10^{-3} \text{ m}^2$, $N_1 = 1000$ turns, and $N_2 = 10$ turns. Then

$$M = \frac{[4\pi \times 10^{-7} \text{ Wb/(A} \cdot \text{m})](1.0 \times 10^{-3} \text{ m}^2)(1000)(10)}{0.50 \text{ m}}$$

$$= 25 \times 10^{-6} \text{ Wb/A} = 25 \times 10^{-6} \text{ H} = 25 \, \mu\text{H}.$$

EXAMPLE 21.9 **The average magnetic flux through a Tesla coil**

Suppose the current i_2 in the smaller coil in Example 21.8 is given by $i_2 = (2.0 \times 10^6 \text{ A/s})t$. At time $t = 3.0 \ \mu s$, what is the average magnetic flux through each turn of the solenoid caused by the current in the smaller coil? What is the induced emf in the solenoid?

SOLUTION

SET UP AND SOLVE At time $t = 3.0 \ \mu s$, the current in coil 2 is

$$i_2 = (2.0 \times 10^6 \text{ A/s})(3.0 \times 10^{-6} \text{ s}) = 6.0 \text{ A}.$$

To find the flux in the solenoid, we use Equation 21.9, with the roles of coils 1 and 2 interchanged, so that $N_1|\Phi_{B1}| = M|i_2|$. Solving for the average magnetic flux Φ_{B1}, we obtain

$$|(\Phi_{B1})_{av}| = \frac{M|i_2|}{N_1} = \frac{(25 \times 10^{-6} \text{ H})(6.0 \text{ A})}{1000} = 1.5 \times 10^{-7} \text{ Wb}.$$

This is an average value; the flux will vary considerably from the center to the ends.

The induced emf $\mathcal{E}_1$ is given by Equation 21.12, with the change in current per unit time, $\Delta i_2/\Delta t$, equal to $2.0 \times 10^6 \text{ A/s}$:

$$\mathcal{E}_1 = \left| M \frac{\Delta i_2}{\Delta t} \right| = (25 \times 10^{-6} \text{ H})(2.0 \times 10^6 \text{ A/s}) = 50 \text{ V}.$$

REFLECT In an operating Tesla coil, $\Delta i_2/\Delta t$ would be alternating much more rapidly, and its magnitude would be much larger than in this example.

Practice Problem: For the given Tesla coil, how many turns (N_1) should coil 1 have in order to get an induced emf magnitude of 650 V? *Answer:* $N_1 = 13,000$ turns.

21.8 Self-Inductance

In our discussion of mutual inductance, we assumed that one circuit acted as the source of a magnetic field and that the emf under consideration was induced in a separate, independent circuit when some of the magnetic flux created by the first circuit passed through the second. However, when a current is present in *any* circuit, this current sets up a magnetic field that links with *the same* circuit and changes when the current changes. Any circuit that carries a varying current has an induced emf in it resulting from the variation in *its own* magnetic field. Such an emf is called a **self-induced emf.**

As an example, consider a coil with N turns of wire, carrying a current i (Figure 21.24). As a result of this current, a magnetic flux Φ_B passes through each turn. In analogy to Equation 21.11, we define the **inductance** L of the circuit (sometimes called **self-inductance**):

$$L = \left| \frac{N\Phi_B}{i} \right|, \quad \text{or} \quad N|\Phi_B| = L|i|. \tag{21.13}$$

If Φ_B and i change with time, then

$$N \left| \frac{\Delta \Phi_B}{\Delta t} \right| = L \left| \frac{\Delta i}{\Delta t} \right|.$$

From Faraday's law, Equation 21.4, the magnitude $\mathcal{E}$ of the self-induced emf is $\mathcal{E} = N|\Delta \Phi_B/\Delta t|$, so we can also state the definition of L as follows:

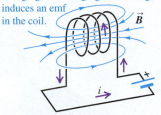

Self-inductance: If the current i in the coil is changing, the changing flux through the coil induces an emf in the coil.

▲ **FIGURE 21.24** A changing current in a coil produces a changing magnetic field that in turn induces an emf in the coil—the phenomenon of self-inductance.

Definition of self-inductance

The self-inductance L of a circuit is the magnitude of the self-induced emf $\mathcal{E}$ per unit rate of change of current, so that:

$$\mathcal{E} = L \left| \frac{\Delta i}{\Delta t} \right|. \tag{21.14}$$

From the definition, the units of self-inductance are the same as those of mutual inductance; the SI unit of self-inductance is *one henry.*

A circuit, or part of a circuit, that is designed to have a particular inductance is called an **inductor,** or a *choke.* Like resistors and capacitors, inductors are among the indispensable circuit elements of modern electronics. In later sections, we'll explore the circuit behavior of inductors. The usual circuit symbol for an inductor is

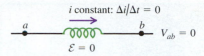

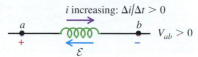

(a) When the current is constant, there is no self-induced emf, so the voltage across the inductor is zero.

(b) and (c) show that a changing current induces an emf and hence a voltage across the inductor.

We can find the direction of the self-induced emf from Lenz's law. The cause of the induced emf is the *changing current* in the conductor, and the emf always acts to oppose this change. Figure 21.25 shows several cases. In Figure 21.25a, the current is constant, Δi is always zero, and $V_{ab} = 0$. In Figure 21.25b, the current is increasing and $\Delta i/\Delta t$ is positive. According to Lenz's law, the induced emf must oppose the increasing current. The emf therefore must be in the sense from b to a; that is, a becomes the terminal with higher potential, and V_{ab} is *positive,* as shown in the figure. The emf opposes the increase in current caused by the external circuit. The direction of the emf is analogous to that of a battery with a as its + terminal.

In Figure 21.25c, the situation is the opposite. The current is decreasing and $\Delta i/\Delta t$ is negative. The induced emf $\mathcal{E}$ opposes this decrease, and V_{ab} is negative. In both cases, the induced emf opposes, not the current itself, but the *rate of change* $\Delta i/\Delta t$ of the current. Thus, the circuit behavior of an inductor is quite different from that of a resistor.

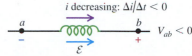

(b) If the current is *increasing,* then, by Lenz's law, the emf points opposite to i.

(c) If the current is *decreasing,* then, by Lenz's law, the emf points in the same direction as i.

▲ **FIGURE 21.25** Relation of current and self-induced emf in an inductor with negligible resistance. The voltage V_{ab} depends on the time rate of change of current.

EXAMPLE 21.10 Inductance of a toroidal solenoid

An air-core toroidal solenoid with cross-sectional area A and mean radius r is closely wound with N turns of wire (Figure 21.26). Determine its self-inductance L. In calculating the flux, assume that B is uniform across a cross section; neglect the variation of B with distance from the toroidal axis.

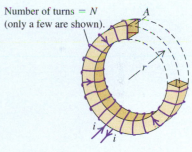

▲ **FIGURE 21.26**

SOLUTION

SET UP From Equation 21.13, which defines inductance,

$$L = \left| \frac{N\Phi_B}{i} \right|.$$

SOLVE To find Φ_B, we first find the field magnitude B. From Equation 20.15, the magnetic field magnitude B at a distance r from the toroidal axis is $B = \mu_0 Ni/2\pi r$. If we assume that the field has this magnitude over the entire cross-sectional area A, then the total flux Φ_B through the cross section is

$$\Phi_B = BA = \left| \frac{\mu_0 NiA}{2\pi r} \right|.$$

All the flux links with each turn, and the self-inductance L is

$$L = \left| \frac{N\Phi_B}{i} \right| = \frac{\mu_0 N^2 A}{2\pi r}.$$

REFLECT As an example, suppose $N = 200$ turns, $A = 5.0 \text{ cm}^2 = 5.0 \times 10^{-4} \text{ m}^2$, and $r = 0.10 \text{ m}$; then

$$L = \frac{[4\pi \times 10^{-7} \text{ Wb}/(\text{A} \cdot \text{m})](200)^2(5.0 \times 10^{-4} \text{ m}^2)}{2\pi(0.10 \text{ m})}$$

$$= 40 \times 10^{-6} \text{ H} = 40 \text{ }\mu\text{H}.$$

This inductor with 200 turns is somewhat bigger than an ordinary doughnut.

Practice Problem: For this toroidal solenoid, how many turns would be required for an inductance of $360 \text{ }\mu\text{H}$? *Answer:* 600 turns.

◀ **Application But I thought it was green!** We're all familiar with the "loops" at intersections that allow lights to respond to the presence of cars (or bicycles). Similar loops within intersections allow cameras to automatically photograph cars that run red lights. A current through the loop generates a magnetic field. The presence of a car above the loop dramatically increases the inductance of the circuit. If this large increase in inductance happens after the light has turned red, it triggers a camera that snaps a photo of the offending vehicle.

The inductance of a circuit depends on its size, shape, and number of turns. For N turns close together, it is always proportional to N^2. The inductance also depends on the magnetic properties of the material enclosed by the circuit. In the preceding examples, we assumed that the conductor was surrounded by vacuum. If the flux is concentrated in a region containing a magnetic material with permeability μ, then, in the expression for B, we must replace μ_0 (the permeability of vacuum) by $\mu = K_m\mu_0$, as we discussed in Section 20.11. If the material is *ferromagnetic,* this difference is of crucial importance: An inductor wound on a soft iron core having $K_m = 5000$ has an inductance approximately 5000 times as great as the same coil with an air core. Iron-core and ferrite-core inductors are widely used in a variety of electric-power and electronic applications.

21.9 Transformers

One of the great advantages of alternating current (ac, usually varying sinusoidally with time) over direct current (dc, not varying with time), for electric-power distribution is that it is much easier to step voltage levels up and down with ac than with dc. For long-distance power transmission, it is desirable to use as high a voltage and as small a current as possible; this approach reduces I^2R losses in the transmission lines, so smaller wires can be used for a given power level, enabling us to save on material costs. Present-day transmission lines routinely operate at voltages on the order of 500 kV. However, safety considerations and insulation requirements dictate relatively low voltages in generating equipment and in household and industrial power distribution. The standard voltage for household wiring is 120 V in the United States and Canada and 240 V in most of Western Europe. The necessary voltage conversion is accomplished by the use of **transformers.**

In the discussion that follows, the symbols for voltage and current usually represent amplitudes—that is, *maximum* magnitudes of quantities that vary sinusoidally with time. In fact, alternating voltages and currents are usually described in terms of rms (root-mean-square) values, which, for sinusoidally varying quantities, are less than the maximum values by a factor of $1/\sqrt{2}$. For example, the standard 120 V household voltage has a *maximum* (peak) value of $(120\text{ V})\sqrt{2} = 170$ V. (We'll discuss this distinction in greater detail in Chapter 22.)

A transformer consists of two coils (usually called *windings*), electrically insulated from each other but wound on the same core. In Section 21.7, we discussed the mutual inductance of such a system. The winding to which power is supplied is called the **primary;** the winding from which power is delivered is called the **secondary.** Transformers used in power-distribution systems have soft iron cores. The circuit symbol for an iron-core transformer is

Here's how a transformer works: An alternating current in either winding sets up an alternating flux in the core, and according to Faraday's law, this induces an

emf in each winding. Energy is transferred from the primary winding to the secondary winding via the core flux and its associated induced emf's.

An idealized transformer is shown in Figure 21.27. We assume that all the flux is confined to the iron core, so at any instant, the magnetic flux Φ_B is the same in the primary and secondary coils. We also neglect the resistance of the windings. The primary winding has N_1 turns, and the secondary winding has N_2 turns. When the magnetic flux changes because of changing currents in the two coils, the resulting induced emf's are

$$\mathcal{E}_1 = N_1 \left| \frac{\Delta \Phi_B}{\Delta t} \right| \quad \text{and} \quad \mathcal{E}_2 = N_2 \left| \frac{\Delta \Phi_B}{\Delta t} \right|.$$

Because the same flux links both primary and secondary, these expressions show that the induced emf *per turn* is the same in each. The ratio of the secondary emf $\mathcal{E}_2$ to the primary emf $\mathcal{E}_1$ is therefore equal to the ratio of secondary to primary turns, often called the *turns ratio:*

$$\frac{\mathcal{E}_2}{\mathcal{E}_1} = \frac{N_2}{N_1}.$$

If the windings have zero resistance, the induced emf's $\mathcal{E}_1$ and $\mathcal{E}_2$ are respectively equal to the corresponding terminal voltages V_1 and V_2, and we find the following:

Relation of voltage to winding turns for a transformer

For an ideal transformer (with zero resistance),

$$\frac{V_2}{V_1} = \frac{N_2}{N_1}. \tag{21.15}$$

By choosing the appropriate turns ratio N_2/N_1, we may obtain any desired secondary voltage from a given primary voltage. If $V_2 > V_1$, we have a *step-up* transformer; if $V_2 < V_1$, we have a *step-down* transformer. The V's can be either both amplitudes or both rms values.

If the secondary circuit is completed by a resistance R, then $I_2 = V_2/R$. From energy considerations, the power delivered to the primary equals that taken out of the secondary, so

$$V_1 I_1 = V_2 I_2. \tag{21.16}$$

We can combine Equations 21.15 and 21.16 with the relation $I_2 = V_2/R$ to eliminate V_2 and I_2:

$$\frac{V_1}{I_1} = \frac{R}{(N_2/N_1)^2}. \tag{21.17}$$

This equation shows that when the secondary circuit is completed through a resistance R, the result is the same as if the *source* had been connected directly to

The induced emf *per turn* is the same in both coils, so we adjust the ratio of terminal voltages by adjusting the ratio of turns:

$$\frac{V_2}{V_1} = \frac{N_2}{N_1}$$

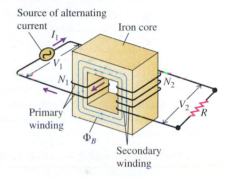

▲ **FIGURE 21.27** Schematic diagram of an idealized step-up transformer.

a resistance equal to R divided by the square of the turns ratio, $(N_2/N_1)^2$. In other words, the transformer "transforms" not only voltages and currents, but resistances as well.

EXAMPLE 21.11 A high-voltage coffee maker

A friend brings back from Europe a device that she claims to be the world's greatest coffee maker. Unfortunately, it was designed to operate from a 240 V line, standard in Europe. At this rms voltage, the coffee draws 960 W of power. (a) If your friend wants to operate the coffee maker in the United States, where the rms line voltage is 120 V, what does she need to do? (b) What current will the coffee maker draw from the 120 V line? (c) What is its resistance? (These voltages and currents are rms values, to be discussed in Chapter 22, so the power quantities are given by Equation 21.16.)

SOLUTION

SET UP Our friend needs to step up the 120 V that comes into her home to the 240 V required to operate the coffee maker, so she has to use a step-up transformer. From Equation 21.15, our friend needs a transformer with a turns ratio equal to the voltage ratio.

SOLVE Part (a): The input (primary) voltage is $V_1 = 120$ V, and we need an output (secondary) voltage $V_2 = 240$ V. From Equation 21.15, the required turns ratio is

$$\frac{N_2}{N_1} = \frac{V_2}{V_1} = \frac{240\ \text{V}}{120\ \text{V}} = 2.$$

Part (b): For a power of 960 W, the current on the 240 V side is 4.0 A. From Equation 21.16, the primary current (from the 120 V source) is 8.0 A. The *power* is the same on the primary and secondary sides:

$$P = (120\ \text{V})(8.0\ \text{A}) = (240\ \text{V})(4.0\ \text{A}) = 960\ \text{W}.$$

Part (c): For a current of 4.0 A on the 240 V side, the resistance R of the coffee maker must be

$$R = \frac{V_2}{I_2} = \frac{240\ \text{V}}{4.0\ \text{A}} = 60\ \Omega.$$

REFLECT The ratio of primary voltage to current is $R' = \dfrac{V_1}{I_1} = \dfrac{120\ \text{V}}{8.0\ \text{A}} = 15\ \Omega$. The primary voltage–current relation is the same as though we had connected a 15 Ω resistor directly to the 120 V source.

Practice Problem: Suppose we want to use the coffee maker on an airplane on which the voltage is 480 V. Find the turns ratio of the required transformer and the ratio R' of primary voltage to current. *Answer:* $N_2/N_1 = 1/2$, $R' = 240\ \Omega$.

Real transformers always have some energy losses. The windings have some resistance, leading to i^2R losses, although superconducting transformers may appear on the horizon in the next few years. There are also energy losses through eddy currents (Section 21.6) in the core, as shown in Figure 21.28. The alternating current in the primary winding sets up an alternating flux within the core and a current in the secondary winding. However, the iron core is also a conductor, and any section, such as AA, can be pictured as several conducting circuits, one within the other (Figure 21.28b). The flux through each of these

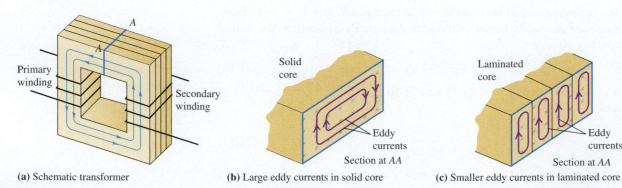

(a) Schematic transformer   (b) Large eddy currents in solid core   (c) Smaller eddy currents in laminated core

▲ **FIGURE 21.28** Eddy currents in a transformer can be minimized by use of a laminated core.

circuits is continually changing, so eddy currents circulate in the entire volume of the core, wasting energy through I^2R heating (due to the resistance of the core material) and setting up unwanted opposing flux.

In actual transformers, the eddy currents are greatly reduced by the use of a *laminated* core—that is, a core built up of thin sheets, or laminae. The large electrical surface resistance of each lamina, due to either a natural coating of oxide or an insulating varnish, effectively confines the eddy currents to individual laminae (Figure 21.28c). The possible eddy-current paths are narrower, the induced emf in each path is smaller, and the eddy currents are greatly reduced. Transformer efficiencies are usually well over 90%; in large installations, they may reach 99%.

In small transformers in which it is important to keep eddy-current losses to an absolute minimum, the cores are sometimes made of *ferrites*, which are complex oxides of iron and other metals. These materials are ferromagnetic, but their resistivity is much greater than that of pure iron.

21.10 Magnetic Field Energy

Establishing a current in an inductor requires an input of energy, so an inductor carrying a current has energy stored in it, associated with its magnetic field. Let's see how this comes about. A changing current in an inductor causes an emf $\mathcal{E}$ between the inductor's terminals. The source that supplies the current must maintain a corresponding potential difference V_{ab} between its terminals while the current is changing; therefore, the source must supply energy to the inductor. We can calculate the total energy input U needed to establish a final current I in an inductor with inductance L if the initial current is zero.

Let the current at some instant be i and its rate of change be $\Delta i/\Delta t$. Then the terminal voltage at that instant is $\mathcal{E} = V_{ab} = L\,\Delta i/\Delta t$, and the average power P supplied by the current source during the small time interval Δt is

$$P = V_{ab}i = Li\frac{\Delta i}{\Delta t}. \tag{21.18}$$

The energy ΔU supplied to the inductor during this time interval is approximately $\Delta U = P\,\Delta t$, so $\Delta U = Li\,\Delta i$.

To find the total energy U supplied while the current increases from zero to a final value I, we note that the *average* value of Li during the entire increase is equal to half of the final value—that is, to $LI/2$. The product of this quantity and the *total* increase in current I gives the total energy U supplied, and we find the following relationship:

Energy stored in an inductor

The energy stored by an inductor with inductance L carrying a current I is

$$U = \frac{1}{2}LI^2. \tag{21.19}$$

This result can also be derived with the use of integral calculus.

After the current has reached its final steady value I, $\Delta i/\Delta t = 0$ and the power input is zero. We can think of the energy U as analogous to a *kinetic energy* associated with the current. This energy is zero when there is no current; when the current is I, the energy is $\frac{1}{2}LI^2$.

When the current decreases from I to zero, the inductor acts as a source, supplying a total amount of energy $\frac{1}{2}LI^2$ to the external circuit. If we interrupt the circuit suddenly by opening a switch, the current decreases very rapidly, the induced

▲ **BIO Application SQUID tales.** The transient electrical signals from excitable animal tissues generate tiny magnetic fields, generally below the detection limit of the usual devices for measuring magnetic fields. However, a specialized instrument called SQUID (for Superconducting Quantum Interference Device) uses a detector coil that is superconducting; that is, it has zero electrical resistance. With careful shielding from stray magnetic fields and other tricks, magnetic fields smaller than 1 fT can be measured. These devices have been used to detect the fields generated by single nerve impulses; the contraction of abdominal smooth muscle in intact, living animals; and the fetal heartbeat. Arrays of SQUID detectors as shown in the figure can be used to pinpoint the electrical storm in the brain that accompanies epilepsy. An important difference between SQUID detection and magnetic resonance (MRI) imaging of brain tissue is that SQUID directly detects electrical activity by way of the tiny magnetic fields, while MRI detects increased blood flow to active areas of the brain—an indirect (and slower) measure of nervous activity.

emf is correspondingly very large, and the energy may be dissipated in an arc across the switch contacts. This large emf is the electrical analog of the large force exerted on a car that runs into a concrete bridge abutment and stops suddenly.

The energy in an inductor is actually stored in the magnetic field within the coil, just as the energy of a capacitor is stored in the electric field between its plates. We can develop a relation for magnetic-field energy analogous to the one we obtained for electric-field energy in Section 18.7 (Equation 18.20). We'll concentrate on one simple case: the ideal toroidal solenoid. This system has the advantage that its magnetic field is confined completely to a finite region of space within its core. As in Example 21.10, we assume that the cross-sectional area A is small enough that we can consider the magnetic field to be uniform over the area. The volume V enclosed by the toroid is approximately equal to the circumference $2\pi r$ multiplied by the area A, or $V = 2\pi rA$. From Example 21.10, the self-inductance of the toroidal solenoid (with vacuum in the core) is

$$L = \frac{\mu_0 N^2 A}{2\pi r},$$

and the stored energy U when the current is I is

$$U = \frac{1}{2}LI^2 = \frac{1}{2}\frac{\mu_0 N^2 A}{2\pi r}I^2. \tag{21.20}$$

The magnetic field, and therefore the energy U, are localized in the volume $V = 2\pi rA$ enclosed by the windings. The energy *per unit volume*, or **energy density,** $u = U/V$, is then

$$u = \frac{U}{2\pi rA} = \frac{1}{2}\mu_0 \frac{N^2 I^2}{(2\pi r)^2}.$$

We can express this energy in terms of the magnetic field B inside the toroid. From Equation 20.15, B is given by

$$B = \frac{\mu_0 NI}{2\pi r}, \quad \text{and} \quad \frac{N^2 I^2}{(2\pi r)^2} = \frac{B^2}{\mu_0^2}.$$

When we substitute the last expression into the preceding equation for u, we finally find that

$$u = \frac{B^2}{2\mu_0}. \tag{21.21}$$

This is the magnetic analog of the energy per unit volume in the *electric* field of an air capacitor, $u = \frac{1}{2}\epsilon_0 E^2$, which we derived in Section 18.7.

EXAMPLE 21.12 Storing energy in an inductor

The electric-power industry would like to find efficient ways to store surplus energy generated during low-demand hours to help meet customer requirements during high-demand hours. Perhaps superconducting coils can be used. What inductance would be needed to store 1.00 kWh of energy in a coil carrying a current of 200 A?

SOLUTION

SET UP AND SOLVE The problem tells us that $U = 1.00$ kWh $= (1.00 \times 10^3 \text{ W})(3600 \text{ s}) = 3.6 \times 10^6$ J and $I = 200$ A. Solving Equation 21.20 for L, we obtain

$$L = \frac{2U}{I^2} = \frac{2(3.60 \times 10^6 \text{ J})}{(200 \text{ A})^2} = 180 \text{ H}.$$

REFLECT A 180 H inductor using conventional wire heavy enough to carry 200 A would be very large (the size of a room), but a superconducting inductor could be much smaller.

Practice Problem: How much energy, in kWh, could be stored in a 200 H inductor carrying a current of 350 A? *Answer:* $U = 3.40$ kWh.

21.11 The *R–L* Circuit

An inductor is primarily a circuit device. A circuit containing a resistor and an inductor is called an *R–L* circuit. Let's look at the behavior of a simple *R–L* circuit. One thing is clear already: We aren't going to see any sudden changes in the current through an inductor. Equation 21.14 shows that the greater the rate of change of current, $\Delta i / \Delta t$, the greater must be the potential difference between the inductor terminals. This equation, together with Kirchhoff's rules, gives us the principles that we need to analyze circuits containing inductors.

Mastering PHYSICS

ActivPhysics 14.1: The *R–L* Circuit

PROBLEM-SOLVING STRATEGY 21.2   **Inductors in circuits**   (MP)

SET UP

1. When an inductor is used as a *circuit* device, all the voltages, currents, and capacitor charges are, in general, functions of time, not constants as they have been in most of our previous circuit analysis. But Kirchhoff's rules, which we studied in Chapter 19, are still valid. When the voltages and currents vary with time, Kirchhoff's rules hold at each instant of time.

2. As in all circuit analysis, getting the signs right is sometimes more challenging than understanding the principles. We suggest that you review the strategy in Section 19.2. In addition, study carefully the sign rule described with Equation 21.14 and Figure 21.25. In Kirchhoff's loop rule, when we go through an inductor in the *same* direction as the assumed current, we encounter a voltage *drop* equal to $L \, \Delta i / \Delta t$, so the corresponding term in the loop equation is $-L \, \Delta i / \Delta t$.

Current Growth in an *R–L* Circuit

We can learn several basic things about inductor behavior by analyzing the circuit of Figure 21.29. The resistor R may be a separate circuit element, or it may be the resistance of the inductor windings; every real-life inductor has some resistance, unless it is made of superconducting wire. By closing the switch, we can connect the *R–L* combination to a source having a constant emf $\mathcal{E}$. (We assume that the source has zero internal resistance, so the terminal voltage equals the emf.) Suppose the switch is initially open, and then, at some initial time $t = 0$, we close it. As we have mentioned, the current cannot change suddenly from zero to some final value, because of the infinite induced emf that would result. Instead, it begins to grow at a definite rate that depends on the value of L in the circuit.

Let i be the current at some time t after the switch is closed, and let $\Delta i / \Delta t$ be the rate of change of i at that time. Then the potential difference v_{bc} across the inductor at that time is

$$v_{bc} = L \frac{\Delta i}{\Delta t},$$

and the potential difference v_{ab} across the resistor is

$$v_{ab} = iR.$$

We apply Kirchhoff's voltage rule, starting at the negative terminal of the source and proceeding counterclockwise around the loop:

$$\mathcal{E} - iR - L \frac{\Delta i}{\Delta t} = 0. \tag{21.22}$$

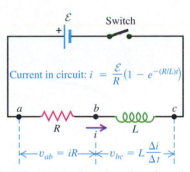

The voltage across the resistor depends on the current; the voltage across the inductor depends on the *rate of change of* the current.

▲ **FIGURE 21.29** An *R–L* circuit.

Solving this equation for $\Delta i/\Delta t$, we find that the rate of increase of current is

$$\frac{\Delta i}{\Delta t} = \frac{\mathcal{E} - iR}{L} = \frac{\mathcal{E}}{L} - \frac{R}{L}i. \tag{21.23}$$

At the instant the switch is first closed, $i = 0$ and the potential drop across R is zero. The initial rate of change of current is

$$\left(\frac{\Delta i}{\Delta t}\right)_{\text{initial}} = \frac{\mathcal{E}}{L}. \tag{21.24}$$

The greater the inductance L, the more slowly the current increases.

As the current increases, the term $(R/L)i$ in Equation 21.23 also increases, and the *rate* of increase of current becomes smaller and smaller. When the current reaches its final *steady-state* value I, its rate of increase is zero. Then Equation 21.23 becomes

$$\frac{\Delta i}{\Delta t} = 0 = \frac{\mathcal{E}}{L} - \frac{R}{L}I,$$

and the final current I is

$$I = \frac{\mathcal{E}}{R}.$$

That is, the *final* current I doesn't depend on the inductance L; it is the same as it would be if the resistance R alone were connected to the source with emf $\mathcal{E}$.

We can use Equation 21.23 to derive an expression for the current i as a function of time. The derivation requires calculus, and we'll omit the details. The final result is

$$i = \frac{\mathcal{E}}{R}\left(1 - e^{-(R/L)t}\right). \tag{21.25}$$

Figure 21.30 is a graph of Equation 21.25, showing the variation of current with time. At time $t = 0$, $i = 0$, and the initial slope of the curve is $\Delta i/\Delta t = \mathcal{E}/L$. As $t \to \infty$, $i \to \mathcal{E}/R$ and the slope $\Delta i/\Delta t$ approaches zero, as we predicted.

As Figure 21.30 shows, the instantaneous current i first rises rapidly, then increases more slowly, and approaches the final value $I = \mathcal{E}/R$ asymptotically. At a time τ equal to L/R, the current has risen to $1 - (1/e)$, or about 0.63 of its final value. The quantity $\tau = L/R$ is called the **time constant** for the circuit:

$$\tau = \frac{L}{R}. \tag{21.26}$$

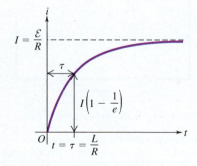

▲ FIGURE 21.30 Graph of i versus t for the growth of current in an R–L circuit. The final current is $I = \mathcal{E}/R$; after one time constant, the current is $1 - (1/e)$ of this value.

In a time equal to 2τ, the current reaches 86% of its final value; in 5τ, 99.3%; and in 10τ, 99.995%.

For a given value of R, the time constant τ is greater for greater values of L. When L is small, the current rises rapidly to its final value; when L is large, it rises more slowly. For example, if $R = 100 \ \Omega$ and $L = 10$ H, then

$$\tau = \frac{L}{R} = \frac{10 \text{ H}}{100 \ \Omega} = 0.10 \text{ s},$$

and the current increases to about 0.632 of its final value in 0.10 s. But if $R = 100 \ \Omega$ and $L = 0.010$ H, then $\tau = 1.0 \times 10^{-4}$ s $= 0.10$ ms, and the rise is much more rapid.

This entire discussion should look familiar; the situation is similar to that of a charging and discharging capacitor, which we analyzed in Section 19.8. It would be a good idea to compare that section with our discussion of the L–R circuit here.

EXAMPLE 21.13 **Current in an *R–L* circuit**

In Figure 21.31, suppose $\mathcal{E} = 120$ V, $R = 200\ \Omega$, and $L = 10.0$ H. Find (a) the final current after the switch has been closed a long time, (b) the initial time rate of change of the current, immediately after the switch has been closed, and (c) the time at which the current has reached 63.2% of its final value.

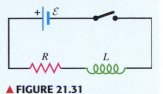

▲ **FIGURE 21.31**

SOLUTION

SET UP AND SOLVE **Part (a):** The final current I, which does not depend on the inductance L, is

$$I = \frac{\mathcal{E}}{R} = \frac{120\ \text{V}}{200\ \Omega} = 0.60\ \text{A.}$$

Part (b): The current can't change discontinuously, so immediately after the switch is closed, $i = 0$ and there is no voltage drop across the resistor. At this time, Kirchhoff's loop rule (Equation 21.22) gives

$$\mathcal{E} - L\left(\frac{\Delta i}{\Delta t}\right)_{\text{initial}} = 0, \quad \text{and} \quad \left(\frac{\Delta i}{\Delta t}\right)_{\text{initial}} = \frac{\mathcal{E}}{L} = \frac{120\ \text{V}}{10.0\ \text{H}} = 12.0\ \text{A/s.}$$

Part (c): From Equation 21.25, the current reaches 0.632 $(=1 - 1/e)$ of its final value when

$$1 - \frac{1}{e} = 1 - e^{-(R/L)t}, \quad -1 = -\frac{R}{L}t, \quad \text{and}$$

$$t = \frac{L}{R} = \frac{10.0\ \text{H}}{200\ \Omega} = 0.050\ \text{s.}$$

Note that this value of t is equal to the time constant τ defined by Equation 21.26.

REFLECT The final current is independent of the inductance because the current is no longer changing with time and there is no induced emf in the inductor. The initial current is independent of the resistance because the initial current is zero and there is no voltage drop across the resistor. The time constant is directly proportional to the inductance: The greater the inductance, the more slowly the current increases.

Practice Problem: Suppose we now want the final current to be 0.300 A, with $\mathcal{E}$ and τ keeping their previous values. What values of R and L should be used? *Answer:* 400 Ω, 20 H.

Energy considerations offer us additional insight into the behavior of an *R–L* circuit. The instantaneous rate at which the source delivers energy to the circuit is $P = \mathcal{E}i$. The instantaneous rate at which energy is dissipated in the resistor is i^2R, and the rate at which energy is stored in the inductor is $iv_{bc} = Li\,\Delta i/\Delta t$. When we multiply Equation 21.23 by Li and rearrange terms, we find that

$$\mathcal{E}i = Li\frac{\Delta i}{\Delta t} + i^2R. \tag{21.27}$$

This shows that part of the power $\mathcal{E}i$ supplied by the source is dissipated (i^2R) in the resistor and part is stored $(Li\,\Delta i/\Delta t)$ in the inductor.

Current Decay in an *R–L* Circuit

In Figure 21.32, we show an *R–L* circuit with two switches. Suppose that switch S_1 has been closed for a long time and that the current has reached a steady value I_0. Resetting our stopwatch to redefine the initial time, we close switch S_2 at time $t = 0$, bypassing the battery. (At the same time, we should open S_1 to save the battery from ruin.) The current through R and L does not instantaneously go to zero, but decays smoothly, as shown in Figure 21.33. The Kirchhoff's-rule loop equation is obtained from Equation 21.23 by simply omitting the $\mathcal{E}$ term. When we retrace the steps in the preceding analysis, we find that the current i varies with time according to the relationship

$$i = I_0 e^{-(R/L)t}, \tag{21.28}$$

where I_0 is the initial current at time $t = 0$. The time constant, $\tau = L/R$, is the time required for the current to decrease to $1/e$, or about 0.37, of its original value. In time 2τ, it has dropped to $1/e^2 = 0.135$, and so on.

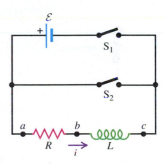

▲ **FIGURE 21.32** An *R–L* circuit with a second switch that constitutes a short circuit, disconnecting the resistor and inductor from the emf source.

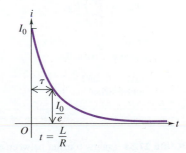

▲ **FIGURE 21.33** Graph of i versus t for decay of current in an *R–L* circuit.

The energy needed to maintain the current during this decay is provided by the energy stored in the magnetic field of the inductor. The detailed energy analysis is simpler this time. In place of Equation 21.27, we have

$$0 = Li\frac{\Delta i}{\Delta t} + i^2 R. \tag{21.29}$$

In this case, $Li\,\Delta i/\Delta t$ is negative, because Δi is negative. Equation 21.29 shows that the energy stored in the inductor *decreases* at a rate equal to the rate of dissipation of energy $i^2 R$ in the resistor.

21.12 The *L–C* Circuit

A circuit containing an inductor and a capacitor (which we'll call an **L–C circuit**) shows an entirely new mode of behavior, characterized by *oscillating* current and charge. This is in sharp contrast to the *exponential* approach to a steady-state situation that we have seen with both *R–C* and *R–L* circuits. In the *L–C* circuit in Figure 21.34, we charge the capacitor to a potential difference V_{max} and an initial charge $Q_{max} = CV_{max}$, as shown in Figure 21.34a, and then close the switch. What happens?

The capacitor begins to discharge through the inductor. Because of the induced emf in the inductor, the current cannot change instantaneously; rather, it starts at zero and eventually builds up to a maximum value I_{max}. During this buildup, the capacitor is discharging. At each instant, the capacitor potential difference equals the induced emf, so as the capacitor discharges, the *rate of change* of current decreases. When the capacitor potential difference becomes zero, the induced emf is also zero, and the current has leveled off at its maximum value I_{max}. Figure 21.34b shows this situation; the capacitor has completely discharged.

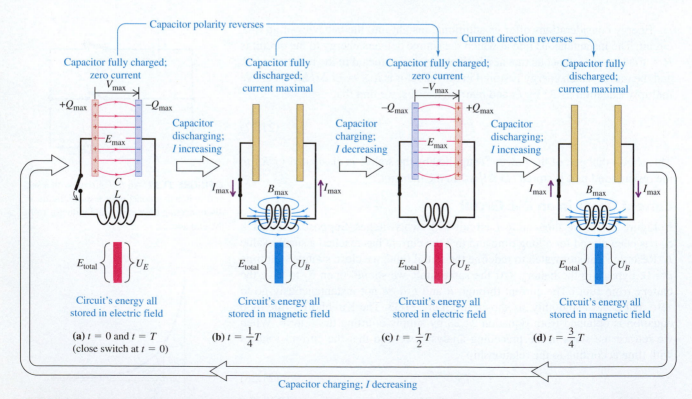

▲ **FIGURE 21.34** Energy transfer between electric and magnetic fields in an oscillating *L–C* circuit. The switch is closed at time $t = 0$, when the capacitor charge is Q_{max}. As in simple harmonic motion, the total energy E_{total} is constant.

The potential difference between its terminals (and those of the inductor) has decreased to zero, and the current has reached its maximum value I_{max}.

During the discharge of the capacitor, the increasing current in the inductor has established a magnetic field in the space around it, and the energy that was initially stored in the capacitor's electric field is now stored in the inductor's magnetic field.

The current cannot drop to zero instantaneously; as it persists, the capacitor begins to charge with polarity opposite to that in the initial state. As the current decreases, the magnetic field also decreases, inducing an emf in the inductor in the same direction as the current. Eventually, the current and the magnetic field reach zero, and the capacitor has been charged in the *opposite* sense to its initial polarity (Figure 21.34c), with a potential difference $-V_{max}$ and charge $-Q_{max}$.

The process now repeats in the reverse direction; a little later, the capacitor has again discharged, and there is a current in the inductor in the opposite direction (Figure 21.34d). The whole process then repeats once more. If there are no energy losses, the charges on the capacitor continue to oscillate back and forth indefinitely. This process is called an **electrical oscillation.**

From an energy standpoint, the oscillations of an electric circuit transfer energy from the capacitor's electric field to the inductor's magnetic field and back. The *total* energy associated with the circuit is constant. This phenomenon is analogous to the transfer of energy in an oscillating, frictionless mechanical system, from potential to kinetic and back, with constant total energy.

Detailed analysis shows that the angular frequency ω of the oscillations we've just described is given by

$$\omega = \sqrt{\frac{1}{LC}}. \qquad (21.30)$$

Thus, the charge and current in the *L−C* circuit oscillate sinusoidally with time, with an angular frequency determined by the values of L and C. In many ways, this behavior is directly analogous to simple harmonic motion in mechanical systems.

▲ **Application Tuning in.** What happens when you turn the tuning knob on this radio? Radios often used *L−C* circuits to tune in on one desired frequency among all the frequencies bombarding the radio's antenna. Radio waves create small oscillating currents in the antenna. A given *L−C* circuit is most responsive to currents oscillating at its own resonance frequency; it tends to pass frequencies close to this and screen out others. When you twist the knob, you change the capacitance or inductance of the circuit and thus its resonant frequency.

SUMMARY

Electromagnetic Induction and Faraday's Law

(Sections 21.1–21.3) A changing magnetic flux through a circuit loop induces an emf in the circuit. **Faraday's law** states that the magnitude $\mathcal{E}$ of the induced emf in a circuit equals the absolute value of the time rate of change of magnetic flux through the circuit: $\mathcal{E} = |\Delta\Phi_B/\Delta t|$ (Equation 21.4). This relation is valid whether the change in flux is caused by a changing magnetic field, motion of the conductor, or both.

The magnet's motion causes a *changing* magnetic field through the coil, inducing a current in the coil.

Lenz's Law

(Section 21.4) **Lenz's law** states that an induced current or emf always acts to oppose the change that caused it. Lenz's law can be derived from Faraday's law and is a convenient way to determine the correct sign for any induced effect.

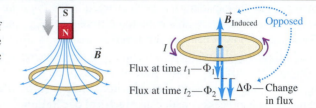

Motional Electromotive Force

(Section 21.5) When a conductor moves in a magnetic field, the charges in the conductor are acted upon by magnetic forces that create a current. When a conductor with length L moves with speed v perpendicular to a uniform magnetic field with magnitude B, the induced emf is $\mathcal{E} = vBL$ (Equation 21.7).

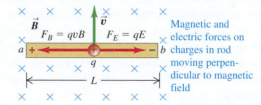

Magnetic and electric forces on charges in rod moving perpendicular to magnetic field

Eddy Currents

(Section 21.6) When a bulk piece of conducting material, such as a metal, is in a changing magnetic field or moves through a nonuniform field, **eddy currents** are induced in it.

Mutual Inductance and Self-Inductance

(Sections 21.7 and 21.8) When changing magnetic flux created by a changing current i_1 in one circuit links a second circuit, an emf with magnitude $\mathcal{E}_2$ is induced in the second circuit. A changing current i_2 in the second circuit induces an emf with magnitude $\mathcal{E}_1$ in the first circuit. The two emfs are given by

$$\mathcal{E}_2 = \left| M\frac{\Delta i_1}{\Delta t}\right| \quad \text{and} \quad \mathcal{E}_1 = \left| M\frac{\Delta i_2}{\Delta t}\right|, \quad (21.12)$$

where M is a constant called the **mutual inductance**. A changing current i in any circuit induces an emf $\mathcal{E}$ in that same circuit, called a self-induced emf, given by

$$\mathcal{E} = \left| L\frac{\Delta i}{\Delta t}\right|, \quad (21.14)$$

where L is a constant called **inductance** or self-inductance.

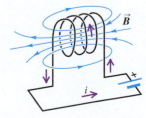

Self-inductance: If the current i in the coil is changing, the changing flux through the coil induces an emf in the coil.

Continued

Transformers

(Section 21.9) A **transformer** is used to transform the voltage and current levels in an ac circuit. An alternating current in either winding results in an alternating flux in the other winding, inducing an emf. In an ideal transformer with no energy losses, if the primary winding has N_1 turns and the secondary winding has N_2 turns, the two voltages are related by

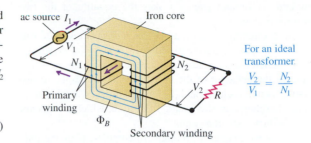

For an ideal transformer.

$$\frac{V_2}{V_1} = \frac{N_2}{N_1}$$

$$\frac{V_2}{V_1} = \frac{N_2}{N_1}. \qquad (21.15)$$

Magnetic Field Energy

(Section 21.10) An inductor with inductance L carrying current I has energy $U = \frac{1}{2}LI^2$ (Equation 21.19). This energy is stored in the magnetic field of the inductor. The energy density u (energy per unit volume) is given by $u = B^2/2\mu_0$ (Equation 21.21).

R–L and *L–C* Circuits

(Sections 21.11 and 21.12) In a circuit containing a resistor R, an inductor L, and a source of emf $\mathcal{E}$, the growth and decay of current are exponential, with a characteristic time τ called the **time constant,** given by $\tau = L/R$ (Equation 21.26). The time constant τ is the time required for the increasing current to approach within a fraction $1 - (1/e)$, or about 63%, of its final value.

A circuit containing an inductance L and a capacitance C undergoes electrical oscillations with angular frequency ω, where $\omega = \sqrt{1/LC}$ (Equation 21.30).

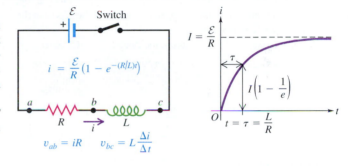

$$i = \frac{\mathcal{E}}{R}\left(1 - e^{-(R/L)t}\right)$$

$$v_{ab} = iR \qquad v_{bc} = L\frac{\Delta i}{\Delta t}$$

$$I = \frac{\mathcal{E}}{R}$$

$$t = \tau = \frac{L}{R}$$

$$I\left(1 - \frac{1}{e}\right)$$

 For instructor-assigned homework, go to www.masteringphysics.com

Conceptual Questions

1. Two circular loops lie adjacent to each other on a tabletop. One is connected to a source that supplies an increasing current; the other is a simple closed ring. Is the induced current in the ring in the same direction as that in the ring connected to the source, or opposite? What if the current in the first ring is decreasing?

2. Small one-cylinder gasoline engines sometimes use a device called a *magneto* to supply current to the spark plug. A permanent magnet is attached to the flywheel, and a stationary coil is mounted adjacent to it. Explain how this device is able to generate current. What happens when the magnet passes the coil?

3. A long, straight current-carrying wire passes through the center of a metal ring, perpendicular to its plane. If the current in the wire increases, is a current induced in the ring? Explain.

4. Two closely wound circular coils have the same number of turns, but one has twice the radius of the other. How are the self-inductances of the two coils related?

5. One of the great problems in the field of energy resources and utilization is the difficulty of storing electrical energy in large quantities economically. Discuss the feasibility of storing large amounts of energy by means of currents in large inductors.

6. Suppose there is a steady current in an inductor. If one attempts to reduce the current to zero instantaneously by opening a switch, a big fat spark appears at the switch contacts. Why? What happens to the induced emf in this situation? Is it physically possible to stop the current instantaneously?

7. Why does a transformer *not* work with dc current?

8. Does Lenz's law say that the induced current in a metal loop always flows to oppose the magnetic flux through that loop? Explain.

9. Does Faraday's law say that a large magnetic flux induces a large emf in a coil? Explain.

10. True or false? Inductors always oppose the current through them. Can you think of any situations in which they actually *help* the current through them?

11. An airplane is in level flight over Antarctica, where the magnetic field of the earth is mostly directed upward away from the ground. As viewed by a passenger facing toward the front of the plane, is the left or the right wingtip at higher potential? Does your answer depend on the direction the plane is flying?

12. Capacitors store energy by accumulating charges on their plates, but how do inductors store energy?

13. A metal ring can be moved in and out of the space between the poles of a horseshoe magnet. Show that you must do work *both* to push it in *and* to pull it out.

14. The ratio of the primary voltage to the secondary voltage in a certain ideal step-down transformer is 2 : 1. The ratio of the

secondary *current* to the primary *current* is also 2:1. Explain how conservation of energy makes the equality of the two ratios inevitable.

15. In an *R–C* circuit, a resistor, an uncharged capacitor, a dc battery, and an open switch are in series. In an *R–L* circuit, a resistor, an inductor, a dc battery, and an open switch are in series. Compare the behavior of the current in these circuits (a) just after the switch is closed and (b) long after the switch has been closed. In other words, compare the way in which a capacitor and an inductor affect a circuit.

16. You have a bar magnet with unidentified poles. Utilizing only simple laboratory equipment, such as conducting wire and an ammeter, show how you could use induction phenomena to determine which are the north and south poles of this magnet.

Multiple-Choice Problems

1. A square loop of wire is pulled upward out of the space between the poles of a magnet, as shown in Figure 21.35. As this is done, the current induced in this loop, as viewed from the N pole of the magnet, will be directed
   A. clockwise.   B. counterclockwise.   C. zero.

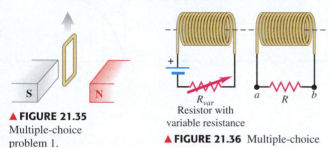

▲ **FIGURE 21.35**
Multiple-choice
problem 1.

▲ **FIGURE 21.36** Multiple-choice problem 2.

2. The two solenoids in Figure 21.36 are coaxial and fairly close to each other. While the resistance of the variable resistor in the left-hand solenoid is *increased* at a constant rate, the induced current through the resistor *R* will
   A. flow from *a* to *b*.   B. flow from *b* to *a*.
   C. be zero because the rate is constant.

3. A metal ring is oriented with the plane of its area perpendicular to a spatially uniform magnetic field that increases at a steady rate. After the radius of the ring is doubled, while the rate of increase of the field is cut in half, the emf induced in the ring
   A. remains the same.
   B. increases by a factor of 2.
   C. increases by a factor of 4.
   D. decreases by a factor of 2.

4. The slide wire of the variable resistor in Figure 21.37 is moved steadily to the right, increasing the resistance in the circuit. While this is being done, the current induced in the small circuit *A* is directed
   A. clockwise.   B. counterclockwise.   C. zero.

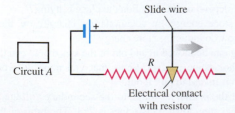

▲ **FIGURE 21.37** Multiple-choice problem 4.

5. The slide wire on the variable resistor in Figure 21.38 is moved steadily to the left. While this is being done, the current induced in the small circuit *A* is directed
   A. clockwise.
   B. counterclockwise.
   C. zero.

6. A metal loop moves at constant velocity toward a long wire carrying a steady current *I*, as shown in Figure 21.39. The current induced in the loop is directed
   A. clockwise.
   B. counterclockwise.
   C. zero.

7. The primary coil of an ideal transformer carries a current of 2.5 A, while the secondary coil carries a current of 5.0 A. The ratio of number of turns of wire in the primary to that in the secondary is
   A. 1:1.
   B. 1:2.
   C. 2:1.

8. A metal loop is held above the S pole of a bar magnet, as shown in Figure 21.40, when the magnet is suddenly dropped from rest. Just after the magnet is dropped, the induced current in the loop, as viewed from above it, is directed
   A. clockwise.
   B. counterclockwise.
   C. zero.

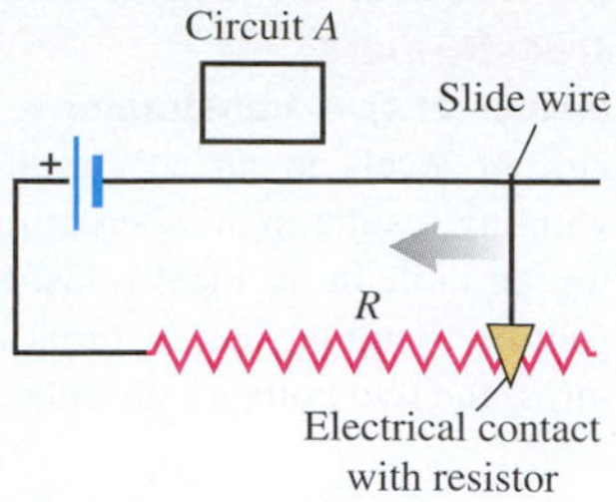

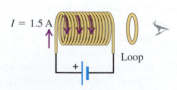

▲ **FIGURE 21.38** Multiple-choice problem 5.

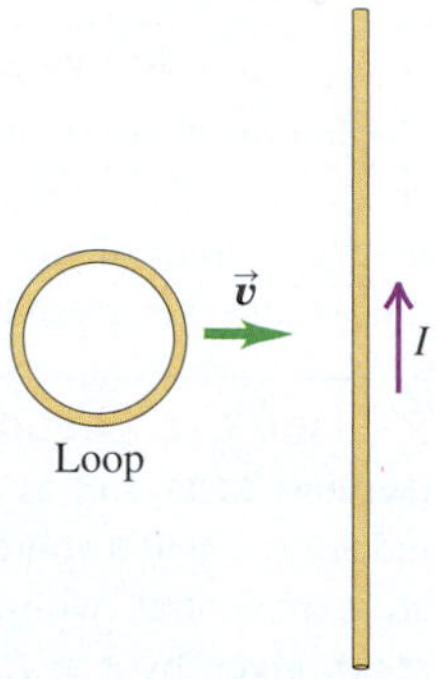

▲ **FIGURE 21.39**
Multiple-choice
problem 6.

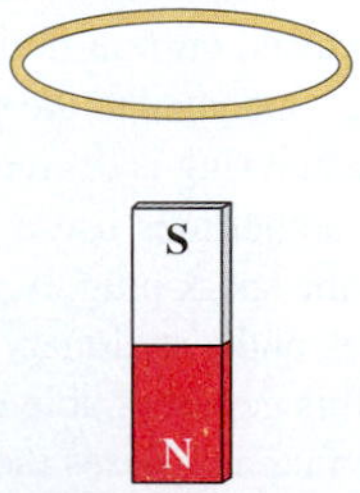

▲ **FIGURE 21.40**
Multiple-choice
problem 8.

▲ **FIGURE 21.41** Multiple-choice problem 9.

9. A steady current of 1.5 A flows through the solenoid shown in Figure 21.41. The current induced in the loop, as viewed from the right, is directed
   A. clockwise.
   B. counterclockwise.
   C. zero.

10. A vertical bar moves horizontally at constant velocity through a uniform magnetic field, as shown in Figure 21.42. We observe that point *b* is at a higher potential than point *a*. We

can therefore conclude that the magnetic field must have a component that is directed

A. vertically downward.
B. vertically upward.
C. perpendicular to the plane of the paper, outward.
D. perpendicular to the plane of the paper, inward.

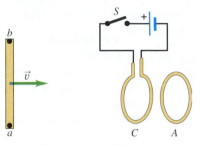

▲ FIGURE 21.42 Multiple-choice problem 10.

▲ FIGURE 21.43 Multiple-choice problem 11.

11. The vertical loops A and C in Figure 21.43 are parallel to each other and are centered on the same horizontal line that is perpendicular to both of them. Just after the switch S is closed, loop A will
   A. not be affected by loop C.   B. be attracted by loop C.
   C. be repelled by loop C.   D. move upward.
   E. move downward.

12. After the switch S in Figure 21.43 has been closed for a very long time, loop A will
   A. not be affected by loop C.
   B. be attracted by loop C.
   C. be repelled by loop C.
   D. move upward.
   E. move downward.

13. After the switch S in the circuit in Figure 21.44 is closed,
   A. The current is zero 1.5 ms (one time constant) later.
   B. The current is zero for a very long time afterward.
   C. The largest current is 5.0 A and it occurs just after S is closed.
   D. The largest current is 5.0 A and it occurs a very long time after S has been closed.

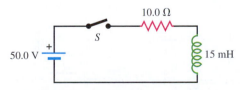

▲ FIGURE 21.44 Multiple-choice problem 13.

14. A square metal loop is pulled to the right at a constant velocity perpendicular to a uniform magnetic field, as shown in Figure 21.45. The current induced in this loop is directed
   A. clockwise.   B. counterclockwise.
   C. zero.

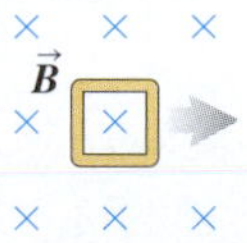

15. A metal loop is being pushed at a constant velocity into a uniform magnetic field, as

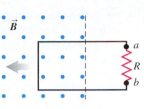

shown in Figure 21.46, but is only partway into the field. As a result of this motion,
   A. End a of the resistor R is at a higher potential than end b.
   B. End b of the resistor R is at a higher potential than end a.
   C. Ends a and b are at the same potential.

▲ FIGURE 21.46 Multiple-choice problem 15.

Problems

21.2 Magnetic Flux

1. • A circular area with a radius of 6.50 cm lies in the x-y plane. What is the magnitude of the magnetic flux through this circle due to a uniform magnetic field $B = 0.230$ T that points (a) in the $+z$ direction? (b) at an angle of 53.1° from the $+z$ direction? (c) in the $+y$ direction?

2. • The magnetic field $\vec{B}$ in a certain region is 0.128 T, and its direction is that of the $+z$ axis in Figure 21.47. (a) What is the magnetic flux across the surface $abcd$ in the figure? (b) What is the magnetic flux across the surface $befc$? (c) What is the magnetic flux across the surface $aefd$? (d) What is the net flux through all five surfaces that enclose the shaded volume?

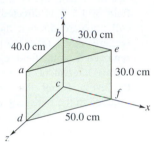

▲ FIGURE 21.47 Problem 2.

3. •• An open plastic soda bottle with an opening diameter of 2.5 cm is placed on a table. A uniform 1.75 T magnetic field directed upward and oriented 25° from vertical encompasses the bottle. What is the total magnetic flux through the plastic of the soda bottle?

21.3 Faraday's Law

4. • A single loop of wire with an area of 0.0900 m² is in a uniform magnetic field that has an initial value of 3.80 T, is perpendicular to the plane of the loop, and is decreasing at a constant rate of 0.190 T/s. (a) What emf is induced in this loop? (b) If the loop has a resistance of 0.600 Ω, find the current induced in the loop.

5. • A coil of wire with 200 circular turns of radius 3.00 cm is in a uniform magnetic field along the axis of the coil. The coil has $R = 40.0$ Ω. At what rate, in teslas per second, must the magnetic field be changing to induce a current of 0.150 A in the coil?

6. • In a physics laboratory experiment, a coil with 200 turns enclosing an area of 12 cm² is rotated from a position where its plane is perpendicular to the earth's magnetic field to one where its plane is parallel to the field. The rotation takes 0.040 s. The earth's magnetic field at the location of the laboratory is 6.0×10^{-5} T. (a) What is the total magnetic flux through the coil before it is rotated? After it is rotated? (b) What is the average emf induced in the coil?

7. • A closely wound rectangular coil of 80 turns has dimensions of 25.0 cm by 40.0 cm. The plane of the coil is rotated from a position where it makes an angle of 37.0° with a magnetic field of 1.10 T to a position perpendicular to the field. The rotation takes 0.0600 s. What is the average emf induced in the coil?

8. •• A very long, straight solenoid with a cross-sectional area of 6.00 cm² is wound with 40 turns of wire per centimeter, and the windings carry a current of 0.250 A. A secondary winding of 2 turns encircles the solenoid at its center. When the primary circuit is opened, the magnetic field of the solenoid becomes zero in 0.0500 s. What is the average induced emf in the secondary coil?

9. •• A 30.0 cm × 60.0 cm rectangular circuit containing a 15 Ω resistor is perpendicular to a uniform magnetic field that starts out at 2.65 T and steadily decreases at 0.25 T/s. (See Figure 21.48.) While this field is changing, what does the ammeter read?

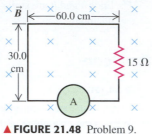

▲ **FIGURE 21.48** Problem 9.

10. •• A circular loop of wire with a radius of 12.0 cm is lying flat on a tabletop. A magnetic field of 1.5 T is directed vertically upward through the loop (Figure 21.49). (a) If the loop is removed from the field region in a time interval of 2.0 ms, find the average emf that will be induced in the wire loop during the extraction process. (b) If the loop is viewed looking down on it from above, is the induced current in the loop clockwise or counterclockwise?

▲ **FIGURE 21.49** Problem 10.

11. •• A flat, square coil with 15 turns has sides of length 0.120 m. The coil rotates in a magnetic field of 0.0250 T. (a) What is the angular velocity of the coil if the maximum emf produced is 20.0 mV? (*Hint:* Look at the motional emf induced across the ends of the segments of the coil.) (b) What is the average emf at this angular velocity?

21.4 Lenz's Law

12. • A cardboard tube is wrapped with two windings of insulated wire, as shown in Figure 21.50. Is the induced current in the resistor R directed from left to right or from right to left in the following circumstances? The current in winding A is directed (a) from a to b and is increasing, (b) from b to a and is decreasing, (c) from b to a and is increasing, and (d) from b to a and is constant.

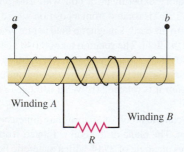

▲ **FIGURE 21.50** Problem 12.

13. • A circular loop of wire is in a spatially uniform magnetic field, as shown in Figure 21.51. The magnetic field is directed into the plane of the figure. Determine the direction (clockwise or counterclockwise) of the induced current in the loop when (a) B is increasing; (b) B is decreasing; (c) B is constant with a value of B_0. Explain your reasoning.

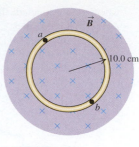

▲ **FIGURE 21.51** Problem 13.

14. • Using Lenz's law, determine the direction of the current in resistor ab of Figure 21.52 when (a) switch S is opened after having been closed for several minutes; (b) coil B is brought closer to coil A with the switch closed; (c) the resistance of R is decreased while the switch remains closed.

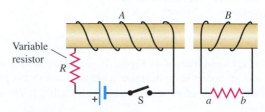

▲ **FIGURE 21.52** Problem 14.

15. • A solenoid carrying a current I is moving toward a metal ring, as shown in Figure 21.53. In what direction, clockwise or counterclockwise (as seen from the solenoid) is a current induced in the ring? In what direction will the induced current be if the solenoid now stops moving toward the ring, but the current in it begins to decrease?

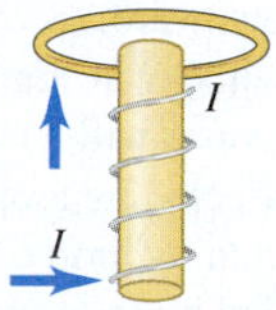

▲ **FIGURE 21.53** Problem 15.

16. • A metal bar is pulled to the right perpendicular to a uniform magnetic field. The bar rides on parallel metal rails connected through a resistor, as shown in Figure 21.54, so the apparatus makes a complete circuit. Find the direction of the current induced in the circuit in two ways: (a) by looking at the magnetic force on the charges in the moving bar and (b) using Lenz's law.

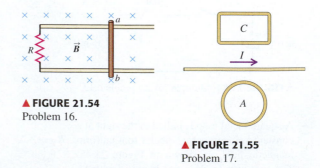

▲ **FIGURE 21.54** Problem 16.

▲ **FIGURE 21.55** Problem 17.

17. • Two closed loops A and C are close to a long wire carrying a current I. (See Figure 21.55.) Find the direction (clockwise or

counterclockwise) of the current induced in each of these loops if I is steadily *increasing*.

18. • A bar magnet is held above a circular loop of wire as shown in Figure 21.56. Find the direction (clockwise or counterclockwise, as viewed from *below* the loop) of the current induced in this loop in each of the following cases. (a) The loop is dropped. (b) The magnet is dropped. (c) Both the loop and magnet are dropped at the same instant.

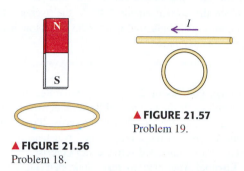

▲ **FIGURE 21.56**
Problem 18.

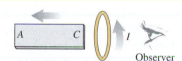

▲ **FIGURE 21.57**
Problem 19.

19. •• The current in Figure. 21.57 obeys the equation $Ie = I_0 e^{-2bt}$, where $b > 0$. Find the direction (clockwise or counterclockwise) of the current induced in the round coil for $t > 0$.

20. • A bar magnet is close to a metal loop. When this magnet is suddenly moved to the left away from the loop, as shown in Figure 21.58, a counterclockwise current is induced in the coil, as viewed by an observer looking through the coil toward the magnet. Identify the north and south poles of the magnet.

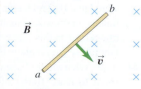

▲ **FIGURE 21.58** Problem 20.

21.5 Motional Electromotive Force

21. • A very thin 15.0 cm copper bar is aligned horizontally along the east–west direction. If it moves horizontally from south to north at 11.5 m/s in a vertically upward magnetic field of 1.22 T, (a) what potential difference is induced across its ends, and (b) which end (east or west) is at a higher potential? (c) What would be the potential difference if the bar moved from east to west instead?

22. • When a thin 12.0 cm iron rod moves with a constant velocity of 4.50 m/s perpendicular to the rod in the direction shown in Figure 21.59, the induced emf across its ends is measured to be 0.450 V. (a) What is the magnitude of the magnetic field? (b) Which point is at a higher potential, a or b? (c) If the bar is rotated clockwise by 90° in the plane of the paper, but keeps the same velocity, what is the potential difference induced across its ends?

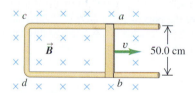

▲ **FIGURE 21.59**
Problem 22.

23. •• You're driving at 95 km/h in a direction 35° east of north, in a region where the earth's magnetic field of 5.5×10^{-5} T is horizontal and points due north. If your car measures 1.5 m from its underbody to its roof, calculate the induced emf between roof and underbody. (You can assume the sides of the car are straight and vertical.) Is the roof of the car at a higher or lower potential than the underbody?

24. •• A 1.41 m bar moves through a uniform, 1.20 T magnetic field with a speed of 2.50 m/s (Figure 21.60). In each case, find the emf induced between the ends of this bar and identify which, if any, end (a or b) is at the higher potential. The bar moves in the direction of (a) the $+x$-axis; (b) the $-y$-axis; (c) the $+z$-axis. (d) How should this bar move so that the emf across its ends has the greatest possible value with b at a higher potential than a, and what is this maximum emf?

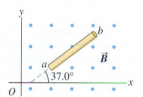

▲ **FIGURE 21.60** Problem 24.

25. •• The conducting rod ab shown in Figure 21.61 makes frictionless contact with metal rails ca and db. The apparatus is in a uniform magnetic field of 0.800 T, perpendicular to the plane of the figure. (a) Find the magnitude of the emf induced in the rod when it is moving toward the right with a speed 7.50 m/s. (b) In what direction does the current flow in the rod? (c) If the resistance of the circuit $abdc$ is a constant 1.50 Ω, find the magnitude and direction of the force required to keep the rod moving to the right with a constant speed of 7.50 m/s.

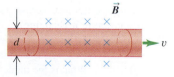

▲ **FIGURE 21.61** Problem 25.

26. •• **Measuring blood flow.**
BIO Blood contains positive and negative ions and therefore is a conductor. A blood vessel, therefore, can be viewed as an electrical wire. We can even picture the flowing blood as a series of parallel conducting slabs whose thickness is the diameter d of the vessel moving with speed v. (See Figure 21.62.) (a) If the blood vessel is placed in a magnetic field B perpendicular to the vessel, as in the figure, show that the motional potential difference induced across it is $\mathcal{E} = vBd$. (b) If you expect that the blood will be flowing at 15 cm/s for a vessel 5.0 mm in diameter, what strength of magnetic field will you need to produce a potential difference of 1.0 mV? (c) Show that the volume rate of flow (R) of the

blood is equal to $R = \pi\mathcal{E}d/4B$. (*Note:* Although the method developed here is useful in measuring the rate of blood flow in a vessel, it is limited to use in surgery because measurement of the potential $\mathcal{E}$ must be made directly across the vessel.)

21.7 Mutual Inductance

27. • A toroidal solenoid has a mean radius of 10.0 cm and a cross-sectional area of 4.00 cm^2 and is wound uniformly with 100 turns. A second coil with 500 turns is wound uniformly on top of the first. What is the mutual inductance of these coils?

28. • A 10.0-cm-long solenoid of diameter 0.400 cm is wound uniformly with 800 turns. A second coil with 50 turns is wound around the solenoid at its center. What is the mutual inductance of the combination of the two coils?

29. • Two coils are wound around the same cylindrical form, like the coils in Example 21.8. When the current in the first coil is decreasing at a rate of 0.242 A/s, the induced emf in the second coil has magnitude 1.65 mV. (a) What is the mutual inductance of the pair of coils? (b) If the second coil has 25 turns, what is the average magnetic flux through each turn when the current in the first coil equals 1.20 A? (c) If the current in the second coil increases at a rate of 0.360 A/s, what is the magnitude of the induced emf in the first coil?

30. • One solenoid is centered inside another. The outer one has a length of 50.0 cm and contains 6750 coils, while the coaxial inner solenoid is 3.0 cm long and 0.120 cm in diameter and contains 15 coils. The current in the outer solenoid is changing at 37.5 A/s. (a) What is the mutual inductance of these solenoids? (b) Find the emf induced in the inner solenoid.

31. •• Two toroidal solenoids are wound around the same form so that the magnetic field of one passes through the turns of the other. Solenoid 1 has 700 turns, and solenoid 2 has 400 turns. When the current in solenoid 1 is 6.52 A, the average flux through each turn of solenoid 2 is 0.0320 Wb. (a) What is the mutual inductance of the pair of solenoids? (b) When the current in solenoid 2 is 2.54 A, what is the average flux through each turn of solenoid 1?

21.8 Self-Inductance

32. • A 4.5 mH toroidal inductor has 125 identical equally spaced coils. (a) If it carries an 11.5 A current, how much magnetic flux passes through each of its coils? (b) If the potential difference across its ends is 1.16 V, at what rate is the current in it changing?

33. • At the instant when the current in an inductor is increasing at a rate of 0.0640 A/s, the magnitude of the self-induced emf is 0.0160 V. What is the inductance of the inductor?

34. • An inductor has inductance of 0.260 H and carries a current that is decreasing at a uniform rate of 18.0 mA/s. Find the self-induced emf in this inductor.

35. • A 2.50 mH toroidal solenoid has an average radius of 6.00 cm and a cross-sectional area of 2.00 cm^2. (a) How many coils does it have? (Make the same assumption as in Example 21.10.) (b) At what rate must the current through it change so that a potential difference of 2.00 V is developed across its ends?

36. •• **Self-inductance of a solenoid.** A long, straight solenoid has N turns, a uniform cross-sectional area A, and length l. Use the definition of self-inductance expressed by Equation 21.13 to show that the inductance of this solenoid is given approximately by the equation $L = \mu_0 AN^2/l$. Assume that the magnetic field is uniform inside the solenoid and zero outside. (Your answer is approximate because B is actually smaller at the ends than at the center of the solenoid. For this reason, your answer is actually an upper limit on the inductance.)

37. •• When the current in a toroidal solenoid is changing at a rate of 0.0260 A/s, the magnitude of the induced emf is 12.6 mV. When the current equals 1.40 A, the average flux through each turn of the solenoid is 0.00285 Wb. How many turns does the solenoid have?

21.9 Transformers

38. • A transformer consists of 275 primary windings and 834 secondary windings. If the potential difference across the primary coil is 25.0 V, (a) what is the voltage across the secondary coil, and (b) what is the effective load resistance of the secondary coil if it is connected across a 125 Ω resistor?

39. • **Off to Europe!** You plan to take your hair blower to Europe, where the electrical outlets put out 240 V instead of the 120 V seen in the United States. The blower puts out 1600 W at 120 V. (a) What could you do to operate your blower via the 240 V line in Europe? (b) What current will your blower draw from a European outlet? (c) What resistance will your blower appear to have when operated at 240 V?

40. •• You need a transformer that will draw 15 W of power from a 220 V (rms) power line, stepping the voltage down to 6.0 V (rms). (a) What will be the current in the secondary coil? (b) What should be the resistance of the secondary circuit? (c) What will be the equivalent resistance of the input circuit?

41. • **A step-up transformer.** A transformer connected to a 120 V (rms) ac line is to supply 13,000 V (rms) for a neon sign. To reduce the shock hazard, a fuse is to be inserted in the primary circuit and is to blow when the rms current in the secondary circuit exceeds 8.50 mA. (a) What is the ratio of secondary to primary turns of the transformer? (b) What power must be supplied to the transformer when the rms secondary current is 8.50 mA? (c) What current rating should the fuse in the primary circuit have?

21.10 Magnetic Field Energy

42. • An air-filled toroidal solenoid has a mean radius of 15.0 cm and a cross-sectional area of 5.00 cm^2. When the current is 12.0 A, the energy stored is 0.390 J. How many turns does the winding have?

43. • **Energy in a typical inductor.** (a) How much energy is stored in a 10.2 mH inductor carrying a 1.15 A current? (b) How much current would such an inductor have to carry to store 1.0 J of energy? Is this a reasonable amount of current for ordinary laboratory circuit elements?

44. •• (a) What would have to be the self-inductance of a solenoid for it to store 10.0 J of energy when a 1.50 A current runs through it? (b) If this solenoid's cross-sectional diameter is 4.00 cm, and if you could wrap its coils to a density of 10 coils/mm, how long would the solenoid be? (See problem 36.) Is this a realistic length for ordinary laboratory use?

45. •• A solenoid 25.0 cm long and with a cross-sectional area of 0.500 cm^2 contains 400 turns of wire and carries a current

of 80.0 A. Calculate: (a) the magnetic field in the solenoid; (b) the energy density in the magnetic field if the solenoid is filled with air; (c) the total energy contained in the coil's magnetic field (assume the field is uniform); (d) the inductance of the solenoid.

46. •• Large inductors have been proposed as energy-storage devices. (a) How much electrical energy is converted to light and thermal energy by a 200 W lightbulb in one day? (b) If the amount of energy calculated in part (a) is stored in an inductor in which the current is 80.0 A, what is the inductance?

47. •• When a certain inductor carries a current I, it stores 3.0 mJ of magnetic energy. How much current (in terms of I) would it have to carry to store 9.0 mJ of energy?

21.11 The R–L Circuit

48. • A 12.0 V dc battery having no appreciable internal resistance, a 150.0 Ω resistor, an 11.0 mH inductor, and an open switch are all connected in series. After the switch is closed, what are (a) the time constant for this circuit, (b) the maximum current that flows through it, (c) the current 73.3 μs after the switch is closed, and (d) the maximum energy stored in the inductor?

49. • An inductor with an inductance of 2.50 H and a resistor with a resistance of 8.00 Ω are connected to the terminals of a battery with an emf of 6.00 V and negligible internal resistance. Find (a) the initial rate of increase of the current in the circuit, (b) the initial potential difference across the inductor, (c) the current 0.313 s after the circuit is closed, and (d) the maximum current.

50. • In Figure 21.63, both switches S_1 and S_2 are initially open. S_1 is then closed and left closed until a constant current is established. Then S_2 is closed just as S_1 is opened, taking the battery out of the circuit. (a) What is the initial current in the resistor just after S_2 is closed and S_1 is opened? (b) What is the time constant of the circuit? (c) What is the current in the resistor after a large number of time constants have elapsed?

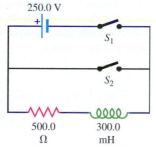

▲ **FIGURE 21.63** Problem 50.

51. •• In the circuit shown in Figure 21.64, the battery and the inductor have no appreciable internal resistance and there is no current in the circuit. After the switch is closed, find the readings

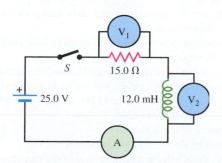

▲ **FIGURE 21.64** Problem 51.

of the ammeter (A) and voltmeters $(V_1$ and $V_2)$ (a) the instant after the switch is closed; (b) after the switch has been closed for a very long time. (c) Which answers in parts (a) and (b) would change if the inductance were 24.0 mH instead?

52. •• A 35.0 V battery with negligible internal resistance, a 50.0 V resistor, and a 1.25 mH inductor with negligible resistance are all connected in series with an open switch. The switch is suddenly closed. (a) How long after closing the switch will the current through the inductor reach one-half of its maximum value? (b) How long after closing the switch will the energy stored in the inductor reach one-half of its maximum value?

53. •• A 1.50 mH inductor is connected in series with a dc battery of negligible internal resistance, a 0.750 kΩ resistor, and an open switch. How long after the switch is closed will it take for (a) the current in the circuit to reach half of its maximum value, (b) the energy stored in the inductor to reach half of its maximum value? (*Hint:* You will have to solve an exponential equation.)

21.12 The L–C Circuit

54. • A 12.0 μF capacitor and a 5.25 mH inductor are connected in series with an open switch. The capacitor is initially charged to 6.20 μC. What is the angular frequency of the charge oscillations in the capacitor after the switch is closed?

55. •• A 5.00 μF capacitor is initially charged to a potential of 16.0 V. It is then connected in series with a 3.75 mH inductor. (a) What is the total energy stored in this circuit? (b) What is the maximum current in the inductor? What is the charge on the capacitor plates at the instant the current in the inductor is maximal?

General Problems

56. •• A 15.0 μF capacitor is charged to 175 μC and then connected across the ends of a 5.00 mH inductor. (a) Find the maximum current in the inductor. At the instant the current in the inductor is maximal, how much charge is on the capacitor plates? (b) Find the maximum potential across the capacitor. At this instant, what is the current in the inductor? (c) Find the maximum energy stored in the inductor. At this instant, what is the current in the circuit?

57. •• An inductor is connected to the terminals of a battery that has an emf of 12.0 V and negligible internal resistance. The current is 4.86 mA at 0.725 ms after the connection is completed. After a long time the current is 6.45 mA. What are (a) the resistance R of the inductor and (b) the inductance L of the inductor?

58. •• A rectangular circuit is moved at a constant velocity of 3.0 m/s into, through, and then out of a uniform 1.25 T magnetic field, as shown in Figure 21.65. The magnetic field region is considerably wider than 50.0 cm. Find the magnitude and direction (clockwise or counterclockwise) of the current induced in the circuit as it is (a) going into the magnetic field, (b) totally

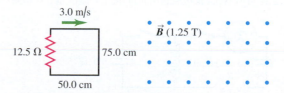

▲ **FIGURE 21.65** Problem 58.

within the magnetic field, but still moving, and (c) moving out of the field. (d) Sketch a graph of the current in this circuit as a function of time, including the preceding three cases.

59. ●● The rectangular loop in Figure 21.66, with area A and resistance R, rotates at uniform angular velocity ω about the y axis. The loop lies in a uniform magnetic field $\vec{B}$ in the direction of the x axis. Sketch graphs of the following quantities, as functions of time, letting $t = 0$ when the loop is in the position shown in the figure: (a) the magnetic flux through the loop, (b) the rate of change of flux with respect to time, (c) the induced emf in the loop, (d) the induced emf if the angular velocity is doubled.

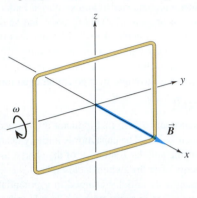

▲ **FIGURE 21.66** Problem 59.

60. ●● A flexible circular loop 6.50 cm in diameter lies in a magnetic field with magnitude 0.950 T, directed into the plane of the page as shown in Figure 21.67. The loop is pulled at the points indicated by the arrows, forming a loop of zero area in 0.250 s. (a) Find the average induced emf in the circuit. (b) What is the direction of the current in R: from a to b or from b to a? Explain your reasoning.

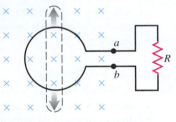

▲ **FIGURE 21.67** Problem 60.

61. ●● **An electromagnetic car alarm.** Your latest invention is a car alarm that produces sound at a particularly annoying frequency of 3500 Hz. To do this, the car-alarm circuitry must produce an alternating electric current of the same frequency. That's why your design includes an inductor and a capacitor in series. The maximum voltage across the capacitor is to be 12.0 V (the same voltage as the car battery). To produce a sufficiently loud sound, the capacitor must store 0.0160 J of energy. What values of capacitance and inductance should you choose for your car-alarm circuit?

62. ●● In the circuit shown in Figure 21.68, S_1 has been closed for a long enough time so that the current reads a steady 3.50 A. Suddenly, S_2 is closed and S_1 is opened at the same instant. (a) What is the maximum charge that the capacitor will receive? (b) What is the current in the inductor at this time?

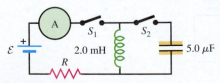

▲ **FIGURE 21.68** Problem 62.

63. ●● Consider the circuit in Figure 21.69. (a) Just after the switch is closed, what is the current through each of the resistors? (b) After the switch has been closed a long time, what is the current through each resistor? (c) After S has been closed a long time, it is opened again. Just after it is opened, what is the current through the 20.0 Ω resistor?

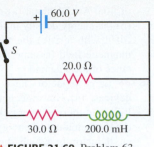

▲ **FIGURE 21.69** Problem 63.

Passage Problems

BIO Stimulating the brain. Communication in the nervous system is based on propagating electrical signals called action potentials that travel along the extended nerve cell processes, the axons. Action potentials are generated when the electrical potential difference across the membrane changes so that the inside of the cell becomes more positive. Researchers in clinical medicine and neurobiology want to stimulate nerves non-invasively at specific locations in conscious subjects. But using electrodes to apply current on the skin is painful and requires large currents.

Anthony Barker and colleagues at the University of Sheffield in England developed a technique that is now widely used called transcranial magnetic stimulation (TMS). In the TMS technique, a coil positioned near the skull produces a time-varying magnetic field, which induces electric currents in the conductive brain tissue sufficient to cause action potentials in nerve cells. For example, if the coil is placed near the motor cortex, the region of the brain that controls voluntary movement, scientists can monitor the contraction of muscles and assess the state of the connections between the brain and the muscle.

▲ **FIGURE 21.70** Problems 64–65.

64. In the diagram of TMS shown in Figure 21.70, a current pulse increases to a peak and then decreases to zero in the direction shown in the stimulating coil. What will be the direction (1 or 2) of the induced current (dotted line) in the brain tissue?
 A. 1   B. 2
 C. 1 while the current increases in the stimulating coil and 2 while the current decreases
 D. 2 while the current increases in the stimulating coil, 1 while the current decreases

65. The brain tissue at the level of the dotted line may be considered as a series of concentric circles, with each circle behaving independently. Where will the induced EMF be the greatest?
 A. At the center of the dotted line
 B. At the periphery of the dotted line
 C. The EMF will be the same in all concentric circles
 D. At the center during the increasing phase of the stimulating current and at the periphery during the decreasing phase

22 Alternating Current

During the 1880s, there was a heated and acrimonious debate over the best method of electric-power distribution. Thomas Edison favored direct current (dc)—that is, steady current that does not vary with time. George Westinghouse favored alternating current (ac)—current (and hence voltage) that varies sinusoidally with time. Westinghouse argued that transformers (which we studied in Section 21.9) could be used to step voltage up and down with ac, but not with dc. Edison claimed that dc was inherently safer.

Eventually, the arguments of Westinghouse prevailed, and most present-day household and industrial power-distribution systems operate with alternating current. Any appliance that you plug into a wall outlet uses ac, and many battery-powered devices, such as portable audio players and cell phones, make use of the dc supplied by the battery to create or amplify alternating currents. Circuits in modern communications equipment, including radios and televisions, make extensive use of ac.

In this chapter, we'll learn how resistors, inductors, and capacitors behave in circuits with sinusoidally varying voltages and currents. Many of the principles that we found useful in Chapters 18, 19, and 21 are applicable, and we'll explore several new concepts related to the circuit behavior of inductors and capacitors. A key idea in this discussion is the concept of *resonance*, which we studied in Chapter 13 in relation to mechanical systems.

As you likely know, good speakers have separate drivers for different parts of the frequency spectrum—small tweeters, larger midrange drivers, and still larger woofers for the low sounds. But the input to the amplifier is a single curve, not three. How can an amplifier sort out the frequencies and send them to the right drivers? You'll find out in this chapter.

22.1 Phasors and Alternating Currents

We've already studied a source of alternating emf (voltage) that can supply an **alternating current** to a circuit. A coil of wire rotating with constant angular

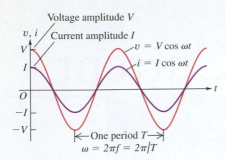

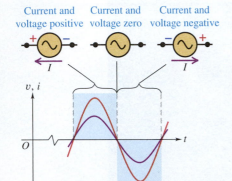

(a) Graphs of sinusoidal current and voltage versus time. The relative heights of the two curves are not significant.

(b) The graphs related to a schematic ac source.

▲ **FIGURE 22.1** Sinusoidal current and voltage graphed as functions of time.

velocity in a magnetic field (Section 21.3) develops a sinusoidally alternating emf and is the prototype of the commercial alternating-current generator, or *alternator.* An *L–C* circuit (Section 21.12) can be used to produce an alternating current with a frequency that may range from a few hertz to many millions.

We'll use the term **ac source** for any device that supplies a sinusoidally varying potential difference v or current i. A sinusoidal voltage might be described by a function such as

$$v = V\cos\omega t. \tag{22.1}$$

In this expression, V is the maximum potential difference, which we call the **voltage amplitude;** v is the *instantaneous* potential difference, and ω is the **angular frequency,** equal to 2π times the frequency f (the number of cycles per second). Figure 22.1 shows graphs of voltage and current that vary sinusoidally with time.

In North America, commercial electric-power distribution systems always use a frequency $f = 60$ Hz, corresponding to an angular frequency $\omega = (2\pi \, \text{rad})(60 \, \text{s}^{-1}) = 377$ rad/s. Similarly, a sinusoidal current i might be described by

$$i = I\cos\omega t, \tag{22.2}$$

where I is the maximum current, or **current amplitude** (Figure 22.1). In the next section, we'll look at the behavior of individual circuit elements when they carry a sinusoidal current. The usual circuit-diagram symbol for an ac source is

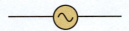

To represent sinusoidally varying voltages and currents, we'll use vector diagrams similar to the diagrams we used with the circle of reference in our study of harmonic motion in Section 11.4. In these diagrams, the instantaneous value of a quantity that varies sinusoidally with time is represented by the *projection* onto a horizontal axis of a vector with a length equal to the amplitude of the quantity. For example, the vector (from point O to point Q in Figure 11.20) has length A, representing the amplitude of the motion, and it rotates counterclockwise with constant angular velocity ω. These rotating vectors are called **phasors,** and diagrams containing them are called **phasor diagrams.** Figure 22.2 shows a phasor diagram for the sinusoidal current described by Equation 22.2. The length I of the phasor is the maximum value of I, and the projection of the phasor onto the horizontal axis at time t is $I\cos\omega t$. At time $t = 0$, $i = I$; this is why we chose to use the cosine function rather than the sine in Equation 22.2.

A phasor isn't a real physical quantity with a direction in space, as are velocity, momentum, and electric field. Rather, it is a *geometric* entity that provides a language for describing and analyzing physical quantities that vary sinusoidally with time. In Section 11.4, we used a single phasor to represent the position and motion of a point mass undergoing simple harmonic motion. In this chapter, we'll use phasors to *add* sinusoidal voltages and currents. Combining sinusoidal quantities with phase differences then becomes a matter of vector addition.

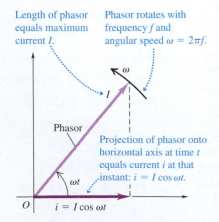

▲ **FIGURE 22.2** A phasor diagram.

Conceptual Analysis 22.1

A current phasor

A sinusoidal current is described by $i = I\cos\omega t$, where $\omega = 1.57$ rad/s and I is the maximum magnitude of the current. At some time $t = t'$, where 1 s $< t' <$ 2 s, the current is -5 amperes. Which of the phasors in Figure 22.3 (A–D) can represent the current at time t'?

SOLUTION Since the current is the phasor's projection onto the *horizontal* axis, the fact that the current at time t' is negative rules out phasors A and D. (They represent *positive* currents.) To decide between B and C, we need to know which quadrant the phasor is in at times between 1 s and 2 s. For that, we need to know the period T—the time taken for the phasor to make one complete revolution (or the time required for one cycle on a graph of i versus t). We get the period from the angular frequency ω: $T = 2\pi/\omega = 2\pi/(1.57$ rad/s$) = 4$ s. Thus, the phasor rotates through one rev-

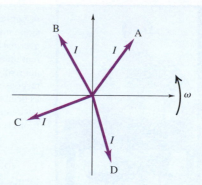

▲ **FIGURE 22.3**

olution in 4 s, or one quadrant each second; during the second second, it is in the second (upper left) quadrant, so B is correct.

We can describe and measure a sinusoidally varying current i in terms of its maximum value, which we can denote as I_{max} or simply I. An alternative notation that is often more useful is the concept of *root-mean-square (rms) value*. We encountered this concept in Section 15.3 in connection with the speeds of molecules in a gas. As shown graphically in Figure 22.4, we *square* the instantaneous current i, calculate the *average* value of i^2, and then take the *square root* of that average. This procedure defines the **root-mean-square (rms) current,** denoted as I_{rms}. Even when i is negative, i^2 is always positive, so I_{rms} is never zero (unless i is zero at every instant).

There's a simple relation between I_{rms} and the maximum current I. If the instantaneous current i is given by $I\cos\omega t$, then

$$i^2 = I^2\cos^2\omega t.$$

From trigonometry, we use the double-angle formula

$$\cos^2 A = \tfrac{1}{2}(1 + \cos 2A)$$

to find that

$$i^2 = \tfrac{1}{2}I^2(1 + \cos 2\omega t)$$
$$= \tfrac{1}{2}I^2 + \tfrac{1}{2}I^2\cos 2\omega t.$$

The average of $\cos 2\omega t$ is zero because it is positive half the time and negative half the time. Thus, the average of i^2 is simply $I^2/2$; the square root of this quantity is I_{rms}:

Meaning of the rms value of a sinusoidal quantity (here, ac current with $I = 3$ A). We:

① Graph current i versus time,

② *square* the instantaneous current i,

③ take the *average* (mean) value of i^2,

④ take the *square root* of that average.

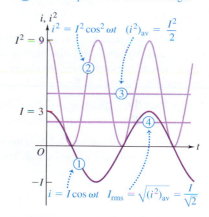

▲ **FIGURE 22.4** Root-mean-square value of a sinusoidally varying quantity.

rms values of sinusoidally varying current and voltage

The rms current for a sinusoidal current with amplitude I is

$$I_{rms} = \frac{I}{\sqrt{2}}. \qquad (22.3)$$

In the same way, the root-mean-square value of a sinusoidal voltage with amplitude (maximum value) V is

$$V_{rms} = \frac{V}{\sqrt{2}}. \qquad (22.4)$$

◀ **Application Going nowhere.** You may wonder how we can power anything by use of alternating current, since the electrons simply jiggle back and forth, going nowhere, and the *average* current is zero. But direct current is not that different. In a dc circuit, there is no current unless the circuit is a closed loop, and then the electrons merely travel in closed loops, not really "going anywhere" either. In both cases, what matters is that the electrons move in response to a potential difference, so energy is available. In a lightbulb, ac is just as effective as dc at heating the filament. A motor that uses ac is constructed differently from a dc motor, but again, the current still interacts with the magnetic field to turn the rotor. Thus, this industrious runner is getting nowhere, thanks to electrons doing the same.

Voltages and currents in power-distribution systems are always described in terms of their rms values, not their amplitudes (maximum values). The usual household power supply in North America, designated 120 volt ac, has an rms voltage of $V_{rms} = 120$ V. The voltage *amplitude V* is

$$V = \sqrt{2}\, V_{rms} = \sqrt{2}\,(120 \text{ V}) = 170 \text{ V}.$$

Meters that are used for ac voltage and current measurements are nearly always calibrated to read rms values, not maximum values.

EXAMPLE 22.1   Current in a personal computer

The plate on the back of a personal computer says that the machine draws 2.7 A from a 120 V, 60 Hz line. For this PC, what are **(a)** the average current, **(b)** the average of the square of the current, and **(c)** the current amplitude?

SOLUTION

SET UP We're given the current, voltage, and frequency (60 Hz), but we actually need only the current. We remember that the stated current is an rms value. To clarify our thinking, we sketch the approximate graph in Figure 22.5.

SOLVE Part (a):   The average of any sinusoidal alternating current over any whole number of cycles is zero.

Part (b):   The rms current, 2.7 A, is the *square root* of the *mean* (average) of the *square* of the current. That is,

$$I_{rms} = \sqrt{(i^2)_{av}}, \quad \text{or}$$
$$(i^2)_{av} = (I_{rms})^2 = (2.7 \text{ A})^2 = 7.3 \text{ A}^2.$$

Part (c):   From Equation 22.3, the current amplitude I is

$$I = \sqrt{2}\, I_{rms} = \sqrt{2}\,(2.7 \text{ A}) = 3.8 \text{ A}.$$

REFLECT We can always describe a sinusoidally varying quantity either in terms of its amplitude or in terms of its rms value. We'll see later that the rms value is usually more convenient for expressing energy and power relations.

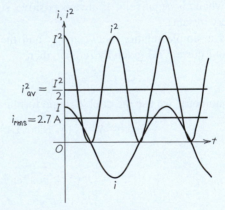

▲ **FIGURE 22.5** Our sketch for this problem.

Practice Problem: A lightbulb draws 0.50 A from a 120 V, 60 Hz line. For this bulb, find the current amplitude. *Answer: I =* 0.71 A.

22.2 Resistance and Reactance

In this section, we'll derive voltage–current relations for an individual resistor, inductor, or capacitor carrying a sinusoidal current.

Resistor in an ac Circuit

First, let's consider a resistor with resistance R, with a sinusoidal current given by Equation 22.2: $i = I\cos\omega t$, as in Figure 22.6a. The current amplitude (maximum current) is I. We denote the instantaneous potential of point a with respect to point b by v_R; from Ohm's law, v_R is given by

$$v_R = iR = IR\cos\omega t. \qquad (22.5)$$

The maximum voltage V_R (the voltage amplitude) is the coefficient of the cosine function; that is,

$$V_R = IR. \qquad (22.6)$$

So we can also write

$$v_R = V_R\cos\omega t. \qquad (22.7)$$

The instantaneous voltage v_R and current i are both proportional to $\cos\omega t$, so the voltage is in step, or *in phase,* with the current. Thus, the voltage and current reach their peak values at the same time, go through zero at the same time, and so on. Equation 22.6 shows that the current and voltage amplitudes are related in the same way as in a dc circuit.

Figure 22.6b shows graphs of i and v_R as functions of time. The vertical scales for i and v_R are different and have different units, so the relative heights of the two curves are not significant. The corresponding phasor diagram is shown in Figure 22.6c. Because i and v_R are *in phase* and have the same frequency, the current and voltage phasors rotate together; they are parallel at each instant. Their projections on the horizontal axis represent the instantaneous current and voltage, respectively.

Inductor in an ac Circuit

Next, suppose we replace the resistor in Figure 22.6a with a pure inductor with self-inductance L and zero resistance (Figure 22.7a). Again, we assume that the current is $i = I\cos\omega t$. The induced emf in the direction of i is given by Equation 21.14, as shown in Figure 21.25: $\mathcal{E} = -L\,\Delta i/\Delta t$. This corresponds to the potential of point b with respect to point a; the potential v_L of point a with respect to point b is the negative of $\mathcal{E}$, and we have

$$v_L = L\,\Delta i/\Delta t.$$

The voltage across the inductor at any instant is proportional to the *rate of change* of the current. The points of maximum voltage on the graph correspond to maximum steepness of the current curve, and the points of zero voltage are the points where the current curve instantaneously levels off at its maximum and minimum values (Figure 22.7b). The voltage and current are "out of step," or *out of phase,*

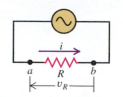

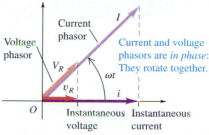

(a) Circuit with ac source and resistor

(b) Graphs of current and voltage versus time

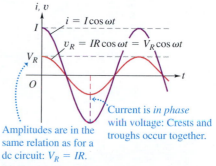

(c) Phasor diagram

▲ **FIGURE 22.6** Current and voltage in a resistance R connected across an ac source.

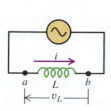

(a) Circuit with ac source and inductor

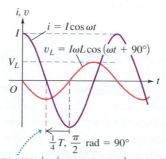

(b) Graphs of current and voltage versus time

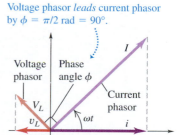

(c) Phasor diagram

▲ **FIGURE 22.7** Current and voltage in an inductance L connected across an ac source.

by a quarter cycle: The voltage peaks occur a quarter cycle *earlier* than the current peaks, and we say that the voltage *leads* the current by 90°. The phasor diagram in Figure 22.7c also shows this relationship; the voltage phasor is ahead of the current phasor by 90°.

The general shape of the graph of v_L as a function of t in Figure 22.7b is that of a *sine* function. Indeed, it can be shown by using calculus that in this case,

$$v_L = -I\omega L\sin\omega t. \tag{22.8}$$

We can also obtain the phase relationship just described by rewriting Equation 22.8, using the identity $\cos(A + 90°) = -\sin A$:

$$v_L = I\omega L\cos(\omega t + 90°). \tag{22.9}$$

This result and the phasor diagram show that the voltage can be viewed as a cosine function with a "head start" of 90°, $\pi/2$ radians, or $\frac{1}{4}$ cycle.

For consistency with later discussions, we'll usually follow this pattern, describing the phase of the *voltage* relative to the *current*, not the reverse. Thus, if the current i in a circuit is

$$i = I\cos\omega t$$

and the voltage v of one point with respect to another is

$$v = V\cos(\omega t + \phi),$$

we call ϕ the **phase angle,** understanding that it gives the phase of the *voltage* relative to the *current*. For a pure inductor, $\phi = 90°$, and the voltage *leads* the current by 90°, $\pi/2$ radians, or $\frac{1}{4}$ cycle. For a pure resistor, $\phi = 0$, and the voltage and current are *in phase*.

From Equation 22.8 or 22.9, the voltage amplitude V_L is

$$V_L = I\omega L. \tag{22.10}$$

Inductive reactance
We define the **inductive reactance** X_L of an inductor as the product of the inductance L and the angular frequency ω:

$$X_L = \omega L. \tag{22.11}$$

Using X_L, we can write Equation 22.10 in the same form as for a resistor $(V_R = IR)$:

$$V_L = IX_L. \tag{22.12}$$

Because X_L is the ratio of a voltage to a current, its unit is the same as for resistance: the ohm.

The inductive reactance of an inductor is directly proportional both to its inductance L and to the angular frequency. The greater the inductance and the higher the frequency, the *larger* is the reactance and the larger is the induced voltage amplitude, for a given current amplitude. In some circuit applications, such as power supplies and radio-interference filters, inductors are used to block high frequencies while permitting lower frequencies or dc to pass through. Thus, inductors are also called *chokes,* because they choke off high-frequency currents.

EXAMPLE 22.2 Inductance in a radio

The current amplitude in an inductor in a radio receiver is to be 250 μA when the voltage amplitude is 3.60 V at a frequency of 1.60 MHz (corresponding to the upper end of the AM broadcast band). What inductive reactance is needed? What inductance is needed?

SOLUTION

SET UP AND SOLVE From Equation 22.12,

$$X_L = \frac{V_L}{I} = \frac{3.60 \text{ V}}{250 \times 10^{-6} \text{ A}} = 1.44 \times 10^4 \ \Omega = 14.4 \text{ k}\Omega.$$

From Equation 22.11, with $\omega = 2\pi f$, we find that

$$L = \frac{X_L}{2\pi f} = \frac{1.44 \times 10^4 \ \Omega}{2\pi(1.60 \times 10^6 \text{ Hz})} = 1.43 \times 10^{-3} \text{ H} = 1.43 \text{ mH}.$$

REFLECT The inductance L is constant, but the inductive reactance X_L is proportional to frequency. If the frequency approaches zero, the inductive reactance also approaches zero.

Practice Problem: A generator contains a 3.5 mH inductor. Find the inductive reactance in the circuit when the generator frequency is 2.5×10^3 Hz. *Answer:* $X_L = 55 \ \Omega$.

Capacitor in an ac Circuit

Finally, suppose we connect a capacitor with capacitance C to the source, as in Figure 22.8a, producing a current $i = I\cos\omega t$ through the capacitor. You may object that charge can't really move *through* the capacitor, because its two plates are insulated from each other. True enough, but at each instant, as the capacitor charges and discharges, we have a current i into one plate and an equal current out of the other plate, just as though the charge were being conducted through the capacitor. For this reason, we often speak about alternating current *through* a capacitor.

To find the voltage v_C of point a with respect to point b, we first note that v_C is the charge q on a capacitor plate, divided by the capacitance C: $v_C = q/C$ (as we learned in Section 18.7). Thus, the rate of change of v_C is equal to $1/C$ times the rate of change of q, where q is in turn equal to the current i in the circuit. That is,

$$\frac{\Delta v_C}{\Delta t} = \frac{1}{C}\frac{\Delta q}{\Delta t} = \frac{i}{C}.$$

If the current i is once again given by $i = I\cos\omega t$, then

$$\frac{\Delta v_C}{\Delta t} = \frac{1}{C}I\cos\omega t.$$

The current is proportional to the rate of change of the voltage. The current is greatest when v_C is increasing most rapidly, and it is zero when v_C instantaneously levels off at a maximum or minimum value. In short, the graph of v_C as a function of time must resemble a sine curve, as shown in Figure 22.8b.

In fact, it can be shown, using methods of calculus, that

$$v_C = \frac{I}{\omega C}\sin\omega t. \qquad (22.13)$$

As Figure 22.8b shows, the current is greatest when the v_C curve is rising or falling most steeply (and the capacitor is charging or discharging most rapidly) and zero when the v_C curve instantaneously levels off at its maximum and minimum values.

The capacitor voltage and current are out of phase by a quarter cycle. The peaks of voltage occur a quarter cycle *after* the corresponding current peaks, and we say that the voltage *lags* the current by 90°. The phasor diagram in Figure 22.8c

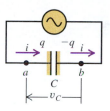

(a) Circuit with ac source and capacitor

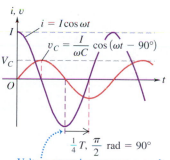

Voltage curve *lags* current curve by a quarter cycle (corresponding to $\phi = \pi/2$ rad = 90°).

(b) Graphs of current and voltage versus time

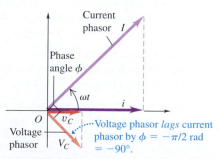

(c) Phasor diagram

▲ **FIGURE 22.8** Current and voltage in a capacitance C connected across an ac source.

shows this relationship; the voltage phasor is *behind* the current phasor by a quarter cycle, 90°, or $\pi/2$.

We can also derive this phase difference by rewriting Equation 22.13, using the identity $\cos(A - 90°) = \sin A$:

$$v_C = \frac{I}{\omega C}\cos(\omega t - 90°). \qquad (22.14)$$

This cosine function has a "late start" of 90° compared with that for the current.

Equations 22.13 and 22.14 show that the *maximum* voltage V_C (the voltage amplitude) is

$$V_C = \frac{I}{\omega C}. \qquad (22.15)$$

To put this expression in the same form as that for a resistor $(V_R = IR)$, we define a quantity X_C, called the **capacitive reactance** of the capacitor, as follows:

> **Capacitive reactance**
> We define the capacitive reactance of a capacitor as the inverse of the product of the angular frequency and the capacitance:
>
> $$X_C = \frac{1}{\omega C}. \qquad (22.16)$$

Then

$$V_C = IX_C. \qquad (22.17)$$

Because X_C is the ratio of a voltage to a current, its unit is the same as that for resistance and inductive reactance: the ohm.

The capacitive reactance of a capacitor is inversely proportional both to the capacitance C and to the angular frequency ω. The greater the capacitance and the higher the frequency, the *smaller* is the reactance X_C. Capacitors tend to pass high-frequency current and to block low-frequency and dc currents, just the opposite of inductors.

Conceptual Analysis 22.2

Reactance and current

In Figure 22.9, the current in circuit 1, containing an inductor, has an angular frequency ω_1, and the current in circuit 2, containing a capacitor, has an angular frequency ω_2. What happens to the brightness of each lightbulb if we keep the source voltage constant while we increase ω_1 and decrease ω_2?

A. Both bulbs get brighter.
B. The brightness of each bulb remains constant.
C. Both bulbs get dimmer.

SOLUTION For the inductor circuit, $X_L = \omega_1 L$, so the inductive reactance increases with increasing angular frequency ω_1. Since $X_C = 1/\omega_2 C$, the reactance of the capacitor circuit increases as its angular frequency ω_2 decreases. The current amplitude

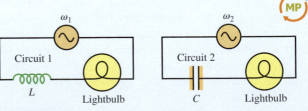

▲ **FIGURE 22.9**

through each element for a given voltage amplitude V is given by $I = V/X_L$ and $I = V/X_C$, respectively. We see that the currents in both circuits decrease as their reactances increase, so both bulbs get dimmer.

EXAMPLE 22.3 **Resistor and capacitor in series**

A 300 Ω resistor is connected in series with a 5.0 μF capacitor, as shown in Figure 22.10. The voltage across the resistor is $v_R = (1.20\text{ V}) \cos(2500\text{ rad/s})t$. **(a)** Derive an expression for the circuit current. **(b)** Determine the capacitive reactance of the capacitor. **(c)** Derive an expression for the voltage v_C across the capacitor.

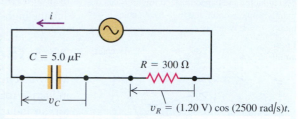

$C = 5.0\ \mu\text{F}$ $R = 300\ \Omega$

$v_R = (1.20\text{ V})\cos(2500\text{ rad/s})t.$

▲ **FIGURE 22.10**

SOLUTION

SET UP AND SOLVE **Part (a):** We recall that circuit elements in series have the same current. We can thus use $v_R = iR$ and solve for i:

$$i = \frac{v_R}{R} = \frac{(1.20\text{ V})\cos(2500\text{ rad/s})t}{300\ \Omega}$$

$$= (4.0 \times 10^{-3}\text{ A})\cos(2500\text{ rad/s})t.$$

Part (b): The expression for v_R tells us that $\omega = 2500$ rad/s. Therefore, we can use Equation 22.16 to find the capacitive reactance of the capacitor:

$$X_C = \frac{1}{\omega C} = \frac{1}{(2500\text{ rad/s})}\frac{1}{(5.0 \times 10^{-6}\text{ F})} = 80\ \Omega.$$

Part (c): First we use Equation 22.17 to find the amplitude V_C of the voltage across the capacitor:

$$V_C = IX_C = (4.0 \times 10^{-3}\text{ A})(80\ \Omega) = 0.32\text{ V}.$$

The capacitor voltage lags the current by 90°. From Equation 22.14, the instantaneous capacitor voltage is

$$v_C = \frac{I}{\omega C}\cos(\omega t - 90°)$$

$$v_C = (0.32\text{ V})\cos[(2500\text{ rad/s})t - \pi/2\text{ rad}].$$

REFLECT We have converted 90° to $\pi/2$ rad, so all the angular quantities have the same units. In ac circuit analysis, phase angles are often given in degrees, so be careful to convert to radians when necessary.

Practice Problem: A 200 Ω resistor is connected in series with a 5.8 μF capacitor. The voltage across the resistor is $v_R = (1.20\text{ V})\cos(2500\text{ rad/s})t$. Determine the capacitive reactance of the capacitor. *Answer:* $X_C = 69\ \Omega$.

TABLE 22.1 Circuit elements with alternating current

Circuit element	Circuit quantity	Amplitude relation	Phase of v
Resistor	R	$V_R = IR$	In phase with i
Inductor	$X_L = \omega L$	$V_L = IX_L$	Leads i by 90°
Capacitor	$X_C = 1/\omega C$	$V_C = IX_C$	Lags i by 90°

Table 22.1 summarizes the relations between voltage and current amplitudes for the three circuit elements that we've discussed.

The graphs in Figure 22.11 show how the resistance of a resistor and the reactances of an inductor and a capacitor vary with angular frequency. As the frequency approaches infinity, the reactance of the inductor approaches infinity and that of the capacitor approaches zero. As the frequency approaches zero, the inductive reactance approaches zero and the capacitive reactance approaches infinity. The limiting case of zero frequency corresponds to a dc circuit; in that case, there is *no* current through a capacitor, because $X_C \to \infty$, and there is no inductive effect, because $X_L \to 0$.

Figure 22.12 shows an application of the preceding discussion to a loudspeaker system. The woofer and tweeter are connected in parallel across the amplifier output. The capacitor in the tweeter branch blocks the low-frequency components of sound, but passes the higher frequencies; the inductor in the woofer

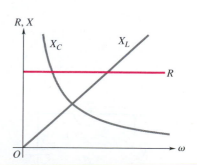

▲ **FIGURE 22.11** Graphs of R, X_L, and X_C as functions of angular frequency ω.

The inductor and capacitor feed low frequencies mainly to the woofer and high frequencies mainly to the tweeter.

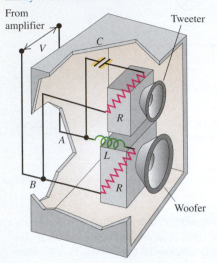

(a) A crossover network in a loudspeaker system

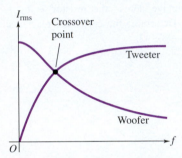

(b) Graphs of rms current as functions of frequency for a given amplifier voltage

▲ **FIGURE 22.12** How capacitance and inductance can be used to direct frequencies to the tweeter and woofer of a speaker.

Mastering PHYSICS®

PhET: Circuit Construction Kit (AC + DC)

branch does the opposite. Thus, the low-frequency sounds are routed to the woofer and the high-frequency sounds to the tweeter.

22.3 The Series R–L–C Circuit

Many ac circuits that are used in practical electronic systems involve resistance, inductive reactance, and capacitive reactance. A series circuit containing a resistor, an inductor, and a capacitor is shown in Figure 22.13a. To analyze this and similar circuits, we'll use a phasor diagram that includes the voltage and current phasors for each of the components. In this circuit, because of Kirchhoff's loop rule, the instantaneous *total* voltage v_{ad} across all three components is equal to the source voltage at that instant. We'll show that the phasor representing this total voltage is the *vector sum* of the phasors for the individual voltages. The complete phasor diagram for the circuit is shown in Figure 22.13b. This diagram may appear complicated; we'll explain it one step at a time.

Let's assume that the source supplies a current i given by $i = I\cos\omega t$. Because the circuit elements are connected in series, the current i at any instant is the same at every point in the circuit. Thus, a *single phasor I*, with length proportional to the current amplitude, represents the current in *all* circuit elements.

We use the symbols v_R, v_L, and v_C for the instantaneous voltages across R, L, and C, respectively, and we use V_R, V_L, and V_C for their respective maximum values (voltage amplitudes). We denote the instantaneous and maximum *source* voltages by v and V. Then $v = v_{ad}$, $v_R = v_{ab}$, $v_L = v_{bc}$, and $v_C = v_{cd}$.

We've shown that the potential difference between the terminals of a resistor is *in phase* with the current in the resistor and that its maximum value V_R is

$$V_R = IR. \tag{22.6}$$

The phasor V_R in Figure 22.13b, in phase with the current phasor I, represents the voltage across the resistor. Its projection onto the horizontal axis at any instant gives the instantaneous potential difference v_R.

The voltage across an inductor *leads* the current by 90°. Its voltage amplitude is

$$V_L = IX_L. \tag{22.12}$$

The phasor V_L in Figure 22.13b represents the voltage across the inductor, and its projection onto the horizontal axis at any instant equals v_L.

The voltage across a capacitor *lags* the current by 90°. Its voltage amplitude is

$$V_C = IX_C. \tag{22.17}$$

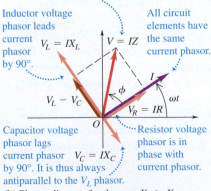

Source voltage phasor is the vector sum of the V_R, V_L, and V_C phasors.

Inductor voltage phasor leads current phasor by 90°. $V_L = IX_L$

All circuit elements have the same current phasor. $V = IZ$

$V_L - V_C$

$V_R = IR$

Capacitor voltage phasor lags current phasor by 90°. It is thus always antiparallel to the V_L phasor. $V_C = IX_C$

Resistor voltage phasor is in phase with current phasor.

If $X_L < X_C$, the source voltage phasor lags the current phasor, $X < 0$, and ϕ is a negative angle between 0 and $-90°$.

$V_R = IR$

$V_L = IX_L$

$V = IZ$

$V_L - V_C$

$V_C = IX_C$

(a) Series R-L-C circuit

(b) Phasor diagram for the case $X_L > X_C$

(c) Phasor diagram for the case $X_L < X_C$

▲ **FIGURE 22.13** Current and voltage in a series R–L–C circuit.

The phasor V_C in Figure 22.13b represents the voltage across the capacitor, and its projection onto the horizontal axis at any instant equals v_C.

The instantaneous potential difference v between terminals a and d is equal at every instant to the (algebraic) sum of the potential differences v_R, v_L, and v_C. That is, it equals the sum of the *projections* of the phasors V_R, V_L, and V_C. But the sum of the projections of these phasors is equal to the *projection* of their *vector* sum. So the vector sum V must be the phasor that represents the source voltage v and the instantaneous total voltage v_{ad} across the series of elements.

To form this vector sum, we first subtract the phasor V_C from the phasor V_L. (These two phasors always lie along the same line and have opposite directions, because V_L is always 90° ahead of the I phasor, and V_C is always 90° behind it.) This gives the phasor $V_L - V_C$, which is always at right angles to the phasor V_R. So, from the Pythagorean theorem, the magnitude of the phasor V is

$$V = \sqrt{V_R^2 + (V_L - V_C)^2} = \sqrt{(IR)^2 + (IX_L - IX_C)^2}$$
$$= I\sqrt{R^2 + (X_L - X_C)^2}. \tag{22.18}$$

The quantity $(X_L - X_C)$ is called the **reactance** of the circuit, denoted by X:

$$X = X_L - X_C. \tag{22.19}$$

Finally, we define the **impedance** Z of the circuit as

$$Z = \sqrt{R^2 + (X_L - X_C)^2} = \sqrt{R^2 + X^2}. \tag{22.20}$$

We can now rewrite Equation 22.18 as

$$V = IZ. \tag{22.21}$$

Equation 22.21 again has the same form as $V_R = IR$, with the impedance Z playing the same role as the resistance R in a dc circuit. Note, however, that the impedance is actually a function of R, L, and C, as well as of the angular frequency ω. The complete expression for Z for a series circuit is as follows:

Impedance of a series *R–L–C* circuit

$$Z = \sqrt{R^2 + X^2}$$
$$= \sqrt{R^2 + (X_L - X_C)^2} \tag{22.22}$$
$$= \sqrt{R^2 + [\omega L - (1/\omega C)]^2}.$$

Impedance is always a ratio of a voltage to a current; the SI unit of impedance is the ohm.

Equation 22.22 gives the impedance Z only for a *series R–L–C* circuit, but, using Equation 22.21, we can *define* the impedance of *any* network as the ratio of the voltage amplitude to the current amplitude.

The angle ϕ shown in Figure 22.13b is the phase angle of the source voltage v with respect to the current i. From the diagram,

$$\tan\phi = \frac{V_L - V_C}{V_R} = \frac{I(X_L - X_C)}{IR} = \frac{X_L - X_C}{R} = \frac{X}{R},$$
$$\phi = \arctan\left(\frac{\omega L - 1/\omega C}{R}\right). \tag{22.23}$$

If $X_L > X_C$, as in Figure 22.13b, the source voltage leads the current by an angle ϕ between 0 and 90°. If the current is $i = I\cos\omega t$, then the source voltage v is

$$v = V\cos(\omega t + \phi).$$

If $X_L < X_C$, as in Figure 22.13c, then vector V lies on the opposite side of the current vector I and the voltage *lags* the current. In this case, $X = X_L - X_C$ is a *negative* quantity, $\tan\phi$ is negative, and ϕ is a negative angle between 0 and $-90°$.

> **NOTE ▶** All the relations that we have developed for a series R–L–C circuit are still valid even if one of the circuit elements is missing. If the resistor is missing, we set $R = 0$; if the inductor is missing, we set $L = 0$. But if the capacitor is missing, we set $C = \infty$, corresponding to zero potential difference ($v = q/C$) or to zero capacitive reactance ($X_C = 1/\omega C$). ◀

Conceptual Analysis 22.3

Limiting behavior of impedance

Figure 22.14 shows a series R–L–C circuit constructed by a student in a lab class. The resistance is due to the filament of a lightbulb. When the student closes the switch, the bulb does not light up. Which of the following is or are possible explanations?

A. The inductance is too small.
B. The frequency is too small.
C. The capacitance is too great.
D. The frequency is too great.

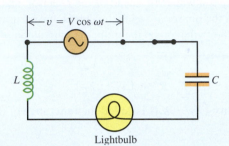

▲ **FIGURE 22.14**

SOLUTION Since the bulb won't light and the current through the bulb is $I = V/Z$, we are looking for causes of high impedance. The impedance $Z = \sqrt{R^2 + [\omega L - (1/\omega C)]^2}$ approaches infinity in both the limits $\omega \to 0$ (from the $1/\omega C$ term) and $\omega \to \infty$ (from the ωL term). Thus, choices B and D are possible explanations.

PROBLEM-SOLVING STRATEGY 22.1 **Alternating-current circuits**

SET UP

1. In ac circuit problems, it is nearly always easiest to work with angular frequency ω. But you may be given the ordinary frequency f, expressed in hertz. Don't forget to convert, using $\omega = 2\pi f$.

2. Keep in mind a few basic facts about phase relationships: For a resistor, voltage and current are always *in phase,* and the two corresponding phasors in a phasor diagram always have the same direction. For an inductor, the voltage always *leads* the current by 90° (that is, $\phi = +90°$) and the voltage phasor is always turned 90° counterclockwise from the current phasor. For a capacitor, the voltage always *lags* the current by 90° (that is, $\phi = -90°$) and the voltage phasor is always turned 90° clockwise from the current phasor.

SOLVE

3. Remember that Kirchhoff's rules are applicable to ac circuits. All the voltages and currents are sinusoidal functions of time instead of being constant, but Kirchhoff's rules hold at each instant. Thus, in a series circuit, the instantaneous current is the same in all circuit elements; in a parallel circuit, the instantaneous potential difference is the same across all circuit elements.

REFLECT

4. Reactance and impedance are analogous to resistance: Each represents the ratio of voltage amplitude V to current amplitude I in a circuit element or combination of elements. But keep in mind that phase relations play an essential role: Resistance and reactance have to be combined by *vector* addition of the corresponding phasors. For example, when you have several circuit elements in series, you can't just *add* all the numerical values of resistance and reactance; doing that would ignore the phase relations.

EXAMPLE 22.4   An *R–L–C* circuit

In the series circuit of Figure 22.13a, suppose $R = 300 \; \Omega$, $L = 60$ mH, $C = 0.50 \; \mu$F, $V = 50$ V, and $\omega = 10{,}000$ rad/s. Find the reactances X_L, X_C, and X, the impedance Z, the current amplitude I, the phase angle ϕ, and the voltage amplitude across each circuit element.

SOLUTION

SET UP If the circuit and phasor diagrams were not already provided in Figure 22.13, we would sketch them as the first step in this problem.

SOLVE From Equations 22.11 and 22.16, the reactances are

$$X_L = \omega L = (10{,}000 \text{ rad/s})(60 \times 10^{-3} \text{ H}) = 600 \; \Omega,$$

$$X_C = \frac{1}{\omega C} = \frac{1}{(10{,}000 \text{ rad/s})(0.50 \times 10^{-6} \text{ F})} = 200 \; \Omega.$$

The reactance X of the circuit is

$$X = X_L - X_C = 600 \; \Omega - 200 \; \Omega = 400 \; \Omega,$$

and the impedance Z is

$$Z = \sqrt{R^2 + X^2} = \sqrt{(300 \; \Omega)^2 + (400 \; \Omega)^2} = 500 \; \Omega.$$

With source voltage amplitude $V = 50$ V, the current amplitude I is

$$I = \frac{V}{Z} = \frac{50 \text{ V}}{500 \; \Omega} = 0.10 \text{ A}.$$

The phase angle ϕ is

$$\phi = \arctan \frac{X_L - X_C}{R} = \arctan \frac{400 \; \Omega}{300 \; \Omega} = 53°.$$

Because the phase angle ϕ is positive, the voltage *leads* the current by 53°. From Equation 22.6, the voltage amplitude V_R across the resistor is

$$V_R = IR = (0.10 \text{ A})(300 \; \Omega) = 30 \text{ V}.$$

From Equation 22.12, the voltage amplitude V_L across the inductor is

$$V_L = IX_L = (0.10 \text{ A})(600 \; \Omega) = 60 \text{ V}.$$

From Equation 22.17, the voltage amplitude V_C across the capacitor is

$$V_C = IX_C = (0.10 \text{ A})(200 \; \Omega) = 20 \text{ V}.$$

REFLECT Note that because of the phase differences between voltages across the separate elements, the source voltage amplitude $V = 50$ V is *not* equal to the sum of the voltage amplitudes across the separate circuit elements. That is, 50 V ≠ 30 V + 60 V + 20 V. These voltages must be combined by *vector* addition of the corresponding phasors, *not* by simple numerical addition.

In this problem, the phase angle ϕ is positive, so the voltage *leads* the current by an angle (between 0 and 90°) equal to ϕ. If ϕ had turned out to be negative, then the voltage would *lag* the current by that angle.

Practice Problem: In a series circuit, suppose $R = 100 \; \Omega$, $L = 200$ mH, $C = 0.60 \; \mu$F, $V = 60$ V, and $\omega = 4000$ rad/s. Find the reactances X_L and X_C, the impedance Z, the current amplitude I, the phase angle ϕ, and the amplitude across each circuit element. *Answers:* $X_L = 800 \; \Omega$; $X_C = 420 \; \Omega$; $Z = 400 \; \Omega$; $I = 0.15$ A, $\phi = 75°$; $V_R = 15$ V, $V_L = 120$ V; $V_C = 63$ V.

In this entire discussion, we've described magnitudes of voltages and currents in terms of their *maximum* values: the voltage and current *amplitudes*. But we remarked at the end of Section 22.1 that these quantities are usually described not in terms of their amplitudes but in terms of rms values. For any sinusoidally varying quantity, the rms value is always $1/\sqrt{2}$ times the amplitude. All the relations between voltage and current that we've derived in this and the preceding sections are still valid if we use rms quantities throughout instead of amplitudes. For example, if we divide Equation 22.21 by $\sqrt{2}$, we get

$$\frac{V}{\sqrt{2}} = \frac{I}{\sqrt{2}} Z,$$

which we can rewrite as

$$V_{\text{rms}} = I_{\text{rms}} Z. \tag{22.24}$$

We can translate Equations 22.6, 22.12, 22.17, and 22.18 in exactly the same way.

Finally, we remark that what we have been describing throughout this section is the *steady-state* condition of a circuit: the state that exists after the circuit has been connected to the source for a long time. When the source is first connected, there may be additional voltages and currents, called *transients*, whose nature

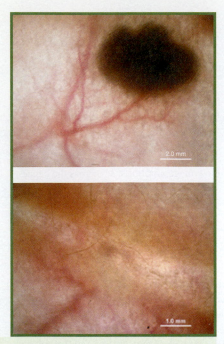

depends on the time in the cycle when the circuit is initially completed. A detailed analysis of transients is beyond our scope. They always die out after a sufficiently long time, and they do not affect the steady-state behavior of the circuit. But they can cause dangerous and damaging surges in power lines, and delicate electronic systems such as computers should always be provided with power-line surge protectors.

22.4 Power in Alternating-Current Circuits

Alternating currents play a central role in systems for distributing, converting, and using electrical energy, so it's important to look at power relationships in ac circuits. When a source with voltage amplitude V and instantaneous potential difference v supplies an instantaneous current i (with current amplitude I) to an ac circuit, the instantaneous power p that it supplies is $p = vi$. Let's first see what this means for individual circuit elements.

We'll assume in each case that $i = I\cos\omega t$. Suppose first that the circuit consists of a *pure resistance R*, as in Figure 22.6; then i and v are *in phase*. We obtain the graph representing p by multiplying the heights of the graphs of v (red curve) and i (purple curve) in Figure 22.6b at each instant. This graph is shown as the black curve in Figure 22.15a. The product vi is always positive, because v and i are always either both positive or both negative. Energy is supplied to the resistor at every instant for both directions of i, although the power is not constant.

The power curve is symmetrical about a value equal to one-half of its maximum value VI, so the *average power P* is

$$P = \tfrac{1}{2}VI. \tag{22.25}$$

An equivalent expression is

$$P = \frac{V}{\sqrt{2}}\frac{I}{\sqrt{2}} = V_{\text{rms}}I_{\text{rms}}. \tag{22.26}$$

Also, $V_{\text{rms}} = I_{\text{rms}}R$, so we can express P in any of these equivalent forms:

$$P = I_{\text{rms}}^2 R = \frac{V_{\text{rms}}^2}{R} = V_{\text{rms}}I_{\text{rms}}. \tag{22.27}$$

Note that the preceding expressions have the same form as the corresponding relations for a dc circuit, Equation 19.10. Note also that they are valid only for pure resistors, not for more complicated combinations of circuit elements.

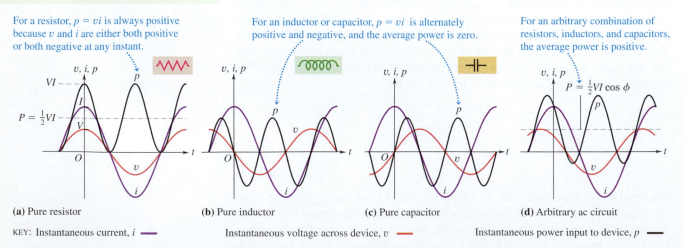

For a resistor, $p = vi$ is always positive because v and i are either both positive or both negative at any instant.

For an inductor or capacitor, $p = vi$ is alternately positive and negative, and the average power is zero.

For an arbitrary combination of resistors, inductors, and capacitors, the average power is positive.

(a) Pure resistor   **(b)** Pure inductor   **(c)** Pure capacitor   **(d)** Arbitrary ac circuit

KEY: Instantaneous current, i ——   Instantaneous voltage across device, v ——   Instantaneous power input to device, p ——

▲ **FIGURE 22.15** Graphs of voltage, current, and power as functions of time for various circuits.

Next, we connect the source to a pure inductor L, as in Figure 22.7. The voltage leads the current by 90°. When we multiply the curves of v and i, the product vi is *negative* during the half of the cycle when v and i have *opposite* signs. We get the power curve in Figure 22.15b, which is symmetrical about the horizontal axis. It is positive half the time and negative the other half, and the *average* power is zero. When p is positive, energy is being supplied to set up the magnetic field in the inductor; when p is negative, the field is collapsing and the inductor is returning energy to the source. The net energy transfer over one cycle is zero.

Finally, we connect the source to a pure capacitor C, as in Figure 22.8. The voltage lags the current by 90°. Figure 22.15c shows the power curve; the average power is again zero. Energy is supplied to charge the capacitor and is returned to the source when the capacitor discharges. The net energy transfer over one cycle is again zero.

In *any* ac circuit, with any combination of resistors, capacitors, and inductors, the voltage v has some phase angle ϕ with respect to the current i, and the instantaneous power p is given by

$$p = vi = [V\cos(\omega t + \phi)][I\cos\omega t]. \tag{22.28}$$

The instantaneous power curve has the form shown in Figure 22.15d. The area under the positive loops is greater than that under the negative loops, and the average power is positive.

To derive an expression for the *average* power, which we'll denote by capital P, we use the identity for the cosine of the sum of two angles:

$$\cos(a + b) = \cos a \cos b - \sin a \sin b.$$

Applying this identity to Equation 22.28, we find:

$$p = [V(\cos\omega t \cos\phi - \sin\omega t \sin\phi)][I\cos\omega t]$$
$$= VI\cos\phi \cos^2\omega t - VI\sin\phi \cos\omega t \sin\omega t.$$

From the discussion leading to Equation 22.3 in Section 22.1, we see that in the first term the average value of $\cos^2\omega t$ (over one cycle) is $\frac{1}{2}$. In the second term, the average value of $\cos\omega t \sin\omega t$ is zero, because this product is equal to $\frac{1}{2}\sin 2\omega t$, whose average over a cycle is zero. So the average power P is

$$P = \frac{1}{2}VI\cos\phi = V_{rms}I_{rms}\cos\phi. \tag{22.29}$$

When v and i are *in phase*, $\phi = 0$, $\cos\phi = 1$, and the average power equals $V_{rms}I_{rms}$ (which also equals $\frac{1}{2}VI$). When v and i are 90° *out of phase*, the average power is zero. In the general case, when v has phase angle ϕ with respect to i, the average power equals $\frac{1}{2}I$ multiplied by $V\cos\phi$, the component of V that is *in phase* with I. The relationship of the current and voltage phasors for this case is shown in Figure 22.16. For the series R–L–C circuit, $V\cos\phi$ is the voltage amplitude for the resistor, and Equation 22.29 is the power dissipated in the resistor. The power dissipation in the inductor and capacitor is zero.

The factor $\cos\phi$ is called the **power factor** of the circuit. For a pure resistance, $\phi = 0$, $\cos\phi = 1$, and $P = V_{rms}I_{rms}$. For a pure (resistanceless) capacitor or inductor, $\phi = \pm90°$, $\cos\phi = 0$, and $P = 0$. For a series R–L–C circuit, the power factor is equal to R/Z. Can you prove this?

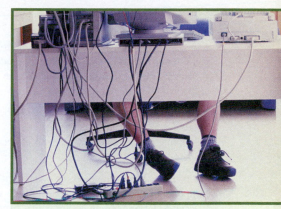

▲ Application **Too much power.** Many types of modern electronic equipment, such as computers and DVD players, can be damaged irreversibly by abrupt changes in electrical current known as power surges. Power surges can be caused by nearby lightning strikes, by malfunctioning transformers, or by switching on a large piece of electrical equipment such as a compressor or elevator that can momentarily disrupt the line voltage. A common type of surge protector has a component called a metal oxide varistor (a variable resistor containing semiconductors connecting the hot wire to the ground wire). At normal line voltage, the varistor resistance is very high and the current travels past it to the outlet. However, if the voltage exceeds a certain limit, the varistor resistance drops dramatically and the current flows through it directly to the ground wire, bypassing the outlet and protecting your computer from being "fried."

Average power $= \frac{1}{2}I(V\cos\phi)$, where $V\cos\phi$ is the component of V in phase with I.

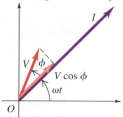

▲ **FIGURE 22.16** Average power in an ac circuit

Conceptual Analysis 22.4

Comparing power

Figure 22.17 shows phasors for two R–L–C circuits. The circuits have the same current amplitude I; circuit 1 has voltage amplitude V_1 and circuit 2 has voltage amplitude V_2. Which of the following statements is correct?

A. Circuit 1 has greater average power because it has a larger power factor: $\cos\phi_1 > \cos\phi_2$.
B. Circuit 2 has greater average power because its resistor dissipates more power.

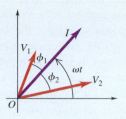

▲ **FIGURE 22.17**

SOLUTION The average power in a circuit is proportional to the component of the voltage that is in phase with the current: $P = \frac{1}{2}(V\cos\phi)I$. For a *single circuit* with a fixed V and I, the average power increases or decreases according to the power factor $\cos\phi$. Circuit 1 has a greater power factor $(\cos\phi_1 > \cos\phi_2)$. However, we're comparing the component of the voltage that is in phase with the current between *two circuits*. By inspection of Figure 22.17, $V_2\cos\phi_2 > V_1\cos\phi_1$, so choice B is correct. In addition, as stated in choice B, it's always true that the average power in an R–L–C circuit depends only on the energy dissipated through its resistance.

EXAMPLE 22.5 An electric hair dryer

An electric hair dryer is rated at 1500 W at 120 V. Calculate **(a)** the resistance, **(b)** the rms current, and **(c)** the maximum instantaneous power of the dryer. Assume pure resistance.

SOLUTION

SET UP AND SOLVE Part (a): We solve Equation 22.27 for R and substitute the given values:

$$R = \frac{V_{rms}^2}{P} = \frac{(120\ \text{V})^2}{1500\ \text{W}} = 9.6\ \Omega.$$

Part (b): From Equation 22.26,

$$I_{rms} = \frac{P}{V_{rms}} = \frac{1500\ \text{W}}{120\ \text{V}} = 12.5\ \text{A}.$$

Or, from Equation 22.6,

$$I_{rms} = \frac{V_{rms}}{R} = \frac{120\ \text{V}}{9.6\ \Omega} = 12.5\ \text{A}.$$

Part (c): The maximum instantaneous power is VI; from Equation 22.29,

$$VI = 2P = 2(1500\ \text{W}) = 3000\ \text{W}.$$

REFLECT To mislead the unwary consumer, some manufacturers of stereo amplifiers state power outputs in terms of the peak value rather than the lower average power. Caveat emptor!

Practice Problem: A toaster is rated at 900 W at 120 V. Calculate the resistance and the rms current. *Answers:* $R = 16\ \Omega$; $I_{rms} = 7.5\ \text{A}$.

EXAMPLE 22.6 Power factor for an *R*–*L*–*C* circuit

For the series R–L–C circuit of Example 22.4, calculate the power factor and the average power to the entire circuit and to each circuit element.

SOLUTION

SET UP AND SOLVE The power factor is $\cos\phi = \cos 53° = 0.60$.
From Equation 22.29, the average power to the circuit is

$$P = \tfrac{1}{2}VI\cos\phi = \tfrac{1}{2}(50\ \text{V})(0.10\ \text{A})(0.60) = 1.5\ \text{W}.$$

REFLECT All of this power is dissipated in the resistor; the average power to a pure inductor or pure capacitor is always zero.

Practice Problem: If the inductance L in this circuit could be changed, what value of L would give a power factor of unity? *Answer:* 20 mH.

A low power factor (large angle of lag or lead) is usually undesirable in power circuits because, for a given potential difference, a large current is needed to supply a given amount of power. This results in large I^2R losses in the transmission lines. Your electric power company may charge a higher rate to a client with a low power factor. Many types of ac machinery draw a lagging current; the power factor can be corrected by connecting a capacitor in parallel with the load. The leading current drawn by the capacitor compensates for the lagging current in the other branch of the circuit. The capacitor itself absorbs no net power from the line.

22.5 Series Resonance

The impedance of a series R–L–C circuit depends on the frequency, as the following equation shows:

$$Z = \sqrt{R^2 + X^2}$$
$$= \sqrt{R^2 + (X_L - X_C)^2} \qquad (22.22)$$
$$= \sqrt{R^2 + [\omega L - (1/\omega C)]^2}.$$

Figure 22.18a shows graphs of R, X_L, X_C, and Z as functions of ω. We have used a logarithmic angular frequency scale so that we can cover a wide range of frequencies. Because X_L increases and X_C decreases with increasing frequency, there is always one particular frequency at which X_L and X_C are equal and $X = X_L - X_C$ is zero. At this frequency, the impedance Z has its *smallest* value, equal to just the resistance R.

Suppose we connect an ac voltage source with constant voltage amplitude V, but variable angular frequency ω, across a series R–L–C circuit. As we vary ω, the current amplitude I varies with frequency as shown in Figure 22.18b; its *maximum* value occurs at the frequency at which the impedance Z attains its *minimum* value. This peaking of the current amplitude at a certain frequency is called **resonance.** The angular frequency ω_0 at which the resonance peak occurs is called the **resonance angular frequency.** This is the angular frequency at which the inductive and capacitive reactances are equal, so

$$X_L = X_C, \qquad \omega_0 L = \frac{1}{\omega_0 C}, \qquad \omega_0 = \frac{1}{\sqrt{LC}}. \qquad (22.30)$$

The **resonance frequency** f_0 is $\omega_0/2\pi$.

Now let's look at what happens to the *voltages* in a series R–L–C circuit at resonance. The current at any instant is the same in L and C. The voltage across an inductor always *leads* the current by 90°, or a quarter cycle, and the voltage across a capacitor always *lags* the current by 90°. Therefore, the instantaneous voltages across L and C always differ in phase by 180°, or a half cycle; they have opposite signs at each instant. If the *amplitudes* of these two voltages are equal, then they add to zero at each instant and the *total* voltage v_{bd} across the L–C combination in Figure 22.13a is exactly zero! This occurs only at the resonance frequency f_0.

NOTE ▶ Depending on the numerical values of R, L, and C, the voltages across L and C individually can be larger than that across R. Indeed, at frequencies close to resonance, the voltages across L and C individually can be *much larger* than the source voltage! ◀

The *phase* of the voltage relative to the current is given by Equation 22.23. At frequencies below resonance, X_C is greater than X_L; the capacitive reactance dominates, the voltage *lags* the current, and the phase angle ϕ is between zero and $-90°$. Above resonance, the inductive reactance dominates; the voltage *leads* the current, and the phase angle is between zero and $+90°$. At resonance, $\phi = 0$ and the power factor is $\cos\phi = 1$. This variation of ϕ with angular frequency is shown in Figure 22.18b.

When we vary the inductance L or the capacitance C of a circuit, we can also vary the resonance frequency. That's how some older radio or television receiving

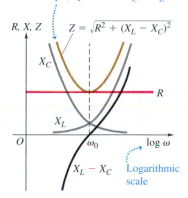

(a) Reactance, resistance, and impedance as functions of angular frequency

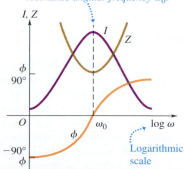

(b) Impedance, current, and phase angle as functions of angular frequency

▲ **FIGURE 22.18** Graphs showing the impedance minimum in a series R–L–C circuit.

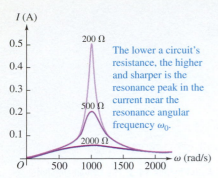

◄ FIGURE 22.19 Resonance peaks for R–L–C circuits with different resistances.

Text in figure: The lower a circuit's resistance, the higher and sharper is the resonance peak in the current near the resonance angular frequency ω_0.

sets were "tuned" to receive particular stations. In the early days of radio, this was accomplished by the use of capacitors with movable metal plates whose overlap could be varied to change C. Alternatively, L could be varied with the use of a coil with a ferrite core that slid in or out.

In a series R–L–C circuit, the impedance reaches its minimum value and the current reaches its maximum value at the resonance frequency. Figure 22.19 shows a graph of rms current as a function of frequency for such a circuit, with $V = 100$ V, $R = 500$ Ω, $L = 2.0$ H, and $C = 0.50\ \mu$F. This curve is called a *response curve,* or a *resonance curve.* The resonance angular frequency is $\omega_0 = 1/\sqrt{LC} = 1000$ rad/s. As we expect, the curve has a peak at this angular frequency.

The resonance frequency is determined by L and C; what happens when we change R? Figure 22.19 also shows graphs of I as a function of ω for $R = 200$ Ω and for $R = 2000$ Ω. The curves are all similar for frequencies far away from resonance, where the impedance is dominated by X_L or X_C. But near resonance, where X_L and X_C nearly cancel each other, the curve is higher and more sharply peaked for small values of R than for larger values. The maximum height of the curve is in fact inversely proportional to R: A small R gives a sharply peaked response curve, and a large value of R gives a broad, flat curve.

In the early days of radio and television, the shape of the response curve was of crucial importance. The sharply peaked curve made it possible to discriminate between two stations broadcasting on adjacent frequency bands. But if the peak was *too* sharp, some of the information in the received signal, such as the high-frequency sounds in music, was lost. A sharply peaked resonance curve corresponds to a small value of R and a lightly damped oscillating system; a broad flat curve goes with a large value of R and a heavily damped system.

Quantitative Analysis 22.5

The resonance peak

We draw the resonance curve for a circuit with electrical components L, R, and C and resonant frequency ω_0. If the values of L, R, and C are all doubled, how does the new resonance curve differ from the original one? Sketch the resonance curve for each case.

A. Peak twice as high, peak frequency twice as great.
B. Peak half as high, peak frequency twice as great.
C. Peak twice as high, peak frequency half as great.
D. Peak half as high, peak frequency half as great.

SOLUTION The greater the resistance, the flatter is the resonance peak. The maximum height is inversely proportional to R: When we double R, the peak height decreases to half its original value. This narrows the possibilities to choices B and D. Since $\omega_0 = 1/\sqrt{LC}$, doubling L and C halves the resonant frequency, so curve D is correct.

EXAMPLE 22.7 **Tuning a radio**

The series circuit in Figure 22.20 is similar to arrangements that are sometimes used in tuning circuits in simple radio receivers (often available in kit form). This circuit is connected to the terminals of an ac source with a constant rms terminal voltage of 1.0 V and a variable frequency. Find **(a)** the resonance frequency; **(b)** the inductive reactance, the capacitive reactance, the reactance, and the impedance at the resonance frequency; **(c)** the rms current at resonance; and **(d)** the rms voltage across each circuit element at resonance.

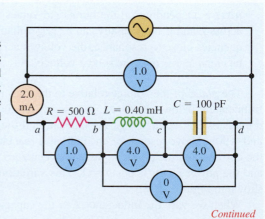

► FIGURE 22.20

Continued

SOLUTION

SET UP AND SOLVE Part (a): The resonance angular frequency is

$$\omega_0 = \frac{1}{\sqrt{LC}} = \frac{1}{\sqrt{(0.40 \times 10^{-3}\,\text{H})(100 \times 10^{-12}\,\text{F})}}$$

$$= 5.0 \times 10^6\,\text{rad/s}.$$

The corresponding frequency is $f = \omega/2\pi = 8.0 \times 10^5\,\text{Hz} = 800\,\text{kHz}$.

Part (b): At this frequency,

$$X_L = \omega L = (5.0 \times 10^6\,\text{rad/s})(0.40 \times 10^{-3}\,\text{H}) = 2000\,\Omega,$$

$$X_C = \frac{1}{\omega C} = \frac{1}{(5.0 \times 10^6\,\text{rad/s})(100 \times 10^{-12}\,\text{F})} = 2000\,\Omega,$$

$$X = X_L - X_C = 2000\,\Omega - 2000\,\Omega = 0.$$

From Equation 22.22, the impedance Z at resonance is equal to the resistance: $Z = R = 500\,\Omega$.

Part (c): At resonance, the rms current is

$$I = \frac{V}{Z} = \frac{V}{R} = \frac{1.0\,\text{V}}{500\,\Omega} = 0.0020\,\text{A} = 2.0\,\text{mA}.$$

Part (d): The rms potential difference across the resistor is

$$V_R = IR = (0.0020\,\text{A})(500\,\Omega) = 1.0\,\text{V}.$$

The rms potential differences across the inductor and capacitor are, respectively,

$$V_L = IX_L = (0.0020\,\text{A})(2000\,\Omega) = 4.0\,\text{V},$$
$$V_C = IX_C = (0.0020\,\text{A})(2000\,\Omega) = 4.0\,\text{V}.$$

The rms potential difference V_{bd} across the inductor–capacitor combination is

$$V_{bd} = IX = I(X_L - X_C) = 0.$$

REFLECT The frequency found in part (a) corresponds to the lower part of the AM radio band. At resonance, the instantaneous potential differences across the inductor and the capacitor have equal amplitudes; but they are 180° out of phase and so add to zero at each instant. Note also that, at resonance, V_R is equal to the source voltage V, but in this example, V_L and V_C are both considerably *larger* than V.

Practice Problem: In a radio tuning circuit, a 300 Ω resistor, a 0.50 mH inductor, and a 150 pF capacitor are connected in series with an rms terminal voltage of 1.5 V. Find the rms voltage across each circuit element at resonance. *Answers:* $V_R = 1.5$ V; $V_L = 9.1$ V; $V_C = 9.1$ V.

Resonance phenomena occur in all areas of physics; we have already seen one example in the forced oscillation of the harmonic oscillator (Section 11.6). In that case, the amplitude of a mechanical oscillation peaked at a driving-force frequency close to the natural frequency of the system. The behavior of the R–L–C circuit is analogous to this behavior. We suggest that you review that discussion now, looking for the analogies. Other important examples of resonance occur in acoustics, in atomic and nuclear physics, and in the study of fundamental particles (high-energy physics).

22.6 Parallel Resonance

A different kind of resonance occurs when a resistor, an inductor, and a capacitor are connected in *parallel,* as shown in Figure 22.21a. This circuit has resonance behavior similar to that of the series R–L–C circuit we analyzed in Section 22.5, but the roles of voltage and current are reversed. This time, the instantaneous potential difference is the same for all three circuit elements and is equal to the source voltage $v = V\cos\omega t$, but the current is different in each of the three elements.

The instantaneous current i_R through the resistor is in phase with the source voltage. Its peak value is $I_R = V/R$. As we discussed in Section 22.2, the instantaneous current i_L through the inductor lags the source voltage by 90° and has peak value $I_L = V/X_L = V/\omega L$. The instantaneous current i_C through the capacitor leads the source voltage by 90° and has peak value $I_C = V/X_C = V\omega C$. The phasor diagram is shown in Figure 22.21b.

Thus, at any frequency, the inductor current and the capacitor current differ in phase by exactly a half cycle (180°) and tend to cancel each other. At one particular frequency ω_0, the two reactances X_L and X_C are equal; at that frequency, the

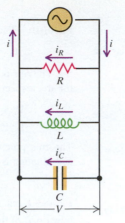

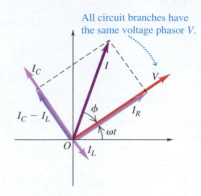

(a) A parallel *R-L-C* circuit

All circuit branches have the same voltage phasor *V*.

(b) Phasor diagram showing current phasors for the three branches

▲ **FIGURE 22.21** Voltage and current in a parallel *R–L–C* circuit.

inductor current and the capacitor current are equal in magnitude and opposite in direction at every instant and therefore add to zero. This occurs when

$$X_L = X_C, \qquad \omega_0 L = \frac{1}{\omega_0 C}, \qquad \text{and} \qquad \omega_0 = \frac{1}{\sqrt{LC}}.$$

We recognize this relation as the same as Equation 22.30, the condition for resonance in a *series R–L–C* circuit. The difference is that, at resonance, the total current through the parallel combination reaches a *minimum* because the total current through *L* and *C* is zero. Thus at resonance the impedance of the parallel circuit reaches a *maximum* value equal simply to $Z = R$.

A detailed analysis of the currents in the three branches shows that, at any frequency, the impedance *Z* of the parallel combination is given by

$$\frac{1}{Z} = \sqrt{\frac{1}{R^2} + \left(\omega C - \frac{1}{\omega L}\right)^2}. \tag{22.31}$$

This result confirms the earlier statement that at resonance $1/Z$ attains its minimum value and therefore *Z* attains its maximum value, $Z = R$.

NOTE ▶ The total current in a *parallel R–L–C* circuit is *minimum* at resonance. When $\omega C = 1/\omega L$, Equation 22.31 becomes simply $I = V/R$. This does *not* mean that there is *no* current in *L* or *C* at resonance, but only that the two currents cancel completely at every instant. If *R* is large, the impedance *Z* of the circuit near resonance is much *larger* than the individual reactances X_L and X_C, and the individual currents in *L* and *C* can be much larger than the total current. ◀

EXAMPLE 22.8 **More current is less current**

In the parallel circuit of Figure 22.21, suppose the circuit elements, the applied voltage, and the angular frequency have the same values as in Example 22.4, in which $R = 300\ \Omega$, $X_L = 600\ \Omega$, and $X_C = 200\ \Omega$. Determine **(a)** the impedance of the parallel combination, **(b)** the current amplitude for each element, and **(c)** the total current amplitude.

Continued

SOLUTION

SET UP AND SOLVE **Part (a):** The impedance Z is given by Equation 22.31. Substituting the values from Example 22.4 into this equation, we get

$$\frac{1}{Z} = \sqrt{\frac{1}{(300\ \Omega)^2} + \left(\frac{1}{200\ \Omega} - \frac{1}{600\ \Omega}\right)^2},$$
$$Z = 212\ \Omega.$$

Part (b): The current through the resistor is given by Equation 22.6:

$$I_R = \frac{V}{R} = \frac{50\ \text{V}}{300\ \Omega} = 0.167\ \text{A}.$$

The current through the inductor is found from Equation 22.12:

$$I_L = \frac{V}{X_L} = \frac{50\ \text{V}}{600\ \Omega} = 0.083\ \text{A}.$$

The current through the capacitor is given by Equation 22.17:

$$I_C = \frac{V}{X_C} = \frac{50\ \text{V}}{200\ \Omega} = 0.25\ \text{A}.$$

Part (c): The amplitude of the total current is

$$I = \frac{V}{Z} = \frac{50\ \text{V}}{212\ \Omega} = 0.24\ \text{A},$$

or

$$I = \sqrt{I_R^2 + (I_L - I_C)^2}$$
$$= \sqrt{(0.167\ \text{A})^2 + (0.083\ \text{A} - 0.25\ \text{A})^2}$$
$$= 0.24\ \text{A}.$$

REFLECT The amplitude I of the total current is less than that of the current through the capacitor; the inductor and capacitor currents partially cancel each other because they are a half cycle out of phase. Make sure you understand that, because of the phase differences of the individual currents, you *cannot* simply add the individual current amplitudes to get the amplitude of the total current.

Practice Problem: Find the resonance angular frequency, and find the circuit impedance and the total current at resonance, if $V = 50\ \text{V}$ as before. *Answers:* $5.8 \times 10^3\ \text{s}^{-1}$, $300\ \Omega$, $0.17\ \text{A}$.

SUMMARY

Phasors and Alternating Currents

(Section 22.1) An ac source produces an emf that varies sinusoidally with time. A sinusoidal voltage or current can be represented by a **phasor**—a vector that rotates counterclockwise with constant angular velocity ω equal to the angular frequency of the sinusoidal quantity. Its projection on the horizontal axis at any instant represents the instantaneous value of the quantity. For a sinusoidal current i with maximum value I, the phasor is given by $i = I\cos\omega t$ (Equation 22.2). In power calculations, it is useful to use the **root-mean-square** (rms) value: $I_{rms} = I/\sqrt{2}$ (Equation 22.3).

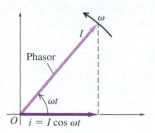

Resistance and Reactance

(Section 22.2) If the current is given by $i = I\cos\omega t$ (Equation 22.2) and the voltage v between two points is $v = V\cos(\omega t + \phi)$, then ϕ is called the **phase angle** of the voltage relative to the current.

The voltage across a resistor R is in phase with the current, and the voltage and current amplitudes are related by $V_R = IR$ (Equation 22.6). The voltage across an inductor L leads the current with a phase angle of $\phi = 90°$; the voltage and current amplitudes are related by $V_L = IX_L$ (Equation 22.12), where $X_L = \omega L$ (Equation 22.11) is the **inductive reactance** of the inductor. The voltage across a capacitor C lags the current with a phase angle $\phi = -90°$; the voltage and current amplitudes are related by $V_C = IX_C$ (Equation 22.17), where $X_C = 1/\omega C$ (Equation 22.16) is the **capacitive reactance** of the capacitor.

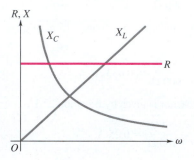

The Series R–L–C Circuit

(Section 22.3) In a series R–L–C circuit, the voltage and current amplitudes are related by $V = IZ$ (Equation 22.21), where Z is the **impedance** of the circuit: $Z = \sqrt{R^2 + [\omega L - (1/\omega C)]^2}$ (Equation 22.22). The phase angle ϕ of the voltage relative to the current is given by

$$\phi = \arctan\frac{\omega L - 1/\omega C}{R}. \qquad (22.23)$$

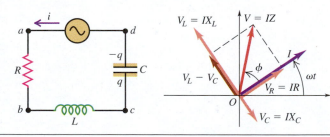

Power in Alternating-Current Circuits

(Section 22.4) The average power input P to an ac circuit is I times one-half the component of the voltage that is in phase with the current, or $P = \frac{1}{2}VI\cos\phi = V_{rms}I_{rms}\cos\phi$ (Equation 22.29), where ϕ is the phase angle of voltage with respect to current. Power is dissipated only through the resistor. For circuits containing only capacitors and inductors, $\phi = \pm90°$ and the average power is zero. The quantity $\cos\phi$ is called the **power factor.**

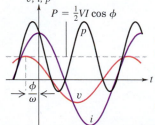

Graphs of p, v, and i versus time for an arbitrary combination of resistors, inductors, and capacitors. The average power is positive.

Series and Parallel Resonance

(Sections 22.5 and 22.6) The current in a series R–L–C circuit reaches a maximum, and the impedance reaches a minimum, at an angular frequency $\omega_0 = 1/(LC)^{1/2}$ known as the **resonance angular frequency.** This phenomenon is called **resonance.** At resonance, the voltage and current are in phase and the impedance Z is equal to the resistance R. The smaller the resistance, the sharper is the **resonance peak.** In an R–L–C parallel circuit, the total current attains a minimum, and the impedance attains a maximum, at the resonance angular frequency ω_0.

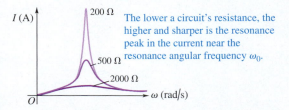

The lower a circuit's resistance, the higher and sharper is the resonance peak in the current near the resonance angular frequency ω_0.

Conceptual Questions

1. For a series ac R–L–C circuit, (a) why do the voltage amplitudes *not* obey the equation $V = V_R + V_L + V_C$; and (b) would the instantaneous voltages obey the equation $v = v_R + v_L + v_C$? Why or why not?
2. In Example 22.5, a hair dryer was treated as a pure resistor. But because there are coils in the heating element and in the motor that drives the blower fan, a hair dryer also has inductance. Qualitatively, does including an inductance increase or decrease the values of R, I_{rms}, and P?
3. Fluorescent lights often use an inductor, called a "ballast," to limit the current through the tubes. Why is it better to use an inductor than a resistor for this purpose?
4. At high frequencies, a capacitor becomes a short circuit. Discuss why this is so.
5. At high frequencies, an inductor becomes an open circuit. Discuss why this is so.
6. The current in an ac power line changes direction 120 times per second, and its average value is zero. So how is it possible for power to be transmitted in such a system?
7. Electric-power companies like to have their power factors as close to unity as possible. Why?
8. Some electrical appliances operate equally well on ac or dc, while others work only on ac or only on dc. Give examples of each, and explain the reasons for the differences.
9. When a series-resonant circuit is connected across a 120 V ac line, the voltage rating of the capacitor may be exceeded even if it is rated at 200 or 400 V. How can this be?
10. Is it possible for the power factor of an R–L–C ac series circuit to be zero? Justify your answer on *physical* grounds.
11. During the last quarter of the 19th century, there was great and acrimonious controversy over whether ac or dc should be used for power transmission. Edison favored dc, while George Westinghouse championed ac. What arguments might each have used to promote his scheme?
12. In an ac circuit, why is the average power delivered to a capacitor or an inductor equal to zero, and why is this not the case for a resistor?
13. dc voltage comes from batteries, but how is ac voltage generated?
14. In what ways is impedance similar to ordinary resistance, and in what ways is it different?
15. Transformers, such as those which plug into the wall socket for use with small electrical appliances, often feel warm to the touch. What is the source of this heat?

Multiple-Choice Problems

1. A piece of electrical equipment in an ac circuit draws a root-mean-square current of 5.00 A. The average current over each cycle is
   A. $5\sqrt{2}A = 7.07$ A.   B. 5.00 A.
   C. $5/\sqrt{2}A = 3.54$ A.   D. 0.
2. A sinusoidal current is described by $i = I\cos\omega t$, where $\omega = 1.57$ rad/s. At some time t', where 2 s $< t' < 4$ s, the current is $+3.0$ A. Which phasor can represent the current at time t'?

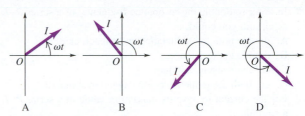

3. A lightbulb is the resistance in a series R–L–C circuit having an ac voltage source $v = V\cos\omega t$. As the frequency of the source is adjusted closer and closer to the value $1/\sqrt{LC}$, what happens to the brightness of the bulb?
   A. It increases.   B. It decreases.
   C. It does not change.
4. A series R–L–C ac circuit with a sinusoidal voltage source of angular frequency ω has a total reactance X. If this frequency is doubled, the reactance becomes
   A. $4X$.   B. $2X$.   C. $X/2$.   D. $X/4$.
   E. none of the above.
5. In a series R–L–C circuit powered by an ac sinusoidal voltage source, which phasor diagram best illustrates the relationship between the current i and the potential drop v_R across the *resistor*?

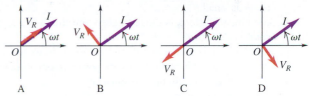

6. In a series R–L–C circuit powered by an ac sinusoidal voltage source, which phase diagram best illustrates the relationship between the current i and the potential drop v_C across the *capacitor*?

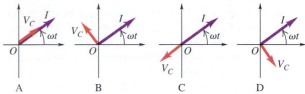

7. In a series R–L–C circuit powered by an ac sinusoidal voltage source, which phase diagram best illustrates the relationship between the current i and the potential drop v_L across the *inductor*?

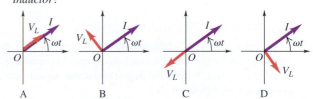

8. A series circuit contains an inductor, a resistor, a capacitor, and a sinusoidal voltage source of angular frequency ω. If we double this frequency (there may be more than one correct choice),
   A. the inductive reactance is doubled.
   B. the capacitive reactance is doubled.
   C. the total reactance is doubled.
   D. the impedance is doubled.

9. In order to double the resonance frequency of a series R–L–C ac circuit, you could
   A. double both the inductance and capacitance.
   B. double the resistance.
   C. cut the resistance in half.
   D. cut both the inductance and capacitance in half.

10. For the current to have its maximum value in a series R–L–C ac circuit,
    A. $\omega = 0$.   B. $\omega \to \infty$.   C. $\omega = 1/\sqrt{LC}$.

11. If the root-mean-square current in an ac circuit is 10.0 A, the current amplitude is approximately
    A. 7.07 A.   B. 5.00 A.   C. 14.1 A.   D. 20.0 A.

12. In a series R–L–C circuit, the voltage source produces an angular frequency ω. If this frequency is *increased* slightly (there may be more than one correct choice),
    A. The total impedance will *necessarily* increase.
    B. The reactance will *necessarily* increase.
    C. The impedance due to the resistor does not change.
    D. The total impedance increases if $\omega \geq 1/\sqrt{LC}$.

13. In an R–L–C ac series circuit, if the resistance is much smaller than the inductive or capacitive reactance and the system is operated at a frequency much higher than the resonance frequency,
    A. the phase angle ϕ is close to zero and the power factor $\cos\phi$ is small.
    B. the phase angle ϕ is close to zero and the power factor $\cos\phi$ is close to 1.
    C. the phase angle is close to 90° and the power factor is small.
    D. the phase angle is close to 90° and the power factor is close to 1.

14. In a series R–L–C ac circuit at resonance,
    A. The impedance is zero.
    B. The impedance has its maximum value.
    C. The reactance is equal to R.
    D. The total impedance has its minimum value, which is equal to R.

15. A circuit consists of a light bulb, a capacitor, and an inductor connected in series to an ac power source. The capacitor and the inductor have equal reactances. If both the capacitor and the inductor are removed from the circuit, what will happen to the brightness of the light bulb?
    A. It will increase.   B. It will decrease.
    C. It will remain the same.

Problems

22.1 Phasors and Alternating Currents

1. • You have a special lightbulb with a *very* delicate wire filament. The wire will break if the current in it ever exceeds 1.50 A, even for an instant. What is the largest root-mean-square current you can run through this bulb?

2. • The plate on the back of a certain computer scanner says that the unit draws 0.34 A of current from a 120 V, 60 Hz line. Find (a) the root-mean-square current, (b) the current amplitude, (c) the average current, and (d) the average square of the current.

22.2 Resistance and Reactance

3. • A capacitance C and an inductance L are operated at the same angular frequency. (a) At what angular frequency will they have the same reactance? (b) If $L = 5.00$ mH and $C = 3.50$ μF, what is the numerical value of the angular frequency in part (a), and what is the reactance of each element?

4. • (a) Compute the reactance of a 0.450 H inductor at frequencies of 60.0 Hz and 600 Hz. (b) Compute the reactance of a 2.50 μF capacitor at the same frequencies. (c) At what frequency is the reactance of a 0.450 H inductor equal to that of a 2.50 μF capacitor?

5. •• **A radio inductor.** You want the current amplitude through a 0.450-mH inductor (part of the circuitry for a radio receiver) to be 2.60 mA when a sinusoidal voltage with amplitude 12.0 V is applied across the inductor. What frequency is required?

6. • A 2.20 μF capacitor is connected across an ac source whose voltage amplitude is kept constant at 60.0 V, but whose frequency can be varied. Find the current amplitude when the angular frequency is (a) 100 rad/s; (b) 1000 rad/s; (c) 10,000 rad/s.

7. • The voltage amplitude of an ac source is 25.0 V, and its angular frequency is 1000 rad/s. Find the current amplitude if the capacitance of a capacitor connected across the source is (a) 0.0100 μF, (b) 1.00 μF, (c) 100 μF.

8. • Find the current amplitude if the self-inductance of a resistanceless inductor that is connected across the source of the previous problem is (a) 0.0100 H, (b) 1.00 H, (c) 100 H.

22.3 The Series R–L–C Circuit

9. • A sinusoidal ac voltage source in a circuit produces a maximum voltage of 12.0 V and an rms current of 7.50 mA. Find (a) the voltage and current amplitudes and (b) the rms voltage of this source.

10. • A 65 Ω resistor, an 8.0 μF capacitor, and a 35 mH inductor are connected in series in an ac circuit. Calculate the impedance for a source frequency of (a) 300 Hz and (b) 30.0 kHz.

11. • In an R–L–C series circuit, the rms voltage across the resistor is 30.0 V, across the capacitor it is 90.0 V, and across the inductor it is 50.0 V. What is the rms voltage of the source?

12. •• A 1500 Ω resistor is connected in series with a 350 mH inductor and an ac power supply. At what frequency will this combination have twice the impedance that it has at 120 Hz?

13. •• (a) Compute the impedance of a series R–L–C circuit at angular frequencies of 1000, 750, and 500 rad/s. Take $R = 200$ Ω, $L = 0.900$ H, and $C = 2.00$ μF. (b) Describe how the current amplitude varies as the angular frequency of the source is slowly reduced from 1000 rad/s to 500 rad/s. (c) What is the phase angle of the source voltage with respect to the current when $\omega = 1000$ rad/s? (d) Construct a phasor diagram when $\omega = 1000$ rad/s.

14. •• A 200 Ω resistor is in series with a 0.100 H inductor and a 0.500 μF capacitor. Compute the impedance of the circuit and draw the phasor diagram (a) at a frequency of 500 Hz, (b) at a frequency of 1000 Hz. In each case, compute the phase angle of the source voltage with respect to the current and state whether the source voltage lags or leads the current.

22.4 Power in Alternating-Current Circuits

15. • The power of a certain CD player operating at 120 V rms is 20.0 W. Assuming that the CD player behaves like a pure resistance, find (a) the maximum instantaneous power, (b) the rms current, and (c) the resistance of this player.

16. • A series R–L–C circuit is connected to a 120 Hz ac source that has $V_{rms} = 80.0$ V. The circuit has a resistance of 75.0 Ω and an impedance of 105 Ω at this frequency. What average power is delivered to the circuit by the source?

17. • The circuit in Problem 13 carries an rms current of 0.250 A with a frequency of 100 Hz. (a) What is the average rate at which electrical energy is converted to heat in the resistor? (b) What average power is delivered by the source? (c) What is the average rate at which electrical energy is dissipated (converted to other forms) in the capacitor? in the inductor?

18. •• A series ac circuit contains a 250 Ω resistor, a 15 mH inductor, a 3.5 μF capacitor, and an ac power source of voltage amplitude 45 V operating at an angular frequency of 360 rad/s. (a) What is the power factor of this circuit? (b) Find the average power delivered to the entire circuit. (c) What is the average power delivered to the resistor, to the capacitor, and to the inductor?

22.5 Series Resonance

19. • An ac series R–L–C circuit contains a 120 Ω resistor, a 2.0 μF capacitor, and a 5.0 mH inductor. Find (a) the resonance angular frequency and (b) the length of time that each cycle lasts at the resonance angular frequency.

20. • (a) At what angular frequency will a 5.00 μF capacitor have the same reactance as a 10.0 mH inductor? (b) If the capacitor and inductor in part (a) are connected in an L–C circuit, what will be the resonance angular frequency of that circuit?

21. •• In an R-L-C series circuit, $R = 150$ Ω, $L = 0.750$ H, and $C = 0.0180$ μF. The source has voltage amplitude $V = 150$ V and a frequency equal to the resonance frequency of the circuit. (a) What is the power factor? (b) What is the average power delivered by the source? (c) The capacitor is replaced by one with $C = 0.0360$ μF and the source frequency is adjusted to the new resonance value. Then what is the average power delivered by the source?

22. • You need to make a series ac circuit having a resonance angular frequency of 1525 rad/s using a 138 Ω resistor, a 10.5 μF capacitor, and an inductor. (a) What should be the inductance of the inductor, and (b) what is the impedance of this circuit when you use it with an ac voltage source having an angular frequency of 1525 Hz?

23. •• A series circuit consists of an ac source of variable frequency, a 115 Ω resistor, a 1.25 μF capacitor, and a 4.50 mH inductor. Find the impedance of this circuit when the angular frequency of the ac source is adjusted to (a) the resonance angular frequency, (b) twice the resonance angular frequency, and (c) half the resonance angular frequency.

24. • In a series R–L–C circuit, $R = 400$ Ω, $L = 0.350$ H, and $C = 0.0120$ μF. (a) What is the resonance angular frequency of the circuit? (b) The capacitor can withstand a peak voltage of 550 V. If the voltage source operates at the resonance frequency, what maximum voltage amplitude can it have if the maximum capacitor voltage is not exceeded?

25. • In a series R–L–C circuit, $L = 0.200$ H, $C = 80.0$ μF, and the voltage amplitude of the source is 240 V. (a) What is the resonance angular frequency of the circuit? (b) When the source operates at the resonance angular frequency, the current amplitude in the circuit is 0.600 A. What is the resistance R of the resistor? (c) At the resonance frequency, what are the peak voltages across the inductor, the capacitor, and the resistor?

26. •• In an R–L–C series circuit, $R = 300$ Ω, $L = 0.400$ H, and $C = 6.00 \times 10^{-8}$ F. When the ac source operates at the resonance frequency of the circuit, the current amplitude is 0.500 A. (a) What is the voltage amplitude of the source? (b) What is the amplitude of the voltage across the resistor, across the inductor, and across the capacitor? (c) What is the average power supplied by the source?

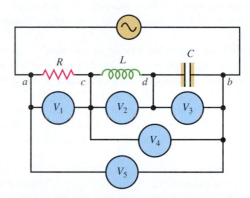

▲ **FIGURE 22.22** Problems 26, 35, and 36.

22.6 Parallel Resonance

27. • A 125 Ω resistor, an 8.50 μF capacitor, and an 11.2 mH inductor are all connected in parallel across an ac voltage source of variable frequency. (a) At what angular frequency will the impedance have its maximum value, and (b) what is that value?

28. • For the circuit in Figure 22.23, $R = 300$ Ω, $L = 0.500$ H, and $C = 0.600$ μF. The voltage amplitude of the source is 120 V. (a) What is the resonance frequency of the circuit? (b) Sketch the phasor diagram at the resonance frequency. (c) At the resonance frequency, what is the current amplitude through the source? (d) At the resonance frequency, what is the current amplitude through the resistor? Through the inductor? Through the branch containing the capacitor?

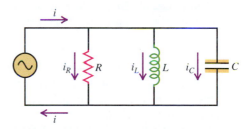

▲ **FIGURE 22.23** Problem 28.

29. • For the circuit in Figure 22.23, $R = 200$ Ω, $L = 0.800$ H, and $C = 5.00$ μF. When the source is operated at the resonance frequency, the current amplitude in the inductor is 0.400 A. Determine the current amplitude (a) in the branch containing the capacitor and (b) through the resistor.

30. •• (a) Use the phasor diagram for a parallel R–L–C circuit (see Figure 22.21) to show that the current amplitude I for the current i through the source is given by $I = \sqrt{I_R^2 + (I_C - I_L)^2}$. (b) Show that the result of part (a) can be written as $I = V/Z$, with $1/Z = \sqrt{1/R^2 + (\omega C - 1/\omega L)^2}$.

General Problems

31. •• A coil has a resistance of 48.0 Ω. At a frequency of 80.0 Hz, the voltage across the coil leads the current in it by 52.3°. Determine the inductance of the coil.

32. •• A large electromagnetic coil is connected to a 120 Hz ac source. The coil has resistance 400 Ω, and at this source frequency the coil has inductive reactance 250 Ω. (a) What is the inductance of the coil? (b) What must the rms voltage of the source be if the coil is to consume an average electrical power of 800 W?

33. •• A parallel-plate capacitor having square plates 4.50 cm on each side and 8.00 mm apart is placed in series with an ac source of angular frequency 650 rad/s and voltage amplitude 22.5 V, a 75.0 Ω resistor, and an ideal solenoid that is 9.00 cm long, has a circular cross section 0.500 cm in diameter, and carries 125 coils per centimeter. What is the resonance angular frequency of this circuit? (See problem 36 in Chapter 21.)

34. •• At a frequency ω_1, the reactance of a certain capacitor equals that of a certain inductor. (a) If the frequency is changed to $\omega_2 = 2\omega_1$, what is the ratio of the reactance of the inductor to that of the capacitor? Which reactance is larger? (b) If the frequency is changed to $\omega_3 = \omega_1/3$, what is the ratio of the reactance of the inductor to that of the capacitor? Which reactance is larger?

35. •• Five voltmeters, calibrated to read rms values, are connected as shown in Figure 22.22. Let $R = 200 \, \Omega$, $L = 0.400$ H, and $C = 6.00 \, \mu$F. The source voltage amplitude is $V = 30.0$ V. What is the reading of each voltmeter if (a) $\omega = 200$ rad/s; (b) $\omega = 1000$ rad/s?

36. •• Consider the circuit sketched in Figure 22.22. The source has a voltage amplitude of 240 V, $R = 150 \, \Omega$, and the reactance of the capacitor is 600 Ω. The voltage amplitude across the capacitor is 720 V. (a) What is the current amplitude in the circuit? (b) What is the impedance? (c) What two values can the reactance of the inductor have?

37. •• In a series R–L–C circuit, the components have the following values: $L = 20.0$ mH, $C = 140$ nF, and $R = 350 \, \Omega$. The generator has an rms voltage of 120 V and a frequency of 1.25 kHz. Determine (a) the power supplied by the generator; and (b) the power dissipated in the resistor.

38. •• (a) Show that for an R–L–C series circuit the power factor is equal to R/Z. (*Hint:* Use the phasor diagram; see Figure 22.13b.) (b) Show that for any ac circuit, not just one containing pure resistance only, the average power delivered by the voltage source is given by $P_{av} = I_{rms}^2 R$.

39. •• In an R–L–C series circuit the magnitude of the phase angle is 54.0°, with the source voltage lagging the current. The reactance of the capacitor is 350 Ω, and the resistor resistance is 180 Ω. The average power delivered by the source is 140 W. Find (a) the reactance of the inductor; (b) the rms current; (c) the rms voltage of the source.

40. •• In a series R–L–C circuit, $R = 300 \, \Omega$, $X_C = 300 \, \Omega$, and $X_L = 500 \, \Omega$. The average power consumed in the resistor is 60.0 W. (a) What is the power factor of the circuit? (b) What is the rms voltage of the source?

41. •• In a series R–L–C circuit, the phase angle is 40.0°, with the source voltage leading the current. The reactance of the capacitor is 400 Ω, and the resistance of the resistor is 200 Ω. The average power delivered by the source is 150 W. Find (a) the reactance of the inductor, (b) the rms current, (c) the rms voltage of the source.

42. •• A 100.0 Ω resistor, a 0.100 μF capacitor, and a 300.0 mH inductor are connected in series to a voltage source with amplitude 240 V. (a) What is the resonance angular frequency? (b) What is the maximum current in the resistor at resonance? (c) What is the maximum voltage across the capacitor at resonance? (d) What is the maximum voltage across the inductor at resonance? (e) What is the maximum energy stored in the capacitor at resonance? in the inductor?

43. •• Consider the same circuit as in the previous problem, with the source operated at an angular frequency of 400 rad/s. (a) What is the maximum current in the resistor? (b) What is the maximum voltage across the capacitor? (c) What is the maximum voltage across the inductor? (d) What is the maximum energy stored in the capacitor? in the inductor?

Passage Problems

BIO Converting dc to ac. Individual cells such as eggs are often organized spatially, as manifested in part by asymmetries in the cell membrane. These asymmetries include non-uniform distributions of ion transport mechanisms that result in net electrical current entering one region of the membrane and leaving another. Because these steady cellular currents may regulate cell polarity, leading (in the case of eggs) to embryonic polarity, scientists are interested in measuring them.

The cellular currents move in loops through the extracellular fluid around the cells. Ohm's Law requires that there be voltage differences between any two points in the fluid near current-producing cells. While the currents may be significant, the extracellular voltage differences are tiny, on the order of nanovolts. If the voltage differences in the medium near a cell can be mapped, the current density can be calculated using Ohm's Law. One way to measure these voltage differences might be to use two electrodes spaced 10 or 20 μm apart, but that fails because the dc impedance (the resistance) of such electrodes is high and the inherent noise in these high-impedance electrodes swamps the cellular voltages.

One successful method uses a platinum ball electrode moved sinusoidally between two points near a cell. The electrical potential that the electrode measures (with respect to a distant reference electrode) also varies sinusoidally, so the dc potential difference between the two extremes of the electrode's excursion is converted to a sine-wave ac potential difference. The platinum electrode behaves as a capacitor in series with the resistance of the fluid, called the access resistance. The access resistance (R_A) has a value of about, $\rho/(10a)$, where ρ is the resistivity of the medium (usually expressed in $\Omega \cdot$ cm) and a is the radius of the electrode. The platinum ball typically has a diameter of 20 μm and a capacitance of 10 nF, and the resistivity of many biological fluids is 100 $\Omega \cdot$ cm.

44. What is the dc impedance of the electrode, assuming that it behaves as an ideal capacitor?
A. 0   B. Infinite   C. $\sqrt{2} \times 10^4 \, \Omega$ D. $\sqrt{2} \times 10^6 \, \Omega$

45. If the electrode is oscillated between two points 20 μ apart with a frequency of $(5000/\pi)$ Hz, what is the impedance of the electrode?
A. 0   B. Infinite   C. $\sqrt{2} \times 10^4 \, \Omega$ D. $\sqrt{2} \times 10^6 \, \Omega$

46. The signal from the oscillating electrode is fed into an amplifier, which reports the measured voltage as an rms value, V_{rms}. However, the number of interest for analyzing currents driven by the cell is the peak-to-peak voltage difference (V_{pp}), that is, the voltage difference between the two extremes of the electrode's excursion. What is the value of V_{pp} in terms of V_{rms}.?
A. $V_{rms}/\sqrt{2}$ B. $V_{rms}/2\sqrt{2}$ C. $\sqrt{2}V_{rms}$ D. $2\sqrt{2}V_{rms}$

23 Electromagnetic Waves

Energy from the sun, an essential requirement for life on earth, reaches us by means of electromagnetic waves (light) that travel through 93 million miles of (nearly) empty space. Electromagnetic waves occur in an astonishing variety of physical situations, including TV and radio transmission, cellular phones, microwave oscillators for ovens and radar, lightbulbs, x-ray machines, and radioactive nuclei. So it's important for us to make a careful study of their properties and behavior.

The existence of electromagnetic waves depends on two facts. First, a time-varying magnetic field acts as a source of an electric field, as implied by Faraday's law of induction. Second (and less familiar), a time-varying electric field acts as a source of a magnetic field, as we'll discuss later. Thus, a time-varying field of either kind gives rise to a time-varying field of the other kind in neighboring regions of space. In this way, time-varying electric and magnetic fields can propagate through space as a *wave*. Such waves carry energy and momentum and have the property of polarization. A wave can be *sinusoidal,* in which case the $\vec{E}$ and $\vec{B}$ fields are sinusoidal functions of time. The spectrum of electromagnetic waves covers an extremely broad range of frequencies and wavelengths. In particular, light consists of electromagnetic waves, and our study of optics in later sections of this chapter will be based in part on the electromagnetic nature of light.

The telescope in the foreground is Gemini North, one of a pair of twin 8.1-m telescopes located on Mauna Kea in Hawaii (Gemini North) and in the Chilean Andes (Gemini South). Together, these two telescopes can see the entire visible universe.

23.1 Introduction to Electromagnetic Waves

In the last several chapters, we've studied various aspects of electric and magnetic fields. When the fields don't vary with time, such as an electric field produced by charges at rest, or the magnetic field of a steady current, we can analyze the electric and magnetic fields independently, without considering interactions between them. But when the fields vary with time, they're no longer independent.

Faraday's law (Section 21.3) tells us that a time-varying magnetic field acts as a source of an electric field, as shown by induced emf's in inductances and transformers. In 1865, the Scottish mathematician and physician James C. Maxwell proposed that a time-varying *electric* field could play the same role as a current, as an additional source of *magnetic* field. This two-way interaction between the two fields can be summarized elegantly in a set of four equations now called *Maxwell's equations.* We won't study these equations here because they require calculus, but you may encounter them in a later course.

Thus, when *either* an electric or a magnetic field is changing with time, a field of the other kind is induced in adjacent regions of space. In this way, time-varying electric and magnetic fields can propagate through space from one region to another, even when there is no matter in the intervening region. Such a propagating disturbance is called an **electromagnetic wave.**

One outcome of Maxwell's analysis was his prediction that an electromagnetic disturbance should propagate in free space with a speed equal to that of light and that light waves were therefore very likely to be electromagnetic in nature. In 1887, Heinrich Hertz used oscillating *L–C* circuits (discussed in Section 21.12) to produce, in his laboratory, electromagnetic waves with wavelengths of the order of a few meters. Hertz also produced electromagnetic *standing waves* and measured the distance between adjacent nodes (one half-wavelength) to determine their wavelength λ. Knowing the resonant frequency f of his circuits, he then determined the speed v of the waves from the wavelength–frequency relation $v = \lambda f$. In this way, Hertz confirmed that the wave speed was the same as that of light, thus verifying Maxwell's theoretical prediction directly. The SI unit of frequency, one cycle per second, is named the *hertz* in honor of the great German scientist.

Maxwell's synthesis, wrapping up the basic principles of electromagnetism neatly and elegantly in four equations, stands as a towering intellectual achievement. It is comparable to the Newtonian synthesis we described at the end of Section 6.5 and to the 20th-century development of relativity, quantum mechanics, and the understanding of DNA. All are beautiful, and all are monuments to the achievements of which the human intellect is capable!

23.2 Speed of an Electromagnetic Wave

We're now ready to introduce the basic ideas of electromagnetic waves and their relation to the principles of electromagnetism. To start, we'll postulate a simple field configuration that has wavelike behavior. We'll assume an electric field that has only a *y* component and a magnetic field with only a *z* component, and we'll assume that both fields move together in the $+x$ direction with a speed c that is initially unknown. Then we'll ask whether these fields are physically possible—that is, whether they are consistent with the laws of electromagnetism we've studied.

A Simple Plane Electromagnetic Wave

Using an *x-y-z* coordinate system (Figure 23.1), we imagine that all space is divided into two regions by a plane perpendicular to the *x* axis (parallel to the *y-z* plane). At every point to the left of this plane there exist a uniform electric field $\vec{E}$ in the $+y$ direction and a uniform magnetic field $\vec{B}$ in the $+z$ direction, as shown. Furthermore, we suppose that the boundary plane, which we call the *wave front,* moves to the right with a constant speed *c,* as yet unknown. Thus, the $\vec{E}$ and $\vec{B}$ fields travel to the right into previously field-free regions with a definite speed. The situation, in short, describes a rudimentary electromagnetic wave.

▲ **Application Making waves** To produce its radio signal, a radio station sends an appropriately modulated alternating current into the antenna of its radio mast. The time-varying current creates changing electric and magnetic fields, and an electromagnetic wave is emitted. When they reach your antenna, the process is reversed: The changing fields create an emf in the antenna; the radio selects and amplifies a narrow band of frequencies emitted by a particular station.

A rudimentary electromagnetic wave. The electric and magnetic fields are uniform behind the advancing wave front and zero in front of it

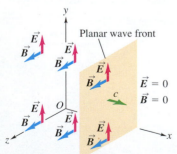

▲ **FIGURE 23.1** A planar electromagnetic wave front.

We won't concern ourselves with the problem of actually *producing* such a field configuration. Instead, we simply ask whether it is consistent with the laws of electromagnetism.

The analysis leading to the answer to this question is somewhat complicated, so we'll simply state the answer: It is: Yes, this primitive electromagnetic wave *is* consistent with the laws of electromagnetism, provided that two conditions are satisfied: (1) When an electromagnetic wave travels in vacuum, the magnitudes of $\vec{E}$ and $\vec{B}$ are in a definite, constant ratio:

$$E = cB. \tag{23.1}$$

(2) The wave front moves with a speed c given by

$$c = \frac{1}{\sqrt{\epsilon_0 \mu_0}}, \tag{23.2}$$

where ϵ_0 and μ_0 are the proportionality constants we encountered in Coulomb's law (Equation 17.1) and the law of Biot and Savart (Equation 20.16), respectively. Inserting the numerical values of these quantities, we find that

$$c = \frac{1}{\sqrt{[8.85 \times 10^{-12}\ C^2/(N \cdot m^2)](4\pi \times 10^{-7}\ N/A^2)}}$$
$$= 3.00 \times 10^8\ m/s.$$

Our assumed wave is consistent with the principles of electromagnetism, provided that the wave front moves with the speed c just given. We recognize this as the speed of light in vacuum! We shouldn't be too surprised by this result; the constant ϵ_0 appears in the equation that relates an electric field to its sources, and μ_0 plays a similar role for magnetic fields. Since electromagnetic waves depend on the interaction between these two fields, we might expect that the two constants should play a central role in their propagation.

We've chosen a simple and primitive wave for our study in order to avoid mathematical complications, but this special case illustrates several important features of *all* electromagnetic waves:

Characteristics of electromagnetic waves in vacuum

1. The wave is **transverse:** Both $\vec{E}$ and $\vec{B}$ are perpendicular to the direction of propagation of the wave and to each other.
2. There is a definite ratio between the magnitudes of $\vec{E}$ and $\vec{B}$: $E = cB$.
3. The wave travels in vacuum with a definite and unchanging speed c.
4. Unlike mechanical waves, which need the oscillating particles of a medium such as water or air to be transmitted, electromagnetic waves require no medium. What's "waving" in an electromagnetic wave are the electric and magnetic fields.

NOTE ▸ The relation $E = cB$ is correct only in the SI unit system. It would be wrong to conclude that E is larger than B. Comparing E and B is like comparing apples and oranges: They are different physical quantities and have different units. ◂

Figure 23.2 shows a "right-hand rule" to determine the directions of $\vec{E}$ and $\vec{B}$: Point the thumb of your right hand in the direction the wave is traveling. Imagine rotating the $\vec{E}$ field vector 90° in the sense your fingers curl; that gives you the direction of the $\vec{B}$ field.

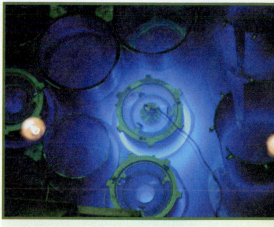

▲ **Application The eerie glow of faster-than-light travel** Nuclear reactors and their radioactive fuel rods are often submerged in pools of water that absorb their radiation. These pools glow with the gorgeous cerulean blue light you see in the photo. This glow, an example of a phenomenon called Cerenkov radiation, is caused by energetic electrons that travel *faster* than the speed of light in water. Nothing can travel faster than the speed of light in vacuum, but there's no law against exceeding the speed of light in a given medium. The fast electrons, which originate in nuclear reactions, interact with atoms in the water, which emit light. Because of the electrons' speed, these interactions are confined to a shock wave like that of a supersonic jet, and the emitted light is mainly blue.

Right-hand rule for an electromagnetic wave:

① Point the thumb of your right hand in the wave's direction of propagation.

② Imagine rotating the $\vec{E}$ field vector 90° in the sense your fingers curl.

That is the direction of the $\vec{B}$ field.

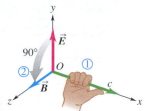

▲ **FIGURE 23.2** The right-hand rule for the directions of $\vec{E}$ and $\vec{B}$ in an electromagnetic wave.

Conceptual Analysis 23.1

The directions of $\vec{E}$ and $\vec{B}$

Consider a simple plane electromagnetic wave that is traveling vertically upward with its electric field pointing eastward. In which direction does the magnetic field point?

A. North
B. South
C. West

SOLUTION Using the right-hand rule just given, we point the thumb of our right hand in the direction of the wave's propagation (upward). We now imagine rotating the $\vec{E}$ vector 90° in the direction our fingers curl. It starts out pointing east and ends up pointing north, so northward is the direction of $\vec{B}$.

EXAMPLE 23.1 **The speed of light in vacuum**

(a) What distance does light travel in 1 second, expressed in units of the circumference of the earth? The equatorial circumference of the earth is 4.01×10^7 m. (b) How much time is required for light to travel 1 foot?

SOLUTION

SET UP AND SOLVE Part (a): In 1 second, light travels a distance

$$x = ct = (3.00 \times 10^8 \text{ m/s})(1 \text{ s}) = 3.00 \times 10^8 \text{ m}$$
$$= (3.00 \times 10^8 \text{ m})\left(\frac{1 \text{ circumference unit}}{4.01 \times 10^7 \text{ m}}\right)$$
$$= 7.48 \text{ circumferences.}$$

Part (b): With 1 ft = 0.3048 m, the time t for light to travel 1 foot is

$$t = \frac{x}{c} = \frac{0.3048 \text{ m}}{3.00 \times 10^8 \text{ m/s}} = 1.02 \times 10^{-9} \text{ s.}$$

REFLECT The straight-line distance light travels in 1 second is about 7.5 circular trips around the earth. The time to travel 1 ft is 1.02×10^{-9} s = 1.02 ns. A useful approximation is that light travels at about 1 ft/ns.

Practice Problem: How far does light travel in a year (a light year)? *Answer:* $x = 9.46 \times 10^{15}$ m.

23.3 The Electromagnetic Spectrum

Electromagnetic waves cover an extremely broad spectrum of wavelengths and frequencies. Radio and TV transmission, visible light, infrared and ultraviolet radiation, x rays, and gamma rays all form parts of the **electromagnetic spectrum.** The extent of this spectrum is shown in Figure 23.3, which gives approximate wavelength and frequency ranges for the various segments. Despite vast

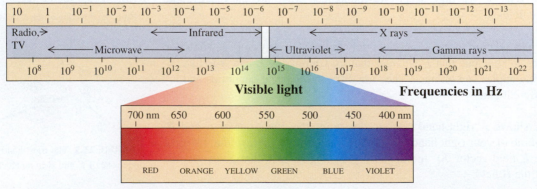

▲ **FIGURE 23.3** The electromagnetic spectrum. The frequencies and wavelengths found in nature extend over such a wide range that we must use a logarithmic scale to graph them. The boundaries between bands are somewhat arbitrary.

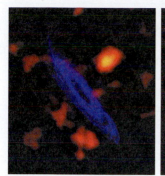   

Radio This image combines a high-resolution radio image of the galaxy disk (blue) with a larger-scale image of the surrounding regions of space (red). Both images are sensitive to hydrogen gas. The clouds of hydrogen gas surrounding the galaxy are invisible in other spectral bands.

Infrared The wavelengths used for this image are particularly sensitive to the warm dust present in star-forming regions. Because these regions occur in galactic arms, the image shows the galaxy's arm structure especially clearly.

Visible In the visible we see mainly the light of stars, blocked in places by dark streamers of dust. Regions containing mostly old stars appear yellow white; star-forming regions containing young stars have a blue cast.

X-ray This x-ray image of the central part of the galaxy's disk shows many point sources of x rays, including a cluster near the galaxy's center. These points are mostly *x-ray binaries* containing a normal star orbiting a neutron star or black hole. The hot disk around the galaxy's central black hole also radiates x rays.

▲ **FIGURE 23.4** Images of the nearby Andromeda galaxy taken in several spectral regions. The Andromeda galaxy is a large spiral galaxy like our own Milky Way; the Andromeda and Milky Way galaxies dominate the local group of galaxies. Except for the visible-light image, these images are all in false color.

differences in their uses and means of production, all these electromagnetic waves have the general characteristics described in the preceding sections, including the common propagation speed $c = 3.00 \times 10^8$ m/s (in vacuum). All are the same in principle; they differ in frequency f and wavelength λ, but the relation $c = \lambda f$ holds for each.

As Figure 23.3 shows, we can detect only a very small segment of this spectrum directly through our sense of sight. Within the visually detectable range, we perceive wavelength (or frequency) in terms of color, from long-wavelength red to short-wavelength violet. Some animals (including bees and birds) can see into the ultraviolet; pit vipers use their pit organs to "see" infrared radiation.

Light from many familiar sources (including the sun) is a mixture of many different wavelengths. By using special sources or filters, we can select a narrow band of wavelengths with a range of, say, from 500 to 501 nm. Such light is approximately *monochromatic* (single-color) light. Absolutely monochromatic light with only a single wavelength is an unattainable idealization. When we use the expression "monochromatic light with wavelength 500 nm" with reference to a laboratory experiment, we really mean a small band of wavelengths *around* 500 nm. One distinguishing characteristic of light from a *laser* is that it is much more nearly monochromatic than light produced in any other way.

Despite the limitations of the human eye, science and technology use all parts of the electromagnetic spectrum. Figure 23.4 shows how wavelengths in the radio, infrared, visible, and x-ray parts of the spectrum are used to explore the Andromeda galaxy, a neighbor of our own Milky Way.

23.4 Sinusoidal Waves

Sinusoidal electromagnetic waves are analogous to sinusoidal transverse mechanical waves on a stretched string. We studied mechanical waves in Chapter 12; we suggest that you review that discussion. In a sinusoidal electromagnetic wave, the

▲ **BIO Application Ultraviolet vision.**
What we call "visible light" is just the part of the electromagnetic spectrum that human eyes see. Many other animals would define "visible" somewhat differently. For instance, many animals, including insects and birds, see into the UV, and the natural world is full of signals that they see and we don't. The left-hand photo shows how a black-eyed Susan looks to us; the right-hand photo (in false color) shows the same flower in UV light. The bees that pollinate these flowers see the prominent central spot that is invisible to us. Similarly, many birds—including bluebirds, budgies, parrots, and even peacocks—have ultraviolet patterns that make them even more vivid to each other than they are to us.

$\vec{E}$ and $\vec{B}$ fields at any point in space are sinusoidal functions of time, and at any instant of time the *spatial* variation of the fields is also sinusoidal.

Some sinusoidal electromagnetic waves share with the primitive wave we described in Section 23.2 the property that at any instant the fields are uniform over any plane perpendicular to the direction of propagation (as shown in Figure 23.1). Such a wave is called a **plane wave.** The entire pattern travels in the direction of propagation with speed c. The directions of $\vec{E}$ and $\vec{B}$ are perpendicular to the direction of propagation (and to each other), so the wave is *transverse.*

The frequency f, the wavelength λ, and the speed of propagation c of any periodic wave are related by the usual wavelength–frequency relation $c = \lambda f$. For visible light, a typical frequency is $f = 5 \times 10^{14}$ Hz; the corresponding wavelength is

$$\lambda = \frac{c}{f} = \frac{3 \times 10^8 \text{ m/s}}{5 \times 10^{14} \text{ Hz}} = 6 \times 10^{-7} \text{ m} = 600 \text{ nm},$$

which is similar in size to some bacteria and about one-hundredth the size of a human hair! If the frequency is 10^8 Hz (100 MHz), typical of commercial FM radio stations, the wavelength is

$$\lambda = \frac{3 \times 10^8 \text{ m/s}}{10^8 \text{ Hz}} = 3 \text{ m}.$$

Figure 23.5 shows a sinusoidal electromagnetic wave traveling in the $+x$ direction. The $\vec{E}$ and $\vec{B}$ vectors are shown only for a few points on the positive side of the x axis. Imagine a plane perpendicular to the x axis at a particular point and a particular time; the fields have the same values at all points in that plane. Of course, the values are different in different planes. In the planes where the $\vec{E}$ vector is in the $+y$ direction, $\vec{B}$ is in the $+z$ direction; where $\vec{E}$ is in the $-y$ direction, $\vec{B}$ is in the $-z$ direction. These directions illustrate the direction relations that we described in Section 23.2.

We can describe electromagnetic waves by means of *wave functions,* just as we did in Section 12.4 for waves on a string. One form of the wave function for a transverse wave traveling to the right along a stretched string is Equation 12.5,

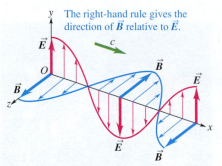

The right-hand rule gives the direction of $\vec{B}$ relative to $\vec{E}$.

▲ **FIGURE 23.5** Representation of the electric and magnetic fields as functions of x at time $t = T/4$ for a sinusoidal electromagnetic wave traveling in the $+x$ direction. The fields are shown only for points on the positive side of the x axis.

$$y(x, t) = A\sin 2\pi\left(\frac{t}{T} - \frac{x}{\lambda}\right) = A\sin(\omega t - kx),$$

where y is the transverse displacement from its equilibrium position at time t of a point with coordinate x on the string. The quantity A is the maximum displacement, or *amplitude,* of the wave; ω is its *angular frequency,* equal to 2π times the frequency f; and k is the **wave number,** or *propagation constant,* equal to $2\pi/\lambda$, where λ is the wavelength.

In Figure 23.5, let E and B represent the instantaneous values of the electric and magnetic fields, respectively, and let E_{max} and B_{max} represent the maximum values, or *amplitudes,* of those fields. The wave functions for the wave are then

$$E = E_{max}\sin 2\pi\left(\frac{t}{T} - \frac{x}{\lambda}\right) = E_{max}\sin(\omega t - kx),$$

$$B = B_{max}\sin 2\pi\left(\frac{t}{T} - \frac{x}{\lambda}\right) = B_{max}\sin(\omega t - kx). \tag{23.3}$$

The sine curves in Figure 23.5 represent instantaneous values of E and B as functions of x at time $t = T/4$. The wave travels to the right with speed c.

Equations 23.3 show that, at any point, the sinusoidal oscillations of $\vec{E}$ and $\vec{B}$ are *in phase*. From Equation 23.1, the amplitudes must be related by

$$E_{max} = cB_{max}. \tag{23.4}$$

Figure 23.6 shows the electric and magnetic fields of a wave traveling in the *negative x* direction. At points where $\vec{E}$ is in the positive y direction, $\vec{B}$ is in the *negative z* direction; where $\vec{E}$ is in the negative y direction, $\vec{B}$ is in the *positive z* direction. The wave functions for this wave are

$$E = -E_{max}\sin 2\pi\left(\frac{t}{T} + \frac{x}{\lambda}\right) = -E_{max}\sin\left(\omega t + kx\right),$$

$$B = B_{max}\sin 2\pi\left(\frac{t}{T} + \frac{x}{\lambda}\right) = B_{max}\sin\left(\omega t + kx\right). \tag{23.5}$$

As with the wave traveling in the $+x$ direction, the sinusoidal oscillations of the $\vec{E}$ and $\vec{B}$ fields at any point are *in phase*.

A wave whose $\vec{E}$ field always lies along the same line is said to be *linearly polarized*. Both of these sinusoidal waves are linearly polarized in the y direction (because the $\vec{E}$ field always lies along the y axis).

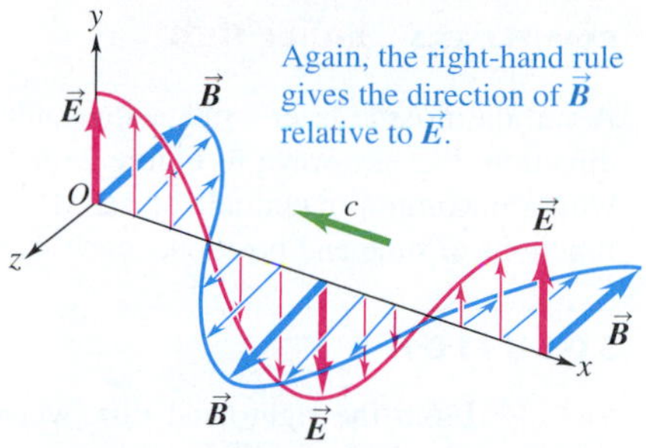

▲ FIGURE 23.6 An electromagnetic wave like the one in Figure 23.5, but traveling in the $-x$ direction, shown at time $t = 3T/4$.

PROBLEM-SOLVING STRATEGY 23.1 **Electromagnetic waves**

SET UP

1. For the problems posed in this chapter, the most important advice that we can give is to concentrate on basic relationships, such as the relation of $\vec{E}$ to $\vec{B}$ (both magnitude and direction), how the wave speed is determined, the transverse nature of the waves, and so on. Don't get sidetracked by mathematical details.

SOLVE

2. In the discussions of sinusoidal waves, both traveling and standing, you need to use the language of sinusoidal waves from Chapter 12. Don't hesitate to go back and review that material, including the problem-solving strategies presented in those chapters. Keep in mind the basic relationships for periodic waves: $v = \lambda f$ and $\lambda = vT$. For electromagnetic waves in vacuum, $v = c$. Be careful to distinguish between ordinary frequency f, usually expressed in hertz, and angular frequency $\omega = 2\pi f$, expressed in rad/s. Remember that the wave number k is $k = 2\pi/\lambda$, and that $\omega = ck$.

EXAMPLE 23.2 **Remote control**

A remote-control unit for a stereo system emits radiation with a frequency of 1.0×10^{14} Hz. Calculate the wavelength of the radiation it emits. Identify the corresponding region of the electromagnetic spectrum by referring to Figure 23.3.

SOLUTION

SET UP AND SOLVE The wavelength of the radiation is

$$\lambda = c/f = (3.00 \times 10^8 \text{ m/s})/(1.0 \times 10^{14} \text{ Hz})$$
$$= 3.0 \times 10^{-6} \text{ m}.$$

REFLECT This is in the infrared (IR) band of the spectrum.

Practice Problem: Calculate the wavelength of a 92.9 MHz FM-station radio wave. *Answer:* $\lambda = 3.23$ m.

Laser light

A carbon dioxide laser emits a sinusoidal electromagnetic wave that travels in vacuum in the negative x direction, like the wave in Figure 23.6. The wavelength is 10.6 μm, and the $\vec{E}$ field is along the z axis, with a maximum magnitude of 1.5 MV/m. Find the equations for the magnitudes of vectors $\vec{E}$ and $\vec{B}$ as functions of time and position.

SOLUTION

SET UP From the right-hand rule, when $\vec{E}$ is in the positive z direction, $\vec{B}$ is in the positive y direction; and when $\vec{E}$ is in the negative z direction, $\vec{B}$ is in the negative y direction (consistent with the right-hand rule). Thus, we don't need the negative sign in the expression for E in Equation 23.5.

SOLVE Since the wave is traveling along the negative x axis, the general equations for the wave are

$$E = E_{max} \sin(\omega t + kx) \qquad B = B_{max} \sin(\omega t + kx).$$

To find B_{max}, we use Equation 23.4: $E_{max} = cB_{max}$. The wavelength is $\lambda = 10.6 \times 10^{-6}$ m, so

$$k = \frac{2\pi}{\lambda} = \frac{2\pi \text{ rad}}{10.6 \times 10^{-6} \text{ m}} = 5.93 \times 10^{5} \text{ rad/m}.$$

Also,

$$\omega = ck = (3.00 \times 10^{8} \text{ m/s})(5.93 \times 10^{5} \text{ rad/m})$$
$$= 1.78 \times 10^{14} \text{ rad/s}.$$

Substituting into the above equations, with

$$B_{max} = \frac{E_{max}}{c} = \frac{1.50 \times 10^{6} \text{ V/m}}{3.00 \times 10^{8} \text{ m/s}} = 5.00 \times 10^{-3} \text{ T},$$

we get

$$E = E_{max} \sin(\omega t + kx)$$
$$= (1.5 \times 10^{6} \text{ V/m}) \sin[(1.78 \times 10^{14} \text{ rad/s})t + (5.93 \times 10^{5} \text{ rad/m})x],$$
$$B = B_{max} \sin(\omega t + kx)$$
$$= (5.0 \times 10^{-3} \text{ T}) \sin[(1.78 \times 10^{14} \text{ rad/s})t + (5.93 \times 10^{5} \text{ rad/m})x].$$

REFLECT Note that no negative sign is needed in the expression for E; the right-hand rule is obeyed without it. At any point, the two fields are in phase.

PhET: Microwaves

23.5 Energy in Electromagnetic Waves

It is a familiar fact that energy is associated with electromagnetic waves. Think of the sun's radiation and the radiation in microwave ovens. To derive detailed relationships for the energy in an electromagnetic wave, we begin with the expressions derived in Sections 18.7 and 21.10 for the **energy densities** (energy per unit volume) in electric and magnetic fields; we suggest that you review those derivations now. Specifically, Equations 18.20 and 21.21 show that the total energy density u in a region of space where $\vec{E}$ and $\vec{B}$ fields are present is given by the following expressions:

Energy density in electric and magnetic fields
The energy density u (energy per unit volume) in a region of empty space where electric and magnetic fields are present is

$$u = \frac{1}{2}\epsilon_0 E^2 + \frac{1}{2\mu_0} B^2. \qquad (23.6)$$

The two field magnitudes are related by Equation 23.1:

$$B = \frac{E}{c} = \sqrt{\epsilon_0 \mu_0} E.$$

Combining this equation with Equation 23.6, we can also express the energy density u as

$$u = \frac{1}{2}\epsilon_0 E^2 + \frac{1}{2\mu_0}(\sqrt{\epsilon_0\mu_0}E)^2 = \epsilon_0 E^2. \qquad (23.7)$$

This result shows that the energy density associated with the $\vec{E}$ field is equal to that of the $\vec{B}$ field.

In the simple wave described in Section 23.2, the $\vec{E}$ and $\vec{B}$ fields advance in the $+x$ direction into regions where originally no fields were present, so it is clear that the wave transports energy from one region to another. We can describe this energy transfer in terms of *energy transferred per unit time per unit cross-sectional area,* or *power per unit area,* for an area perpendicular to the direction of wave travel. The average value of this quantity, for any wave, is called the *intensity* of the wave.

To see how the energy flow is related to the fields, consider a stationary plane, perpendicular to the x axis, that coincides with the wave front at a certain time. In a time Δt after this, the wave front moves a distance $\Delta x = c\,\Delta t$ to the right of the plane. Considering an area A on the stationary plane (Figure 23.7), we note that the energy in the space to the right of this area must have passed through it to reach the new location. The volume ΔV of the relevant region is the base area A times the length $c\,\Delta t$. The energy ΔU in this region is the energy density u times this volume:

$$\Delta U = u\,\Delta V = (\epsilon_0 E^2)(Ac\,\Delta t).$$

This energy passes through the area A in time Δt. The energy flow per unit time per unit area, which we'll denote by S, is

$$S = \frac{1}{A}\frac{\Delta U}{\Delta t} = \epsilon_0 c E^2.$$

Using Equations 23.1 $(E = cB)$ and 23.2 $(c = 1/\sqrt{\epsilon_0\mu_0})$, we can derive the alternative forms

$$S = \epsilon_0 c E^2 = \frac{\epsilon_0}{\sqrt{\epsilon_0\mu_0}}E^2 = \sqrt{\frac{\epsilon_0}{\mu_0}}E^2 = \frac{EB}{\mu_0} = cu. \qquad (23.8)$$

The units of S are energy per unit time per unit area, or power per unit area. The SI unit of S is $1\ \mathrm{J/(s\cdot m^2)}$, or $1\ \mathrm{W/m^2}$. That is, S is power per unit area.

In all the preceding equations, E and B are the *instantaneous* values of the electric and magnetic field magnitudes, respectively. For a wave in which the fields vary with time, S also varies. The *average* value of S is the average energy transmitted across a given area perpendicular to the direction of propagation, per unit area and per unit of time—that is, the average power per unit area.

For a *sinusoidal* wave, the average value of E^2 is one-half the square of the amplitude E_{max}. In this case, we can find the average value of S simply by replacing E^2 in Equation 23.8 with $E_{\mathrm{max}}^2/2$. Equation 23.8 then becomes

$$S_{\mathrm{av}} = \frac{1}{2}\sqrt{\frac{\epsilon_0}{\mu_0}}E_{\mathrm{max}}^2 = \frac{1}{2}\epsilon_0 c E_{\mathrm{max}}^2 = \frac{E_{\mathrm{max}}B_{\mathrm{max}}}{2\mu_0}. \qquad \text{(sinusoidal wave)} \qquad (23.9)$$

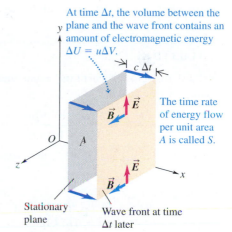

At time Δt, the volume between the plane and the wave front contains an amount of electromagnetic energy $\Delta U = u\Delta V$.

The time rate of energy flow per unit area A is called S.

▲ **FIGURE 23.7** A wave front at a time Δt after it passes through the stationary plane with area A.

Making the same substitution in Equation 23.7, we obtain $u_{av} = \frac{1}{2}\epsilon_0 E_{max}^2$. Comparing this equation with Equation 23.9, we find that S_{av} and u_{av} are simply related:

$$S_{av} = \frac{u_{av}}{\sqrt{\epsilon_0 \mu_0}} = u_{av}c. \quad \text{(sinusoidal wave)} \quad (23.10)$$

The average power per unit area in an electromagnetic wave is also called the **intensity** of the wave, denoted by I. That is, $I = S_{av}$, and we can write

$$I = S_{av} = \frac{1}{2}\sqrt{\frac{\epsilon_0}{\mu_0}}E_{max}^2 = \frac{1}{2}\epsilon_0 c E_{max}^2 = \frac{E_{max}B_{max}}{2\mu_0}. \quad (23.11)$$

EXAMPLE 23.4 **Laser cutter**

A laser cutter used for cutting thin sheets of material emits a beam with electric-field amplitude $E_{max} = 2.76 \times 10^5$ V/m over an area of 2.00 mm². Find **(a)** the maximum magnetic field B_{max}, **(b)** the maximum energy density u_{max}, **(c)** the intensity $S_{av} = I$ of the beam, and **(d)** the average power of the beam.

SOLUTION

SET UP AND SOLVE **Part (a):** From Equation 23.1, the maximum magnetic field B_{max} is

$$B_{max} = \frac{E_{max}}{c} = \frac{2.76 \times 10^5 \text{ V/m}}{3.00 \times 10^8 \text{ m/s}} = 9.20 \times 10^{-4} \text{ T}.$$

Part (b): From Equation 23.7, the maximum energy density u_{max} is

$$u_{max} = \epsilon_0 E_{max}^2$$
$$= [8.85 \times 10^{-12} \text{ C}^2/(\text{N} \cdot \text{m}^2)](2.76 \times 10^5 \text{ N/C})^2$$
$$= 0.674 \text{ N/m}^2 = 0.674 \text{ J/m}^3.$$

The *average* energy density is half of this:

$$u_{av} = \frac{1}{2}\epsilon_0 E_{max}^2 = 0.337 \text{ J/m}^3.$$

Part (c): The intensity $I = S_{av}$ is given by Equation 23.10:

$$I = S_{av} = u_{av}c = (0.337 \text{ J/m}^3)(3.00 \times 10^8 \text{ m/s})$$
$$= 1.01 \times 10^8 \text{ J/(m}^2 \cdot \text{s)} = 1.01 \times 10^8 \text{ W/m}^2.$$

Alternatively, from Equation 23.9,

$$I = S_{av} = \frac{1}{2}\epsilon_0 c E_{max}^2$$
$$= \frac{1}{2}[8.85 \times 10^{-12} \text{ C}^2/(\text{N} \cdot \text{m}^2)](3.00 \times 10^8 \text{ m/s})$$
$$\cdot (2.76 \times 10^5 \text{ N/C})^2$$
$$= 1.01 \times 10^8 \text{ W/m}^2.$$

Part (d): The average total power P_{av} is the intensity S_{av} multiplied by the cross-sectional area A of the beam:

$$P_{av} = (1.01 \times 10^8 \text{ W/m}^2)(2.00 \times 10^{-6} \text{ m}^2)$$
$$= 202 \text{ W}.$$

REFLECT A power of 200 W is enough power to cut cardboard and thin wood. A typical laser pointer has an output power on the order of a few milliwatts.

Practice Problem: Find the maximum energy density for a 2000 W laser with an electric-field amplitude $E_{max} = 8.68 \times 10^5$ N/C. *Answer:* $u_{max} = 6.67 \text{ J/m}^3$.

Radiation Pressure

The fact that electromagnetic waves transport energy follows directly from the fact that energy is required to establish electric and magnetic fields. It can also be shown that electromagnetic waves carry *momentum* p, with a corresponding momentum density (momentum p per volume V) of magnitude

$$\frac{p}{V} = \frac{\epsilon_0 E^2}{c} = \frac{EB}{\mu_0 c^2} = \frac{S}{c^2}.$$

For a sinusoidal wave, the average value of E^2 is $E_{max}^2/2$; from Equation 23.9, the average momentum density is

$$\frac{p_{av}}{V} = \frac{\epsilon_0 E_{max}^2}{2c} = \frac{S_{av}}{c^2}. \quad (23.12)$$

The momentum p of an electromagnetic wave is a property of the field; it is not associated with the mass of a moving particle in the usual sense.

There is a corresponding momentum *flow rate*, equal to the momentum per unit volume (Equation 23.12) multiplied by the wave speed c. For a sinusoidal wave, the average momentum flow Δp in a time interval Δt, per unit area A, is

$$\frac{1}{A}\frac{\Delta p}{\Delta t} = \frac{1}{2}\epsilon_0 E^2 = \left(\frac{S_{av}}{c^2}\right)c = \frac{S_{av}}{c} = \frac{I}{c}. \quad \text{(sinusoidal wave)} \quad (23.13)$$

This is the average momentum transferred per unit surface area per unit time.

This momentum transfer is responsible for the phenomenon of **radiation pressure.** When an electromagnetic wave is completely absorbed by a surface perpendicular to the direction of propagation of the wave, the rate of change of momentum with respect to time equals the *force* on the surface. Thus, the average force per unit area, or, simply, the pressure, is equal to I/c. If the wave is totally reflected, the change in momentum is twice as great, and the pressure is $2I/c$. For example, the value of I (or S_{av}) for direct sunlight before it passes through the earth's atmosphere is about 1.4 kW/m². The corresponding radiation pressure on a completely absorbing surface is

$$\frac{I}{c} = \frac{1.4 \times 10^3 \text{ W/m}^2}{3.0 \times 10^8 \text{ m/s}} = 4.7 \times 10^{-6} \text{ Pa}.$$

The average radiation pressure on a totally *reflecting* surface is twice this, $2I/c$, or 9.4×10^{-6} Pa. These are very small pressures, on the order of 10^{-10} atmosphere, but they can be measured with sensitive instruments.

EXAMPLE 23.5   Solar sails

Suppose a spacecraft with a mass of 2.50×10^4 kg has a solar sail made of perfectly reflective aluminized Kapton® film with an area of 2.59×10^6 m² (about 1 square mile). If the spacecraft is launched into earth orbit and then deploys its sail at right angles to the sunlight, what is the acceleration due to sunlight? (At the earth's distance from the sun, the pressure exerted by sunlight on an absorbing surface is 4.70×10^{-6} Pa.)

SOLUTION

SET UP We sketch the situation in Figure 23.8. Because our sail is perfectly reflective, the pressure exerted on it is twice the value given in the statement of the problem: $p = 9.40 \times 10^{-6}$ Pa. Before the sail is deployed, the spacecraft's net radial velocity relative to the sun is zero.

SOLVE The radiation pressure p is the magnitude of force per unit area, so we start by finding the magnitude F of force exerted on the sail:

$$F = pA = (9.40 \times 10^{-6} \text{ N/m}^2)(2.59 \times 10^6 \text{ m}^2) = 24.3 \text{ N}.$$

We now find the magnitude a of the spacecraft's acceleration due to the radiation pressure:

$$a = F/m = (24.3 \text{ N})/(2.50 \times 10^4 \text{ kg}) = 9.72 \times 10^{-4} \text{ m/s}^2.$$

REFLECT For this spacecraft, a square mile of sail provides an acceleration of only about $10^{-4}g$, small compared with the accelerations provided by chemical rockets. However, rockets burn out; sunlight keeps pushing. Even on its first day, the craft travels

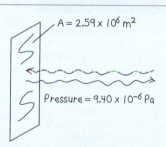

▲ **FIGURE 23.8** Our diagram for this problem.

more than 3000 km in the radial direction, and on day 12 its radial speed passes 1 km/s.

Practice Problem: A communications satellite has solar-energy-collecting panels with a total area of 4.0 m². What is the average magnitude of total force on these panels associated with radiation pressure, if the radiation is completely absorbed? *Answer:* $F = 1.9 \times 10^{-5}$ N.

Dust tail: Consists of fine dust accelerated by light pressure. It points away from the sun, but curves slightly because light accelerates the dust particles only gradually.

Ion tail: Consists of gas molecules ionized by the sun's ultraviolet light; it is quickly entrained by the sun's "wind" of charged particles and points straight away from the sun.

▲ **FIGURE 23.9** A comet actually has two tails. Both point away from the sun, but one is accelerated quickly by electric interactions and the other more slowly by light pressure.

The pressure of the sun's radiation is partially responsible for pushing the tail of a comet away from the sun (Figure 23.9). Also, while stars the size of our sun are supported against gravitational collapse mainly by the pressure of their hot gas, for some massive stars, radiation pressure dominates, and gravitational collapse of the star is prevented mainly by the light radiating outward from its core. Figure 23.10 shows an extreme example.

23.6 Nature of Light

The remainder of this chapter is devoted to optics. We'll lay some of the foundation needed for the study of many recent developments in this area of physics, including optical fibers, holograms, optical computers, and new techniques in medical imaging. We begin with a study of the laws of reflection and refraction and the concepts of dispersion, polarization, and scattering of light. Along the way, we'll compare the various possible descriptions of light in terms of *rays* and *waves,* and we'll look at Huygens's principle, an important connecting link between these two viewpoints.

Until the time of Isaac Newton (1642–1727), most scientists thought that light consisted of streams of particles (called *corpuscles*) emitted by visible objects. Galileo and others tried (unsuccessfully) to measure the speed of light. Around 1665, evidence of *wave* properties of light began to emerge. By the early 19th century, evidence that light is a wave had grown very persuasive. The picture of light as an electromagnetic wave isn't the whole story, however. Several effects associated with the emission and absorption of light reveal that it also has a particle aspect, in that the energy carried by light waves is packaged in discrete bundles called *photons* or *quanta.* These apparently contradictory wave and particle properties have been reconciled since 1930 with the development of quantum electrodynamics, a comprehensive theory that includes *both* wave and particle properties. The *propagation* of light is best described by a wave model, but understanding emission and absorption by atoms and nuclei requires a particle approach.

The fundamental sources of all electromagnetic radiation are electric charges in accelerated motion. All objects emit electromagnetic radiation as a result of thermal motion of their molecules; this radiation, called *thermal radiation,* is a

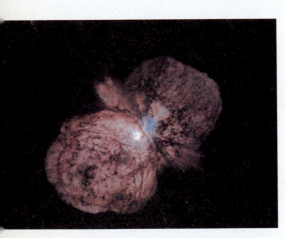

▲ **FIGURE 23.10 Massive star Eta Carinae.** The blue-white monster at the heart of this cloud may be the most massive star presently alive in our galaxy. It has 100–150 times the mass of our sun, but it is about 4 *million* times more luminous, and light pressure makes it very unstable. Gravity has a hard time holding it together. The strange lobed cloud probably dates from an episode around 1840 during which this star temporarily became the second brightest in earth's sky. During such flare-ups, light blows vast quantities of material off the star's surface. Even when relatively quiescent, this star loses matter at a high rate owing to light pressure. Indeed, light pressure limits how massive a star can be: A star's luminosity depends on its mass, but luminosity increases much faster than mass. Eta Carinae is probably close to the limit of stellar stability.

▶ **FIGURE 23.11** We use incandescent bulbs because of the visible light the hot filament emits. However, most of the filament's radiation is in the infrared. That is why incandescent bulbs are hot, and it is also why they are energy inefficient.

mixture of different wavelengths. At sufficiently high temperatures, all matter emits enough visible light to be self-luminous; a very hot body appears "red hot" or even "white hot." Thus, hot matter in any form is a source of light. Familiar examples are a candle flame (hot gas), hot coals in a campfire, the coils in an electric room heater, and an incandescent lamp filament (which usually operates at a temperature of about 3000°C) (Figure 23.11).

Light is also produced during electrical discharges through ionized gases. The bluish light of mercury-arc lamps, the orange-yellow of sodium-vapor lamps, and the various colors of "neon" signs are familiar. A variation of the mercury-arc lamp is the *fluorescent* lamp. This light source uses a material called a *phosphor* to convert the ultraviolet radiation from a mercury arc into visible light. This conversion makes fluorescent lamps more efficient than incandescent lamps in converting electrical energy into light.

A special light source that has attained prominence in the last 50 years is the *laser,* which can produce a very narrow beam of enormously intense radiation. High-intensity lasers are used to cut steel, fuse high-melting-point materials, carry out microsurgery, and in many other applications. A significant characteristic of laser light is that it is much more nearly *monochromatic,* or single frequency, than light from any other source. (Figure 23.12)

The speed of light in vacuum is a fundamental constant of nature. As we discussed in Section 1.3, the speed of light in vacuum is *defined* to be precisely 299,792,458 m/s, and 1 meter is defined to be the distance traveled by light in vacuum in a time of 1/299,792,458 s. The second is defined by the cesium clock, which can measure time intervals with a precision of one part in 10^{13}. If future work results in greater precision in measuring the speed of light, the value just cited won't change, but a small adjustment will be made in the definition of the meter.

Wave Fronts

We often use the concept of a **wave front** to describe wave propagation. We define a wave front as *the locus of all adjacent points at which the phase of vibration of the wave is the same.* That is, at any instant, all points on a wave front are at the same part of the cycle of their periodic variation. During wave propagation, the wave fronts all move with the same speed in the direction of propagation of the wave.

A familiar example of a wave front is a crest of a water wave. When we drop a pebble in a calm pool, the expanding circles formed by the wave crests are wave fronts. Similarly, when sound waves spread out in still air from a pointlike source, any spherical surface concentric with the source is a wave front, as shown in Figure 23.13. The "pressure crests"—the surfaces over which the pressure is maximum—form sets of expanding spheres as the wave travels outward from the source. In diagrams of wave motion, we usually draw only parts of a few wave fronts, often choosing consecutive wave fronts that have the same phase, such as crests of a water wave. These consecutive wave crests are separated from each other by one wavelength.

For a light wave (or any other electromagnetic wave), the quantity that corresponds to the displacement of the surface in a water wave or the pressure in a sound wave is the electric or magnetic field. We'll often use diagrams that show the shapes of the wave fronts or their cross sections. For example, when electromagnetic waves are radiated by a small light source, we can represent the wave fronts as spherical surfaces concentric with the source or, as in Figure 23.14a, by

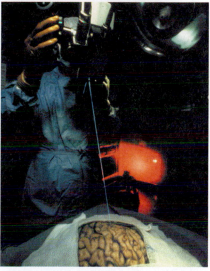

▲ **FIGURE 23.12** A laser being used for brain surgery. Lasers can be used as ultraprecise, bloodless "scalpels" to reach and remove tumors with minimal damage to neighboring healthy tissues.

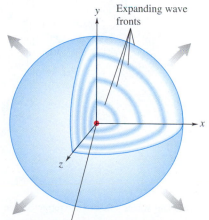

▲ **FIGURE 23.13** Spherical wave fronts, such as those from a point source of sound, spread out uniformly in all directions (provided that the medium is uniform and isotropic).

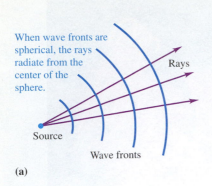

When wave fronts are spherical, the rays radiate from the center of the sphere.

Rays

Source

Wave fronts

(a)

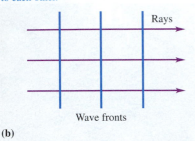

When wave fronts are planar, the rays are perpendicular to the wave fronts and parallel to each other.

Rays

Wave fronts

(b)

▲ **FIGURE 23.14** Spherical and planar wave fronts and rays.

the intersections of these surfaces with the plane of the diagram. Far away from the source, where the radii of the spheres have become very large, a section of a spherical surface can be considered as a plane, and we have a *plane* wave (Figure 23.14b).

It's often convenient to represent a light wave by **rays** rather than by wave fronts. Rays were used to describe light long before its wave nature was firmly established, and in a particle theory of light, rays are the paths of the particles. From the wave viewpoint, *a ray is an imaginary line along the direction of travel of the wave.* In Figure 23.14a, the rays are the radii of the spherical wave fronts; in Figure 23.14b, they are straight lines perpendicular to the wave fronts. When waves travel in a homogeneous, isotropic material (a material with the same properties in all of its regions and in all directions), the rays are always straight lines normal to the wave fronts. At a boundary surface between two materials, such as the surface of a glass plate in air, the wave speed and the direction of a ray usually change, but the ray segments in each material (the air and the glass) are straight lines.

In the remainder of this chapter and in the next three, we'll have many opportunities to see the interplay among the ray, wave, and particle descriptions of light. The branch of optics for which the ray description is adequate is called **geometric optics** (Chapters 24 and 25); the branch dealing specifically with wave behavior is called **physical optics** (Chapter 26).

23.7 Reflection and Refraction

In this section, we'll explore the basic elements of the *ray* model of light. When a light wave strikes a smooth interface (a surface separating two transparent materials, such as air and glass or water and glass), the wave is, in general, partly reflected and partly *refracted* (transmitted) into the second material, as shown in Figure 23.15a. For example, when you look into a store window from the street

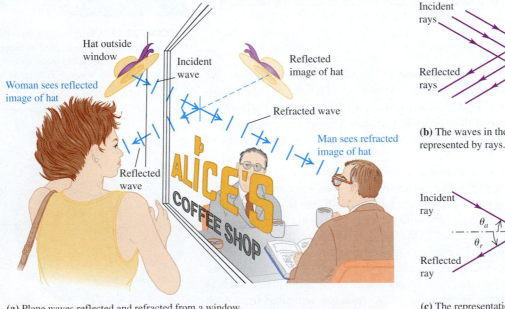

Hat outside window

Incident wave

Woman sees reflected image of hat

Reflected wave

Reflected image of hat

Refracted wave

Man sees refracted image of hat

(a) Plane waves reflected and refracted from a window.

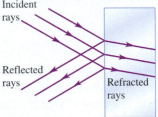

Incident rays

Reflected rays

Refracted rays

(b) The waves in the outside air and glass represented by rays.

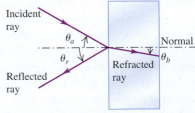

Incident ray

θ_a

θ_r

Reflected ray

Normal

θ_b

Refracted ray

(c) The representation simplified to show just one set of rays.

▲ **FIGURE 23.15** A plane wave is in part reflected and in part refracted at the boundary between two media (in this case, air and glass).

and see a reflection of the street scene, a person inside the store can look out *through* the window at the same scene, as light reaches him by refraction.

The segments of plane waves shown in Figure 23.15b can be represented by bundles of rays forming *beams* of light. For simplicity, we often draw only one ray in each beam (Figure 23.15c). Representing these waves in terms of rays is the basis of *geometric optics.* We begin our study with the behavior of an individual ray.

We describe the directions of the incident, reflected, and refracted rays at a smooth interface between two optical materials in terms of the angles they make with the *normal* to the surface at the point of incidence, as shown in Figure 23.15c. If the interface is rough, both the transmitted light and the reflected light are scattered in various directions, and there is no single angle of transmission or reflection. Reflection at a definite angle from a very smooth surface is called *specular reflection;* scattered reflection from a rough surface is called *diffuse reflection.* This distinction is illustrated in Figure 23.16. Specular reflection also occurs at a very smooth opaque surface, such as one made of highly polished metal or plastic.

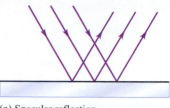

(a) Specular reflection

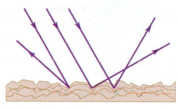

(b) Diffuse reflection

▲ **FIGURE 23.16** Two types of reflection.

Conceptual Analysis 23.2

Laser beam demonstration

When a laser beam is aimed at the wall of a lecture hall, every student in the class observes a red dot on the wall. What can you infer about the direction of the beam as it leaves the wall?

A. The beam is reflected in all directions, an example of diffuse reflection.
B. The beam is reflected at a definite angle, an example of specular reflection.
C. Nothing can be inferred from this experiment. Other results are needed.

SOLUTION The fact that *everyone* in the class can see the dot means that the beam is reflected diffusely, in all directions. If the beam were pointed at a mirror, it would reflect in only one direction and be seen by only one person (or nobody).

The **index of refraction** of an optical material (also called the *refractive index*), denoted by *n*, plays a central role in geometric optics.

Definition of index of refraction

The index of refraction of an optical material, denoted as *n*, is the ratio of the speed of light in vacuum (c) to the speed of light in the material (v):

$$n = \frac{c}{v}. \tag{23.14}$$

Light always travels *more slowly* in a material than in vacuum, so *n* for any material is always greater than one. For vacuum, $n = 1$ by definition.

Experimental studies of the directions of the incident, reflected, and refracted rays at an interface between two optical materials lead to the following conclusions (see Figure 23.17):

Principles of geometric optics

1. **The incident, reflected, and refracted rays, and the normal to the surface, all lie in the same plane.** If the incident ray is in the plane of the diagram and the boundary surface between the two materials is perpendicular to this plane, then the reflected and refracted rays are in the plane of the diagram.

1. The incident, reflected, and refracted rays and the normal to the surface all lie in the same plane.

   Angles θ_a, θ_b, and θ_r are measured *from the normal.*

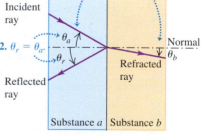

2. $\theta_r = \theta_a$.

3. When a monochromatic light ray crosses the interface between two given substances *a* and *b*, the angles θ_a and θ_b are related to the indexes of refraction of *a* and *b* by

$$\frac{\sin\theta_a}{\sin\theta_b} = \frac{n_b}{n_a}$$

▲ **FIGURE 23.17** The principles of geometric optics.

A ray entering a material of *larger* index of refraction bends *toward* the normal.

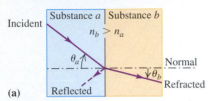

(a)

A ray oriented perpendicular to the surface does not bend, regardless of the materials.

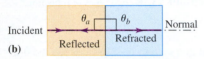

(b)

A ray entering a material of *smaller* index of refraction bends *away from* the normal.

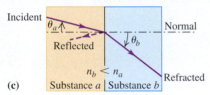

(c)

▲ **FIGURE 23.18** Refraction on crossing an interface to a material of larger or smaller index of refraction.

The path of a refracted ray is reversible (the same from either direction).

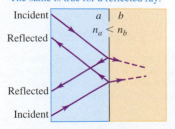

The same is true for a reflected ray.

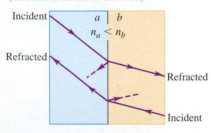

▲ **FIGURE 23.19** Reversibility of refraction and reflection.

2. **The angle of reflection θ_r is equal to the angle of incidence θ_a for all wavelengths and for any pair of substances.** That is, in Figure 23.17,

$$\theta_r = \theta_a. \tag{23.15}$$

This relationship, together with the fact that the incident and reflected rays and the normal all lie in the same plane, is called the **law of reflection.**

3. For monochromatic light and for a given pair of substances a and b on opposite sides of the interface, **the ratio of the sines of the angles θ_a and θ_b, where both angles are measured from the normal to the surface, is equal to the inverse ratio of the two indexes of refraction:**

$$\frac{\sin\theta_a}{\sin\theta_b} = \frac{n_b}{n_a}, \quad \text{or} \quad n_a\sin\theta_a = n_b\sin\theta_b. \tag{23.16}$$

These experimental results, together with the fact that the incident and refracted rays and the normal to the surface all lie in the same plane (with the incident and refracted rays always on opposite sides of the normal), is called the **law of refraction,** or **Snell's law,** after Willebrord Snell (1591–1626), Dutch mathematician and physicist. There is some doubt that Snell actually discovered the law named after him. The discovery that $n = c/v$ came much later.

Equation 23.16 shows that when a ray passes from one material (a) into another material (b) having a larger index of refraction and a smaller wave speed $(n_b > n_a)$, the angle θ_b with the normal is *smaller* in the second material than the angle θ_a in the first, and the ray is bent *toward* the normal (Figure 23.18a). This is the case in Figure 23.15, where light passes from air into glass. When the second index is *less than* the first $(n_b < n_a)$, the ray is bent *away from* the normal (Figure 23.18c). The index of refraction of vacuum is 1, by definition. When a ray passes from vacuum into a material, it is always bent *toward* the normal; when passing from a material into vacuum, it is always bent *away from* the normal. When the incident ray is perpendicular to the interface, $\theta_a = 0$ and $\sin\theta_a = 0$. In this special case, $\sin\theta_b = 0$, $\theta_b = 0$, and the transmitted ray is not bent at all (Figure 23.18b).

When a ray of light approaches the interface from the opposite side (Figure 23.19), there are again reflected and refracted rays; these two rays—the incident ray and the normal to the surface—again lie in the same plane. The laws of reflection and refraction apply regardless of whether the incident ray is in material a or material b in the figure. The path of a refracted ray is *reversible:* The ray follows the same path when going from b to a as when going from a to b. The path of a ray *reflected* from any surface is also reversible.

The *intensities* of the reflected and refracted rays depend on the angle of incidence, the two indexes of refraction, and the polarization of the incident ray. For unpolarized light, the fraction reflected is smallest at *normal* incidence $(0°)$, where it is about 4% for an air–glass interface, and it increases with increasing angle of incidence up to 100% at grazing incidence, when $\theta_a = 90°$.

Most glasses used in optical instruments have indexes of refraction between about 1.5 and 2.0. A few substances have larger indexes; two examples are diamond, with 2.42, and rutile (a crystalline form of titanium dioxide), with 2.62. The index of refraction depends not only on the substance, but also on the wavelength of the light. The dependence on wavelength is called *dispersion;* we'll discuss it in Section 23.9. Indexes of refraction for several solids and liquids are given in Table 23.1.

The index of refraction of *air* at standard temperature and pressure is about 1.0003, and we will usually take it to be exactly 1. The index of refraction of a gas increases in proportion to its density.

TABLE 23.1 **Index of refraction for yellow sodium light ($\lambda_0 = 589$ nm)**

Substance	Index of refraction, n	Substance	Index of refraction, n
Solids		Medium flint	1.62
Ice (H_2O)	1.309	Dense flint	1.66
Fluorite (CaF_2)	1.434	Lanthanum flint	1.80
Polystyrene	1.49	*Liquids at 20°C*	
Rock salt (NaCl)	1.544	Methanol (CH_3OH)	1.329
Quartz (SiO_2)	1.544	Water (H_2O)	1.333
Zircon ($ZrO_2 \cdot SiO_2$)	1.923	Ethanol (C_2H_5OH)	1.36
Fabulite ($SrTiO_3$)	2.409	Carbon tetrachloride (CCl_4)	1.460
Diamond (C)	2.417	Turpentine	1.472
Rutile (TiO_2)	2.62	Glycerine	1.473
Glasses (typical values)		Benzene	1.501
Crown	1.52	Carbon disulfide (CS_2)	1.628
Light flint	1.58		

When light passes from one material into another, the frequency f of the wave doesn't change. The boundary surface cannot create or destroy waves; the number arriving per unit time must equal the number leaving per unit time; otherwise, incident and transmitted waves couldn't have a definite phase relationship. In any material, $v = \lambda f$. Because f is the same in any material as in vacuum and v is always less than the wave speed c in vacuum by the factor n, λ is also correspondingly reduced. Thus, the wavelength λ of the light in a material is *less than* its wavelength λ_0 in vacuum by the factor n. That is,

$$\lambda = \frac{\lambda_0}{n}. \tag{23.17}$$

 Conceptual Analysis 23.3

Properties of a refracted wave

A monochromatic beam of light passes from air into a block of clear plastic. Which of the following properties may differ between the part of the beam in the air and the part in the plastic?

A. Frequency.
B. Wavelength.
C. Speed.

SOLUTION As we've noted, the frequency cannot change when the wave crosses an interface, because that would necessitate creating or destroying waves at the interface. The speed of light is less in any material than that in vacuum and may differ between different materials. (We expect it to differ between air and plastic.) Because wavelength depends on frequency and speed, a change in speed requires a corresponding change in wavelength.

 Conceptual Analysis 23.4

Propagating from air to glass

Figure 23.20 shows a light ray passing from air into glass. Which of the choices A–D represents the ray within the glass?

SOLUTION Glass has a greater index of refraction than air, so we expect the refracted ray in the glass to bend toward the normal. Ray A bends *away from* the normal; this is the result we would get if the index of refraction of glass were less than that of air. Ray B doesn't bend at all; this would be the result if glass and air had exactly the same index of refraction. Ray C is a correct choice. What about ray D? For this ray, $\sin\theta_{\text{glass}} = 0$, so it doesn't satisfy Snell's law (Equation 23.16) $n_{\text{air}}\sin\theta_{\text{air}} = n_{\text{glass}}\sin\theta_{\text{glass}}$.

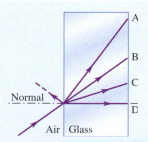

▲ **FIGURE 23.20**

Mastering PHYSICS®

ActivPhysics 15.1: Reflection and Refraction
ActivPhysics 15.3: Refraction Applications

PROBLEM-SOLVING STRATEGY 23.2 **Reflection and refraction**

SET UP

1. In geometric optics problems involving rays and angles, *always* start by drawing a large, neat diagram. Label all known angles and indexes of refraction.
2. Don't forget that, by convention, we always measure the angles of incidence, reflection, and refraction from the *normal* to the surface, *never* from the surface itself.

SOLVE

3. You'll often have to use some simple geometry or trigonometry in working out angular relations: The sum of the interior angles in a triangle is 180° and so on. (See Appendix A.6 for a review.) It often helps to think through the problem, asking yourself, "What information am I given?" "What do I need to know in order to find this angle?" or "What other angles or other quantities can I compute, using the information given in the problem?"

REFLECT

4. Refracted light is always bent toward the normal when the second index is greater than the first, away from the normal when the second index is less than the first. Check whether your results are consistent with this rule.

EXAMPLE 23.6 **A fishpond**

You kneel beside the fishpond in your backyard and look at one of the fish. You see it by sunlight that reflects off the fish and refracts at the water–air interface. If the light from the fish to your eye strikes the water–air interface at an angle of 60.0° to the interface, what is its angle of refraction of the ray in the air?

SOLUTION

SET UP Figure 23.21 shows our sketch. We take the water as medium *a* and the air as medium *b*; from Table 23.1, $n_a = 1.33$ and $n_b = 1.00$. Note that θ_a is 30.0°, not 60.0°! We use Snell's law to find θ_b.

SOLVE From Snell's law,

$$n_a \sin\theta_a = n_b \sin\theta_b,$$

$$\theta_b = \sin^{-1}\frac{n_a \sin\theta_a}{n_b} = \sin^{-1}\frac{1.33 \sin 30.0°}{1.00} = 41.7°.$$

REFLECT Because the difference in refractive index between water and air is substantial, the actual position of the fish is quite different from its apparent position to you.

Practice Problem: You are spearfishing from a boat and eye a large bass swimming below. It is *apparently* at an angle of 40.0° from the normal. At what angle should you aim your spear? *Answer:* $\theta_b = 28.9°$, measured from the normal.

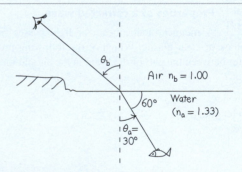

▲ **FIGURE 23.21** Our diagram for this problem.

Looking through the glass

Figure 23.22 shows a setup in which two pairs of pins are separated by a block of glass. You observe the pins so that the ray reaching your eye passes through all the pins (Figure 23.22a). Which of the choices in Figure 23.22b represents the *apparent* position of the pins on the far side of the glass block?

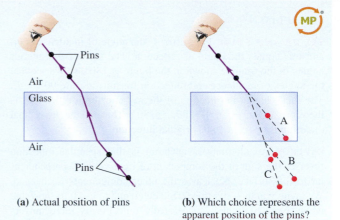

SOLUTION The actual position of the pins below the glass is B. But the eyes can see only the direction of the ray emerging from the top surface of the glass. Nothing about that light ray tells you that it refracted twice as it crossed the two air–glass interfaces. Thus, the apparent position of the pins is A. Notice that the light refracts through the same angle as it crosses from air into glass and then from glass into air. This illustrates the point made in Figure 23.19.

(a) Actual position of pins

(b) Which choice represents the apparent position of the pins?

▲ **FIGURE 23.22**

EXAMPLE 23.7 **Index of refraction in the eye**

The wavelength of the red light from a helium–neon laser is 633 nm in air, but 474 nm in the jellylike fluid inside your eyeball, called the *vitreous humor* (Figure 23.23). Calculate the index of refraction of the vitreous humor and the speed and frequency of light passing through it.

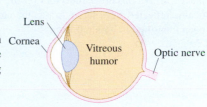

▲ **FIGURE 23.23**

SOLUTION

SET UP AND SOLVE We don't need to sketch this problem, because we're interested only in the index of refraction of a material and the speed of light in it, not in the path a particular ray follows. The index of refraction of air is very close to one, so we assume that the wavelength of the laser light is the same in vacuum as in air. Then the wavelength λ in the material is given by Equation 23.17:

$$n = \frac{\lambda_0}{\lambda} = \frac{633 \text{ nm}}{474 \text{ nm}} = 1.34,$$

which is about the same as for water. Then, to find the speed of light in the vitreous humor, we use Equation 23.14, $n = c/v$:

$$v = \frac{c}{n} = \frac{3.00 \times 10^8 \text{ m/s}}{1.34} = 2.24 \times 10^8 \text{ m/s}.$$

The frequency of the light is

$$f = \frac{v}{\lambda} = \frac{2.24 \times 10^8 \text{ m/s}}{474 \times 10^{-9} \text{ m}} = 4.73 \times 10^{14} \text{ Hz}.$$

REFLECT Note that while the speed and wavelength have different values in air and in the vitreous humor, the *frequency* f_0 in air is the same as the frequency f in the vitreous humor (and the frequency in vacuum):

$$f_0 = \frac{c}{\lambda_0} = \frac{3.00 \times 10^8 \text{ m/s}}{633 \times 10^{-9} \text{ m}} = 4.73 \times 10^{14} \text{ Hz}.$$

This result confirms the general rule that when a light wave passes from one material into another, the frequency doesn't change.

Practice Problem: The newest laser pointers emit green light with a wavelength of $\lambda_0 = 532$ nm in air. What is the wavelength of this light in the vitreous humor of the eyeball? *Answer:* $\lambda = 397$ nm.

EXAMPLE 23.8 **Reflected light rays**

Two mirrors are perpendicular to each other. A ray traveling in a plane perpendicular to both mirrors is reflected from one mirror and then the other, as shown in Figure 23.24. What is the ray's final direction relative to its original direction?

Continued

SOLUTION

SET UP AND SOLVE For mirror 1, the angle of incidence is θ_1, and this equals the angle of reflection. The sum of the interior angles in the triangle shown in the figure is 180°, so we see that the angles of incidence and reflection for mirror 2 are $90° - \theta_1$. The total change in direction of the ray after both reflections is therefore $(180° - 2\theta_1) + 2\theta_1 = 180°$. That is, the ray's final direction is opposite to its original direction.

REFLECT An alternative viewpoint is that specular reflection reverses the sign of the component of light velocity perpendicular to the surface, but leaves the other components unchanged. We invite you to verify this fact in detail and to use it to show that when a ray of light is successively reflected by three mirrors forming a corner of a cube (a "corner reflector"), its final direction is again opposite to its original direction. The principle of a corner reflector is widely used in taillight lenses and highway signs to improve their nighttime visibility. Apollo astronauts placed arrays of corner reflectors on the moon. By use of laser beams reflected from these arrays, the earth–moon distance has been measured to within 0.15 m.

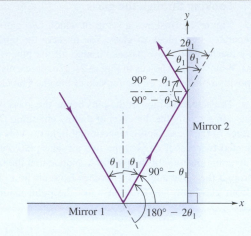

▲ **FIGURE 23.24**

23.8 Total Internal Reflection

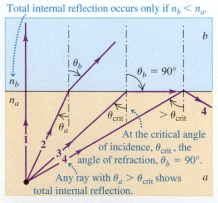

(a) Total internal reflection

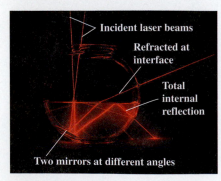

(b) Total internal reflection demonstrated with a laser, mirrors, and water in a fishbowl

▲ **FIGURE 23.25** Total internal reflection.

Figure 23.25a shows several rays diverging from a point source P in a material a with index of refraction n_a. The rays strike the surface of a second material b with index n_b, where $n_b < n_a$. From Snell's law,

$$\sin\theta_b = \frac{n_a}{n_b}\sin\theta_a.$$

Because $n_a/n_b > 1$, $\sin\theta_b$ is larger than $\sin\theta_a$, so the ray is bent *away from* the normal. Thus, there must be some value of θ_a *less than* 90° for which Snell's law gives $\sin\theta_b = 1$ and $\theta_b = 90°$. This is shown by ray 3 in the diagram, which emerges just grazing the surface, at an angle of refraction of 90°.

The angle of incidence for which the refracted ray emerges tangent to the surface is called the **critical angle,** denoted by θ_{crit}. If the angle of incidence is *greater than* the critical angle, then the sine of the angle of refraction, as computed by Snell's law, has to be greater than unity, which is impossible. Hence, for angles of incidence greater than the critical angle, the ray *cannot* pass into the upper material; it is trapped in the lower material and is completely reflected internally at the boundary surface, as shown in Figure 23.25. This situation, called **total internal reflection,** occurs only when a ray is incident on an interface with a second material whose index of refraction is *smaller* than that of the material in which the ray is traveling.

We can find the critical angle for two given materials by setting $\theta_b = 90°$ and $\sin\theta_b = 1$ in Snell's law. We then have the following result:

Total internal reflection

When a ray traveling in a material a with index of refraction n_a reaches an interface with a material b having index n_b, where $n_b < n_a$, it is totally reflected back into material a if the angle incidence is greater than the critical angle given by

$$\sin\theta_{crit} = \frac{n_b}{n_a}. \tag{23.18}$$

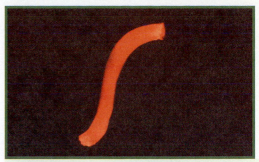

If the incident beam is oriented as shown, total internal reflection occurs on the 45° faces (because, for a glass–air interface, $\theta_{crit} = 41°$).

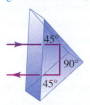

(a) Total internal reflection in a Porro prism.

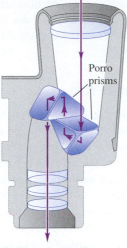

(b) Binoculars use Porro prisms to reflect the light to each eyepiece.

▲ **FIGURE 23.26** Total internal reflection in Porro prisms.

For a glass–air surface, with $n = 1.52$ for the glass,

$$\sin\theta_{crit} = \frac{1}{1.52} = 0.658, \qquad \theta_{crit} = 41.1°.$$

The fact that this critical angle is slightly less than 45° makes it possible to use a triangular prism with angles of 45°, 45°, and 90° as a totally reflecting surface. As reflectors, totally reflecting prisms have some advantages over metallic surfaces such as ordinary coated-glass mirrors. Light is *totally* reflected by a prism, but no metallic surface reflects 100% of the light incident on it. Also, the reflecting properties are permanent and not affected by tarnishing.

A 45°–45°–90° prism, used as in Figure 23.26a, is called a *Porro prism.* Light enters and leaves at right angles to the hypotenuse and is totally reflected at each of the shorter faces. The total change in direction of the rays is 180°. Binoculars often use combinations of two Porro prisms, as shown in Figure 23.26b.

When a beam of light enters at one end of a transparent rod (Figure 23.27), the light is totally reflected internally and is "trapped" within the rod even if the rod is curved, provided that the curvature is not too great. Such a rod is sometimes called a *light pipe.* A bundle of fine plastic fibers behaves in the same way and has the advantage of being flexible. A bundle may consist of thousands of individual fibers, each on the order of 0.002–0.01 mm in diameter. If the fibers are assembled in the bundle so that the relative positions of the ends are the same (or mirror images) at both ends, the bundle can transmit an image, as shown in Figure 23.28.

Fiber-optic devices have found a wide range of medical applications in instruments called *endoscopes,* which can be inserted directly into the bronchial tubes, the knee joint, the colon, and so on, for direct visual examination. A bundle of fibers can be enclosed in a hypodermic needle for the study of tissues and blood vessels far beneath the skin.

Fiber optics are also widely used in communication systems, where they are used to transmit a modulated laser beam. The number of binary digits that can be transmitted per unit time is proportional to the frequency of the wave. Infrared and visible-light waves have much higher frequencies than radio waves, so a modulated laser beam can transmit an enormous amount of information through a single fiber-optic cable. Another advantage of fiber-optic cables is that they are

▲ **BIO Application Let the light shine in.** In addition to photosynthesis, plants and algae use light to regulate many aspects of their physiology, including seed germination, stem elongation, and growth direction. These responses are tuned to specific parts of the visible spectrum, and a variety of molecules are used as photoreceptors. Some of these responses take place in seedlings underground. How is the light transmitted to the responding portions? Scientists found that the columns of cells in seedlings act as fibers that can guide light from near the surface of the soil to portions of the seedling farther below. The photograph shows light piping in an isolated, curved stem of an oat seedling.

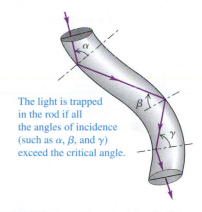

The light is trapped in the rod if all the angles of incidence (such as α, β, and γ) exceed the critical angle.

▲ **FIGURE 23.27** The principle of a light pipe, the basis for fiber optics.

▲ **FIGURE 23.28** An image transmitted by a bundle of optical fibers.

electrical insulators. They are thus immune to electrical interference from lightning and other sources, and they don't allow unwanted currents between a source and a receiver. They are secure and difficult to "bug," but they are also difficult to splice and tap into.

23.9 Dispersion

Ordinarily, white light is a superposition of waves with wavelengths extending throughout the visible spectrum. The speed of light *in vacuum* is the same for all wavelengths, but the speed in a material substance is different for different wavelengths. Therefore, the index of refraction of a material depends on wavelength. The dependence of wave speed and index of refraction on wavelength is called **dispersion.** Figure 23.29 shows the variation of index of refraction with wavelength for a few common optical materials. The value of *n* usually *decreases* with increasing wavelength and thus *increases* with increasing frequency. Hence, light of longer wavelength usually has greater speed in a given material than light of shorter wavelength.

Figure 23.30 shows a ray of white light incident on a prism. The deviation (change in direction) produced by the prism increases with increasing index of refraction and frequency and with decreasing wavelength. Violet light is deviated most and red least, and other colors show intermediate deviations. When it comes out of the prism, the light is spread out into a fan-shaped beam, as shown. The light is said to be *dispersed* into a spectrum. The amount of dispersion depends on the *difference* between the refractive indexes for violet light and for red light. From Figure 23.29, we can see that for a substance such as fluorite, whose refractive index for yellow light is small, the difference between the indexes for red and violet is also small. For silicate flint glass, both the index for yellow light and the difference between extreme indexes are larger. (The values of index of refraction in Table 23.1 are values for a wavelength of 589 nm, near the center of the visible range of wavelengths.)

The brilliance of diamond is due in part to its large dispersion and in part to its unusually large refractive index. Crystals of rutile and of strontium titanate, which can be produced synthetically, have about eight times the dispersion of diamond!

When you experience the beauty of a rainbow, as in Figure 23.31a, you are seeing the combined effects of dispersion and internal reflection. The light comes from behind you and is refracted into many small water droplets in the air. Each ray undergoes internal reflection from the back surface of the droplet and is reflected back to you (Figure 23.31b). Dispersion causes different colors to be refracted preferentially at different angles, so you see the various colors as coming

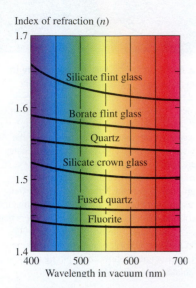

▲ FIGURE 23.29 Variation of index of refraction with wavelength for several materials.

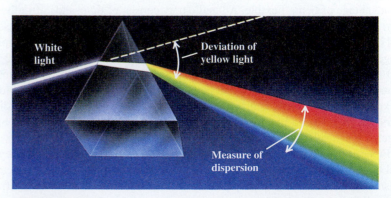

▲ FIGURE 23.30 Schematic representation of dispersion by a prism.

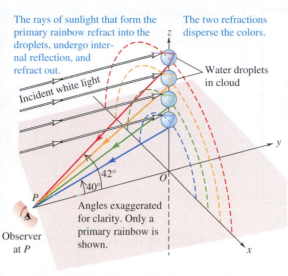

The rays of sunlight that form the primary rainbow refract into the droplets, undergo internal reflection, and refract out.

The two refractions disperse the colors.

Incident white light

Water droplets in cloud

42°
40°

Angles exaggerated for clarity. Only a primary rainbow is shown.

Observer at P

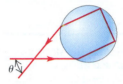

(c) An incoming ray undergoes two refractions and one internal reflection. The angle θ is greater for red light than for violet.

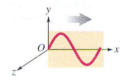

(d) A secondary rainbow is formed by rays that undergo two refractions and two internal reflections; the angle θ is greater for violet light than for red.

Secondary rainbow (note reversed colors)

Primary rainbow

(a) A double rainbow.

(b) How refraction and reflection in cloud droplets forms a rainbow. The x-y plane is horizontal, the z axis vertical.

▲ **FIGURE 23.31** How refraction and internal reflection in water droplets create a double rainbow.

from different regions of the sky, forming concentric circular arcs (Figure 23.31b). When you see a second, slightly larger rainbow with its colors reversed, you are seeing the results of dispersion and *two* internal reflections (Figure 23.31d).

23.10 Polarization

Polarization occurs with all transverse waves. This chapter is mainly about light, but to introduce basic polarization concepts, let's go back to some of the ideas presented in Chapter 12 about transverse waves on a string. For a string whose equilibrium position is along the x axis, the displacements may be along the y direction, as in Figure 23.32a. In this case, the string always vibrates in the x-y plane. But the displacements might instead be along the z axis, as in Figure 23.32b; then the string vibrates in the x-z plane.

When a wave has only y displacements, we say that it is **linearly polarized** in the y direction; similarly, a wave with only z displacements is linearly polarized in the z direction. For mechanical waves, we can build a **polarizing filter** that permits only waves with a certain polarization direction to pass. In Figure 23.32c, the string can slide vertically in the slot without friction, but no horizontal motion is possible. This filter passes waves polarized in the y direction but blocks those polarized in the z direction.

This same language can be applied to electromagnetic waves, which also have polarization. As we learned in Section 23.2, an electromagnetic wave is a *transverse* wave: The fluctuating electric and magnetic fields are perpendicular to the direction of propagation and to each other. We always define the direction of polarization of an electromagnetic wave to be the direction of the *electric*-field vector, not the magnetic-field vector, because most common electromagnetic-wave detectors (including the human eye) respond to the *electric* forces on electrons in materials, not the magnetic forces.

Polarizing Filters

Polarizing filters can be made for electromagnetic waves; the details of their construction depend on the wavelength. For microwaves with a wavelength of a few centimeters, a grid of closely spaced, parallel conducting wires that are insulated

MasteringPHYSICS

ActivPhysics 16.9: Polarization

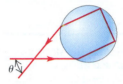

(a) Transverse wave linearly polarized in the y direction

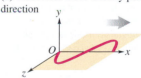

(b) Transverse wave linearly polarized in the z direction

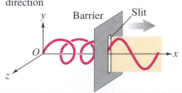

(c) The slit functions as a polarizing filter, passing only motion in the y direction

▲ **FIGURE 23.32** The concept of wave polarization applied to a transverse wave in a string.

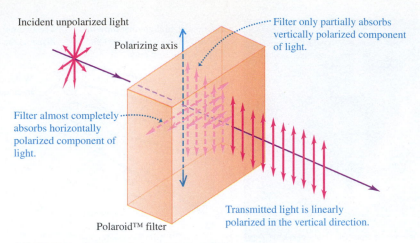

▲ **FIGURE 23.33** How a Polaroid™ filter produces polarized light.

from each other passes waves whose $\vec{E}$ fields are perpendicular to the wires, but not those with $\vec{E}$ fields parallel to the wires. The most common polarizing filter for light is a material known by the trade name Polaroid™, widely used for sunglasses and polarizing filters for camera lenses. This material, developed originally by Edwin H. Land, incorporates substances that exhibit **dichroism,** the selective absorption of one of the polarized components much more strongly than the other (Figure 23.33). A Polaroid™ filter transmits 80% or more of the intensity of waves polarized parallel to a certain axis in the material (called the **polarizing axis**), but only 1% or less of waves polarized perpendicular to this axis.

Waves emitted by radio transmitter antennas are usually linearly polarized. Vertical-rod antennas emit waves that, in a horizontal plane around the antenna, are polarized in the vertical direction (parallel to the antenna). Rooftop TV antennas have horizontal elements in the United States and vertical elements in Great Britain because the transmitted waves have different polarizations.

Light from ordinary sources doesn't have a definite polarization. The "antennas" that radiate light waves are the molecules that make up the sources. The waves emitted by any one molecule may be linearly polarized, like those from a radio antenna. But any actual light source contains a tremendous number of molecules with random orientations, so the light emitted is a random mixture of waves that are linearly polarized in all possible transverse directions.

An ideal polarizing filter, or **polarizer,** passes 100% of the incident light polarized in the direction of the filter's *polarizing axis,* but blocks all light polarized perpendicular to that axis. Such a device is an unattainable idealization, but the concept is useful in clarifying the basic ideas. In the discussion that follows, we'll assume that all polarizing filters are ideal. In Figure 23.34, unpolarized light (a random mixture of all polarization states) is incident on a polarizer in the form of a flat plate. The polarizing axis is represented by the blue line. The $\vec{E}$ vector of the incident wave can be represented in terms of components parallel and perpendicular to the polarizing axis. The polarizer transmits only the components of $\vec{E}$ parallel to that axis. The light emerging from the polarizer is linearly polarized parallel to the polarizing axis.

When we measure the intensity (power per unit area) of the light transmitted through an ideal polarizer, using the photocell in Figure 23.34, we find that it is exactly half that of the incident light, no matter how the polarizing axis is oriented. Here's why: We can resolve the $\vec{E}$ field of the incident wave into a component parallel to the polarizing axis and a component perpendicular to it. Because the incident light is a random mixture of all states of polarization, these two components are, on average, equal. The ideal polarizer transmits only the component parallel to the polarizing axis, so half the incident intensity is transmitted.

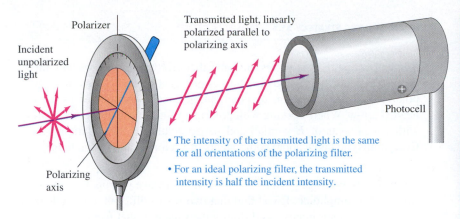

- The intensity of the transmitted light is the same for all orientations of the polarizing filter.
- For an ideal polarizing filter, the transmitted intensity is half the incident intensity.

▲ **FIGURE 23.34** The effect of a polarizing filter on unpolarized incident light.

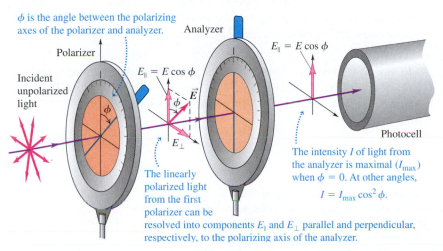

ϕ is the angle between the polarizing axes of the polarizer and analyzer.

$E_\parallel = E\cos\phi$

The linearly polarized light from the first polarizer can be resolved into components $E_\parallel$ and $E_\perp$ parallel and perpendicular, respectively, to the polarizing axis of the analyzer.

The intensity I of light from the analyzer is maximal (I_{max}) when $\phi = 0$. At other angles,

$$I = I_{max}\cos^2\phi.$$

▲ **FIGURE 23.35** The effect of passing unpolarized light through two polarizers.

▲ **BIO Application Polarized communication.** Numerous marine invertebrates have optical arrangements in their eyes that permit the perception of the orientation of polarized light. Some of these animals also can vary the degree of polarization of the light reflected from their bodies as a method of camouflage or communication. The accompanying photographs show a true color image (top) of a cuttlefish (not a fish at all, but an invertebrate related to squids and octopi) and an image in which the degree of polarization of the reflected light is represented by color. Horizontal polarization (in red) is seen in the vertical stripes on the animal. This pattern of polarization changes during mating and response to prey.

Now suppose we insert a second polarizer between the first polarizer and the photocell (Figure 23.35). The polarizing axis of the second polarizer, or *analyzer,* is vertical, and the axis of the first polarizer makes an angle ϕ with the vertical. That is, ϕ is the angle between the polarizing axes of the two polarizers. We can resolve the linearly polarized light transmitted by the first polarizer into two components, as shown in Figure 23.35, one parallel and the other perpendicular to the vertical axis of the analyzer. Only the parallel component, with amplitude $E\cos\phi$, is transmitted by the analyzer. The transmitted intensity is greatest when $\phi = 0$; it is zero when $\phi = 90°$—that is, when polarizer and analyzer are *crossed* (Figure 23.36).

To find the transmitted intensity at intermediate angles, we recall from our discussion in Section 23.5 that the intensity of an electromagnetic wave is proportional to the *square* of the amplitude of the wave. The ratio of the transmitted to the incident *amplitude* is $\cos\phi$, so the ratio of the transmitted to the incident *intensity* is $\cos^2\phi$. Thus, we obtain the following result:

Light transmitted by polarizing filter

When linearly polarized light strikes a polarizing filter with its axis at an angle ϕ to the direction of polarization, the intensity of the transmitted light is

$$I = I_{max}\cos^2\phi, \qquad (23.19)$$

where I_{max} is the maximum intensity of light transmitted (at $\phi = 0$) and I is the amount transmitted at angle ϕ. This relationship, discovered experimentally by Etienne Louis Malus in 1809, is called **Malus's law.**

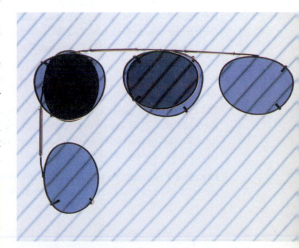

▲ **FIGURE 23.36** Polarizing sunglasses oriented with their polarizing axes parallel (top center) and perpendicular (top left). The crossed polarizers in the top left transmit no light.

PROBLEM-SOLVING STRATEGY 23.3 **Polarization**

SET UP

1. Remember that in light waves (and all other electromagnetic waves), the $\vec{E}$ field is perpendicular to the direction of propagation and is the direction of polarization. The direction of polarization can be thought of as a two-headed arrow. When working with polarizing filters, you are really dealing with components of $\vec{E}$ parallel and perpendicular to the polarizing axis. Everything you know about components of vectors is applicable here.

SOLVE

2. The intensity (average power per unit area) of a wave is proportional to the *square* of its amplitude. If you find that two waves differ in amplitude by a certain factor, their intensities differ by the square of that factor.

3. Unpolarized light is a random mixture of all possible polarization states, so, on average, it has equal components in any two perpendicular directions, with each component having half the total light intensity. Partially linearly polarized light is a superposition of linearly polarized and unpolarized light.

REFLECT

4. In any arrangement of filters, the total intensity of the outgoing light can never exceed that of the incoming light, because of conservation of energy. Check to make sure that your results satisfy this requirement.

EXAMPLE 23.9 **Linear polarization**

You and a friend each have a pair of polarizing sunglasses and decide to test Malus's law by using the light sensors in your physics lab. You orient the sunglasses so that the angle between the polarizing axes of two of the lenses is 30°; then you direct a narrow beam of unpolarized light through both lenses. Relative to the intensity I_0 of the incident unpolarized beam, what intensity do you expect to find after the beam passes through the first lens? After it passes through the second?

SOLUTION

SET UP Figure 23.37 shows our sketch.

SOLVE The intensity of the light transmitted through an ideal polarizer is exactly half that of the unpolarized incident light I_0, regardless of the polarizer's orientation. Therefore, the intensity I_1 after the first lens is $I_1 = I_0/2$. For the intensity after the second lens, with $\phi = 30°$, Malus's law gives

$$I_2 = I_1 \cos^2 30° = \left(\frac{I_0}{2}\right)\cos^2 30° = \left(\frac{I_0}{2}\right)\left(\frac{\sqrt{3}}{2}\right)^2 = \frac{3}{8}I_0.$$

REFLECT If the polarizing axes of the two lenses are parallel, then *all* the light emerging from the first lens is transmitted through the second lens. In this case, the transmitted intensity is $I_0/2$, which is also equal to $(I_0/2)\cos^2 0°$.

Practice Problem: Now you rotate the polarizing axes so they are at a 45° angle relative to each other. What is the intensity of

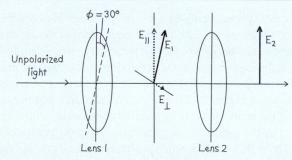

▲ **FIGURE 23.37** Our sketch for this problem.

the light emerging from the second polarizing lens? *Answer:* $I_2 = I_0/4$.

Polarization by Reflection

Unpolarized light can be partially polarized by *reflection*. When unpolarized light strikes a reflecting surface between two optical materials, preferential reflection occurs for those waves in which the electric-field vector is parallel to the reflect-

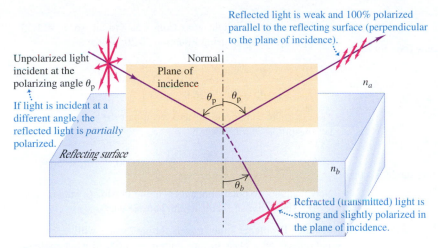

Unpolarized light incident at the polarizing angle θ_p

If light is incident at a different angle, the reflected light is *partially* polarized.

Reflected light is weak and 100% polarized parallel to the reflecting surface (perpendicular to the plane of incidence).

Normal

Plane of incidence

θ_p θ_p

n_a

Reflecting surface

n_b

θ_b

Refracted (transmitted) light is strong and slightly polarized in the plane of incidence.

▲ **FIGURE 23.38** When light is incident at the polarizing angle, the reflected light is 100% linearly polarized.

ing surface. In Figure 23.38, the plane containing the incident and reflected rays and the normal to the surface is called the **plane of incidence.** At one particular angle of incidence, called the **polarizing angle** θ_p, only the light for which the $\vec{E}$ vector is perpendicular to the plane of incidence (and parallel to the reflecting surface) is reflected. The reflected light is therefore linearly polarized perpendicular to the plane of incidence (parallel to the reflecting surface), as shown in Figure 23.38.

When light is incident at the polarizing angle θ_p, *none* of the $\vec{E}$-field component *parallel* to the plane of incidence is reflected; this component is transmitted 100% in the *refracted* beam. So the *reflected* light is *completely* polarized. The *refracted* light is a mixture of the component parallel to the plane of incidence, all of which is refracted, and the remainder of the perpendicular component; it is therefore *partially* polarized.

In 1812, Sir David Brewster noticed that when the angle of incidence is equal to the polarizing angle θ_p, the reflected ray and the refracted ray are perpendicular to each other, as shown in Figure 23.39. In this case, the angle of refraction θ_b is equal to $90° - \theta_p$ so $\sin\theta_b = \cos\theta_p$. From the law of refraction, $n_a\sin\theta_p = n_b\sin\theta_b$, so we find that $n_a\sin\theta_p = n_b\cos\theta_p$ and

$$\frac{\sin\theta_p}{\cos\theta_p} = \tan\theta_p = \frac{n_b}{n_a}. \qquad (23.20)$$

This relation is known as **Brewster's law.** It states that when the tangent of the angle of incidence equals the ratio of the two indexes of refraction, the reflected light is completely polarized.

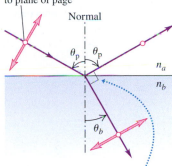

Component perpendicular to plane of page

Normal

θ_p θ_p

n_a

n_b

θ_b

When light strikes a surface at the polarizing angle, the reflected and refracted rays are perpendicular to each other and

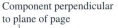

$$\tan\theta_p = \frac{n_b}{n_a}$$

▲ **FIGURE 23.39** Brewster's law. This diagram shows a side view of the scene in Figure 23.38.

EXAMPLE 23.10 **Reflection from a swimming pool's surface**

Sunlight reflects off the smooth surface of an unoccupied swimming pool. **(a)** At what angle of reflection is the light completely polarized? **(b)** What is the corresponding angle of refraction for the light that is transmitted (refracted) into the water? **(c)** At night, an underwater floodlight is turned on in the pool. Repeat parts (a) and (b) for rays from the floodlight that strike the smooth surface from below.

Continued

SOLUTION

SET UP Figure 23.40 shows the sketches we draw (one for each situation).

SOLVE Part (a): We're looking for the polarizing angle for light that passes from air into water, so $n_a = 1.00$ (air) and $n_b = 1.33$ (water). From Equation 23.20,

$$\theta_p = \tan^{-1}\frac{n_b}{n_a} = \tan^{-1}\frac{1.33}{1.00} = 53.1°.$$

Part (b): The incident light is at the polarizing angle, so the reflected and refracted rays are perpendicular; hence,

$$\theta_p + \theta_b = 90°,$$
$$\theta_b = 90° - 53.1° = 36.9°.$$

Part (c): Now the light goes from water to air, so $n_a = 1.33$ and $n_b = 1.00$. Again using Equation 23.20, we have

$$\theta_p = \tan^{-1}\frac{n_b}{n_a} = \tan^{-1}\frac{1.00}{1.33} = 36.9°,$$
$$\theta_b = 90° - 36.9° = 53.1°.$$

REFLECT We can check our answer in part (b) by using Snell's law, $n_a \sin\theta_p = n_b \sin\theta_b$, or

$$\sin\theta_b = \frac{n_a \sin\theta_p}{n_b} = \frac{1.00 \sin 53.1°}{1.33} = 0.600, \qquad \theta_b = 36.9°.$$

Note that the two polarizing angles found in parts (a) and (c) add to 90°. This is *not* an accident; can you see why?

Practice Problem: Light travels through water with $n_a = 1.33$ and reflects off a glass surface with $n_b = 1.52$. At what angle of reflection is the light completely polarized? *Answer:* $\theta_p = 48.8°$.

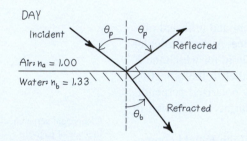 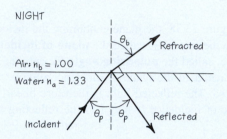

▲ **FIGURE 23.40** Our sketches for this problem.

Polarizing filters are widely used in sunglasses. When sunlight is reflected from a horizontal surface, the reflected light contains a preponderance of light polarized in the horizontal direction. When the reflection occurs at a smooth asphalt road surface or the surface of a lake, it causes unwanted glare. Vision can be improved by eliminating this glare. The manufacturer makes the polarizing axis of the lens material vertical, so very little of the horizontally polarized light is transmitted to the eyes. The glasses also reduce the overall intensity in the transmitted light to somewhat less than 50% of the intensity of the unpolarized incident light.

Photoelasticity

Some optical materials, when placed under mechanical stress, develop the property that their index of refraction is different for different planes of polarization. The result is that the plane of polarization of incident light can be rotated, by an amount that depends on the stress. This effect is the basis of the science of **photoelasticity.** Stresses in girders, boiler plates, gear teeth, and cathedral pillars can be analyzed by constructing a transparent model of the object, usually of a plastic material, subjecting it to stress, and examining it between a polarizer and an analyzer in the crossed position. Very complicated stress distributions can be studied by these optical methods. Figure 23.41 shows photographs of photoelastic models under stress.

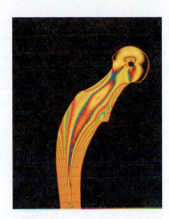

▲ **FIGURE 23.41** In photoelastic stress analysis, a structure such as a cathedral or an artificial hip is modeled in a transparent photoelastic material, subjected to stress, and analyzed in polarized light. The resulting images show where stress is concentrated in the object, allowing engineers to design it appropriately. In the case of cathedrals, such models have explained spectacular historical collapses by showing that inadequate buttressing and high winds can cause tensile stresses in masonry members. Masonry is strong in compression, but not in tension—which is one reason that modern large buildings use both steel and concrete in their frames.

23.11 Huygens's Principle

The laws of reflection and refraction of light rays that we introduced in Section 23.7 were discovered experimentally long before the wave nature of light was firmly established. However, we can *derive* these laws from wave considerations and show that they are consistent with the wave nature of light.

We begin with a principle called **Huygens's principle,** stated originally by Christiaan Huygens in 1678. Huygens's principle offers a geometrical method for finding, from the known shape of a wave front at some instant, the shape of the wave front at some later time. Huygens made the following hypothesis:

> **Huygens's principle:**
>
> **Every point of a wave front may be considered the source of secondary wavelets that spread out in all directions with a speed equal to the speed of propagation of the wave.**
>
> The new wave front at a later time is then found by constructing a surface *tangent* to the secondary wavelets, which is called the *envelope* of the wavelets.

All the results that we obtain from Huygens's principle can also be obtained from Maxwell's equations. Thus, it is not an independent principle, but it is very helpful in demonstrating the close relationship between the wave and ray models of light.

Huygens's principle is illustrated in Figure 23.42. The original wave front AA' is traveling outward from a source, as indicated by the small arrows. We want to find the shape of the wave front after a time interval t. Let v be the speed of propagation of the wave; then, in time t, it travels a distance vt. We construct several circles (traces of spherical wavelets) with radius $r = vt$, centered at points along AA'. The trace of the envelope of these wavelets, which is the new wave front, is the curve BB'. Throughout this discussion we're assuming that the speed v is the same at all points and in all directions.

The law of reflection can be derived from Huygens's principle; here's a brief sketch of the derivation: We consider a plane wave approaching a plane reflecting surface. In Figure 23.43a, the lines AA', BB', and CC' represent successive positions of a wave front approaching the reflecting surface MM'. As the points on

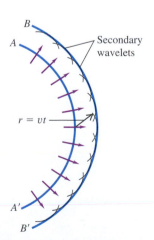

▲ **FIGURE 23.42** Applying Huygens's principle to wave front AA' to construct a new wave front BB'.

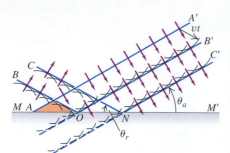

(a) Successive positions of a plane wave AA' as it is reflected from a plane surface

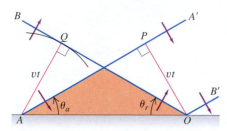

(b) Magnified portion of part (a)

▲ **FIGURE 23.43** Reflection of a plane wave from a surface.

this wave front successively reach points on surface MM', secondary wavelets are produced, as shown. The effect of the reflecting surface is to *change the direction of travel* of the wavelets that strike it, so part of a wavelet that would have penetrated the surface actually lies above it.

Figure 23.43b is an enlarged view of the colored area in Figure 23.43a. The right triangles OAP and AOQ are equal, so $\theta_a = \theta_r$. The angle the *wave front* makes with the surface equals the angle the *ray* makes with the *normal* to the surface, so the angle of reflection of the ray equals the angle of incidence, showing the relation between the law of reflection and Huygens's principle.

We can derive the law of *refraction* from Huygens's principle by a similar procedure. In Figure 23.44a, we consider a wave front, represented by line AA', for which point A has just arrived at the boundary surface SS' between two transparent materials a and b with wave speeds v_a and v_b, respectively. (The *reflected* waves are not shown in the figure; they proceed exactly as in Figure 23.43.) As successive points on the wavefront AA', arrive at the interface, secondary wavelets originate at these points.

Figure 23.44b is an enlarged view of a portion of the interface. Note that AO is the hypotenuse of triangle AOQ and also of triangle AOB. The angles θ_a and θ_b between the surface and the incident and refracted wave fronts are the angle of incidence and the angle of refraction, respectively. In time t, point A' on the wave front moves a distance $OQ = v_a t$, and point A moves a distance $AB = v_b t$. For the right triangles AOQ and AOB, we have

$$\sin\theta_a = \frac{v_a t}{AO} \quad \text{and} \quad \sin\theta_b = \frac{v_b t}{AO}.$$

Combining these two equations, we obtain

$$\frac{\sin\theta_a}{\sin\theta_b} = \frac{v_a}{v_b}. \tag{23.21}$$

We've defined the index of refraction n of a material as the ratio of the speed of light in vacuum (c) to its speed v in the material: $n_a = c/v_a$ and $n_b = c/v_b$. Thus,

$$\frac{n_b}{n_a} = \frac{c/v_b}{c/v_a} = \frac{v_a}{v_b},$$

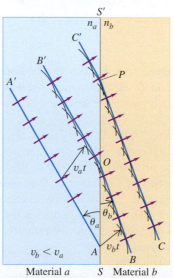

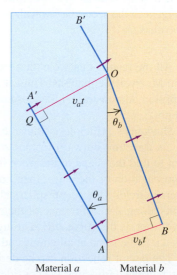

(a) Successive positions of a plane wave AA' as it is refracted by a plane surface

(b) Magnified portion of part (a)

▲ **FIGURE 23.44** Refraction of a plane wave from a surface for the case $v_b < v_a$.

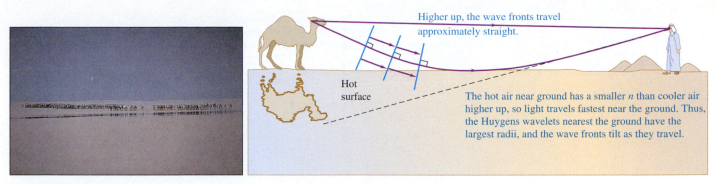

Higher up, the wave fronts travel approximately straight.

Hot surface

The hot air near ground has a smaller n than cooler air higher up, so light travels fastest near the ground. Thus, the Huygens wavelets nearest the ground have the largest radii, and the wave fronts tilt as they travel.

▲ **FIGURE 23.45** The cause of a mirage.

and we can rewrite Equation 23.21 as

$$\frac{\sin\theta_a}{\sin\theta_b} = \frac{n_b}{n_a},$$

or

$$n_a\sin\theta_a = n_b\sin\theta_b,$$

which we recognize as Snell's law, Equation 23.16. So we have derived Snell's law from a wave theory. Alternatively, we may choose to regard Snell's law as an experimental result that defines the index of refraction of a material; in that case, the preceding analysis helps to confirms the relationship $v = c/n$ for the speed of light in a material.

Mirages offer an interesting demonstration of Huygens's principle in action. When a surface of pavement or desert sand is heated intensely by the sun, a hot, less dense layer of air with smaller n forms near the surface. The speed of light is slightly greater in this hotter air near the ground, and the Huygens wavelets have slightly larger radii (because they move slightly faster and travel farther in a given time interval). As a result, the wave fronts tilt somewhat, and rays that were headed slightly toward the ground (with an incident angle near 90°) can be bent upwards, as shown in Figure 23.45. Light farther from the ground is bent less and travels nearly in a straight line. The observer sees the object in its natural position, with an inverted image below it, as though seen in a horizontal reflecting surface. Even when the turbulence of the heated air prevents a clear inverted image from being formed, the mind of the thirsty traveler can interpret the apparent reflecting surface as a sheet of water.

It's important to keep in mind that Maxwell's equations are the fundamental relations for electromagnetic wave propagation. But it is a remarkable fact that Huygens's principle anticipated Maxwell's analysis by two centuries! Indeed, Maxwell provided the theoretical underpinning for Huygens's principle. Every point in an electromagnetic wave, with its time-varying electric and magnetic fields, acts as a source of the continuing wave, as predicted by Ampère's and Faraday's laws.

23.12 Scattering of Light

The sky is blue. Sunsets are red. Skylight is partially polarized; you can see this by looking at the sky directly overhead through a polarizing filter. It turns out that one phenomenon is responsible for all three of these effects.

In Figure 23.46, sunlight (unpolarized) comes from the left along the x axis and passes over an observer looking vertically upward along the y axis. (We are viewing the situation from the side.) Molecules of the earth's atmosphere are located at point O. The electric field in the beam of sunlight sets the electric

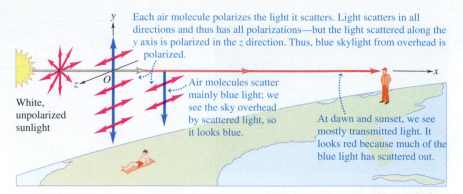

Each air molecule polarizes the light it scatters. Light scatters in all directions and thus has all polarizations—but the light scattered along the y axis is polarized in the z direction. Thus, blue skylight from overhead is polarized.

White, unpolarized sunlight

Air molecules scatter mainly blue light; we see the sky overhead by scattered light, so it looks blue.

At dawn and sunset, we see mostly transmitted light. It looks red because much of the blue light has scattered out.

▲ **FIGURE 23.46** The sky is blue because light scattered by the atmosphere is primarily blue (and linearly polarized). At sunset and sunrise, we see primarily transmitted light, which is red because the blue scattered light has been subtracted from it.

charges in the molecules into vibration. Light is a transverse wave; the direction of the electric field in any component of the sunlight lies in the y-z plane, and the motion of the charges takes place in this plane. There is no field, and therefore no vibration, in the direction of the x axis.

An incident light wave whose $\vec{E}$ field has y and z components causes the electric charges in the molecules to vibrate along the line of $\vec{E}$. We can resolve this vibration into two components, one along the y axis (blue arrows), the other along the z axis (red arrows). The two components in the incident light thus produce the equivalent of two molecular "antennas," oscillating with the same frequency as the incident light and lying along the y and z axes, respectively.

Such an antenna doesn't radiate in the direction of its own length. The antenna along the y axis doesn't send any light to the observer directly below it, although it does emit light in other directions. Therefore, the only light reaching that observer comes from the other antenna, corresponding to the component of vibration along the z axis. This light is linearly polarized, with its electric field parallel to the antenna. The vectors on the y axis below point O show the direction of polarization of the light reaching the observer.

The process that we've just described is called **scattering.** The energy of the scattered light is removed from the original beam, reducing its intensity. Detailed analysis of the scattering process shows that the intensity of the light scattered from air molecules increases in proportion to the fourth power of the frequency (inversely to the fourth power of the wavelength). Thus, the intensity ratio for the two ends of the visible spectrum is $(700 \text{ nm}/400 \text{ nm})^4 = 9.4$. Roughly speaking, scattered light contains nine times as much blue light as red, and that's why the sky is blue.

Because skylight is partially polarized, polarizers are useful in photography. The sky in a photograph can be darkened by appropriate orientation of the polarizer axis. The effect of atmospheric haze can be reduced in exactly the same way, and unwanted reflections can be controlled just as with polarizing sunglasses, discussed in Section 23.10.

Toward evening, when sunlight has to travel a long distance through the earth's atmosphere, a substantial fraction of the blue light is removed by scattering. White light minus blue light appears yellow or red. Thus, when sunlight with the blue component removed is incident on a cloud, the light reflected from the cloud to the observer has the yellow or red hue so often observed at sunset. If the earth had no atmosphere, we would receive *no* skylight at the earth's surface, and the sky would appear as black in the daytime as it does at night. Thus, to an astronaut in a spacecraft or on the moon, the sky appears black, not blue.

SUMMARY

Electromagnetic Waves, the Speed of Light, and the Electromagnetic Spectrum

(Sections 23.1–23.3) When either an electric or a magnetic field changes with time, a field of the other kind is induced in the adjacent regions of space. This electromagnetic disturbance, or **electromagnetic wave,** can travel through space even when there is no matter in the intervening region. In vacuum, electromagnetic waves travel at the speed of light $c = 3.00 \times 10^8$ m/s.

The **electromagnetic spectrum** covers a range of frequencies from at least 1 to 10^{24} Hz and a correspondingly broad range of wavelengths. Radio waves have low frequencies (long wavelengths), and gamma rays have high frequencies (short wavelengths). Visible light is a very small part of this spectrum, with wavelengths in vacuum from 400 to 700 nm.

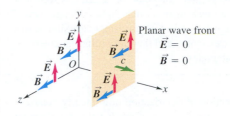

Sinusoidal Waves and Energy

(Sections 23.4 and 23.5) In a sinusoidal electromagnetic plane wave, the $\vec{E}$ and $\vec{B}$ fields vary sinusoidally in space and time. For a sinusoidal plane wave traveling in the $+x$ direction, $E = E_{max} \sin(\omega t - kx)$ and $B = B_{max} \sin(\omega t - kx)$ (Equations 23.3)

The electric and magnetic fields each contain energy. The maximum magnitudes of the electric and magnetic fields are related by the equation $E_{max} = cB_{max}$ (Equation 23.4); the intensity I can be expressed as

$$I = \frac{1}{2}\sqrt{\frac{\epsilon_0}{\mu_0}} E_{max}^2 = \frac{1}{2}\epsilon_0 c E_{max}^2. \qquad (23.11)$$

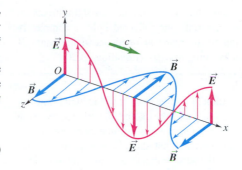

The Nature of Light

(Section 23.6) Light is an electromagnetic wave. When emitted or absorbed, it also shows particle properties. Light is emitted by accelerated electric charges that have been given excess energy by heat or electrical discharge. The speed of light, c, is a fundamental physical constant.

A **wave front** is a surface of constant phase; wave fronts move with a speed equal to the propagation speed of the wave. A ray is a line along the direction of propagation, perpendicular to the wave fronts.

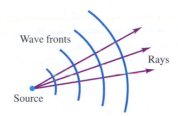

Reflection and Refraction; Total Internal Reflection

(Sections 23.7 and 23.8) The **index of refraction,** n, of a material is the ratio of the speed of light in vacuum, c, to the speed v in the material: $n = c/v$ (Equation 23.14). The incident, reflected, and refracted rays and the normal to the interface all lie in a single plane called the **plane of incidence.** Angles of incidence, reflection, and refraction are always measured from the normal to the interface. The law of reflection states that the angles of incidence and reflection are equal: $\theta_r = \theta_a$ (Equation 23.15). The law of refraction is $n_a \sin\theta_a = n_b \sin\theta_b$ (Equation 23.16).

When a ray travels within a material of greater index of refraction n_a, toward an interface with one of smaller index n_b, total internal reflection occurs when the angle of incidence exceeds a critical value θ_{crit} given by $\sin\theta_{crit} = n_b/n_a$ (Equation 23.18).

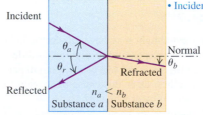

• Incident, reflected, and refracted rays and the normal lie in the same plane.
• $\theta_r = \theta_a$.
• $\dfrac{\sin\theta_a}{\sin\theta_b} = \dfrac{n_b}{n_a}$

Continued

Dispersion

(Section 23.9) The variation of index of refraction, n, with wavelength λ is called **dispersion**. Usually, n increases with decreasing λ. Thus, refraction is greater for light of shorter wavelength (higher frequency) in a given material than light of longer wavelength. When white light is incident on a prism, dispersion causes light with differing wavelengths, and therefore differing colors, to be refracted at different angles.

Polarization

(Section 23.10) The direction of polarization of a linearly polarized electromagnetic wave is the direction of the $\vec{E}$ field. A **polarizing filter** passes radiation that is linearly polarized along its polarizing axis and blocks radiation polarized perpendicular to that axis. **Malus's law** states that when polarized light of intensity I_{max} is incident on an analyzer and ϕ is the angle between the polarizing axes of the polarizer and analyzer, the transmitted intensity I is $I = I_{max} \cos^2 \phi$ (Equation 23.19).

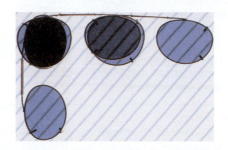

Brewster's law states that when unpolarized light strikes an interface between two materials, the reflected light is completely polarized perpendicular to the plane of incidence if the angle of incidence, θ_p, is given by $\tan \theta_p = n_b/n_a$ (Equation 23.20).

Huygens's Principle

(Section 23.11) If the position of a wave front at one instant is known, the position of the front at a later time can be constructed by using **Huygens's principle:** Every point of a wave front may be considered the source of secondary wavelets that spread out in all directions, and the new wave front is the surface that is tangent to the wavelets.

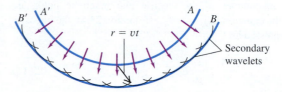

Secondary wavelets

Scattering of Light

(Section 23.12) Light is scattered by air molecules, and the scattered light is partially polarized. Light with higher frequencies is scattered more than light with lower frequencies. The sky is blue because air molecules scatter into view more higher-frequency blue light than lower-frequency red light.

 For instructor-assigned homework, go to www.masteringphysics.com

Conceptual Questions

1. The light beam from a searchlight may have an electric-field magnitude of 1000 V/m, corresponding to a potential difference of 1500 V between the head and feet of a 1.5-m-tall person on whom the light shines. Does this cause the person to feel a strong electric shock? Why or why not?
2. If a light beam carries momentum, should a person holding a flashlight feel a recoil analogous to the recoil of a rifle when it is fired? Why is this recoil not actually observed?
3. Why is the average radiation pressure on a perfectly reflecting surface twice as great as on a perfectly absorbing surface?
4. When an electromagnetic wave is reflected from a moving reflector, the frequency of the reflected wave is different from that of the initial wave. Explain physically how this happens. Also, show why a higher-than-normal frequency results if the reflector is moving toward the observer and a lower frequency if it is moving away. (Some radar systems used for highway-speed control operate on this principle.)
5. When hot air rises around a radiator or from a heating duct, objects behind it appear to shimmer or waver. What is happening?
6. How does the refraction of light account for the twinkling of starlight? (*Hint:* The earth's atmosphere consists of layers of varying density.)
7. Light requires about 8 min to travel from the sun to the earth. Is it delayed appreciably by the earth's atmosphere?
8. Sometimes when looking at a window, one sees two reflected images, slightly displaced from each other. What causes this phenomenon?
9. A student claimed that, because of atmospheric refraction, the sun can be seen after it has set and that the day is therefore longer than it would be if the earth had no atmosphere. First, what does the student mean by saying the sun can be seen after

it has set? Second, comment on the validity of the conclusion. Does the same effect also occur at sunrise?

10. If you look at your pet fish through the corner of your aquarium, you may see a double image of the fish, one image on each side of the corner. Explain how this could happen.

11. How could you determine the direction of the polarizing axis of a single polarizer? (*Hint:* Is there any naturally occurring polarized light you could use?)

12. In three-dimensional movies, two images are projected on the screen, and the viewers wear special glasses to sort them out. How does the polarization of light allow this effect to work?

13. Can sound waves be reflected? Refracted? Give examples to back up your answer.

14. Why should the wavelength, but not the frequency, of light change in passing from one material into another?

15. When light is incident on an interface between two materials, the angle of the refracted ray depends on the wavelength, but the angle of the reflected ray does not. Why should this be?

Multiple-Choice Problems

1. Light having a certain frequency, wavelength, and speed is traveling through empty space. If the frequency of this light were doubled, then
   A. its wavelength would remain the same, but its speed would double.
   B. its wavelength would remain the same, but its speed would be halved.
   C. its wavelength would be halved, but its speed would double.
   D. its wavelength would be halved, but its speed would remain the same.
   E. both its speed and its wavelength would be doubled.

2. Unpolarized light with an original intensity I_0 passes through two ideal polarizers having their polarizing axes turned at $120°$ to each other. After passing through both polarizers, the intensity of the light is

   A. $\dfrac{\sqrt{3}}{2}I_0$.   B. $\dfrac{1}{2}I_0$.   C. $\dfrac{\sqrt{3}}{4}I_0$.

   D. $\dfrac{1}{4}I_0$.   E. $\dfrac{1}{8}I_0$.

3. Light travels from water (with index of refraction 1.33) into air (index of refraction 1.00). Which of the following statements about this light is true? (There may be more than one correct choice.)
   A. The light has the same frequency in the air as it does in the water.
   B. The light travels faster in the air than in the water.
   C. The light has the same wavelength in the air as it does in the water.
   D. The light has the same speed in the air as in the water.
   E. The wavelength of the light in the air is greater than the wavelength in the water.

4. If a sinusoidal electromagnetic wave with intensity 10 W/m² has an electric field of amplitude E, then a 20 W/m² wave of the same wavelength will have an electric field of amplitude
   A. $4E$.   B. $2\sqrt{2}E$.   C. $2E$.   D. $\sqrt{2}E$.

5. A plane electromagnetic wave is traveling vertically downward with its magnetic field pointing northward. Its electric field must be pointing
   A. toward the south.   B. toward the east.
   C. toward the west.   D. vertically upward.
   E. vertically downward.

6. Suppose that a reflective solar sail (see Example 23.5) is deployed not perpendicular to the sun's rays but at some other angle. In what direction will the sail accelerate?
   A. In the direction the sun's rays are moving.
   B. Perpendicular to the surface of the sail.
   C. At an angle somewhere between that of the sun's rays and the perpendicular to the surface of the sail.
   D. The sail will not accelerate unless it is perpendicular to the sun's rays.

7. The index of refraction, n, has which of the following range of values?
   A. $n \geq 1$.   B. $0 \leq n \leq 1$.   C. $n \geq 0$.

8. A ray of light going from one material into another follows the path shown in Figure 23.47. What can you conclude about the relative indexes of refraction of these two materials?
   A. $n_a \geq n_b$.   B. $n_a > n_b$.
   C. $n_a < n_b$.   D. $n_a \leq n_b$.

   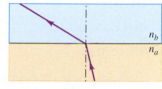
   ▲ **FIGURE 23.47** Multiple-choice problem 8.

9. Which of the following statements about radio waves, infrared radiation, and x rays are correct? (There may be more than one correct choice.)
   A. They all have the same wavelength in vacuum.
   B. They all have the same frequency in vacuum.
   C. They all have exactly the same speed as visible light in vacuum.
   D. The short-wavelength x rays travel faster through vacuum than the long-wavelength radio waves.

10. Two lasers each produce 2 mW beams. The beam of laser B is wider, having twice the cross-sectional area as the beam of laser A. Which of the following statements about these two laser beams are correct? (There may be more than one correct choice.)
   A. Both of the beams have the same average power.
   B. Beam A has twice the intensity of beam B.
   C. Beam B has twice the intensity of beam A.
   D. Both beams have the same intensity.

11. A ray of light follows the path shown in Figure 23.48 as it reaches the boundary between two *transparent* materials. What can you conclude about the relative indexes of refraction of these two materials?
   A. $n_1 \geq n_2$.   B. $n_1 > n_2$.
   C. $n_1 < n_2$.   D. $n_1 \leq n_2$.

   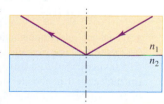
   ▲ **FIGURE 23.48** Multiple-choice problem 11.

12. A light beam has a wavelength of 300 nm in a material of refractive index 1.5. In a material of refractive index 3.0, its wavelength will be
   A. 450 nm.   B. 300 nm.   C. 200 nm.
   D. 150 nm.   E. 100 nm.

13. A light beam has a frequency of 300 MHz in a material of refractive index 1.5. In a material of refractive index 3.0, its frequency will be
    A. 450 MHz. B. 300 MHz. C. 200 MHz.
    D. 150 MHz. E. 100 MHz.
14. You are sunbathing in the late afternoon when the sun is relatively low in the western sky. You are lying flat on your back, looking straight up through Polaroid sunglasses (see Figure 23.46 and the discussion of sunglasses in Section 23.10). To minimize the amount of light reaching your eyes, you should lie with your feet pointing in what direction? (There may be more than one correct answer.)
    A. north.   B. east.   C. south.
    D. west.   E. some other direction.
15. A beam of light takes time t to travel a distance L in a certain liquid. If we now add water to the liquid to reduce its index of refraction by half, the time for the beam to travel the same distance will be
    A. $2t$.   B. $\sqrt{2}t$.   C. $t/\sqrt{2}$.   D. $t/2$.

Problems

23.2 Speed of an Electromagnetic Wave

1. ● When a solar flare erupts on the surface of the sun, how many minutes after it occurs does its light show up in an astronomer's telescope on earth? (Consult Appendix E.)
2. ● **TV ghosting.** In a TV picture, faint, slightly offset ghost images are formed when the signal from the transmitter travels to the receiver both directly and indirectly after reflection from a building or some other large metallic mass. In a 25 inch set, the ghost is about 1.0 cm to the right of the principal image if the reflected signal arrives 0.60 μs after the principal signal. In this case, what is the difference in the distance traveled by the two signals?
3. ● (a) How much time does it take light to travel from the moon to the earth, a distance of 384,000 km? (b) Light from the star Sirius takes 8.61 years to reach the earth. What is the distance to Sirius in kilometers?
4. ●● A geostationary communications satellite orbits the earth directly above the equator at an altitude of 35,800 km. Calculate the time it takes for a signal to travel from a point on the equator to the satellite and back to the ground at another point on the equator exactly halfway around the earth. (See Appendix E.)

23.3 The Electromagnetic Spectrum

23.4 Sinusoidal Waves

5. ● Consider electromagnetic waves propagating in air. (a) Determine the frequency of a wave with a wavelength of (i) 5.0 km, (ii) 5.0 μm, (iii) 5.0 nm. (b) What is the wavelength (in meters and nanometers) of (i) gamma rays of frequency 6.50×10^{21} Hz, (ii) an AM station radio wave of frequency 590 kHz?
6. ● Most people perceive light having a wavelength between 630 nm and 700 nm as red and light with a wavelength between 400 nm and 440 nm as violet. Calculate the approximate frequency ranges for (a) violet light and (b) red light.
7. ● The electric field of a sinusoidal electromagnetic wave obeys the equation $E = -(375 \text{ V/m}) \sin[(5.97 \times 10^{15} \text{ rad/s})t + (1.99 \times 10^{7} \text{ rad/m})x]$. (a) What are the amplitudes of the electric and magnetic fields of this wave? (b) What are the frequency, wavelength, and period of the wave? Is this light visible to humans? (c) What is the speed of the wave?

8. ● A sinusoidal electromagnetic wave having a magnetic field of amplitude 1.25 μT and a wavelength of 432 nm is traveling in the $+x$ direction through empty space. (a) What is the frequency of this wave? (b) What is the amplitude of the associated electric field? (c) Write the equations for the electric and magnetic fields as functions of x and t in the form of Equations (23.3).
9. ● **Visible light.** The wavelength of visible light ranges from 400 nm to 700 nm. Find the corresponding ranges of this light's (a) frequency, (b) angular frequency, (c) wave number.
10. ● **Ultraviolet radiation.** There are two categories of ultraviolet light. Ultraviolet A (UVA) has a wavelength ranging from 320 nm to 400 nm. It is not so harmful to the skin and is necessary for the production of vitamin D. UVB, with a wavelength between 280 nm and 320 nm, is much more dangerous, because it causes skin cancer. (a) Find the frequency ranges of UVA and UVB. (b) What are the ranges of the wave numbers for UVA and UVB?
    BIO
11. ● **Medical x rays.** Medical x rays are taken with electromagnetic waves having a wavelength around 0.10 nm. What are the frequency, period, and wave number of such waves?
    BIO
12. ● Radio station WCCO in Minneapolis broadcasts at a frequency of 830 kHz. At a point some distance from the transmitter, the magnetic-field amplitude of the electromagnetic wave from WCCO is 4.82×10^{-11} T. Calculate (a) the wavelength, (b) the wave number, (c) the angular frequency, and (d) the electric-field amplitude.
13. ●● A sinusoidal electromagnetic wave of frequency 6.10×10^{14} Hz travels in vacuum in the $+x$-direction. The magnetic field is parallel to the y-axis and has amplitude 5.80×10^{-4} T. (a) Find the magnitude and direction of the electric field. (b) Write the wave functions for the electric and magnetic fields in the form of Equations (23.3).
14. ● Consider each of the electric- and magnetic-field orientations given next. In each case, what is the direction of propagation of the wave? (a) $\vec{E}$ in the $+x$ direction, $\vec{B}$ in the $+y$ direction. (b) $\vec{E}$ in the $-y$ direction, $\vec{B}$ in the $+x$ direction. (c) $\vec{E}$ in the $+z$ direction, $\vec{B}$ in the $-x$ direction. (d) $\vec{E}$ in the $+y$ direction, $\vec{B}$ in the $-z$ direction.
15. ●● An electromagnetic wave has a magnetic field given by $B = (8.25 \times 10^{-9} \text{ T}) \sin[(\omega t + 1.38 \times 10^{4} \text{ rad/m})x]$, with the magnetic field in the $+y$ direction. (a) In which direction is the wave traveling? (b) What is the frequency f of the wave? (c) Write the wave function for the electric field.

23.5 Energy in Electromagnetic Waves

16. ● **Laboratory lasers.** He–Ne lasers are often used in physics demonstrations. They produce light of wavelength 633 nm and a power of 0.500 mW spread over a cylindrical beam 1.00 mm in diameter (although these quantities can vary). (a) What is the intensity of this laser beam? (b) What are the maximum values of the electric and magnetic fields? (c) What is the average energy density in the laser beam?
17. ● **Fields from a lightbulb.** We can reasonably model a 75 W incandescent lightbulb as a sphere 6.0 cm in diameter. Typically, only about 5% of the energy goes to visible light; the rest

goes largely to nonvisible infrared radiation. (a) What is the visible light intensity (in W/m^2) at the surface of the bulb? (b) What are the amplitudes of the electric and magnetic fields at this surface, for a sinusoidal wave with this intensity?

18. • **Threshold of vision.** Under controlled darkened condi- **BIO** tions in the laboratory, a light receptor cell on the retina of a person's eye can detect a single photon (more on photons in Chapter 28) of light of wavelength 505 nm and having an energy of 3.94×10^{-19} J. We shall assume that this energy is absorbed by a single cell during one period of the wave. Cells of this kind are called *rods* and have a diameter of approximately 0.0020 mm. What is the intensity (in W/m^2) delivered to a rod?

19. • **High-energy cancer treatment.** Scientists are working on **BIO** a new technique to kill cancer cells by zapping them with ultrahigh-energy (in the range of 10^{12} W) pulses of light that last for an extremely short time (a few nanoseconds). These short pulses scramble the interior of a cell without causing it to explode, as long pulses would do. We can model a typical such cell as a disk 5.0 μm in diameter, with the pulse lasting for 4.0 ns with an average power of 2.0×10^{12} W. We shall assume that the energy is spread uniformly over the faces of 100 cells for each pulse. (a) How much energy is given to the cell during this pulse? (b) What is the intensity (in W/m^2) delivered to the cell? (c) What are the maximum values of the electric and magnetic fields in the pulse?

20. • At the floor of a room, the intensity of light from bright overhead lights is 8.00 W/m^2. Find the radiation pressure on a totally absorbing section of the floor.

21. • The intensity at a certain distance from a bright light source is 6.00 W/m^2. Find the radiation pressure (in pascals and in atmospheres) on (a) a totally absorbing surface and (b) a totally reflecting surface.

22. •• A sinusoidal electromagnetic wave from a radio station passes perpendicularly through an open window that has area 0.500 m^2. At the window, the electric field of the wave has rms value 0.0200 V/m. How much energy does this wave carry through the window during a 30.0 s commercial?

23. •• Two sources of sinusoidal electromagnetic waves have average powers of 75 W and 150 W and emit uniformly in all directions. At the same distance from each source, what is the ratio of the maximum electric field for the 150 W source to that of the 75 W source?

24. •• Radiation falling on a perfectly reflecting surface produces an average pressure p. If radiation of the same intensity falls on a perfectly absorbing surface and is spread over twice the area, what is the pressure at that surface in terms of p?

25. •• A sinusoidal electromagnetic wave emitted by a cellular phone has a wavelength of 35.4 cm and an electric field amplitude of 5.40×10^{-2} V/m at a distance of 250 m from the antenna. Calculate: (a) the frequency of the wave; (b) the magnetic-field amplitude; (c) the intensity of the wave.

23.7 The Reflection and Refraction of Light

26. • Two plane mirrors intersect at right angles. A laser beam strikes the first of them at a point 11.5 cm from their point of intersection, as shown in Figure 23.49. For what angle of inci- dence at the first mirror will this ray strike the midpoint of the second mirror (which is 28.0 cm long) after reflecting from the first mirror?

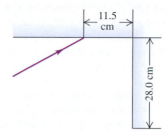

▲ **FIGURE 23.49** Problem 26.

27. • Three plane mirrors intersect at right angles. A beam of laser light strikes the first of them at an angle θ with respect to the normal. (See Figure 23.50.) (a) Show that when this ray is reflected off of the other two mirrors and crosses the original ray, the angle α between these two rays will be $\alpha = 180° - 2\theta$. (b) For what angle θ will the two rays be per- pendicular when they cross?

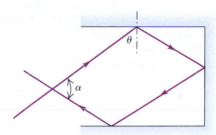

▲ **FIGURE 23.50** Problem 27.

28. • Two plane mirrors A and B intersect at a 45° angle. Three rays of light leave point P (see Figure 23.51) and strike one of the mirrors. What is the subsequent path of each of the following rays until they no longer strike either of the mir- rors? (a) Ray 1, which strikes A at 45° with respect to the normal. (b) Ray 2, which strikes B traveling perpendi- cular to mirror A. (c) Ray 3, which strikes B perpendicular to its surface.

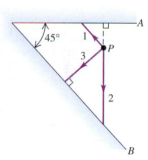

▲ **FIGURE 23.51** Problem 28.

29. •• Prove that when a ray of light travels at any angle into the corner formed by two mirrors placed at right angles to each other, the reflected ray emerges parallel to the origi- nal ray (see Figure 23.52).

30. •• A light beam travels at 1.94×10^8 m/s in quartz. The wavelength of the light in quartz is 355 nm. (a) What is the index of refraction of quartz at this wavelength? (b) If this same light travels through air, what is its wavelength there?

▲ **FIGURE 23.52** Problem 29.

31. •• Using a fast-pulsed laser and electronic timing circuitry, you find that light travels 2.50 m within a plastic rod in 11.5 ns. What is the refractive index of the plastic?

32. • Light with a frequency of 5.80×10^{14} Hz travels in a block of glass that has an index of refraction of 1.52. What is the wavelength of the light (a) in vacuum and (b) in the glass?

33. • The speed of light with a wavelength of 656 nm in heavy flint glass is 1.82×10^8 m/s. What is the index of refraction of the glass at this wavelength?

34. • **Light inside the eye.** The vitreous humor, a transparent,
BIO gelatinous fluid that fills most of the eyeball, has an index of refraction of 1.34. Visible light ranges in wavelength from 400 nm (violet) to 700 nm (red), as measured in air. This light travels through the vitreous humor and strikes the rods and cones at the surface of the retina. What are the ranges of (a) the wavelength, (b) the frequency, and (c) the speed of the light just as it approaches the retina within the vitreous humor?

35. •• Light of a certain frequency has a wavelength of 438 nm in water. What is the wavelength of this light (a) in benzene, (b) in air? (See Table 23.1.)

36. •• A 1.55-m-tall fisherman stands at the edge of a lake, being watched by a suspicious trout who is 3.50 m from the fisherman in the horizontal direction and 45.0 cm below the surface of the water. At what angle from the vertical does the fish see the top of the fisherman's head?

37. • Show that when a light ray travels from air through a sheet of glass with parallel surfaces and back into the air, it emerges traveling parallel to its original direction, although slightly displaced. (*Note:* The result of this problem is important, because it shows us that a sheet with parallel faces does not change the direction of a ray. It can also be shown that a very thin sheet does not displace the beam significantly. This is the case with a thin lens, since its opposite faces near its center are essentially parallel to each other. Rays that strike the center of such a lens go essentially straight through.)

38. • A glass plate having parallel faces and a refractive index of 1.58 lies at the bottom of a liquid of refractive index 1.70. A ray of light in the liquid strikes the top of the glass at an angle of incidence of 62.0°. Compute the angle of refraction of this light in the glass.

39. • A beam of light in air makes an angle of 47.5° with the *surface* (*not* the normal) of a glass plate having a refractive index of 1.66. (a) What is the angle between the reflected part of the beam and the *surface* of the glass? (b) What is the angle between the refracted beam and the *surface* (*not* the normal) of the glass?

40. •• A laser beam shines along the surface of a block of transparent material. (See Figure 23.53.) Half of the beam goes straight to a detector, while the other half travels through the block and then hits the detector. The time delay between the arrival of the two light beams at the detector is 6.25 ns. What is the index of refraction of this material?

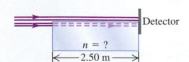

▲ **FIGURE 23.53** Problem 40.

41. •• **Reversibility of rays.** Ray 1 of light in medium *A* (see Figure 23.54) strikes the surface at 51.0° with respect to the normal. (a) Show that the angle of refraction of ray 1 with respect to the normal in medium *B* is 35.8°. (b) Now suppose that ray 2 is the reverse of ray 1, so that it strikes the surface at 35.8° with the normal in *B*. Show that ray 2 will come out in *A* at 51.0° with the normal. In other words, show that rays 1 and 2 follow the same path, except reversed from each other.

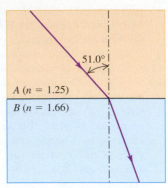

▲ **FIGURE 23.54** Problem 41.

42. •• You (height of your eyes above the water, 1.75 m) are standing 2.00 m from the edge of a 2.50-m-deep swimming pool. You notice that you can barely see your cell phone, which went missing a few minutes before, on the bottom of the pool. How far from the side of the pool is your cell phone?

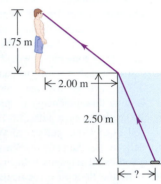

▲ **FIGURE 23.55** Problem 42.

43. •• A parallel-sided plate of glass having a refractive index of 1.60 is in contact with the surface of water in a tank. A ray coming from above makes an angle of incidence of 32.0° with the top surface of the glass. What angle does this ray make with the normal in the water?

44. •• As shown in Figure 23.56, a layer of water covers a slab of material *X* in a beaker. A ray of light traveling upwards follows the path indicated. Using the information on the figure, find (a) the index of refraction of material *X* and (b) the angle the light makes with the normal in the *air*.

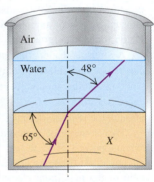

▲ **FIGURE 23.56** Problem 44.

23.8 Total Internal Reflection

45. • A ray of light in diamond (index of refraction 2.42) is incident on an interface with air. What is the *largest* angle the ray can make with the normal and not be totally reflected back into the diamond?

46. • The critical angle for total internal reflection at a liquid–air interface is 42.5°. (a) If a ray of light traveling in the liquid has an angle of incidence of 35.0° at the interface, what angle does the refracted ray in the air make with the normal? (b) If a ray of light traveling in air has an angle of incidence of 35.0° at the interface, what angle does the refracted ray in the liquid make with the normal?

47. •• At the very end of Wagner's series of operas *The Ring of the Nibelung*, Brünnhilde takes the golden ring from the finger of the dead Siegfried and throws it into the Rhine, where it

sinks to the bottom of the river. Assuming that the ring is small enough to be treated as a point compared with the depth of the river and that the Rhine is 10.0 m deep where the ring goes in, what is the *area* of the largest circle at the surface of the water over which light from the ring could escape from the water?

48. •• A ray of light is traveling in a glass cube that is totally immersed in water. You find that if the ray is incident on the glass–water interface at an angle to the normal greater than 48.7°, no light is refracted into the water. What is the refractive index of the glass?

49. •• Light is incident along the normal to face AB of a glass prism of refractive index 1.52, as shown in Figure 23.57. Find the largest value the angle α can have without any light refracted out of the prism at face AC if (a) the prism is immersed in air and (b) the prism is immersed in water.

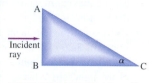

▲ **FIGURE 23.57** Problem 49.

50. •• **Light pipe.** Light enters a solid tube made of plastic having an index of refraction of 1.60. The light travels parallel to the upper part of the tube. (See Figure 23.58.) You want to cut the face AB so that all the light will reflect back into the tube after it first strikes that face. (a) What is the largest that θ can be if the tube is in air? (b) If the tube is immersed in water of refractive index 1.33, what is the largest that θ can be?

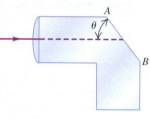

▲ **FIGURE 23.58** Problem 50.

51. •• An optical fiber consists of an outer "cladding" layer and an inner core with a slightly higher index of refraction. Light rays entering the core are trapped inside by total internal reflection and forced to travel along the fiber (see Figure 23.59). Suppose the cladding has an index of refraction of 1.46 and the core has an index of refraction of 1.48. Calculate the largest angle θ between a light ray and the longitudinal axis of the fiber (see the figure) for which the ray will be totally internally reflected at the core/cladding boundary.

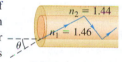

▲ **FIGURE 23.59** Problem 51.

23.9 Dispersion

52. • A beam of light strikes a sheet of glass at an angle of 57.0° with the normal in air. You observe that red light makes an angle of 38.1° with the normal in the glass, while violet light makes a 36.7° angle. (a) What are the indexes of refraction of this glass for these colors of light? (b) What are the speeds of red and violet light in the glass?

53. • Use the information from the graph in Figure 23.29 to construct a graph of the index of refraction of silicate flint glass as a function of the frequency of light.

54. •• A narrow beam of white light strikes one face of a slab of silicate flint glass. The light is traveling parallel to the two adjoining faces, as shown in Figure 23.60. For the transmitted light inside the glass, through what angle $\Delta\theta$ is the complete visible spectrum of light dispersed? (Consult the graph in Figure 23.29.)

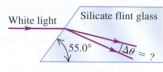

▲ **FIGURE 23.60** Problem 54.

55. •• Use the graph in Figure 23.29 for silicate flint glass. (a) What are the indexes of refraction of this glass for extreme violet light of wavelength 400 nm and for extreme red light of wavelength 700 nm? (b) What are the wavelengths of 400 nm violet light and 700 nm red light in this glass? (c) Calculate the *ratio* of the speed of extreme red light to that of extreme violet light in the glass. Which of these travels faster in the glass? (d) If a beam of white light in air strikes a sheet of this glass at 65.0° with the normal in air, what will be the angle of *dispersion* between the extremes of visible light in the glass? In other words, what will be the angle between extreme red and extreme violet light in the glass?

56. • The indices of refraction for violet light ($\lambda = 400$ nm) and red light ($\lambda = 700$ nm) in diamond are 2.46 and 2.41, respectively. A ray of light traveling through air strikes the diamond surface at an angle of 53.5° to the normal. Calculate the angular separation between these two colors of light in the refracted ray.

23.10 Polarization

57. • Unpolarized light with intensity I_0 is incident on an ideal polarizing filter. The emerging light strikes a second ideal polarizing filter whose axis is at 41.0° to that of the first. Determine (a) the intensity of the beam after it has passed through the second polarizer and (b) its state of polarization.

58. • Two ideal polarizing filters are oriented so that they transmit the *maximum* amount of light when unpolarized light is shone on them. To what fraction of its maximum value I_0 is the intensity of the transmitted light reduced when the second filter is rotated through (a) 22.5°, (b) 45.0°, and (c) 67.5°?

59. •• A beam of unpolarized light of intensity I_0 passes through a series of ideal polarizing filters with their polarizing directions turned to various angles as shown in Figure 23.61. (a) What is the light intensity (in terms of I_0) at points A, B, and C? (b) If we remove the middle filter, what will be the light intensity at point C?

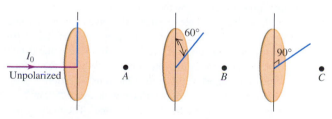

▲ **FIGURE 23.61** Problem 59.

60. •• Three ideal polarizing filters are stacked, with the polarizing axis of the second and third filters at 23.0° and 62.0°, respectively, to that of the first. If unpolarized light is incident on the stack, the light has intensity 75.0 W/cm² after it passes through the stack. If the incident intensity is kept constant, what is the intensity of the light after it has passed through the stack if the second polarizer is removed?

61. •• Light of original intensity I_0 passes through two ideal polarizing filters having their polarizing axes oriented as shown in Figure 23.62. You want to adjust the angle ϕ so that the intensity at point P is equal to $I_0/10$. (a) If the original light is unpolarized, what should ϕ be? (b) If the original light is linearly polarized in the same direction as the polarizing axis of the first polarizer the light reaches, what should ϕ be?

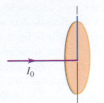 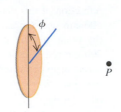

▲ **FIGURE 23.62** Problem 61.

62. • The polarizing angle for light in air incident on a glass plate is 57.6°. What is the index of refraction of the glass?

63. •• A beam of polarized light passes through a polarizing filter. When the angle between the polarizing axis of the filter and the direction of polarization of the light is θ, the intensity of the emerging beam is I. If you instead want the intensity to be $I/2$, what should be the angle (in terms of θ) between the polarizing angle of the filter and the original direction of polarization of the light?

64. •• A beam of unpolarized light in air is incident at an angle of 54.5° (with respect to the normal) on a plane glass surface. The reflected beam is completely linearly polarized. (a) What is the refractive index of the glass? (b) What is the angle of refraction of the transmitted beam?

65. •• Plane-polarized light passes through two polarizers whose axes are oriented at 35.0° to each other. If the intensity of the original beam is reduced to 15.0%, what was the polarization direction of the original beam, relative to the first polarizer?

General Problems

66. •• The energy flow to the earth from sunlight is about 1.4 kW/m². (a) Find the maximum values of the electric and magnetic fields for a sinusoidal wave of this intensity. (b) The distance from the earth to the sun is about 1.5×10^{11} m. Find the total power radiated by the sun.

67. •• A plane sinusoidal electromagnetic wave in air has a wavelength of 3.84 cm and an $\vec{E}$ field amplitude of 1.35 V/m. (a) What is the frequency of the wave? (b) What is the $\vec{B}$ field amplitude? (c) What is the intensity? (d) What average force does this radiation exert perpendicular to its direction of propagation on a totally absorbing surface with area 0.240 m²?

68. •• A powerful searchlight shines on a man. The man's cross-sectional area is 0.500 m² perpendicular to the light beam, and the intensity of the light at his location is 36.0 kW/m². He is wearing black clothing, so that the light incident on him is *totally* absorbed. What is the magnitude of the force the light beam exerts on the man? Do you think he could sense this force?

69. •• **Laser surgery.** Very short pulses of high-intensity laser
BIO beams are used to repair detached portions of the retina of the eye. The brief pulses of energy absorbed by the retina welds the detached portion back into place. In one such procedure, a laser beam has a wavelength of 810 nm and delivers 250 mW of power spread over a circular spot 510 μm in diameter. The vitreous humor (the transparent fluid that fills most of the eye) has an index of refraction of 1.34. (a) If the laser pulses are each 1.50 ms long, how much energy is delivered to the retina with each pulse? (b) What average pressure does the pulse of the laser beam exert on the retina as it is fully absorbed by the circular spot? (c) What are the wavelength and frequency of the laser light inside the vitreous humor of the eye? (d) What are the maximum values of the electric and magnetic fields in the laser beam?

70. •• A small helium-neon laser emits red visible light with a power of 3.20 mW in a beam that has a diameter of 2.50 mm. (a) What are the amplitudes of the electric and magnetic fields of the light? (b) What are the average energy densities associated with the electric field and with the magnetic field? (c) What is the total energy contained in a 1.00 m length of the beam?

71. •• Radio receivers can comfortably pick up a broadcasting station's signal when the electric field strength of the signal is about 10.0 mV/m. If a radio station broadcasts in all directions with an average power of 50.0 kW, what would be the maximum distance at which you could easily pick up its transmissions? (Atmospheric conditions can have major effects on this distance.)

72. •• The 19th-century inventor Nikola Tesla proposed to transmit electric power via sinusoidal electromagnetic waves. Suppose power is to be transmitted in a beam of cross-sectional area 100 m². What electric- and magnetic-field amplitudes are required to transmit an amount of power comparable to that handled by modern transmission lines (which carry voltages and currents of the order of 500 kV and 1000 A)?

73. •• **Solar sail.** NASA is doing research on the concept of *solar sailing.* A solar sailing craft uses a large, low-mass sail and the energy and momentum of sunlight for propulsion. (a) Should the sail be absorptive or reflective? Why? (b) The total power output of the sun is 3.9×10^{26} W. How large a sail is necessary to propel a 10,000 kg spacecraft against the gravitational force of the sun? Express your result in square kilometers. (c) Explain why your answer to part (b) is independent of the distance from the sun.

74. • A thick layer of oil is floating on the surface of water in a tank. A beam of light traveling in the oil is incident on the water interface at an angle of 30.0° from the normal. The refracted beam travels in the water at an angle of 45.0° from the normal. What is the refractive index of the oil?

75. •• A thin beam of light in air is incident on the surface of a lanthanum flint glass plate having a refractive index of 1.80. What is the angle of incidence, θ_a, of the beam with this plate, for which the angle of refraction is $\theta_a/2$? Both angles are measured relative to the normal.

76. •• You want to support a sheet of fireproof paper horizontally, using only a vertical upward beam of light spread uniformly over the sheet. There is no other light on this paper. The sheet measures 22.0 cm by 28.0 cm and has a mass of 1.50 g. (a) If the paper is black and hence absorbs all the light that hits it, what must be the intensity of the light beam? (b) For the light in part (a), what are the maximum values of its electric and magnetic fields? (c) If the paper is white and hence reflects all the light that hits it, what intensity of light beam is needed to support it? (d) To see if it is physically reasonable to expect

to support a sheet of paper this way, calculate the intensity in a typical 0.500 mW laser beam that is 1.00 mm in diameter and compare this value with your answer in part (a).

77. •• A light ray in air strikes the right-angle prism shown in Figure 23.63. This ray consists of two different wavelengths. When it emerges at face *AB,* it has been split into two different rays that diverge from each other by 8.50°. Find the index of refraction of the prism for each of the two wavelengths.

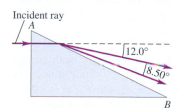

Incident ray

▲ **FIGURE 23.63** Problem 77.

78. •• A ray of light is incident in air on a block of a transparent solid whose index of refraction is *n*. If $n = 1.38$, what is the *largest* angle of incidence, θ_a, for which total internal reflection will occur at the vertical face (point *A* shown in Figure 23.64)?

79. •• A light beam is directed parallel to the axis of a hollow cylindrical tube. When the tube contains only air, it takes the light 8.72 ns to travel the length of the tube, but when the tube is filled with a transparent jelly, it takes the light 2.04 ns longer to travel its length. What is the refractive index of this jelly?

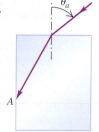

▲ **FIGURE 23.64** Problem 78.

80. •• **Heart sonogram.** Physicians use high-frequency ($f =$ **BIO** 1 MHz to 5 MHz) sound waves, called ultrasound, to image internal organs. The speed of these ultrasound waves is 1480 m/s in muscle and 344 m/s in air. We define the index of refraction of a material for sound waves to be the ratio of the speed of sound in air to the speed of sound in the material. Snell's law then applies to the refraction of sound waves. (a) At what angle from the normal does an ultrasound beam enter the heart if it leaves the lungs at an angle of 9.73° from the normal to the heart wall? (Assume that the speed of sound in the lungs is 344 m/s.) (b) What is the critical angle for sound waves in air incident on muscle?

81. •• The prism shown in Figure 23.65 has a refractive index of 1.66, and the angles *A* are 25.0°. Two light rays *m* and *n* are parallel as they enter the prism. What is the angle between them after they emerge?

82. •• A 45°–45°–90° prism is immersed in water. A ray of light is incident normally on one of the prism's shorter faces. What is the minimum index of refraction that the prism must have if this ray is to be totally reflected within the glass at the long face of the prism?

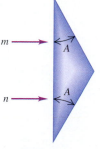

▲ **FIGURE 23.65** Problem 81.

83. •• A beaker with a mirrored bottom is filled with a liquid whose index of refraction is 1.63. A light beam strikes the top surface of the liquid at an angle of 42.5° from the normal. At what angle from the normal will the beam exit from the liquid after traveling down through it, reflecting from the mirrored bottom, and returning to the surface?

84. •• A ray of light traveling *in* a block of glass ($n = 1.52$) is incident on the top surface at an angle of 57.2° with respect to the normal in the glass. If a layer of oil is placed on the top surface of the glass, the ray is totally reflected. What is the maximum possible index of refraction of the oil?

85. •• A block of glass has a polarizing angle of 60.0° for red light and 70.0° for blue light, for light traveling in air and reflecting from the glass. (a) What are the indexes of refraction for red light and for blue light? (b) For the same angle of incidence, which color is refracted more on entering the glass?

86. •• In a physics lab, light with wavelength 490 nm travels in air from a laser to a photocell in 17.0 ns. When a slab of glass 0.840 m thick is placed in the light beam, with the beam incident along the normal to the parallel faces of the slab, it takes the light 21.2 ns to travel from the laser to the photocell. What is the wavelength of the light in the glass?

87. •• (a) Light passes through three parallel slabs of different thicknesses and refractive indexes. The light is incident in the first slab and finally refracts into the third slab. Show that the middle slab has no effect on the final direction of the light. That is, show that the direction of the light in the third slab is the same as if the light had passed directly from the first slab into the third slab. (b) Generalize this result to a stack of *N* slabs. What determines the final direction of the light in the last slab?

88. • The refractive index of a certain glass is 1.66. For what angle of incidence is light that is reflected from the surface of this glass completely polarized if the glass is immersed in (a) air or (b) water?

89. •• A thin layer of ice ($n = 1.309$) floats on the surface of water ($n = 1.333$) in a bucket. A ray of light from the bottom of the bucket travels upward through the water. (a) What is the largest angle with respect to the normal that the ray can make at the ice–water interface and still pass out into the air above the ice? (b) What is this angle after the ice melts?

90. •• **Optical activity of biological molecules.** Many biologi-**BIO** cally important molecules are optically active. When linearly polarized light traverses a solution of compounds containing these molecules, its plane of polarization is rotated. Some compounds rotate the polarization clockwise; others rotate the polarization counterclockwise. The amount of rotation depends on the amount of material in the path of the light. The following data give the amount of rotation through two amino acids over a path length of 100 cm:

Rotation (degrees) *l*-leucine	*d*-glutamic acid	Concentration (g/100 mL)
−0.11	0.124	1.0
−0.22	0.248	2.0
−0.55	0.620	5.0
−1.10	1.24	10.0
−2.20	2.48	20.0
−5.50	6.20	50.0
−11.0	12.4	100.0

From these data, find the relationship between the concentration *C* (in grams per 100 mL) and the rotation of the polarization (in degrees) of each amino acid. (*Hint:* Graph the concentration as a function of the rotation angle for each amino acid.)

91. •• A horizontal cylindrical tank 2.20 m in diameter is half full of water. The space above the water is filled with a pressurized gas of unknown refractive index. A small laser can move along the curved bottom of the water and aims a light beam toward the center of the water surface (Figure 23.66). You observe that when the laser has moved a distance $S = 1.09$ m or more (measured along the curved surface) from the lowest point in the water, no light enters the gas. (a) What is the index of refraction of the gas? (b) How long does it take the light beam to travel from the laser to the rim of the tank when (i) $S > 1.09$ m and (ii) $S < 1.09$ m?

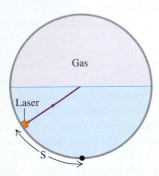

▲ **FIGURE 23.66** Problem 91.

Passage Problems

Reflection and refraction. We can see a transparent object that is illuminated because of the incident light that is reflected from its surfaces and the refracted light that is reflected by small imperfections within the object itself.

When a light ray is incident on a transparent surface we can easily predict the *direction* of the reflected and refracted rays by using the laws of reflection and refraction. However, the *amount* of light reflected from a surface is more difficult to determine since this depends on the direction and polarization of the incident ray, and the refractive indices for both surfaces. For example, when light strikes the boundary between two surfaces (with refractive indices of n_1 and n_2) at an angle of incidence that is near 90°, the fractional intensity of light reflected from the boundary is given by

$$\left(\frac{n_1 - n_2}{n_1 + n_2}\right)^2$$

According to this result, typical optical glass (with an index of refraction of 1.5) should reflect about 4% (0.04) of the light normally incident from the surrounding air, which is in fact the case.

92. If a light beam strikes a 10 cm thick slab of glass (which is immersed in air) at an angle of 30° from the normal to the surface, what will be its angle to the normal when it leaves the back side of the slab? Assume that the slab has parallel sides and an index of refraction of 1.5.
    A. 0°   B. 30°   C. 60°   D. 58.3°

93. If a layer of oil, with an index of refraction of 1.8, is placed on the top of the glass, what will be the new angle at which the light beam leaves the glass?
    A. 0°   B. 16°   C. 30°   D. 46°

94. If the entire slab (without the oil) is submerged in a fluid with an index of refraction of 1.5, what will be the effect?
    A. The slab will appear to change color.
    B. Light striking the slab could be totally reflected.
    C. The slab will be very difficult to see.
    D. Light exiting the slab could be totally reflected.

24 Geometric Optics

When you look at yourself in a flat mirror, you appear your actual size and right side up. But in a curved mirror you may appear larger or smaller, or even upside down. Why is this? Mirrors, magnifying glasses, and telescopes are a familiar part of everyday life. We can use the ray model of light, introduced in the preceding chapter, to understand the behavior of these and similar instruments. All we need are the laws of reflection and refraction and some simple geometry and trigonometry.

In this chapter and the next, we'll make frequent use of the concept of *image*. When a politician worries about his image, he is thinking of how he looks to the voters. We'll use the term in a related, but more precise, way, having to do with the behavior of a collection of rays that converge toward or appear to diverge from a point called an *image point*. In this chapter, we analyze the formation of images by a single reflecting or refracting surface and by a thin lens. This discussion lays the foundation for analyses of many familiar optical instruments, including camera lenses, magnifiers, the human eye, microscopes, and telescopes. We'll study these instruments in Chapter 25.

24.1 Reflection at a Plane Surface

The concept of **image** is the most important new idea in this chapter and the next. Consider Figure 24.1. Several rays diverge from point P and are reflected at the plane mirror, according to the law of reflection. After they are reflected, their final directions are the same as though they had come from point P′. We call point P an *object point* and point P′ the corresponding *image point,* and we say that the mirror forms an *image* of point P. The outgoing rays (those going away from the mirror) don't really come from point P′, but their directions are the same as though they had come from that point.

The lens in this photo has the same fundamental shape as the lenses used for nearsightedness. The lens is held with the convex side facing us, and the objects seen through it look shrunken. You will see the same effect if you look at a nearsighted person's eyes through his or her eyeglasses.

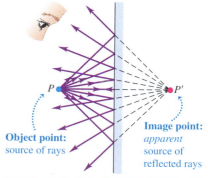

Object point: source of rays

Image point: *apparent* source of reflected rays

Plane mirror

▲ **FIGURE 24.1** The rays entering the eye after reflection from a plane mirror look as though they had come from point P′, the image point for object point P.

Something similar happens at a plane refracting surface, as shown in Figure 24.2. Rays coming from object point P are refracted at the interface between two optical materials. When the angles of incidence are small, the final directions of the rays after refraction are the same as though they had come from point P', and again we call P' an *image point*.

In both of these cases, the rays spread out from point P, whether it is an actual point source of light or a point that scatters the light shining on it. An observer who can see only the rays spreading out from the surface (after they are reflected or refracted) *thinks* that they come from the image point P'. This image point is therefore a convenient way to describe the directions of the various reflected or refracted rays, just as the object point P describes the directions of the rays arriving at the surface *before* reflection or refraction.

To find the precise location of the image point P' that a plane mirror forms of an object point P, we use the construction shown in Figure 24.3. The figure shows two rays diverging from an object point P at a distance s to the left of a plane mirror. We call s the **object distance.** The ray PV is incident normally on the mirror (that is, it is perpendicular to the surface of the mirror), and it returns along its original path.

The ray PB makes an angle θ with PV. It strikes the mirror at an angle of incidence θ and is reflected at an equal angle with the normal. When we extend the two reflected rays backward, they intersect at point P', at a distance s' behind the mirror. We call s' the **image distance.** The line between P and P' is perpendicular to the mirror. The two triangles have equal angles, so P and P' are at equal distances from the mirror, and s and s' have equal magnitudes.

We can repeat the construction of Figure 24.3 for each ray diverging from P. The directions of *all* the outgoing rays are the same *as though* they had originated at point P', confirming that P' is the *image* of P. The rays do not actually pass through point P'. In fact, for an ordinary mirror, there is no light at all on the back side. In cases like this, and like that of Figure 24.2, the outgoing rays don't actually come from P', and we call the image a **virtual image.** Later we will see cases in which the outgoing rays really *do* pass through an image point, and we will call the resulting image a **real image.** The images formed on a projection screen or the photographic film (or digital sensor array) in a camera are real images.

Sign Rules

Before we go further, let's introduce some general sign rules. These may seem unnecessarily complicated for the simple situations discussed so far, but we want to state the rules in a form that will be applicable to *all* the situations that we'll encounter. These will include image formation by a plane or spherical reflecting or refracting surface and by a pair of refracting surfaces forming a lens. Here are the rules:

When $n_a > n_b$, P' is closer to the surface than P; for $n_a < n_b$, the reverse is true.

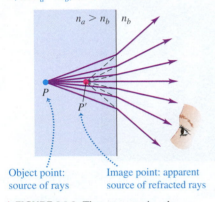

▲ **FIGURE 24.2** The rays entering the eye after refraction at the plane interface look as though they had come from point P', the image point for object P. The angles of incidence have been exaggerated for clarity.

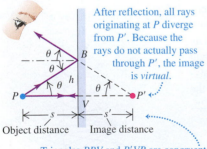

After reflection, all rays originating at P diverge from P'. Because the rays do not actually pass through P', the image is *virtual*.

Triangles PBV and $P'VB$ are congruent, so $|s| = |s'|$.

▲ **FIGURE 24.3** Reflected rays originating at the object point P appear to come from the virtual image point P'.

Sign rules for object and image distances

Object distance: When the object is on the same side of the reflecting or refracting surface as the incoming light, the object distance s is positive; otherwise, it is negative.

Image distance: When the image is on the same side of the reflecting or refracting surface as the outgoing light, the image distance s' is positive; otherwise, it is negative.

Figure 24.4 shows how these rules are applied to a plane mirror and to a plane refracting surface. For a mirror, the incoming and outgoing sides are always the same.

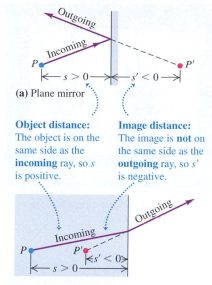

Object distance: The object is on the same side as the **incoming** ray, so s is positive.

Image distance: The image is **not** on the same side as the **outgoing** ray, so s' is negative.

(a) Plane mirror

(b) Plane refracting interface

▲ **FIGURE 24.4** The sign rules for the object and image distances as applied to a plane reflecting mirror and a plane refracting interface.

In Figure 24.4, the object distance s is *positive* in both cases because the object point P is on the incoming side of the reflecting or refracting surface. The image distance s' is *negative* because the image point P' is *not* on the outgoing side of the surface.

Let's now consider just the case of a plane mirror (Figure 24.4a). For a plane mirror, the object and image distances s and s' are related simply by

$$s = -s'. \qquad \text{(plane mirror)} \qquad (24.1)$$

Next we consider an object with finite size, parallel to the mirror. In Figure 24.5, this object is the blue arrow, which extends from P to Q and has height y. Two of the rays from Q are shown; *all* the rays from Q appear to diverge from its image point Q' after reflection. The image of the arrow, represented by the dashed magenta arrow, extends from P' to Q' and has height y'. Other points of the arrow PQ have image points between P' and Q'. The triangles PQV and $P'Q'V$ have equal angles, so the object and image have the same size and orientation, and $y = y'$.

The ratio of the image height to the object height, y'/y, in *any* image-forming situation is called the **lateral magnification** m; that is,

Definition of lateral magnification

For object height y and image height y', the lateral magnification m is

$$m = \frac{y'}{y}. \qquad (24.2)$$

For a plane mirror, PQV and P'Q'V are congruent, so $y = y'$ and the lateral magnification is 1 (the object and image are the same size).

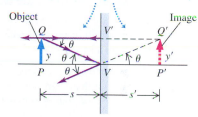

▲ **FIGURE 24.5** Construction for determining the height of an image formed by reflection from a plane mirror.

For a plane mirror, the lateral magnification m is unity. In other words, when you look at yourself in a plane mirror, your image is the same size as the real you.

We'll often represent an object by an arrow. Its image may be an arrow pointing in the *same* direction as the object or in the *opposite* direction. When the directions are the same, as in Figure 24.5, we say that the image is **erect;** when they are opposite, the image is **inverted.** The image formed by a plane mirror is always erect. A positive value of lateral magnification m corresponds to an erect image; a negative value corresponds to an inverted image. That is, for an erect image, y and y' always have the *same* sign, while for an inverted image, they always have *opposite* signs.

Figure 24.6 shows a three-dimensional virtual image of a three-dimensional object—a hand—formed by a plane mirror. The index and middle fingers of the

The object and image are *reversed*: The object and image thumbs point in opposite directions (toward each other).

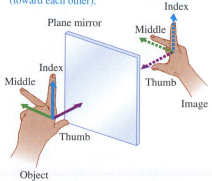

▲ **FIGURE 24.6** The image formed by the mirror is virtual, erect, and reversed. It is the same size as the object.

▲ **FIGURE 24.7** A plane mirror forms a reversed image of a hand.

image point in the same direction as those of the actual hand, but the object and image *thumbs* point in opposite directions: The actual thumb points toward the mirror, but the image thumb points "out of the mirror." The image of a three-dimensional object formed by a plane mirror is the same *size* as the object in all its dimensions, but the image and object are *not* identical. They are related in the same way as a right hand and a left hand, and indeed we speak of a pair of objects with this relationship as mirror-image objects. To verify this object–image relationship, try arranging your left and right hands to match the object and image hands in Figure 24.6. When an object and its image are related in this way, the image is said to be **reversed.** When the transverse dimensions of object and image are in the same direction, the image is erect. A plane mirror always forms an erect, but reversed, image. Figure 24.7 illustrates this point.

An important property of all images formed by reflecting or refracting surfaces is that an image formed by one surface or optical device can serve as the object for a second surface or device. Figure 24.8 shows a simple example. Mirror 1 forms an image P_1' of the object point P, and mirror 2 forms another image P_2', each in the way we have just discussed. But in addition, the image P_1' formed by mirror 1 can serve as an object for mirror 2, which then forms an image of this object at point P_3' as shown. Similarly, mirror 1 can use the image P_2' formed by mirror 2 as an object and form an image of it. We leave it to you to show that this image point is also at P_3'. Later in this chapter, we'll use this principle to locate the image formed by two successive curved-surface refractions in a lens.

Conceptual Analysis 24.1

Mirror, mirror on the wall

You want to install the shortest possible wall mirror that will let you see all of yourself when you stand straight. The bottom edge of the mirror will be

A. halfway between the floor and the level of your eyes.
B. nearer to the floor than to the level of your eyes.
C. nearer to the level of your eyes than to the floor.
D. at floor level.

SOLUTION Because you want the mirror to be as short as possible, the rays that go from your toes to your eyes should reflect from the bottom edge of the mirror. Because the angles of incidence and reflection are the same, the bottom of the mirror should be halfway between your toes and your eyes.

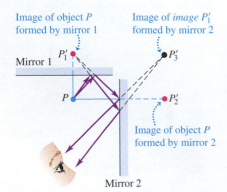

Image of object *P* formed by mirror 1 — P_1'

Mirror 1

Image of *image* P_1' formed by mirror 2 — P_3'

P

Image of object *P* formed by mirror 2 — P_2'

Mirror 2

▲ **FIGURE 24.8** An image formed by the reflection of an image, treated as an object. This figure shows mirror 2 reflecting the image from mirror 1 to form image P_3'. The same image is obtained when mirror 1 reflects the image P_2' from mirror 2.

24.2 Reflection at a Spherical Surface

Continuing our analysis of reflecting surfaces, we consider next the formation of an image by a *spherical* mirror. Figure 24.9a shows a spherical mirror with radius of curvature R, with its concave side facing the incident light. The **center of curvature** of the surface (the center of the sphere of which the surface is a part) is at C. Point P is an object point; for the moment, we assume that the distance from P to V is greater than R. The ray PV, passing through C, strikes the mirror normally and is reflected back on itself. Point V, at the center of the mirror surface, is called the **vertex** of the mirror, and the line PCV is called the **optic axis.**

Ray PB, at an angle α with the axis, strikes the mirror at B, where the angles of incidence and reflection are θ. The reflected ray intersects the axis at point P'. We'll show that *all* rays from P intersect the axis at the *same* point P', as in Figure 24.9b, no matter what α is, provided that it is a *small* angle. Point P' is therefore the *image* of object point P. The object distance, measured from the vertex V, is s, and the image distance is s'. The object point P is on the same side as the incident light, so, according to the sign rule in Section 24.1, the object distance s is positive. The image point P' is on the same side as the reflected light, so the image distance s' is also positive.

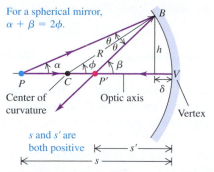

For a spherical mirror, $\alpha + \beta = 2\phi$.

Center of curvature

Optic axis

Vertex

s and s' are both positive

(a) Construction for finding the position P' of an image formed by a concave spherical mirror.

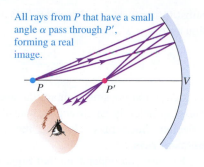

All rays from P that have a small angle α pass through P', forming a real image.

(b) The paraxial approximation, which holds for rays with small α.

▲ **FIGURE 24.9** Reflection from a concave spherical mirror.

Unlike the reflected rays in Figure 24.1, the reflected rays in Figure 24.9b actually do intersect at point P'; then they diverge from P' *as if* they had originated at that point. The image point P' is thus a *real* image point.

A theorem from plane geometry states that an exterior angle of a triangle equals the sum of the two opposite interior angles. Using this theorem with triangles PBC and $P'BC$ in Figure 24.9a, we have

$$\phi = \alpha + \theta, \qquad \beta = \phi + \theta.$$

Eliminating θ between these equations gives

$$\alpha + \beta = 2\phi. \tag{24.3}$$

Now we need a sign rule for the radii of curvature of spherical surfaces:

Sign rule for the radius of curvature

When the center of curvature C is on the same side as the outgoing (reflected) light, the radius of curvature R is positive; otherwise, it is negative.

In Figure 24.10a, R is positive because the center of curvature C is on the same side of the mirror as the reflected light. This is always the case when reflection occurs at the *concave* side of a surface. For a *convex* surface (Figure 24.10b), the center of curvature is on the opposite side from the reflected light, and R is negative.

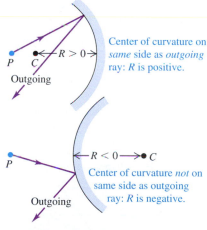

$R > 0$

Center of curvature on *same* side as *outgoing* ray: R is positive.

Outgoing

$R < 0$

Center of curvature *not* on same side as outgoing ray: R is negative.

Outgoing

▲ **FIGURE 24.10** Sign rule for the radius of curvature of a spherical surface.

We may now compute the image distance s'. In Figure 24.9a, let h represent the height of point B above the axis, and let δ denote the short distance from V to the foot of this vertical line. We now write expressions for the tangents of α, β, and ϕ, remembering that s, s', and R are all positive quantities:

$$\tan\alpha = \frac{h}{s - \delta}, \qquad \tan\beta = \frac{h}{s' - \delta}, \qquad \tan\phi = \frac{h}{R - \delta}.$$

These trigonometric equations cannot be solved as simply as the corresponding algebraic equations for a plane mirror. However, *if the angle α is small,* then the angles β and ϕ are also small. Now, the tangent of a small angle is nearly equal to the angle itself (in radians), so we can replace $\tan\alpha$ by α, and so on, in the previous equations. Also, if α is small, we can neglect the distance δ compared with s', s, and R. So, for small angles, we have the approximate relations

$$\alpha = \frac{h}{s}, \qquad \beta = \frac{h}{s'}, \qquad \phi = \frac{h}{R}.$$

Substituting these into Equation 24.3 and dividing by h, we obtain a general relation among s, s', and R:

$$\frac{1}{s} + \frac{1}{s'} = \frac{2}{R}. \qquad \text{(spherical mirror)} \qquad (24.4)$$

This equation does not contain the angle α; this means that *all* rays from P that make sufficiently small angles with the optic axis intersect at P' after they are reflected. Such rays, close to the axis and nearly parallel to it, are called **paraxial rays.**

Be sure you understand that Equation 24.4, as well as many similar relations that we will derive later in this chapter and the next, is only *approximately* correct. It results from a calculation containing approximations, and it is valid only for paraxial rays. The term **paraxial approximation** is often used for the approximations that we've just described. As the angle α increases, the point P' moves somewhat closer to the vertex; a spherical mirror, unlike a plane mirror, does not form a precise point image of a point object. This property of a spherical mirror is called *spherical aberration.*

If $R = \infty$, the mirror becomes *plane,* and Equation 24.4 reduces to Equation 24.1 ($s' = -s$), which we derived earlier for a plane reflecting surface.

Focal Point

When the object point P is very far from the mirror ($s = \infty$), the incoming rays are parallel. From Equation 24.4, the image distance s' is then given by

$$\frac{1}{\infty} + \frac{1}{s'} = \frac{2}{R}, \qquad s' = \frac{R}{2}.$$

The situation is shown in Figure 24.11a. A beam of incident rays parallel to the axis converges, after reflection, to a point F at a distance $R/2$ from the vertex of the mirror. Point F is called the **focal point,** or the *focus,* and its distance from the vertex, denoted by f, is called the **focal length.** We see that f is related to the radius of curvature R by

$$f = \frac{R}{2}. \qquad (24.5)$$

We can discuss the opposite situation, shown in Figure 24.11b. When the *image* distance s' is very large, the outgoing rays are parallel to the optic axis. The object distance s is then given by

$$\frac{1}{s} + \frac{1}{\infty} = \frac{2}{R}, \qquad s = \frac{R}{2}.$$

In Figure 24.11b, all the rays coming to the mirror from the focal point are reflected parallel to the optic axis. Again, we see that $f = R/2$.

Thus, the focal point F of a concave spherical mirror has the following properties:

Focal point of a concave spherical mirror
1. Any incoming ray parallel to the optic axis is reflected through the focal point.
2. Any incoming ray that passes through the focal point is reflected parallel to the optic axis.

For spherical mirrors, these statements are true only for paraxial rays; for parabolic mirrors, they are *exactly* true.

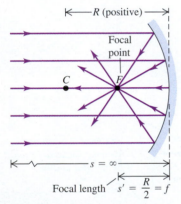

(a) All parallel rays incident on a spherical mirror reflect through the focal point.

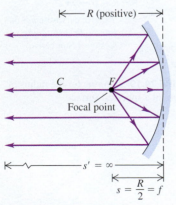

(b) Rays diverging from the focal point reflect to form parallel outgoing rays.

▲ FIGURE 24.11 Reflection and production of parallel rays by a concave spherical mirror. The angles are exaggerated for clarity.

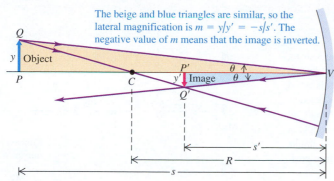

The beige and blue triangles are similar, so the lateral magnification is $m = y/y' = -s/s'$. The negative value of m means that the image is inverted.

▲ **FIGURE 24.12** Construction for determining the position, orientation, and height of an image formed by a concave spherical mirror.

We'll usually express the relationship between object and image distances for a mirror, Equation 24.4, in terms of the focal length f:

$$\frac{1}{s} + \frac{1}{s'} = \frac{1}{f}. \qquad \text{(spherical mirror)} \qquad (24.6)$$

Now suppose we have an object with finite height, represented by the blue arrow PQ in Figure 24.12, perpendicular to the axis PV. The image of P formed by paraxial rays is at P'. The object distance for point Q is very nearly equal to that for point P, so the image $P'Q'$ is nearly straight and is perpendicular to the axis. Note that object and image have different heights, y and y', respectively, and that they have opposite orientations. We've defined the *lateral magnification m* as the ratio of the image height y' to the object height y:

$$m = \frac{y'}{y}.$$

Because triangles PVQ and $P'VQ'$ in Figure 24.12 are *similar,* we also have the relation $y/s = -y'/s'$ The negative sign is needed because object and image are on opposite sides of the optic axis; if y is positive, y' is negative. Therefore, the lateral magnification m is given by

$$m = \frac{y'}{y} = -\frac{s'}{s}. \qquad \text{(spherical mirror)} \qquad (24.7)$$

A negative value of m indicates that the image is *inverted* relative to the object, as the figure shows. In cases that we'll consider later, in which m may be either positive or negative, a positive value always corresponds to an erect image and a negative value to an inverted one. For a *plane* mirror, $s = -s'$, so $y' = y$ and the image is erect, as we have already shown.

Although the ratio of the image height to the object height is called the *lateral magnification,* the image formed by a mirror or lens may be either larger or smaller than the object. If it is smaller, then the magnification is less than unity in absolute value. For instance, the image formed by an astronomical telescope mirror or a camera lens is usually *much* smaller than the object. For three-dimensional objects, the ratio of image distances to object distances measured *along* the optic axis is different from the ratio of *lateral* distances (the lateral magnification). In particular, if m is a small fraction, the three-dimensional image of a three-dimensional object is reduced *longitudinally* much more than it is reduced *transversely,* and the image appears squashed along the optic axis. Also, the image formed by a spherical mirror, like that of a plane mirror, is always reversed.

Conceptual Analysis 24.2

An image problem

Three students stand in front of a large concave mirror at a carnival (Figure 24.13). An apple hangs from a string as shown, at the same level as the center of curvature, but farther from the mirror than its center of curvature. Each student is asked to report to a bystander where the image of the apple (modeled as a point) is located. The bystander suspects that they aren't all telling the truth. Chandra claims to see the image of the apple at position (1). Joe says he sees it at position (2). Michi thinks they're both fibbing; she believes that all of them see it at position (3), where she sees it. Who is telling the truth?

A. Chandra and Joe are telling the truth; the position of the image depends on the observer's position.
B. Michi is correct; all observers see the image at the same position.

SOLUTION As shown in Figure 24.9b, all the reflected rays converge at the image point; it follows that all three observers must

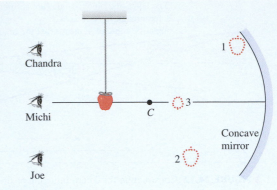

▲ **FIGURE 24.13**

see the image as located at the same point. The object and the center of curvature are on the optic axis, so the image is also on the optic axis. Michi is right; Chandra and Joe are fibbing.

EXAMPLE 24.1 **Image from a concave mirror**

A lamp is placed 10 cm in front of a concave spherical mirror that forms an image of the filament on a screen placed 3.0 m from the mirror. What is the radius of curvature of the mirror? If the lamp filament is 5.0 mm high, how tall is its image? What is the lateral magnification?

SOLUTION

SET UP Figure 24.14 shows our diagram.

SOLVE Both object distance and image distance are positive; we have $s = 10$ cm and $s' = 300$ cm. To find the radius of curvature, we use Equation 24.4:

$$\frac{1}{s} + \frac{1}{s'} = \frac{2}{R},$$

$$\frac{1}{10 \text{ cm}} + \frac{1}{300 \text{ cm}} = \frac{2}{R},$$

and $R = 19.4$ cm. To find the height of the image, we use Equation 24.7:

$$m = \frac{y'}{y} = -\frac{s'}{s},$$

$$\frac{y'}{5.0 \text{ mm}} = -\frac{300 \text{ cm}}{10 \text{ cm}},$$

and $y' = -150$ mm. The lateral magnification m is

$$m = \frac{y'}{y} = \frac{-150 \text{ mm}}{5 \text{ mm}} = -30.$$

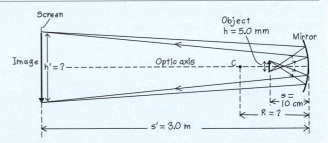

▲ **FIGURE 24.14** Our diagram for this problem (not to scale).

REFLECT The image is inverted (as we know because $m = -30$ is negative) and is 30 times taller than the object. Notice that the filament is *not* located at the mirror's focal point; the image is not formed by rays parallel to the optic axis. (The focal length of this mirror is $f = R/2 = 9.7$ cm.)

Practice Problem: A concave mirror has a radius of curvature $R = 25$ cm. An object of height 2 cm is placed 15 cm in front of the mirror. What is the image distance? What is the height of the image? *Answers: $s' = 75$ cm, $y' = -10$ cm.*

Convex Mirrors

In Figure 24.15a, the *convex* side of a spherical mirror faces the incident light. The center of curvature is on the opposite side from the outgoing rays, so, according to our sign rule, R is negative. Ray *PB* is reflected, with the angles of incidence and reflection both equal to θ. The reflected ray, projected backward, intersects the axis at *P'*. *All* rays from *P* that are reflected by the mirror appear to

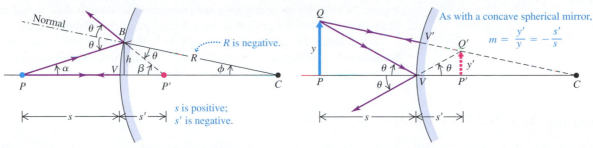

(a) Construction for finding the position of an image formed by a convex mirror.

(b) Construction for finding the magnification of an image formed by a convex mirror.

▲ **FIGURE 24.15** Reflection from a convex spherical mirror. As with a concave spherical mirror, rays from P for which α is small form an image at P'.

diverge from the same point P', provided that the angle α is small. Therefore, P' is the image point of P. The object distance s is positive, the image distance s' is negative, and the radius of curvature R is negative.

Figure 24.15b shows two rays diverging from the head of the arrow PQ and the virtual image $P'Q'$ of this arrow. We can show, by the same procedure that we used for a concave mirror, that

$$\frac{1}{s} + \frac{1}{s'} = \frac{2}{R},$$

and that the lateral magnification m is

$$m = \frac{y'}{y} = -\frac{s'}{s}.$$

These expressions are exactly the same as those for a concave mirror.

When R is negative (as with a convex mirror), incoming rays parallel to the optic axis are not reflected through the focal point F. Instead, they diverge as though they had come from the point F at a distance f behind the mirror, as shown in Figure 24.16a. In this case, f is the focal length and F is called a **virtual focal point**. The corresponding image distance s' is negative, so both f and R are negative, and Equation 24.5 holds for convex as well as concave mirrors. In

▲ **Application A wide, wide world.**
Professional truck drivers rely heavily on their mirrors, and you have probably seen the phrase "If you can't see my mirrors, I can't see you" on a truck or two rolling down the highway. In addition to regular plane mirrors to see the traffic behind them, truckers usually have a smaller, convex mirror that gives them a "wide-angle" view backwards. As we've learned, a convex mirror gives an upright virtual image with a lateral magnification less than one. Because the image size is reduced, a convex mirror has a larger field of view than a plane mirror, so the driver can see more lanes of traffic. Surveillance mirrors are convex for the same reason. *Concave* mirrors are not used for such purposes, because the image they produce is upside down.

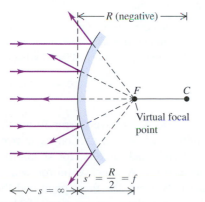

(a) Paraxial rays incident on a convex spherical mirror diverge from a virtual focal point.

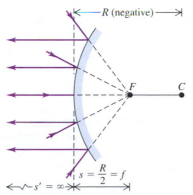

(b) Rays aimed at the virtual focal point are parallel to the axis after reflection.

▲ **FIGURE 24.16** Reflection and production of paraxial rays by a convex spherical mirror. The angles are exaggerated for clarity.

Figure 24.16b, the incoming rays are converging as though they would meet at the virtual focal point F, and they are reflected parallel to the optic axis.

In summary, Equations 24.4 through 24.7, the basic relationships for image formation by a spherical mirror, are valid for *both* concave and convex mirrors, provided that we use the sign rules consistently.

EXAMPLE 24.2 **Santa's image problem**

Santa checks himself for soot, using his reflection in a shiny, silvered Christmas tree ornament 0.750 m away (Figure 24.17). The diameter of the ornament is 7.20 cm. Standard reference works state that Santa is a "right jolly old elf"; we estimate his height as 1.6 m. Where and how tall is the image of Santa formed by the ornament? Is it erect or inverted?

▶ **FIGURE 24.17**

SOLUTION

SET UP Figure 24.18 shows our diagram. (To limit its size, we drew it not to scale; the angles are exaggerated, and Santa would actually be much taller and farther away.) The surface of the ornament closest to Santa acts as a convex mirror with radius $R = -(7.20 \text{ cm})/2 = -3.60 \text{ cm}$ and focal length $f = R/2 = -1.80 \text{ cm}$. The object distance is $s = 0.750 \text{ m} = 75.0 \text{ cm}$.

SOLVE From Equation 24.6,

$$\frac{1}{s'} = \frac{1}{f} - \frac{1}{s} = \frac{1}{-1.80 \text{ cm}} - \frac{1}{75.0 \text{ cm}},$$
$$s' = -1.76 \text{ cm}.$$

The lateral magnification m is given by Equation 24.7:

$$m = \frac{y'}{y} = -\frac{s'}{s} = -\frac{-1.76 \text{ cm}}{75.0 \text{ cm}} = 2.35 \times 10^{-2}.$$

Because m is positive, the image is erect, and it is only about 0.0235 as tall as Santa himself. Thus, the image height y' is

$$y' = my = (0.0235)(1.6 \text{ m}) = 3.8 \times 10^{-2} \text{ m} = 3.8 \text{ cm}.$$

REFLECT The object is on the same side of the mirror as the incoming light, so the object distance s is positive. Because s' is negative, the image is behind the mirror—that is, in Figure 24.18 it is on the side opposite to that of the outgoing light—and it is

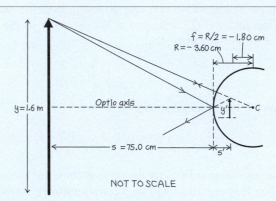

▲ **FIGURE 24.18** Our diagram for this problem.

virtual. The image is about halfway between the front surface of the ornament and its center. Thus, this convex mirror forms an erect, virtual, diminished, reversed image. In fact, when the object distance s is positive, a convex mirror *always* forms an erect, virtual, diminished, reversed image.

Practice Problem: One of Santa's elves scoots in halfway between Santa and the ornament to check and see that his hat is on straight. His image height in the ornament is 2.00 cm. What is the height of the elf? *Answer: $y = 43.7$ cm.*

Spherical mirrors have many important uses. A concave mirror forms the light from a flashlight or headlight bulb into a parallel beam. Convex mirrors are used to give a wide-angled view to car and truck drivers, for shoplifting surveillance in stores, and at "blind" intersections. A concave mirror with a focal length long enough so that your face is between the focal point and the mirror functions as a magnifier. Some solar-power plants use an array of plane mirrors to simulate an approximately spherical concave mirror. This array is used to collect and direct the sun's radiation to the focal point, where a steam boiler is placed. You can probably think of other examples from your everyday experience.

24.3 Graphical Methods for Mirrors

We can find the position and size of the image formed by a mirror by a simple graphical method. This method consists of finding the point of intersection of a few particular rays that diverge from a point of the object (such as point Q in Figure 24.19) and are reflected by the mirror. Then (neglecting aberrations) *all* rays from this point that strike the mirror will intersect at the same point. For this construction, we always choose an object point that is *not* on the optic axis. Four rays that we can usually draw easily are shown in Figure 24.19. These are called **principal rays.**

Definitions of principal rays for spherical mirrors

1. A ray parallel to the axis, after reflection, passes through the focal point F of a concave mirror or appears to come from the (virtual) focal point of a convex mirror.
2. A ray through, away from, or proceeding toward the focal point F is reflected parallel to the axis.
3. A ray along the radius through, away from, or proceeding toward the center of curvature C strikes the surface normally and is reflected back along its original path.
4. A ray to *the vertex V* is reflected forming equal angles with the optic axis.

   Once we have found the position of the real or virtual image point by means of the (real or virtual) intersection of any two of these four principal rays, we can draw the path of any other ray from the object point to the same image point.

① Ray parallel to axis reflects through focal point.
② Ray through focal point reflects parallel to axis.
③ Ray through center of curvature intersects the surface normally and reflects along its original path.
④ Ray to vertex reflects symmetrically around optic axis.

① Reflected parallel ray appears to come from focal point.
② Ray toward focal point reflects parallel to axis.
③ As with concave mirror: Ray radial to center of curvature intersects the surface normally and reflects along its original path.
④ As with concave mirror: Ray to vertex reflects symmetrically around optic axis.

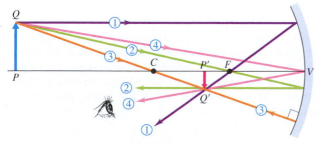

(a) Principal rays for concave mirror

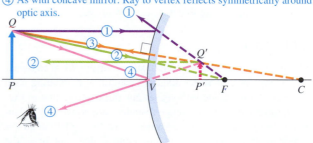

(b) Principal rays for convex mirror

▲ **FIGURE 24.19** Principal-ray diagrams for concave and convex mirrors. To find the image point Q', we draw any two of these rays; the image point is located at their intersection.

PROBLEM-SOLVING STRATEGY 24.1   **Image formation by mirrors**

SET UP

1. The principal-ray diagram is to geometric optics what the free-body diagram is to mechanics! When you attack a problem involving image formation by a mirror, *always* draw a principal-ray diagram first if you have enough information. (And apply the same advice to lenses in the sections that follow.) It's usually best to orient your diagrams consistently, with the incoming rays traveling from left to right. Don't draw a lot of other rays at random; stick with the principal rays—the ones that you know something about.

2. If your principal rays don't converge at a real image point, you may have to extend them straight backward to locate a virtual image point. We recommend drawing the extensions with broken lines. Another useful aid is to color-code your principal rays consistently; for example, referring to the preceding definitions of principal rays, Figure 24.19 uses purple for 1, green for 2, orange for 3, and pink for 4.

SOLVE

3. Identify the known and unknown quantities, such as s, s', R, and f. Make lists of the known and unknown quantities, and identify the relationships among them; then substitute the known values and solve for the unknowns. Pay careful attention to *signs* of object and image distances, radii of curvature, focal lengths, lateral magnifications, and object and image heights.

REFLECT

4. Make certain you understand that the same sign rules work for all four cases in this chapter: reflection and refraction from both plane and spherical surfaces. A negative sign on any one of the quantities mentioned in item 3 *always* has significance; use the equations and the sign rules carefully and consistently, and they will tell you the truth!

EXAMPLE 24.3   **Graphical construction of an image from a mirror**

A concave mirror has a radius of curvature with absolute value 20 cm. Find graphically the image of a real object in the form of an arrow perpendicular to the axis of the mirror at each of the following object distances: **(a)** 30 cm; **(b)** 20 cm; **(c)** 10 cm; and **(d)** 5 cm. Check the construction by *computing* the image distance and lateral magnification of each image.

SOLUTION

SET UP AND SOLVE Figure 24.20 shows the graphical constructions. Study each of these diagrams carefully, comparing each numbered ray to the definitions set out earlier. Note that in Figure 24.20b the object and image distances are equal. Ray 3 cannot be drawn in this case because a ray from Q through (or proceeding from) the center of curvature C does not strike the mirror. For a similar reason, ray 2 cannot be drawn in Figure 24.20c. In this case, the outgoing rays are parallel, corresponding to an infinite image distance. In Figure 24.20d, the outgoing rays have no real intersection point; they must be extended backward to find the point from which they appear to diverge—that is, the virtual image point Q'.

Measurements of the figures, with appropriate scaling, give the following approximate image distances: (a) 15 cm, (b) 20 cm, (c) ∞ or −∞ (because the outgoing rays are parallel and do not converge at any finite distance), (d) −10 cm. To *compute* these distances, we first note that $f = R/2 = 10$ cm; then we use Equation 24.6:

$$\frac{1}{s} + \frac{1}{s'} = \frac{1}{f}.$$

The lateral magnifications measured from the figures are approximately (a) $-\frac{1}{2}$, (b) -1, (c) $\pm\infty$ (because the image distance is infinite), (d) $+2$. To *compute* the lateral magnifications, use Equation 24.7, $m = -s'/s$. Here are the computations:

Continued

Part (a): With $s = 30$ cm and $f = 10$ cm, we obtain

$$\frac{1}{30 \text{ cm}} + \frac{1}{s'} = \frac{1}{10 \text{ cm}}, \qquad s' = 15 \text{ cm},$$

$$m = -\frac{15 \text{ cm}}{30 \text{ cm}} = -\frac{1}{2}.$$

Part (b): With $s = 20$ cm and $f = 10$ cm, we get

$$\frac{1}{20 \text{ cm}} + \frac{1}{s'} = \frac{1}{10 \text{ cm}}, \qquad s' = 20 \text{ cm},$$

$$m = -\frac{20 \text{ cm}}{20 \text{ cm}} = -1.$$

Part (c): With $s = 10$ cm and $f = 10$ cm, we have

$$\frac{1}{10 \text{ cm}} + \frac{1}{s'} = \frac{1}{10 \text{ cm}}, \qquad s' = \pm\infty,$$

$$m = -\frac{\pm\infty}{10 \text{ cm}} = \mp\infty.$$

Part (d): With $s = 5$ cm and $f = 10$ cm, we find that

$$\frac{1}{5 \text{ cm}} + \frac{1}{s'} = \frac{1}{10 \text{ cm}}, \qquad s' = -10 \text{ cm},$$

$$m = -\frac{-10 \text{ cm}}{5 \text{ cm}} = +2.$$

REFLECT When the object is at the focal point, the outgoing rays are parallel and the image is at infinity. When the object distance is greater than the focal length, the image is inverted and real; when the object distance is less than the focal length, the image is erect and virtual.

Practice Problem: If an object is placed 15 cm in front of this mirror, where is the image? *Answer:* $s' = 30$ cm; light rays are reversible!

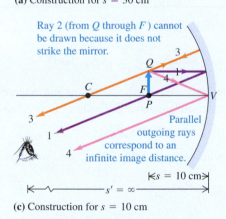

All principal rays can be drawn. The image is inverted.

(a) Construction for $s = 30$ cm

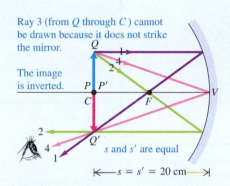

Ray 3 (from Q through C) cannot be drawn because it does not strike the mirror.

The image is inverted.

s and s' are equal

(b) Construction for $s = 20$ cm

Ray 2 (from Q through F) cannot be drawn because it does not strike the mirror.

Parallel outgoing rays correspond to an infinite image distance.

(c) Construction for $s = 10$ cm

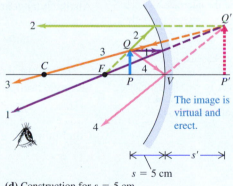

The image is virtual and erect.

(d) Construction for $s = 5$ cm

▲ **FIGURE 24.20**

24.4 Refraction at a Spherical Surface

Our next topic is refraction at a spherical surface—that is, at a spherical interface between two optical materials with different indexes of refraction. This analysis is directly applicable to some real optical systems, such as the human eye. It also provides a stepping-stone for the analysis of lenses, which usually have *two* spherical surfaces.

In Figure 24.21, a spherical surface with radius R forms an interface between two materials with indexes of refraction n_a and n_b. The surface forms an image

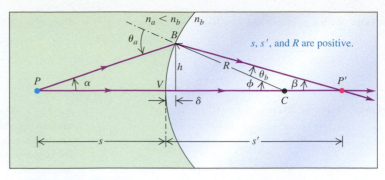

▲ FIGURE 24.21 Construction for finding the position of the image point P' of an object point P, formed by refraction at a spherical surface.

▲ BIO Application Seeing in focus. To see clearly, your eye must focus the light reaching it to form a crisp image on the retina (the light-receptive tissue at the back of the eye). Most of this focusing happens through refraction at the interface between the air and the cornea—the clear outer element of your eye. The surface of the cornea is approximately spherical. Corneal focusing works only in air, because it depends on the difference in refractive index between air and corneal tissue. To see clearly under water, you need to wear a face mask that creates a pocket of air around your eyes.

point P' of an object point P; we want to find how the object and image distances (s and s') are related. We'll use the same sign rules that we used for spherical mirrors. The center of curvature C is on the outgoing side of the surface, so R is positive. Ray PV strikes the vertex V and is perpendicular to the surface at that point. It passes into the second material without deviation. Ray PB, making an angle α with the axis, is incident at an angle θ_a with the normal and is refracted at an angle θ_b. These rays intersect at P' at a distance s' to the right of the vertex. The figure is drawn for the case where n_b is greater than n_a. The object and image distances are both positive.

We are going to prove that if the angle α is small, *all* rays from P intersect at the same point P', so P' is the *real image* of P. From the triangles PBC and $P'BC$, we have

$$\theta_a = \alpha + \phi, \qquad \phi = \beta + \theta_b. \tag{24.8}$$

From the law of refraction,

$$n_a \sin\theta_a = n_b \sin\theta_b.$$

Also, the tangents of α, β, and ϕ are

$$\tan\alpha = \frac{h}{s+\delta}, \qquad \tan\beta = \frac{h}{s'-\delta}, \qquad \tan\phi = \frac{h}{R-\delta}. \tag{24.9}$$

For paraxial rays, we may approximate both the sine and tangent of an angle by the angle itself and neglect the small distance δ. The law of refraction then becomes

$$n_a \theta_a = n_b \theta_b.$$

Combining this equation with the first of Equations 24.8, we obtain

$$\theta_b = \frac{n_a}{n_b}(\alpha + \phi).$$

When we substitute this expression into the second of Equations 24.8, we get

$$n_a \alpha + n_b \beta = (n_b - n_a)\phi. \tag{24.10}$$

Now we use the approximations $\tan\alpha = \alpha$, and so on, in Equations 24.9, and we neglect δ; those equations then become

$$\alpha = \frac{h}{s}, \qquad \beta = \frac{h}{s'}, \qquad \phi = \frac{h}{R}.$$

Finally, we substitute these equations into Equation 24.10 and divide out the common factor h; we finally obtain

$$\frac{n_a}{s} + \frac{n_b}{s'} = \frac{n_b - n_a}{R}. \tag{24.11}$$

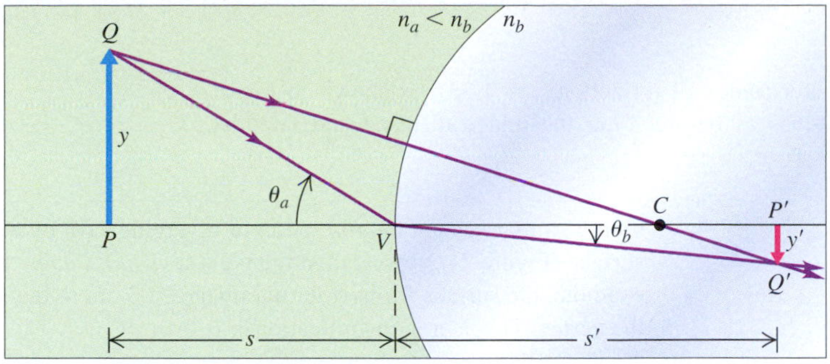

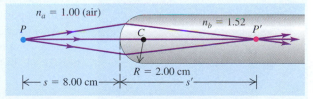

▲ FIGURE 24.22 Construction for determining the height of an image formed by refraction at a spherical surface.

This equation does not contain the angle α, so the image distance is the same for *all* paraxial rays from P.

To obtain the lateral magnification for this situation, we use the construction in Figure 24.22. We draw two rays from point Q, one through the center of curvature C and the other through the vertex V. From the triangles PQV and $P'Q'V$,

$$\tan\theta_a = \frac{y}{s}, \qquad \tan\theta_b = \frac{-y'}{s'},$$

and from the law of refraction,

$$n_a \sin\theta_a = n_b \sin\theta_b.$$

For small angles,

$$\tan\theta_a = \sin\theta_a, \qquad \tan\theta_b = \sin\theta_b,$$

so, finally,

$$\frac{n_a y}{s} = -\frac{n_b y'}{s'},$$

or

$$m = \frac{y'}{y} = -\frac{n_a s'}{n_b s}. \tag{24.12}$$

▲ Application Sometimes the sun is green. As the sun approaches the horizon, a number of interesting phenomena can be observed, especially if the view of the horizon is unrestricted, such as at sea. One of these is the "green flash" in which the last portion of the setting sun appears to be green. While the observation and analysis of this phenomenon is complicated by atmospheric conditions, the phenomenon is essentially a result of refraction along the long path the light must take through the atmosphere at sunset. Under the right conditions, the atmosphere acts as a lens to refract the setting sun's rays into their component parts. The blue-green rays of the spectrum are refracted more than the red, and thus as the sun sinks below the horizon, we can see the shorter wavelengths for a bit longer than the longer wavelengths. As the blue light is scattered strongly by the atmosphere, we see the rim of the sinking sun as a green segment. This may be followed by a rarely observed vertical green ray. The green ray has been the subject of stories and myths, including an 1882 novel by Jules Verne, *The Green Ray*.

EXAMPLE 24.4 **Glass rod in air**

The cylindrical glass rod in Figure 24.23 has index of refraction 1.52. One end is ground to a hemispherical surface with radius $R = 2.00$ cm. The rod is surrounded by air. **(a)** Find the image distance of a small object on the axis of the rod and 8.00 cm to the left of the vertex. **(b)** Find the lateral magnification.

▲ FIGURE 24.23

SOLUTION

SET UP AND SOLVE We are given that

$$n_a = 1.00, \qquad n_b = 1.52,$$
$$R = +2.00 \text{ cm}, \qquad s = +8.00 \text{ cm}.$$

Part (a): From Equation 24.11,

$$\frac{1.00}{8.00 \text{ cm}} + \frac{1.52}{s'} = \frac{1.52 - 1.00}{+2.00 \text{ cm}},$$
$$s' = +11.3 \text{ cm}.$$

Part (b): From Equation 24.12, the lateral magnification is

$$m = -\frac{n_a s'}{n_b s} = -\frac{(1.00)(11.3 \text{ cm})}{(1.52)(8.00 \text{ cm})} = -0.929.$$

REFLECT Our result for part (a) tells us that the image is formed to the right of the vertex (because s' is positive) at a distance of 11.3 cm from it. From part (b), we know that the image is somewhat smaller than the object, and it is inverted. If the object is an arrow 1.00 mm high pointing upward, the image is an arrow 0.929 mm high pointing downward.

EXAMPLE 24.5 **Glass rod in water**

The glass rod of Example 24.4 is immersed in water (index of refraction $n = 1.33$), as shown in Figure 24.24. The other quantities have the same values as before. Find the image distance and lateral magnification.

SOLUTION

SET UP AND SOLVE From Equation 24.11,

$$\frac{1.33}{8.00 \text{ cm}} + \frac{1.52}{s'} = \frac{1.52 - 1.33}{+2.00 \text{ cm}},$$
$$s' = -21.3 \text{ cm}.$$

The fact that s' is negative means that after the rays are refracted by the surface, they are not converging, but *appear* to diverge from a point 21.3 cm to the *left* of the vertex. We have

seen a similar case in the refraction of diverging rays by a plane surface (Figure 24.2); we called the point a *virtual image*. In this example, the surface forms a virtual image 21.3 cm to the left of the vertex. The lateral magnification m is then

$$m = -\frac{(1.33)(-21.3 \text{ cm})}{(1.52)(8.00 \text{ cm})} = +2.33.$$

REFLECT In this case, the image is erect (because m is positive) and 2.33 times as large as the object.

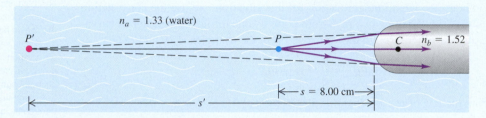

$n_a = 1.33$ (water)

P' P C $n_b = 1.52$

$s = 8.00 \text{ cm}$

s'

▶ **FIGURE 24.24**

Equations 24.11 and 24.12 can be applied to both convex and concave refracting surfaces, provided that you use the sign rules consistently. It doesn't matter whether n_b is greater or less than n_a. We suggest that you construct diagrams like Figures 24.21 and 24.22 when R is negative and when n_b is less than n_a, and use them to derive Equations 24.11 and 24.12 for these cases.

Here's a final note on the sign rule for the radius of curvature R of a surface: For the convex reflecting surface in Figure 24.15 we considered R negative, but in Figure 24.21 the refracting surface with the same orientation has a *positive* value of R. This may seem inconsistent, but it isn't. Both cases are consistent with the rule that R is positive when the center of curvature is on the outgoing side of the surface and negative when it is *not* on the outgoing side. When both reflection and refraction occur at a spherical surface, R has one sign for the reflected light and the opposite sign for the refracted light.

An important special case of a spherical refracting surface is a *plane* surface between two optical materials. This corresponds to setting $R = \infty$ in Equation 24.11. In this case,

$$\frac{n_a}{s} + \frac{n_b}{s'} = 0. \tag{24.13}$$

To find the lateral magnification m for this case, we combine Equation 24.13 with the general relation Equation 24.12, obtaining the simple result

$$m = 1. \tag{24.14}$$

That is, the image formed by a *plane* refracting surface is always the same size as the object, and it is always erect. For a real object, the image is always virtual.

A familiar example of image formation by a plane refracting surface is the appearance of a partly submerged drinking straw or canoe paddle (Figure 24.25). When viewed from some angles, the object appears to have a sharp bend at the water surface because the submerged part appears to be only about three-quarters of its actual distance below the surface.

▲ **FIGURE 24.25** An object partially submerged in water appears to bend at the surface, owing to the refraction of the light rays.

EXAMPLE 24.6 How deep is the pool?

Swimming-pool owners know that the pool always looks shallower than it really is and that it is important to identify clearly the deep parts so that people who can't swim won't jump into water that's over their heads. If a nonswimmer looks down into water that is actually 2.00 m (about 6 ft, 7 in) deep, how deep does it appear to be?

SOLUTION

SET UP Figure 24.26 shows a diagram for the problem. We pick an arbitrary object point P on the bottom of the pool; the image distance s' from the surface of the pool to the image point P' is the apparent depth of the pool. (Note that the image is virtual and that s' is negative.) The rays we draw confirm that the pool does indeed appear shallower than it actually is. (P' is above the actual bottom.)

SOLVE To find s', we use Equation 24.13:

$$\frac{n_a}{s} + \frac{n_b}{s'} = 0,$$

$$\frac{1.33}{2.00 \text{ m}} + \frac{1.00}{s'} = 0,$$

$$s' = -1.50 \text{ m}.$$

REFLECT The apparent depth is only about three-quarters of the actual depth, or about 4 ft, 11 in. A 6 ft nonswimmer who didn't allow for this effect would be in trouble. The negative sign shows that the image is virtual and on the incoming side of the refracting surface.

Practice Problem: An ice fisherman would like to check the thickness of the ice to see whether it is safe to walk on. He knows

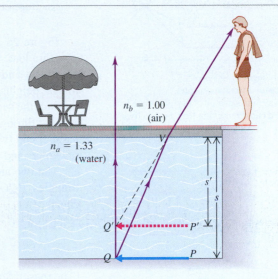

▲ **FIGURE 24.26**

that the ice ($n_{\text{ice}} = 1.309$) should be 15.0 cm thick to be safe. If he looks straight down through the ice, how thick will the ice appear to be if it is just thick enough for safe walking? *Answer:* $s' = -11.5$ cm.

24.5 Thin Lenses

The most familiar and widely used optical device (after the plane mirror) is the *lens.* A lens is an optical system with two refracting surfaces. The simplest lens has two *spherical* surfaces close enough together that we can neglect the distance between them (the thickness of the lens); we call this a **thin lens.** We can analyze such a system in detail by using the results of Section 24.4 for refraction by a single spherical surface. However, we postpone this analysis until later in this section so that we can first discuss the properties of thin lenses.

Focal Points of a Lens

Figure 24.27 shows a lens with two spherical surfaces; this type of lens is thickest at its center. The central horizontal line is called the *optic axis,* as with spherical mirrors. The centers of curvature of the two spherical surfaces lie on and define the optic axis.

When a beam of rays parallel to the optic axis passes through the lens, the rays converge to a point F_2 (Figure 24.27a). Similarly, rays passing through point F_1 emerge from the lens as a beam of rays parallel to the optic axis (Figure 24.27b). The points F_1 and F_2 are called the first and second *focal points,* and the distance f (measured from the center of the lens) is called the *focal length.* We've already used the concepts of focal point and focal length for spherical mirrors in Section 24.2. The two focal lengths in Figure 24.27, both labeled f, *are always equal* for a thin lens, even when the two sides have different curvatures. We will derive this somewhat surprising result later in this section, when we

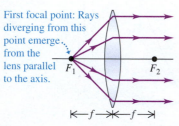

Second focal point: the point to which incoming parallel rays converge

F_1 F_2

$\overleftarrow{-f}\overrightarrow{}\overleftarrow{}\overrightarrow{-f}$

Focal length
• Measured from lens center.
• Always the same on both sides of the lens.
• For a converging thin lens, f is positive.

(a)

First focal point: Rays diverging from this point emerge from the lens parallel to the axis.

F_1 F_2

$\overleftarrow{-f}\overrightarrow{}\overleftarrow{}\overrightarrow{-f}$

(b)

▲ **FIGURE 24.27** The focal points of a converging thin lens. The numerical value of f is positive in this case.

derive the relationship of f to the index of refraction of the lens and the radii of curvature of its surfaces.

NOTE ▶ A ray passing through a thin lens is refracted at both spherical surfaces, but we'll usually draw the rays as bent at the midplane of the lens rather than at the spherical surfaces. This scheme is consistent with the assumption that the lens is very thin. ◀

Figure 24.28 shows how we can determine the position of the image formed by a thin lens. Using the same notation and sign rules as before, we let s and s' be the object and image distances, respectively, and let y and y' be the object and image heights. Ray QA, parallel to the optic axis before refraction, passes through the second focal point F_2 after refraction. Ray QOQ' passes undeflected straight through the center of the lens, because at the center the two surfaces are parallel and (we have assumed) very close together. There is refraction where the ray enters and leaves the material, but no net change in direction.

The two angles labeled α in Figure 24.28 are equal. Therefore, the two right triangles PQO and $P'Q'O$ are *similar*, and ratios of corresponding sides are equal. Thus,

$$\frac{y}{s} = -\frac{y'}{s'}, \qquad \text{or} \qquad \frac{y'}{y} = -\frac{s'}{s}. \tag{24.15}$$

(The reason for the negative sign is that the image is below the optic axis and y' is negative.) Also, the two angles labeled β are equal, and the two right triangles OAF_2 and $P'Q'F_2$ are similar. Hence, we have

$$\frac{y}{f} = -\frac{y'}{s'-f},$$

or

$$\frac{y'}{y} = -\frac{s'-f}{f}. \tag{24.16}$$

We now equate Equations 24.15 and 24.16, divide by s', and rearrange the resulting equation to obtain

$$\frac{1}{s} + \frac{1}{s'} = \frac{1}{f}. \tag{24.17}$$

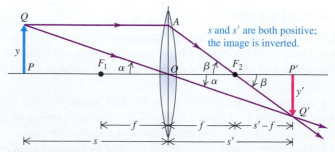

▲ **FIGURE 24.28** Construction used to find the image position of a thin lens. To emphasize that the lens is assumed to be very thin, the upper ray is shown bent at the midplane and the lower ray is shown as a straight line.

▶ **Application** **Coming soon to a disc drive near you . . . .** Data stored on optical media such as DVDs is read with a laser. To increase the density of optical data storage, it was necessary to devise very small lenses that were capable of accurately focusing light of short wavelengths (in the blue and ultraviolet). Such lenses are often made out of diamond, which has an exceptionally high index of refraction (enabling the lens to be small) and is transparent to the desired wavelengths. The lens shown is about 1 mm across and is made from pure synthetic diamond.

This analysis also gives us the lateral magnification $m = y'/y$ of the system; from Equation 24.15,

$$m = \frac{y'}{y} = -\frac{s'}{s}. \tag{24.18}$$

The negative sign tells us that when s and s' are both positive, the image is *inverted,* and y and y' have opposite signs.

Equations 24.17 and 24.18 are the basic equations for thin lenses. It is pleasing to note that their *form* is exactly the same as that of the corresponding equations for spherical mirrors, Equations 24.6 and 24.7. As we'll see, the same sign rules that we used for spherical mirrors are also applicable to lenses.

Figure 24.29 shows how a lens forms a three-dimensional image of a three-dimensional object. The real thumb points toward the lens, and the thumb's image points away from the lens. Thus, in contrast to the image produced by a plane mirror (Figure 24.6), the image is *not* reversed: The image of a left hand is a left hand. However, the real index finger points up, while its image points down, so the image produced by the lens is inverted. To make a real hand match the image produced by this lens, you would simply rotate the hand by 180° around the lens axis. Inversion of an image is equivalent to a rotation of 180° about the lens axis.

A bundle of parallel rays incident on the lens shown in Figure 24.27 converges to a real image point after passing through the lens. This lens is called a **converging lens.** Its focal length is a positive quantity, so it is also called a *positive lens.*

A bundle of parallel rays incident on the lens in Figure 24.30 *diverges* after refraction, and this lens is called a **diverging lens.** Its focal length is a negative quantity, and the lens is also called a *negative lens.* The focal points of a negative lens are *virtual* focal points, and they are reversed relative to the focal points of a positive lens. The second focal point, F_2, of a negative lens is the point from which rays that are originally parallel to the axis *appear to diverge*

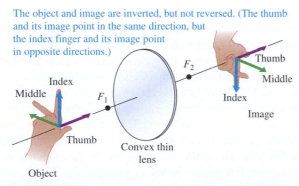

▲ **FIGURE 24.29** A lens does not produce a reversed image.

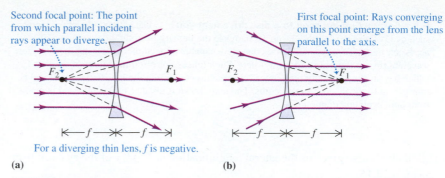

Second focal point: The point from which parallel incident rays appear to diverge.

First focal point: Rays converging on this point emerge from the lens parallel to the axis.

For a diverging thin lens, f is negative.

(a)   (b)

▲ **FIGURE 24.30** F_2 and F_1 are, respectively, the second and first focal points of a diverging thin lens.

after refraction, as in Figure 24.30a. Incident rays converging toward the first focal point F_1, as in Figure 24.30b, emerge from the lens parallel to its axis.

Equations 24.17 and 24.18 apply to *both* negative and positive lenses. Various types of lenses, both converging and diverging, are shown in Figure 24.31. Any lens that is thicker in the center than at the edges is a converging lens with positive f, and any lens that is thinner at the center than at the edges is a diverging lens with negative f (provided that the lens material has a greater index of refraction than the surrounding material has). We can prove this statement using Equation 24.20, which we haven't yet derived.

 Conceptual Analysis 24.3

Half of a lens

A lens projects an image on a screen. If the left half of the lens is covered, then

A. the left half of the image disappears.
B. the right half of the image disappears.
C. the entire image disappears.
D. the image becomes fainter.

SOLUTION When you look at nearly any object, you can usually see it from many different positions of your eye. This means that light rays are leaving the object and traveling out in many directions. If the object is in front of a lens, then the many rays from a single point on the object leave that point and hit every point of the lens. These rays are all refracted, and they all intersect at the position of the image. When you block half the lens, you block only half the rays leaving a given point on the object, so the resulting image is dimmer, but complete.

The Thin-Lens Equation

Now we proceed to derive Equation 24.17 in more detail. At the same time, we'll derive the relationship between the focal length f of the lens, its index of refraction, n, and the radii of curvature, R_1 and R_2, of its surfaces. We use the principle that an image formed by one reflecting or refracting surface can serve as the object for a second reflecting or refracting surface.

We begin with the somewhat more general problem of two spherical interfaces separating three materials with indexes of refraction n_a, n_b, and n_c, as shown in Figure 24.32. The object and image distances for the first surface are s_1 and s_1', and those for the second surface are s_2 and s_2'. We assume that the distance t between the two surfaces is small enough to be neglected in comparison with the object and image distances. Then s_2 and s_1' have the same magnitude, but opposite sign. For example, if the first image is on the outgoing side of the first surface, s_1' is positive. But when viewed as an object for the second surface, it is *not* on the incoming side of that surface. So we can say that $s_2 = -s_1'$.

We need to use the single-surface equation, Equation 24.11, twice, once for each surface. The two resulting equations are

$$\frac{n_a}{s_1} + \frac{n_b}{s_1'} = \frac{n_b - n_a}{R_1},$$

$$\frac{n_b}{s_2} + \frac{n_c}{s_2'} = \frac{n_c - n_b}{R_2}.$$

Ordinarily, the first and third materials are air or vacuum, so we set $n_a = n_c = 1$. The second index, n_b, is that of the lens material; we now call that index simply n. Substituting these values and the relation $s_2 = -s_1'$ into the preceding equations, we get

$$\frac{1}{s_1} + \frac{n}{s_1'} = \frac{n-1}{R_1},$$

$$-\frac{n}{s_1'} + \frac{1}{s_2'} = \frac{1-n}{R_2}.$$

To get a relation between the initial object position s_1 and the final image position s_2', we add these two equations. This eliminates the term n/s_1', and we obtain

$$\frac{1}{s_1} + \frac{1}{s_2'} = (n-1)\left(\frac{1}{R_1} - \frac{1}{R_2}\right).$$

Finally, thinking of the lens as a single unit, we call the object distance simply s instead of s_1 and the final image distance s' instead of s_2'. Making these substitutions, we have

$$\frac{1}{s} + \frac{1}{s'} = (n-1)\left(\frac{1}{R_1} - \frac{1}{R_2}\right). \tag{24.19}$$

Now we compare this equation with the other thin-lens equation, Equation 24.17. We see that the object and image distances s and s' appear in exactly the same places in both equations and that the focal length f is given by

$$\frac{1}{f} = (n-1)\left(\frac{1}{R_1} - \frac{1}{R_2}\right). \tag{24.20}$$

This relation is called the **thin-lens equation,** or the *lensmaker's equation.* In the process of rederiving the thin-lens equation, we have also derived an expression for the focal length f of a lens in terms of its index of refraction n and the radii of curvature R_1 and R_2 of its surfaces. Combining Equations 24.19 and 24.20 gives the general form of the thin-lens equation:

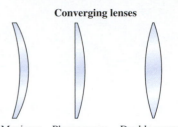

Converging lenses

Meniscus   Planoconvex   Double convex

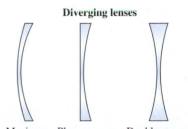

Diverging lenses

Meniscus   Planoconcave   Double concave

▲ **FIGURE 24.31** A selection of lenses.

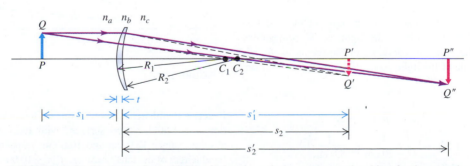

▲ **FIGURE 24.32** The image formed by the first surface of a lens serves as the object for the second surface.

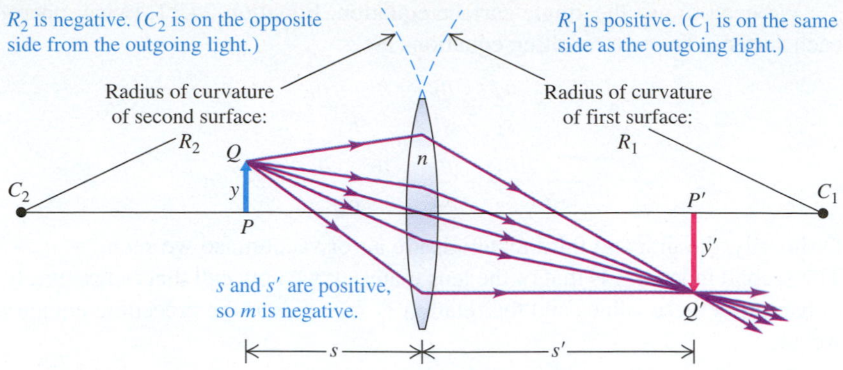

R₂ is negative. (C₂ is on the opposite side from the outgoing light.)

R₁ is positive. (C₁ is on the same side as the outgoing light.)

Radius of curvature of second surface: R_2

Radius of curvature of first surface: R_1

C_2

Q

y

P

n

P'

C_1

y'

Q'

s and s' are positive, so m is negative.

$$\overleftarrow{}\;\; s \;\;\overrightarrow{} \times \overleftarrow{}\;\; s' \;\;\overrightarrow{}$$

▲ **FIGURE 24.33** A thin lens. This lens is converging and thus has a positive focal length f.

Thin-lens equation

$$\frac{1}{s} + \frac{1}{s'} = \frac{1}{f} = (n - 1)\left(\frac{1}{R_1} - \frac{1}{R_2}\right). \qquad (24.21)$$

We use all our previous sign conventions with Equations 24.19, 24.20, and 24.21. For example, in Figure 24.33, s, s', and R_1 are positive, but R_2 is negative.

Quantitative Analysis 24.4

Object distance and magnification

A certain lens forms a real image of a very distant object on a screen that is 10 cm from the lens. As the object is moved closer to the lens, at what distance of the object from the lens are the object and image the same height?

A. only at 20 cm.

B. somewhere between 20 cm and 40 cm.

C. at 40 cm.

D. only at 10 cm.

SOLUTION If the object distance s is infinite, the image distance s' is equal to the focal length f. Thus, the focal length of our lens is 10 cm. For the object and image to be the same height, the object and image distances must be the same; hence, $s = s'$. The thin-lens equation (Equation 24.17 or 24.21) then gives us

$$\frac{1}{s} + \frac{1}{s} = \frac{1}{f} = \frac{1}{10 \text{ cm}}, \qquad s = 20 \text{ cm}.$$

The object and image distances (and therefore their heights) are equal only when $s = s' = 20$ cm.

EXAMPLE 24.7 Double-concave diverging lens

The two surfaces of the lens shown in Figure 24.34 have radii of curvature with absolute values of 20 cm and 5.0 cm, respectively. The index of refraction is 1.52. What is the focal length f of the lens?

SOLUTION

SET UP AND SOLVE The center of curvature of the first surface is on the incoming side of the light, so R_1 is negative: $R_1 = -20$ cm. The center of curvature of the second surface is on the outgoing side of the light, so R_2 is positive: $R_2 = 5.0$ cm. Then, from Equation 24.20,

$$\frac{1}{f} = (n - 1)\left(\frac{1}{R_1} - \frac{1}{R_2}\right)$$

$$= (1.52 - 1)\left(\frac{1}{-20 \text{ cm}} - \frac{1}{5.0 \text{ cm}}\right),$$

$$f = -7.7 \text{ cm}.$$

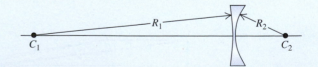

R_1

R_2

C_1

C_2

▲ **FIGURE 24.34**

Practice Problem: A double-concave lens with index of refraction $n = 1.50$ has two surfaces with radii of curvature with absolute values 12.0 cm and 10.0 cm, respectively. What is the focal length of the lens? *Answer: f = −10.9 cm.*

REFLECT This lens is a diverging lens—a negative lens—with a negative focal length.

EXAMPLE 24.8 **The eye's lens**

When light enters your eye, most of the focusing happens at the interface between the air and the cornea (the outermost element of the eye). The eye also has a doubly-convex *lens*, lying behind the cornea, that completes the job of forming an image on the retina. (The lens also enables us to shift our distance of focus; it gets rounder for near vision and flatter for far vision.) The lens has an index of refraction of about 1.40. **(a)** For the lens in Figure 24.35, find the focal length. **(b)** If you could consider this lens in isolation from the rest of the eye, what would the image distance be for an object 0.20 m in front of this crystalline lens?

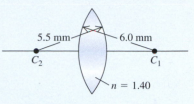

▲ **FIGURE 24.35**

SOLUTION

SET UP The center of curvature of the first surface of the lens is on the outgoing side, so $R_1 = +6.0$ mm. The center of curvature for the second surface is *not* on the outgoing side, so $R_2 = -5.5$ mm. We solve for f, then use the result in Equation 24.21.

SOLVE **Part (a):** Using Equation 24.20, we find that

$$\frac{1}{f} = (1.40 - 1)\left(\frac{1}{+6.0 \text{ mm}} - \frac{1}{-5.5 \text{ mm}}\right),$$
$$f = 7.2 \text{ mm}.$$

Part (b): The object distance is $s = 0.20$ m $= 200$ mm. Using Equation 24.21, we obtain

$$\frac{1}{200 \text{ mm}} + \frac{1}{s'} = \frac{1}{7.2 \text{ mm}}, \qquad s' = 7.5 \text{ mm}.$$

REFLECT The image is slightly farther from the lens than it would be for an infinitely distant object. As expected for a converging lens, the focal length is positive.

It's not hard to generalize Equation 24.20 or 24.21 to the situation in which the lens is immersed in a material with an index of refraction greater than unity. We invite you to work out the thin-lens equation for this more general situation.

24.6 Graphical Methods for Lenses

We can determine the position and size of an image formed by a thin lens by using a graphical method that is similar to the one we used in Section 24.3 for spherical mirrors. Again, we draw a few special rays called *principal rays* that diverge from a point of the object that is *not* on the optic axis. The intersection of these rays, after they pass through the lens, determines the position and size of the image. In using this graphical method, as in Section 24.5, we consider the entire deviation of a ray as occurring at the midplane of the lens, as shown in Figure 24.36; this is consistent with the assumption that the distance between the lens surfaces is negligible.

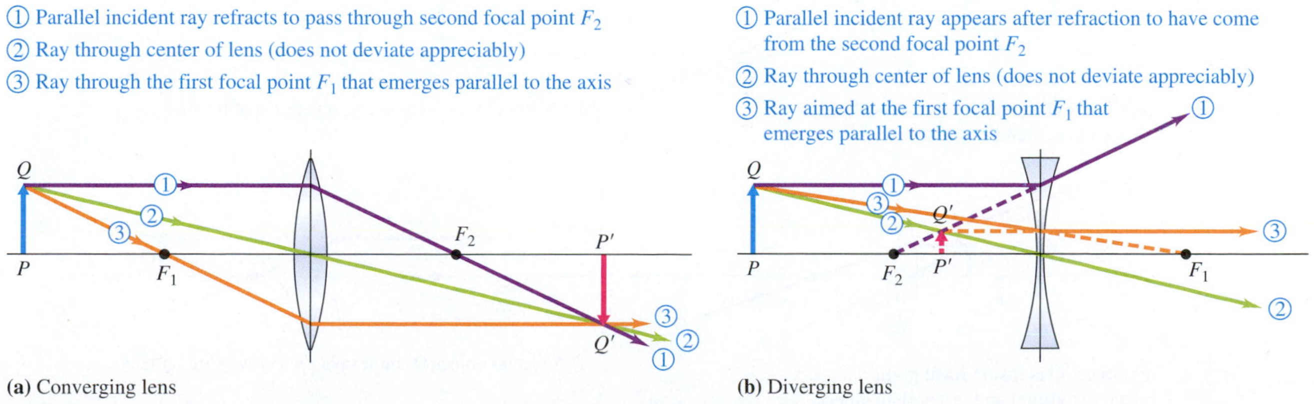

① Parallel incident ray refracts to pass through second focal point F_2

② Ray through center of lens (does not deviate appreciably)

③ Ray through the first focal point F_1 that emerges parallel to the axis

① Parallel incident ray appears after refraction to have come from the second focal point F_2

② Ray through center of lens (does not deviate appreciably)

③ Ray aimed at the first focal point F_1 that emerges parallel to the axis

(a) Converging lens

(b) Diverging lens

▲ **FIGURE 24.36** Principal-ray diagrams showing the graphical method for locating an image produced by a thin lens.

The three principal rays whose paths are usually easy to trace for lenses are shown in Figure 24.36. They are as follows:

> **Principal rays for thin lenses**
> 1. *A ray parallel to the axis,* after refraction by the lens, passes through the second focal point F_2 of a converging lens or appears to come from the second focal point of a diverging lens (purple in diagrams).
> 2. *A ray through the center of the lens* is not appreciably deviated, because, at the center of the lens, the two surfaces are parallel and close together (green in diagrams).
> 3. *A ray through, away from, or proceeding toward the first focal point F_1* emerges parallel to the axis (orange in diagrams).
>
> When the image is real, the position of the image point is determined by the intersection of any two of the three principal rays (Figure 24.36a). When the image is virtual, the outgoing rays diverge. In this case, we extend the diverging outgoing rays backward to their intersection point (Figure 24.36b). Once the image position is known, we can draw any other ray from the same point. Usually, nothing is gained by drawing a lot of additional rays.

Figure 24.37 shows several principal-ray diagrams for a converging lens for several object distances. We suggest that you study each of these diagrams very carefully, comparing each numbered ray with the preceding description. Several points are worth noting. In Figure 24.37d, the object is at the focal point, and ray 3 cannot be drawn because it does not pass through the lens. In Figure 24.37e, the object distance is less than the focal length. The outgoing rays are divergent, and the *virtual image* is located by extending the outgoing rays backward. In this case, the image distance s' is negative. Figure 24.37f corresponds to a *virtual*

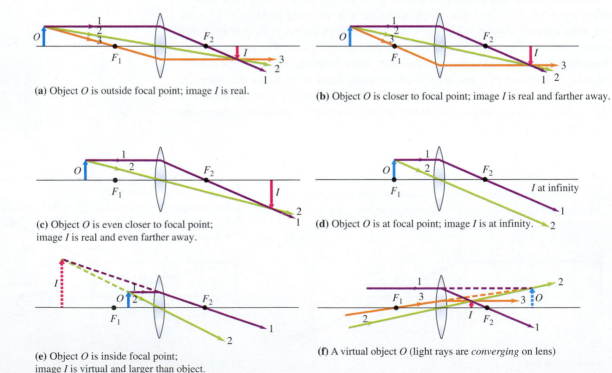

(a) Object O is outside focal point; image I is real.

(b) Object O is closer to focal point; image I is real and farther away.

(c) Object O is even closer to focal point; image I is real and even farther away.

(d) Object O is at focal point; image I is at infinity.

(e) Object O is inside focal point; image I is virtual and larger than object.

(f) A virtual object O (light rays are *converging* on lens)

▲ **FIGURE 24.37** Formation of images by a thin converging lens for various object distances. The principal rays are numbered.

object. The incoming rays do not diverge from a real object point O, but are *converging* as though they would meet at the virtual object point O on the right side. The object distance s is negative in this case. The image is real; the image distance s' is positive and less than f.

Conceptual Analysis 24.5   **Possible rays**

In Figure 28.38, which of the rays (A–D) *did not* originate from point Q at the top of the object?

SOLUTION   Because the lens is a diverging lens, rays spread apart after passing through it. Ray C comes from point Q and passes in a straight line through the lens, crossing the optic axis at its center. When rays A and D are projected backward, they pass through the focal point F_2; thus, before they passed through the lens, they were parallel to the axis of the lens. Ray A therefore came from Q; ray D did not. Ray B is parallel to the axis after refraction, so it was directed toward focal point F_1 before being refracted. Thus, rays D and B do not come from Q.

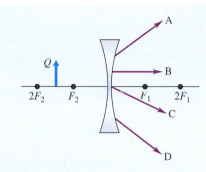

▲ **FIGURE 24.38**

PROBLEM-SOLVING STRATEGY 24.2   **Image formation by a thin lens**

SET UP

1. The strategy outlined in Section 24.3 is equally applicable here, and we suggest that you review it now. Always begin with a principal-ray diagram if you have enough information. Orient your diagrams consistently so that light travels from left to right. For a lens, there are only three principal rays, compared to four for a mirror. Don't just sketch these diagrams; draw the rays with a ruler and measure the distances carefully. Draw the rays so that they bend at the midplane of the lens, as shown in Figure 24.37. Be sure to draw *all three* principal rays whenever possible. The intersection of any two determines the image, but if the third doesn't pass through the same intersection point, you know that you've made a mistake. Redundancy can be useful in spotting errors.

SOLVE

2. If the outgoing principal rays don't converge at a real image point, the image is virtual. Then you have to extend the outgoing rays backward to find the virtual image point, which lies on the *incoming* side of the lens.
3. The same sign rules that we've used for mirrors and single refracting surfaces are also applicable to thin lenses. Be extremely careful to get your signs right and to interpret the signs of results correctly.

REFLECT

4. Always determine the image position and size *both* graphically and by calculating. Then compare the results. This gives an extremely valuable consistency check.
5. Remember that the *image* from one lens or mirror may serve as the *object* for another. In that case, be careful in finding the object and image *distances* for this intermediate image; be sure you include the distance between the two elements (lenses and/or mirrors) correctly.

EXAMPLE 24.9 **Image formed by a converging lens**

A converging lens has a focal length of 20 cm. Find graphically the image location for an object at each of the following distances from the lens: **(a)** 50 cm; **(b)** 20 cm; **(c)** 15 cm; **(d)** −40 cm. Determine the lateral magnification in each case. Check your results by calculating the image distance and lateral magnification.

SOLUTION

SET UP AND SOLVE The principal-ray diagrams are shown in Figures 24.37a, d, e, and f. The approximate image distances, from measurements of these diagrams, are 35 cm, ∞, −40 cm, and 15 cm, respectively, and the approximate lateral magnifications are $-\frac{2}{3}$, ∞, +3, and $+\frac{1}{3}$. To calculate the image positions, we use Equation 24.17:

$$\frac{1}{s} + \frac{1}{s'} = \frac{1}{f}.$$

To calculate the lateral magnifications, we use Equation 24.18, $m = -s'/s$.

Part (a): $s = 50$ cm and $f = 20$ cm:

$$\frac{1}{50\text{ cm}} + \frac{1}{s'} = \frac{1}{20\text{ cm}}, \qquad s' = 33.3\text{ cm},$$

$$m = -\frac{33.3\text{ cm}}{50\text{ cm}} = -\frac{2}{3}.$$

Part (b): $s = 20$ cm and $f = 20$ cm:

$$\frac{1}{20\text{ cm}} + \frac{1}{s'} = \frac{1}{20\text{ cm}}, \qquad s' = \pm\infty,$$

$$m = -\frac{\pm\infty}{20\text{ cm}} = \mp\infty.$$

Part (c): $s = 15$ cm and $f = 20$ cm:

$$\frac{1}{15\text{ cm}} + \frac{1}{s'} = \frac{1}{20\text{ cm}}, \qquad s' = -60\text{ cm},$$

$$m = -\frac{-60\text{ cm}}{15\text{ cm}} = +4.$$

Part (d): $s = -40$ cm and $f = 20$ cm:

$$\frac{1}{-40\text{ cm}} + \frac{1}{s'} = \frac{1}{20\text{ cm}}, \qquad s' = 13.3\text{ cm},$$

$$m = -\frac{13.3\text{ cm}}{-40\text{ cm}} = +\frac{1}{3}.$$

REFLECT The graphical results for the image distances are fairly close to the calculated results, except for those from Figure 24.37e, where the precision of the diagram is limited because the rays extended backward have nearly the same direction.

Practice Problem: If an object is placed at a point that is 5.0 cm to the left of the lens in this problem, where is the image and what is the lateral magnification of the image? *Answer:* 6.7 cm to the left of the lens; $s' = -6.7$ cm, $m = 1.3$.

EXAMPLE 24.10 **Image formed by a diverging lens**

You are given a thin diverging lens. You find that a beam of parallel rays spreads out after passing through the lens, as though all the rays came from a point 20.0 cm from the center of the lens. You want to use this lens to form an erect virtual image that is one-third the height of a real object. **(a)** Where should the object be placed? **(b)** Draw a principal-ray diagram.

SOLUTION

SET UP AND SOLVE Part (a): The behavior of the parallel incident rays indicates that the focal length is negative: $f = -20.0$ cm. We want the lateral magnification to be $m = +\frac{1}{3}$ (positive because the image is to be erect). From Equation 24.18, $m = \frac{1}{3} = -s'/s$. So we use $s' = -s/3$ in Equation 24.17:

$$\frac{1}{s} + \frac{1}{s'} = \frac{1}{f},$$

$$\frac{1}{s} + \frac{1}{-s/3} = \frac{1}{-20.0\text{ cm}},$$

$$s = 40.0\text{ cm},$$

$$s' = -\frac{40.0\text{ cm}}{3} = -13.3\text{ cm}.$$

Part (b): Figure 24.39 shows our principal-ray diagram for this problem, with the rays numbered the same way as in Figure 24.36b.

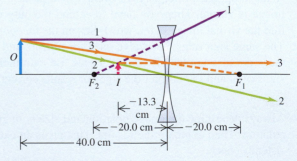

▲ **FIGURE 24.39**

REFLECT In part (a), the image distance is negative, so the real object and the virtual image are on the same side of the lens.

Practice Problem: For a diverging lens with a focal length of $f = -10$ cm, where should an object be placed to form an erect virtual image that is one-fourth the height of the object? *Answer:* $s = 30$ cm.

SUMMARY

Reflection at a Plane Surface

(Section 24.1) When rays diverge from an **object point** Q and are reflected at a plane surface, the directions of the outgoing rays are the same as though they had diverged from a point Q' called the **image point.** Since the rays don't actually converge at Q', the image formed is a **virtual image.** An image formed by a plane mirror is always **reversed;** for example, the image of a right hand is a left hand.

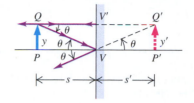

The **lateral magnification** m in any reflecting or refracting situation is defined as the ratio of the image height y' to the object height y:

$$m = \frac{y'}{y}. \qquad (24.2)$$

When m is positive, the image is erect; when m is negative, the image is inverted.

Reflection at a Spherical Surface

(Section 24.2) The **focal point** of a mirror is the point at which parallel rays converge after reflection from a concave mirror or the point from which they appear to diverge after reflection from a convex mirror. Rays diverging from the focal point of a concave mirror are parallel after reflection; rays converging toward the focal point of a convex mirror are parallel after reflection. The distance from the focal point to the vertex is called the focal length f, where $f = R/2$ (Equation 24.5). The object distance, image distance, and focal length are related by

$$\frac{1}{s} + \frac{1}{s'} = \frac{1}{f}. \qquad (24.6)$$

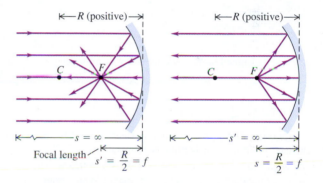

Graphical Methods for Mirrors

(Section 24.3) Four principal rays can be drawn to find the size and position of the image formed by a mirror:

1. *A ray parallel to the axis,* after reflection, passes through the focal point F of a concave mirror or appears to come from the (virtual) focal point of a convex mirror.
2. *A ray through (or proceeding toward) the focal point F is* reflected parallel to the axis.
3. *A ray along the radius through or away from the center of curvature C intersects the surface normally and is reflected back along its original path.*
4. *A ray to the vertex V is reflected and forms equal angles with the optic axis.*

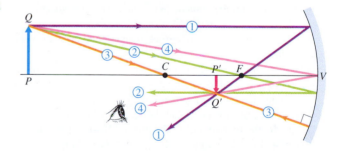

Refraction at a Spherical Surface and Thin Lenses

(Sections 24.4 and 24.5) When a spherical surface forms the interface between two materials, the magnification of the image is given by

$$m = -\frac{n_a s'}{n_b s}. \qquad (24.12)$$

A **thin lens** has two spherical surfaces close enough together that we can ignore the distance between them. The magnification of a thin lens is given by $m = -s'/s$ (Equation 24.18). The focal length

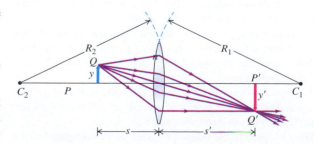

Continued

is the same on both sides of a thin lens, even when the two sides have different curvatures. The behavior of a thin lens is described by the **thin-lens equation:**

$$\frac{1}{s} + \frac{1}{s'} = \frac{1}{f} = (n - 1)\left(\frac{1}{R_1} - \frac{1}{R_2}\right). \qquad (24.19, 24.20, 24.21)$$

Graphical Methods for Lenses

(Section 24.6) Three principal rays can be drawn to find the size and position of the image formed by a thin lens:

1. *A ray parallel to the axis,* after refraction by the lens, passes through the second focal point F_2 of a converging lens or appears to come from the second focal point of a diverging lens.
2. *A ray through the center of the lens* is not appreciably deviated, because, at the center of the lens, the two surfaces are parallel and close together.
3. *A ray through (or proceeding toward) the first focal point F_1* emerges parallel to the axis.

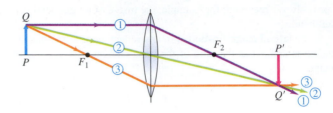

 For instructor-assigned homework, go to www.masteringphysics.com

Conceptual Questions

1. If a spherical mirror is immersed in water, does its focal length change? What if a lens is immersed in water?
2. For what range of object positions does a concave spherical mirror of focal length f form a real image? What about a convex spherical mirror?
3. If a screen is placed at the location of a real image, will the image appear on the screen? What happens if the image is a virtual one?
4. Is it possible to view a virtual image directly with your eye? How?
5. When a room has mirrors on two opposite walls, an infinite series of images can be seen. Show why this happens. Why do the distant images appear progressively darker?
6. A spherical mirror is cut in half horizontally. Will an image be formed by the bottom half of the mirror? If so, where will the image be formed, and how will its appearance compare to the image formed by the full mirror?
7. A concave mirror (sometimes surrounded by lights) is often used as an aid for applying cosmetics to the face. Why is such a mirror always concave rather than convex? What considerations determine its radius of curvature?
8. On a sunny day, you can use the sun's rays and a concave mirror to start a fire. How is this done? Could you do the same thing with a convex mirror? with a double concave lens? with a double convex lens?
9. A person looks at her reflection in the concave side of a shiny spoon. Is this image right side up or inverted? What does she see if she looks in the convex side?
10. What happens to the image produced by a converging lens as the object is slowly moved *through* the focal point? Does the same thing happen for a diverging lens?

11. Without measuring its radius of curvature (which is not so easy to do), explain how you can experimentally determine the focal length of (a) a concave mirror, (b) a convex mirror. Your apparatus consists of viewing screens and an optical bench on which to mount the mirrors and measure distances.
12. Without measuring its radii of curvature (which is not so easy to do), explain how you can experimentally determine the focal length of (a) a converging lens, (b) a diverging lens. Your apparatus consists of viewing screens and an optical bench on which to mount the lenses and measure distances.
13. What happens to the focal length of a thin lens if you turn it around? Which things change and which ones stay the same?
14. A spherical air bubble in water can function as a lens. Is it a converging or a diverging lens? How is its focal length related to its radius?
15. You have a curved spherical mirror about a foot across. You find that when your eye is very close to the mirror, you see an erect image of your face. But as you move farther and farther from the mirror, your face suddenly looks upside down. (a) What kind of mirror (convex or concave) is this? (b) Are the images you see real or virtual?

Multiple-Choice Problems

1. A ray from an object passes through a thin lens, as shown in Figure 24.40. What can we conclude about the lens from this ray?
   A. It is a converging lens.
   B. It is a diverging lens.
   C. It is not possible to tell which type of lens it is.

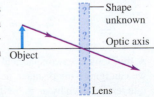

▲ **FIGURE 24.40** Multiple-choice problem 1.

2. If a single lens forms a real image, we can conclude that
   A. It is a converging lens.
   B. It is a diverging lens.
   C. It could be either type of lens.

3. If a single lens forms a virtual image, we can conclude that
   A. It is a converging lens.
   B. It is a diverging lens.
   C. It could be either type of lens.

4. An object lies outside the focal point of a converging lens. Which of the following statements about the image formed by this lens must be true? (There may be more than one correct choice.)
   A. The image is always real and inverted.
   B. The image could be real or virtual, depending on how far the object is past the focal point.
   C. The image could be erect or inverted, depending on how far the object is past the focal point.
   D. The image is always on the opposite side of the lens from the object.

5. An object lies outside the focal point of a diverging lens. Which of the following statements about the image formed by this lens must be true? (There may be more than one correct choice.)
   A. The image is always virtual and inverted.
   B. The image could be real or virtual, depending on how far the object is past the focal point.
   C. The image could be erect or inverted, depending on how far the object is past the focal point.
   D. The image is always virtual and on the same side of the lens as the object.

6. A spherical mirror is shown in Figure 24.41, and the object is placed in front of the curved front surface. Which of the following statements about this mirror must be true? (There may be more than one correct choice.)
   A. Its radius of curvature is negative.
   B. The image it produces is inverted.
   C. A ray parallel to the optic axis is reflected from the mirror so that it passes through the focal point of the mirror.
   D. It always produces a virtual image on the opposite side of the mirror from the object.

Front face

▲ FIGURE 24.41
Multiple-choice problem 6.

7. An object is beyond the focal point of a converging lens. If you want to bring its image closer to the lens, you should move the object
   A. away from the lens.
   B. toward the lens.
   C. to the focal point of the lens.

8. A thin glass lens has a focal length f in air. If you now make a lens of identical shape, using glass having twice the refractive index of the original glass, the focal length of the new lens will be
   A. $f/2$.
   B. less than $f/2$.
   C. $2f$.
   D. greater than $2f$.

9. A ray from an object passes through a thin lens, as shown in Figure 24.42. What can we conclude about the lens from this ray?
   A. It must be a converging lens.
   B. It must be a diverging lens.
   C. It could be either a converging or a diverging lens.

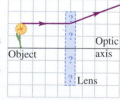

▲ FIGURE 24.42
Multiple-choice problem 9.

10. As you move an object from just outside to just inside the focal point of a converging lens, its image
    A. goes from real to virtual and from inverted to erect.
    B. goes from inverted to erect, but remains real.
    C. goes from inverted to erect, but remains virtual.
    D. goes from real to virtual, but remains inverted.
    E. remains both erect and virtual.

11. As you move an object from just outside to just inside the focal point of a diverging lens, its image
    A. goes from real to virtual and from inverted to erect.
    B. goes from inverted to erect, but remains real.
    C. goes from inverted to erect, but remains virtual.
    D. goes from real to virtual, but remains inverted.
    E. remains both erect and virtual.

12. You have a shiny salad bowl with a spherical shape. If you hold the bowl at arm's length with the inside of the bowl facing you, the image of your face that you see will be:
    A. upside down and smaller than your face.
    B. right side up and smaller than your face.
    C. right side up and bigger than your face.
    D. upside down and bigger than your face.

13. A very long tube of transparent glass has a hemispherical end, as shown in Figure 24.43, and is surrounded by air. A small object is placed to the left of the curved glass surface. Which of the following statements must be true about the image of this object formed by the glass? (There may be more than one correct choice.)
    A. It will lie within the glass.
    B. It will lie in the air to the left of the glass.
    C. It will be real.
    D. It will be virtual.

Glass
Air

▲ FIGURE 24.43
Multiple-choice problem 13.

14. A very long glass tube, with index of refraction 1.50, has a convex hemispherical end (the reverse of the tube in question 13). Suppose the tube is immersed in a fluid whose index of refraction is 1.60, and a small object is placed to the left of the curved glass surface. Which of the following statements must be true about the image of this object formed by the glass? (There may be more than one correct choice.)
    A. It will lie within the glass.
    B. It will lie in the fluid to the left of the glass.
    C. It will be real.
    D. It will be virtual.

15. A certain thin lens has a focal length f. If you double both of its radii of curvature, but change nothing else, its focal length will now be
    A. $4f$.
    B. $2f$.
    C. $f/2$.
    D. $f/4$.

Problems

24.1 Reflection at a Plane Surface

1. • A candle 4.85 cm tall is 39.2 cm to the left of a plane mirror. Where is the image formed by the mirror, and what is the height of this image?

2. • What is the size of the smallest vertical plane mirror in which a 10 ft tall giraffe standing erect can see her full-length image? (*Hint:* Locate the image by drawing a number of rays from the giraffe's body that reflect off the mirror and go to her eye. Then eliminate that part of the mirror for which the reflected rays do not reach her eye.)

3. •• An object is placed between two plane mirrors arranged at right angles to each other at a distance d_1 from the surface of one mirror and a distance d_2 from the surface of the other. (a) How many images are formed? Show the location of the images in a diagram. (b) Draw the paths of rays from the object to the eye of an observer.

4. •• If you run away from a plane mirror at 2.40 m/s, at what speed does your image move away from you?

24.2 Reflection at a Spherical Surface

5. • A concave spherical mirror has a radius of curvature of 10.0 cm. Calculate the location and size of the image formed of an 8.00-mm-tall object whose distance from the mirror is (a) 15.0 cm, (b) 10.0 cm, (c) 2.50 cm, and (d) 10.0 m.

6. • Repeat the previous problem, except use a convex mirror with the same magnitude of focal length.

7. • The diameter of Mars is 6794 km, and its minimum distance from the earth is 5.58×10^7 km. (a) When Mars is at this distance, find the diameter of the image of Mars formed by a spherical, concave telescope mirror with a focal length of 1.75 m. (b) Where is the image located?

8. • A concave mirror has a radius of curvature of 34.0 cm. (a) What is its focal length? (b) A ladybug 7.50 mm tall is located 22.0 cm from this mirror along the principal axis. Find the location and height of the image of the insect. (c) If the mirror is immersed in water (of refractive index 1.33), what is its focal length?

9. • **Rearview mirror.** A mirror on the passenger side of your car is convex and has a radius of curvature with magnitude 18.0 cm. (a) Another car is seen in this side mirror and is 13.0 m behind the mirror. If this car is 1.5 m tall, what is the height of its image? (b) The mirror has a warning attached that objects viewed in it are closer than they appear. Why is this so?

10. •• Examining your image in a convex mirror whose radius of curvature is 25.0 cm, you stand with the tip of your nose 10.0 cm from the surface of the mirror. (a) Where is the image of your nose located? What is its magnification? (b) Your ear is 10.0 cm behind the tip of your nose; where is the image of your ear located, and what is its magnification? Do your answers suggest reasons for your strange appearance in a convex mirror?

11. •• A coin is placed next to the convex side of a thin spherical glass shell having a radius of curvature of 18.0 cm. An image of the 1.5-cm-tall coin is formed 6.00 cm behind the glass shell. Where is the coin located? Determine the size, orientation, and nature (real or virtual) of the image.

12. •• (a) Show that when an object is *outside* the focal point of a concave mirror, its image is always *inverted* and *real*. Is there any limitation on the magnification? (b) Show that when an object is *inside* the focal point of a concave mirror, its image is always *erect* and *virtual*. Is there any limitation on the magnification?

24.3 Graphical Methods for Mirrors

13. • A spherical, concave shaving mirror has a radius of curvature of 32.0 cm. (a) What is the magnification of a person's face when it is 12.0 cm to the left of the vertex of the mirror? (b) Where is the image? Is the image real or virtual? (c) Draw a principal-ray diagram showing the formation of the image.

14. • An object 0.600 cm tall is placed 16.5 cm to the left of the vertex of a concave spherical mirror having a radius of curvature of 22.0 cm. (a) Draw a principal-ray diagram showing the formation of the image. (b) Calculate the position, size, orientation (erect or inverted), and nature (real or virtual) of the image.

15. • Repeat the previous problem for the case in which the mirror is *convex*.

16. • The stainless steel rear end of a tanker truck is convex, shiny, and has a radius of curvature of 2.0 m. You're tailgating the truck, with the front end of your car only 5.0 m behind it. Making the not very realistic assumption that your car is on the axis of the mirror formed by the tank, (a) determine the position, orientation, magnification, and type (real or virtual) of the image of your car's front end that forms in this mirror; (b) draw a principal-ray diagram of the situation to check your answer.

17. • The thin glass shell shown in Figure 24.44 has a spherical shape with a radius of curvature of 12.0 cm, and both of its surfaces can act as mirrors. A seed 3.30 mm high is placed 15.0 cm from the center of the mirror along the optic axis, as shown in the figure. (a) Calculate the location and height of the image of this seed. (b) Suppose now that the shell is reversed. Find the location and height of the seed's image.

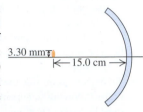

▲ **FIGURE 24.44** Problem 17.

18. •• **Dental mirror.** A dentist uses a curved mirror to view **BIO** teeth on the upper side of the mouth. Suppose she wants an erect image with a magnification of 2.00 when the mirror is 1.25 cm from a tooth. (Treat this problem as though the object and image lie along a straight line.) (a) What kind of mirror (concave or convex) is needed? Use a ray diagram to decide, without performing any calculations. (b) What must be the focal length and radius of curvature of this mirror? (c) Draw a principal-ray diagram to check your answer in part (b).

24.4 Refraction at a Spherical Surface

19. • The left end of a long glass rod 6.00 cm in diameter has a convex hemispherical surface 3.00 cm in radius. The refractive index of the glass is 1.60. Determine the position of the image if an object is placed in air on the axis of the rod at the following distances to the left of the vertex of the curved end: (a) infinitely far, (b) 12.0 cm, and (c) 2.00 cm.

20. • The rod of the previous problem is immersed in a liquid. An object 90.0 cm from the vertex of the left end of the rod and on its axis is imaged at a point 1.60 m inside the rod. What is the refractive index of the liquid?

21. • The left end of a long glass rod 8.00 cm in diameter and with an index of refraction of 1.60 is ground and polished to a convex hemispherical surface with a radius of 4.00 cm. An object in the form of an arrow 1.50 mm tall, at right angles to the axis of the rod, is located on the axis 24.0 cm to the left of the vertex of the convex surface. Find the position and height of the image of the arrow formed by paraxial rays incident on the convex surface. Is the image erect or inverted?

22. •• A large aquarium has portholes of thin transparent plastic with a radius of curvature of 1.75 m and their convex sides facing into the water. A shark hovers in front of a porthole, sizing

up the dinner prospects outside the tank. (a) If one of the shark's teeth is exactly 45.0 cm from the plastic, how far from the plastic does it appear to be to observers outside the tank? (You can ignore refraction due to the plastic.) (b) Does the shark appear to be right side up or upside down? (c) If the tooth has an actual length of 5.00 cm, how long does it appear to the observers?

23. • **A spherical fishbowl.** A small tropical fish is at the center of a water-filled spherical fishbowl 28.0 cm in diameter. (a) Find the apparent position and magnification of the fish to an observer outside the bowl. The effect of the thin walls of the bowl may be ignored. (b) A friend advised the owner of the bowl to keep it out of direct sunlight to avoid blinding the fish, which might swim into the focal point of the parallel rays from the sun. Is the focal point actually within the bowl?

24. •• **Focus of the eye.** The cornea of the eye has a radius of
BIO curvature of approximately 0.50 cm, and the aqueous humor behind it has an index of refraction of 1.35. The thickness of the cornea itself is small enough that we shall neglect it. The depth of a typical human eye is around 25 mm. (a) What would have to be the radius of curvature of the cornea so that it alone would focus the image of a distant mountain on the retina, which is at the back of the eye opposite the cornea? (b) If the cornea focused the mountain correctly on the retina as described in part (a), would it also focus the text from a computer screen on the retina if that screen were 25 cm in front of the eye? If not, where would it focus that text, in front of or behind the retina? (c) Given that the cornea has a radius of curvature of about 5.0 mm, where does it actually focus the mountain? Is this in front of or behind the retina? Does this help you see why the eye needs help from a lens to complete the task of focusing?

25. • A speck of dirt is embedded 3.50 cm below the surface of a sheet of ice having a refractive index of 1.309. What is the apparent depth of the speck, when viewed from directly above?

26. • A skin diver is 2.0 m below the surface of a lake. A bird flies overhead 7.0 m above the surface of the lake. When the bird is directly overhead, how far above the diver does it appear to be?

27. •• A zoo aquarium has transparent walls, so that spectators on both sides of it can watch the fish. The aquarium is 5.50 m across, and the spectators on both sides of it are standing 1.20 m from the wall. How far away do spectators on one side of the aquarium appear to those on the other side? (Ignore any refraction in the walls of the aquarium.)

28. • To a person swimming 0.80 m beneath the surface of the water in a swimming pool, the diving board directly overhead appears to be a height of 5.20 m above the swimmer. What is the actual height of the diving board above the surface of the water?

24.5 The Thin Lens

29. • A converging lens with a focal length of 7.00 cm forms an image of a 4.00-mm-tall real object that is to the left of the lens. The image is 1.30 cm tall and erect. Where are the object and image located? Is the image real or virtual?

30. • A converging lens with a focal length of 90.0 cm forms an image of a 3.20-cm-tall real object that is to the left of the lens. The image is 4.50 cm tall and inverted. Where are the

object and image located in relation to the lens? Is the image real or virtual?

31. • You are standing in front of a lens that projects an image of you onto a wall 1.80 m on the other side of the lens. This image is three times your height. (a) How far are you from the lens? (b) Is your image erect or inverted? (c) What is the focal length of the lens? Is the lens converging or diverging?

32. •• Figure 24.45 shows an object and its image formed by a thin lens. (a) What is the focal length of the lens and what type of lens (converging or diverging) is it? (b) What is the height of the image? Is it real or virtual?

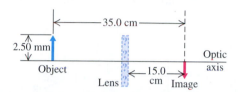

▲ **FIGURE 24.45** Problem 32.

33. •• Figure 24.46 shows an object and its image formed by a thin lens. (a) What is the focal length of the lens and what type of lens (converging or diverging) is it? (b) What is the height of the image? Is it real or virtual?

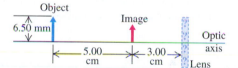

▲ **FIGURE 24.46** Problem 33.

34. •• Figure 24.47 shows an object and its image formed by a thin lens. (a) What is the focal length of the lens and what type of lens (converging or diverging) is it? (b) What is the height of the image? Is it real or virtual?

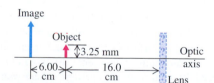

▲ **FIGURE 24.47** Problem 34.

35. • The two surfaces of a plastic converging lens have equal radii of curvature of 22.0 cm, and the lens has a focal length of 20.0 cm. Calculate the index of refraction of the plastic.

36. • The front, convex, surface of a lens made for eyeglasses has a radius of curvature of 11.8 cm, and the back, concave, surface has a radius of curvature of 6.80 cm. The index of refraction of the plastic lens material is 1.67. Calculate the focal length of the lens.

37. • For each of the thin lenses (L_1 and L_2) shown in Figure 24.48, the index of refraction of the lens' glass is 1.50, and the object is to the left of the lens. The radii of curvature indicated are just the magnitudes. Calculate the focal length of each lens.

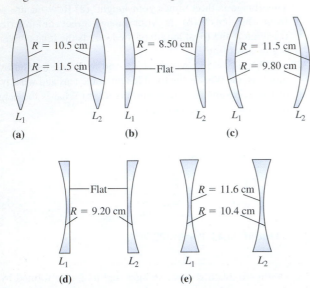

▲ **FIGURE 24.48** Problem 37.

38. • For each thin lens shown in Figure 24.49, calculate the location of the image of an object that is 18.0 cm to the left of the lens. The lens material has a refractive index of 1.50, and the radii of curvature shown are only the magnitudes.

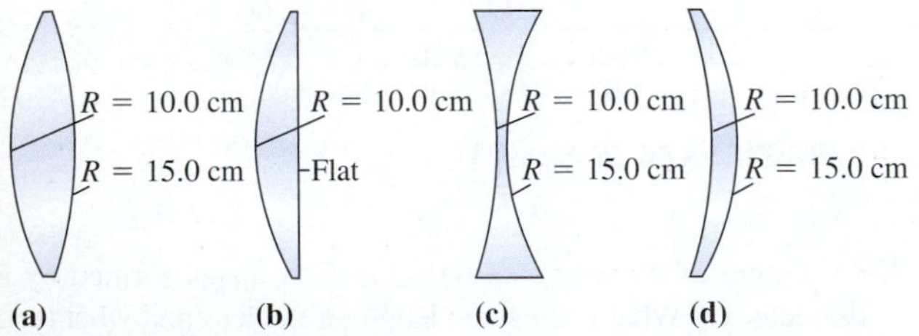

▲ **FIGURE 24.49** Problem 38.

39. • **The lens of the eye.** The crystalline lens of the human eye is
BIO a double-convex lens made of material having an index of refraction of 1.44 (although this varies). Its focal length in air is about 8.0 mm, which also varies. We shall assume that the radii of curvature of its two surfaces have the same magnitude. (a) Find the radii of curvature of this lens. (b) If an object 16 cm tall is placed 30.0 cm from the eye lens, where would the lens focus it and how tall would the image be? Is this image real or virtual? Is it erect or inverted? (*Note:* The results obtained here are not strictly accurate, because the lens is embedded in fluids having refractive indexes different from that of air.)

40. • **The cornea as a simple lens.** The cornea behaves as a
BIO thin lens of focal length approximately 1.8 cm, although this varies a bit. The material of which it is made has an index of refraction of 1.38, and its front surface is convex, with a radius of curvature of 5.0 mm. (a) If this focal length is in air, what is the radius of curvature of the back side of the cornea? (b) The closest distance at which a typical person can focus on an object (called the near point) is about 25 cm,

although this varies considerably with age. Where would the cornea focus the image of an 8.0-mm-tall object at the near point? (c) What is the height of the image in part (b)? Is this image real or virtual? Is it erect or inverted? (*Note:* The results obtained here are not strictly accurate, because, on one side, the cornea has a fluid with a refractive index different from that of air.)

41. • An insect 3.75 mm tall is placed 22.5 cm to the left of a thin planoconvex lens. The left surface of this lens is flat, the right surface has a radius of curvature of magnitude 13.0 cm, and the index of refraction of the lens material is 1.70. (a) Calculate the location and size of the image this lens forms of the insect. Is it real or virtual? erect or inverted? (b) Repeat part (a) if the lens is reversed.

42. •• A double-convex thin lens has surfaces with equal radii of curvature of magnitude 2.50 cm. Looking through this lens, you observe that it forms an image of a very distant tree, at a distance of 1.87 cm from the lens. What is the index of refraction of the lens?

43. •• A converging meniscus lens (see Fig. 24.31) with a refractive index of 1.52 has spherical surfaces whose radii are 7.00 cm and 4.00 cm. What is the position of the image if an object is placed 24.0 cm to the left of the lens? What is the magnification?

44. •• A converging lens with a focal length of 12.0 cm forms a virtual image 8.00 mm tall, 17.0 cm to the right of the lens. Determine the position and size of the object. Is the image erect or inverted? Are the object and image on the same side or opposite sides of the lens?

45. •• **Combination of lenses, I.** When two lenses are used in combination, the first one forms an image that then serves as the object for the second lens. The magnification of the combination is the ratio of the height of the final image to the height of the object. A 1.20-cm-tall object is 50.0 cm to the left of a converging lens of focal length 40.0 cm. A second converging lens, this one having a focal length of 60.0 cm, is located 300.0 cm to the right of the first lens along the same optic axis. (a) Find the location and height of the image (call it I_1) formed by the lens with a focal length of 40.0 cm. (b) I_1 is now the object for the second lens. Find the location and height of the image produced by the second lens. This is the final image produced by the combination of lenses.

46. •• (a) You want to use a lens with a focal length of 35.0 cm to produce a real image of an object, with the image twice as long as the object itself. What kind of lens do you need, and where should the object be placed? (b) Suppose you want a virtual image of the same object, with the same magnification—what kind of lens do you need, and where should the object be placed?

47. •• **Combination of lenses, II.** Two thin lenses with a focal length of magnitude 12.0 cm, the first diverging and the second converging, are located 9.00 cm apart. An object 2.50 mm tall is placed 20.0 cm to the left of the first (diverging) lens. (a) How far from this first lens is the final image formed? (b) Is the final image real or virtual? (c) What is the height of the final image? Is it erect or inverted? (*Hint:* See Problem 45.)

48. •• A lens forms a real image, which is 214 cm away from the object and 1⅗ times its length. What kind of lens is this, and what is its focal length?

24.6 Graphical Methods for Lenses

49. • A converging lens has a focal length of 14.0 cm. For each of two objects located to the left of the lens, one at a distance of 18.0 cm and the other at a distance of 7.00 cm, determine (a) the image position, (b) the magnification, (c) whether the image is real or virtual, and (d) whether the image is erect or inverted. Draw a principal-ray diagram in each case.

50. • A converging lens forms an image of an 8.00-mm-tall real object. The image is 12.0 cm to the left of the lens, 3.40 cm tall, and erect. (a) What is the focal length of the lens? (b) Where is the object located? (c) Draw a principal-ray diagram for this situation.

51. • A diverging lens with a focal length of −48.0 cm forms a virtual image 8.00 mm tall, 17.0 cm to the right of the lens. (a) Determine the position and size of the object. Is the image erect or inverted? Are the object and image on the same side or opposite sides of the lens? (b) Draw a principal-ray diagram for this situation.

52. • When an object is 16.0 cm from a lens, an image is formed 12.0 cm from the lens on the same side as the object. (a) What is the focal length of the lens? Is the lens converging or diverging? (b) If the object is 8.50 mm tall, how tall is the image? Is it erect or inverted? (c) Draw a principal-ray diagram.

53. •• Figure 24.50 shows a small plant near a thin lens. The ray shown is one of the principal rays for the lens. Each square is 2.0 cm along the horizontal direction, but the vertical direction is not to the same scale. Use information from the diagram to answer the following questions: (a) Using only the ray shown, decide what type of lens (converging or diverging) this is. (b) What is the focal length of the lens? (c) Locate the image by drawing the other two principal rays. (d) Calculate where the image should be, and compare this result with the graphical solution in part (c).

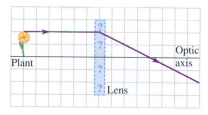

▲ **FIGURE 24.50** Problem 53.

54. •• Figure 24.51 shows a small plant near a thin lens. The ray shown is one of the principal rays for the lens. Each square is 2.0 cm along the horizontal direction, but the vertical direction is not to the same scale. Use information from the diagram to answer the following questions: (a) Using only the ray shown, decide what type of lens (converging or diverging) this is. (b) What is the focal length of the lens? (c) Locate the image by drawing the other two principal rays. (d) Calculate where the image should be, and compare this result with the graphical solution in part (c).

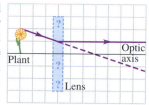

▲ **FIGURE 24.51** Problem 54.

55. •• Figure 24.52 shows a small plant near a thin lens. The ray shown is one of the principal rays for the lens. Each square is 2.0 cm along the horizontal direction, but the vertical direction is not to the same scale. Use information from the diagram to answer the following questions: (a) Using only the ray shown, decide what type of lens (converging or diverging) this is. (b) What is the focal length of the lens? (c) Locate the image by drawing the other two principal rays. (d) Calculate where the image should be, and compare this result with the graphical solution in part (c).

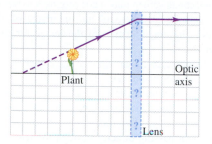

▲ **FIGURE 24.52** Problem 55.

56. •• Figure 24.53 shows a small plant near a thin lens. The ray shown is one of the principal rays for the lens. Each square is 2.0 cm along the horizontal direction, but the vertical direction is not to the same scale. Use information from the diagram to answer the following questions: (a) Using only the ray shown, decide what type of lens (converging or diverging) this is. (b) What is the focal length of the lens? (c) Locate the image by drawing the other two principal rays. (d) Calculate where the image should be, and compare this result with the graphical solution in part (c).

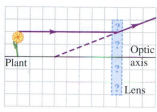

▲ **FIGURE 24.53** Problem 56.

General Problems

57. •• A layer of benzene ($n = 1.50$) 2.60 cm deep floats on water ($n = 1.33$) that is 6.50 cm deep. What is the apparent distance from the upper benzene surface to the bottom of the water layer when it is viewed at normal incidence?

58. •• Where must you place an object in front of a concave mirror with radius R so that the image is erect and $2\frac{1}{2}$ times the size of the object? Where is the image?

59. •• A luminous object is 4.00 m from a wall. You are to use a concave mirror to project an image of the object on the wall, with the image 2.25 times the size of the object. How far should the mirror be from the wall? What should its radius of curvature be?

60. •• A concave mirror is to form an image of the filament of a headlight lamp on a screen 8.00 m from the mirror. The filament is 6.00 mm tall, and the image is to be 36.0 cm tall. (a) How far in front of the vertex of the mirror should the filament be placed? (b) To what radius of curvature should you grind the mirror?

61. •• A plastic lens ($n = 1.67$) has one convex surface of radius 12.2 cm and one concave surface of radius 15.4 cm. If an object is placed 35.0 cm from the lens, (a) where is the image located and (b) what is its magnification?

62. • A 3.80-mm-tall object is 24.0 cm from the center of a silvered spherical glass Christmas tree ornament 6.00 cm in diameter. What are the position and height of its image?

63. •• A lensmaker wants to make a magnifying glass from glass with $n = 1.55$ and with a focal length of 20.0 cm. If the two surfaces of the lens are to have equal radii, what should that radius be?

64. •• An object is placed 18.0 cm from a screen. (a) At what two points between object and screen may a converging lens with a 3.00 cm focal length be placed to obtain an image on the screen? (b) What is the magnification of the image for each position of the lens?

65. •• As shown in Figure 24.54, a candle is at the center of curvature of a concave mirror whose focal length is 10.0 cm. The converging lens has a focal length of 32.0 cm and is 85.0 cm to the right of the candle. The candle is viewed through the lens from the right. The lens forms two images of the candle. The first is formed by light passing directly through the lens. The second image is formed from the light that goes from the candle to the mirror, is reflected, and then passes through the lens. (a) For each of these two images, draw a principal-ray diagram that locates the image. (b) For each image, answer the following questions: (i) Where is the image? (ii) Is the image real or virtual? (iii) Is the image erect or inverted with respect to the original object?

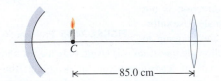

▲ **FIGURE 24.54** Problem 65.

66. •• In the text, Equations (24.4) and (24.7) were derived for the case of a concave mirror. Give a similar derivation for a convex mirror, and show that the same equations result if you use the sign convention established in the text.

67. •• **A lens in a liquid.** A lens obeys Snell's law, bending light rays at each surface an amount determined by the index of refraction of the lens and the index of the medium in which the lens is located. (a) Equation (24.20) assumes that the lens is surrounded by air. Consider instead a thin lens immersed in a liquid with refractive index n_{liq}. Prove that the focal length f' is then given by Eq. (24.20), with n replaced by n/n_{liq}. (b) A thin lens with index n has focal length f in vacuum. Use the result of part (a) to show that when this lens is immersed in a liquid of index n_{liq}, it will have a new focal length given by

$$f' = \left[\frac{n_{\text{liq}}(n - 1)}{n - n_{\text{liq}}} \right] f.$$

Passage Problems

Refraction of liquids. The focal length of a mirror can be determined entirely from the shape of the mirror. In contrast, to determine the focal length of a lens we must know both the shape of the lens and its index of refraction—and the index of refraction of the surrounding medium. For instance, when a thin lens is immersed in a liquid we must modify the thin-lens equation to take into account the refractive properties of the surrounding liquid:

$$\frac{1}{f} = \left(\frac{n}{n_{\text{liq}}} - 1 \right) \left(\frac{1}{R_1} - \frac{1}{R_2} \right)$$

where n_{liq} is the index of refraction of the liquid and n is the index of refraction of the glass.

68. If you place a glass lens ($n = 1.5$), which has a focal length of 0.5 meters in air, into a tank of water ($n = 1.33$), what will happen to its focal length?
   A. Nothing will happen.
   B. The focal length of the lens will be reduced.
   C. The focal length of the lens will be increased.
   D. There is not enough information to answer the question.

69. If you place a concave glass lens into a tank of a liquid that has an index of refraction that is greater than that of the lens, what will happen?
   A. The lens will no longer be able to create any images.
   B. The focal length of the lens will become longer.
   C. The focal length of the lens will become shorter.
   D. The lens will become a converging lens.

70. If you place a concave mirror with a focal length of 1 m into a liquid that has an index of refraction of 3, what will happen?
   A. The mirror will no longer be able to focus light.
   B. The focal length of the mirror will decrease.
   C. The focal length of the mirror will increase.
   D. Nothing will happen.

25 Optical Instruments

How does a camera resemble the human eye? What are the significant differences? What does a projector operator have to do to "focus" the picture on the screen? Why do large telescopes usually include curved mirrors as well as lenses? This chapter is concerned with these and other questions having to do with a number of familiar optical systems, some of which use several lenses or combinations of lenses and mirrors. In answering such questions, we can apply the basic principles of mirror and lens behavior that we studied in Chapter 24.

The concept of *image* plays an important role in the analysis of optical instruments. We continue to base our analysis on the *ray* model of light, so the content of this chapter comes under the general heading *geometric optics.*

The eyes and the camera viewing this landscape form similar images by similar means. Both have lenses that focus light on a recording surface (retina or CCD array). Both can adjust their focus for near or distant objects. In this chapter we'll study these and similar optical devices.

25.1 The Camera

The essential elements of a **camera** are a lens equipped with a shutter, a light-tight enclosure, and a light-sensitive film (or an electronic sensor, which we'll discuss later) that records an image (Figure 25.1). The lens forms an inverted, usually reduced, real image, in the plane of the film or sensor, of the object being photographed. To provide proper image distances for various object distances, the lens is moved closer to or farther from the film or sensor, often by being turned in a threaded mount. Most lenses have several elements, permitting partial correction of various *aberrations,* including the dependence of index of refraction on wavelength and the limitations imposed by the paraxial approximation. (We'll discuss lens aberrations in Section 25.7).

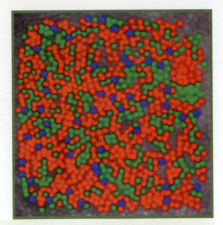

As shown in Figure 25.7, the lens of the human eye focuses light on the retina, which comprises an array of photoreceptor cells, each of which generates an electrical signal when it absorbs light. The eyes of primates contain three types of cone cells responsible for color vision, each containing pigment molecules tuned to absorb different portions of the visible spectrum. The protein portion of the pigment is different in the three types and determines the peak wavelength response of each cell. The three molecular types have peak sensitivities in the red, green, and blue portions of the visible spectrum, but their sensitivities overlap considerably. Our perception of a color depends on the ratio of activation among the three, which explains why two pigments can be mixed to provide an impression of a color that is quite different from either color of the pigments. The photograph shows an actual image of a human retina in which the three cone types have been identified and artificially colored red, green, and blue to show the distribution of the three types.

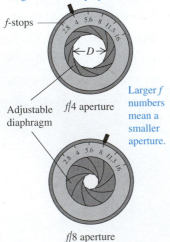

Changing the diameter by a factor of 2 changes the intensity by a factor of 4.

f-stops

Adjustable diaphragm $f/4$ aperture

Larger f numbers mean a smaller aperture.

$f/8$ aperture

▲ **FIGURE 25.2** The diaphragm that controls a camera's aperture.

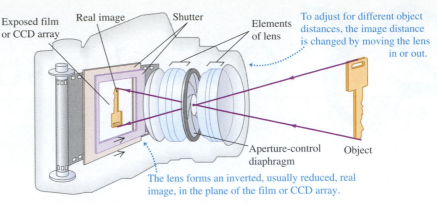

The lens forms an inverted, usually reduced, real image, in the plane of the film or CCD array.

▲ **FIGURE 25.1** Camera with aperture control.

In order for the film or sensor to record the image properly, the total light energy per unit area reaching the film (the exposure) must fall within certain limits. This is controlled by the *shutter* and the *lens aperture*. The shutter controls the time interval during which light enters the lens. The interval is usually adjustable in steps corresponding to factors of about two, often from 1 s to $\frac{1}{1000}$ s.

The intensity of light reaching the film is proportional to the effective area of the lens, which may be varied by means of an adjustable aperture, or *diaphragm*—a nearly circular hole with variable diameter D. The aperture size is usually described as a fraction of the focal length f of the lens. A lens with a focal length $f = 50$ mm and an aperture diameter of 25 mm has an aperture of $f/2$; many photographers would call this an f-number of 2, or simply "$f/2$." In general,

$$f\text{-number} = \frac{\text{Focal length}}{\text{Aperture diameter}} = \frac{f}{D}. \qquad (25.1)$$

Because the light intensity at the film or sensor is proportional to the area of the lens aperture and thus to the *square* of its diameter, changing the diameter by a factor of $\sqrt{2}$ changes the intensity by a factor of 2. Adjustable apertures (often called *f-stops*) usually have scales labeled with successive numbers related by factors of approximately $\sqrt{2}$, such as $f/2$, $f/2.8$, $f/4$, $f/5.6$, $f/8$, $f/11$, $f/16$, $f/22$, and so on. The larger numbers represent smaller apertures and intensities, and each step corresponds to a factor of 2 in intensity (Figure 25.2). The actual exposure is proportional to both the aperture area and the time of exposure. Thus, $\left(f/4 \text{ and } \frac{1}{500} \text{ s}\right)$, $\left(f/5.6 \text{ and } \frac{1}{250} \text{ s}\right)$, and $\left(f/8 \text{ and } \frac{1}{125} \text{ s}\right)$ all correspond to the same exposure.

The choice of focal length for a camera lens depends on the film size and the desired angle of view, or *field*. The popular 35 mm camera has an image size of 24×36 mm on the film. The normal lens for such a camera usually has a focal length of about 50 mm, permitting an angle of view of about 45°. A lens with a longer focal length, often 135 mm or 200 mm, provides a *smaller* angle of view and a larger image of *part* of the object, compared with a normal lens. This gives the impression that the camera is *closer* to the object than it really is, and such a lens is called a *telephoto lens*. A lens with a shorter focal length, such as 35 mm or 28 mm, permits a wider angle of view and is called a *wide-angle lens*. An extreme wide-angle, or "fish-eye," lens may have a focal length as small as 6 mm. Figure 25.3 shows a scene photographed from the same point with lenses of various focal lengths.

(a) $f = 28$ mm

(b) $f = 105$ mm

(c) $f = 300$ mm

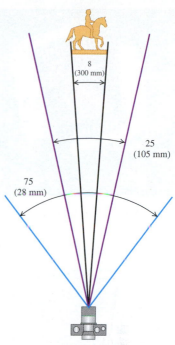

(d) The angles of view for the photos in (a)–(c)

▲ **FIGURE 25.3** As f increases, the image size increases proportionally.

EXAMPLE 25.1 Photographic exposures

A common telephoto lens for a 35 mm camera has a focal length of 200 mm and a range of f-stops from $f/5.6$ to $f/45$. **(a)** What is the corresponding range of aperture diameters? **(b)** What is the corresponding range of intensities of the film image?

SOLUTION

SET UP AND SOLVE Part (a): From Equation 25.1, the diameters range from

$$D = \frac{f}{f\text{-number}} = \frac{200 \text{ mm}}{5.6} = 36 \text{ mm}$$

to

$$D = \frac{200 \text{ mm}}{45} = 4.4 \text{ mm}.$$

Part (b): Because the intensity is proportional to the square of the diameter, the ratio of the intensity at $f/5.6$ to that at $f/45$ is approximately

$$\left(\frac{36 \text{ mm}}{4.4 \text{ mm}}\right)^2 \cong \left(\frac{45}{5.6}\right)^2 \cong 65 \quad \text{(about } 2^6\text{)}.$$

REFLECT If the correct exposure time at $f/5.6$ is $(1/1000 \text{ s})$, then at $f/45$ it is $(65)(1/1000 \text{ s}) = 1/15 \text{ s}$.

Practice Problem: If the correct exposure at $f/5.6$ is $1/1000$ s, what is the correct exposure at $f/11$? At $f/16$? *Answers: 1/250 s, 1/125 s.*

FIGURE 25.4 A CCD chip on a circuit board.

The Digital Camera

In a digital camera, the light-sensitive film is replaced by an array of tiny photo-cells fabricated on a semiconductor chip called a charge-coupled device (CCD) chip. This device covers the image plane, dividing it into many rectangular areas called *pixels*. The total number of pixels is typically 2 to 10 million; a 2 megapixel camera has about 2 million pixels. The intensity of light of each primary color striking each pixel is recorded and stored digitally. Once stored, the image can be processed digitally to crop or enlarge it, change the color balance, and so on. Figure 25.4 shows a CCD chip.

The exposure time is controlled electronically, so a mechanical shutter is not needed. The dimensions of the CCD chip can be made smaller than the image size in a typical 35 mm camera by a factor on the order of $\frac{1}{5}$ in each dimension. The focal length of the lens can be correspondingly smaller, so the entire camera can be made much more compact than the usual 35 mm camera. This rapidly developing technology is used in applications such as cell-phone cameras.

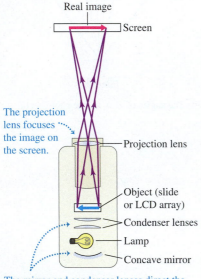

The projection lens focuses the image on the screen.

Real image

Screen

Projection lens

Object (slide or LCD array)

Condenser lenses

Lamp

Concave mirror

The mirror and condenser lenses direct the lamp's light onto the slide or LCD array.

FIGURE 25.5 The optical components of a slide projector.

25.2 The Projector

A **projector** for viewing slides or motion pictures operates very much like a camera in reverse. The essential elements are shown in Figure 25.5. Light from the source (an incandescent bulb or, in large motion-picture projectors, a carbon-arc lamp) shines through the film, and the projection lens forms a real, enlarged, inverted image of the film on the projection screen. Additional lenses called *condenser lenses* are placed between lamp and film. Their function is to direct the light from the source so that most of it enters the projection lens after passing through the film. A concave mirror behind the lamp also helps direct the light. The condenser lenses must be large enough to cover the entire area of the film. The image on the screen is always real and inverted; this is why slides have to be put into a projector upside down.

The position and size of the image projected on the screen are determined by the position and focal length of the projection lens.

EXAMPLE 25.2   A slide projector

An ordinary 35 mm color slide has a picture area of 24 × 36 mm. What focal length would a projection lens need in order to project a 1.2 m × 1.8 m image of this picture on a screen 5.0 m from the lens?

SOLUTION

SET UP Figure 25.6 diagrams the situation.

SOLVE We use the thin-lens analysis from Section 24.5. We need a lateral magnification (disregarding the sign) of

$$|m| = \frac{y'}{y} = \frac{1.2 \text{ m}}{24 \times 10^{-3} \text{ m}} = 50.$$

From Equation 24.18,

$$m = \frac{y'}{y} = -\frac{s'}{s},$$

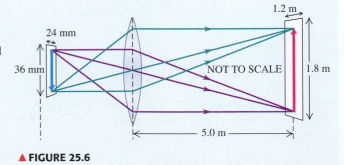

FIGURE 25.6

Continued

so the ratio s'/s must also be 50. (The image is real, so s' is positive.) We are given that $s' = 5.0$ m. Thus,

$$s = \frac{5.0 \text{ m}}{50} = 0.10 \text{ m.}$$

The object and image distances are related by Equation 24.17, which we solve for the focal length f:

$$\frac{1}{f} = \frac{1}{s} + \frac{1}{s'} = \frac{1}{0.10 \text{ m}} + \frac{1}{5.0 \text{ m}},$$
$$f = 0.098 \text{ m} = 98 \text{ mm.}$$

REFLECT A popular focal length for home slide projectors is 100 mm; such lenses are readily available and would be the appropriate choice in this situation.

Practice Problem: For this same projector, what size image would appear on a screen 1.5 m away? *Answer:* 34 cm × 51 cm.

▶ **BIO** Application **The original digital image?** Insects view the world through compound eyes, which are arrays of individual light-sensing units packed together over the surface of the eye. (Honeybees, with fairly good vision, have about 7000 units; dragonflies have about 30,000.) Each unit records the intensity and color of the light entering it, producing a mosaic image. The modern digital camera uses a similar approach: A CCD is an array of individual photoreceptors (pixels) that record intensity at specific colors. However, with several million pixels, typical digital cameras now have better resolution than any insect. Unlike the honeybee photoreceptors, which respond to yellow, blue, and ultraviolet light, the pixels in a digital camera record red, green, and blue light, creating a color image that looks natural to our eyes.

Digital Projectors

Digital projectors (also called *data projectors*) use the same optical principles as the projectors just described, but the film or slide is replaced by a rectangular array of pixels consisting of liquid-crystal diodes (LCDs). Each diode can be made transparent or opaque by imposing appropriate electrical signals on it, and the resulting image is projected onto a screen. This scheme is widely used in computer monitors and TV display systems.

25.3 The Eye

The optical behavior of the eye is similar to that of a camera. The essential parts of the human eye, considered as an optical system, are shown in Figure 25.7. The eye is nearly spherical in shape and about 2.5 cm in diameter. The front portion is somewhat more sharply curved and is covered by a tough, transparent membrane called the *cornea*. The region behind the cornea contains a liquid called the

MasteringPHYSICS

PhET: Color Vision

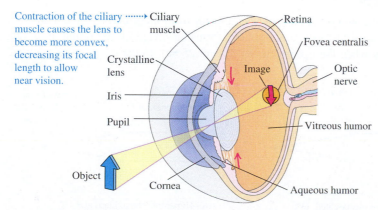

(a) Diagram of the eye

▲ **FIGURE 25.7** The human eye.

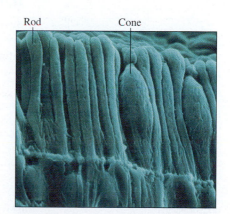

(b) Scanning electron micrograph showing retinal rods and cones in cross section

▲ **BIO** Application **Focus-o-rama.** The crystalline lens and ciliary muscle found in humans and other mammals is only one of a number of focusing mechanisms used by animals. Birds, for instance, can change the shape not only of their lens, but also of the corneal surface. In aquatic animals, the corneal surface is not very useful for focusing, because its refractive index is close to that of water. Thus, focusing is accomplished entirely by the lens, which is nearly spherical. Fishes change their focal length by using a muscle to move the lens either inward or outward. Whales and dolphins achieve the same effect, but hydraulically: They fill or empty a fluid chamber behind the lens to move the lens in or out. In the compound eyes of insects, each unit has its own lens, but adaptive focusing is not needed.

TABLE 25.1 Receding of near point with age

Age (years)	Near point (cm)
10	7
20	10
30	14
40	22
50	40
60	200

aqueous humor. Next comes the *crystalline lens,* a capsule containing a fibrous jelly, hard at the center and progressively softer at the outer portions. The crystalline lens is held in place by ligaments that attach it to the ciliary muscle, which encircles it. Behind the lens, the eye is filled with a thin, watery jelly called the *vitreous humor.* The indexes of refraction of both the aqueous humor and the vitreous humor are nearly equal to that of water, about 1.336. The crystalline lens, although not homogeneous, has an average index of refraction of 1.437. This is not very different from the indexes of the aqueous and vitreous humors; most of the bending of light rays entering the eye occurs at the outer surface of the cornea.

Refraction at the cornea and the surfaces of the lens produces a *real image* of the object being viewed; the image is formed on the light-sensitive *retina* that lines the rear inner surface of the eye. The retina plays the same role as the CCD in a digital camera. The *rods* and *cones* in the retina act like an array of photocells, sensing the image and transmitting it via the *optic nerve* to the brain. Vision is most acute in a small central region called the *fovea centralis,* about 0.25 mm in diameter.

In front of the lens is the *iris.* This structure contains an aperture with variable diameter called the *pupil,* which opens and closes to adapt to changing light intensity. The receptors of the retina also change their sensitivity in response to light intensity (as when your eyes adapt to darkness).

For an object to be seen sharply, the image must be formed exactly on the retina. The lens-to-retina distance, corresponding to s', does not change, but the eye accommodates to different object distances s by changing the focal length of its lens. When the ciliary muscle surrounding the lens contracts, the lens bulges and the radii of curvature of its surfaces *decrease* thereby decreasing the focal length. For the normal eye, an object at infinity is sharply focused when the ciliary muscle is relaxed. With increasing tension, the focal length decreases to permit sharp imaging on the retina of closer objects. This process is called *accommodation.*

The extremes of the range over which distinct vision is possible are known as the *far point* and the *near point* of the eye. The far point of a normal eye is at infinity. The position of the near point depends on the amount by which the ciliary muscle can increase the curvature of the crystalline lens. The range of accommodation gradually diminishes with age as the crystalline lens loses its flexibility. For this reason, the near point gradually recedes as one grows older. This recession of the near point is called *presbyopia;* it is the reason that people need reading glasses when they get older, even if their vision is good otherwise. Table 25.1 shows the approximate position of the near point for an average person at various ages. For example, an average person 50 years of age cannot focus on an object closer than about 40 cm.

Defects of Vision

Several common defects of vision result from incorrect distance relations in the eye. As we just saw, a normal eye forms an image on the retina of an object at infinity when the eye is relaxed (Figure 25.8a). In the *myopic* (nearsighted) eye,

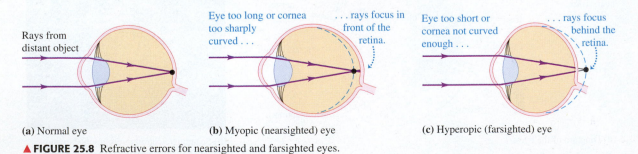

(a) Normal eye

(b) Myopic (nearsighted) eye

(c) Hyperopic (farsighted) eye

▲ **FIGURE 25.8** Refractive errors for nearsighted and farsighted eyes.

the eyeball is too long from front to back in comparison to the radius of curvature of the cornea (or the cornea is too sharply curved), and rays from an object at infinity are focused in front of the retina (Figure 25.8b). The most distant object for which an image can be formed on the retina is then nearer than infinity. The *hyperopic* (farsighted) eye has the opposite problem: The eyeball is too short or the cornea is not curved enough, and the image of an infinitely distant object is behind the retina (Figure 25.8c). The myopic eye produces *too much* convergence in a parallel bundle of rays for an image to be formed on the retina; the hyperopic eye produces *not enough* convergence.

Astigmatism refers to a defect in which the surface of the cornea is not spherical, but is more sharply curved in one plane than another. As a result, horizontal lines may be imaged in a different plane from vertical lines (Figure 25.9a). Astigmatism may make it impossible, for example, to focus clearly on the horizontal and vertical bars of a window at the same time.

All these defects can be corrected by the use of corrective lenses (eyeglasses or contact lenses) or, in recent years, by refractive surgery in which the cornea itself is reshaped. The near point of either a presbyopic or a hyperopic eye is *farther* from the eye than normal. To see an object clearly at normal reading distance (often assumed to be 25 cm), such an eye needs an eyeglass lens that forms a virtual image of the object at or beyond the near point. This can be accomplished by a converging (positive) lens, as shown in Figure 25.10. In effect, the lens moves the object farther away from the eye, to a point where a sharp retinal image can be formed. Similarly, eyeglasses for myopic eyes use diverging (negative) lenses to move the image closer to the eye than the actual object, as shown in Figure 25.11.

Astigmatism is corrected by use of a lens with a *cylindrical* surface. For example, suppose the curvature of the cornea in a horizontal plane is correct for focusing rays from infinity on the retina, but the curvature in the vertical plane is not great enough to form a sharp retinal image. Then, when a cylindrical lens with its axis horizontal is placed before the eye, the rays in a horizontal plane are unaffected, but the additional divergence of the rays in a vertical plane causes these to be sharply imaged on the retina, as shown in Figure 25.9b.

Lenses for correcting vision are usually described in terms of the **power,** defined as the reciprocal of the focal length, expressed in meters. The unit of power is the **diopter.** Thus, a lens with $f = 0.50$ m has a power of 2.0 diopters, $f = -0.25$ m corresponds to -4.0 diopters, and so on. The numbers on a prescription for glasses are usually powers expressed in diopters. When the

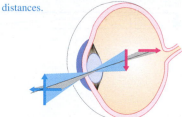

Shape of eyeball or lens causes vertical and horizontal elements to focus at different distances.

(a) Vertical lines are imaged in front of the retina.

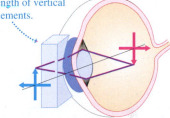

This cylindrical lens is curved in the vertical, but not the horizontal, direction; it changes the focal length of vertical elements.

(b) A cylindrical lens corrects for astigmatism.

▲ **FIGURE 25.9** (a) An uncorrected astigmatic eye. (b) Correction of the astigmatism by a cylindrical lens.

▲ **Application Near- or farsighted?** Can you tell by looking at these eyeglasses whether their owner is myopic (nearsighted) or hyperopic (farsighted)? Yes, easily: Just notice whether the glasses seem to magnify or shrink the person's face. Nearsighted people need eyeglasses with diverging lenses, which make their eyes look smaller from the outside; farsighted people wear glasses with converging lenses, which make their eyes look larger. This man is nearsighted.

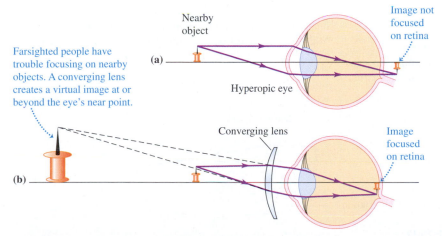

Farsighted people have trouble focusing on nearby objects. A converging lens creates a virtual image at or beyond the eye's near point.

(a) Nearby object — Hyperopic eye — Image not focused on retina

(b) Converging lens — Image focused on retina

▲ **FIGURE 25.10** (a) An uncorrected farsighted (hyperopic) eye. (b) A positive lens gives the extra convergence needed to focus the image on the retina.

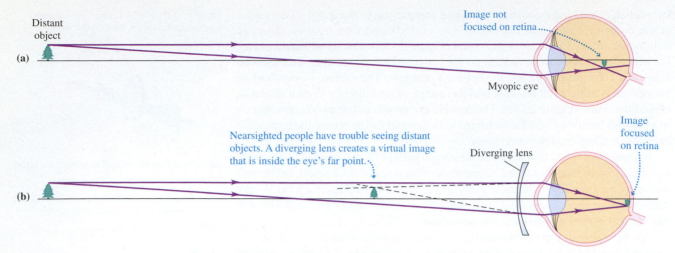

Distant object

Image not focused on retina

(a)

Myopic eye

Nearsighted people have trouble seeing distant objects. A diverging lens creates a virtual image that is inside the eye's far point.

Image focused on retina

Diverging lens

(b)

▲ **FIGURE 25.11** (a) An uncorrected nearsighted (myopic) eye. (b) A negative lens spreads the rays farther apart to compensate for the eye's excessive convergence.

correction involves both astigmatism and myopia or hyperopia, there are three numbers for each lens: one for the spherical power, one for the cylindrical power, and an angle to describe the orientation of the cylinder axis.

NOTE ▶ The use of the term *power* in reference to lenses is unfortunate; it has nothing to do with the familiar meaning of "energy per unit time" in various other areas of physics. The unit of power of a lens is $(1/\text{meter})$, *not* $(1 \text{ joule}/\text{second})$. ◀

With ordinary eyeglasses, the corrective lens is typically 1 to 2 cm from the eye. In the examples that follow, we'll ignore this small distance and assume that the corrective lens and the eye coincide.

EXAMPLE 25.3 **Correcting for farsightedness**

The near point of a certain hyperopic eye is 100 cm in front of the eye. What lens should be used to enable the eye to see clearly an object that is 25 cm in front of the eye? (Neglect the small distance from the lens to the eye.)

SOLUTION

SET UP Figure 25.12 shows the relevant diagram; it is essentially the same as Figure 25.10b, except that it emphasizes the role of the corrective lens. We want the lens to form a virtual image of the object at a location corresponding to the near point of the eye, 100 cm from the lens. That is, when $s = 25$ cm, we want s' to be -100 cm.

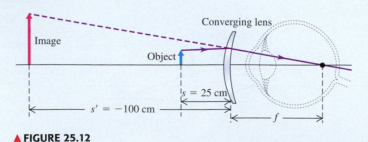

Image

Converging lens

Object

$s = 25$ cm

$s' = -100$ cm

f

▲ **FIGURE 25.12**

Continued

SOLVE From the basic thin-lens equation (Equation 24.17),

$$\frac{1}{f} = \frac{1}{s} + \frac{1}{s'} = \frac{1}{+25 \text{ cm}} + \frac{1}{-100 \text{ cm}},$$
$$f = +33 \text{ cm}.$$

REFLECT We need a converging lens with focal length $f = 33$ cm. The corresponding power is $1/(0.33 \text{ m})$, or $+3.0$ diopters.

Practice Problem: A 60-year-old man has a near point of 200 cm. What lens should he have to see clearly an object that is 25 cm in front of the lens? *Answer:* $f = 29$ cm (about 3.5 diopters).

EXAMPLE 25.4 **Correcting for nearsightedness**

The far point of a certain myopic eye is 50 cm in front of the eye. What lens should be used to focus sharply an object at infinity? (Assume that the distance from the lens to the eye is negligible.)

SOLUTION

SET UP Figure 25.13 shows the relevant diagram. The far point of a *myopic* eye is nearer than infinity. To see clearly objects that are beyond the far point, such an eye needs a lens that forms a virtual image of the object no farther from the eye than the far point. We assume that the virtual image is formed at the far point. Then, when $s = \infty$, we want s' to be -50 cm.

SOLVE From the basic thin-lens equation,

$$\frac{1}{f} = \frac{1}{s} + \frac{1}{s'} = \frac{1}{\infty} + \frac{1}{-50 \text{ cm}},$$
$$f = -50 \text{ cm}.$$

REFLECT Because all rays originating at an object distance of infinity are parallel to the axis of the lens, the image distance s' equals the focal length f. We need a *diverging* lens with focal length $f = -50$ cm $= -0.50$ m. The power is $-1/(0.50 \text{ m}) = -2.0$ diopters.

Practice Problem: If the far point of an eye is 75 cm, what lens should be used to see clearly an object that is at infinity? *Answer:* $f = -75$ cm (about -1.3 diopters).

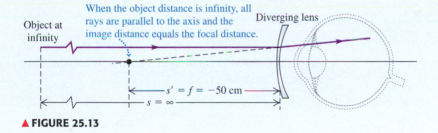

When the object distance is infinity, all rays are parallel to the axis and the image distance equals the focal distance.

Object at infinity

Diverging lens

$s' = f = -50 \text{ cm}$

$s = \infty$

▲ **FIGURE 25.13**

25.4 The Magnifier

The apparent size of an object is determined by the size of its image on the retina. If the eye is unaided, this size depends upon the *angle θ* subtended by the object at the eye, called its **angular size** (Figure 25.14a).

To look closely at a small object, such as an insect or a crystal, you bring it close to your eye, making the subtended angle and the retinal image as large as possible. But your eye cannot focus sharply on objects closer than the near point, so the angular size of an object is greatest (subtends the largest possible viewing angle) when it is placed at the near point. In the discussion that follows, we'll assume an average viewer with a near point 25 cm from the eye.

A converging lens can be used to form a virtual image that is larger and farther from the eye than the object itself, as shown in Figure 25.14. Then the object can be moved closer to the eye, and the angular size of the image may be substantially larger than the angular size of the object at 25 cm without the lens. A

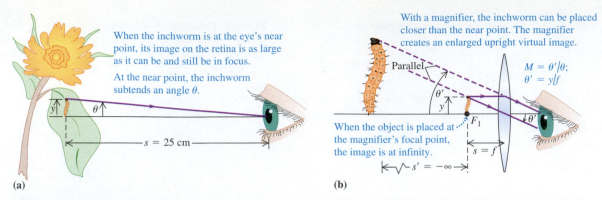

When the inchworm is at the eye's near point, its image on the retina is as large as it can be and still be in focus.

At the near point, the inchworm subtends an angle θ.

$s = 25$ cm

(a)

With a magnifier, the inchworm can be placed closer than the near point. The magnifier creates an enlarged upright virtual image.

Parallel

$M = \theta'/\theta$;
$\theta' = y/f$

When the object is placed at the magnifier's focal point, the image is at infinity.

F_1

$s = f$

$s' = -\infty$

(b)

▲ **FIGURE 25.14** (a) The subtended angle θ is largest when an object is at the near point. A simple magnifier produces a virtual image at infinity, which acts as a real object subtending a larger angle θ' for the eye.

lens used in this way is called a *magnifying glass,* or simply a **magnifier.** The virtual image is most comfortable to view when it is placed at infinity, and in the discussion that follows we'll assume that this is done.

In Figure 25.14a, the object is at the near point, where it subtends an angle θ at the eye. In Figure 25.14b, a magnifier in front of the eye forms an image at infinity, and the angle subtended at the magnifier is θ'. We define the **angular magnification** M as follows:

> **Definition of angular magnification**
>
> Angular magnification M is the ratio of the angle θ' subtended by an object at the eye when the magnifier is used to the angle θ subtended without the magnifier:
>
> $$M = \frac{\theta'}{\theta}. \tag{25.2}$$

To find the value of M, we first assume that the angles are small enough that each angle (in radians) is equal to its sine and its tangent. From Figure 25.14, θ and θ' are given (in radians) by

$$\theta = \frac{y}{25 \text{ cm}}, \qquad \theta' = \frac{y}{f}.$$

Combining these expressions with Equation 25.2, we find that

$$M = \frac{\theta'}{\theta} = \frac{y/f}{y/25 \text{ cm}} = \frac{25 \text{ cm}}{f}. \tag{25.3}$$

NOTE ▶ Be careful not to confuse *angular magnification M* (a ratio of two angles) with *lateral magnification m* (the ratio of image to object height, which we defined in Chapter 24). In some of the examples we'll discuss later, M is the more relevant quantity. Be on the lookout for this distinction, and make sure you understand why one or the other is relevant in a specific situation. ◀

It may seem that we can make the angular magnification as large as we like by decreasing the focal length f. However, the aberrations of a simple double-convex lens (to be discussed in Section 25.7) set an upper limit on M of about $3\times$ to $4\times$. If these aberrations are corrected, the angular magnification may be made as great as $20\times$. When greater magnification than this is needed, we usually use a compound microscope, discussed in the next section.

Comparing magnifying lenses

A simple magnifier with a shorter focal length gives a greater angular magnification than one with a longer focal length primarily because

A. the image is the same angular size for both lenses, but the image can be seen clearly closer to the eye with the lens having smaller focal length;
B. The angular image size is larger for the lens having smaller focal length;
C. The object can be held farther from the eye and still be seen clearly;
D. the magnification depends, *not* on the focal length, but on the location of your eye relative to the lens.

SOLUTION The purpose of any magnifying device is to create an image with larger angular size than is possible with the unaided eye, in order to produce the largest possible image on the retina. The minimum eye-to-object distance for the normal unaided eye is about 25 cm, so an object with height y has maximum angular size $y/25$ cm. When a converging lens with focal length f is placed closer to the eye than 25 cm, the lens forms a virtual image. If this image is at infinity, its angular size is y/f, as discussed earlier. The angular magnification is the quotient of these two angular sizes, which, from Equation 25.3, is $(25\text{ cm}/f)$. If the focal length is anything less than 25 cm, the angular size of the image is *greater than* the angular size of the object. The object must be at or inside the focal point to create a virtual image. Shortening the focal length means that the object must be closer, creating a larger image size and greater angular magnification. The correct answer is B.

EXAMPLE 25.5 A simple magnifier

You have two plastic lenses, one double convex and the other double concave, each with a focal length with absolute value 10.0 cm. **(a)** Which lens can you use as a simple magnifier? **(b)** What is the angular magnification?

SOLUTION

SET UP AND SOLVE Part (a): To act as a magnifier, a lens must produce an upright virtual image that is larger than the object. Figure 25.14 shows that a double-convex lens does this.

REFLECT A double-*concave* lens would produce an upright virtual image *smaller* than the object.

Part (b): From Equation 25.3, the angular magnification M is

$$M = \frac{25\text{ cm}}{f} = \frac{25\text{ cm}}{10\text{ cm}} = 2.5.$$

Practice Problem: A simple magnifier has a focal length of 14 cm. What is the angular magnification of this magnifier? *Answer: M = 1.8.*

25.5 The Microscope

When we need greater magnification than we can get with a simple magnifier, the instrument we usually use is the **microscope,** sometimes called a *compound microscope.* The essential elements of a microscope are shown in Figure 25.15a. To analyze this system, we use the principle that an image formed by one optical element, such as a lens or mirror, can serve as the object for a second element. We used this principle in Section 24.5 when we derived the thin-lens equation by repeated application of the single-surface refraction equation.

The object O to be viewed is placed just beyond the first focal point F_1 of the **objective,** a lens that forms a real, enlarged, inverted image I (Figure 25.15b). In a properly designed instrument, this image lies just inside the first focal point F_2 of the **eyepiece** (also called the *ocular*), which forms a final virtual image of I at I'. The position of I' may be anywhere between the near and far points of the eye. Both the objective and the eyepiece of an actual microscope are highly corrected compound lenses, but for simplicity, we show them here as simple thin lenses.

The notation in Figure 25.15b merits careful study. The objective lens has focal length f_1 and focal points F_1 and F_1'; the eyepiece lens has focal length f_2 and focal points F_2 and F_2'. The object and image distances for the objective lens are s_1 and s_1', respectively, and the object and image distances for the eyepiece lens are s_2 and s_2', respectively (not shown in the figure).

MasteringPHYSICS

ActivPhysics 15.12: Two-Lens Optical Systems

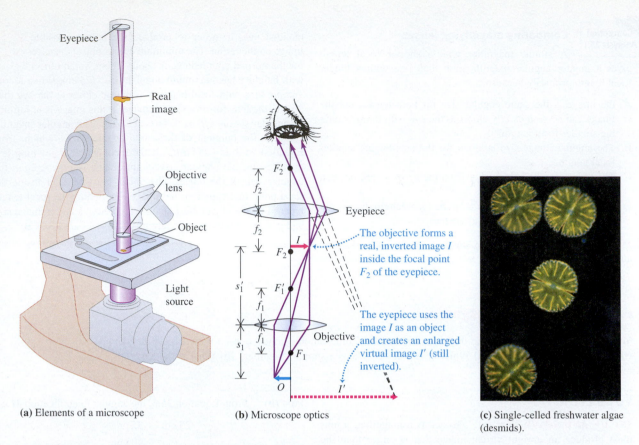

(a) Elements of a microscope

(b) Microscope optics

The objective forms a real, inverted image I inside the focal point F_2 of the eyepiece.

The eyepiece uses the image I as an object and creates an enlarged virtual image I' (still inverted).

(c) Single-celled freshwater algae (desmids).

▲ **FIGURE 25.15** A light microscope. Typically, light microscopes can resolve details as small as 200 nm, comparable to the wavelength of light.

As with a simple magnifier, the most significant quantity for a microscope is the *angular* magnification M. The objective forms an enlarged, real, inverted image that is viewed through the eyepiece. Thus, the overall angular magnification M of the compound microscope is the product of the *lateral* magnification m_1 of the objective and the *angular* magnification M_2 of the eyepiece. The first is given by

$$m_1 = -\frac{s_1'}{s_1},$$

where s_1 and s_1' are the object and image distances, respectively, for the objective. Ordinarily, the object is very close to the focal point, and the resulting image distance s_1' is very great in comparison to the focal length f_1 of the objective. Thus, s_1 is approximately equal to f_1, and $m_1 \simeq -s_1'/f_1$.

The eyepiece functions as a simple magnifier, as discussed in Section 25.4. From Equation 25.3, its angular magnification is $M_2 = (25 \text{ cm})/f_2$, where f_2 is the focal length of the eyepiece, considered as a simple lens. The overall magnification M of the compound microscope (apart from a negative sign, which is customarily ignored) is the product of the two magnifications; that is,

$$M = m_1 M_2 = \frac{(25 \text{ cm}) s_1'}{f_1 f_2}, \tag{25.4}$$

where s_1', f_1, and f_2 are measured in centimeters. The final image is inverted with respect to the object (Figure 25.15b). Microscope manufacturers usually specify the values of m_1 and M_2 for microscope components, rather than the focal lengths of the objective and eyepiece.

Conceptual Analysis 25.2

Lenses in a compound microscope

The basic compound microscope consists of

A. two diverging lenses, both of short focal length;
B. two converging lenses, both of short focal length;
C. two converging lenses, one of short and one of long focal length;
D. one diverging lens and one converging lens.

SOLUTION The eyepiece of a compound microscope serves as a simple magnifier to view the image created by the objective lens. (The telescope, discussed in the next section, uses the same arrangement.) Simple magnifiers are always converging lenses. As we discussed in Conceptual Analysis 25.1, the magnifier (the eyepiece in this case) requires a focal length less than 25 cm. The total length of the microscope must be small enough so that the observer can reach the sample while looking into the eyepiece. The focal length of the objective must be small enough to get a large lateral magnification and still keep the objective's image inside the microscope. Thus, the focal length of the objective must be much less than the barrel length of the microscope; the correct answer is B.

EXAMPLE 25.6 **Watching bacteria**

The maximum magnification attainable with a compound microscope is about 2000×. The smallest bacteria are roughly 0.5 micron in size. **(a)** For an eyepiece with 15×, what magnification must the objective lens have to reach an overall magnification of 2000×? **(b)** How big does a half-micron bacterium appear? Is this larger or smaller than the period at the end of this sentence?

SOLUTION

SET UP We are given the angular magnification of the eyepiece: $M_2 = 15\times$. We want an overall magnification $M = 2000\times$. The size of the bacterium is $0.5\ \mu m = 0.5 \times 10^{-6}$ m. We want to find the lateral magnification m_1 of the objective.

SOLVE **Part (a):** From Equation 25.4, $m_1 = M/M_2 = 2000\times/15\times = 130\times$.

Part (b): The apparent size of the bacterium is

$$(2000)(0.5 \times 10^{-6}\,\text{m}) = 1.0 \times 10^{-3}\,\text{m} = 1.0\,\text{mm}.$$

REFLECT The apparent size is a few times larger than the period at the end of this sentence. Thus, the bacterium is visible with 2000× magnification, but without much detail.

Practice Problem: If the only available eyepiece has an angular magnification of 10×, what is the overall magnification of this microscope? *Answer:* 1300×.

25.6 Telescopes

The optical system of a refracting **telescope** is similar to that of a compound microscope. In both instruments the image formed by an objective is viewed through an eyepiece. The difference is that the telescope is used to view large objects at large distances, while the microscope is used to view small objects close at hand.

An *astronomical telescope* is shown in Figure 25.16. The objective forms a reduced real image *I* of the object, and the eyepiece forms an enlarged virtual image of *I*. As with the microscope, the image *I'* may be formed anywhere between the near and far points of the eye. Objects viewed with a telescope are usually so far away from the instrument that the first image *I* is formed very nearly at the second focal point of the objective. This image is the object for the eyepiece. If the final image *I'* formed by the eyepiece is at infinity (for the most comfortable viewing by a normal eye), the first image must be at the first focal point of the eyepiece. The distance between the objective and the eyepiece is the length of the telescope, which is therefore the *sum* $f_1 + f_2$ of the focal lengths of the objective and the eyepiece.

The angular magnification *M* of a telescope is defined as the ratio of the angle subtended at the eye by the final image *I'* to the angle subtended at the (unaided) eye by the object. We can express this ratio in terms of the focal lengths of the objective and the eyepiece. In Figure 25.16, the ray passing through F_1, the first

ActivPhysics 15.12: Two-Lens Optical Systems

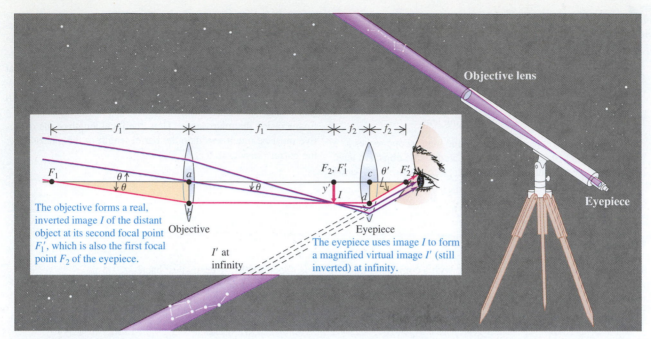

▲ **FIGURE 25.16** Optical system of an astronomical telescope.

▲ **Application X-ray telescope.** It's not easy to make a telescope that works on x rays, because these rays simply pass through the surface of a normal mirror. They can be reflected only by a metal surface that they strike at a grazing angle of incidence. Therefore, an x-ray telescope consists of a sheaf of concentric, barrel-shaped mirrors, each shaped to reflect grazing-incidence x rays toward a focus. Because x rays do not penetrate earth's atmosphere, x-ray telescopes must be placed in space. This photo shows the mirrors of the Chandra x-ray telescope being assembled.

focal point of the objective, and through F_2', the second focal point of the eyepiece, has been emphasized. The object subtends an angle θ at the objective and would subtend essentially the same angle at the unaided eye. Also, since the observer's eye is placed just to the right of the focal point F_2', the angle subtended at the eye by the final image is very nearly equal to the angle θ'. Because bd is parallel to the optic axis, the distances ab and cd are equal to each other and also to the height y' of the image I. Because θ and θ' are small, they may be approximated by their tangents. From the right triangles F_1ab and $F_2'cd$, we obtain

$$\theta = \frac{-y'}{f_1}, \quad \theta' = \frac{y'}{f_2},$$

and the angular magnification M is

$$M = \frac{\theta'}{\theta} = -\frac{y'/f_2}{y'/f_1} = -\frac{f_1}{f_2}. \tag{25.5}$$

The angular magnification M of a telescope is equal to the ratio of the focal length of the objective to that of the eyepiece. The negative sign shows that the final image is inverted.

An inverted image is no particular disadvantage for astronomical observations. When we use a telescope or binoculars on earth, though, we want the image to be right side up. Inversion of the image is accomplished in *prism binoculars* by a pair of 45°–45°–90° totally reflecting prisms called *Porro prisms* (introduced in Section 23.8). Porro prisms are inserted between objective and eyepiece, as shown in Figure 25.17. The image is inverted by the four internal reflections from the faces of the prisms at 45°. The prisms also have the effect of folding the optical path and making the instrument shorter and more compact than it would otherwise be. (See Figure 23.26.) Binoculars are usually described by two numbers separated by a multiplication sign, such as 7 × 50. The first number is the angular magnification M, and the second is the diameter of the objective lenses (in millimeters). The diameter determines the light-gathering capacity of the objective lenses and thus the brightness of the image.

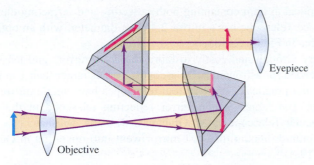

▲ **FIGURE 25.17** Inversion of an image in prism binoculars.

 Conceptual Analysis 25.3

The objective lens

The purpose of the objective lens of a telescope is to

A. produce an image whose angular size is greater than that of the object;
B. take parallel light rays and spread them apart;
C. produce an image that is larger than the object;
D. form an image that can be looked at closely with the aid of an eyepiece.

SOLUTION When we look at a distant object (for example, a person 100 m away), the angular size is the ratio of the height of the person to the distance to the person. The image created by the objective lens is much smaller than the person. However, using an eyepiece, the observer can get very close to the objective's image to view that image. Viewed through the eyepiece, the angular size of the objective's image is much larger than the angular size of the distant object. The correct answer is D.

Reflecting Telescopes

In the *reflecting telescope* (Figure 25.18a), the objective lens is replaced by a concave mirror. In large telescopes, this scheme has many advantages, both theoretical and practical. Mirrors are inherently free of chromatic aberrations (the dependence of focal length on wavelength), and spherical aberrations (associated with the paraxial approximation) are easier to correct than for a lens. The reflecting surface is sometimes parabolic rather than spherical. The material of the mirror need not be transparent, and the mirror can be made more rigid than a lens, which can be supported only at its edges.

Because the image in a reflecting telescope is formed in a region traversed by incoming rays, it can be observed directly with an eyepiece only by blocking off part of the incoming beam (Figure 25.18a), an arrangement that is practical only for the very largest telescopes. Alternative schemes use a mirror to reflect the image through a hole in the mirror or out the side, as shown in Figures 25.18b and 25.18c. This scheme is also used in some long-focal-length telephoto lenses for cameras. In the context of photography, such a system is called a *catadioptric*

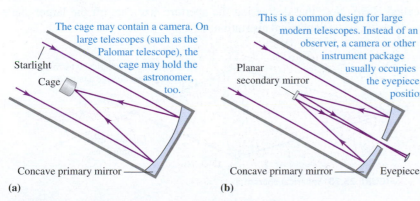
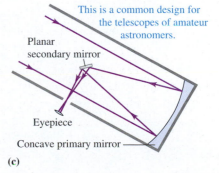

(a) (b) (c)

▲ **FIGURE 25.18** Three optical systems for reflecting telescopes.

(a) One of the 8.4 m single primary mirrors of the Large Binocular Telescope being shaped by a computer-controlled polishing machine. (The mirror is not yet silvered.)

(b) A worker in front of the last segment of the Southern African Large Telescope (SALT).

▲ **FIGURE 25.19** Examples of single and composite telescope mirrors.

lens—an optical system containing both reflecting and refracting elements. The Hubble Space Telescope has a mirror 2.4 m in diameter, with an optical system similar to that in Figure 25.18b.

It is very difficult and costly to fabricate large mirrors and lenses with the accuracy needed for astronomical telescopes. Lenses larger than 1 m in diameter are usually not practical. A few current telescopes have single mirrors over 8 m in diameter. The largest single-mirror reflecting telescope in the world, the Large Binocular Telescope on Mt Graham in Arizona, will have two 8.4 m mirrors on a common mount; the first mirror went into operation in October 2005 (Figure 25.19a).

Another technology for creating very large telescope mirrors is to construct a composite mirror of individual hexagonal segments and use computer control for precise alignment. The twin Keck telescopes, atop Mauna Kea in Hawaii, have mirrors with an overall diameter of 10 m, made up of 36 separate 1.8 m hexagonal mirrors. The orientation of each mirror is controlled by computer to within one millionth of an inch. In the Southern Hemisphere, the Southern African Large Telescope (SALT) has an 11 m mirror composed of 91 hexagonal segments (Figure 25.19b). It went into operation in September 2005.

25.7 Lens Aberrations

An **aberration** is any failure of a mirror or lens to behave precisely according to the simple formulas that we've derived. Aberrations can be classified as **chromatic aberrations,** which involve wavelength-dependent imaging behavior, and **monochromatic aberrations,** which occur even with monochromatic (single-wavelength) light. Lens aberrations are not caused by faulty construction of the lens, such as irregularities in its surfaces, but rather are inevitable consequences of the laws of refraction at spherical surfaces.

Monochromatic aberrations are all related to the *paraxial approximation.* Our derivations of equations for object and image distances, focal lengths, and magnification have all been based upon this approximation. We have assumed that all rays are *paraxial*—that is, that they are very close to the optic axis and make very small angles with it. Real lenses never conform exactly to this condition.

For any lens with an aperture of finite size, the cone of rays that forms the image of any point has a finite size. In general, nonparaxial rays that proceed from a given object point *do not* all intersect at precisely the same point after they are refracted by a lens. For this reason, the image formed by these rays is never perfectly sharp. *Spherical aberration* is the failure of rays from a point object on the optic axis to converge to a point image. Instead, the rays converge within a circle of minimum radius, called *the circle of least confusion,* and then diverge again, as shown in Figure 25.20. The corresponding effect for points off the optic axis produces images that are comet-shaped figures rather than circles; this effect is called *coma.* Note that decreasing the aperture size cuts off the larger-angle rays, thus decreasing spherical aberration.

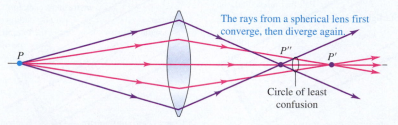

The rays from a spherical lens first converge, then diverge again.

Circle of least confusion

▲ **FIGURE 25.20** Spherical aberration.

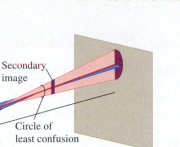

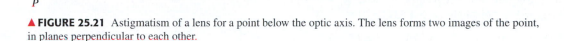

▲ **FIGURE 25.21** Astigmatism of a lens for a point below the optic axis. The lens forms two images of the point, in planes perpendicular to each other.

Spherical aberration is also present in spherical mirrors. Mirrors used in astronomical reflecting telescopes are usually paraboloidal rather than spherical; paraboloidal shapes eliminate spherical aberration for points on the axis, but are much more difficult to fabricate precisely than are spherical shapes.

Astigmatism is the imaging of a point off the axis as two perpendicular lines in different planes. In this aberration, the rays from a point object converge, at a certain distance from the lens, to a line called the *primary image,* perpendicular to the plane defined by the optic axis and the object point. At a somewhat different distance from the lens, they converge to a second line called the *secondary image,* which is *parallel* to that plane. This effect is shown in Figure 25.21. The circle of least confusion (greatest convergence) appears between these two positions.

The location of the circle of least confusion depends on the object point's *transverse* distance from the axis, as well as its *longitudinal* distance from the lens. As a result, object points lying in a plane are, in general, imaged not in a plane, but in some curved surface; this effect is called *curvature of field.*

Finally, the image of a straight line that does not pass through the optic axis may be curved. As a result, the image of a square with the axis through its center may resemble a barrel (sides bent outward) or a pincushion (sides bent inward). This effect, called *distortion,* is not related to lack of sharpness of the image, but results from a change in lateral magnification with distance from the axis.

Chromatic aberrations are a result of *dispersion*—the variation of index of refraction with wavelength that we discussed in Section 23.9. Dispersion causes the focal length of a lens to be somewhat different for different wavelengths. When an object is illuminated with light containing a mixture of wavelengths, different wavelengths are imaged at different points. The magnification of a lens also varies with wavelength; this effect is responsible for the rainbow-fringed images that are seen with inexpensive binoculars or telescopes. Reflectors are inherently free of chromatic aberrations; this is one of the reasons for their usefulness in large astronomical telescopes.

It is impossible to eliminate all these aberrations from a single lens, but in a compound lens with several optical elements, the aberrations of one element may partially cancel those of another. The design of such lenses is an extremely complex undertaking, aided greatly in recent years by the use of computers. It is still impossible to eliminate all aberrations, but it *is* possible to decide which ones are most troublesome for a particular application and to design accordingly.

EXAMPLE 25.7 **Chromatic aberration**

A glass planoconvex lens has its flat side toward the object, and the other side has a radius of curvature with magnitude 30.0 cm. The index of refraction of the glass for violet light (wavelength 400 nm) is 1.537, and for red light (700 nm) it is 1.517. The color purple is a mixture of red and violet. A purple object is placed 80.0 cm from this lens. Where are the red and violet images formed?

SOLUTION

SET UP AND SOLVE We use the thin-lens equation in the form given by Equation 24.19:

$$\frac{1}{s} + \frac{1}{s'} = (n-1)\left(\frac{1}{R_1} - \frac{1}{R_2}\right).$$

In this case, using the usual sign rules, we have $R_1 = \infty$ and $R_2 = -30.0$ cm. For violet light $(n = 1.537)$,

$$\frac{1}{80.0 \text{ cm}} + \frac{1}{s'} = (1.537 - 1)\left(\frac{1}{\infty} - \frac{1}{-30.0 \text{ cm}}\right)$$

$$s' = 185 \text{ cm}.$$

For red light $(n = 1.517)$, we find that $s' = 211$ cm.

REFLECT The violet light is refracted more than the red, and its image is formed closer to the lens. We see that a rather small variation in index of refraction causes a substantial displacement of the image.

SUMMARY

The Camera and the Projector

(Sections 25.1 and 25.2) A camera forms a real, inverted, usually reduced image of the object being photographed. For cameras using film, the amount of light striking the film is controlled by the shutter speed and the aperture. A projector is essentially a camera in reverse: A lens forms a real, inverted, enlarged image on a screen of the slide or motion-picture film.

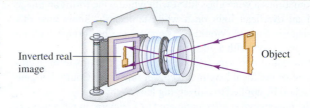

Inverted real image

Object

The Eye

(Section 25.3) In the eye, a real image is formed on the retina and transmitted to the optic nerve. Adjustment for various object distances is made by the ciliary muscle; for close vision, the lens becomes more convex, decreasing its focal length. For sharp vision, the image must form exactly on the retina. In a nearsighted (myopic) eye, the image is formed in front of the retina. In a farsighted (hyperopic) eye, the image is formed behind the retina. The power of a corrective lens, in diopters, is the reciprocal of the focal length, in meters.

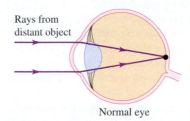

Rays from distant object

Normal eye

Magnifiers, Microscopes, and Telescopes

(Section 25.4–25.6) The apparent size of an object is determined by the size of its image on the retina. A simple magnifier creates a virtual image whose angular size is larger than that of the object itself. The angular magnification is the ratio of the angular size of the virtual image to that of the object.

In a compound microscope, the objective lens forms a first image in the barrel of the instrument, and this image becomes the object for the second lens, called the eyepiece. The eyepiece forms a final virtual image, often at infinity, of the first image. A telescope operates on the same principle, but the object is far away. In a reflecting telescope, the objective lens is replaced by a concave mirror, which eliminates chromatic aberrations.

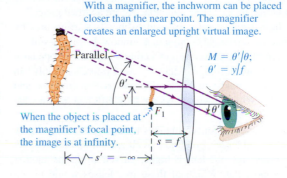

With a magnifier, the inchworm can be placed closer than the near point. The magnifier creates an enlarged upright virtual image.

Parallel

$M = \theta'/\theta;$
$\theta' = y/f$

When the object is placed at the magnifier's focal point, the image is at infinity.

F_1

$s = f$

$s' = -\infty$

Lens Aberrations

(Section 25.7) Lens aberrations account for the failure of a lens to form a perfectly sharp image of an object. Monochromatic aberrations occur because of limitations of the paraxial approximation. Chromatic aberrations result from the dependence of index of refraction on wavelength, causing the focal length of a lens to be somewhat different for different wavelengths.

Conceptual Questions

1. Sometimes a wine glass filled with white wine forms an image of an overhead light on a white tablecloth. Show how this image is formed. Would the same image be formed with an empty glass? With a glass of water?

2. How could one very quickly make an approximate measurement of the focal length of a converging lens? Could the same method be applied to a diverging lens? Why?

3. The human eye is often compared to a camera. In what ways is it similar to a camera? In what ways does it differ?

4. How could one make a lens for sound waves?

5. A diver proposed using a plastic bag full of air, immersed in water, as an underwater lens. Is this possible? If the lens is to be a converging lens, what shape should the air pocket have?

6. When a converging lens is immersed in water, does its focal length increase or decrease in comparison to its focal length in air?

7. You are marooned on a desert island and want to use your eyeglasses to start a fire. Can this be done if you are nearsighted? If you are farsighted?

8. If the sensor array of a digital camera is placed at the location of a real image, the sensor will record this image. Can this be done with a virtual image? How might one record a virtual image?

9. There have been reports of round fishbowls starting fires by focusing the sun's rays coming in a window. Is this possible?

10. Since a refracting telescope and a compound microscope have the same basic design, can they be used interchangeably? In other words, could you use a telescope as a microscope and vice versa? Why? In what ways are they different and in what ways similar?

11. You are selecting a converging lens for a magnifier. By simply feeling several lenses, you find that all are flat on one side, yet on the other side one lens is highly curved while the another one is almost flat. Which of these two lenses would give a greater angular magnification?

12. While choosing between two refracting astronomical telescopes, you notice that one is twice as long as the other one. Will the longer telescope necessarily produce a greater angular magnification than the shorter one? Why?

13. When choosing between two refracting astronomical telescopes, you notice that the tube of one is twice as wide as the tube of the other one because the objective lens is twice as wide. Will the telescope with the wider tube necessarily produce a greater angular magnification than the narrower one? Why? Would there be *any* advantages to the wider-tube telescope? If so, what would they be?

14. You've entered a survival contest that will include building a crude telescope. You are given a large box of lenses. Which two lenses do you pick? How do you quickly identify them?

15. If a person is severely nearsighted, can the optometrist prescribe a *single* lens that will correct this defect and allow him to see clearly at all the usual distances?

16. Ads for amateur telescopes sometimes contain statements such as "maximum magnification 600×." Are such statements meaningful? Why can they be misleading? Is there really a maximum magnification for a telescope?

17. The focal length of a simple lens depends on the color (wavelength) of light passing through it. Why? Is it possible for a lens to have a positive focal length for some colors and negative for others? Explain.

Multiple-Choice Problems

1. The focusing mechanism of the human eye most closely resembles that of a
   A. telescope.
   B. microscope.
   C. camera.
   D. magnifier.

2. Which of the following statements are true about the lenses used in eyeglasses to correct nearsightedness and farsightedness? (There may be more than one correct choice.)
   A. They produce a real image.
   B. They produce a virtual image.
   C. Both nearsightedness and farsightedness are corrected with a converging lens.
   D. Both nearsightedness and farsightedness are corrected with a diverging lens.

3. The discussion of simple magnifiers in Sec. 25.4 was based on the assumption of a viewer with a normal near point of 25 cm. *Compared to the angular size they can see with their own unaided eyes,* how does the angular magnification obtained by a myopic person compare with that obtained by a person with normal vision, if they look at the same object with the same magnifying glass?
   A. The myopic and the normal person obtain the same magnification.
   B. The myopic person obtains a greater magnification.
   C. The myopic person obtains less magnification.

4. If, without changing anything else, we double the focal lengths of both of the lenses in a refracting telescope that had an original angular magnification M, its new angular magnification will be
   A. $4M$.
   B. $2M$.
   C. M.
   D. $M/2$.
   E. $M/4$.

5. If a person's eyeball is 2.7 cm deep instead of the usual 2.5 cm, he most likely suffers from
   A. astigmatism.
   B. color blindness.
   C. nearsightedness.
   D. farsightedness.

6. Which of the following statements are true about the eye? (There may be more than one correct choice.)
   A. When focusing on a very distant object, the lens is curved the most.
   B. When focusing on a nearby object, the lens is curved the most.
   C. Most of the bending of light is accomplished by the lens.
   D. Most of the bending of light is accomplished by the cornea.
   E. Most of the focusing is due to variations in the diameter of the pupil.

7. If a camera lens gives the proper exposure for a photograph at a shutter speed of $1/200$ s at an f-stop of $f/2.8$, the proper shutter speed at $f/5.6$ is
   A. $1/800$ s.
   B. $1/400$ s.
   C. $1/100$ s.
   D. $1/50$ s.

8. A slide projector produces
   A. a real erect image of the slide.
   B. a virtual erect image of the slide.
   C. a real inverted image of the slide.
   D. a virtual inverted image of the slide.

9. Which of the following operations would increase the angular magnification of a refracting telescope? (There may be more than one correct choice.)
   A. Increase the focal length of the objective lens.
   B. Increase the focal length of the eyepiece.
   C. Decrease the focal length of the objective.
   D. Decrease the focal length of the eyepiece.
   E. Increase the focal length of both the objective lens and the eyepiece by the same factor.

10. Which of the following statements are true about a compound microscope? (There may be more than one correct choice.)
    A. The objective lens produces an inverted virtual image of the object.
    B. The image produced by the objective lens is just inside the focal point of the eyepiece.
    C. The final image is inverted compared with the original object.
    D. The object to be viewed is placed just inside the focal point of the objective lens.
    E. The final image is virtual.

11. Laser eye surgery can correct various vision defects by altering the shape of the cornea's outer surface. Suppose a surgeon performs an operation that decreases the radius of curvature of the cornea's surface; this would be an appropriate procedure for what kind of vision defect?
    A. Astigmatism
    B. Presbyopia
    C. Nearsightedness
    D. Farsightedness

12. A camera is focusing on an animal. As the creature moves *closer* to the lens, what must be done to keep the animal in focus?
    A. The lens must be moved closer to the film (or light sensors in a digital camera).
    B. The lens must be moved farther from the film (or light sensors in a digital camera).
    C. The *f*-number of the lens must be increased.
    D. The *f*-number of the lens must be decreased.

13. Your eye is focusing on a person. As he walks toward you, what must be done to keep him in focus?
    A. The pupil of your eye must open up.
    B. The distance from the eye lens to the retina must decrease.
    C. The distance from the eye lens to the retina must increase.
    D. The focal length of your eye must decrease.
    E. The focal length of your eye must increase.

14. An astronomical telescope is made with an objective lens of focal length 150 cm and an eyepiece with focal length of 2.0 cm. The magnification of this instrument is
    A. 75×.   B. 300×.
    C. 152×.   D. 148×.

15. A simple magnifying glass produces a
    A. real inverted image.
    B. real erect image.
    C. virtual inverted image.
    D. virtual erect image.

Problems

25.1 The Camera

1. • The focal length of an $f/4$ camera lens is 300 mm. (a) What is the aperture diameter of the lens? (b) If the correct exposure of a certain scene is $\frac{1}{250}$ s at $f/4$, what is the correct exposure at $f/8$?

2. • Camera A has a lens with an aperture diameter of 8.00 mm. It photographs an object, using the correct exposure time of $\frac{1}{30}$ s. What exposure time should be used with camera B in photographing the same object with the same film if camera B has a lens with an aperture diameter of 23.1 mm?

3. • (a) A small refracting telescope designed for individual use has an objective lens with a diameter of 6.00 cm and a focal length of 1.325 m. What is the *f*-number of this instrument? (b) The 200-inch-diameter objective mirror of the Mt. Palomar telescope has an *f*-number of 3.3. Calculate its focal length. (c) The distance between lens and retina for a normal human eye is about 2.50 cm, and the pupil can vary in size from 2.0 mm to 8.0 mm. What is the range of *f*-numbers for the human eye?

4. •• A 135 mm telephoto lens for a 35 mm camera has *f*-stops that range from $f/2.8$ to $f/22$. (a) What are the smallest and largest aperture diameters for this lens? What is the diameter at $f/11$? (b) If a 50 mm lens had the same *f*-stops as the telephoto lens, what would be the smallest and largest aperture diameters for that lens? (c) At a given shutter speed, what is the ratio of the greatest to the smallest light intensity of the film image? (d) If the shutter speed for correct exposure at $f/22$ is 1/30 s, what shutter speed is needed at $f/2.8$?

5. • A camera lens has a focal length of 200 mm. How far from the lens should the subject for the photo be if the lens is 20.4 cm from the film?

6. •• A camera with a 90-mm-focal-length lens is focused on an object 1.30 m from the lens. To refocus on an object 6.50 m from the lens, by how much must the distance between the lens and the film be changed? To refocus on the more distant object, is the lens moved toward or away from the film?

7. • A certain digital camera having a lens with focal length 7.50 cm focuses on an object 1.85 m tall that is 4.25 m from the lens. (a) How far must the lens be from the sensor array? (b) How tall is the image on the sensor array? Is it erect or inverted? Real or virtual? (c) A SLR digital camera often has pixels measuring 8.0 μm × 8.0 μm. How many such pixels does the height of this image cover?

8. •• Your digital camera has a lens with a 50 mm focal length and a sensor array that measures 4.82 mm × 3.64 mm. Suppose you're at the zoo, and want to take a picture of a 4.50-m-tall giraffe. If you want the giraffe to exactly fit the longer dimension of your sensor array, how far away from the animal will you have to stand?

9. •• You want to take a full-length photo of your friend who is 2.00 m tall, using a 35 mm camera having a 50.0-mm-focal-length lens. The image dimensions of 35 mm film are 24 mm × 36 mm, and you want to make this a vertical photo in which your friend's image completely fills the image area. (a) How far should your friend stand from the lens? (b) How far is the lens from the film?

10. •• **Zoom lens, I.** A *zoom lens* is a lens that varies in focal length. The zoom lens on a certain digital camera varies in focal length from 6.50 mm to 19.5 mm. This camera is focused

on an object 2.00 m tall that is 1.50 m from the camera. Find the distance between the lens and the photo sensors and the height of the image (a) when the zoom is set to 6.50 mm focal length and (b) when it is at 19.5 mm. (c) Which is the tele-photo focal length, 6.50 mm or 19.5 mm?

25.2 The Projector

11. • A slide projector uses a lens of focal length 115 mm to focus a 35 mm slide (having dimensions 24 mm × 36 mm) on a screen. The slide is placed 12.0 cm in front of the lens. (a) Where should you place the screen to view the image of this slide? (b) What are the dimensions of the slide's image on the screen?

12. •• An LCD projector (see Sec. 25.2) has a projection lens with *f*-number of 1.8 and a diameter of 46 mm. The LCD array measures 3.30 cm × 3.30 cm and will be projected on a screen 8.00 m from the lens. If the array is 800 × 600 pixels, what will be the dimensions of a single pixel on the screen?

13. • The dimensions of the picture on a 35 mm color slide are 24 mm × 36 mm. An image of the slide is projected onto a screen 9.00 m from the projector lens with a lens of focal length 150 mm. (a) How far is the slide from the lens? (b) What are the dimensions of the image on the screen?

14. •• You are designing a projection system for a hall having a screen measuring 4.00 m square. The lens of a 35 mm slide projector in the projection booth is 15.0 m from this screen. You want to focus the image of 35 mm slides (which are 24 mm × 36 mm) onto this screen so that you can fill as much of the screen as possible without any part of the image extending beyond the screen. (a) What focal-length lens should you use in the projector? (b) How far from the lens should the slide be placed? (c) What are the dimensions of the slide's image on the screen?

15. •• In a museum devoted to the history of photography, you are setting up a projection system to view some historical 4.0 inch × 5.0 inch color slides. Your screen is 6.0 m from the projector lens, and you want the image to be 4.0 ft × 5.0 ft on the screen. (a) What focal-length lens do you need? (b) How far from the lens should you put the slide?

25.3 The Eye

16. • **The cornea as a thin lens.**
BIO Measurements on the cornea of a person's eye reveal that the mag-nitude of the front surface radius of curvature is 7.80 mm, while the magnitude of the rear surface radius of curvature is 7.30 mm (see Figure 25.22), and that the index of refraction of the cornea

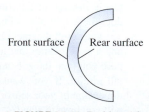
▲ **FIGURE 25.22** Problem 16.

is 1.38. If the cornea were simply a thin lens in air, what would be its focal length and its power in diopters? What type of lens would it be?

17. • **Range of the focal length of the eye.** We can model the eye
BIO as a sphere 2.50 cm in diameter with a thin lens in the front and the retina along the back surface. A 20-year-old person should be able to focus on objects from her near point (see Table 25.1) up to infinity. What is the range of the effective focal length of the lens of the eye for this person?

18. • A 40-year-old optometry patient focuses on a 6.50-cm-tall
BIO photograph at his near point. (See Table 25.1.) We can model his eye as a sphere 2.50 cm in diameter, with a thin lens at the front and the retina at the rear. (a) What is the effective focal length of his eye and its power in diopters when he focuses on the photo? (b) How tall is the image of the photo on his retina? Is it erect or inverted? Real or virtual? (c) If he views the pho-tograph from a distance of 2.25 m, how tall is its image on his retina?

19. •• **Crystalline lens of the eye.** The crystalline lens of the eye
BIO is double convex and has a typical index of refraction of 1.43. At minimum power, the front surface has a radius of 10.0 mm and the back surface has a radius of 6.0 mm; at maximum power, these radii are 6.0 mm and 5.5 mm, respectively (although the values do vary from person to person). (a) Find the maximum and minimum power (in diopters) of the crys-talline lens if it were in air. (b) What is the range of focal lengths the eye can achieve? (c) At minimum power, where does it focus the image of a very distant object? (d) At maxi-mum power, where does it focus the image of an object at the near point of 25 cm?

20. • **Contact lenses.** Contact lenses are placed right on the eye-
BIO ball, so the distance from the eye to an object (or image) is the same as the distance from the lens to that object (or image). A certain person can see distant objects well, but his near point is 45.0 cm from his eyes instead of the usual 25.0 cm. (a) Is this person nearsighted or farsighted? (b) What type of lens (con-verging or diverging) is needed to correct his vision? (c) If the correcting lenses will be contact lenses, what focal length lens is needed and what is its power in diopters?

21. •• **Ordinary glasses.** Ordinary glasses are worn in front of
BIO the eye and usually 2.0 cm in front of the eyeball. Suppose that the person in the previous problem prefers ordinary glasses to contact lenses. What focal length lenses are needed to correct his vision, and what is their power in diopters?

22. • A person can see clearly up close, but cannot focus on
BIO objects beyond 75.0 cm. She opts for contact lenses to correct her vision. (a) Is she nearsighted or farsighted? (b) What type of lens (converging or diverging) is needed to correct her vision? (c) What focal-length contact lens is needed, and what is its power in diopters?

23. •• In one form of cataract surgery the person's natural lens,
BIO which has become cloudy, is replaced by an artificial lens. The refracting properties of the replacement lens can be chosen so that the person's eye focuses on distant objects. But there is no accommodation, and glasses or contact lenses are needed for close vision. What is the power, in diopters, of the corrective contact lenses that will enable a person who has had such sur-gery to focus on the page of a book at a distance of 24 cm?

24. •• **Bifocals, I.** A person can focus clearly only on objects
BIO between 35.0 cm and 50.0 cm from his eyes. Find the focal length and power of the correcting contact lenses needed to correct (a) his closeup vision, (b) his distant vision.

25. •• A student's far point is at 17.0 cm, and she needs glasses to
BIO view her computer screen comfortably at a distance of 45.0 cm. What should be the power of the lenses for her glasses?

26. •• (a) Where is the near point of an eye for which a contact
BIO lens with a power of +2.75 diopters is prescribed? (b) Where is the far point of an eye for which a contact lens with a power of −1.30 diopters is prescribed for distant vision?

27. •• **Corrective lenses.** Determine the power of the corrective
BIO contact lenses required by (a) a hyperopic eye whose near point is at 60.0 cm, and (b) a myopic eye whose far point is at 60.0 cm.

25.4 The Magnifier

28. • You want to view an insect 2.00 mm in length through a magnifier. If the insect is to be at the focal point of the magnifier, what focal length will give the image of the insect an angular size of 0.025 radian?

29. • A simple magnifier for viewing postage stamps and other pieces of paper consists of a thin lens mounted on a stand. When this device is placed on a stamp, the stand holds the lens at the proper distance from the stamp for viewing. (See Figure 25.23.) If the lens has a focal length of 11.5 cm, and the image is to be at infinity, (a) what is the angular magnification of the device, and (b) how high should the stand hold the lens above the stamp?

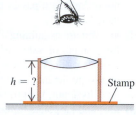

▲ **FIGURE 25.23** Problem 29.

30. • A thin lens with a focal length of 6.00 cm is used as a simple magnifier. (a) What angular magnification is obtainable with the lens if the object is at the focal point? (b) When an object is examined through the lens, how close can it be brought to the lens? Assume that the image viewed by the eye is at infinity and that the lens is very close to the eye.

31. •• The focal length of a simple magnifier is 8.00 cm. Assume the magnifier to be a thin lens placed very close to the eye. (a) How far in front of the magnifier should an object be placed if the image is formed at the observer's near point, 25.0 cm in front of her eye? (b) If the object is 1.00 mm high, what is the height of its image formed by the magnifier?

25.5 The Microscope

32. • A microscope has an objective lens with a focal length of
BIO 12.0 mm. A small object is placed 0.8 mm beyond the focal point of the objective lens. (a) At what distance from the objective lens does a real image of the object form? (b) What is the magnification of the real image? (c) If an eyepiece with a focal length of 2.5 cm is used, with a final image at infinity, what will be the overall angular magnification of the object?

33. • A compound microscope has an objective lens of focal
BIO length 10.0 mm with an eyepiece of focal length 15.0 mm, and it produces its final image at infinity. The object to be viewed is placed 2.0 mm beyond the focal point of the objective lens. (a) How far from the objective lens is the first image formed? (b) What is the overall magnification of this microscope?

34. •• An insect 1.2 mm tall is placed 1.0 mm beyond the focal
BIO point of the objective lens of a compound microscope. The objective lens has a focal length of 12 mm, the eyepiece a focal length of 25 mm. (a) Where is the image formed by the objective lens and how tall is it? (b) If you want to place the eyepiece so that the image it produces is at infinity, how far should this lens be from the image produced by the objective lens? (c) Under the conditions of part (b), find the overall magnification of the microscope.

35. •• The objective lens and the eyepiece of a microscope are
BIO 16.5 cm apart. The objective lens has a magnification of 62× and the eyepiece has a magnification of 10×. Calculate (a) the overall magnification of the microscope, (b) the focal length of each lens, and (c) where an object should be placed in order for a normal eye to focus comfortably on the image.

36. •• The focal length of the eyepiece of a certain microscope is
BIO 18.0 mm. The focal length of the objective is 8.00 mm. The distance between objective and eyepiece is 19.7 cm. The final image formed by the eyepiece is at infinity. Treat all lenses as thin. (a) What is the distance from the objective to the object being viewed? (b) What is the magnitude of the linear magnification produced by the objective? (c) What is the overall angular magnification of the microscope?

37. •• A certain microscope is provided with objectives that have
BIO focal lengths of 16 mm, 4 mm, and 1.9 mm and with eyepieces that have angular magnifications of 5× and 10×. Each objective forms an image 120 mm beyond its second focal point. Determine (a) the largest overall angular magnification obtainable and (b) the smallest overall angular magnification obtainable.

38. •• **Resolution of a microscope.** The image formed by a
BIO microscope objective with a focal length of 5.00 mm is 160 mm from its second focal point. The eyepiece has a focal length of 26.0 mm. (a) What is the angular magnification of the microscope? (b) The unaided eye can distinguish two points at its near point as separate if they are about 0.10 mm apart. What is the minimum separation that can be resolved with this microscope?

25.6 Telescopes

39. • A refracting telescope has an objective lens of focal length 16.0 in and eyepieces of focal lengths 15 mm, 22 mm, 35 mm, and 85 mm. What are the largest and smallest angular magnifications you can achieve with this instrument?

40. • The eyepiece of a refracting astronomical telescope (see Figure 25.16) has a focal length of 9.00 cm. The distance between objective and eyepiece is 1.80 m, and the final image is at infinity. What is the angular magnification of the telescope?

41. • **Galileo's telescopes, I.** While Galileo did not invent the telescope, he was the first known person to use it astronomically, beginning around 1609. Five of his original lenses have survived (although he did work with others). Two of these have focal lengths of 1710 mm and 980 mm. (a) For greatest magnification, which of these two lenses should be the eyepiece and which the objective? How long would this telescope be between the two lenses? (b) What is the greatest angular magnification that Galileo could have obtained with these lenses? (*Note:* Galileo actually obtained magnifications up to about 30×, but by using a diverging lens as the eyepiece.) (c) The Moon subtends an angle of $\frac{1}{2}°$ when viewed with the naked eye. What angle would it subtend when viewed through this telescope (assuming that all of it could be seen)?

42. •• The objective mirror of the Hubble Space telescope has a focal length of 57.6 meters. The planet Mars's closest approach to the earth is about 35 million miles. Use data from Appendix E to help you calculate the size of the real image of Mars formed by the Hubble's objective mirror when the planet is closest to earth.

43. •• The largest refracting telescope in the world is at Yerkes Observatory in Wisconsin. The objective lens is 1.02 m in diameter and has a focal length of 19.4 m. Suppose you want to magnify Jupiter, which is 138,000 km in diameter, so that its image subtends an angle of $\frac{1}{2}°$ (about the same as the moon) when it is 6.28×10^8 km from earth. What focal-length eyepiece do you need?

25.7 Lens Aberrations

44. • A double-concave lens having radii of curvature of magnitudes 32.0 cm and 24.0 cm is made of a glass for which the refractive index for red light of wavelength 700 nm is 1.44 and for violet light of wavelength 400 nm is 1.57. A white object is placed 50.0 cm in front of this lens. Where will the red light and violet light be focused?

45. • A thin double-convex lens has radii of curvature of magnitudes 25.0 cm and 35.0 cm and is made of silicate flint glass. (See Figure 23.29.) (a) What is its focal length for (i) red light of wavelength 700 nm and (ii) violet light of wavelength 400 nm? (b) A color chart is placed 30.0 cm from the front of this lens. How far from the lens will the red and violet light be focused?

46. • A thin planoconvex lens has a radius of curvature of magnitude 22.5 cm on the curved side. When a color chart is placed 48.0 cm from the lens, green light of wavelength 550 nm is focused 277 cm from the lens and blue light of wavelength 450 nm is focused 171 cm from the lens. What are the indices of refraction for these two wavelengths of light?

47. •• A thin planoconvex lens with a radius of curvature of magnitude 28.0 cm on the curved face is made of silicate flint glass. (See Figure 23.29.) When a colorful object is placed 65.0 cm from this lens, the yellow light of wavelength 550 nm is perfectly focused on a screen. (a) How far is the screen from the lens? (b) Where will red light of wavelength 700 nm and violet light of wavelength 400 nm be focused?

General Problems

48. • A photographer takes a photograph of a Boeing 747 airliner (length 70.7 m) when it is flying directly overhead at an altitude of 9.50 km. The lens has a focal length of 5.00 m. How long is the image of the airliner on the film?

49. •• **Curvature of the cornea.** In a simplified model of the
BIO human eye, the aqueous and vitreous humors and the lens all have a refractive index of 1.40, and all the refraction occurs at the cornea, whose vertex is 2.60 cm from the retina. What should be the radius of curvature of the cornea such that the image of an object 40.0 cm from the cornea's vertex is focused on the retina? (See Sec. 24.4.)

50. •• **A nearsighted eye.** A certain very nearsighted person cannot
BIO focus on anything farther than 36.0 cm from the eye. Consider the simplified model of the eye described in problem 49. If the radius of curvature of the cornea is 0.75 cm when the eye is focusing on an object 36.0 cm from the cornea vertex and the indexes of refraction are as described in problem 49, what is the distance from the cornea vertex to the retina? What does this tell you about the shape of the nearsighted eye?

51. •• **What is the smallest thing we can see?** The smallest
BIO object we can resolve with our eye is limited by the size of the light receptor cells on the retina. In order to distinguish any detail in an object, its image cannot be any smaller than a single retinal cell. Although the size depends on the type of cell (rod or cone), a diameter of a few microns (μm) is typical near the center of the eye. We shall model the eye as a sphere 2.50 cm in diameter with a single thin lens at the front and the retina at the rear, with light receptor cells 5.0 μm in diameter. (a) What is the smallest object you can resolve at a near point of 25 cm? (b) What angle is subtended by this object at the eye? Express your answer in units of minutes ($1° = 60$ min), and compare it with the typical experimental value of about 1.0 min. (*Note:* There are other limitations, such as the bending of light as it passes through the pupil, but we shall ignore them here.)

52. •• You are examining a flea with a converging lens that has a focal length of 4.00 cm. If the image of the flea is 6.50 times the size of the flea, how far is the flea from the lens? Where, relative to the lens, is the image?

53. •• **Physician, heal thyself!** (a) Experimentally determine the
BIO near and far points for both of your own eyes. Are these points the same for both eyes? (All you need is a tape measure or ruler and a cooperative friend.) (b) Design correcting lenses, as needed, for your closeup and distant vision in one of your eyes. If you prefer contact lenses, design that type of lens. Otherwise design lenses for ordinary glasses, assuming that they will be 2.0 cm from your eye. Specify the power (in diopters) of each correcting lens.

54. •• **Laser eye surgery.** The distance from the vertex of
BIO the cornea to the retina for a certain nearsighted person is 2.75 cm, and the radius of curvature of her cornea is 0.700 cm. She decides to get laser surgery to correct her vision. Using the simplified model of the eye described in problem 49, calculate the radius of curvature for her cornea that the surgeon should aim for, in order to allow her to view distant objects with a relaxed eye.

55. •• **It's all done with mirrors.** A photographer standing 0.750 m in front of a plane mirror is taking a photograph of her image in the mirror, using a digital camera having a lens with a focal length of 19.5 mm. (a) How far is the lens from the light sensors of the camera? (b) If the camera is 8.0 cm high, how high is its image on the sensors?

56. •• During a lunar eclipse, a picture of the moon (which has a diameter of 3.48×10^6 m and is 3.86×10^8 m from the earth) is taken with a camera whose lens has a focal length of 300 mm. (a) What is the diameter of the image on the film? (b) What percent is this of the width of a 24 mm × 36 mm color slide?

57. •• A person with a digital camera uses a lens of focal length 15.0 mm to take a photograph of the image of a 1.40-cm-tall seedling located 10.0 cm behind a double-

▲ **FIGURE 25.24** Problem 57.

convex lens of 17.0 cm focal length. (See Figure 25.24.) The camera's lens is 5.00 cm from the convex lens. (a) How far is the camera's lens from the photoreceptors? (b) What is the height of the seedling on the photoreceptors? (c) Will the camera actually record *two* images, one of the seedling directly and the other of the seedling as viewed through the lens?

58. •• A microscope with an objective of focal length 8.00 mm
BIO and an eyepiece of focal length 7.50 cm is used to project an image on a screen 2.00 m from the eyepiece. Let the image

distance of the objective be 18.0 cm. (a) What is the lateral magnification of the image? (b) What is the distance between the objective and the eyepiece?

59. •• A person with a near point of 85 cm, but excellent distant
BIO vision, normally wears corrective glasses. But he loses them while traveling. Fortunately, he has his old pair as a spare. (a) If the lenses of the old pair have a power of +2.25 diopters, what is his near point (measured from his eye) when he is wearing the old glasses if they rest 2.0 cm in front of his eye? (b) What would his hear point be if his old glasses were contact lenses instead?

60. •• A telescope is constructed from two lenses with focal lengths of 95.0 cm and 15.0 cm, the 95.0-cm lens being used as the objective. Both the object being viewed and the final image are at infinity. (a) Find the angular magnification of the telescope. (b) Find the height of the image formed by the objective of a building 60.0 m tall and 3.00 km away. (c) What is the angular size of the final image as viewed by an eye very close to the eyepiece?

61. •• **Galileo's telescopes, II.** The characteristics that follow are characteristics of two of Galileo's surviving double-convex lenses. The numbers given are *magnitudes* only; you must supply the correct signs. L_1: front radius = 950 mm, rear radius = 2700 mm, refractive index = 1.528; L_2: front radius = 535 mm, rear radius = 50,500 mm, refractive index = 1.550. (a) What is the largest angular magnification that Galileo could have obtained with these two lenses? (b) How long would this telescope be between the two lenses?

62. •• **Water drop magnifier.** You can make a pretty good magnifying lens by putting a small drop of water on a piece of transparent kitchen wrap. Suppose your drop has an upper surface with a radius of curvature of 1.6 cm and the side on the kitchen wrap is essentially flat. (a) Calculate the focal length of your water lens. (b) What's the angular magnification of the lens? (c) Suppose you place this planoconvex water lens directly onto the surface of a table, so that the tabletop is in effect about half the thickness of the drop, or 1.0 mm, away from the lens. Where does the image of the tabletop form, what type is it, and what is its magnification? (Use the thin lens equation here, even though the small object distance relative to the thickness of the lens makes it a poor approximation in this case.) What does this result tell you about how a simple magnifier works?

26 Interference and Diffraction

I f you've ever blown soap bubbles, you know that part of the fun is watching the multicolored reflections from the bubbles. An ugly black oil spot on the pavement can become a thing of beauty after a rain, when it reflects a rainbow of colors. These familiar sights give us a hint that there are aspects of light that we haven't yet explored. In our discussion of lenses, mirrors, and optical instruments, we've used the model of *geometric optics,* which represents light as *rays*—straight lines that are bent at a reflecting or refracting surface.

But many aspects of the behavior of light (including colors in soap bubbles and oil films) *can't* be understood on the basis of rays. We have already learned that light is fundamentally a *wave,* and in some situations we have to consider its wave properties explicitly. In this chapter, we'll study *interference* and *diffraction* phenomena. When light passes through apertures or around obstacles, the patterns that are formed are a result of the *wave* nature of light; they cannot be understood on the basis of rays. Such effects are grouped under the heading *physical optics.* We'll look at several practical applications of physical optics, including diffraction gratings, x-ray diffraction, and holography.

A soap bubble is just a film of soapy water. Why is it iridescent? And how is this effect related to the iridescence on a CD or a hummingbird's throat? In this chapter we'll study these and related optical phenomena.

26.1 Interference and Coherent Sources

In our discussions of mechanical waves in Chapter 12 and electromagnetic waves in Chapter 23, we often talked about *sinusoidal* waves with a single frequency and a single wavelength. In optics, such a wave is characteristic of **monochromatic light** (light of a single color). Common sources of light, such as an incandescent lightbulb or a flame, *do not* emit monochromatic light; rather, they give off a continuous distribution of wavelengths.

A precisely monochromatic light wave is an idealization, but monochromatic light can be *approximated* in the laboratory. For example, some optical filters block all but a very narrow range of wavelengths. Gas discharge lamps, such as a mercury-vapor lamp, emit light with a discrete set of colors, each having a narrow band of wavelengths. The bright green line in the spectrum of a mercury-vapor lamp has a wavelength of about 546.1 nm, with a spread on the order of ± 0.001 nm. By far the most nearly monochromatic light source available at present is the *laser.* The familiar helium–neon laser, inexpensive and readily available, emits red light at 632.8 nm, with a line width (wavelength range) on the order of ± 0.000001 nm, or about 1 part in 10^9. As we analyze interference and diffraction effects in this chapter and the next, we'll often assume that we're working with monochromatic light.

The term **interference** refers to any situation in which two or more waves overlap in space. When this occurs, the total displacement at any point at any instant of time is governed by the **principle of linear superposition.** We introduced these ideas in Sections 12.5 and 12.8 in relation to mechanical waves; you may want to review those sections. Linear superposition is the most important principle in all of physical optics, so make sure you understand it well.

▲ **Application Waves are waves.** All waves exhibit the phenomena of diffraction and interference that we discuss in this chapter for light waves. This photo shows water waves diffracting and interfering as they pass between offshore barriers. Indeed, the enormous tsunami waves created by the Sumatra–Andaman earthquake of December 2004, which spread around the whole globe, exhibited interference and diffraction as they passed through the gaps between islands and continents.

The principle of superposition

When two or more waves overlap, the resultant displacement at any point and at any instant may be found by adding the instantaneous displacements that would be produced at that point by the individual waves if each were present alone.

We use the term *displacement* in a general sense. For waves on the surface of a liquid, we mean the actual displacement of the surface above or below its normal level. For sound waves, the term refers to the excess or deficiency of pressure. For electromagnetic waves, we usually mean a specific component of the electric or magnetic field.

Another term we'll often use is **phase,** introduced in Chapter 22. When we say that two periodic motions are *in phase,* we mean that they are in step; they reach their maximum values at the same time, their minimum values at the same time, and so on. When two periodic motions are *one-half cycle out of phase,* the positive peaks of one occur at the same times as the negative peaks of the other, and so on.

To introduce the essential ideas of interference, let's consider two identical sources of monochromatic waves separated in space by a certain distance. The two sources are permanently *in phase;* they vibrate in unison. They might be two agitators in a ripple tank, two loudspeakers driven by the same amplifier, two radio antennas powered by the same transmitter, or two small holes or slits in an opaque screen illuminated by the same monochromatic light source.

We position the sources at points S_1 and S_2 along the y axis, equidistant from the origin, as shown in Figure 26.1a. Let P_0 be any point on the x axis. By symmetry, the distance from S_1 to P_0 is *equal* to the distance from S_2 to P_0; waves from the two sources thus require equal times to travel to P_0. Accordingly, waves that leave S_1 and S_2 in phase arrive at P_0 in phase. The two waves add, and the total amplitude at P_0 is *twice* the amplitude of each individual wave.

Similarly, the distance from S_2 to point P_1 is exactly one wavelength *greater* than the distance from S_1 to P_1. Hence, a wave crest from S_1 arrives at P_1 exactly one cycle earlier than a crest emitted at the same time from S_2, and again the two waves arrive *in phase.* For point P_2, the path difference is *two* wavelengths, and again the two waves arrive in phase, and so on.

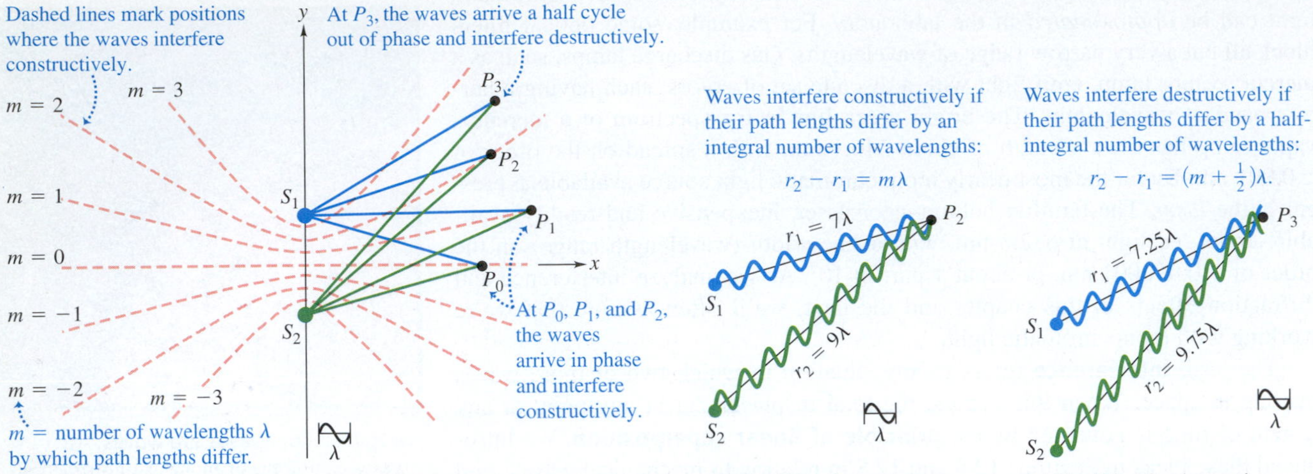

Dashed lines mark positions where the waves interfere constructively.

At P_3, the waves arrive a half cycle out of phase and interfere destructively.

$m = 2$ $m = 3$

$m = 1$ S_1

$m = 0$

$m = -1$ S_2

$m = -2$ $m = -3$

m = number of wavelengths λ by which path lengths differ.

P_3

P_2

P_1

P_0

At P_0, P_1, and P_2, the waves arrive in phase and interfere constructively.

(a) Interference of waves from two monochromatic sources S_1 and S_2 that are in phase and equidistant from the origin

Waves interfere constructively if their path lengths differ by an integral number of wavelengths:

$$r_2 - r_1 = m\lambda$$

$r_1 = 7\lambda$ P_2

S_1

$r_2 = 9\lambda$

S_2

λ

(b) Condition for constructive interference

Waves interfere destructively if their path lengths differ by a half-integral number of wavelengths:

$$r_2 - r_1 = (m + \tfrac{1}{2})\lambda.$$

$r_1 = 7.25\lambda$ P_3

S_1

$r_2 = 9.75\lambda$

S_2

λ

(c) Condition for destructive interference

▲ **FIGURE 26.1** Interference of monochromatic light waves from two sources that are in phase. In this example, the distance between the sources is 4λ.

The addition of amplitudes that results when waves from two or more sources arrive at a point *in phase* is called **constructive interference,** or *reinforcement* (Figure 26.1b). Let the distance from S_1 to any point P be r_1 and the distance from S_2 to P be r_2. Then the condition that must be satisfied for constructive interference to occur at P is that the path difference $r_2 - r_1$ for the two sources must be an integral multiple of the wavelength λ:

Constructive interference

Constructive interference of two waves arriving at a point occurs when the path difference from the two sources is an integer number of wavelengths:

$$r_2 - r_1 = m\lambda \qquad (m = 0, \pm 1, \pm 2, \pm 3, \cdots). \tag{26.1}$$

For our example, all points satisfying this condition lie on the dashed curves in Figure 26.1a.

Intermediate between the dashed curves in Figure 26.1a is another set of curves (not shown) for which the path difference for the two sources is a *half-integral* number of wavelengths. Waves from the two sources arrive at a point on one of these lines (such as point P_3 in Figure 26.1a) exactly a half cycle out of phase. A crest of one wave arrives at the same time as a "trough" (a crest in the opposite direction) from the other wave (Figure 26.1c), and the resultant amplitude is the *difference* between the two individual amplitudes. If the individual amplitudes are equal, then the *total* amplitude is zero! This condition is called **destructive interference,** or *cancellation.* For our example, the condition for destructive interference is as follows:

Destructive interference

Destructive interference of two waves arriving at a point occurs when the path difference from the two sources is a half-integer number of wavelengths:

$$r_2 - r_1 = (m + \tfrac{1}{2})\lambda \qquad (m = 0, \pm 1, \pm 2, \pm 3, \cdots). \tag{26.2}$$

Figures 26.1b and 26.1c show the phase relationships for constructive and destructive interference of two waves.

An example of this interference pattern is the familiar ripple-tank pattern shown in Figure 26.2. The two wave sources are two agitators driven by the same vibrating mechanism. The regions of both maximum and zero amplitude are clearly visible.

For Equations 26.1 and 26.2 to hold, the two sources must *always* be in phase. With light waves, there is no practical way to achieve such a relationship with two *separate* sources because of the way light is emitted. In ordinary light sources, atoms gain excess energy by thermal agitation or by impact with accelerated electrons. An atom thus "excited" begins to radiate energy and continues until it has lost all the energy it can, typically in a time on the order of 10^{-8} s. The many atoms in a source ordinarily radiate in an unsynchronized and random phase relationship, and the separate beams of light emitted from *two* such sources have no definite phase relation to each other.

However, the light from a single source can be split so that parts of it emerge from two or more regions of space, forming two or more *secondary sources.* Then any random phase change in the source affects these secondary sources equally and does not change their *relative* phase. Light from two such sources, derived from a single primary source and with a definite, constant phase relation, is said to be **coherent.** We'll consider the interference of light from two secondary sources in the next section.

The distinguishing feature of light from a *laser* is that the emission of light from many atoms is *synchronized* in frequency and phase. As a result, the random phase changes mentioned in the paragraphs above occur *much* less frequently. Definite phase relations are preserved over correspondingly much greater lengths in the beam, and laser light is much more *coherent* than ordinary light.

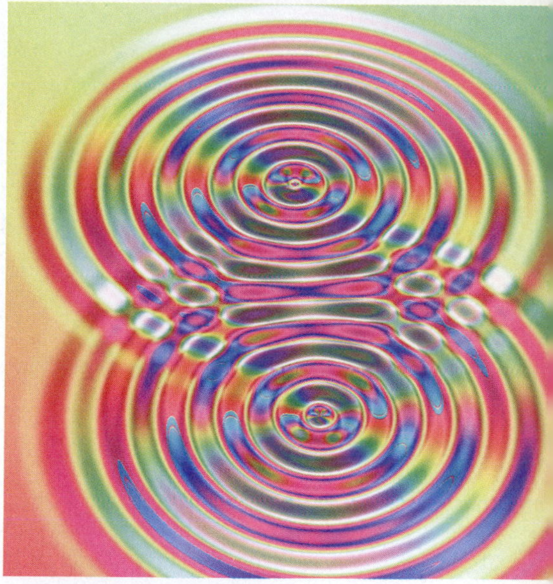

▲ **FIGURE 26.2** Photograph of ripple pattern on water. The ripples are produced by two objects moving up and down in phase just under the water surface.

26.2 Two-Source Interference of Light

One of the earliest quantitative experiments involving the interference of light was performed in 1800 by the English scientist Thomas Young. His experiment involved interference of light from two sources, which we discussed in Section 26.1. Young's apparatus is shown in Figure 26.3a. Monochromatic light emerging from a narrow slit S_0 (1.0 μm or so wide) falls on a screen with two other narrow slits S_1 and S_2, each 1.0 μm or so wide and a few micrometers apart. According to Huygens's principle, cylindrical wave fronts spread out from slit S_0 and reach slits S_1 and S_2 *in phase* because they travel equal distances from S_0. The waves emerging from slits S_1 and S_2 are therefore always in phase, so S_1 and S_2 are *coherent* sources (Figure 26.3b). But the waves from these sources do not necessarily arrive at point P in phase, because of the path difference $(r_2 - r_1)$.

To simplify the analysis that follows, we assume that the distance R from the slits to the screen is so large in comparison to the distance d between the slits that the lines from S_1 and S_2 to P are very nearly parallel, as shown in Figure 26.3c. This is usually the case for experiments with light. The difference in path length is then given by

$$r_2 - r_1 = d\sin\theta. \qquad (26.3)$$

We found in Section 26.1 that constructive interference (reinforcement) occurs at a point P (in a brightly illuminated region of the screen) when the path difference $d\sin\theta$ is an integral number of wavelengths, $m\lambda$, where $m = 0, \pm1, \pm2, \pm3, \cdots$. So we have the following principles:

MasteringPHYSICS

PhET: Wave Interference
ActivPhysics 16.1: Two-Source Interference: Introduction
ActivPhysics 16.2: Two-Source Interference: Qualitative Questions
ActivPhysics 16.3: Two-Source Interference: Problems

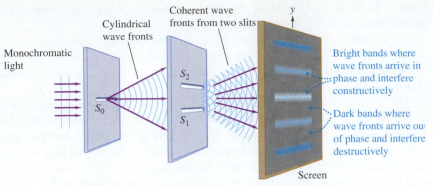

(a) Interference of light waves passing through two slits

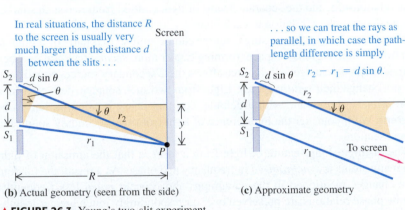

(b) Actual geometry (seen from the side)

(c) Approximate geometry

▲ **FIGURE 26.3** Young's two-slit experiment.

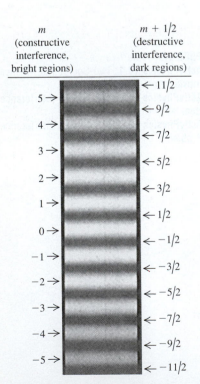

m (constructive interference, bright regions)	$m + 1/2$ (destructive interference, dark regions)

▲ **FIGURE 26.4** Interference fringes produced by Young's two-slit experiment.

Constructive and destructive interference, two slits

Constructive interference occurs at angles θ for which

$$d\sin\theta = m\lambda \qquad (m = 0, \pm 1, \pm 2, \cdots). \qquad (26.4)$$

Similarly, destructive interference (cancellation) occurs, forming dark regions on the screen, at points for which the path difference is a half-integral number of wavelengths, $(m + \frac{1}{2})\lambda$:

$$d\sin\theta = (m + \tfrac{1}{2})\lambda \qquad (m = 0, \pm 1, \pm 2, \cdots). \qquad (26.5)$$

Thus, the pattern on the screen of Figure 26.3a is a succession of bright and dark bands, often called *fringes*. A photograph of such a pattern is shown in Figure 26.4.

We can derive an expression for the positions of the centers of the bright bands on the screen. In Figure 26.3b, y is measured from the center of the pattern, corresponding to the distance from the center of Figure 26.4. Let y_m be the distance from the center of the pattern $(\theta = 0)$ to the center of the mth bright band. We denote the corresponding value of θ as θ_m; then, when $R \gg d$,

$$y_m = R\tan\theta_m.$$

In experiments such as this, the y_m's are often much smaller than R. This means that θ_m is very small, $\tan\theta_m$ is very nearly equal to $\sin\theta_m$, and

$$y_m = R\sin\theta_m.$$

Combining this relationship with Equation 26.4, we obtain the equation for constructive interference in Young's experiment:

Constructive interference, Young's experiment

$$y_m = R\frac{m\lambda}{d} \qquad (m = 0, \pm1, \pm2, \cdots).$$
(26.6)

We can measure R and d, as well as the positions y_m of the bright fringes, so this experiment provides a direct measurement of the wavelength λ. Young's experiment was in fact the first direct measurement of wavelengths of light.

With this same approximation, the condition for formation of a *dark* fringe is

$$y_m = R\frac{(m + \frac{1}{2})\lambda}{d} \qquad (m = 0, \pm1, \pm2, \cdots).$$

NOTE ▶ The distance between adjacent bright bands in the pattern is *inversely* proportional to the distance d between the slits. The closer together the slits are, the more the pattern spreads out. When the slits are far apart, the bands in the pattern are closer together. ◀

Quantitative Analysis 26.1   **Constructive interference**

The experimental arrangement in Figure 26.5 is used to observe a two-slit interference pattern. Which of the following changes would increase the separation between the bright bands? (A lightbulb filament emits all the colors of the visible spectrum.) More than one answer may be correct.

A. Move the screen closer to the slits.
B. Replace the green filter by a red one.
C. Increase the width of each slit.
D. Decrease the separation of the slits.
E. Move the filament closer to the slits.

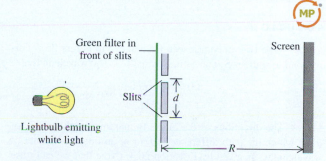

▲ **FIGURE 26.5**

SOLUTION Equation 26.6, $y_m = R(m\lambda/d)$, predicts the results of this experiment. The separation of the bright lines (fringes) decreases with the distance R to the screen, so A won't work. Red light has a greater wavelength λ than green light, so B will work. The width of each slit has an effect on the pattern, but not on the separation, of the bright bands, so C won't work. The separation of the fringes is inversely related to the separation d of the slits, so D will work. The distance from the source to the slits has no effect on any part of the pattern, so E won't work.

EXAMPLE 26.1   Determining wavelength

A two-slit interference experiment is used to determine the unknown wavelength of a laser light source, as shown in Figure 26.6. With the slits 0.200 mm apart and a screen at a distance of 1.00 m, the third bright band out from the central bright band is found to be 9.49 mm from the center of the screen. **(a)** What is the wavelength of the light? **(b)** How far apart would the slits have to be so that the fourth minimum (dark band) would occur at 9.49 mm from the center of the screen?

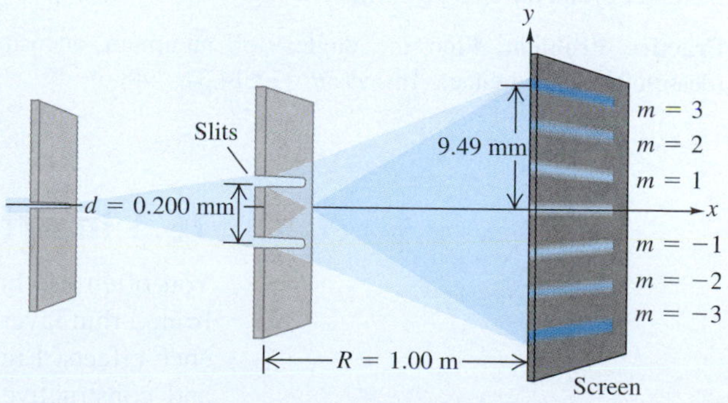

▶ **FIGURE 26.6**

Continued

SOLUTION

SET UP AND SOLVE Part (a): From Equation 26.6, with $m = 3$,

$$\lambda = \frac{y_m d}{mR} = \frac{(9.49 \times 10^{-3}\ \text{m})(0.200 \times 10^{-3}\ \text{m})}{(3)(1.00\ \text{m})}$$

$$= 633 \times 10^{-9}\ \text{m} = 633\ \text{nm}.$$

Part (b): For destructive interference, and therefore a minimum, $m = 0$ gives the first minimum, $m = 1$ the second, $m = 2$ the third, and $m = 3$ the fourth. Thus,

$$d = \frac{\left(m + \frac{1}{2}\right)\lambda R}{y_m} = \frac{\left(3 + \frac{1}{2}\right)(633 \times 10^{-9}\ \text{m})(1.00\ \text{m})}{9.49 \times 10^{-3}\ \text{m}}$$

$$= 2.33 \times 10^{-4}\ \text{m} = 0.233\ \text{mm}.$$

REFLECT This wavelength corresponds to the red light from a helium–neon laser.

Practice Problem: If a green laser pointer with $\lambda = 523$ nm is used, where would the screen have to be placed in order to find the third bright fringe 9.49 mm from the central fringe? *Answer: R = 1.21 m.*

EXAMPLE 26.2 **Broadcast pattern of a radio station**

A radio station operating at a frequency of 1500 kHz $= 1.5 \times 10^6$ Hz (near the top end of the AM broadcast band) has two identical vertical dipole antennas spaced 400 m apart, oscillating in phase. At distances much greater than 400 m, in what directions is the intensity greatest in the resulting radiation pattern?

SOLUTION

SET UP The two antennas, seen from above in Figure 26.7, correspond to sources S_1 and S_2 in Figure 26.3. The wavelength is

$$\lambda = \frac{c}{f} = \frac{3.0 \times 10^8\ \text{m/s}}{1.5 \times 10^6\ \text{Hz}} = 200\ \text{m}.$$

SOLVE The directions of the intensity maxima are the values of θ for which the path difference is zero or an integral number of wavelengths, as given by Equation 26.4. (Note that we can't use Equation 26.6, because the angles θ_m aren't necessarily small.) Inserting the numerical values, with $m = 0, \pm 1, \pm 2$, we find that

$$\sin\theta = \frac{m\lambda}{d} = \frac{m(200\ \text{m})}{400\ \text{m}} = \frac{m}{2}, \qquad \theta = 0, \pm 30°, \pm 90°.$$

REFLECT In this example, values of m greater than 2 or less than -2 give values of $\sin\theta$ greater than 1 or less than -1, which is impossible. There is no direction for which the path difference is three or more wavelengths. Thus, in this example, values of m of ± 3 and beyond have no physical meaning.

Practice Problem: Find the angles for minimum intensity (destructive interference). *Answer: $\theta = \pm 14.5°, \pm 48.6°$.*

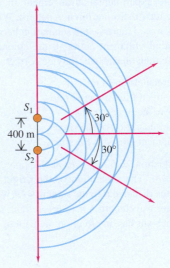

▲ **FIGURE 26.7**

26.3 Interference in Thin Films

You often see bright bands of color when light reflects from a soap bubble or from a thin layer of oil floating on water. These bands are the results of interference effects. Light waves are reflected from opposite surfaces of the thin films, and constructive interference between the two reflected waves (with different path lengths) occurs in different places for different wavelengths. The situation is shown schematically in Figure 26.8a. Light shining on the upper surface of a thin film with thickness t is partly reflected at the upper surface (path *abc*). Light *transmitted* at the upper surface is partly reflected at the lower surface (path *abdef*). The two reflected waves come together at point P on the retina of the eye. Depending on the phase relationship, they may interfere constructively or destructively. Different colors have different wavelengths, so the interference may be constructive for some colors and destructive for others. That's why you see rainbow-colored rings or fringes, like the ones in Figure 26.8b.

Here's an example involving *monochromatic* light reflected from two nearly parallel surfaces at nearly normal incidence. Figure 26.9 shows two plates of glass separated by a thin wedge of air. We want to consider interference between the two light waves reflected from the surfaces adjacent to the air wedge, as shown. (Reflections also occur at the top surface of the upper plate and the bottom surface of the lower plate; to keep our discussion simple, we won't include those reflections.) The situation is the same as in Figure 26.8, except that the thickness of the film isn't uniform. The path difference between the two waves is just twice the thickness t of the air wedge at each point. At points where $2t$ is an integral number of wavelengths, we expect to see constructive interference and a bright area; where $2t$ is a half-integral number of wavelengths, we expect to see destructive interference and a dark area. Along the line where the plates are in contact, there is practically *no* path difference, and we expect a bright area.

When we carry out the experiment, the bright and dark fringes appear, but they are interchanged! Along the line of contact, we find a *dark* fringe, not a bright one. This suggests that one or the other of the reflected waves has undergone a half-cycle phase shift during its reflection. In that case, the two waves reflected at the line of contact are a half cycle out of phase even though they have the same path length.

This unexpected phase shift illustrates a general principle that we can state as follows: When a wave traveling in medium a is reflected at an interface between this material and a different material b, there may or may not be an additional phase shift associated with the reflection, depending on the refractive indexes n_a and n_b of the two materials. If the second material (b) has *greater* refractive index than the first (a) (that is, when $n_b > n_a$), the reflected wave undergoes a half-cycle phase shift during reflection. When the second material (b) has *smaller* refractive index than the first (a) (that is, when $n_b < n_a$), there is *no* phase shift.

In fact, this phase shift behavior is predicted by Maxwell's equations from a detailed description of electromagnetic waves. The derivation is beyond the scope of this text, but let's check the prediction with the situation shown in Figure 26.9. For the wave reflected from the upper surface of the air wedge, n_b (air) is less than n_a (glass), so this wave has zero phase shift. For the wave reflected from the lower surface, n_b (glass) is greater than n_a (air), so this wave has a half-cycle phase shift. Waves reflected from the line of contact have no path difference to give additional phase shifts, and they interfere destructively; this is what we observe. We invite you to use the foregoing principle to show that, in Figure 26.8, the wave reflected at point b is shifted by a half cycle and the wave reflected at d is not (assuming the index of the bottom layer is less than n).

We can summarize this discussion symbolically: For a film with thickness t and light at normal (perpendicular) incidence, the reflected waves from the two surfaces interfere *constructively* if neither or both have a half-cycle reflection phase shift whenever the condition

$$2t = m\lambda \qquad (m = 0, 1, 2, \cdots),\qquad (26.7)$$

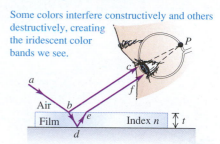

Light reflected from the upper and lower surfaces of the film comes together in the eye at P and undergoes interference.

Some colors interfere constructively and others destructively, creating the iridescent color bands we see.

(a) Interference between rays reflected from the two surfaces of a thin film

(b) The rainbow fringes of an oil slick on water

▲ **FIGURE 26.8** Interference patterns produced by reflection from a thin film.

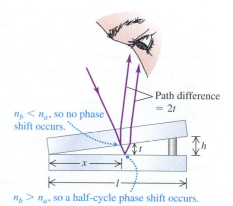

$n_b < n_a$, so no phase shift occurs.

Path difference $= 2t$

$n_b > n_a$, so a half-cycle phase shift occurs.

▲ **FIGURE 26.9** Interference between two light waves reflected from the two sides of an air wedge separating two glass plates.

(where λ is the wavelength *in the film*), is satisfied. However, when *one* of the two waves has a half-cycle reflection phase shift, the same equation is the condition for *destructive* interference.

Similarly, if neither wave, or both, has a half-cycle phase shift, the condition for *destructive* interference in the reflected waves is

$$2t = \left(m + \tfrac{1}{2}\right)\lambda \qquad (m = 0, 1, 2, \cdots).\qquad(26.8)$$

However, if one wave has a half-cycle phase shift, the same equation is the condition for *constructive* interference.

Quantitative Analysis 26.2

Destructive interference

Two very flat planes of glass touch at one end, but are held apart at the other end by a tiny filament of nylon placed between them. When the planes of glass are viewed by reflection from a sodium lamp, 23 evenly spaced dark bands are counted. The width of the filament

A. is 23 wavelengths of sodium light.
B. is 11 wavelengths of sodium light.
C. cannot be determined without a knowledge of the length of the glass plates.

SOLUTION Equation 26.7 gives the thickness of the film of air between the two glass plates at the locations of the dark bands. The first dark band occurs where the thickness is zero—that is, where the glass planes touch. The second occurs where the air thickness is $\lambda/2$, the third where it is $2\lambda/2$, and the fourth where it is $3\lambda/2$. Continuing this pattern, we see that, at the 23rd dark line, the thickness is $22\lambda/2$, or 11 wavelengths of light. If the filament is just beyond the 23rd dark line, it will actually be just a little thicker than 11 wavelengths of light.

EXAMPLE 26.3   Interference fringes

Suppose the two glass plates in Figure 26.10 are two microscope slides 10 cm long. At one end, they are in contact; at the other end, they are separated by a piece of paper 0.020 mm thick. What is the spacing of the interference fringes seen by reflection? Is the fringe at the line of contact bright or dark? Assume monochromatic light with $\lambda_0 = 500$ nm.

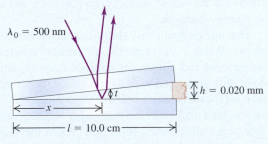

▶ **FIGURE 26.10**

SOLUTION

SET UP AND SOLVE For the wave reflected from the upper surface of the air wedge, n_b (air) is less than n_a (glass), so this wave has zero phase shift. For the wave reflected from the lower surface of the air wedge, n_b (glass) is greater than n_a (air), so this wave has a half-cycle wave shift. Thus, the fringe at the line of contact is dark.

The condition for *destructive* interference (a dark fringe) is Equation 26.7:

$$2t = m\lambda \qquad (m = 0, 1, 2, \cdots).$$

From similar triangles in Figure 26.10, the thickness t at each point is proportional to the distance x from the line of contact:

$$\frac{t}{x} = \frac{h}{l}.$$

Combining this equation with Equation 26.7, we find that

$$\frac{2xh}{l} = m\lambda;$$

$$x = m\frac{l\lambda}{2h} = m\frac{(0.10\text{ m})(500 \times 10^{-9}\text{ m})}{(2)(0.020 \times 10^{-3}\text{ m})} = m(1.25\text{ mm}).$$

Successive dark fringes, corresponding to successive integer values of m, are spaced 1.25 mm apart.

REFLECT Both glass slides have refractive indexes greater than unity. The indexes don't need to be equal; the results would be the same if the bottom slide had $n = 1.50$ and the top slide had $n = 1.60$.

Practice Problem: Suppose we use two layers of paper, so that $h = 0.040 \times 10^{-3}$ m. What is the fringe spacing? *Answer:* 0.62 mm.

EXAMPLE 26.4   A thin film of water

Suppose the glass plates in Example 26.3 have $n = 1.52$ and the space between the plates contains water $(n = 1.33)$ instead of air. What happens now?

SOLUTION

SET UP AND SOLVE The glass has a greater refractive index than either water or air, so the phase shifts are the same as in Example 26.3. Thus, the fringe at the line of contact is still dark.

The wavelength in water is

$$\lambda = \frac{\lambda_0}{n} = \frac{500 \text{ nm}}{1.33} = 376 \text{ nm}.$$

The fringe spacing is reduced by a factor of 1.33 and is therefore equal to 0.94 mm.

REFLECT If the space between the glass plates were filled with a liquid, such as carbon disulfide, that had a *greater* index of refraction than glass, a phase shift would occur at the top surface, and not at the bottom surface, of the liquid wedge. However, the waves from the contact point would still be out of phase by a half cycle, and the fringe at the line of contact would still be dark.

EXAMPLE 26.5   Grease as a thin film

Suppose the upper of the two plates in Example 26.3 is made of a plastic material with $n = 1.40$, the wedge is filled with a silicone grease having $n = 1.50$, and the bottom plate is of a dense flint glass with $n = 1.60$ (Figure 26.11). What is the spacing of the inter-ference fringes seen by reflection? Is the fringe at the line of contact bright or dark?

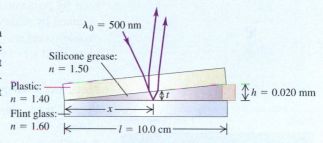

▶ **FIGURE 26.11**

SOLUTION

SET UP For the wave reflected from the upper surface of the silicone grease wedge, n_b (silicone grease) is greater than n_a (plastic), so this wave has a half-cycle phase shift. For the wave reflected from the lower surface of the silicone grease wedge, n_b (flint glass), is greater than n_a (silicone grease), so this wave also has a half-cycle wave shift. The fringe spacing is again determined by the wavelength in the film (that is, in the silicone grease), which is

$$\lambda = \frac{\lambda_0}{n} = \frac{500 \text{ nm}}{1.50} = 333 \text{ nm}.$$

With half-cycle phase shifts at both surfaces, the condition for constructive interference (a bright fringe) is $2t = m\lambda$ $(m = 0, 1, 2, \cdots)$.

SOLVE With half-cycle phase shifts at both surfaces, the fringe at the line of contact is bright. From the similar triangles in Fig-ure 26.11,

$$\frac{t}{x} = \frac{h}{l}.$$

Combining this equation with Equation 26.7, we find that

$$\frac{2xh}{l} = m\lambda,$$

$$x = m\frac{l\lambda}{2h} = m\frac{(0.10 \text{ m})(333 \times 10^{-9} \text{ m})}{(2)(0.020 \times 10^{-3} \text{ m})} = m(0.833 \text{ mm}).$$

Successive bright fringes, corresponding to successive integer values of m, are spaced 0.833 mm apart.

REFLECT In this case, both reflected waves undergo half-cycle phase shifts on reflection, so the *relative* phase doesn't change. Thus, the line of contact is a bright fringe.

Practice Problem: Suppose we repeat this experiment with light having a wavelength in vacuum of $\lambda_0 = 600$ nm. What is the fringe spacing? *Answer:* 1.0 mm.

The results of Examples 26.4 and 26.5 show that the fringe spacing is propor-tional to the wavelength of the light used. Thus, for a given choice of materials, the fringes are farther apart with red light (larger λ) than with blue light (smaller λ). If we use white light, the reflected light at any point is a mixture of wave-lengths for which constructive interference occurs; the wavelengths that interfere

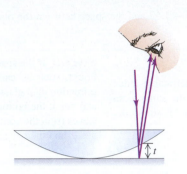

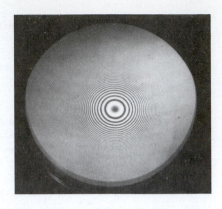

(a) A convex lens in contact with a glass plane

(b) Newton's rings: circular interference fringes

▲ **FIGURE 26.12** Newton's rings.

destructively are weak or absent in the reflected light. But the colors that are weak in the *reflected* light are strong in the *transmitted* light. At any point, the color of the wedge as viewed by reflected light is the *complement* of its color as seen by transmitted light! Roughly speaking, the complement of any color is obtained by removing that color from white light (a mixture of all colors in the visible spectrum). For example, the complement of blue is yellow, and the complement of green is magenta.

Newton's Rings

Figure 26.12a shows the convex surface of a lens in contact with a plane glass plate. A thin film of air is formed between the two surfaces. When you view the setup with monochromatic light, you see circular interference fringes (Figure 26.12b). These fringes were studied by Newton and are called *Newton's rings*. When viewed by reflected light, the center of the pattern is black. Can you see why a black center should be expected?

We can compare the surfaces of two optical parts by placing the two in contact and observing the interference fringes that are formed. Figure 26.13 is a photograph made during the grinding of a telescope objective lens. The lower, larger-diameter, thicker disk is the correctly shaped master, and the smaller upper disk is the lens under test. The "contour lines" are Newton's interference fringes; each one indicates an additional distance of a half-wavelength between the specimen and the master. At 10 lines from the center spot, the distance between the two surfaces is 5 wavelengths, or about 0.003 mm. This isn't very good; high-quality lenses are routinely ground with a precision of less than 1 wavelength. The surface of the primary mirror of the Hubble Space Telescope was ground to a precision of better than $\frac{1}{50}$ wavelength.

Fringes map lack of fit between lens and master.

Master   Lens being tested

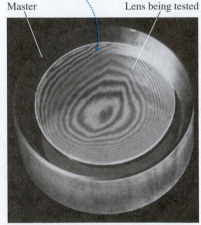

▲ **FIGURE 26.13** The surface of a telescope objective lens under inspection.

◀ **Application See my smile.** The Mona Lisa by Leonardo da Vinci may be the most famous painting of all time; millions of visitors come to the Louvre in Paris to see her every year. For security, the painting is protected by a bulletproof glass case. If this case did not have a nonreflective coating, visitors would perpetually be bobbing their heads trying to see the painting without the glare of obscuring reflections.

▶ **BIO** Application **Structural color.** Many of the most brilliant colors in the animal world are created by *structures* rather than pigments. Iridescence is always structural, involving thin films or regularly spaced structures that cause interference. These photos show the shining wings of the butterfly *Morpho rhetenor* and the structure that creates this effect. Butterfly wings are covered with small scales that can be intricately sculpted. In this butterfly, the scales have a profusion of tiny ridges; these carry regularly spaced flanges that function as reflectors, interacting constructively for blue light. The multilayered structure reflects very efficiently, giving the wings an almost mirrorlike brilliance.

Nonreflective Coatings

Nonreflective coatings for lens surfaces make use of thin-film interference. A thin layer or film of hard transparent material with an index of refraction smaller than that of the lens material is deposited on the lens surface, as shown in Figure 26.14. Light is reflected from both surfaces of the layer. In both reflections, the light is reflected from a medium of greater index than that in which it is traveling, so the same phase change occurs. If the film thickness is a quarter (one-fourth) of the wavelength *in the film* (assuming normal incidence), the total path difference is a half-wavelength. Light reflected from the first surface is then a half cycle out of phase with light reflected from the second surface, and there is destructive interference.

The thickness of the nonreflective coating can be a quarter-wavelength for only one particular wavelength, usually chosen in the central yellow-green portion of the spectrum (550 nm), where the eye is most sensitive. Then there is somewhat more reflection at both longer and shorter wavelengths, and the reflected light has a purple hue. The overall reflection from a lens or prism surface can be reduced in this way from 4–5% to less than 1%. This treatment is particularly important in eliminating stray reflected light in highly corrected lenses with many air–glass surfaces. The same principle is used to minimize reflection from silicon photovoltaic solar cells by means of a thin surface layer of silicon monoxide (SiO).

Reflective Coatings

If a material with an index of refraction *greater* than that of glass is deposited on glass, forming a film with a thickness of a quarter-wavelength, then the reflectivity of the glass is *increased*. In this case, there is a half-cycle phase shift at the air–film interface, but none at the film–glass interface, and reflections from the two sides of the film interfere constructively. For example, a coating with an index of refraction of 2.5 allows 38% of the incident energy to be reflected, compared with 4% or so with no coating. By the use of multiple-layer coatings, it is possible to achieve nearly 100% transmission or reflection for particular wavelengths. These coatings are used for infrared "heat reflectors" in motion-picture projectors, solar cells, and astronauts' visors and for color separation in color television cameras, to mention only a few applications.

26.4 Diffraction

According to *geometric* optics, when an opaque object is placed between a point light source and a screen, as in Figure 26.15, the shadow of the object should form a perfectly sharp line. No light at all should strike the screen at points within the shadow. But we've already seen in this chapter that the *wave* nature of light causes things to happen that can't be understood with the simple model of geometric optics. The edge of the shadow is never perfectly sharp. Some light appears in the area that we expect to be in the shadow, and we find alternating bright and dark fringes in the illuminated area. More generally, light emerging from apertures doesn't behave precisely according to the predictions of the

Destructive interference occurs when
• the film is about $\frac{1}{4}\lambda$ thick and
• the light undergoes a phase change at both reflecting surfaces,
so that the two reflected waves emerge from the film about $\frac{1}{2}$ cycle out of phase.

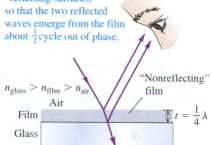

▲ **FIGURE 26.14** How a nonreflective coating works.

Geometric optics predicts that this situation should produce a sharp boundary between illumination and solid shadow. That's NOT what really happens!

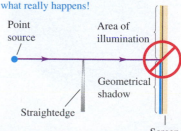

▲ **FIGURE 26.15** Setup for the shadow produced when a sharp edge obstructs monochromatic light from a point source.

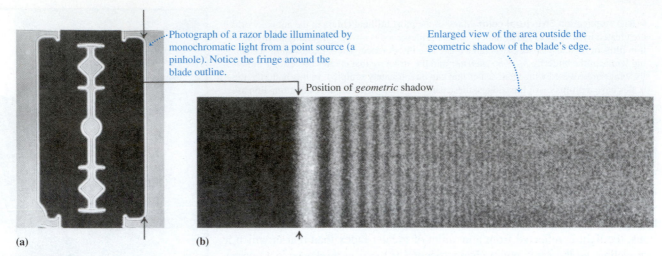

Photograph of a razor blade illuminated by monochromatic light from a point source (a pinhole). Notice the fringe around the blade outline.

Enlarged view of the area outside the geometric shadow of the blade's edge.

Position of *geometric* shadow

(a)　　　(b)

▲ **FIGURE 26.16** Actual shadow of a razor blade produced in a situation like that of Figure 26.15.

straight-line ray model of geometric optics. An important class of such effects occurs when light strikes a barrier with an aperture or an edge. The interference patterns formed in such a situation are grouped under the heading **diffraction.**

Here's an example: The photograph in Figure 26.16a was made by placing a razor blade halfway between a pinhole, illuminated by monochromatic light, and a photographic film. The film recorded the shadow cast by the blade. Figure 26.16b is an enlargement of a region near the shadow of the right edge of the blade. The position of the *geometric* shadow line is indicated by arrows. The area outside the geometric shadow is bordered by alternating bright and dark bands. There is some light in the shadow region, although this is not very visible in the photograph. The first bright band, just outside the geometric shadow, is actually *brighter* than in the region of uniform illumination to the extreme right. This simple experiment gives us some idea of the richness and complexity of a seemingly simple phenomenon: the casting of a shadow by an opaque object.

We don't often observe diffraction patterns such as Figure 26.16 in everyday life, because most ordinary light sources are neither monochromatic nor point sources. If a frosted lightbulb were used instead of a point source in Figure 26.15, each wavelength of the light from every point of the bulb would form its own diffraction pattern, and the patterns would overlap to such an extent that we wouldn't see any individual pattern.

Diffraction is sometimes described as "the bending of light around an obstacle." But the process that causes diffraction effects is present in the propagation of *every* wave. When part of the wave is cut off by some obstacle, we observe diffraction effects that result from interference of the remaining parts of the wave fronts. Every optical instrument uses only a limited portion of a wave; for example, a telescope uses only the part of a wave admitted by its objective lens. Thus, diffraction plays a role in nearly all optical phenomena.

Figure 26.17 shows a diffraction pattern formed by a steel ball about 3 mm in diameter. Note the rings in the pattern, both outside and inside the geometric shadow area, and the bright spot at the very center of the shadow. The existence of this spot was predicted in 1818, on the basis of a wave theory of light, by the French mathematician Siméon-Denis Poisson, during an extended debate in the French Academy of Sciences concerning the nature of light. Ironically, Poisson himself was *not* a believer in the wave theory of light, and he published this *apparently* absurd prediction hoping to deal a death blow to the wave theory. But the prize committee of the Academy arranged for an experimental test, and soon the bright spot was actually observed. (It had in fact been seen as early as 1723, but those experiments had gone unnoticed.)

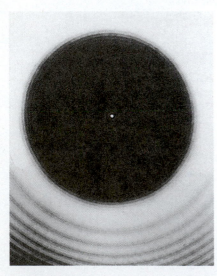

▲ **FIGURE 26.17** Fresnel diffraction pattern formed by a steel ball 3 mm in diameter. The Poisson bright spot is seen at the center of the area in shadow.

The fringe patterns formed by diffraction effects can be analyzed using Huygens's principle (Section 23.11). Let's review that principle briefly. Every point of a wave front can be considered the source of secondary wavelets that spread out in all directions with a speed equal to the speed of propagation of the wave. The position of the wave front at any later time is the *envelope* of the secondary wavelets at that time. To find the resultant displacement at any point, we combine all the individual displacements produced by these secondary wavelets, using the superposition principle and taking into account their amplitudes and relative phases.

In Figure 26.15, both the point source and the screen are at finite distances from the obstacle forming the diffraction pattern. This situation is described as *near-field diffraction,* or **Fresnel diffraction** (after Augustin Jean Fresnel, 1788–1827). If the source, obstacle, and screen are far enough away that all lines from the source to the obstacle can be considered parallel and all lines from the obstacle to a point in the pattern can be considered parallel, the phenomenon is called *far-field diffraction,* or **Fraunhofer diffraction** (after Joseph von Fraunhofer, 1787–1826). Fraunhofer diffraction is usually simpler to analyze in detail than is Fresnel diffraction, so we'll confine our discussion in this chapter to Fraunhofer diffraction.

Finally, we emphasize that there is no fundamental distinction between *interference* and *diffraction.* In preceding sections of this chapter, we used the term *interference* for effects involving waves from a small number of sources, usually two. *Diffraction* usually involves a *continuous* distribution of Huygens's wavelets across the area of an aperture or a very large number of sources or apertures. But the same physical principles—superposition and Huygens's principle—govern both categories.

26.5 Diffraction from a Single Slit

In this section, we'll discuss the diffraction pattern formed by plane-wave (parallel-ray) monochromatic light when it emerges from a long, narrow slit, as shown in Figure 26.18. (We call the narrow dimension the *width,* even though, in this figure, it is a vertical dimension. We'll assume that the slit width a is much greater than the wavelength λ of the light.) According to geometric optics, the transmitted beam should have the same cross section as the slit, as in Figure 26.18a. What is *actually* observed, however, is the pattern shown in Figure 26.18b. The beam spreads out vertically after passing through the slit. The diffraction pattern consists of a central bright band, which may be much broader than the width of the vertical slit, bordered by alternating dark and

MasteringPHYSICS

PhET: Wave Interference
ActivPhysics 16.6: Single-Slit Diffraction

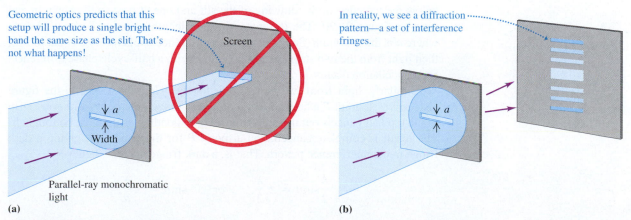

Geometric optics predicts that this setup will produce a single bright band the same size as the slit. That's not what happens!

Screen

$\downarrow a$

Width

Parallel-ray monochromatic light

(a)

In reality, we see a diffraction pattern—a set of interference fringes.

$\downarrow a$

(b)

▲ **FIGURE 26.18** Comparison of the result predicted by geometric optics and the actual result when monochromatic light passes through a narrow slit and illuminates a screen.

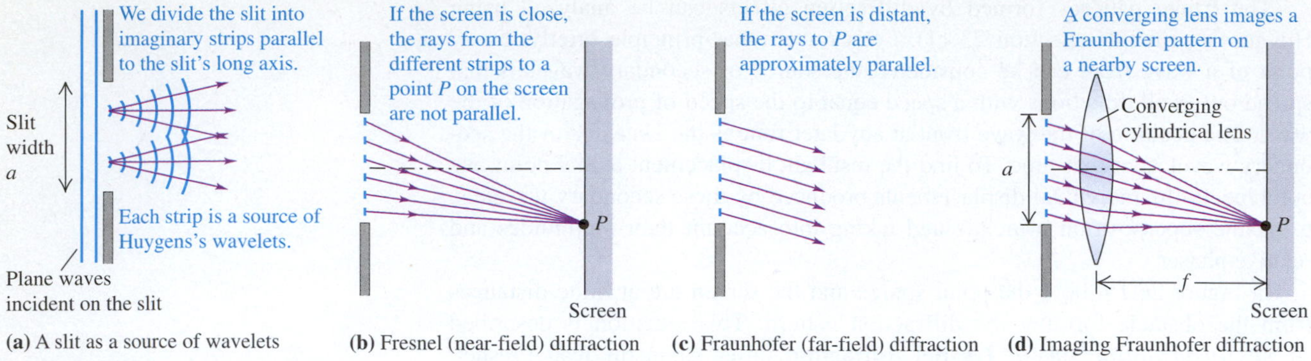

(a) A slit as a source of wavelets

We divide the slit into imaginary strips parallel to the slit's long axis.

Slit width a

Each strip is a source of Huygens's wavelets.

Plane waves incident on the slit

(b) Fresnel (near-field) diffraction

If the screen is close, the rays from the different strips to a point P on the screen are not parallel.

Screen

(c) Fraunhofer (far-field) diffraction

If the screen is distant, the rays to P are approximately parallel.

(d) Imaging Fraunhofer diffraction

A converging lens images a Fraunhofer pattern on a nearby screen.

Converging cylindrical lens

a

f

P

Screen

▲ **FIGURE 26.19** Analysis of diffraction by a single rectangular slit.

bright bands with rapidly decreasing intensity. When a is much greater than λ, about 85% of the total energy is in this central bright band, whose width is found to be *inversely* proportional to the width of the slit. You can easily observe a similar diffraction pattern by looking at a point source, such as a distant street light, through a narrow slit formed between two fingers in front of your eye. The retina of your eye then corresponds to the screen.

Figure 26.19 shows a side view of the same setup; the long sides of the slit are perpendicular to the figure. According to Huygens's principle, each element of area of the slit opening can be considered as a source of secondary wavelets. In particular, imagine dividing the slit into several narrow strips with equal width, parallel to the long edges and perpendicular to the page in Figure 26.19a. Cylindrical secondary wavelets spread out from each strip, as shown in cross section.

In Figure 26.19b, a screen is placed to the right of the slit. We can calculate the resultant intensity at a point P on the screen by adding the contributions from the individual wavelets, taking proper account of their various phases and amplitudes. We assume that the screen is far enough away that all the rays from various parts of the slit to a particular point P on the screen are parallel, as in Figure 26.19c. An equivalent situation is Figure 26.19d, in which the rays to the lens are parallel and the lens forms a reduced image of the same pattern that would be formed on an infinitely distant screen without the lens. We might expect that the various light paths through the lens would introduce additional phase shifts, but in fact it can be shown that all the paths have *equal* phase shifts, so this is not a problem.

The situation of Figure 26.19b is Fresnel diffraction; those of Figures 26.19c and 26.19d, in which the outgoing rays are considered parallel, are Fraunhofer diffraction. We can derive quite simply the most important characteristics of the diffraction pattern from a single slit. First consider two narrow strips, one just below the top edge of the drawing of the slit and one at its center, shown in end view in Figure 26.20. The difference in path length to point P is $(a/2)\sin\theta$, where a is the slit width. Suppose this path difference happens to be equal to $\lambda/2$; then light from the two strips arrives at point P with a half-cycle phase difference, and cancellation occurs.

Similarly, light from two strips immediately *below* those two in the figure also arrives at point P a half-cycle out of phase. In fact, the light from *every* strip in the top half cancels out the light from a corresponding strip in the bottom half. The result is complete cancellation at point P for the entire slit, giving a dark fringe in the interference pattern. That is, a dark fringe occurs whenever

$$\frac{a}{2}\sin\theta = \pm\frac{\lambda}{2}, \qquad \text{or} \qquad \sin\theta = \pm\frac{\lambda}{a}.$$

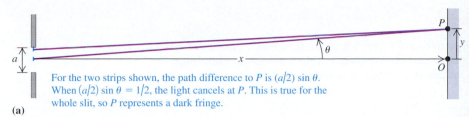

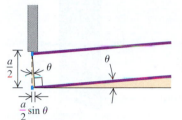

(a)

For the two strips shown, the path difference to P is $(a/2) \sin \theta$.
When $(a/2) \sin \theta = 1/2$, the light cancels at P. This is true for the whole slit, so P represents a dark fringe.

θ is usually very small, so we can use the approximations $\sin \theta = \theta$ and $\tan \theta = \theta$. Then the condition for a dark band is

$$y_m = R\frac{m\lambda}{a}.$$

(b) Enlarged view of the top half of the slit

▲ **FIGURE 26.20** Geometry of rays passing through a single slit.

We may also divide the screen into quarters, sixths, and so on and use the preceding argument to show that a dark fringe occurs whenever $\sin\theta = \pm 2\lambda/a$, $\pm 3\lambda/a$, and so on. Thus, the condition for a *dark* fringe is

$$\sin\theta = \frac{m\lambda}{a} \qquad (m = \pm 1, \pm 2, \pm 3, \cdots). \qquad (26.9)$$

For example, if the slit width is equal to 10 wavelengths $(a = 10\lambda)$, dark fringes occur at $\sin\theta = \pm\frac{1}{10}, \pm\frac{2}{10}, \pm\frac{3}{10}, \cdots$. Between the dark fringes are bright fringes.

NOTE ▶ In the "straight ahead" direction $(\sin\theta = 0)$ is a *bright* band; in this case, light from the entire slit arrives at P in phase. It would be wrong to put $m = 0$ in Equation 26.9. The central bright fringe is wider than the others, as Figure 26.18 shows. In the small-angle approximation we'll use in the following discussion, it is exactly *twice* as wide. ◀

With light, the wavelength λ is on the order of 500 nm $= 5 \times 10^{-7}$ m. This is often much smaller than the slit width a; a typical slit width is 10^{-2} cm $= 10^{-4}$ m. Therefore, the values of θ in Equation 26.9 are often so small that the approximation $\sin\theta = \theta$ is very good. In that case, we can rewrite that equation as

$$\theta = \frac{m\lambda}{a} \qquad (m = \pm 1, \pm 2, \pm 3, \cdots).$$

Also, if the distance from slit to screen is R, and the vertical distance of the mth dark band from the center of the pattern is y_m, then $\tan\theta = y_m/R$. For small θ, we may approximate $\tan\theta$ by θ, and we then find that

$$y_m = R\frac{m\lambda}{a} \qquad (m = \pm 1, \pm 2, \pm 3, \cdots). \qquad (26.10)$$

NOTE ▶ This equation has the same form as the equation for the two-slit pattern, Equation 26.6, but here it gives the positions of the *dark* fringes in a *single-slit* pattern rather than the *bright* fringes in a *double-slit* pattern. Also, $m = 0$ is *not* a dark fringe. Be careful! ◀

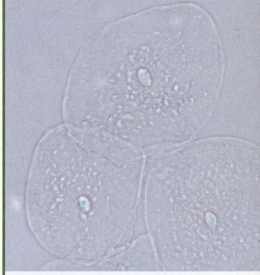

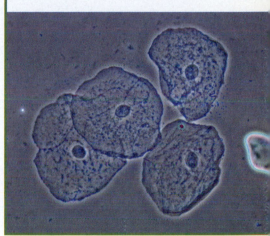

▲ **BIO Application Enhancing the image.** A major problem facing the biological or medical microscopist is the generation of contrast in the object under study. If the object strongly absorbs light, then the outline of the object will be clearly seen; however, individual cells are often thin and transparent, and the cells are nearly invisible in the microscope. One solution is to stain the cells, but this generally kills them. A far better solution was achieved by Frits Zernike, who applied the principles of wave optics to create a microscope in which the small modifications in the phase of the illuminating light due to slight differences in the index of refraction between the cell and the surrounding medium were converted to differences in light amplitude. Unlike phase differences, amplitude differences can be detected by the eye. The upper photograph shows living human epithelial cells in ordinary bright-field view, and the lower one shows the same cells in phase-contrast microscopy. Zernike was awarded the Nobel Prize in Physics in 1953, and the phase-contrast technique is still used routinely in research laboratories.

EXAMPLE 26.6 **A single-slit experiment**

You pass 633 nm (helium–neon) laser light through a narrow slit and observe the diffraction pattern on a screen 6.0 m away. You find that the distance between the centers of the first minima (dark fringes) on either side of the central bright fringe in the pattern is 27 mm. How wide is the slit?

SOLUTION

SET UP The angle θ in this situation is very small, so we can use the approximate relation of Equation 26.10. Figure 26.21 shows the various distances. The distance y_1 from the central maximum to the first minimum on either side is half the distance between the two first minima, so $y_1 = (27 \text{ mm})/2$.

SOLVE Solving Equation 26.10 for the slit width a and substituting $m = 1$, we find that

$$a = \frac{R\lambda}{y_1} = \frac{(6.0 \text{ m})(633 \times 10^{-9} \text{ m})}{(27 \times 10^{-3} \text{ m})/2}$$

$$= 2.8 \times 10^{-4} \text{ m} = 0.28 \text{ mm}.$$

REFLECT It can also be shown that the distance between the *second* minima on the two sides is $2(27 \text{ mm})$, and so on.

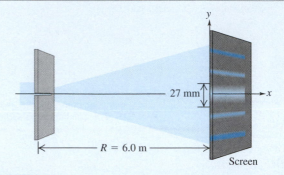

▲ **FIGURE 26.21**

Practice Problem: What is the distance between the fifth minima on the two sides of the central bright fringe? *Answer:* $2y_5 = 5(27 \text{ mm}) = 135 \text{ mm}$.

In the preceding analysis, we've located the maxima and minima in the diffraction pattern formed by a single slit. It's also possible to calculate the intensity at any point in the pattern, using Huygens's principle and some nontrivial mathematical analysis. We'll omit the details, but Figure 26.22a is a graph of intensity as a function of position for a single-slit pattern. Figure 26.22b is a photograph of the same pattern. Figure 26.23 shows how the width of the diffraction pattern varies inversely with that of the slit: The narrower the slit, the wider is the diffraction pattern.

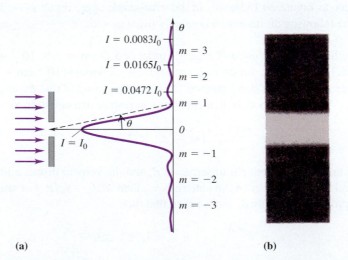

(a)   (b)

▲ **FIGURE 26.22** (a) Intensity distribution for diffraction from a single slit. (b) Photograph of the Fraunhofer diffraction pattern of a single slit.

If the slit width is equal to or narrower than the wavelength, only one broad maximum forms.

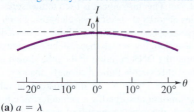

(a) $a = \lambda$

The wider the slit (or the shorter the wavelength), the narrower and sharper is the central peak.

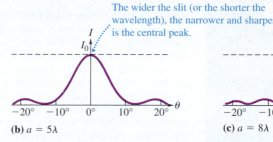

(b) $a = 5\lambda$

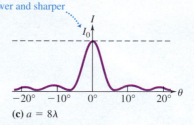

(c) $a = 8\lambda$

▲ **FIGURE 26.23** Effect of slit width or wavelength on the diffraction pattern formed by a single slit.

26.6 Multiple Slits and Diffraction Gratings

In Sections 26.1 and 26.2, we analyzed interference from two point sources or from two very thin slits. In Section 26.5, we carried out a similar analysis for a single slit with finite width. Now let's consider patterns produced by *several* very narrow slits. Assume that each slit is narrow in comparison to the wavelength, so its diffraction pattern spreads out nearly uniformly. Figure 26.24 shows an array of several narrow slits, with distance d between adjacent slits. Constructive interference occurs for those rays that are at an angle θ to the normal and that arrive at point P when the path difference between adjacent slits is an integral number of wavelengths:

$$d\sin\theta = m\lambda \qquad (m = 0, \pm1, \pm2, \cdots). \qquad (26.11)$$

That is, the maxima in the pattern occur at the same positions as for a two-slit pattern with the same spacing (Equation 26.4). In this respect, the pattern resembles the two-slit pattern.

But what happens *between* the maxima? In the two-slit pattern, there is exactly one intensity minimum between each pair of maxima, corresponding to angles for which

$$d\sin\theta = \left(m + \tfrac{1}{2}\right)\lambda \qquad (m = 0, 1, 2, \cdots),$$

or for which the phase difference between waves from the two sources is $\frac{1}{2}$ cycle, $\frac{3}{2}$ cycle, $\frac{5}{2}$ cycle, and so on. (See Figure 26.25a.) In the eight-slit pattern, these are also minima, because the light from adjacent slits cancels out in pairs. Detailed

Maxima occur where the path difference for adjacent slits is a whole number of wavelengths: $d\sin\theta = m\lambda$.

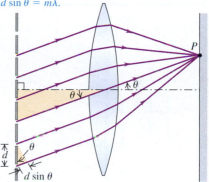

▲ **FIGURE 26.24** In multiple-slit diffraction, rays from every slit arrive in phase to give a sharp maximum if the path difference between adjacent slits is a whole number of wavelengths.

Two slits produce one minimum between adjacent maxima.

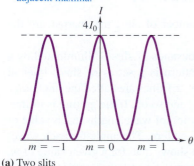

(a) Two slits

Eight slits produce larger, narrower maxima in the same locations, separated by seven minima.

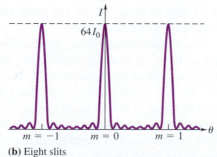

(b) Eight slits

With sixteen slits, the maxima are still taller and narrower, with more intervening minima.

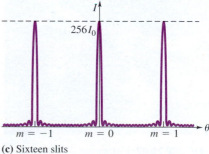

(c) Sixteen slits

▲ **FIGURE 26.25** Effect of the number of slits on a diffraction pattern, for a given slit width and spacing.

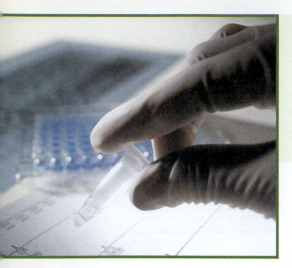

calculation shows that the interference pattern is as shown in Figure 26.25b. The maxima are in the same positions as for the two-slit pattern of Figure 26.25a, but they are much narrower. Figure 26.25c shows the corresponding pattern with 16 slits. Increasing the number of slits in an interference experiment while keeping the spacing of adjacent slits constant gives interference patterns with the maxima in the same positions as with two slits, but progressively sharper and narrower.

Diffraction Gratings

An array of a large number of parallel slits, all with the same width a and spaced equal distances d between centers, is called a **diffraction grating.** The first one was constructed out of fine wires by Fraunhofer. Gratings can be made by using a diamond point to scratch many equally spaced grooves on a glass or metal surface or by photographically reducing a pattern of black and white stripes drawn on paper with a pen or a computer. For a grating, what we have been calling *slits* are often called *rulings* or *lines*.

In Figure 26.26, GG' is a cross section of a grating; the slits are perpendicular to the plane of the page. Only six slits are shown in the diagram, but an actual grating may contain several thousand. The spacing d between centers of adjacent slits is called the *grating spacing;* typically, it is about 0.002 mm. A plane wave of monochromatic light is incident normally on the grating from the left side. We assume that the pattern is formed on a screen far enough away that all rays emerging from the grating in a particular direction are parallel.

We noted earlier that the principal intensity maxima occur in the same directions as for the two-slit pattern, directions for which the path difference for adjacent slits is an integer number of wavelengths. So the positions of the maxima are once again given by Equation 26.11:

$$d\sin\theta = m\lambda \qquad (m = 0, \pm1, \pm2, \pm3, \cdots).$$

As we saw in Figure 26.25, the larger the number of slits, the sharper are the peaks of the diffraction pattern.

When a grating containing hundreds or thousands of slits is illuminated by a parallel beam of monochromatic light, the pattern is a series of sharp lines at angles determined by Equation 26.11. The $m = \pm1$ lines are called the *first-order lines,* the $m = \pm2$ lines the *second-order* lines, and so on. If the grating is illuminated by white light with a continuous distribution of wavelengths, each value of m corresponds to a continuous spectrum in the pattern. The angle for each wavelength is determined by Equation 26.11, which shows that, for a given value of m, long wavelengths (at the red end of the spectrum) lie at larger angles (i.e., are deviated more from the straight-ahead direction) than the shorter wavelengths (at the violet end of the spectrum). Figure 26.27 shows a familiar example.

As Equation 26.11 shows, the sines of the deviation angles of the maxima are proportional to the ratio λ/d. For substantial deviation to occur, the grating

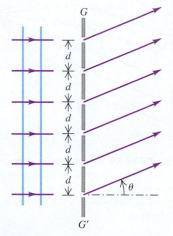

▲ **FIGURE 26.26** A portion of a transmission diffraction grating.

▲ **FIGURE 26.27** Unless you download all your music electronically, you probably have a few diffraction gratings around the house. The tracks in a commercial CD act as a reflection diffraction grating for visible light, producing rainbow patterns.

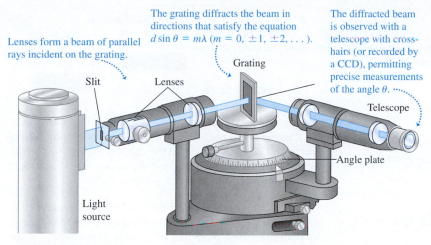

The grating diffracts the beam in directions that satisfy the equation $d \sin \theta = m\lambda$ ($m = 0, \pm 1, \pm 2, \ldots$).

The diffracted beam is observed with a telescope with cross-hairs (or recorded by a CCD), permitting precise measurements of the angle θ.

Lenses form a beam of parallel rays incident on the grating.

Grating

Slit

Lenses

Telescope

Angle plate

Light source

▲ **FIGURE 26.28** A diffraction grating spectrometer.

spacing d should be of the same order of magnitude as the wavelength λ. Gratings for use in the visible spectrum usually have about 500 to 1500 slits per millimeter, so d (which equals the reciprocal of the number of slits per unit width) is on the order of 1000 nm.

Grating Spectrometers

Diffraction gratings are widely used in spectrometry as a means of dispersing a light beam into spectra. If the grating spacing is known, then we can measure the angles of deviation and use Equation 26.11 to compute the wavelength. A typical setup is shown in Figure 26.28. A prism can also be used to disperse the various wavelengths through different angles, because the index of refraction always varies with wavelength. But there is no simple relationship that describes this variation, so a spectrometer using a prism has to be calibrated with known wavelengths that are determined in some other way. Another difference is that a prism deviates red light the least and violet the most, while a grating does the opposite.

Mastering**PHYSICS**

ActivPhysics 16.4: The Grating: Introduction and Qualitative Questions
ActivPhysics 16.5: The Grating: Problems

EXAMPLE 26.7 **Width of a grating spectrum**

The wavelengths of the visible spectrum are approximately 400 nm (violet) through 700 nm (red). Find the angular width of the first-order visible spectrum produced by a plane grating with 600 lines per millimeter when white light falls normally onto the grating.

SOLUTION

SET UP The first-order spectrum corresponds to $m = 1$. The grating has 600 lines per millimeter, so the grating spacing (the distance d between adjacent lines is

$$d = \frac{1}{600} \text{ mm} = 1.67 \times 10^{-6} \text{ m}.$$

SOLVE From Equation 26.11, with $m = 1$, the angular deviation θ_v of the violet light is given by

$$\sin\theta_v = \frac{m\lambda}{d} = \frac{(1)400 \times 10^{-9} \text{ m}}{1.67 \times 10^{-6} \text{ m}} = 0.240,$$

so that

$$\theta_v = 13.9°.$$

The angular deviation of the red light is given by

$$\sin\theta_r = \frac{700 \times 10^{-9} \text{ m}}{1.67 \times 10^{-6} \text{ m}} = 0.419;$$

thus,

$$\theta_r = 24.8°.$$

Therefore, the angular width of the first-order visible spectrum is

$$24.8° - 13.9° = 10.9°.$$

REFLECT Increasing the number of lines per millimeter in the grating increases the angular width of the spectrum, but the grating spacing d can't be less than the wavelengths of the spectrum being observed. Do you see why?

Practice Problem: If the angular width of the first-order visible spectrum is instead $20.5° - 11.5° = 9.0°$, how many lines per millimeter are there in the grating? *Answer:* 500 lines/mm.

EXAMPLE 26.8 **Spectra that overlap**

Show that, in the situation of Example 26.7, the violet end of the third-order spectrum overlaps the red end of the second-order spectrum (Figure 26.29).

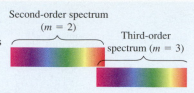

Second-order spectrum
(m = 2)

Third-order
spectrum (m = 3)

▶ **FIGURE 26.29**

SOLUTION

SET UP AND SOLVE From Equation 26.11, with $m = 3$, the angular deviation of the third-order violet end is given by

$$\sin\theta_v = \frac{(3)(400 \times 10^{-9}\,\text{m})}{d} = \frac{1.20 \times 10^{-6}\,\text{m}}{d}.$$

The deviation of the second-order red end, with $m = 2$, is given by

$$\sin\theta_r = \frac{(2)(700 \times 10^{-9}\,\text{m})}{d} = \frac{1.40 \times 10^{-6}\,\text{m}}{d}.$$

These two equations show that no matter what the grating spacing d is, the largest angle θ_r (at the red end) for the second-order spectrum is always greater than the smallest angle θ_v (at the violet end) for the third-order spectrum, so the second and third orders *always* overlap, as shown in Figure 26.29.

REFLECT If we could separate the third-order and second-order spectra, then we *would* be able to distinguish the third-order violet from the second-order red. In actuality, they overlap, so the separate spectra can't be viewed clearly.

26.7 X-Ray Diffraction

X rays were discovered by Wilhelm Röntgen (1845–1923) in 1895, and early experiments suggested that they were electromagnetic waves with wavelengths on the order of 10^{-10} m. At about the same time, the idea began to emerge that the atoms in a crystalline solid are arranged in a lattice in a regular repeating pattern, with spacing between adjacent atoms also on the order of 10^{-10} m. Putting these two ideas together, Max von Laue (1879–1960) proposed in 1912 that a crystal might serve as a kind of three-dimensional diffraction grating for x rays. That is, a beam of x rays might be scattered (absorbed and re-emitted) by the individual atoms in a crystal, and the scattered waves might interfere just like waves from a diffraction grating.

The first experiments in **x-ray diffraction** were performed in 1912 by Walther Friederich, Paul Knipping, and von Laue, using the experimental setup sketched in Figure 26.30a. The scattered x rays *did* form an interference pattern, which the three scientists recorded on photographic film. Figure 26.30b is a photograph of such a pattern. These experiments verified that x rays *are* waves, or at least have wavelike properties, and also that the atoms in a crystal *are*

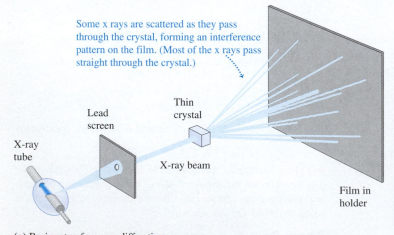

Some x rays are scattered as they pass through the crystal, forming an interference pattern on the film. (Most of the x rays pass straight through the crystal.)

Lead screen

Thin crystal

X-ray tube

X-ray beam

Film in holder

(a) Basic setup for x-ray diffraction

(b) Laue diffraction pattern for a thin section of quartz crystal

▲ **FIGURE 26.30** X-ray diffraction.

arranged in a regular pattern (Figure 26.31). Since that time, x-ray diffraction has proved an invaluable research tool, both for measuring wavelengths of x rays and for studying crystal structure.

To introduce the basic ideas, we consider first a two-dimensional scattering situation, as shown in Figure 26.32a, in which a plane wave is incident on a rectangular array of scattering centers. The situation might involve a ripple tank with an array of small posts, or 3 cm microwaves striking an array of small conducting spheres, or x rays incident on an array of atoms. In the case of x rays, the wave induces vibrations in the individual atoms, which then act like little antennas, emitting scattered waves. The resulting interference pattern is the superposition of all these scattered waves. The situation is different from that obtained with a diffraction grating, in which the waves from all the slits are emitted *in phase* (for a plane wave at normal incidence). With x rays, the scattered waves are *not* all in phase, because their distances from the *source* are different. To compute the interference pattern, we have to consider the *total* path differences for the scattered waves, including the distances both from source to scatterer and from scatterer to observer.

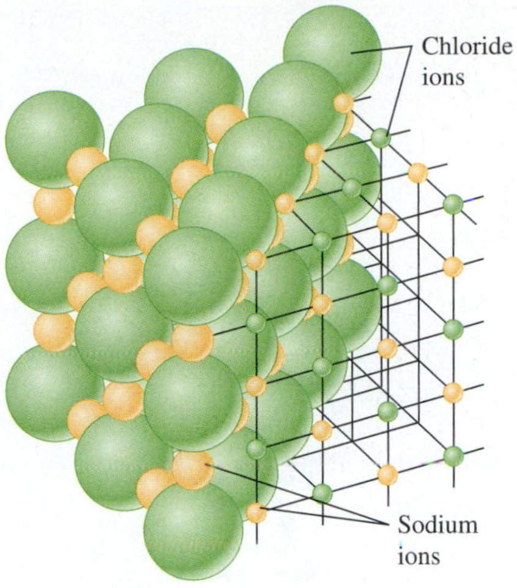

As Figure 26.32b shows, the path length from source to observer is the same for all the scatterers in a single row if the angles θ_a and θ_r are equal. Scattered radiation from *adjacent* rows is *also* in phase if the path difference for adjacent rows is an integral number of wavelengths. Figure 26.32c shows that this path difference is $2d\sin\theta$. Therefore, the conditions for radiation from the *entire* array to reach the observer in phase are that (1) the angle of incidence must equal the angle of scattering and (2) the path difference for adjacent rows must equal $m\lambda$, where m is an integer. We can express the second condition as

$$2d\sin\theta = m\lambda \qquad (m = 1, 2, 3, \cdots).\qquad (26.12)$$

In directions for which this condition is satisfied, we see a strong maximum in the interference pattern. We can describe this interference in terms of *reflections* of the wave from the horizontal rows of scatterers in Figure 26.32a. Strong reflection (constructive interference) occurs at angles such that the angles of the incident and scattered x rays are equal and Equation 26.12 is satisfied.

NOTE ▶ The angle θ is customarily measured with respect to the surface of the crystal. This approach is different from our usual one, in which we measure θ with respect to the normal to the plane of an array of slits or a grating. Also, Equation 26.12 is *not* the same as Equation 26.11. Be careful! ◀

We can extend this discussion to a three-dimensional array by considering *planes* of scatterers instead of *rows*. Figure 26.33 shows two different sets of parallel planes that pass through all the scatterers. Waves from all the scatterers in a

▲ **FIGURE 26.31** Model of the arrangement of ions in a crystal of NaCl (table salt). The spacing of adjacent ions is 0.282 nm. For convenience, the ions are represented as spheres, although their electron clouds actually overlap somewhat.

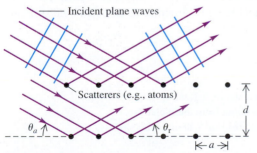

(a) Scattering of waves from a rectangular array

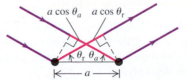

(b) Scattering from adjacent atoms in row

(c) Scattering from atoms in adjacent rows

▲ **FIGURE 26.32** Scattering of waves from rows of atoms (or other scatterers) in a two-dimensional rectangular array.

Spacing of planes is $d = a/\sqrt{2}$.

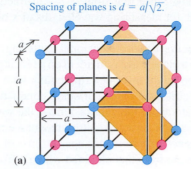

Spacing of planes is $d = a/\sqrt{3}$.

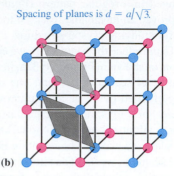

(a) (b)

▲ **FIGURE 26.33** Cubic crystal lattice showing two different families of crystal planes. There are also three sets of planes parallel to the cube faces, with spacing a.

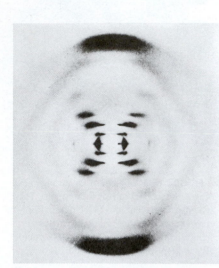

The historic x-ray diffraction image of DNA fibers obtained by Rosalind Franklin in 1953. This image was central to the elucidation of the double-helix structure of DNA.

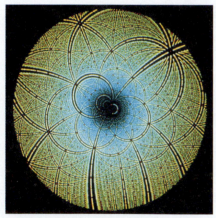

The x-ray diffraction pattern of the enzyme rubisco, which plants use to "fix" atmospheric carbon dioxide into carbohydrate.

▲ **FIGURE 26.34** Historic and modern examples of x-ray diffraction spectra.

given plane interfere constructively if the angles of incidence and scattering are equal. There is also constructive interference between planes when Equation 26.12 is satisfied, where d is now the distance between adjacent planes. Because there are many different sets of parallel planes, there are many values of d and many sets of angles that give constructive interference for the whole crystal lattice. This phenomenon is called **Bragg reflection,** and Equation 26.12 is called the **Bragg condition,** in honor of Sir William Bragg and his son Laurence Bragg, two pioneers in x-ray analysis.

NOTE ▶ Don't let the term *reflection* obscure the fact that we are dealing with an *interference* effect. In fact, the reflections from various planes are closely analogous to interference effects in thin films (Section 26.3). ◀

As Figure 26.30b shows, nearly complete cancellation occurs for all but certain very specific directions, where constructive interference occurs and forms bright spots. Such a pattern is usually called an x-ray *diffraction* pattern, although *interference* pattern might be more appropriate. This particular type of pattern is called a *Laue pattern*.

If the crystal lattice spacing is known, we can determine the wavelength of the x rays (just as we determined wavelengths of visible light in Section 26.6 by measuring diffraction patterns from slits or gratings). For example, we can determine the crystal lattice spacing for sodium chloride from its density and Avogadro's number. Then, once we know the x-ray wavelength, we can use x-ray diffraction to explore the structure and lattice spacing of crystals with unknown structure.

X-ray diffraction is by far the most important experimental tool in the investigation of the crystal structure of solids. Atomic spacings in crystals can be measured precisely, and the lattice structure of complex crystals can be determined. X-ray diffraction also plays an important role in studies of the structures of liquids and of organic molecules. Indeed, it was one of the chief experimental techniques used in working out the double-helix structure of DNA and the structures of proteins (Figure 26.34).

EXAMPLE 26.9 **X-ray diffraction with silicon crystal**

Suppose you direct an x-ray beam with wavelength 0.154 nm at certain planes of a silicon crystal. As you increase the angle of incidence from zero, you find the first strong interference maximum from these planes when the beam makes an angle of 34.5° with them. **(a)** How far apart are the planes? **(b)** Will you find other interference maxima from these planes at larger angles?

Continued

SOLUTION

SET UP AND SOLVE **Part (a):** To find the plane spacing d, we solve the Bragg equation (Equation 26.12) for d and set $m = 1$:

$$d = \frac{m\lambda}{2\sin\theta} = \frac{(1)(0.154 \text{ nm})}{2\sin 34.5°} = 0.136 \text{ nm}.$$

This is the distance between adjacent planes.

Part (b): To calculate other angles, we solve Equation 26.12 for $\sin\theta$:

$$\sin\theta = \frac{m\lambda}{2d} = m\frac{0.154 \text{ nm}}{2(0.136 \text{ nm})} = m(0.566).$$

REFLECT Values of m of 2 or greater give values of $\sin\theta$ greater than unity, which is impossible. Therefore, there are no other angles for interference maxima for this particular set of crystal planes.

26.8 Circular Apertures and Resolving Power

We've studied in detail the diffraction patterns formed by long, thin slits or arrays of such slits. But an aperture of *any* shape forms a diffraction pattern. The diffraction pattern formed by a *circular* aperture is of special interest because of its role in limiting the resolving power of optical instruments. In principle, we could compute the intensity at any point P in the diffraction pattern by dividing the area of the aperture into small elements, finding the resulting wave amplitude and phase at P, and then summing all these elements to find the resultant amplitude and intensity at P. In practice, this calculation requires the use of integral calculus with numerical approximations, and we won't pursue it further here. We'll simply *describe* the pattern and quote a few relevant numbers.

The diffraction pattern formed by a circular aperture consists of a central bright spot surrounded by a series of bright and dark rings, as shown in Figure 26.35. We can describe the pattern in terms of the angle θ, representing the angular size of each ring. If the aperture diameter is D and the wavelength is λ, then the angular size θ_1 of the first *dark* ring is given by the following expression:

MasteringPHYSICS®

ActivPhysics 16.7: Circular Hole Diffraction
ActivPhysics 16.8: Resolving Power

First dark ring from a circular aperture

$$\sin\theta_1 = 1.22\frac{\lambda}{D}. \qquad (26.13)$$

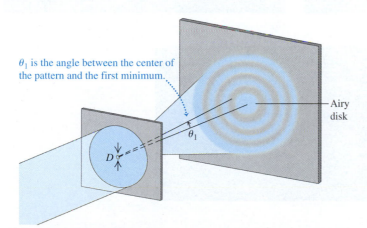

θ_1 is the angle between the center of the pattern and the first minimum.

Airy disk

D

θ_1

▲ **FIGURE 26.35** Diffraction pattern formed by a circular aperture of diameter D. The pattern consists of a central bright spot and alternating bright and dark rings. (Not to scale.)

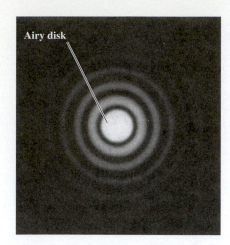

▲ FIGURE 26.36 Diffraction pattern formed by a circular aperture. The Airy disk is over-exposed so that the other rings may be seen.

The angular sizes $(\theta_2 \text{ and } \theta_3)$ of the next two dark rings are given, respectively, by

$$\sin\theta_2 = 2.23\frac{\lambda}{D}, \qquad \sin\theta_3 = 3.24\frac{\lambda}{D}.$$

The central bright spot is called the **Airy disk,** in honor of Sir George Airy (1801–1892), Astronomer Royal of England, who first derived the expression for the intensity in the pattern. The angular size of the Airy disk is that of the first dark ring, given by Equation 26.13.

The *intensities* in the bright rings drop off very quickly. When D is much larger than the wavelength λ, which is the usual case for optical instruments, the peak intensity in the center of the first bright ring is only 1.7% of the value at the center of the Airy disk, and at the center of the second bright ring it is only 0.4%. Most (85%) of the total energy falls within the Airy disk. Figure 26.36 shows a diffraction pattern from a circular aperture 1.0 mm in diameter.

This analysis has far-reaching implications for image formation by lenses and mirrors. In our study of optical instruments in Chapter 25, we assumed that a lens with focal length f focuses a parallel beam (plane wave) to a *point* at a distance f from the lens. This assumption ignored diffraction effects. We now see that what we get is not a point, but the diffraction pattern just described. If we have two point objects, their images are not two points, but two diffraction patterns. When the objects are close together, their diffraction patterns overlap; if they are close enough, their patterns overlap almost completely and cannot be distinguished. The effect is shown in Figure 26.37, which shows the patterns for four (very small) "point" objects. In Figure 26.37a, the images of objects 1 and 2 are well separated, but the images of objects 3 and 4 on the right have merged together. In

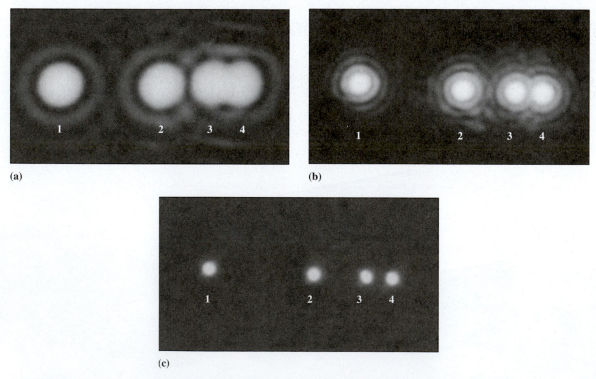

(a)   (b)

(c)

▲ FIGURE 26.37 Diffraction pattern of four "point" sources as seen through a circular opening whose size increases from (a) to (c). In (a), the aperture is so small that the patterns of sources 3 and 4 overlap and are barely resolved by Rayleigh's criterion (discussed later in text). Increasing the aperture size decreases the size of the diffraction patterns, as shown in (b) and (c).

Figure 26.37b, with a larger aperture diameter and resulting smaller Airy disks, images 3 and 4 are better resolved. In Figure 26.37c, with a still larger aperture, they are well resolved.

A widely used criterion for the resolution of two point objects, proposed by Lord Rayleigh (1887–1905) and called **Rayleigh's criterion,** is that the objects are just barely resolved (that is, distinguishable) if the center of one diffraction pattern coincides with the first minimum of the other. In that case, the angular separation of the image centers is given by Equation 26.13. The angular separation of the *objects* is the same as that of the *images,* so two point objects are barely resolved, according to Rayleigh's criterion, when their angular separation is given by Equation 26.13. The angles are usually very small, so $\sin\theta \simeq \theta$, and we have

Rayleigh's criterion

The minimum angular separation of two objects that can barely be resolved by an optical instrument is called the **limit of resolution, θ_{res},** of the instrument:

$$\theta_{res} = 1.22\frac{\lambda}{D}. \tag{26.14}$$

The smaller the limit of resolution, the greater the *resolution,* or **resolving power,** of the instrument. Diffraction sets the ultimate limits on the resolution of lenses and mirrors. *Geometric* optics may make it seem that we can make images as large as we like. Eventually, though, we always reach a point at which the image becomes larger, but does not gain in detail. The images in Figure 26.37 would not become sharper with further enlargement.

EXAMPLE 26.10 Resolving power of the human eye

The iris of the eye is a circular aperture that allows light to pass into the eye. **(a)** For an iris with radius 0.25 cm, and for visible light with a wavelength of 550 nm, what is the resolving angle, or limiting resolution, of the eye, based on Rayleigh's criterion? **(b)** In fact, the actual limiting resolution of the human eye is about four times poorer: $\theta_{res} = 4\theta_{Rayleigh}$. What is the farthest distance s from a tree that you could stand and resolve two birds sitting on a limb, separated transversely by a distance $y = 10$ cm?

SOLUTION

SET UP AND SOLVE Part (a): Figure 26.38 shows the situation. We use Equation 26.14, with the diameter $D = 2 \times$ radius $= 2(0.25\text{ cm}) = 0.50\text{ cm} = 5.0 \times 10^{-3}$ m:

$$\theta_{Rayleigh} = 1.22\frac{\lambda}{D} = (1.22)\frac{(550 \times 10^{-9}\text{ m})}{(5.0 \times 10^{-3}\text{ m})}$$
$$= 1.34 \times 10^{-4}\text{ rad.}$$

Part (b): Using $\theta_{res} = 4\theta_{Rayleigh}$, we find that the actual limiting resolution of the human eye is

$$\theta_{res} = 4\theta_{Rayleigh} = 4(1.34 \times 10^{-4}\text{ rad}) = 5.4 \times 10^{-4}\text{ rad.}$$

To find the farthest distance s from the tree that you could stand and still resolve two birds separated by 10 cm, we use the small-angle approximation

$$\sin\theta \simeq \tan\theta \simeq \theta \simeq \frac{y}{s}, \text{ and}$$

$$s = \frac{y}{\theta} = \frac{10\text{ cm}}{5.4 \times 10^{-4}} = 19,000\text{ cm} = 190\text{ m.}$$

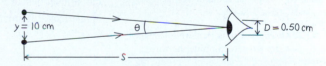

▲ **FIGURE 26.38**

REFLECT This result is a rather optimistic estimate of the resolving power of the human eye. Many other factors, including the illumination level and small defects of vision, also act to limit the actual resolution.

Practice Problem: For a person whose vision has angular resolution eight times that given by the Rayleigh criterion, what is the maximum distance at which the person can distinguish two birds sitting 10 cm apart on a tree limb? *Answer:* 93 m.

Radio interferometry. The Very Large Array in Soccoro, New Mexico, consists of 27 radio dishes that can be moved on tracks; at their greatest separation, their resolution equals that of a single dish 36 km across.

Optical interferometry. The four 8.2 m telescopes of the European Southern Observatory's Very Large Telescope in Cerro Paranal, Chile, can be combined optically in pairs. Functioning together, the outer two telescopes have the resolution of a single much larger telescope.

▲ **FIGURE 26.39** Some modern telescopes use *interferometry* to combine the waves from several source telescopes, producing an image that has the resolving power of a telescope with the diameter of the whole array.

An important lesson to be learned from this analysis is that resolution improves with shorter wavelengths. Thus, ultraviolet microscopes have higher resolution than visible-light microscopes. In electron microscopes, the resolution is limited by the wavelengths associated with the wavelike behavior of electrons. These wavelengths can be made 100,000 times smaller than wavelengths of visible light, with a corresponding gain in resolution. Finally, one reason for building very large reflecting telescopes is to increase the aperture diameter and thus minimize diffraction effects. Such telescopes also provide a greater light-gathering area for viewing very faint stars.

The Hubble Space Telescope, launched April 25, 1990, from the space shuttle *Discovery,* has a mirror diameter of 2.4 m. The telescope was designed to resolve objects 2.8×10^{-7} rad apart with 550 nm light. This was at least a factor of six better than the much larger earth-based telescopes of the time, whose resolving power was limited primarily by the distorting effects of atmospheric turbulence. Many of the current generation of earth-based telescopes use *adaptive optics* to counteract atmospheric distortion. Adaptive optics is a technique in which computer-controlled actuators distort the telescope mirror in real time to compensate for the effects of the atmosphere, allowing such telescope to achieve resolutions closer to their native diffraction limits.

Diffraction is an important consideration for satellite "dishes"—parabolic reflectors designed to receive satellite transmission. Satellite dishes have to be able to pick up transmissions from two satellites that are only a few degrees apart and are transmitting at the same frequency. The need to resolve two such transmissions determines the minimum diameter of the dish. As higher frequencies are used, the diameter needed decreases. For example, when two satellites 5.0° apart broadcast 7.5 cm microwaves, the minimum dish diameter required to resolve them (by Rayleigh's criterion) is about 1.0 m.

The effective diameter of a telescope can be increased in some cases by using arrays of smaller telescopes. The Very Long Baseline Array (VLBA), a group of 10 radio telescopes scattered at locations from Mauna Kea to the Virgin Islands, has a maximum separation of about 8000 km and can resolve radio signals to 10^{-8} rad. This astonishing resolution is comparable, in the optical realm, to seeing a parked car on the moon. The same technique is harder to apply to optical telescopes because of the shorter wavelengths involved. Nevertheless, large optical telescopes are now achieving this goal (Figure 26.39). The use of widely spaced satellite arrays will increase resolution even more in the future.

26.9 Holography

Holography is a technique for recording and reproducing an image of an object without the use of lenses. Unlike the two-dimensional images recorded by an ordinary photograph or television system, a holographic image is truly three dimensional. Such an image can be viewed from different directions to reveal different sides and from various distances to reveal changing perspective. If you had never seen a hologram, you wouldn't believe it was possible!

The basic procedure for making a hologram is shown in Figure 26.40a. We illuminate the object to be holographed with monochromatic light, and we place a photographic film so that it is struck by scattered light from the object (the object beam) and also by direct light from the source (the reference beam). In practice, the source must be a laser, for reasons that we'll discuss later. Interference between the direct and scattered light leads to the formation and recording of a complex interference pattern on the film.

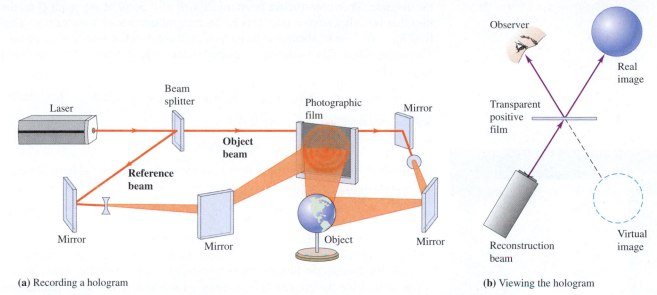

(a) Recording a hologram

(b) Viewing the hologram

▲ **FIGURE 26.40** (a) A hologram is the record on film of the interference pattern formed with light from a coherent source and light scattered from an object. (b) Images are formed when light is projected through the hologram.

To form the images, we simply project light (the reconstruction beam) through the developed film, as shown in Figure 26.40b. Two images are formed: a virtual image on the side of the film nearer the source and a real image on the opposite side.

A complete analysis of holography is beyond our scope, but we can gain some insight into the process by looking at how a single point is imaged to form a hologram. Consider the interference pattern formed on a photographic film by the superposition of an incident plane wave and a spherical wave, as shown in Figure 26.41a. The spherical wave originates at a point source P at a distance b_0 from the film; P may in fact be a small object that scatters part of the incident plane wave. We assume that the two waves are monochromatic and coherent and that the phase relation is such that constructive interference occurs at point O on

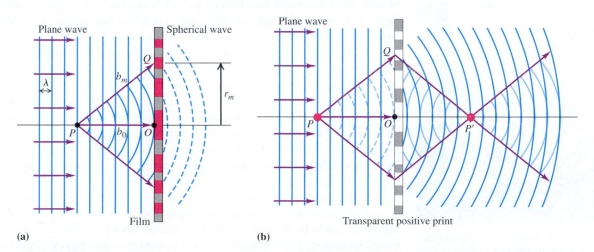

(a)

(b)

▲ **FIGURE 26.41** (a) Constructive interference of plane and spherical waves occurs in the plane of the film at every point Q for which the distance b_m from P is greater than the distance b_0 from P to O by an integral number of wavelengths $m\lambda$. For the point shown, $m = 2$. (b) When a plane wave strikes the developed film, the diffracted wave consists of a wave converging to P' and then diverging again and a diverging wave that appears to originate at P. These waves form real and virtual images, respectively.

the diagram. Then constructive interference will *also* occur at any point Q on the film that is farther from P than O is by an integral number of wavelengths. That is, if $b_m - b_0 = m\lambda$, where m is an integer, then constructive interference occurs. The points where this condition is satisfied form circles centered at O, with radii r_m given by

$$b_m - b_0 = \sqrt{b_0^2 + r_m^2} - b_0 = m\lambda \qquad (m = 1, 2, 3, \cdots). \quad (26.15)$$

Solving this equation for r_m^2, we find that

$$r_m^2 = \lambda(2mb_0 + m^2\lambda).$$

Ordinarily, b_0 is very much larger than λ, so we neglect the second term in parentheses, obtaining

$$r_m = \sqrt{2m\lambda b_0} \qquad (m = 1, 2, 3, \cdots). \quad (26.16)$$

The interference pattern consists of a series of concentric bright circular fringes with radii given by Equation 26.16. Between these bright fringes are dark fringes.

Now we develop the film and make a transparent positive print, so the bright-fringe areas have the greatest transparency on the film. Then we illuminate the print with monochromatic plane-wave light of the same wavelength λ that we used initially. In Figure 26.41b, consider a point P' at a distance b_0 along the axis from the film. The centers of successive bright fringes differ in their distances from P' by an integral number of wavelengths; therefore, a strong *maximum* in the diffracted wave occurs at P'. That is, light converges to P' and then diverges from it on the opposite side. Hence, P' is a *real image* of point P.

This is not the entire diffracted wave, however. The interference of the wavelets that spread out from all the transparent areas form a second spherical wave that is diverging rather than converging. When traced back behind the film, this wave appears to be spreading out from point P. Thus, the total diffracted wave from the hologram is a superposition of a spherical wave converging to form a real image at P' and a spherical wave that diverges as though it had come from the virtual image point P.

Because of the principle of linear superposition, what is true for the imaging of a single point is also true for the imaging of any number of points. The film records the superposed interference pattern from the various points, and when light is projected through the film, the various image points are reproduced simultaneously. Thus, the images of an extended object can be recorded and reproduced just as they can for a single point object. Figure 26.42 shows photographs of a holographic image from two different angles, revealing the changing perspective in this three-dimensional image.

In making a hologram, we have to overcome several practical problems. First, the light used must be *coherent* over distances that are large in comparison to the dimensions of the object and its distance from the film. Ordinary light sources *do not* satisfy this requirement, for reasons that we discussed in Section 26.1. Therefore, laser light is essential for making a hologram. (However, many common kinds of holograms can be *viewed* with ordinary light.)

Second, extreme mechanical stability is needed. If any relative motion of the source, object, or film occurs during exposure, even by as much as a wavelength, the interference pattern on the film is blurred enough to prevent satisfactory image formation. These obstacles are not insurmountable, however, and holography promises to become increasingly important in research, entertainment, and a wide variety of technological applications.

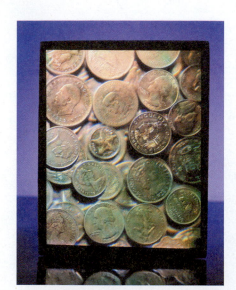

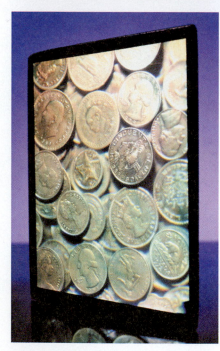

▲ **FIGURE 26.42** Two views of a hologram, showing how the perspective of the image changes with the angle from which it is viewed.

SUMMARY

Interference and Coherent Sources

(Section 26.1) The overlap of waves from two sources of monochromatic light forms an **interference pattern.** The principle of linear superposition states that the total wave disturbance at any point is the sum of the disturbances from the separate waves. **Constructive interference** results when two waves arrive at a point in phase; **destructive interference** results when two waves arrive at a point exactly a half cycle out of phase.

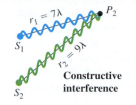
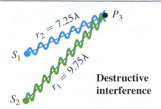

Two-Source Interference of Light

(Section 26.2) When two light sources are in phase, constructive interference occurs at a point when the difference in path length from the two sources is zero or an integral number of wavelengths; destructive interference occurs when the path difference is a half-integral number of wavelengths. When the lines from the sources to a distant point P make an angle θ with the horizontal line in the figure, the condition for constructive interference is $d \sin \theta = m\lambda$ (Equation 26.4) and the condition for destructive interference is $d \sin \theta = \left(m + \frac{1}{2}\right)\lambda$ (Equation 26.5), where $m = 0, \pm 1, \pm 2, \cdots$ for both cases.

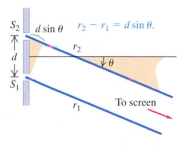

Interference in Thin Films

(Section 26.3) When light is reflected from both sides of a thin film with thickness t, constructive interference between the reflected waves occurs when $2t = m\lambda$ $(m = 0, 1, 2, \cdots)$ (Equation 26.7), unless a half-cycle phase shift occurs at only one surface; then this is the condition for destructive interference. A half-cycle phase shift occurs during reflection whenever the index of refraction of the second material is greater than that of the first.

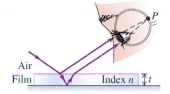

Diffraction; Single and Multiple Slits

(Sections 26.4–26.6) Diffraction occurs when light passes through an aperture or around an edge. For a single narrow slit with width a, the condition for destructive interference (a dark fringe) at a point P at an angle θ from the direction of the incident light is $\sin \theta = m\lambda/a$, where $m = \pm 1, \pm 2, \pm 3, \cdots$ (Equation 26.9).

A **diffraction grating** consists of a large number of thin parallel slits spaced a distance d apart. The condition for maximum intensity in the interference pattern is $d \sin \theta = m\lambda$, where $m = 0, \pm 1, \pm 2, \pm 3, \cdots$ (Equation 26.11). This is the same condition as for the two-source pattern, but for the grating, the maxima are very sharp and narrow.

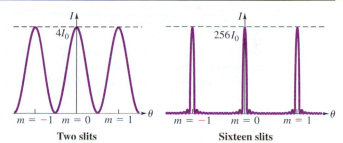

X-Ray Diffraction

(Section 26.7) A crystal having its atoms arranged in a lattice in a regularly repeating pattern can serve as a three-dimensional diffraction grating for x rays with wavelengths of the same order of magnitude as the lattice spacing, typically 10^{-10} m. The scattered x rays emerging from the crystal form an interference pattern.

Continued

Circular Apertures and Resolving Power

(**Section 26.8**) The diffraction pattern from a circular aperture with diameter D consists of a central bright spot, called the **Airy disk,** and a series of concentric dark and bright rings. The limit of resolution, determined by the angular size of the first dark ring, is $\theta_{res} = 1.22\lambda/D$ (Equation 26.14). Diffraction sets the ultimate limit on the resolution (sharpness of the image) of optical instruments. According to Rayleigh's criterion, two point objects are just barely resolved when their angular separation θ is given by Equation 26.14.

Holography

(**Section 26.9**) A hologram is a photographic record of an interference pattern formed by light scattered from an object and light coming directly from the source. When properly illuminated, a hologram forms a true three-dimensional image of the object.

 For instructor-assigned homework, go to www.masteringphysics.com

Conceptual Questions

1. Could an experiment similar to Young's two-slit experiment be performed with sound? How might it be carried out? Does it matter that sound waves are longitudinal and electromagnetic waves are transverse?

2. One refracting astronomical telescope has a tube twice as wide as another one because the objective lens is twice the diameter. Are there any advantages to the wide telescope over the narrow one? What are they?

3. A two-slit interference experiment is set up, and the fringes are displayed on a screen. Then the whole apparatus is immersed in the nearest swimming pool. How does the fringe pattern change?

4. Would the headlights of a distant car form a two-source interference pattern? If so, how might it be observed? If not, why not?

5. Coherent red light illuminates two narrow slits that are 25 cm apart. Will a two-slit interference pattern be observed when the light from the slits falls on a screen? Explain.

6. If a two-slit interference experiment were done with white light, what would be seen?

7. Could x ray diffraction effects with crystals be observed by using visible light instead of x rays? Why or why not?

8. Does a microscope have better resolution with red light or blue light? Why?

9. Around harbors, where oil from boat engines is on the water, you often see patterns of closed colored fringes, like the ones in Figure 26.8b. Why do the patterns make *closed* fringes? Why are the fringes different colors?

10. What happens to the width of the central bright *region* (not just the central point) of a single-slit diffraction pattern as you make the slit thinner and thinner?

11. What are some advantages to making a telescope with a large-diameter objective lens (or mirror)? Does the large diameter contribute to the magnification of the telescope?

12. A *very* thin soap film $(n = 1.33)$, whose thickness is much less than a wavelength of visible light, looks black; it appears to reflect no light at all. Why? By contrast, an equally thin layer of soapy water $(n = 1.33)$ on glass $(n = 1.50)$ appears quite shiny. Why is there a difference?

13. When monochromatic light passes through two thin slits, what characteristics of the light and of the slits limit the number of bright spots (interference maxima) that will occur, or are there an infinite number of them? Assume that you can detect all the bright spots, even if they are very dim.

14. If we view a double-slit interference pattern on a fairly distant screen, are the bright spots necessarily equally spaced on the screen? Consider the cases of small and large angles from the central spot.

15. Optical telescopes having a principal mirror only a few meters in diameter can produce extremely sharp images, yet radiotelescopes need to be hundreds (or even thousands) of meters in diameter to make sharp images. Why do they need to be so much larger than optical telescopes?

Multiple-Choice Problems

1. Two sources of waves are at A and B in Figure 26.43. At point P, the path difference for waves from these two sources is
   A. $x + y$.
   B. $\dfrac{x + y}{2}$.
   C. $x - y$.
   D. $\dfrac{x - y}{2}$.

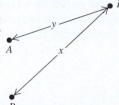

▲ **FIGURE 26.43** Multiple-choice problem 1.

2. If the sources A and B in the previous problem are emitting waves of wavelength λ that are in

phase with each other, *constructive* interference will occur at point P if (there may be more than one correct choice):

A. $x = y$.

B. $x + y = \lambda$.

C. $x - y = 2\lambda$.

D. $x - y = 5\lambda$.

3. To obtain the greatest resolution from a microscope,

A. You should view the object in long-wavelength visible light.

B. You should view the object in short-wavelength visible light.

C. You can use any wavelength of visible light, since the resolution is determined only by the diameter of the microscope lenses.

4. A monochromatic beam of laser light falls on a thin slit and produces a series of bright and dark spots on a screen. If this apparatus (slit and screen) is submerged in water, the dark spots

A. will not change their location on the screen.

B. will move away from the center spot.

C. will move toward the center spot.

5. A person is standing at a distance x from a stereo speaker that is emitting a continuous tone. She hears the sound directly from the speaker, as well as the sound reflected from a wall a distance y ($y > x$) away. (See Figure 26.44.) The path difference between these two sound waves as they reach the listener is

A. $2y$.   B. $2x$.   C. $y - x$.

D. $2y - x$.   E. $x + y$.

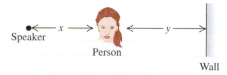

▲ **FIGURE 26.44** Multiple-choice problem 5.

6. When a thin oil film spreads out on a puddle of water, the thinnest part of the film looks dark in the resulting interference pattern. This implies that

A. the oil has a higher index of refraction than water.

B. water has a higher index of refraction than the oil.

C. the oil and water have identical indexes of refraction.

7. A laser beam of wavelength 500 nm is shone through two different diffraction gratings A and B, and the pattern is viewed on a distant screen. The pattern from grating A consists of many closely spaced bright dots, while that from B contains few dots spaced widely apart. What can you conclude about the line densities N_A and N_B (the number of lines/cm) of these two gratings?

A. $N_A > N_B$.   B. $N_A < N_B$.

C. You cannot conclude anything about the line densities without knowing the order of the bright spots.

8. A film contains a single thin slit of width a. When monochromatic light passes through this slit, the first two dark fringes on either side of the center on a distant screen are a distance x apart. If you increase the width of the slit, these two dark fringes will

A. move closer together.

B. move farther apart.

C. remain the same distance apart.

9. Light of wavelength λ strikes a pane of glass of thickness T and refractive index n, as shown in Figure 26.45. Part of the beam is reflected off the upper surface of the glass, and part is transmitted and then reflected off the lower surface of the glass. *Destructive* interference between these two beams will occur if

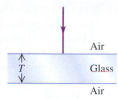

▲ **FIGURE 26.45** Multiple-choice problem 9.

A. $T = \dfrac{\lambda}{2}$.   B. $2T = \dfrac{\lambda}{2}$.

C. $T = \dfrac{\lambda}{2n}$.   D. $2T = \dfrac{\lambda}{2n}$.

10. Two thin parallel slits are a distance d apart. Monochromatic light passing through them produces a series of interference bright spots on a distant screen. If you *decrease* the distance between these slits, the bright spots will

A. move closer to the center spot.

B. move farther from the center spot.

C. not change position.

11. Laser light of wavelength λ passes through a thin slit of thickness a and produces its first dark fringes at angles of $\pm 45°$ with the original direction of the beam. The slit is then reduced in size to a *circle* of diameter a. When the same laser light is passed through the circle, the first dark fringe occurs at

A. $\pm 59.6°$.   B. $\pm 54.9°$.   C. $\pm 36.9°$.   D. $\pm 35.4°$.

12. The formula $y_m = R\dfrac{m\lambda}{d}$ for the location of the points of constructive interference from two slits is valid

A. only for large angles θ.

B. only for small angles θ.

C. for all angles θ, because it is a general formula.

13. A light beam strikes a pane of glass as shown in Figure 26.46. Part of it is reflected off the air–glass surface, and part is transmitted and then reflected off the glass–water surface. Which statements are true about this light? (There may be more than one correct choice.)

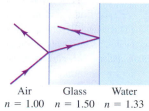

▲ **FIGURE 26.46** Multiple-choice problem 13.

A. Half-cycle phase shifts occur at both the air–glass and glass–water surfaces.

B. Half-cycle phase shifts occur at neither of the surfaces.

C. A half-cycle phase shift occurs at the glass–water surface, but not at the air–glass surface.

D. A half-cycle phase shift occurs at the air–glass surface, but not at the glass–water surface.

E. The phase shifts at both surfaces cancel each other out.

14. Light of wavelength λ and frequency f passes through a single slit of width a. The diffraction pattern is observed on a screen a distance x from the slit. Which of the following will decrease the width of the central maximum? (There may be more than one correct answer.)

A. Decrease the slit width.

B. Decrease the frequency of the light.

C. Decrease the wavelength of the light.

D. Decrease the distance x of the screen from the slit.

15. Both CDs and DVDs will flash a rainbow spectrum when viewed from certain angles. The "track pitch," or distance between rows of pits, is 1600 nm on a CD but only 740 nm on a DVD. This is part of the reason why a DVD can store much more data. Which of the two disks should produce a wider separation between red light and violet light in the reflected spectrum?
    A. The CD.
    B. The DVD.
    C. The same for both.

Problems

26.1 Interference and Coherent Sources

1. • Two small stereo speakers A and B that are 1.40 m apart are sending out sound of wavelength 34 cm in all directions and all in phase. A person at point P starts out equidistant from both speakers and walks (see Figure 26.47) so that he is always 1.50 m from speaker B. For what values of x will the sound this person hears be (a) maximally reinforced, (b) cancelled? Limit your solution to the cases where $x \leq 1.50$ m.

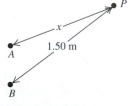
▲ **FIGURE 26.47** Problem 1.

2. • A person with a radio-wave receiver starts out equidistant from two FM radio transmitters A and B that are 11.0 m apart, each one emitting in-phase radio waves at 92.0 MHz. She then walks so that she always remains 50.0 m from transmitter B. (See Figure 26.48.) For what values of x will she find the radio signal to be (a) maximally enhanced, (b) cancelled? Limit your solution to the cases where $x \geq 50.0$ m.

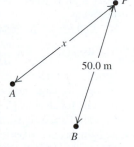
▲ **FIGURE 26.48** Problem 2.

3. • **Radio interference.** Two radio antennas A and B radiate in phase. Antenna B is 120 m to the right of antenna A. Consider point Q along the extension of the line connecting the antennas, a horizontal distance of 40 m to the right of antenna B. The frequency, and hence the wavelength, of the emitted waves can be varied. (a) What is the longest wavelength for which there will be destructive interference at point Q? (b) What is the longest wavelength for which there will be constructive interference at point Q?

4. •• Two speakers that are 15.0 m apart produce in-phase sound waves of frequency 250.0 Hz in a room where the speed of sound is 340.0 m/s. A woman starts out at the midpoint between the two speakers. The room's walls and ceiling are covered with absorbers to eliminate reflections, and she listens with only one ear for best precision. (a) What does she hear, constructive or destructive interference? Why? (b) She now walks slowly toward one of the speakers. How far from the center must she walk before she first hears the sound reach a minimum intensity? (c) How far from the center must she walk before she first hears the sound maximally enhanced?

5. •• Suppose that the situation is the same as in the previous problem, except that the two speakers are 180° out of phase. Repeat parts (a), (b), and (c) of that problem.

26.2 Two-Source Interference of Light

6. • Coherent light of wavelength 525 nm passes through two thin slits that are 0.0415 mm apart and then falls on a screen 75.0 cm away. How far away from the central bright fringe on the screen is (a) the fifth bright fringe (not counting the central bright fringe); (b) the eighth dark fringe?

7. • Coherent light from a sodium-vapor lamp is passed through a filter that blocks everything except for light of a single wavelength. It then falls on two slits separated by 0.460 mm. In the resulting interference pattern on a screen 2.20 m away, adjacent bright fringes are separated by 2.82 mm. What is the wavelength of the light that falls on the slits?

8. • Young's experiment is performed with light of wavelength 502 nm from excited helium atoms. Fringes are measured carefully on a screen 1.20 m away from the double slit, and the center of the 20th fringe (not counting the central bright fringe) is found to be 10.6 mm from the center of the central bright fringe. What is the separation of the two slits?

9. • Coherent light of frequency 6.32×10^{14} Hz passes through two thin slits and falls on a screen 85.0 cm away. You observe that the third bright fringe occurs at ± 3.11 cm on either side of the central bright fringe. (a) How far apart are the two slits? (b) At what distance from the central bright fringe will the third dark fringe occur?

10. •• Coherent light with wavelength 600 nm passes through two very narrow slits and the interference pattern is observed on a screen 3.00 m from the slits. The first-order bright fringe is at 4.84 mm from the center of the central bright fringe. For what wavelength of light will the first-order dark fringe be observed at this same point on the screen?

11. • Two slits spaced 0.450 mm apart are placed 75.0 cm from a screen. What is the distance between the second and third dark lines of the interference pattern on the screen when the slits are illuminated with coherent light with a wavelength of 500 nm?

12. •• Coherent light that contains two wavelengths, 660 nm (red) and 470 nm (blue), passes through two narrow slits separated by 0.300 mm, and the interference pattern is observed on a screen 5.00 m from the slits. What is the distance on the screen between the first-order bright fringes for the two wavelengths?

13. •• Two thin parallel slits that are 0.0116 mm apart are illuminated by a laser beam of wavelength 585 nm. (a) On a very large distant screen, what is the *total* number of bright fringes (those indicating complete constructive interference), including the central fringe and those on both sides of it? Solve this problem without calculating all the angles! (*Hint:* What is the largest that $\sin \theta$ can be? What does this tell you is the largest value of m?) (b) At what angle, relative to the original direction of the beam, will the fringe that is most distant from the central bright fringe occur?

14. •• Two small loudspeakers that are 5.50 m apart are emitting sound in phase. From both of them, you hear a singer singing C# (frequency 277 Hz), while the speed of sound in the room is 340 m/s. Assuming that you are rather far from these speakers, if you start out at point P equidistant from both of them and walk around the room in front of them, at what angles (measured relative to the line from P to the midpoint between the speakers) will you hear the sound (a) maximally enhanced, (b) cancelled? Neglect any reflections from the walls.

26.3 Interference in Thin Films

15. • The walls of a soap bubble have about the same index of refraction as that of plain water, $n = 1.33$. There is air both inside and outside the bubble. (a) What wavelength (in air) of visible light is most strongly reflected from a point on a soap bubble where its wall is 290 nm thick? To what color does this correspond (see Figure 23.3)? (b) Repeat part (a) for a wall thickness of 340 nm.

16. • What is the thinnest soap film (excluding the case of zero thickness) that appears black when viewed by reflected light with a wavelength of 480 nm? The index of refraction of the film is 1.33, and there is air on both sides of the film.

17. • A thin film of polystyrene of refractive index 1.49 is used as a nonreflecting coating for Fabulite (strontium titanate) of refractive index 2.409. What is the minimum thickness of the film required? Assume that the wavelength of the light in air is 480 nm.

18. • **Conserving energy.** You want to coat the *inner* surfaces of your windows (which have refractive index of 1.51) with a film in order to enhance the reflection of light back into the room so that you can use bulbs of lower wattage than usual. You find that MgF_2, with $n = 1.38$, is not too expensive, so you decide to use it. Since incandescent home lightbulbs emit reddish light with a peak wavelength of approximately 650 nm, you decide that this wavelength is the one to enhance in the light reflected back into the room. (a) What is the minimum thickness of film that you will need? (b) If this layer seems too thin to be able to put on accurately, what other thicknesses would also work? Give only the three thinnest ones.

19. • **Nonglare glass.** When viewing a piece of art that is behind glass, one often is affected by the light that is reflected off the front of the glass (called *glare*), which can make it difficult to see the art clearly. One solution is to coat the outer surface of the glass with a thin film to cancel part of the glare. (a) If the glass has a refractive index of 1.62 and you use TiO_2, which has an index of refraction of 2.62, as the coating, what is the minimum film thickness that will cancel light of wavelength 505 nm? (b) If this coating is too thin to stand up to wear, what other thicknesses would also work? Find only the three thinnest ones.

20. • The lenses of a particular set of binoculars have a coating with index of refraction $n = 1.38$, and the glass itself has $n = 1.52$. If the lenses reflect a wavelength of 525 nm the most strongly, what is the minimum thickness of the coating?

21. • A plate of glass 9.00 cm long is placed in contact with a second plate and is held at a small angle with it by a metal strip 0.0800 mm thick placed under one end. The space between the plates is filled with air. The glass is illuminated from above with light having a wavelength in air of 656 nm. How many interference fringes are observed per centimeter in the reflected light?

22. • Two rectangular pieces of plane glass are laid one upon the other on a table. A thin strip of paper is placed between them at one edge, so that a very thin wedge of air is formed. The plates are illuminated at normal incidence by 546 nm light from a mercury-vapor lamp. Interference fringes are formed, with 15.0 fringes per centimeter. Find the angle of the wedge.

23. •• A researcher measures the thickness of a layer of benzene $(n = 1.50)$ floating on water by shining monochromatic light onto the film and varying the wavelength of the light. She finds that light of wavelength 575 nm is reflected most strongly from the film. What does she calculate for the minimum thickness of the film?

24. •• **Compact disc player.** A compact disc (CD) is read from the bottom by a semiconductor laser beam with a wavelength of 790 nm that passes through a plastic substrate of refractive index 1.8. When the beam encounters a pit, part of the beam is reflected from the pit and part from the flat region between the pits, so these two beams interfere with each other. (See Figure 26.49.) What must the minimum pit depth be so that the part of the beam reflected from a pit cancels the part of the beam reflected from the flat region? (It is this cancellation that allows the player to recognize the beginning and end of a pit. For a fuller explanation of the physics behind CD technology, see the article "The Compact Disc Digital Audio System," by Thomas D. Rossing, in the December 1987 issue of *The Physics Teacher.*)

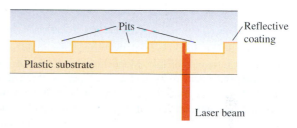

▲ **FIGURE 26.49** Problem 24.

26.5 Diffraction from a Single Slit

25. • A beam of laser light of wavelength 632.8 nm falls on a thin slit 0.00375 mm wide. After the light passes through the slit, at what angles relative to the original direction of the beam is it completely cancelled when viewed far from the slit?

26. • Parallel rays of green mercury light with a wavelength of 546 nm pass through a slit with a width of 0.437 mm. What is the distance from the central maximum to the first minimum on a screen 1.75 m away from the slit?

27. • Parallel light rays with a wavelength of 600 nm fall on a single slit. On a screen 3.00 m away, the distance between the first dark fringes on either side of the central maximum is 4.50 mm. What is the width of the slit?

28. • Monochromatic light from a distant source is incident on a slit 0.750 mm wide. On a screen 2.00 m away, the distance from the central maximum of the diffraction pattern to the first minimum is measured to be 1.35 mm. Calculate the wavelength of the light.

29. •• Red light of wavelength 633 nm from a helium–neon laser passes through a slit 0.350 mm wide. The diffraction pattern is observed on a screen 3.00 m away. Define the width of a bright fringe as the distance between the minima on either side. (a) What is the width of the central bright fringe? (b) What is the width of the first bright fringe on either side of the central one?

30. • Light of wavelength 633 nm from a distant source is incident on a slit 0.750 mm wide, and the resulting diffraction pattern is observed on a screen 3.50 m away. What is the distance between the two dark fringes on either side of the central bright fringe?

31. •• **Doorway diffraction.** Diffraction occurs for all types of waves, including sound waves. Suppose sound of frequency 1250 Hz leaves a room through a 1.00-m-wide doorway. At which angles relative to the centerline perpendicular to the doorway will someone outside the room hear no sound? Use

344 m/s for the speed of sound in air and assume that the source and listener are both far enough from the doorway for Fraunhofer diffraction to apply. You can ignore effects of reflections.

32. •• Light of wavelength 585 nm falls on a slit 0.0666 mm wide. (a) On a very large distant screen, how many *totally* dark fringes (indicating complete cancellation) will there be, including both sides of the central bright spot? Solve this problem *without* calculating all the angles! (*Hint:* What is the largest that $\sin\theta$ can be? What does this tell you is the largest that m can be?) (b) At what angle will the dark fringe that is most distant from the central bright fringe occur?

33. •• A glass sheet measuring 10.0 cm × 25.0 cm is covered by a very thin opaque coating. In the middle of this sheet is a thin, straight scratch 0.00125 mm thick, as shown in Figure 26.50. The

▲ **FIGURE 26.50** Problem 33.

sheet is totally immersed beneath the surface of a liquid having an index of refraction of 1.45. Monochromatic light strikes the sheet perpendicular to its surface and passes through the scratch. A screen is placed under water a distance 30.0 cm away from the sheet and parallel to it. You observe that the first dark fringes on either side of the central bright fringe on this screen are 22.4 cm apart. What is the wavelength of the light in air?

26.6 Multiple Slits and Diffraction Gratings

34. • A laser beam of unknown wavelength passes through a diffraction grating having 5510 lines/cm after striking it perpendicularly. Taking measurements, you find that the first pair of bright spots away from the central maximum occurs at ±15.4° with respect to the original direction of the beam. (a) What is the wavelength of the light? (b) At what angle will the next pair of bright spots occur?

35. • A laser beam of wavelength 600.0 nm is incident normally on a transmission grating having 400.0 lines/mm. Find the angles of deviation in the first, second, and third orders of bright spots.

36. • When laser light of wavelength 632.8 nm passes through a diffraction grating, the first bright spots occur at ±17.8° from the central maximum. (a) What is the line density (in lines/cm) of this grating? (b) How many additional bright spots are there beyond the first bright spots, and at what angles do they occur?

37. • A diffraction grating has 5580 lines/cm. When a beam of monochromatic light goes through it, the *second* pair of bright spots occurs at ±26.3 cm from the central spot on a screen 42.5 cm past the grating. (a) What is the wavelength of this light? (b) How far from the central spot does the next pair of bright spots occur on the screen?

38. • Monochromatic light is at normal incidence on a plane transmission grating. The first-order maximum in the interference pattern is at an angle of 8.94°. What is the angular position of the fourth-order maximum?

39. •• **CDs and DVDs as diffraction gratings.** A laser beam of wavelength $\lambda = 632.8$ nm shines at normal incidence on the reflective side of a compact disc. (a) The tracks of tiny pits in which information is coded onto the CD are 1.60 μm apart. For what angles of reflection (measured from the normal) will the intensity of light be maximum? (b) On a DVD, the tracks are only 0.740 μm apart. Repeat the calculation of part (a) for the DVD.

40. •• Light of wavelength 631 nm passes through a diffraction grating having 485 lines/mm. (a) What is the *total* number of bright spots (indicating complete constructive interference) that will occur on a large distant screen? Solve this problem *without* finding the angles. (*Hint:* What is the largest that $\sin\theta$ can be? What does this imply for the largest value of m?) (b) What is the angle of the bright spot farthest from the center?

41. •• If a diffraction grating produces a third-order bright spot for red light (of wavelength 700 nm) at 65.0° from the central maximum, at what angle will the second-order bright spot be for violet light (of wavelength 400 nm)?

26.7 X-Ray Diffraction

42. • X-rays of wavelength 0.0850 nm are scattered from the atoms of a crystal. The second-order maximum in the Bragg reflection occurs when the angle θ in Figure 26.32 is 21.5°. What is the spacing between adjacent atomic planes in the crystal?

43. • Monochromatic x rays are incident on a crystal for which the spacing of the atomic planes is 0.440 nm. The first-order maximum in the Bragg reflection occurs when the incident and reflected x rays make an angle of 39.4° with the crystal planes. What is the wavelength of the x rays?

44. • Electromagnetic waves of wavelength 0.173 nm fall on a crystal surface. As the angle from the plane is gradually increased, starting at 0°, you find that the first strong interference maximum occurs when the beam makes an angle of 22.4° with the surface of the crystal planes in the Bragg reflection. (a) What is the distance between the crystal planes? (b) At what other angles will interference maxima occur?

26.8 Circular Apertures and Resolving Power

45. • A converging lens 7.20 cm in diameter has a focal length of 300 mm. If the resolution is diffraction limited, how far away can an object be if points on it transversely 4.00 mm apart are to be resolved (according to Rayleigh's criterion) by means of light of wavelength 550 nm?

46. • A telescope is used to observe two distant point sources transversely 2.50 m apart with light of wavelength 600 nm. The objective of the telescope is covered with a slit of width 0.350 mm. What is the maximum distance in meters at which the two sources may be distinguished if the resolution is diffraction limited and Rayleigh's criterion is used?

47. • Two satellites at an altitude of 1200 km are separated by 28 km. If they broadcast 3.6-cm microwaves, what minimum receiving-dish diameter is needed to resolve (by Rayleigh's criterion) the two transmissions?

48. •• **Resolution of telescopes.** Due to blurring caused by atmospheric distortion, the best resolution that can be obtained by a normal, earth-based, visible-light telescope is about 0.3 arcsecond (there are 60 arcminutes in a degree and 60 arcseconds in an arcminute). (a) Using Rayleigh's criterion, calculate the diameter of an earth-based telescope that gives this resolution with 550-nm light. (b) Increasing the telescope diameter beyond the value found in part (a) will increase the light-gathering power of the telescope, allowing more distant and dimmer astronomical objects to be studied, but it will not improve the resolution. In what ways are the Keck telescopes (each of 10-m diameter) atop Mauna Kea in Hawaii superior to the Hale Telescope (5-m diameter) on Palomar Mountain in California? In what ways are they *not* superior? Explain.

49. •• **Resolution of the eye, I.** Even if the lenses of our eyes
BIO functioned perfectly, our vision would still be limited due to diffraction of light at the pupil. Using Rayleigh's criterion, what is the smallest object a person can see clearly at his near point of 25.0 cm with a pupil 2.00 mm in diameter and light of wavelength 550 nm? (To get a reasonable estimate without having to go through complicated calculations, we'll ignore the effect of the fluid in the eye.) Based upon your answer, does it seem that diffraction plays a significant role in limiting our visual acuity?

50. •• **Resolution of the eye, II.** The maximum resolution of the
BIO eye depends on the diameter of the opening of the pupil (a diffraction effect) and the size of the retinal cells, as illustrated in problem 51 in Chapter 25. In that problem, we saw that the size of the retinal cells (about 5.0 μm in diameter) limits the size of an object at the near point (25 cm) of the eye to a height of about 50 μm. (To get a reasonable estimate without having to go through complicated calculations, we shall ignore the effect of the fluid in the eye.) (a) Given that the diameter of the human pupil is about 2.0 mm, does the Rayleigh criterion allow us to resolve a 50-μm-tall object at 25 cm from the eye with light of wavelength 550 nm? (b) According to the Rayleigh criterion, what is the shortest object we could resolve at the 25 cm near point with light of wavelength 550 nm? (c) What angle would the object in part (b) subtend at the eye? Express your answer in minutes (60 min = 1°), and compare it with the experimental value of about 1 min. (d) Which effect is more important in limiting the resolution of our eyes, diffraction or the size of the retinal cells?

51. •• **Spy satellites?** Rumor has it that the U.S. military has spy satellites in orbit carrying telescopes that can resolve objects on the ground as small as the width of a car's license plate. If we assume that such satellites orbit about 400 km above the ground (which is typical for orbiting telescopes) and that they focus light of wavelength 500 nm, what would have to be the minimum diameter of the mirror (or objective lens) of this kind of a telescope in order to resolve such small objects? (Measure the width of a car's license plate.)

General Problems

52. •• Two identical audio speakers connected to the same amplifier produce in-phase sound waves with a single frequency that can be varied between 300 and 600 Hz. The speed of sound is 340 m/s. You find that where you are standing, you hear minimum-intensity sound. (a) Explain why you hear minimum-intensity sound. (b) If one of the speakers is moved 39.8 cm toward you, the sound you hear has maximum intensity. What is the frequency of the sound? (c) How much closer to you from the position in part (b) must the speaker be moved to the next position where you hear maximum intensity?

53. •• Suppose you illuminate two thin slits by monochromatic coherent light in air and find that they produce their first interference *minima* at ±35.20° on either side of the central bright spot. You then immerse these slits in a transparent liquid and illuminate them with the same light. Now you find that the first minima occur at ±19.46° instead. What is the index of refraction of this liquid?

54. • **Coating eyeglass lenses.** Eyeglass lenses can be coated on
BIO the *inner* surfaces to reduce the reflection of stray light to the eye. If the lenses are medium flint glass of refractive index

1.62 and the coating is fluorite of refractive index 1.432, (a) what minimum thickness of film is needed on the lenses to cancel light of wavelength 550 nm reflected toward the eye at normal incidence, and (b) will any other wavelengths of visible light be cancelled or enhanced in the reflected light?

55. •• **Sensitive eyes.** After an eye examination, you put some
BIO eyedrops on your sensitive eyes. The cornea (the front part of the eye) has an index of refraction of 1.38, while the eyedrops have a refractive index of 1.45. After you put in the drops, your friends notice that your eyes look red, because red light of wavelength 600 nm has been reinforced in the reflected light. (a) What is the minimum thickness of the film of eyedrops on your cornea? (b) Will any other wavelengths of visible light be reinforced in the reflected light? Will any be cancelled? (c) Suppose you had contact lenses, so that the eyedrops went on them instead of on your corneas. If the refractive index of the lens material is 1.50 and the layer of eyedrops has the same thickness as in part (a), what wavelengths of visible light will be reinforced? What wavelengths will be cancelled?

56. •• A wildlife photographer uses a moderate telephoto lens of focal length 135 mm and maximum aperture *f*/4.00 to photograph a bear that is 11.5 m away. Assume the wavelength is 550 nm. (a) What is the width of the smallest feature on the bear that this lens can resolve if it is opened to its maximum aperture? (b) If, to gain depth of field, the photographer stops the lens down to *f*/22.0, what would be the width of the smallest resolvable feature on the bear?

57. •• **Thickness of human hair.** Although we have discussed
BIO single-slit diffraction only for a slit, a similar result holds when light bends around a straight, thin object, such as a strand of hair. In that case, *a* is the width of the strand. From actual laboratory measurements on a human hair, it was found that when a beam of light of wavelength 632.8 nm was shone on a single strand of hair, and the diffracted light was viewed on a screen 1.25 m away, the first dark fringes on either side of the central bright spot were 5.22 cm apart. How thick was this strand of hair?

58. •• An oil tanker spills a large amount of oil $(n = 1.45)$ into the sea $(n = 1.33)$. (a) If you look down onto the oil spill from overhead, what predominant wavelength of light do you see at a point where the oil is 380 nm thick? (b) In the water under the slick, what visible wavelength (as measured in air) is predominant in the transmitted light at the same place in the slick as in part (a)?

59. •• A glass plate $(n = 1.53)$ that is 0.485 μm thick and surrounded by air is illuminated by a beam of white light normal to the plate. What wavelengths (in air) within the limits of the visible spectrum ($\lambda = 400$ nm to 700 nm) (a) are intensified in the reflected beam, (b) are cancelled in the reflected light?

60. •• The radius of curvature of the convex surface of a planoconvex lens is 95.2 cm. The lens is placed convex side down on a perfectly flat glass plate that is illuminated from above with red light having a wavelength of 580 nm. Find the diameter of the second bright ring in the interference pattern.

61. • **X-ray diffraction of salt.** X rays with a wavelength of 0.125 nm are scattered from a cubic array (of a sodium chloride crystal), for which the spacing of adjacent atoms is $a = 0.282$ nm. (a) If diffraction from planes parallel to a cube face is considered, at what angles θ of the incoming beam relative

to the crystal planes will maxima be observed? (b) Repeat part (a) for diffraction produced by the planes shown in Fig. 26.33a, which are separated by $a/\sqrt{2}$.

62. •• **Searching for planets around other stars.** If an optical telescope focusing light of wavelength 550 nm had a perfectly ground mirror, what would have to be the minimum diameter of its mirror so that it could resolve a Jupiter-size planet around our nearest star, Alpha Centauri, which is about 4.3 light years from earth? (Consult Appendix E.)

63. •• You need a diffraction grating that will disperse the visible spectrum (400 nm to 700 nm) through 30.0° in the first-order pattern. What must be the line density of this grating? (*Hint:* Use the fact that $\sin(A + B) = \sin A \cos B + \cos A \sin B$.)

64. •• A uniform thin film of material of refractive index 1.40 coats a glass plate of refractive index 1.55. This film has the proper thickness to cancel normally incident light of wavelength 525 nm that strikes the film surface from air, but it is somewhat greater than the minimum thickness to achieve this cancellation. As time goes by, the film wears away at a steady rate of 4.20 nm per year. What is the minimum number of years before the reflected light of this wavelength is now enhanced instead of cancelled?

65. •• A diffraction grating has 650 slits/mm. What is the highest order that contains the entire visible spectrum? (The wavelength range of the visible spectrum is approximately 400–700 nm.)

Passage Problems

Interference and sound waves. The phenomenon of interference occurs not only with light waves, but also with all frequencies of electromagnetic waves and all other types of waves, such as sound and water waves. Suppose that your professor sets up two sound speakers in the front of your classroom and uses an electronic oscillator to produce sound waves of a single frequency. When the professor begins, you and many other students hear a loud tone, while other students hear nothing. The speed of sound in air is 340 m/s.

66. The professor then does something to the apparatus. The frequency that you hear does not change, but the loudness decreases and now all your fellow student can also hear the tone. What did the professor do?
  A. She did nothing.
  B. She turned down the volume of the speakers.
  C. She changed the phase relationship of the speakers.
  D. She disconnected one speaker.

67. The professor now returns the apparatus to the same condition that it was in at the start of lecture. She then does something else to the speakers and all the students who heard nothing before now hear a loud tone, while you and those who heard the tone hear nothing. What did the professor do?
  A. She did nothing.
  B. She turned down the volume of the speakers.
  C. She changed the phase relationship of the speakers.
  D. She disconnected one speaker.

68. The professor now returns the apparatus to its original configuration (the one it had at the start of the lecture) so you again hear the original loud tone. She now slowly moves one of the speakers away from you until it first reaches a point where you can no longer hear the tone. If the speaker was moved 0.34 meters further from you, what is the frequency of the tone?
  A. 1000 Hz
  B. 2000 Hz
  C. 500 Hz
  D. 250 Hz

27 Relativity

When the year 1905 began, 25-year-old Albert Einstein was an unknown clerk in the Swiss patent office. By the end of that year, he had published three papers of extraordinary importance. The first was an analysis of Brownian motion; the second (for which he was later awarded the Nobel Prize) was on the photoelectric effect. In the third, Einstein introduced his **special theory of relativity,** in which he proposed drastic revisions in the classical concepts of space and time.

The concept of an **inertial frame of reference** plays a central role in our discussion. This concept was introduced in Section 4.2, and we strongly recommend that you review that section now. Briefly, an inertial frame of reference is a frame of reference in which Newton's laws are valid. Any frame of reference that moves with constant velocity with respect to an inertial frame is also an inertial frame.

The special theory of relativity is based on the simple statement that all the laws of physics are the same in every inertial frame of reference. This innocent-sounding proposition has far-reaching implications. Here are three that we'll explore in this chapter:

1. When two observers who are moving relative to each other measure a time interval or a length, they may not get the same results.
2. Two events that appear to one observer to occur at the same time may appear to another observer to occur at different times.
3. For the conservation principles of momentum and energy to be valid in all inertial systems, Newton's second law and the definitions of momentum and kinetic energy have to be revised.

A recurring theme in our discussions will be the role of the observer in the formulation of physical laws.

Einstein in 1932. At this point the former patent clerk is Director of the great Kaiser Wilhelm Physical Institute in Berlin. In the following year he will leave Nazi Germany for the United States, where he will settle at Princeton.

In studying this material, you'll confront some ideas that at first sight may seem too strange to be believed. You'll find that your intuition is often unreliable when you consider phenomena far removed from everyday experience, especially when objects are moving with speeds much faster than those we encounter in everyday life. But the theory rests on a solid foundation of experimental evidence. It has far-reaching consequences in *all* areas of physics (and beyond), including thermodynamics, electromagnetism, optics, atomic and nuclear physics, and high-energy physics. It doesn't *refute* Newton's laws, but it generalizes them to a vastly expanded realm of physical phenomena.

27.1 Invariance of Physical Laws

Although the implications of Einstein's theory were revolutionary, they are based on simple principles. What's more, we can work out several important results using only straightforward algebra.

Principle of Relativity

Einstein's **principle of relativity** is as follows:

> **Fundamental Postulate of Special Relativity**
> All the laws of physics, including mechanics and electromagnetism, are the same in all inertial frames of reference.

In other words, all inertial frames are equivalent; any one inertial frame is as good as any other.

Here are two examples: Suppose you watch two children playing catch with a ball while the three of you are aboard a train that moves with constant velocity with respect to the earth, assumed to be an inertial frame of reference. No matter how carefully you study the motion of the ball, you can't tell how fast (or, in fact, whether) the train is moving. This is because the laws of mechanics (Newton's laws) are the same in every inertial system (in this case, a system stationary on the ground and a system moving with the train).

Another example is the electromotive force (emf) induced in a coil of wire by a nearby moving permanent magnet (Figure 27.1a). In the frame of reference in which the coil is stationary, the moving magnet causes a change in magnetic flux through the coil, and this change induces an emf. In a frame of reference where the magnet is stationary (Figure 27.1b), the motion of the coil through a magnetic field causes magnetic forces to act upon the mobile charges in the conductor, inducing an emf. According to the principle of relativity, both of these points of view have equal validity and both must predict the same induced emf. As we saw in Chapter 21, Faraday's law of electromagnetic induction can be applied to either description, and it does indeed satisfy this requirement. If the moving-magnet and moving-coil situations *did not* give the same results, we could use this experiment to distinguish one inertial frame from another, and that would contradict the principle of relativity.

Equally significant is the fact that Maxwell's equations predicted the speed of electromagnetic radiation, as discussed in Chapter 23. The laws of electromagnetism show that light and all other electromagnetic waves travel with a constant speed $c = 299{,}792{,}458$ m/s. (We'll often use the approximate value $c = 3.00 \times 10^8$ m/s, which is within one part in 1000 of the exact value.) If these laws are the same in all inertial frames, then the principle of relativity requires that this speed must be the same in all inertial frames.

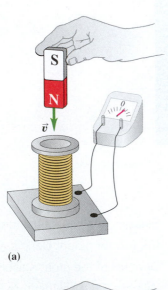

(a)

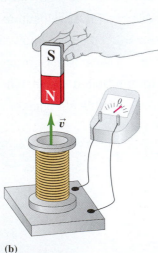

(b)

▲ **FIGURE 27.1** Induction experiments using (a) a magnet that is moving relative to the coil and (b) a stationary magnet.

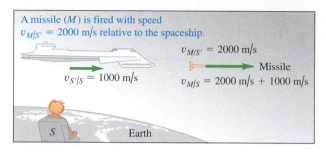

A missile (M) is fired with speed
$v_{M/S'} = 2000$ m/s relative to the spaceship.

$v_{M/S'} = 2000$ m/s

$v_{S'/S} = 1000$ m/s

Missile

$v_{M/S} = 2000$ m/s $+$ 1000 m/s

S Earth

▲ **FIGURE 27.2** A spaceship (S') moves with speed $v_{S'/S} =$ 1000 m/s relative to an observer on earth (S).

Speed of Light

During the nineteenth century, most physicists believed that light traveled through a hypothetical medium called the *ether,* just as sound waves travel through air. If so, the speed of light would depend on the motion of the observer relative to the ether and would therefore be different in different directions. The Michelson–Morley experiment, an interference experiment similar to those described in Chapter 26, was an effort to detect the motion of the earth relative to the ether. Einstein's conceptual leap was to recognize that if Maxwell's equations for the electromagnetic field are to be valid in all inertial frames, then the speed of light in vacuum should also be the same in all inertial frames and in all directions. In fact, Michelson and Morley detected *no* motion of their laboratory relative to the supposed ether. Indeed, there is no experimental evidence for the existence of an ether, and the concept has been discarded.

Thus, a direct consequence of Einstein's principle of relativity, sometimes called the second postulate of relativity, is this statement:

Invariance of speed of light
The speed of light in vacuum is the same in all inertial frames of reference and is independent of the motion of the source.

Let's think about what this statement means. Suppose two observers measure the speed of light. One is at rest with respect to the light source, and the other is moving away from it. Both are in inertial frames of reference. According to the principle of relativity, the two observers must obtain the same result for the speed of light, despite the fact that one is moving with respect to the other.

To explore this statement further, let's consider the following situation: A spaceship moving past earth at 1000 m/s fires a missile straight ahead with a speed of 2000 m/s (relative to the spacecraft) (Figure 27.2). What is the missile's speed relative to earth? This looks like an elementary problem in relative velocity. The correct answer, according to Newtonian mechanics, is 3000 m/s. But now suppose there is a searchlight in the spaceship, pointing in the same direction that the missile was fired. An observer on the spaceship measures the speed of light emitted by the searchlight and obtains the value c. According to our previous discussion, the observer on earth who measures the speed of this same light must *also* obtain the value c, not $c + 1000$ m/s. This contradicts our elementary notion of relative velocities; it may not appear to agree with common sense, and it certainly represents a break from our Newtonian notions of relative velocity. But "common sense" is intuition based on everyday experience, and such experience does not usually include measurements of the speed of light.

Let's restate this argument symbolically, using the two inertial frames of reference, labeled S for the observer on earth and S' for the moving spaceship,

▲ **Application It's all relative.** If this child throws the ball upward, it lands back in his hands, regardless of whether he is standing still or running at constant velocity. In *his* inertial frame of reference, the two events are the same. To a person standing still, they aren't—the ball moves along a line in one case and along a parabolic path in the other. The ball's path depends on the frame in which it is observed. Similarly, in this chapter we'll see that measurements of distances and time intervals also vary depending on the relative motion of the observer and the observed, especially at speeds approaching the speed of light.

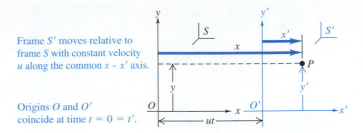

Frame S' moves relative to frame S with constant velocity u along the common x - x' axis.

Origins O and O' coincide at time $t = 0 = t'$.

▲ **FIGURE 27.3** A moving coordinate system, showing that the position of particle P can be described by the earth coordinates x and y in S or by the spaceship coordinates x' and y' in S'.

shown in Figure 27.3. To keep things as simple as possible, we have omitted the z axes. The x axes of the two frames lie along the same line, but the origin O' of frame S' moves relative to the origin O of frame S with constant velocity u along the common x-x' axis. We set our clocks so that the two origins coincide at time $t = 0$. Then their separation at a later time t is ut.

Now think about how we describe the motion of a particle P. This particle might represent an exploratory vehicle launched from the spaceship or a flash of light from a searchlight. We can describe the *position* of point P by using the earth coordinates (x, y) in S or the spaceship coordinates (x', y') in S'. The figure shows that these coordinates are related by the transformation equations $x = x' + ut$, $y = y'$. For completeness, we add the assumption that all time measurements are the same in the two frames of reference, so $t = t'$. These three equations, based on the familiar Newtonian notions of space and time, are called the **Galilean coordinate transformation.**

Galilean coordinate transformation:
$$x = x' + ut, \qquad y = y', \qquad t = t'. \tag{27.1}$$

As we'll see in the sections that follow, these equations aren't consistent with the principle of relativity. Developing the needed modifications will yield the fundamental equations of relativity.

Relative Velocity

If point P moves in the x direction, its velocity v as measured by an observer who is stationary in S is given by $v = \Delta x/\Delta t$. Its velocity v' measured by an observer at rest in S' is $v' = \Delta x'/\Delta t$. From our discussion of relative velocities in Sections 2.7 and 3.5, we know that these velocities are related by

$$v = v' + u, \tag{27.2}$$

which agrees with the relative velocity equation we derived at the end of Chapter 2. We can also derive this relation from Equations 27.1: Suppose that the particle is at a point described by coordinate x_1 or x_1' at time t_1 and by x_2 or x_2' at time t_2. Then $\Delta t = t_2 - t_1$. From Equation 27.1,

$$\Delta x = x_2 - x_1 = (x_2' - x_1') + u(t_2 - t_1) = \Delta x' + u\,\Delta t,$$

$$\frac{\Delta x}{\Delta t} = \frac{\Delta x'}{\Delta t} + u,$$

and

$$v = v' + u,$$

in agreement with Equation 27.2.

Now here's the fundamental problem: If we apply Equation 27.2 to the speed of light, using c for the speed seen by the observer in S and c' for the speed seen by the observer in S', then Equation 27.2 says that $c = c' + u$. Einstein's principle of relativity, supported by experimental evidence, says that $c = c'$. This is a genuine inconsistency, not an illusion, and it demands resolution. If we accept the principle of relativity, we are forced to conclude that Equations 27.1 and 27.2, intuitively appealing as they are, *cannot* be correct for the speed of light and hence aren't consistent with the principle of relativity. They have to be modified to bring them into harmony with that principle.

The resolution involves some fundamental modifications in our kinematic concepts. First is the seemingly obvious assumption that the observers in frames S and S' use the same time scale. We stated this assumption formally by including in Equations 27.1 the equation $t = t'$. But we are about to show that the assumption $t = t'$ *cannot* be correct; the two observers must have different time scales. The difficulty lies in the concept of simultaneity, which is our next topic. A careful analysis of simultaneity will help us to develop the appropriate modifications of our notions about space and time. In this analysis, we'll make frequent use of hypothetical experiments, often called thought experiments or, in Einstein's original German, *gedanken* experiments, to help clarify concepts and relationships.

Conceptual Analysis 27.1

Two spaceships and a star

Two spaceships can travel in any direction. Each ship carries equipment to measure the speed of light coming from any targeted star. In one experiment, each ship travels at a constant speed and they measure the speed of light from the same star. Which of the following statements is correct?

A. If the spaceships travel in a direction perpendicular to the direction of the incoming light, they both measure a light speed of zero.
B. No matter in what direction each spaceship travels, they measure a light speed of $c = 3.00 \times 10^8$ m/s.
C. For each ship to measure a light speed $c = 3.00 \times 10^8$ m/s, both ships must travel in the same direction, but they can have unequal speeds.
D. If the spaceships travel parallel to the incoming light, but in opposite directions, the ship traveling towards the star measures a greater speed of light.

SOLUTION In all of the choices, each ship travels at a constant speed in a constant direction. Therefore, each ship is an inertial frame of reference. The invariance of the speed of light means that the speed of light is the same in all inertial frames of reference, regardless of the direction of travel. So choice B is correct. Note that choice D would be correct if we were measuring the speed of an object that has mass (such as a proton emitted by the star) instead of measuring light.

27.2 Relative Nature of Simultaneity

When two events occur at the same time, we say that they are *simultaneous*. If you awoke at seven o'clock this morning, then two events (your awakening and the arrival of the hour hand of your clock at the number seven) occurred simultaneously. The fundamental problem in measuring time intervals is that, in general, two events that are simultaneous when seen by an observer in one frame of reference are *not* simultaneous as seen by an observer in a second frame if it is moving in relation to the first, *even if both are inertial frames.* This is the relative nature of **simultaneity.**

This idea may seem to be contrary to common sense. But here is a thought experiment, devised by Einstein, that illustrates the point: Consider a long train moving with uniform velocity, as shown in Figure 27.4. Two lightning bolts strike the train, one at each end. Each bolt leaves a mark on the train and one on the ground at the same instant. The points on the ground are labeled A and B in the figure, and the corresponding points on the train are A' and B'. An observer is standing on the ground at O, midway between A and B. Another observer is at O' at the middle of the train, midway between A' and B', moving with the train.

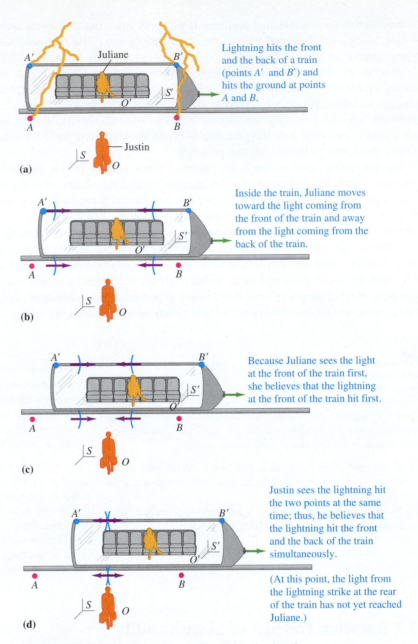

Lightning hits the front and the back of a train (points A' and B') and hits the ground at points A and B.

(a)

Inside the train, Juliane moves toward the light coming from the front of the train and away from the light coming from the back of the train.

(b)

Because Juliane sees the light at the front of the train first, she believes that the lightning at the front of the train hit first.

(c)

Justin sees the lightning hit the two points at the same time; thus, he believes that the lightning hit the front and the back of the train simultaneously.

(At this point, the light from the lightning strike at the rear of the train has not yet reached Juliane.)

(d)

▲ **FIGURE 27.4** A thought experiment illustrating simultaneity.

Both observers see both light flashes emitted from the points where the lightning strikes.

Suppose the two light flashes reach the observer at O simultaneously. He knows he is the same distance from A as from B, so he concludes that the two bolts struck A and B simultaneously. But the observer at O' is moving to the right, with the train, with respect to the observer at O, and the light flash from B' reaches her before the light flash from A' does. Because the observer at O' is the same distance from A' as from B', she concludes that the lightning bolt at the front of the train struck *earlier* than the one at the rear. The two events appear simultaneous to observer O, but not to observer O'.

Relative nature of simultaneity
Whether two events at different space points are simultaneous depends on the state of motion of the observer.

Furthermore, there is no basis for saying that O is right and O' is wrong, or the reverse. According to the principle of relativity, no inertial frame of reference is preferred over any other in the formulation of physical laws. Each observer is correct in his or her own frame of reference. In other words, simultaneity is not an absolute concept. Whether two events are simultaneous depends on the frame of reference. Because of the essential role of simultaneity in measuring time intervals, it also follows that *the time interval between two events depends on the frame of reference in which it is measured.* So our next task is to learn how to compare time intervals in different frames of reference.

27.3 Relativity of Time

We need to learn how to relate time intervals between events that are observed in two different frames of reference. If a train passes through a certain station at 12:00 and another at 12:05, the stationmaster sees a time interval of 5 minutes between trains. How would the conductor on the front train measure this time interval, and what result would he get? To answer these questions, let's consider another thought experiment. As before, a frame of reference S' (perhaps the moving train) moves along the common x-x' axis with constant speed u relative to a frame S (the station platform). For reasons that will become clear later, we assume that u is always less than the speed of light c. A passenger Juliane at O', moving with frame S', directs a flash of light at a mirror a distance d away, as shown in Figure 27.5a. She measures the time interval Δt_0 for light to make the round-trip to the mirror and back. (We use the subscript zero as a reminder that the timing apparatus is at rest in frame S'.) The total distance is $2d$, and the light flash moves with constant speed c, so the time interval Δt_0 is the distance $2d$ that the light flash travels, divided by the speed c:

$$\Delta t_0 = \frac{2d}{c}. \tag{27.3}$$

We call S' Juliane's **rest frame,** because she is not moving with respect to S'.

The time for the round-trip, as measured by Justin in frame S, is a different interval Δt. During this time, the light source moves at constant speed u in relation to S. In time Δt, it moves a distance $u \, \Delta t$. The total round-trip distance, as seen in S, is not just $2d$, but $2l$. Figure 27.5 shows that d and $u \, \Delta t / 2$ are the two legs of a right triangle, with l as the hypotenuse, so

$$l = \sqrt{d^2 + \left(\frac{u \, \Delta t}{2}\right)^2}.$$

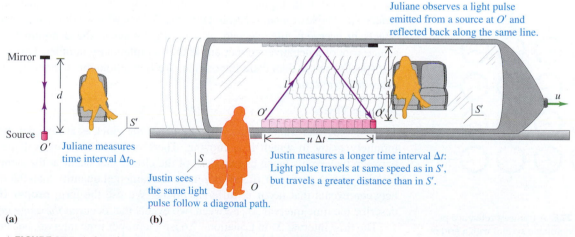

(a) Juliane measures time interval Δt_0.

(b) Justin sees the same light pulse follow a diagonal path.

Juliane observes a light pulse emitted from a source at O' and reflected back along the same line.

Justin measures a longer time interval Δt: Light pulse travels at same speed as in S', but travels a greater distance than in S'.

▲ **FIGURE 27.5** A thought experiment illustrating the relativity of time.

▲ **Application Not so fast.** Is the measured time for this runner to complete the race absolute or does it depend on who has the clock? A clock in the pocket of a runner crossing the finish line at constant velocity would be running at a very slightly slower rate from the viewpoint of an official with a stationary clock at the finish line. This would make the official measured time infinitesimally longer than that measured by the runner's clock. Fortunately, this time dilation effect would be significant only if this runner were approaching the speed of light, in which case we probably wouldn't need a clock to tell who won the race.

The grid is three dimensional; there are identical planes of clocks that are parallel to the page, in front of and behind it, connected by grid lines perpendicular to the page.

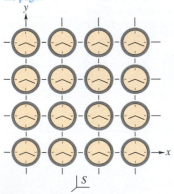

▲ **FIGURE 27.6** A frame of reference pictured as a coordinate system with a grid of synchronized clocks.

(In writing this expression, we've used the fact that the distance d is the same for both observers. We'll discuss this point later.) The speed of light is the same for both observers, so the relation in S analogous to Equation 27.3 is

$$\Delta t = \frac{2l}{c} = \frac{2}{c}\sqrt{d^2 + \left(\frac{u\,\Delta t}{2}\right)^2}. \tag{27.4}$$

We'd like to have a relation between the two time intervals Δt and Δt_0 that doesn't contain d. (We don't want the time intervals to depend on the dimensions of our clock.) To get this relation, we solve Equation 27.3 for d and substitute the result into Equation 27.4, obtaining

$$\Delta t = \frac{2}{c}\sqrt{\left(\frac{c\,\Delta t_0}{2}\right)^2 + \left(\frac{u\,\Delta t}{2}\right)^2}. \tag{27.5}$$

Now we square both sides of this equation and solve for Δt; the result is

$$\Delta t = \frac{\Delta t_0}{\sqrt{1 - u^2/c^2}}.$$

We may generalize this important result: If two events occur at the same point in space in a particular frame of reference, and if the time interval between them, as measured by an observer at rest in this frame (which we call the **rest frame** of this observer) is Δt_0, then an observer in a second frame moving with constant velocity u relative to the first frame measures the time interval to be Δt:

Time dilation:

$$\Delta t = \frac{\Delta t_0}{\sqrt{1 - u^2/c^2}}. \tag{27.6}$$

The denominator is always smaller than unity, so Δt is always larger than Δt_0. Think of an old-fashioned pendulum clock that ticks once a second, as observed in its rest frame; this is Δt_0. When the time between ticks is measured by an observer in a frame that is moving with velocity u with respect to the clock, the time interval Δt that he observes is *longer* than 1 second, and he thinks the clock is running slow. This effect is called **time dilation.** A clock moving with respect to an observer appears to run more slowly than a clock that is at rest in the observer's frame. Note that this conclusion is a direct result of the fact that the speed of light is the same in both frames of reference.

Two special cases are worth mentioning. First, if the velocity u of frame S' relative to S is much less than c, then u^2/c^2 is much smaller than one; in that case, the denominator in Equation 27.6 is nearly equal to one, and that equation reduces to the Newtonian relation $\Delta t = \Delta t_0$. That's why we don't see relativistic effects in everyday life. Second, if u is larger than c, the denominator is the square root of a negative number and is thus an imaginary number. This suggests that speeds greater than c are impossible. We'll examine this conclusion in more detail in later sections.

Proper Time

There is only one frame of reference in which the clock is at rest, but there are infinitely many in which it is moving. Therefore, the time interval measured between two events (such as two ticks of the clock) that occur at the same point (as viewed in a particular frame) is a more fundamental quantity than the interval between events that occur at different points. We use the term **proper time** to describe the time interval Δt_0 between two events that occur *at the same point.*

The time interval Δt in Equation 27.6 isn't a proper time interval, because it is the interval between two events that occur *at different spatial points* in the frame of reference S. Therefore, it can't really be measured by a single observer at rest in S. But we can use *two* observers, one stationary at the location of the first event and the other at the second, each with his or her own clock. We can synchronize these two clocks without difficulty, as long as they are at rest in the same frame of reference. For example, we could send a light pulse simultaneously to the two clocks from a point midway between them. When the pulses arrive, the observers set their clocks to a prearranged time. (But note that clocks synchronized in one frame of reference are, in general, *not* synchronized in any other frame.)

In thought experiments, it's often helpful to imagine *many* observers with synchronized clocks at rest at various points in a particular frame of reference. We can picture a frame of reference as a coordinate grid with lots of synchronized clocks distributed around it, as suggested by Figure 27.6. Only when a clock is *moving* relative to a given frame of reference do we have to watch for ambiguities of synchronization or simultaneity.

Conceptual Analysis 27.2

Pendulums in relative motion

Two identical pendulums have a period of τ_0 when measured in the factory. While one pendulum swings on earth, the other is taken by astronauts on a spaceship that travels at 99% the speed of light. Which statement is correct about observations of the pendulums? (There may be more than one correct choice.)

A. When observed from the earth, the pendulum on the ship has a period less than τ_0.
B. When observed from the earth, the pendulum on the ship has a period greater than τ_0.

C. The astronauts observe the pendulum on earth with a period greater than τ_0.
D. The astronauts observe the pendulum on the ship with a period greater than τ_0.

SOLUTION Recall that a clock moving with respect to an observer appears to run more slowly than a clock that is at rest in the observer's frame. Therefore, people on earth will observe the pendulum on the ship with a period greater than τ_0 (choice B). In addition, the pendulum on earth is moving with respect to the astronauts, so the astronauts observe the pendulum on earth with a period greater than τ_0 (choice C).

EXAMPLE 27.1   Time dilation at 0.99c

A spaceship flies past earth with a speed of $0.990c$ (about 2.97×10^8 m/s) relative to earth (Figure 27.7). A high-intensity signal light (perhaps a pulsed laser) blinks on and off; each pulse lasts $2.20\ \mu s$ ($= 2.20 \times 10^{-6}$ s), as measured on the spaceship. At a certain instant, the ship is directly above an observer on earth and is traveling perpendicular to the line of sight. **(a)** What is the duration of each light pulse, as measured by the observer on earth? **(b)** How far does the ship travel in relation to the earth during each pulse?

SOLUTION

SET UP First we label the frames of reference. Let S be the earth's frame of reference and S' that of the spaceship. Next, we note that the duration of a laser pulse, measured by an observer on the spaceship, is a proper time in S', because the laser is at rest in S' and the two events (the starting and stopping of the pulse) occur at the same point relative to S'. In Equation 27.6, we denote this interval by Δt_0. The time interval Δt between these events, as seen in any other frame of reference, is greater than Δt_0. The speed of the spaceship (frame S') relative to earth (frame S) is $u = 0.990c = 2.97 \times 10^8$ m/s.

SOLVE Part (a): In the preceding discussion, the proper time is $\Delta t_0 = 2.20\ \mu s$. According to Equation 27.6, the corresponding interval Δt measured by an observer on the earth (S) is

$$\Delta t = \frac{\Delta t_0}{\sqrt{1 - u^2/c^2}} = \frac{2.20\ \mu s}{\sqrt{1 - (0.990)^2}} = 15.6\ \mu s.$$

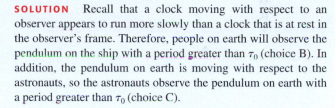
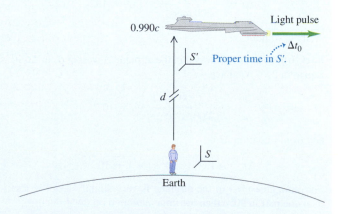

▲ **FIGURE 27.7** A spaceship traveling perpendicular to earth transmits a high-intensity light signal.

Continued

Part (b): During this interval, the spaceship travels a distance d relative to earth, given by

$$d = u \, \Delta t = (0.990)(3.00 \times 10^8 \text{ m/s})(15.6 \times 10^{-6} \text{ s})$$
$$= 4600 \text{ m} = 4.6 \text{ km}.$$

REFLECT The time dilation in S is about a factor of 7. To an observer on earth, each pulse appears to last about seven times as long as it does to an observer on the spaceship.

Practice Problem: Suppose the spaceship changes velocity relative to earth, so that the duration of the laser pulses as seen on earth is 4.40×10^{-6} s. What is the spaceship's velocity relative to earth? *Answer:* 2.60×10^8 m/s.

An experiment that is similar in principle to this spaceship example, although different in detail, provided the first direct experimental confirmation of Equation 27.6. μ leptons, or *muons,* are unstable particles, first observed in cosmic rays. They decay with a mean lifetime of 2.20 μs, as measured in a frame of reference in which the particles are at rest. But in cosmic-ray showers, the particles are moving very fast. The mean lifetime of a muon with speed of $0.99c$ has been measured to be 15.6 μs. Note that the numbers are identical to those in the spaceship example; the duration of the light pulse is replaced by the mean lifetime of the muon. These measurements provide a direct experimental confirmation of Equation 27.6.

EXAMPLE 27.2 Time dilation at jetliner speeds

An airplane flies from San Francisco to New York (about 4800 km, or 4.80×10^6 m) at a steady speed of 300 m/s (about 670 mi/h) (Figure 27.8). How much time does the trip take, as measured **(a)** by observers on the ground; **(b)** by an observer in the plane?

SOLUTION

SET UP We denote the airplane's speed relative to the ground as u. The time between the plane's departure and arrival, as measured by an observer on the plane, is a proper time, measured between two events at the same point in the plane's frame of reference, and we denote it as Δt_0. The corresponding time interval as seen by observers who are stationary on the ground is Δt, and the distance d traveled during this interval is $d = u \, \Delta t$.

SOLVE Part (a): The time Δt measured by the ground observers is simply the distance divided by the speed:

$$\Delta t = \frac{4.80 \times 10^6 \text{ m}}{300 \text{ m/s}} = 1.60 \times 10^4 \text{ s},$$

or about $4\frac{1}{2}$ hours.

Part (b): The time interval in the airplane corresponds to Δt_0 in Equation 27.6. We have

$$\frac{u^2}{c^2} = \frac{(300 \text{ m/s})^2}{(3.00 \times 10^8 \text{ m/s})^2} = 10^{-12},$$

and from Equation 27.6,

$$\Delta t_0 = (1.60 \times 10^4 \text{ s})\sqrt{1 - 10^{-12}}.$$

The time measured in the airplane is very slightly less (by less than one part in 10^{12}) than the time measured on the ground.

REFLECT To measure the time interval Δt seen by the ground observers, we need *two* observers with synchronized clocks, one in San Francisco and one in New York, because the two events

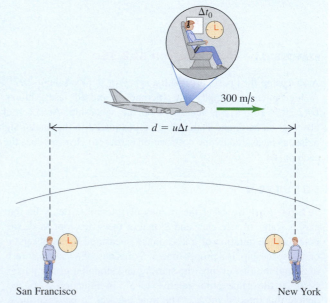

▲ **FIGURE 27.8** An airplane flying from San Francisco to New York, observed from the ground and from the plane.

(takeoff and landing) occur at different spatial points in the ground frame of reference. In the airplane's frame, they occur at the *same* point, so Δt_0 can be measured by a single observer. (But note that if the plane develops engine trouble and has to return to San Francisco, it must be in an accelerated, non-inertial frame of reference during its turnaround. In this case, the preceding

Continued

analysis isn't valid, because we have used only inertial frames in our derivations.)

We don't notice such tiny effects in everyday life. But as we mentioned in Section 1.3, present-day atomic clocks can attain a precision of about one part in 10^{13}, and measurements similar to the one in this example have been carried out, verifying Equation 27.6 directly.

Practice Problem: The π^+ meson (or positive pion) is an unstable particle that has a lifetime (measured in its rest frame) of about 2.60×10^{-8} s. If the particle is moving with a speed of $0.992c$ with respect to your laboratory, what is its lifetime as measured in the laboratory? What distance does it travel in the laboratory during this time? *Answer:* 2.06×10^{-7} s; 61.3 m.

Space Travel and the Twin Paradox

Time dilation has some interesting consequences for space travel. Suppose we want to explore a planet in a distant solar system 200 light years away. Could a person possibly live long enough to survive the trip? One light year is the distance light travels in 1 year. In ordinary units, this distance is $(3.00 \times 10^8 \text{ m/s})(3.15 \times 10^7 \text{ s}) = 9.45 \times 10^{15}$ m. Even if the spaceship can attain a speed of $0.999c$ relative to earth, the trip requires nearly 200 years, as observed in earth's reference frame.

But the aging of the astronauts is determined by the time measured in the spaceship's frame of reference. With reference to Equation 27.6, Δt is the time interval measured on earth, and Δt_0 is the corresponding proper time interval measured in the moving spaceship. If the ship's speed relative to earth is $0.999c$, then Equation 27.6 gives

$$\sqrt{1 - (0.999c)^2/c^2} = \sqrt{1 - (0.999)^2} = 0.0447 \qquad \text{and}$$
$$\Delta t_0 = \Delta t \sqrt{1 - (0.999)^2} = (200 \text{ y})(0.0447) = 8.94 \text{ years}.$$

Observers on earth think the trip lasts 200 years, but to the astronauts on the spaceship, the interval is only 8.94 years.

But there's another problem. Suppose the astronauts want to come back home to earth. The time-dilation equation (Equation 27.6) suggests an apparent paradox called the **twin paradox.** Consider two identical-twin astronauts named Eartha and Astrid. Eartha remains on earth, while Astrid takes off on a high-speed trip through the galaxy. Because of time dilation, Eartha sees all of Astrid's life processes, such as her heartbeat, proceeding more slowly than her own. Thus, Eartha thinks that Astrid ages more slowly, so when Astrid returns to earth, she is younger than Eartha.

Now here's the paradox: All inertial frames are equivalent. Can't Astrid make exactly the same arguments to conclude that Eartha is in fact the younger? Then each twin thinks that the other is younger, and that's a paradox.

To resolve the paradox, we recognize that the twins are *not* identical in all respects. If Eartha remains in an inertial frame at all times, Astrid must have an acceleration with respect to inertial frames during part of her trip in order to turn around and come back. Eartha remains always at rest in the same inertial frame; Astrid does not. Thus, there is a real physical difference between the circumstances of the twins. Careful analysis shows that Eartha is correct: When Astrid returns, she *is* younger than Eartha.

▲ Application **Who's the grandmother?** The answer to this question may seem obvious, but to the student of relativity, it could depend on which person had traveled to a distant planet at 99.9% of the speed of light. Imagine that a 20-year-old woman had given birth to a child and then immediately left on a 100-light-year trip at nearly the speed of light. Because of time dilation for the traveler, only about 10 years would pass, and she would appear to be about 30 years old when she returned, even though 100 years had passed by for people on Earth. Meanwhile, the child she left behind at home could have had a baby 20 years after her departure, and this grandchild would now be 80 years old!

27.4 Relativity of Length

Just as the time interval between two events depends on the observer's frame of reference, the distance between two points also depends on the observer's frame of reference. The concept of simultaneity is involved. Suppose you want to measure the length of a moving car. One way is to have two friends make marks on the pavement *at the same time* at the positions of the front and rear bumpers. Then you measure the distance between the marks. If you mark the position of the front bumper at one time and that of the rear bumper half a second later, you won't get the car's true length. But we've learned that simultaneity isn't an absolute concept,

MasteringPHYSICS®

ActivPhysics 17.2: Relativity of Length

so we have to be very careful to treat times and time intervals correctly in various reference frames.

To develop a relationship between lengths in various coordinate systems, we consider another thought experiment. We attach a source of light pulses to one end of a ruler and a mirror to the other end, as shown in Figure 27.9. The ruler is at rest in reference frame S', and its length in that frame is l_0. Then the time Δt_0 required for a light pulse to make the round-trip from source to mirror and back (a distance $2l_0$) is given by

$$\Delta t_0 = \frac{2l_0}{c}. \tag{27.7}$$

This is a proper time interval in S', because departure and return occur at the same point in S'.

In reference frame S, the ruler is moving to the right with speed u during the time the light pulse is traveling. The length of the ruler in S is l, and the time of travel from source to mirror, as measured in S, is Δt_1. During this interval, the ruler, with source and mirror attached, moves a distance $u\,\Delta t_1$. The total length of path d from source to mirror is not l, but

$$d = l + u\,\Delta t_1.$$

The light pulse travels with speed c, so it is also true that

$$d = c\,\Delta t_1.$$

Combining these two equations to eliminate d, we find that

$$c\,\Delta t_1 = l + u\,\Delta t_1 \qquad \text{or} \qquad \Delta t_1 = \frac{l}{c - u}.$$

In the same way, we can show that the time Δt_2 for the return trip from mirror to source, as measured in S, is

$$\Delta t_2 = \frac{l}{c + u}.$$

The *total* time $\Delta t = \Delta t_1 + \Delta t_2$ for the round-trip, as measured in S, is the sum of these two time intervals:

$$\Delta t = \frac{l}{c - u} + \frac{l}{c + u} = \frac{2l}{c(1 - u^2/c^2)}. \tag{27.8}$$

The ruler is stationary in Juliane's frame of reference S'. The light pulse travels a distance l_0 from the light source to the mirror.

(a)

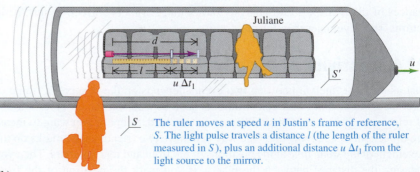

The ruler moves at speed u in Justin's frame of reference, S. The light pulse travels a distance l (the length of the ruler measured in S), plus an additional distance $u\,\Delta t_1$ from the light source to the mirror.

(b)

▶ **FIGURE 27.9** A thought experiment illustrating the relativity of length.

We also know that Δt and Δt_0 are related by Equation 27.6, because Δt_0 is a proper time in S'. When we solve Equation 27.6 for Δt_0 and substitute the result into Equation 27.7, we get

$$\Delta t \sqrt{1 - \frac{u^2}{c^2}} = \frac{2l_0}{c}.$$

Finally, combining this equation with Equation 27.8 to eliminate Δt and simplifying, we obtain the following relationship:

Length contraction parallel to motion:

$$l = l_0 \sqrt{1 - \frac{u^2}{c^2}}. \tag{27.9}$$

Thus, the length l measured in S, in which the ruler is moving, is *shorter* than the length l_0 measured in S', where the ruler is at rest. A length measured in the rest frame of the body is called a **proper length;** thus, in Equation 27.9, l_0 is a proper length in S', and the length measured in any other frame is *less than* l_0. This effect is called **length contraction.**

EXAMPLE 27.3   The shrinking spaceship

In Example 27.1 (Section 27.3), suppose a crew member on the spaceship measures its length, obtaining the value 400 m. What is the length measured by observers on earth?

SOLUTION

SET UP The length of the spaceship in the frame in which it is at rest (400 m) is a proper length in this frame, corresponding to l_0 in Equation 27.9. We want to find the length l measured by observers on earth, who see the ship moving with speed u.

SOLVE From Equation 27.9,

$$l = l_0 \sqrt{1 - \frac{u^2}{c^2}} = (400 \text{ m}) \sqrt{1 - (0.990)^2} = 56.4 \text{ m}.$$

REFLECT To measure l, we need two observers, because we have to observe the positions of the two ends of the spaceship simultaneously in the earth's reference frame. We could use two observers with synchronized clocks (Figure 27.10). These two observations are simultaneous in the earth's reference frame, but they are *not* simultaneous as seen by an observer in the spaceship.

Practice Problem: Suppose your car is 5.00 m long when it is sitting at rest in your driveway. How fast would it have to move for its length (as seen by someone at rest in the driveway) to be 4.00 m? *Answer:* 1.80×10^8 m/s.

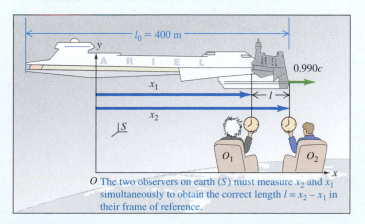

▲ **FIGURE 27.10**

When u is very small compared with c, the square-root factor in Equation 27.9 approaches unity, and in the limit of small speeds we recover the Newtonian

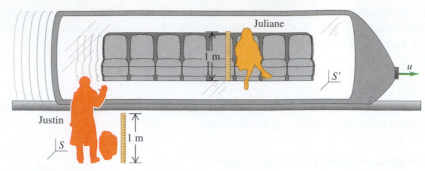

▲ **FIGURE 27.11** The rulers are perpendicular to the relative velocity, so, for any value of u, the observers measure each ruler to have a length of 1 meter.

relation $l = l_0$. This and the corresponding result for time dilation show that Equations 27.1 retain their validity when all speeds are much smaller than c; only at speeds comparable to c are modifications needed.

Lengths Perpendicular to Relative Motion

We have derived Equation 27.13 for lengths measured in the direction *parallel* to the relative motion of the two frames of reference. Lengths measured *perpendicular* to the direction of motion are *not* contracted. To prove this, let's consider two identical rulers (Figure 27.11). One ruler is at rest in frame S and lies along the y axis with one end at O, the origin of S. The other ruler is at rest in frame S' and lies along the y' axis with one end at O', the origin of S'. At the instant the two origins coincide, observers in the two frames of reference S and S' simultaneously make marks at the positions of the upper ends of the rulers. Because the observations occur at the same space point for both observers, they agree that they are simultaneous.

If the marks made by the two observers *do not* coincide, then one must be higher than the other. This would imply that the two frames aren't equivalent. But any such asymmetry would contradict the basic postulate of relativity that all inertial frames of reference are equivalent. Consistency with the postulates of relativity requires that both observers see the two rulers as having the same length, even though, to each observer, one of them is stationary and one is moving.

> **Length contraction perpendicular to motion**
> There is no length contraction perpendicular to the direction of relative motion.

Finally, let's consider the visual appearance of a moving three-dimensional body. If we could see the positions of all points of the body simultaneously, it would appear just to shrink in the direction of motion. But we *don't* see all the points simultaneously; light from points farther from us takes longer to get to us than light from points near to us, so we see the positions of farther points as they were at earlier times.

Suppose we have a cube with its faces parallel to the coordinate planes. When we look straight-on at the closest face of such a cube at rest, we see only that face. But when the cube is moving past us toward the right, we can also see the left side because of the effect just described. More generally, we can see some points that we couldn't see when the body was at rest, because it moves out of the way of the light rays. Conversely, some light that can get to us when the body is at rest is blocked by the moving body. Because of all this, the cube appears rotated and distorted. Figure 27.12 shows a computer-generated image of a more complicated body, moving at a relativistic speed relative to the observer, that illustrates this effect.

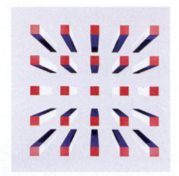

(a) Array at rest

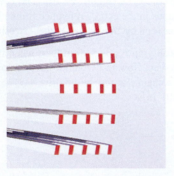

(b) Array moving to the right at $0.2c$

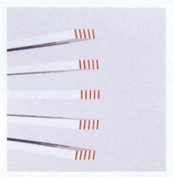

(c) Array moving to the right at $0.9c$

▲ **FIGURE 27.12** Computer-generated images of 25 rods with square cross sections.

Quantitative Analysis 27.3 **A moving square**

An engineer in space measures the area of a square of girders with sides of length l_0. If she is moving with constant velocity parallel to one side of the square, which of the following is true of the area $A_{observer}$ that she measures?

A. $A_{observer} > l_0^2$.
B. $A_{observer} = l_0^2$.
C. $A_{observer} < l_0^2$.

SOLUTION Because the engineer moves parallel to one side of the square, she sees the square shortened by length contraction in her direction of motion—to her, it looks like a rectangle that is narrower than it is tall, rather than a square. Since the measured *height* of the square is not affected by her motion, the area she measures is less than the area she would measure if she were at rest relative to the square (so $A_{observer} < l_0^2$). Being a space engineer, she naturally takes this length contraction into account and computes the correct area.

27.5 The Lorentz Transformation

In Section 27.1, we discussed the Galilean coordinate transformation equations (Equations 27.1), which relate the coordinates (x, y) of a point in frame of reference S to the coordinates (x', y') of the same point in a second frame S' when S' moves with constant velocity u relative to S along the common x-x' axis (Figure 27.13).

As we have seen, this Galilean transformation is valid only when u is much smaller than c. The relativistic equations for time dilation and length contraction (Equations 27.6 and 27.9) can be combined to form a more general set of transformation equations that are consistent with the principle of relativity. The more general relations are called the **Lorentz transformation.** In the limit of very small u, they reduce to the Galilean transformation, but they may also be used when u is comparable to c.

The basic question is this: When an event occurs at point (x, y) at time t, as observed in a frame of reference S, what are the coordinates (x', y') and time t' of the event as observed in a second frame S' moving relative to S with constant velocity u along the x direction?

As before, we assume that the origins of the two frames S and S' coincide at the initial time $t = t' = 0$. Deriving the transformation equations involves several intermediate steps. We'll omit the details; here are the results:

Lorentz transformation equations

$$x' = \frac{x - ut}{\sqrt{1 - u^2/c^2}}, \qquad y' = y, \qquad t' = \frac{t - ux/c^2}{\sqrt{1 - u^2/c^2}}. \qquad (27.10)$$

The Lorentz transformation equations are the relativistic generalization of the Galilean transformation, Equations 27.1. When u is much smaller than c, the square roots in the denominators approach one, and the second term in the numerator of the t' equation approaches zero. In this limit, Equations 27.10 become

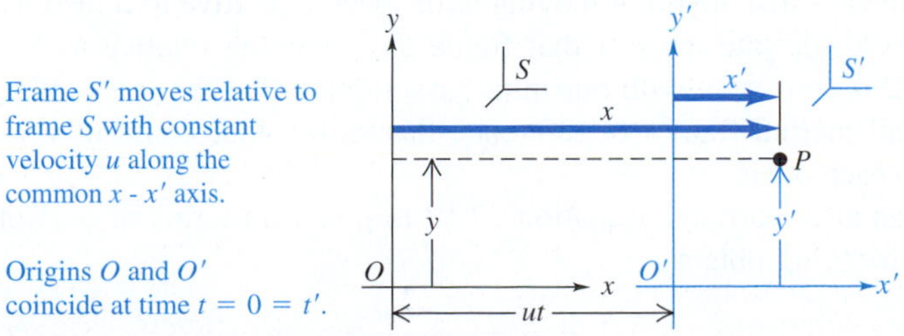

Frame S' moves relative to frame S with constant velocity u along the common x - x' axis.

Origins O and O' coincide at time $t = 0 = t'$.

▲ **FIGURE 27.13** As measured in frame of reference S, x' is contracted to $x'\sqrt{1 - u^2/c^2}$ so $x = ut + x'\sqrt{1 - u^2/c^2}$ and $x' = (x - ut)/\sqrt{1 - u^2/c^2}$.

identical to Equations 27.1: $x' = x - ut$ and $t' = t$. In general, though, both the coordinates and the time of an event in one frame depend on both its coordinates and time in another frame. *Space and time have become intertwined; we can no longer say that length and time have absolute meanings, independent of a frame of reference.*

Relativistic Velocity Transformation

We can use Equations 27.10 to derive the relativistic generalization of the Galilean velocity transformation, Equation 27.2. First we use Equations 27.10 to obtain relations for distance and time *intervals* between two events in the two coordinate systems. Suppose a body is moving along the x axis. As observed in S, it arrives at point x_1 at time t_1 and at point x_2 at time t_2. The distance interval as observed in S is $\Delta x = x_2 - x_1$, the time interval is $\Delta t = t_2 - t_1$, and the velocity as seen in S (strictly speaking, the x component of velocity) is $v = \Delta x/\Delta t$. The corresponding intervals as observed in the moving frame S' (moving with x component of velocity u relative to S) are $\Delta x' = x_2' - x_1'$ and $\Delta t = t_2' - t_1'$. The x component of velocity as seen in S' is $v' = \Delta x'/\Delta t'$. Using the first and third of Equations 27.10 for each event and subtracting, we find that

$$\Delta x' = \frac{\Delta x - u\,\Delta t}{\sqrt{1 - u^2/c^2}}, \qquad \Delta t' = \frac{\Delta t - u\,\Delta x/c^2}{\sqrt{1 - u^2/c^2}}. \qquad (27.11)$$

Finally, we divide the first of Equations 27.11 by the second:

$$v' = \frac{\Delta x'}{\Delta t'} = \frac{\Delta x - u\,\Delta t}{\Delta t - u\,\Delta x/c^2} = \frac{(\Delta x/\Delta t) - u}{1 - (u/c^2)(\Delta x/\Delta t)}.$$

But $\Delta x/\Delta t$ equals v, the x component of velocity as measured in S, so we finally obtain the relation between v and v':

Relativistic velocity transformation

$$v' = \frac{v - u}{1 - uv/c^2}. \qquad (27.12)$$

This surprisingly simple result says that if a body is moving with x component of velocity v' relative to a frame S', and if S' is moving with x component of velocity u relative to another frame S, then the x components of velocity are related by Equation 27.12.

When u and v are much smaller than c, the second term in the denominator in Equation 27.12 is very small, and we obtain the nonrelativistic result $v' = v - u$. The opposite extreme is the case $v = c$; then we find that

$$v' = \frac{c - u}{1 - uc/c^2} = c.$$

This result says that anything moving with speed c relative to S also has speed c relative to S', despite the fact that frame S' is moving relative to S. So Equation 27.12 is consistent with our initial assumption that the speed of light is the same in all inertial frames of reference that move with constant velocity with respect to each other.

We can also rearrange Equation 27.12 to give v in terms of v'. Solving that equation for v, we obtain

$$v = \frac{v' + u}{1 + uv'/c^2}. \qquad (27.13)$$

Note that this equation has the same form as Equation 27.12, with v and v' interchanged and the sign of u reversed. Indeed, this resemblance is in fact *required*

▲ **Application Space and time are what??**
This may be how Newton would feel after talking to Einstein and finding out that neither distances in space nor intervals of time have absolute meanings, independent of the frame of reference. Both space and time, according to relativity theory, have meaning only when considering the relative motion of the observer in relation to what is being observed and measured. Much like Maxwell's equations of electromagnetism, which described the interdependence of electricity and magnetism, Einstein's theory of relativity demonstrated the intimate association between space and time. For example, length contraction can be compensated for by time dilation. In fact, time is often considered to be a fourth dimension, and relativistic events may be described in terms of the four-dimensional space-time continuum.

by the principle of relativity, which insists that there is no basic distinction between the two inertial frames S and S'.

PROBLEM-SOLVING STRATEGY 27.1   **Lorentz transformation**

SET UP

1. Observe carefully the roles of the concepts of proper time and proper length. A time interval between two events that happen at the same space point in a particular frame of reference is a proper time in that frame, and the time interval between the same two events is greater in any other frame. The length of a body measured in a frame in which it is at rest is a proper length in that frame, and the length is less in any other frame.

SOLVE

2. The Lorentz transformation equations tell you how to relate measurements made in different inertial frames of reference. When you use them, it helps to make a list of coordinates and times of events in the two frames, such as x_1, x_1', t_2, and so on. List and label carefully what you know and don't know. Do you know the coordinates in one frame? The time of an event in one frame? What are your knowns and unknowns?

3. In velocity-transformation problems, if you have two observers measuring the motion of a body, decide which you want to call S and which S', identify the velocities v and v' clearly, and make sure you know the velocity u of S' relative to S. Use either form of the velocity transformation equation, Equation 27.12 or 27.13, whichever is more convenient.

REFLECT

4. Don't be surprised if some of your results don't seem to make sense or if they disagree with your intuition. Reliable intuition about relativity takes time to develop; keep trying!

EXAMPLE 27.4   **Relative velocities in space travel**

A spaceship moving away from the earth with speed $0.90c$ fires a robot space probe in the same direction as its motion, with speed $0.70c$ relative to the spaceship (Figure 27.14). What is the probe's speed relative to the earth?

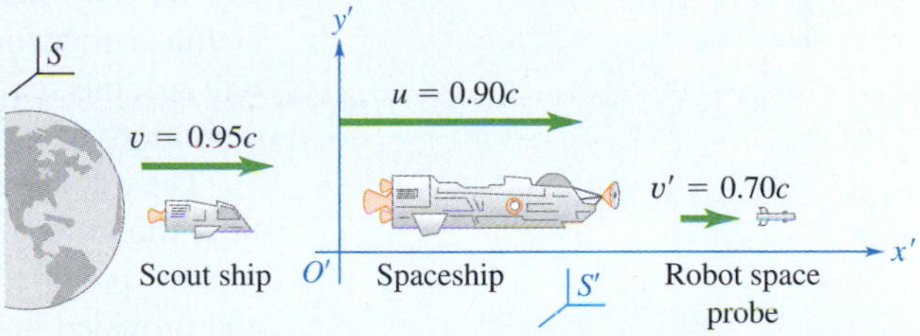

▶ **FIGURE 27.14**

SOLUTION

SET UP We resist the Newtonian temptation to simply add the two speeds, obtaining the impossible result $1.60c$. Instead, we use the relativistic formulation given by Equations 27.12 and 27.13, which you may want to review before proceeding. Let the earth's frame of reference be S, and let the spaceship's be S'. Then u is the spaceship's velocity relative to earth, and v' is the velocity of the probe relative to the spaceship. We want to find v, the velocity of the probe relative to earth.

SOLVE From the statement of the problem and the discussion in the preceding paragraph, $u = 0.90c$, $v' = 0.70c$, and v is to be determined. We use Equation 27.13:

$$v = \frac{v' + u}{1 + uv'/c^2} = \frac{0.70c + 0.90c}{1 + (0.90c)(0.70c)/c^2} = 0.98c.$$

REFLECT We can check that the *incorrect* value obtained from the Galilean velocity-addition formula $(v = v' + u)$ is a velocity of $1.60c$ relative to earth; this velocity is larger than the speed of light, and larger than the correct relativistic value by a factor of about 1.63.

Practice Problem: Suppose the robot space probe is aimed directly toward the earth instead of away from it. What is its velocity relative to earth? *Answer:* $0.54c$.

EXAMPLE 27.5 **Space travel again**

A scout ship from the earth tries to catch up with the spaceship of Example 27.4 by traveling at $0.95c$ relative to the earth. What is its speed relative to the spaceship?

SOLUTION

SET UP Again we let the earth's frame of reference be S and the spaceship's frame be S'. Again we have $u = 0.90c$, but now $v = 0.95c$.

SOLVE According to nonrelativistic velocity addition, the scout ship's velocity relative to the spaceship would be $0.05c$. We get the correct result from Equation 27.12:

$$v' = \frac{v - u}{1 - uv/c^2} = \frac{0.95 - 0.90c}{1 - (0.90c)(0.95c)} = 0.34c.$$

REFLECT Are you surprised that v' is greater than $0.05c$? Try to think of an argument as to why this might have been expected.

Practice Problem: Referring to Equation 27.12, verify that when both v and u are much smaller than c, the expression for v' reduces to the nonrelativistic expression.

Equations 27.12 and 27.13 can be used to show that when the relative velocity u of two frames is less than c, an object moving with a speed less than c in one frame of reference also has a speed less than c in *every other* frame of reference. This is one reason for thinking that no material object may travel with a speed greater than that of light relative to *any* frame of reference. The relativistic generalizations of energy and momentum, which we'll consider in the sections that follow, give further support to this hypothesis.

27.6 Relativistic Momentum

Newton's laws of motion have the same form in all inertial frames of reference. When we transform coordinates from one inertial frame to another using the Galilean coordinate transformation, the laws should be *invariant* (unchanging). But we have just learned that the principle of relativity forces us to replace the Galilean transformation with the more general Lorentz transformation. As we will see, this requires corresponding generalizations in the laws of motion and the definitions of momentum and energy.

The principle of conservation of momentum states that **when two objects collide, the total momentum is constant,** provided that they are an isolated system (that is, provided that they interact only with each other, not with anything else) and provided that the velocities of the objects are measured in the same inertial reference frame. If conservation of momentum is a valid physical law, then, according to the principle of relativity, it must be valid in *all* inertial frames of reference.

But suppose we look at a collision of two particles in one inertial coordinate system S and find that the total momentum is conserved. Then we use the Lorentz transformation to obtain the velocities of the particles in a second inertial system S'. It turns out that if we use the Newtonian definition of momentum, $\vec{p} = m\vec{v}$, momentum is *not* conserved in the second system. If we're confident that the principle of relativity and the Lorentz transformation are correct, the only way to save momentum conservation is to generalize the Newtonian definition of momentum.

We'll omit the detailed derivation of the correct relativistic generalization of momentum, but here is the result: If a particle has mass m when it is at rest (or moving with a speed much smaller than c), then the **relativistic momentum $\vec{p}$** of the particle when it is moving with velocity $\vec{v}$ is given by the following expression:

Relativistic momentum

$$\vec{p} = \frac{m\vec{v}}{\sqrt{1 - v^2/c^2}}. \qquad (27.14)$$

When the particle's speed v is much less than c, Equation 27.14 is approximately equal to the Newtonian expression $\vec{p} = m\vec{v}$, but in general, the momentum is greater in magnitude than mv (Figure 27.15). In fact, as v approaches c, the magnitude of the momentum approaches infinity.

It's often convenient to use the abbreviation

$$\gamma = \frac{1}{\sqrt{1 - v^2/c^2}}. \qquad (27.15)$$

Using this abbreviation, we can rewrite Equation 27.14 as

$$\vec{p} = \gamma m\vec{v}. \qquad (27.16)$$

When v is much smaller than c, γ is approximately equal to one; when v is almost as large as c, γ approaches infinity. Thus, γ provides a measure of the amount by which Equation 27.14 and similar relations differ from their Newtonian counterparts. Figure 27.16 shows how γ increases as v approaches c.

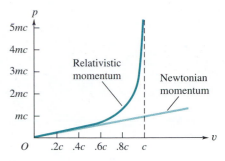

▲ **FIGURE 27.15** The magnitude of momentum as a function of speed.

Quantitative Analysis 27.4

The possible range of γ

Two objects, 1 and 2, are shot away from an observer with equal speeds and in opposite directions. Which of the following statements are correct about their respective values of γ (that is, γ_1 and γ_2)?

A. If $\gamma_1 > 0$, then $\gamma_2 < 0$.
B. If $\gamma_1 > 1$, then $\gamma_2 < 1$.
C. $1 \leq \gamma_1 < \infty$ and $1 \leq \gamma_2 < \infty$.
D. $0 \leq \gamma_1 < 1$ and $0 \leq \gamma_2 < 1$.

SOLUTION In Equation 27.15, the function γ depends on v^2, so objects with $+v$ and $-v$ have the same γ, since $(+v)^2 = (-v)^2 = v^2$. Therefore, $\gamma_1 = \gamma_2$, which is one way to rule out choices A and B. In addition, the speed of each object must be somewhere in the range $0 \leq v < c$. If we consider γ in the limit as $v \to 0$, then $\gamma \to 1$; likewise, in the limit as $v \to c$, $\gamma \to \infty$. Thus, for both objects, $1 \leq \gamma < \infty$.

In Equation 27.14, m is a *constant* that describes the inertial properties of a particle. Because $\vec{p} = m\vec{v}$ is still valid in the limit of very small velocities, m must be the same quantity we used (and learned to measure) in our study of Newtonian mechanics. In relativistic mechanics, m is often called the **rest mass** of a particle.

What about the relativistic generalization of Newton's second law? In Newtonian mechanics, one form of the second law, as we discussed in Section 8.1, is

$$\vec{F} = \frac{\Delta\vec{p}}{\Delta t}.$$

That is, force equals rate of change of momentum with respect to time. Experiments show that this result is still valid in relativistic mechanics, provided that we use the *relativistic* momentum given by Equation 27.14. An interesting aspect of this relation is that, because momentum is no longer directly proportional to velocity, the rate of change of momentum is no longer directly proportional to acceleration. As a result, **a constant force does not cause constant acceleration.** Furthermore, if the force has components both parallel and perpendicular to the object's instantaneous velocity, the relation between force and acceleration turns out to be different for the two components. In this case, the force and the acceleration are in different directions!

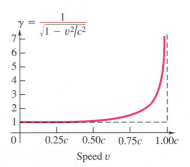

▲ **FIGURE 27.16** As v approaches c, $\gamma = 1/\sqrt{1 - v^2/c^2}$ grows larger without bound.

If the force and velocity have the same direction, it turns out that the acceleration a is given by

$$F = \frac{ma}{(1 - v^2/c^2)^{\frac{3}{2}}} = \gamma^3 ma. \tag{27.17}$$

When v is much smaller than c, γ is approximately one, and this expression reduces to the familiar Newtonian $F = ma$.

As Figure 27.16 shows, the factor $\gamma = 1/\sqrt{1 - v^2/c^2}$ in the relativistic momentum expression approaches infinity as the object's speed increases. Therefore, the acceleration caused by a given constant force continuously *decreases*. As the speed approaches c, the magnitude of the momentum approaches infinity and the acceleration approaches zero, no matter how great a force is applied. **It is impossible to accelerate a particle from a state of rest to a speed equal to or greater than c.** The speed of light is sometimes called **the ultimate speed;** no material object can travel faster than light.

EXAMPLE 27.6 **Relativistic dynamics of a proton**

A proton (rest mass 1.67×10^{-27} kg, charge 1.60×10^{-19} C) is moving parallel to an electric field that has magnitude $E = 5.00 \times 10^5$ N/C. Find the magnitudes of momentum and acceleration at the instants when $v = 0.010c$, $0.90c$, and $0.99c$.

SOLUTION

SET UP The relativistic definition of momentum, Equation 27.14, differs from its Newtonian counterpart by the factor γ. As Equation 27.15 shows, γ is only slightly greater than unity when v is much smaller than c, but γ becomes very large when v approaches c. Thus, our first step is to compute the value of γ for each speed given in the problem. Then we can find the corresponding momentum values from Equation 27.16. We'll also need to know that the force (magnitude F) exerted by the electric field on the proton is

$$F = qE = (1.60 \times 10^{-19} \text{ C})(5.00 \times 10^5 \text{ N/C})$$
$$= 8.00 \times 10^{-14} \text{ N}.$$

SOLVE Using Equation 27.15, we find that $\gamma = 1.00$, 2.29, and 7.09. From Equation 27.16, the corresponding values of p are

$$p_1 = \gamma mv = (1.00)(1.67 \times 10^{-27} \text{kg})(0.010)(3.00 \times 10^8 \text{m/s})$$
$$= 5.01 \times 10^{-21} \text{ kg} \cdot \text{m/s},$$

$$p_2 = (2.29)(1.67 \times 10^{-27} \text{ kg})(0.90)(3.00 \times 10^8 \text{ m/s})$$
$$= 1.03 \times 10^{-18} \text{ kg} \cdot \text{m/s},$$

$$p_3 = (7.09)(1.67 \times 10^{-27} \text{ kg})(0.99)(3.00 \times 10^8 \text{ m/s})$$
$$= 3.52 \times 10^{-18} \text{ kg} \cdot \text{m/s}.$$

From Equation 27.17,

$$a = \frac{F}{\gamma^3 m}.$$

When $v = 0.01c$ and $\gamma = 1.00$,

$$a_1 = \frac{8.00 \times 10^{-14} \text{ N}}{(1)^3(1.67 \times 10^{-27} \text{ kg})} = 4.79 \times 10^{13} \text{ m/s}^2.$$

The other accelerations are smaller by factors of γ^3:

$$a_2 = 3.99 \times 10^{12} \text{ m/s}^2, \qquad a_3 = 1.34 \times 10^{11} \text{ m/s}^2.$$

▲ FIGURE 27.17 The Stanford Linear Accelerator (SLAC).

These are only 8.33% and 0.280% of the values predicted by nonrelativistic mechanics.

REFLECT As the proton's speed increases, the relativistic values of momentum differ more and more from the nonrelativisitic values computed from $p = mv$. The momentum at $0.99c$ is more than three times as great as that at $0.90c$ because of the increase in the factor γ. We also note that as v approaches c, the acceleration drops off very quickly. In the Stanford Linear Accelerator (Figure 27.17), a path length of 3 km is needed to give electrons the speed that, according to classical physics, they could acquire in 1.5 cm.

Practice Problem: At what speed is the momentum of a proton twice as great as the nonrelativistic result (from $p = mv$)? Ten times as great? Would your results be different for an electron? *Answers:* $0.866c$; $0.995c$; no.

Equation 27.14 is sometimes interpreted to mean that a rapidly moving particle undergoes an increase in mass. If the mass at zero velocity (the rest mass) is denoted by m, then the "relativistic mass" m_{rel} is given by

$$m_{\text{rel}} = \frac{m}{\sqrt{1 - v^2/c^2}} = \gamma m. \qquad (27.18)$$

Indeed, when we consider the motion of a system of particles (such as gas molecules in a moving container), the total mass of the system is the sum of the relativistic masses of the particles, not the sum of their rest masses.

The concept of relativistic mass also has its pitfalls, however. As Equation 27.17 shows, it is *not* correct to say that the relativistic generalization of Newton's second law is $\vec{F} = m_{\text{rel}}\vec{a}$, and it is *not* correct that the relativistic kinetic energy of an object is $K = \frac{1}{2}m_{\text{rel}}v^2$. Thus, this concept must be approached with great caution. We prefer to think of m as a constant, unvarying quantity for any given object and to incorporate the correct relativistic relationships into the definitions of momentum (as we have done) and kinetic energy (the subject of the next section).

27.7 Relativistic Work and Energy

When we developed the relationship between work and kinetic energy in Chapter 7, we used Newton's laws of motion. But these laws have to be generalized to bring them into harmony with the principle of relativity, so we also need to generalize the definition of kinetic energy. It turns out that the classical definition of work $(W = F\,\Delta x)$ can be retained. Einstein showed that when a force does work W on a moving body, causing acceleration and changing the object's speed from v_1 to v_2, these quantities are related by the equation

$$W = \frac{mc^2}{\sqrt{1 - v_2^2/c^2}} - \frac{mc^2}{\sqrt{1 - v_1^2/c^2}}. \qquad (27.19)$$

This result suggests that we might define kinetic energy as

$$K \overset{?}{=} \frac{mc^2}{\sqrt{1 - v^2/c^2}}. \qquad (27.20)$$

But this expression isn't zero when the particle is at rest. Instead, when $v = 0$, K becomes equal to mc^2. Thus, a more reasonable relativistic definition of kinetic energy K is

Relativistic kinetic energy

$$K = \frac{mc^2}{\sqrt{1 - v^2/c^2}} - mc^2. \qquad (27.21)$$

This looks rather different from the Newtonian expression $K = \frac{1}{2}mv^2$, but the two must agree whenever v is much smaller than c. Indeed, it can be shown that as v approaches zero, the Newtonian and relativistic expressions give the same result, and we obtain the classical $\frac{1}{2}mv^2$. Figure 27.18 shows a graph of relativistic kinetic energy as a function of v, compared with the Newtonian expression.

But what is the significance of the term mc^2 that we had to subtract in Equation 27.21? Although Equation 27.20 does not give the *kinetic energy* of the particle, perhaps it represents some kind of *total* energy, including both the kinetic energy and an additional energy mc^2 that the particle possesses even when it is not moving. We'll call this additional energy the **rest energy** of the particle,

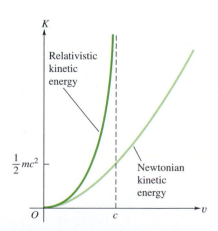

▲ **FIGURE 27.18** Comparison of relativistic and classical expressions for kinetic energy, as functions of speed v.

denoted by $E_{rest} = mc^2$. Then when we denote the *total* energy by E and use Equation 27.21, we obtain this result:

Total energy of an object

$$E = K + mc^2 = K + E_{rest} = \frac{mc^2}{\sqrt{1 - v^2/c^2}}. \qquad (27.22)$$

Rest energy of an object

$$E_{rest} = mc^2. \qquad (27.23)$$

There is, in fact, direct experimental evidence that rest energy really does exist. The simplest example is the decay of the π^0 meson (also called the *neutral pion*). This is an unstable particle, produced in high-energy collisions of other particles. When it decays, it disappears and electromagnetic radiation appears. When the particle, with mass m, is at rest (and therefore has no kinetic energy) before its decay, the total energy of the radiation produced is found to be exactly equal to mc^2. There are many other examples of fundamental particle transformations in which the sum of the rest masses of the particles in the system changes. In every case, a corresponding energy change occurs, consistent with the assumption of a rest energy mc^2 associated with a rest mass m. We'll study such particle transformations in greater detail in Chapter 30.

Historically, the principles of conservation of mass and of energy developed quite independently. The theory of relativity now shows that they are actually two special cases of a single broader conservation principle: the **principle of conservation of mass and energy.** In some physical phenomena, neither the sum of the rest masses of the particles nor the total energy (other than rest energy) is separately conserved, but there is a more general conservation principle: **In an isolated system, when the sum of the rest masses changes, there is always an equal and opposite change in the total energy other than the rest energy.**

The conversion of mass into energy is the fundamental principle involved in the generation of power through nuclear reactions, a subject we will discuss in Chapter 30. When a uranium nucleus undergoes fission in a nuclear reactor, the total mass of the resulting fragments is less than that of the parent nucleus, and the total kinetic energy of the fragments is equal to this mass deficit multiplied by c^2. This kinetic energy can be used in a variety of ways, such as producing steam to operate turbines for electric-power generators.

Energy–Momentum Relations

In Chapter 8, we learned that in Newtonian mechanics, the kinetic energy K and the magnitude of momentum p for a particle with mass m are related by $K = p^2/2m$. In relativistic mechanics, there is an analogous (and equally simple) relation between the total energy E (kinetic energy plus rest energy) of an object and its magnitude of momentum p.

Relativistic energy–momentum relation

$$E^2 = (mc^2)^2 + (pc)^2. \qquad (27.24)$$

This equation shows again the existence of rest energy; for a particle at rest ($p = 0$), the total energy is $E = mc^2$.

Equation 27.24 also suggests the possible existence of particles that have energy and momentum even when they have zero rest mass. In such a case, $m = 0$ and the relation between E and p is even simpler:

Energy and momentum of massless particles
$$E = pc. \qquad (27.25)$$

In fact, massless particles *do* exist. One example is the *photon*—the quantum of electromagnetic radiation. Photons always travel with the speed of light; they are emitted and absorbed during changes of state of an atomic or nuclear system when the energy and momentum of the system change.

EXAMPLE 27.7 Rest energy of the electron

(a) Find the rest energy of an electron ($m = 9.09 \times 10^{-31}$ kg, $e = 1.60 \times 10^{-19}$ C) in joules and in electronvolts. (b) Find the speed of an electron that has been accelerated by an electric field from rest through a potential difference of 20.0 kV (typical of TV picture tubes in the pre-flat-panel era) and through 5.00 MV (a high-voltage x-ray machine).

SOLUTION

SET UP We need the expression for rest energy, $E_{rest} = mc^2$, the definition of the electronvolt (which we suggest you review in Section 18.3), and the conversion factor from electronvolts to joules. We find the total energy E in Equation 27.22 by adding the kinetic energy (from the accelerating potential difference) to the rest energy, and then we solve Equation 27.22 for v.

SOLVE Part (a): The electron's rest energy is
$$mc^2 = (9.09 \times 10^{-31}\ \text{kg})(2.998 \times 10^8\ \text{m/s})^2$$
$$= 8.187 \times 10^{-14}\ \text{J}.$$

From Section 18.3, 1 eV = 1.602×10^{-19} J, so
$$mc^2 = (8.187 \times 10^{-14}\ \text{J})(1\ \text{eV}/1.602 \times 10^{-19}\ \text{J})$$
$$= 5.11 \times 10^5\ \text{eV} = 0.511\ \text{MeV}.$$

Part (b): An electron accelerated through a potential difference of 20.0 kV gains a kinetic energy of $K = 20.0$ keV, equal to about 4% of the rest energy found in part (a). Combining these values with Equation 27.22, we get
$$20.0 \times 10^3\ \text{eV} + 5.11 \times 10^5\ \text{eV} = \frac{5.11 \times 10^5\ \text{eV}}{\sqrt{1 - v^2/c^2}}.$$

The rest is arithmetic; we first simplify this equation by dividing both sides by 5.11×10^5 eV and inverting both sides, obtaining
$$\sqrt{1 - v^2/c^2} = 0.962 \qquad \text{and} \qquad v = 0.272c.$$

REFLECT When $K = 20$ keV, the kinetic energy is a small fraction of the rest energy $E_{rest} = 511$ keV, and the speed is about one-fourth the speed of light. We invite you to use Equation 27.15 to show that in this case $\gamma = 1.04$. For this energy, the electron's behavior is not very different from what Newtonian mechanics would predict. But when we repeat the calculation with a potential difference of 5.00 MV and kinetic energy $K = 5.00$ MeV = 5.00×10^6 eV, we find that $v = 0.996c$ and $\gamma = 10.8$. In this case, the kinetic energy is much *larger* than the rest energy, the speed is very close to c, and the particle's behavior is emphatically non-Newtonian. Such a speed is said to be in the *extreme relativistic range*.

Practice Problem: In Example 27.7, suppose the particle is a proton accelerated through the same potential differences as was the electron in that example. Find the speed in each case. Could nonrelativistic approximations be used? *Answers:* 1.96×10^6 m/s, 3.08×10^7 m/s; yes (to 0.4%).

EXAMPLE 27.8 Colliding protons and meson production

Two protons, each with rest mass $M = 1.67 \times 10^{-27}$ kg, have equal speeds in opposite directions. They collide head-on, producing a π^0 meson (also called a *neutral pion*) with mass $m = 2.40 \times 10^{-28}$ kg (Figure 27.19). If all three particles are at rest after the collision, find the initial speed of each proton and find its kinetic energy, expressed as a fraction of its rest energy.

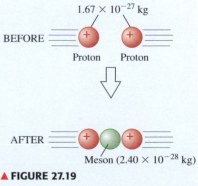

▲ **FIGURE 27.19**

Continued

SOLUTION

SET UP The key is to understand that the total energy of the system (including the rest energy) is conserved, so we can write the "before-and-after" equation:

$$\frac{2Mc^2}{\sqrt{1 - v^2/c^2}} = 2Mc^2 + mc^2.$$

SOLVE Substituting numerical values, taking care to use consistent units, we find that

$$\frac{2(1.67 \times 10^{-27}\,\text{kg})}{\sqrt{1 - v^2/c^2}} = 2(1.67 \times 10^{-27}\,\text{kg})$$
$$+ (2.40 \times 10^{-28}\,\text{kg}).$$

We divide by the proton mass M and invert, obtaining

$$\sqrt{1 - v^2/c^2} = 0.933 \qquad \text{and} \qquad v = 0.360c.$$

The initial kinetic energy K of each proton is half the pion rest energy, $K = mc^2/2$, so the ratio of K to the proton rest energy Mc^2 is

$$\frac{mc^2/2}{Mc^2} = \frac{m}{2M} = \frac{2.40 \times 10^{-28}\,\text{kg}}{2(1.67 \times 10^{-27}\,\text{kg})} = 0.0719.$$

The rest energy of a proton is 938 MeV, so its initial kinetic energy is $(0.0719)(938\,\text{MeV}) = 67.4\,\text{MeV}$.

REFLECT We invite you to verify that the rest energy of the π^0 meson is twice this result, or 135 MeV. Also, note that the initial kinetic energies of the protons are not much different from the values they would have according to Newtonian mechanics. Thus, the speeds might be called nonrelativistic, but the direct conversion of kinetic energy into mass is definitely a relativistic process.

Practice Problem: In this example, suppose the colliding particles are electrons instead of protons. Find the initial energy of each electron, and show that the energies and speeds are in the extreme relativistic range. *Answer:* 67.4 MeV (electron rest energy is 0.511 MeV).

27.8 Relativity and Newtonian Mechanics

The sweeping changes required by the principle of relativity go to the very roots of Newtonian mechanics, including the concepts of length and time, the equations of motion, and the conservation principles. Thus it may appear that we have destroyed the foundations on which Newtonian mechanics is built. In one sense this is true, and yet the Newtonian formulation is still valid whenever speeds are small in comparison to the speed of light. In such cases, time dilation, length contraction, and the modifications of the laws of motion are so small that they are unobservable. In fact, every one of the principles of Newtonian mechanics survives as a special case of the more general relativistic formulation.

So the laws of Newtonian mechanics are not *wrong;* they are *incomplete.* They are a limiting case of relativistic mechanics; they are approximately correct when all speeds are small in comparison to c and they become exactly correct in the limit when all speeds approach zero. Thus, relativity does not destroy the laws of Newtonian mechanics, but rather generalizes them. Newton's laws rest on a solid base of experimental evidence, and it would be strange to advance a new theory that is inconsistent with this evidence. There are many situations for which Newtonian mechanics is clearly inadequate, including all phenomena in which particle speeds are comparable to that of light or in which the direct conversion of mass to energy occurs. But there is still a large area, including nearly all the behavior of macroscopic bodies in mechanical systems, in which Newtonian mechanics is perfectly adequate. It's interesting to speculate how different the experiences of everyday life would be if the speed of light were 10 m/s; we would be living in a very different world indeed!

At this point, we may ask whether relativistic mechanics is the final word on this subject or whether *further* generalizations are possible or necessary. For example, inertial frames of reference have occupied a privileged position in our

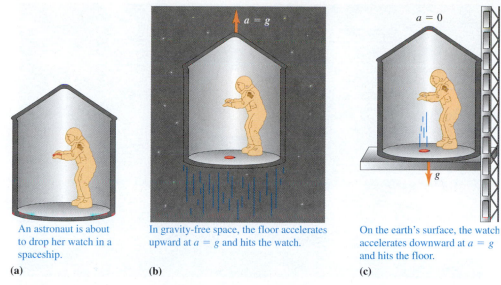

An astronaut is about to drop her watch in a spaceship.

(a)

In gravity-free space, the floor accelerates upward at $a = g$ and hits the watch.

(b)

On the earth's surface, the watch accelerates downward at $a = g$ and hits the floor.

(c)

▲ **FIGURE 27.20** Without information from outside the spaceship, the astronaut cannot distinguish situation (b) from situation (c).

discussion. Should the principle of relativity be extended to noninertial frames as well?

Here's an example that illustrates some implications of this question. A student is standing in an elevator. The elevator cables have all broken, and the safety devices have all failed at once, so the elevator is in free fall with an acceleration of 9.8 m/s^2 relative to earth. Temporarily losing her composure, the student loses her grasp of her physics textbook. The book doesn't fall to the floor, because it, the student, and the elevator all have the same free-fall acceleration. But the student can view the situation in two ways: Is the elevator really in free fall, with her in it, or is it possible that somehow the earth's gravitational attraction has suddenly vanished? As long as she remains in the elevator and it remains in free fall, she can't tell whether she is indeed in free fall or whether the gravitational interaction has vanished.

A more realistic illustration is an astronaut in a space station in orbit around the earth. Objects in the spaceship *seem* to be weightless because both they and the station are in circular or elliptical orbits and are constantly accelerating toward the earth. But without looking outside the ship, the astronauts have no way to determine whether gravitational interactions have disappeared or whether the spaceship is in a noninertial frame of reference accelerating toward the center of the earth (Figure 27.20).

These considerations form the basis of Einstein's **general theory of relativity.** If we can't distinguish experimentally between a gravitational field at a particular location and an accelerated reference system, then there can't be any real distinction between the two. Pursuing this concept, we may try to represent *any* gravitational field in terms of special characteristics of the coordinate system. This turns out to require even more sweeping revisions of our space–time concepts than did the special theory of relativity, and we find that, in general, the geometric properties of space are non-Euclidean (Figure 27.21) and gravitational fields are closely related to the geometry of space.

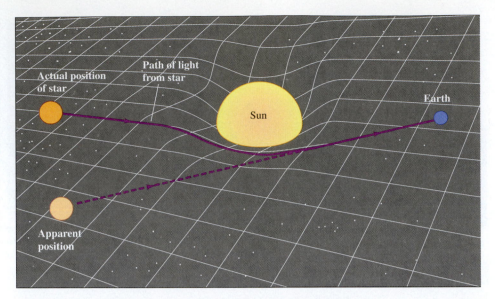

▲ **FIGURE 27.21** Curved space. The change of position is greatly exaggerated.

The general theory of relativity has passed several experimental tests, including three proposed by Einstein. One test has to do with understanding the rotation of the axes of the planet Mercury's elliptical orbit, called the *precession of the perihelion* of Mercury. (The perihelion is the point of closest approach to the sun.) Another test concerns the apparent bending of light rays from distant stars when they pass near the sun, and the third test is the *gravitational red shift,* the increase in wavelength of light proceeding outward from a massive source. Some details of the general theory are more difficult to test and remain speculative in nature, but this theory has played a central role in cosmological investigations of the structure of the universe, the formation and evolution of stars, and related matters.

SUMMARY

Relativity and Simultaneity

(Sections 27.1 and 27.2) All the fundamental laws of physics have the same form in all inertial frames of reference. The speed of light in vacuum is the same in all inertial frames of reference and is independent of the motion of the source. Simultaneity is not an absolute concept: two events that appear to be simultaneous in one inertial frame may not appear simultaneous in a second frame moving relative to the first.

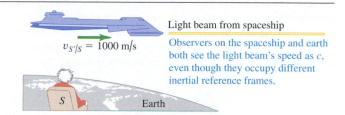

Light beam from spaceship

Observers on the spaceship and earth both see the light beam's speed as c, even though they occupy different inertial reference frames.

$v_{S'/S} = 1000$ m/s

S

Earth

Relativity of Time

(Section 27.3) If Δt_0 is the time interval between two events that occur at the same space point in a particular frame of reference, it is called a **proper time**. If the first frame moves with a constant velocity u relative to a second frame, the time interval Δt between the events, as observed in the second frame, is longer than Δt_0:

$$\Delta t = \frac{\Delta t_0}{\sqrt{1 - u^2/c^2}}. \quad (27.6)$$

This effect is called **time dilation:** A clock moving with respect to an observer appears to run more slowly (Δt) than a clock that is at rest in the observer's frame (Δt_0).

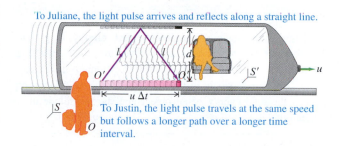

To Juliane, the light pulse arrives and reflects along a straight line.

To Justin, the light pulse travels at the same speed but follows a longer path over a longer time interval.

Relativity of Length

(Section 27.4) If l_0 is the distance between two points that are at rest in a particular frame of reference, it is called a **proper length**. If this first frame is moving with a constant velocity u relative to a second frame, and distances in each frame are measured parallel to that frame's velocity, then the distance l between the points, as observed in the second frame, is shorter than l_0, in accordance with the formula.

$$l = l_0\sqrt{1 - \frac{u^2}{c^2}}. \quad (27.9)$$

This effect is called **length contraction:** A ruler oriented parallel to the frame's velocity and moving with respect to an observer appears shorter (l) than when measured in a frame at rest (l_0) in the observer's frame.

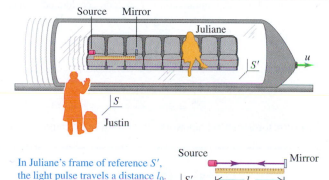

Source   Mirror

Juliane

Justin

In Juliane's frame of reference S', the light pulse travels a distance l_0.

In Justin's frame of reference S, the light pulse travels a distance $l + u\,\Delta t_1$.

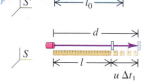

Source

Mirror

The Lorentz Transformation

(Section 27.5) The **Lorentz transformation** relates the coordinates and time of an event in an inertial coordinate system S to the coordinates and time of the same event as observed in a second inertial frame S' moving with constant velocity u relative to the first. The Lorentz transformation equations are

$$x' = \frac{x - ut}{\sqrt{1 - u^2/c^2}}, \qquad y' = y, \qquad t' = \frac{t - ux/c^2}{\sqrt{1 - u^2/c^2}}. \quad (27.10)$$

For one-dimensional motion, the velocity v' in S' is related to the velocity v in S by

$$v' = \frac{v - u}{1 - uv/c^2}. \quad (27.12)$$

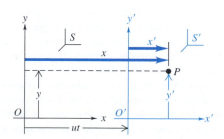

Continued

Relativistic Momentum and Energy

(Sections 27.6 and 27.7) In order for the principle of momentum conservation to hold in all inertial frames, the classical definition of momentum must be modified. Similarly, the definition of kinetic energy K must be modified in order to generalize the relationship between work and energy. The new, relativistic equations are

$$\vec{p} = \frac{m\vec{v}}{\sqrt{1 - v^2/c^2}} \qquad (27.14)$$

and

$$K = \frac{mc^2}{\sqrt{1 - v^2/c^2}} - mc^2 \qquad (27.21)$$

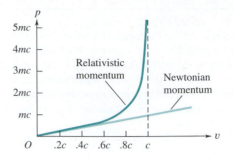

The latter form suggests assigning a **rest energy** $E_{\text{rest}} = mc^2$ (Equation 27.23) to a particle so that the total energy is $E = K + mc^2$ (Equation 27.22). In Newtonian mechanics, $K = p^2/2m$. In relativistic mechanics, the analogous relationship between the total energy of an object and the magnitude of its momentum p is $E^2 = (mc^2)^2 + (pc)^2$ (Equation 27.24).

Relativity and Newtonian Mechanics

(Section 27.8) The special theory of relativity is a generalization of Newtonian mechanics. All the principles of Newtonian mechanics are present as limiting cases when all the speeds are small in comparison to c. Further generalization to include accelerated frames of reference and their relation to gravitational fields leads to the general theory of relativity.

 For instructor-assigned homework, go to www.masteringphysics.com

Conceptual Questions

1. Suppose the speed of light were 30 m/s instead of its actual value. Describe various ways in which the behavior of your everyday world would be different than it is now.

2. The average life span in the United States is about 70 years. Does this mean that it is impossible for an average person to travel a distance greater than 70 light years away from the earth? (A light year is the distance light travels in a year.)

3. Two events occur at the same spatial point in a particular frame of reference and appear to be simultaneous in that frame. Is it possible that they will not appear to be simultaneous in another frame?

4. Does the fact that simultaneity is not an absolute concept also destroy the concept of *causality?* If event A is to *cause* event B, A must occur first. Is it possible that in some frames A will appear to cause B and in others B will appear to cause A?

5. You are standing on a train platform watching a high-speed train pass by. A light inside one of the train cars is turned on and then a little later it is turned off. (a) Who can measure the proper time interval for the duration of the light: you or a passenger on the train? (b) Who can measure the proper length of the train car, you or a passenger on the train? (c) Who can measure the proper length of a sign attached to a post on the train platform, you or a passenger on the train? In each case, explain your answer.

6. According to the twin paradox mentioned in Section 27.3, if one twin stays on earth while the other takes off in a spaceship at a relativistic speed and then returns, one will be older than the other. Can you think of a practical experiment, perhaps using two very precise atomic clocks, that would test this conclusion?

7. Photons are considered to be "massless" particles. Yet since they have energy, you might expect that they also have mass. So what does it mean to call a particle "massless"?

8. A student asserted that a massless particle must always travel at exactly the speed of light. Why do you think this is correct? If so, how do massless particles such as photons acquire that speed? Can they start from rest and accelerate?

9. The theory of relativity sets an upper limit on the speed a particle can have. Does this mean that there are also upper limits on its energy and momentum?

10. Why do you think the development of Newtonian mechanics preceded the more refined relativistic mechanics by so many years?

11. You're approaching the star Betelgeuse in your spaceship at a speed of 0.2c. At what speed does the light from the star reach you?
12. Does relativity say that we can travel fast enough to actually get younger than when we started? Explain.
13. Discuss several good reasons for believing that no matter can reach the speed of light. (*Hint:* Look at what happens to several physical quantities, such as kinetic energy and momentum, as $v \rightarrow c$.)
14. People sometimes interpret the theory of relativity as saying that "everything is relative." Is this really what the theory says? Can you think of any physical quantities that are *not* relative in this theory? (In fact, Einstein thought of his theory as a theory of *absolutes* rather than *relativity.*)
15. Some people have expressed doubt about the theory of relativity, dismissing it as little more than a *theory.* What *experimental evidence* can you cite that this theory is actually correct?

Multiple-Choice Problems

1. A rocket flies toward the earth at $\frac{1}{2}c$, and the captain shines a laser light beam in the forward direction. Which of the following statements about the speed of this light are correct? (There may be more than one correct answer.)
   A. The captain measures speed c for the light.
   B. An observer on earth measures speed c for the light.
   C. An observer on earth measures speed $\frac{3}{2}c$ for the light.
   D. The captain measures speed $\frac{1}{2}c$ for the light.
2. A rocket is traveling at $\frac{1}{3}c$ relative to earth when a lightbulb in the center of a cubical room is suddenly turned on. An astronaut at rest in the rocket and a person at rest on earth both observe the light hit the opposite walls of the room. (See Figure 27.22.) We shall call these events A and B. Which statements about the two events are true? (There may be more than one true statement.)

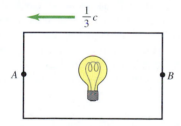

▲ **FIGURE 27.22** Multiple-choice problem 2.

   A. To the observer on earth, both events happen at the same time.
   B. To the astronaut, both events happen at the same time.
   C. To the observer on earth, event B happens before event A.
   D. To the astronaut, event A happens after event B.
3. According to the principles of special relativity (there may be more than one correct choice),
   A. The speed of light is the same for all observers in inertial reference frames, no matter how fast those frames are moving.
   B. All physical quantities are relative, their values depending on the motion of the observer.

C. Two events that occur at the same time for one observer must occur at the same time for all other observers.
D. The laws of physics are the same in all inertial reference frames, no matter how fast they are moving.
4. A square measuring 1 m by 1 m is moving away from observer A along a direction parallel to one of its sides at a speed such that γ is equal to 2. The *area* of this square, as measured by observer A, is
   A. 4 m². B. 2 m². C. 1 m². D. $\frac{1}{2}$ m². E. $\frac{1}{4}$ m².
5. To an observer moving along with the square in the previous question, the area of the square is
   A. 4 m². B. 2 m². C. 1 m². D. $\frac{1}{2}$ m². E. $\frac{1}{4}$ m².
6. To the observer moving along with the square in Question 4, the time interval between consecutive blinks of her eyes is 1 s. To the stationary observer A, this time interval is
   A. 2 s. B. 1 s. C. $\frac{1}{2}$ s. D. $\frac{1}{4}$ s.
7. A high-speed train passes a train platform. Anthony is a passenger on the train, Miguel is standing on the train platform, and Carolyn is riding a bicycle toward the platform in the same direction as the train is traveling. Choose the proper order of how long each of these observers measures the train to be, from longest to shortest.
   A. Carolyn, Miguel, Anthony.
   B. Miguel, Carolyn, Anthony.
   C. Miguel, Anthony, Carolyn.
   D. Anthony, Carolyn, Miguel.
   E. Anthony, Miguel, Carolyn.
8. For a material object, such as a rocket ship, the possible range of γ is
   A. $0 \leq \gamma \leq 1$. B. $\gamma \geq 1$. C. $0 \leq \gamma < \infty$.
9. If it requires energy U to accelerate a rocket from rest to $\frac{1}{2}c$, the energy needed to accelerate that rocket from $\frac{1}{2}c$ to c would be
   A. $\frac{1}{2}U$. B. U. C. $2U$. D. infinite.
10. The reason we do not observe relativistic effects (such as time dilation or length contraction) at ordinary speeds on earth is that
    A. Special relativity is valid at all speeds, but the effects are normally too small to observe at ordinary speeds on earth.
    B. Special relativity is valid only when the speed of an object approaches that of light.
    C. We do readily observe relativistic effects for objects such as jet planes.
11. A rocket is traveling toward the earth at $\frac{1}{2}c$ when it ejects a missile forward at $\frac{1}{2}c$ relative to the rocket. According to *Galilean* velocity addition, the speed of this missile as measured by an observer on earth would be
    A. 0. B. $\frac{4}{5}c$. C. c.
12. For the missile in the previous question, the *correct* value for its speed measured by an observer on earth would be
    A. 0. B. $\frac{4}{5}c$. C. c.
13. Suppose a rocket traveling at 99.99% of the speed of light measured relative to the earth makes a trip to a star 100 light years from earth (meaning that it would take light 100 years to make the trip). During this rocket trip (there may be more than one correct choice),
    A. People on earth would age essentially 100 years.
    B. The astronauts in the rocket would age more than 100 years.
    C. The astronauts in the rocket would age less than 100 years.
    D. The astronauts in the rocket would age the same as the people on earth.
14. A rocket ship is moving toward earth at $\frac{2}{3}c$. The crew is using a telescope to watch a Cubs baseball game in Chicago. The

batter hits the ball (event *A*), which is soon caught (event *B*) by a player 175 ft away, as measured in the ball park. Which one of the following is the proper length of the distance the ball traveled?

A. the 175 ft measured in the ball park.
B. the distance measured by the rocket's crew.
C. Both distances are equal, and hence both are the proper length.

15. A large constant force is used to accelerate an object from rest to a high speed. In which form of Newton's second law—relativistic or classical nonrelativistic—does the object take a longer time to reach a speed of $0.9c$?

A. Relativistic.   B. Nonrelativistic.   C. Same for both.

Problems

27.1 Invariance of Physical Laws
27.2 Relative Nature of Simultaneity

1. •• A spaceship is traveling toward earth from the space colony on Asteroid 1040A. The ship is at the halfway point of the trip, passing Mars at a speed of $0.9c$ relative to Mars's frame of reference. At the same instant, a passenger on the spaceship receives a radio message from her boyfriend on 1040A and another from her hairdresser on earth. According to the passenger on the ship, were these messages sent simultaneously or at different times. If at different times, which one was sent first? Explain your reasoning.

2. •• A rocket is moving to the right at half the speed of light relative to the earth. A lightbulb in the center of a room inside the rocket suddenly turns on. Call the light hitting the front end of the room event *A* and the light hitting the back of the room event *B*. (See Figure 27.23.) Which event occurs first, *A* or *B*, or are they simultaneous, as viewed by (a) an astronaut riding in the rocket and (b) a person at rest on the earth?

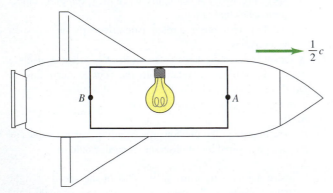

▲ **FIGURE 27.23** Problem 2.

27.3 Relativity of Time

3. • A futuristic spaceship flies past Pluto with a speed of $0.964c$ relative to the surface of the planet. When the spaceship is directly overhead at an altitude of 1500 km, a very bright signal light on the surface of Pluto blinks on and then off. An observer on Pluto measures the signal light to be on for 80.0 μs. What is the duration of the light pulse as measured by the pilot of the spaceship?

4. • Inside a spaceship flying past the earth at three-fourths the speed of light, a pendulum is swinging. (a) If each swing takes 1.50 s as measured by an astronaut performing an experiment inside the spaceship, how long will the swing take as measured by a person at mission control on earth who is watching the experiment? (b) If each swing takes 1.50 s as measured by a person at mission control on earth, how long will it take as measured by the astronaut in the spaceship?

5. • You take a trip to Pluto and back (round trip 11.5 billion km), traveling at a constant speed (except for the turnaround at Pluto) of 45,000 km/h. (a) How long does the trip take, in hours, from the point of view of a friend on earth? About how many years is this? (b) When you return, what will be the difference between the time on your atomic wristwatch and the time on your friend's? (*Hint:* Assume the distance and speed are highly precise, and carry a lot of significant digits in your calculation!)

6. • The negative pion (π^-) is an unstable particle with an average lifetime of 2.60×10^{-8} s (measured in the rest frame of the pion). (a) If the pion is made to travel at very high speed relative to a laboratory, its average lifetime is measured in the laboratory to be 4.20×10^{-7} s. Calculate the speed of the pion expressed as a fraction of c. (b) What distance, as measured in the laboratory, does the pion travel during its average lifetime?

7. • An alien spacecraft is flying overhead at a great distance as you stand in your backyard. You see its searchlight blink on for 0.190 s. The first officer on the craft measures the searchlight to be on for 12.0 ms. (a) Which of these two measured times is the proper time? (b) What is the speed of the spacecraft relative to the earth, expressed as a fraction of the speed of light, c?

8. •• How fast must a rocket travel relative to the earth so that time in the rocket "slows down" to half its rate as measured by earth-based observers? Do present-day jet planes approach such speeds?

9. •• A spacecraft flies away from the earth with a speed of 4.80×10^6 m/s relative to the earth and then returns at the same speed. The spacecraft carries an atomic clock that has been carefully synchronized with an identical clock that remains at rest on earth. The spacecraft returns to its starting point 365 days (1 year) later, as measured by the clock that remained on earth. What is the difference in the elapsed times on the two clocks, measured in hours? Which clock, the one in the spacecraft or the one on earth, shows the smaller elapsed time?

27.4 Relativity of Length

10. • You measure the length of a futuristic car to be 3.60 m when the car is at rest relative to you. If you measure the length of the car as it zooms past you at a speed of $0.900c$, what result do you get?

11. • A meterstick moves past you at great speed. Its motion relative to you is parallel to its long axis. If you measure the length of the moving meterstick to be 1.00 ft (1 ft $= 0.3048$ m)—for example, by comparing it with a 1-foot ruler that is at rest relative to you, at what speed is the meterstick moving relative to you?

12. • In the year 2084, a spacecraft flies over Moon Station III at a speed of $0.800c$. A scientist on the moon measures the length of the moving spacecraft to be 140 m. The spacecraft later lands on the moon, and the same scientist measures the length of the now stationary spacecraft. What value does she get?

13. • A rocket ship flies past the earth at 85.0% of the speed of light. Inside, an astronaut who is undergoing a physical examination is having his height measured while he is lying down parallel to the direction the rocket ship is moving. (a) If his height is measured to be 2.00 m by his doctor inside the ship, what height would a person watching this from earth measure for his height? (b) If the earth-based person had measured 2.00 m, what would the doctor in the spaceship have measured for the astronaut's height? Is this a reasonable height? (c) Suppose the astronaut in part (a) gets up after the examination and stands with his body perpendicular to the direction of motion. What would the doctor in the rocket and the observer on earth measure for his height now?

14. •• A spaceship makes the long trip from earth to the nearest star system, Alpha Centauri, at a speed of 0.955c. The star is about 4.37 light years from earth, as measured in earth's frame of reference (1 light year is the distance light travels in a year). (a) How many years does the trip take, according to an observer on earth? (b) How many years does the trip take according to a passenger on the spaceship? (c) How many light years distant is Alpha Centauri from earth, as measured by a passenger on the speeding spacecraft? (Note that, in the ship's frame of reference, the passengers are at rest, while the space between earth and Alpha Centauri goes rushing past at 0.955c.) (d) Use your answer from part (c) along with the speed of the spacecraft to calculate another answer for part (b). Do your two answers for that part agree? Should they?

15. •• A muon is created 55.0 km above the surface of the earth (as measured in the earth's frame). The average life-time of a muon, measured in its own rest frame, is 2.20 μs, and the muon we are considering has this lifetime. In the frame of the muon, the earth is moving toward the muon with a speed of 0.9860c. (a) In the muon's frame, what is its initial height above the surface of the earth? (b) In the muon's frame, how much closer does the earth get during the lifetime of the muon? What fraction is this of the muon's original height, as measured in the muon's frame? (c) In the earth's frame, what is the lifetime of the muon? In the earth's frame, how far does the muon travel during its life-time? What fraction is this of the muon's original height in the earth's frame?

27.5 The Lorentz Transformation

16. • An enemy spaceship is moving toward your starfighter with a speed of 0.400c, as measured in your reference frame. The enemy ship fires a missile toward you at a speed of 0.700c relative to the enemy ship. (See Figure 27.24.) (a) What is the speed of the missile relative to you? Express your answer in terms of the speed of light. (b) If you measure the enemy ship to be 8.00 × 10⁶ km away from you when the missile is fired, how much time, measured in your frame, will it take the missile to reach you?

▲ FIGURE 27.24 Problem 16.

17. •• An imperial spaceship, moving at high speed relative to the planet Arrakis, fires a rocket toward the planet with a speed of 0.920c relative to the spaceship. An observer on Arrakis measures the rocket to be approaching with a speed of 0.360c. What is the speed of the spaceship relative to Arrakis? Is the spaceship moving toward or away from Arrakis?

18. • Two particles in a high-energy accelerator experiment are approaching each other head-on, each with a speed of 0.9520c as measured in the laboratory. What is the magnitude of the velocity of one particle relative to the other?

19. • A pursuit spacecraft from the planet Tatooine is attempting to catch up with a Trade Federation cruiser. As measured by an observer on Tatooine, the cruiser is traveling away from the planet with a speed of 0.600c. The pursuit ship is traveling at a speed of 0.800c relative to Tatooine, in the same direction as the cruiser. What is the speed of the cruiser relative to the pursuit ship?

20. • Two particles are created in a high-energy accelerator and move off in opposite directions. The speed of one particle, as measured in the laboratory, is 0.650c, and the speed of each particle relative to the other is 0.950c. What is the speed of the second particle, as measured in the laboratory?

21. •• Neutron stars are the remains of exploded stars, and they rotate at very high rates of speed. Suppose a certain neutron star has a radius of 10.0 km and rotates with a period of 1.80 ms. (a) Calculate the surface rotational speed at the equator of the star as a fraction of c. (b) Assuming the star's surface is an inertial frame of reference (which it isn't, because of its rotation), use the Lorentz velocity transformation to calculate the speed of a point on the equator with respect to a point directly opposite it on the star's surface.

27.6 Relativistic Momentum

22. • At what speed is the momentum of a particle three times as great as the result obtained from the nonrelativistic expression mv?

23. •• (a) At what speed does the momentum of a particle differ by 1.0% from the value obtained with the nonrelativistic expression mv? (b) Is the correct relativistic value greater or less than that obtained from the nonrelativistic expression?

24. • **Relativistic baseball.** Calculate the magnitude of the force required to give a 0.145 kg baseball an acceleration of $a = 1.00$ m/s² in the direction of the baseball's initial velocity, when this velocity has a magnitude of (a) 10.0 m/s; (b) 0.900c; (c) 0.990c.

25. •• Sketch a graph of (a) the nonrelativistic Newtonian momentum as a function of speed v and (b) the relativistic momentum as a function of v. In both cases, start from $v = 0$ and include the region where $v \rightarrow c$. Does either of these graphs extend beyond $v = c$?

26. • An electron is acted upon by a force of 5.00 × 10⁻¹⁵ N due to an electric field. Find the acceleration this force produces in each case: (a) The electron's speed is 1.00 km/s. (b) The electron's speed is 2.50 × 10⁸ m/s and the force is parallel to the velocity.

27.7 Relativistic Work and Energy

27. • Using both the nonrelativistic and relativistic expressions, compute the kinetic energy of an electron and the ratio of the two results (relativistic divided by nonrelativistic), for speeds of (a) 5.00 × 10⁷ m/s, (b) 2.60 × 10⁸ m/s.

28. • What is the speed of a particle whose kinetic energy is equal to (a) its rest energy, (b) five times its rest energy?

29. • **Particle annihilation.** In proton–antiproton annihilation, a proton and an antiproton (a negatively charged particle with the mass of a proton) collide and disappear, producing electromagnetic radiation. If each particle has a mass of 1.67×10^{-27} kg and they are at rest just before the annihilation, find the total energy of the radiation. Give your answers in joules and in electron volts.

30. • The sun produces energy by nuclear fusion reactions, in which matter is converted into energy. By measuring the amount of energy we receive from the sun, we know that it is producing energy at a rate of 3.8×10^{26} W. (a) How many kilograms of matter does the sun lose each second? Approximately how many tons of matter is this? (b) At this rate, how long would it take the sun to use up all its mass? (See Appendix E.)

31. •• A proton (rest mass 1.67×10^{-27} kg) has total energy that is 4.00 times its rest energy. What are (a) the kinetic energy of the proton; (b) the magnitude of the momentum of the proton; (c) the speed of the proton?

32. • In a hypothetical nuclear-fusion reactor, two deuterium nuclei combine or "fuse" to form one helium nucleus. The mass of a deuterium nucleus, expressed in atomic mass units (u), is 2.0136 u; that of a helium nucleus is 4.0015 u. $(1 \text{ u} = 1.661 \times 10^{-27} \text{ kg}.)$ (a) How much energy is released when 1.0 kg of deuterium undergoes fusion? (b) The annual consumption of electrical energy in the United States is on the order of 1.0×10^{19} J. How much deuterium must react to produce this much energy?

33. •• **An antimatter reactor.** When a particle meets its antiparticle (more about this in Chapter 30), they annihilate each other and their mass is converted to light energy. The United States uses approximately 1.0×10^{20} J of energy per year. (a) If all this energy came from a futuristic antimatter reactor, how much mass would be consumed yearly? (b) If this antimatter fuel had the density of Fe (7.86 g/cm^3) and were stacked in bricks to form a cubical pile, how high would it be? (Before you get your hopes up, antimatter reactors are a *long* way in the future—if they ever will be feasible.)

34. • A particle has a rest mass of 6.64×10^{-27} kg and a momentum of 2.10×10^{-18} kg · m/s. (a) What is the total energy (kinetic plus rest energy) of the particle? (b) What is the kinetic energy of the particle? (c) What is the ratio of the kinetic energy to the rest energy of the particle?

35. •• (a) Through what potential difference does an electron have to be accelerated, starting from rest, to achieve a speed of $0.980c$? (b) What is the kinetic energy of the electron at this speed? Express your answer in joules and in electronvolts.

36. •• Sketch a graph of (a) the nonrelativistic Newtonian kinetic energy as a function of speed v, (b) the relativistic kinetic energy as a function of speed v. In both cases, start from $v = 0$ and include the region where $v \to c$. Does either of these graphs extend beyond $v = c$?

General Problems

37. •• The starships of the Solar Federation are marked with the symbol of the Federation, a circle, while starships of the Denebian Empire are marked with the Empire's symbol, an ellipse whose major axis is 1.40 times its minor axis

($a = 1.40b$ in Figure 27.25). How fast, relative to an observer, does an Empire ship have to travel for its markings to be confused with those of a Federation ship?

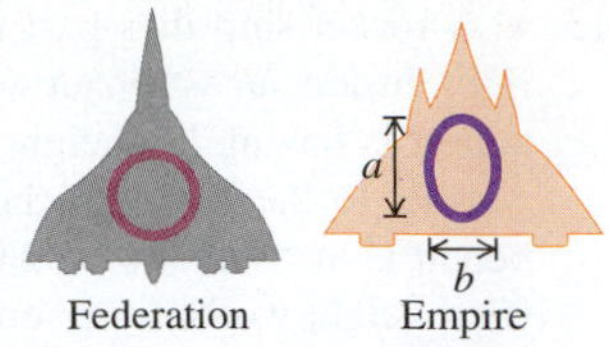

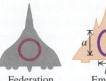

▲ **FIGURE 27.25** Problem 37.

38. • A space probe is sent to the vicinity of the star Capella, which is 42.2 light years from the earth. (A light year is the distance light travels in a year.) The probe travels with a speed of $0.9910c$ relative to the earth. An astronaut recruit on board is 19 years old when the probe leaves the earth. What is her biological age when the probe reaches Capella, as measured by (a) the astronaut and (b) someone on earth?

39. •• Two events are observed in a frame of reference S to occur at the same space point, the second occurring 1.80 s after the first. In a second frame S' moving relative to S, the second event is observed to occur 2.35 s after the first. What is the difference between the positions of the two events as measured in S'?

40. •• **Why are we bombarded by muons?** Muons are unstable subatomic particles (more on them in Chapter 30) that decay to electrons with a mean lifetime of 2.2 μs. They are produced when cosmic rays bombard the upper atmosphere about 10 km above the earth's surface, and they travel very close to the speed of light. The problem we want to address is why we see any of them at the earth's surface. (a) What is the greatest distance a muon could travel during its 2.2 μs lifetime? (b) According to your answer in part (a), it would seem that muons could never make it to the ground. But the 2.2 μs lifetime is measured in the frame of the muon, and they are moving very fast. At a speed of $0.999c$, what is the mean lifetime of a muon as measured by an observer at rest on the earth? How far could the muon travel in this time? Does this result explain why we find muons in cosmic rays? (c) From the point of view of the muon, it still lives for only 2.2 μs, so how does it make it to the ground? What is the thickness of the 10 km of atmosphere through which the muon must travel, as measured by the muon? Is it now clear how the muon is able to reach the ground?

41. •• How fast does a muon (see the previous problem) have to move (according to an outside observer) in order to travel 1.0 km during its brief lifetime of 2.2 μs?

42. • A cube of metal with sides of length a sits at rest in the laboratory with one edge parallel to the x axis. Therefore, in the laboratory frame, its volume is a^3. A rocket ship flies past the laboratory parallel to the x axis with a velocity v. To an observer in the rocket, what is the volume of the metal cube?

43. •• In an experiment, two protons are shot directly toward each other, each moving at half the speed of light relative to the laboratory. (a) What speed does one proton measure for the other proton? (b) What would be the answer to part (a) if we used only nonrelativistic Newtonian mechanics? (c) What is the kinetic energy of each proton as measured by (i) an observer at rest in the laboratory and (ii) an observer riding along with one of the protons? (d) What would be the answers to part (c) if we used only nonrelativistic Newtonian mechanics?

44. •• A 0.100 μg speck of dust is accelerated from rest to a speed of $0.900c$ by a constant 1.00×10^6 N force. (a) If the nonrelativistic form of Newton's second law $(\sum F = ma)$ is used, how far does the object travel to reach its final speed?

(b) Using the correct relativistic form of Equation 27.19, how far does the object travel to reach its final speed? (c) Which distance is greater? Why?

45. •• By what minimum amount does the mass of 4.00 kg of ice increase when the ice melts at 0.0°C to form water at that same temperature? (The heat of fusion of water is 3.34×10^5 J/kg.)

46. •• In certain radioactive beta decay processes (more about these in Chapter 30), the beta particle (an electron) leaves the atomic nucleus with a speed of 99.95% the speed of light relative to the decaying nucleus. If this nucleus is moving at 75.00% the speed of light, find the speed of the emitted electron relative to the laboratory reference frame if the electron is emitted (a) in the same direction that the nucleus is moving, (b) in the opposite direction from the nucleus's velocity. (c) In each case in parts (a) and (b), find the kinetic energy of the electron as measured in (i) the laboratory frame and (ii) the reference frame of the decaying nucleus.

47. •• Starting from Equation 27.24, show that in the classical limit $(pc \ll mc^2)$ the energy approaches the classical kinetic energy plus the rest mass energy. (*Hint:* If $x \ll 1$, $\sqrt{1 + x} \approx 1 + x/2$.)

48. •• **Space travel?** Travel to the stars requires hundreds or thousands of years, even at the speed of light. Some people have suggested that we can get around this difficulty by accelerating the rocket (and its astronauts) to very high speeds so that they will age less due to time dilation. The fly in this ointment is that it takes a *great deal* of energy to do this. Suppose you want to go to the immense red giant Betelgeuse, which is about 500 light years away. (A light year is the distance that light travels in one year.) You plan to travel at constant speed in a 1000 kg rocket ship (a little over a ton), which, in reality, is far too small for this purpose. In each case that follows, calculate the time for the trip, as measured by people on earth and by astronauts in the rocket ship, the energy needed in joules, and the energy needed as a percent of U.S. yearly use (which is 1.0×10^{20} J). For comparison, arrange your results in a table showing v_{Rocket}, t_{Earth}, t_{Rocket}, E (in J), and E (as % of U.S. use). The rocket ship's speed is (a) $0.50c$, (b) $0.99c$, and (c) $0.9999c$. On the basis of your results, does it seem likely that any government will invest in such high-speed space travel any time soon?

49. •• A nuclear device containing 8.00 kg of plutonium explodes. The rest mass of the products of the explosion is less than the original rest mass by one part in 10^4. (a) How much energy is released in the explosion? (b) If the explosion takes place in 4.00 μs, what is the average power developed by the bomb? (c) What mass of water could the released energy lift to a height of 1.00 km?

50. •• Electrons are accelerated through a potential difference of 750 kV, so that their kinetic energy is 7.50×10^5 eV. (a) What is the ratio of the speed v of an electron having this energy to the speed of light, c? (b) What would the speed be if it were computed from the principles of classical mechanics?

51. •• The distance to a particular star, as measured in the earth's frame of reference, is 7.11 light years (1 light year is the distance light travels in 1 year). A spaceship leaves earth headed for the star, and takes 3.35 years to arrive, as measured by passengers on the ship. (a) How long does the trip take, according to observers on earth? (b) What distance for the trip do passengers on the spacecraft measure? (*Hint:* What is the speed of light in units of ly/y?)

52. •• **Čerenkov radiation.** The Russian physicist P. A. Čerenkov discovered that a charged particle traveling in a solid with a speed exceeding the speed of light *in that material* radiates electromagnetic radiation. (This phenomenon is analogous to the sonic boom produced by an aircraft moving faster than the speed of sound in air.) Čerenkov shared the 1958 Nobel Prize for this discovery. What is the minimum kinetic energy (in electronvolts) that an electron must have while traveling inside a slab of crown glass $(n = 1.52)$ in order to create Čerenkov radiation?

53. •• Scientists working with a particle accelerator determine that an unknown particle has a speed of 1.35×10^8 m/s and a momentum of 2.52×10^{-19} kg · m/s. From the curvature of its path in a magnetic field they also deduce that it has a positive charge. Using this information, identify the particle.

Passage Problems

Speed of light. Our universe has properties that are determined by the values of the fundamental physical constants, and it would be a much different place if the charge of the electron, the mass of the proton, or the speed of light were substantially different from what they actually are. For instance, the speed of light is so large that the effects of relativity usually go unnoticed in everyday events. Let's imagine an alternate universe where the speed of light is 1,000,000 times smaller than it is in our universe to see what would happen.

54. What is the speed of light in the alternate universe?
   A. 3×10^8 m/s
   B. 3×10^6 m/s
   C. 3000 m/s
   D. 300 m/s

55. An airplane has a length of 60 m when measured at rest. When the airplane is moving at 180 m/s (400 mph) in the alternate universe, how long would it appear to be to a stationary observer?
   A. 24 m   B. 36 m
   C. 48 m   D. 60 m
   E. 75 m

56. If the airplane has a rest mass of 20,000 kg, what is its relativistic mass when moving at 180 m/s?
   A. 8000 kg
   B. 12,000 kg
   C. 16,000 kg
   D. 25,000 kg
   E. 33,300 kg

57. In our universe the rest energy of an electron is approximately 8.2×10^{-14} J. What would it be in the alternate universe?
   A. 8.2×10^{-8} J
   B. 8.2×10^{-26} J
   C. 8.2×10^{-2} J
   D. 0.82 J

28 Photons, Electrons, and Atoms

W hat is light? The work of Maxwell, Hertz, and others established firmly that light is an electromagnetic wave. Interference, diffraction, and polarization phenomena show convincingly the wave nature of light and other electromagnetic radiation.

But there are also many phenomena, particularly those involving the emission and absorption of electromagnetic radiation, that show a completely different aspect of the nature of light, in which it seems to behave as a stream of *particles*. In such phenomena, the energy of light is emitted and absorbed in packages with a definite size, called *photons* or *quanta*. The energy of a single photon is proportional to the frequency of the radiation, and we say that the energy is *quantized*.

The energy associated with the internal motion within atoms is also quantized. Each kind of atom has a set of possible energy values called *energy levels*. The internal energy of an atom must be equal to one of these values; an atom cannot have an energy between two values. Understanding the internal structure of atoms requires a new language in which electrons sometimes behave like waves rather than particles.

The three common threads woven through this chapter are the quantization of electromagnetic radiation, the existence of discrete energy levels in atoms, and the dual wave–particle nature of both particles and electromagnetic radiation. These three basic concepts take us a long way toward understanding a wide variety of otherwise puzzling phenomena, including the photoelectric effect (the emission of electrons from a surface when light strikes it), the line spectra emitted by gaseous elements, the operation of lasers, and the production and scattering of x rays. Analysis of these phenomena and their relation to atomic structure takes us to the threshold of quantum mechanics, which involves some radical changes in our views of the nature of radiation and of matter itself.

When you take a snapshot, you use light. However, this false-color scanning electron micrograph of a fruit fly was made using a beam of electrons instead. Up to now, we've treated electrons as particles and light as a wave. Now we'll see that both of them can actually be treated in both ways.

28.1 The Photoelectric Effect

The **photoelectric effect,** first observed by Heinrich Hertz in 1887, is the emission of electrons from the surface of a conductor when light strikes the surface. The liberated electrons absorb energy from the incident radiation and are thus able to overcome the potential-energy barrier that normally confines them inside the material. The process is analogous to **thermionic emission,** discovered in 1883 by Edison, in which the escape energy is supplied by heating the material to a high temperature, liberating electrons by a process analogous to boiling a liquid. The minimum amount of energy an individual electron has to gain in order to escape from a particular surface is called the **work function** for that surface; it is denoted by ϕ.

The photoelectric effect was investigated in detail by Wilhelm Hallwachs and Philipp Lenard during the years 1886–1900. These two researchers used an apparatus shown schematically in Figure 28.1a. Two conducting electrodes are enclosed in an evacuated glass tube. The negatively charged electrode is called the **cathode,** and the positively charged electrode is called the **anode.** The battery, or other source of potential difference, creates an electric field (red arrows) in the direction from anode to cathode. Monochromatic (single-frequency) light (purple arrows) falls on the surface of the cathode, causing electrons to be emitted from the cathode. A high vacuum, with residual pressure of 0.01 Pa $(10^{-7}$ atm$)$ or less, is needed to avoid collisions of electrons with gas molecules.

Once emitted, the electrons are pushed toward the anode by the electric field, causing a current i in the external circuit; the current is measured by the galvanometer G. Hallwachs and Lenard studied how this current varies with voltage and with the frequency and intensity of the light.

Hallwachs and Lenard found that when monochromatic light falls on the cathode, *no* electrons are emitted unless the frequency of the light is greater than some minimum value f, called the **threshold frequency,** that depends on the material of the cathode. For most metals, this frequency corresponds to that of light in the ultraviolet range (wavelengths of 200 to 300 nm), but for potassium and cesium oxides, it is in the visible spectrum (400 to 700 nm). This experimental result is consistent with the idea that each liberated electron absorbs an amount of energy E proportional to the frequency f of the light. When the frequency f is not great enough, the energy E is not great enough for the electron to surmount the potential-energy barrier ϕ at the surface.

When f is *greater than* the threshold value, it is found that some electrons are emitted from the cathode with substantial initial speeds. The evidence for this conclusion is the fact that, even with *no* potential difference between anode and cathode, a few electrons reach the anode, causing a small current in the external circuit. Indeed, even when the polarity of the potential difference V is reversed (Figure 28.1b) and the associated electric-field force on the electrons is back toward the cathode, some electrons still reach the anode. The electron flow stops completely only when the reversed potential difference V is made large enough that the corresponding potential energy eV is greater than the maximum kinetic energy $\frac{1}{2}mv_{max}^2$ of the emitted electrons.

Light causes the cathode to emit electrons, which are pushed toward the anode by the electric field force.

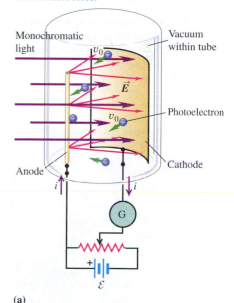

(a)

Even when the direction of $\vec{E}$ field is reversed, some electrons still reach the anode.

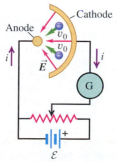

(b) Overhead view with $\vec{E}$ field reversed

▲ **FIGURE 28.1** A phototube demonstrating the photoelectric effect. The stopping potential V_0 is the minimum absolute value of the reverse potential difference that gives zero current.

Stopping potential

The reversed potential difference required to stop the electron flow completely is called the stopping potential, denoted by V_0. From the preceding discussion,

$$\frac{1}{2}mv_{max}^2 = eV_0. \qquad (28.1)$$

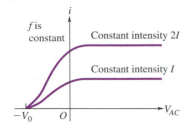

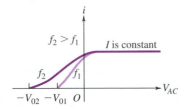

▲ **FIGURE 28.2** Photocurrent I as a function of V_{AC}.

▲ **FIGURE 28.3** Variation of stopping potential with frequency. The potential of the anode with respect to the cathode is V_{AC}.

▲ **Application Fire!** Countless lives have been saved from tragic fires by smoke detectors, thanks to our understanding and use of the photoelectric effect, in which light shining on a surface causes ejection of electrons and a resulting electric current. Some types of smoke detectors make use of this effect to determine whether the air in a room contains particles of smoke. A tiny beam of light inside the detector passes by a photosensitive cell placed at a right angle to the light beam. Normally the light bypasses the photocell, but smoke particles scatter the light, causing some of it to hit the photocell and produce a small current via the photoelectric effect. This current results in activation of a very loud alarm, giving any unwary occupants of a burning building advance warning of a potentially life-threatening fire.

Measuring the stopping potential V_0 therefore gives us a direct measurement of the maximum kinetic energy electrons have when they leave the cathode.

A classical wave theory of light would predict that when we increase the *intensity* of the light striking the cathode, the electrons should come off with greater energy and the stopping potential should be greater. But that isn't what actually happens. Figure 28.2 shows how the photocurrent varies with voltage for two different intensity levels. When the potential V_{AC} of the anode with respect to the cathode is sufficiently large and positive, the curve levels off, showing that *all* the emitted electrons are being collected by the anode. When the light intensity is increased (say, from I to $2I$), the maximum current increases, but the stopping potential is found to be the same. However, when the frequency f of the light is increased, the stopping potential V_0 increases linearly (Figure 28.3). These results suggest that the maximum kinetic energy of an emitted electron *does not* depend on the *intensity* of the incident light, but it *does* depend on the *frequency* or *wavelength*.

The correct analysis of the photoelectric effect was developed by Albert Einstein in 1905. Extending a proposal made five years earlier by Max Planck, Einstein postulated that a beam of light consists of small bundles of energy called **quanta** or **photons.**

Energy of a photon
The energy E of an individual photon is equal to a constant times the frequency f of the photon; that is,

$$E = hf, \qquad (28.2)$$

where h is a universal constant called **Planck's constant.** The numerical value of h, to the accuracy presently known, is

$$h = 6.6260693(11) \times 10^{-34} \text{ J} \cdot \text{s}.$$

In Einstein's analysis, a photon striking the surface of a conductor is absorbed by an electron. The energy transfer is an "all-or-nothing" process: The electron gets either *all* the photon's energy or *none* of it. If this energy is greater than the surface potential-energy barrier (the work function ϕ), the electron can escape from the surface.

It follows that the maximum kinetic energy $\frac{1}{2}mv_{max}^2$ for an emitted electron is the energy hf it gains by absorbing a photon, minus the work function ϕ:

$$\frac{1}{2}mv_{max}^2 = hf - \phi. \qquad (28.3)$$

Combining this relationship with Equation 28.1, we find that

$$eV_0 = hf - \phi, \qquad \text{or} \qquad V_0 = \frac{h}{e}f - \frac{\phi}{e}. \qquad (28.4)$$

That is, if we measure the stopping potential V_0 for each of several values of frequency f, we expect V_0 to be a linear function of f. A graph of V_0 (on the vertical axis) as a function of f (on the horizontal axis) should then be a straight line. Measurements of V_0 and f confirm this prediction. Furthermore, the slope of the line is h/e, and the intercept on the vertical axis (corresponding to $f = 0$) is at $V_0 = -\phi/e$. Thus, we can determine both the work function ϕ (in electronvolts) for the material and the value of the quantity h/e. (Example 28.3 shows in detail how this can be done.) After the electron charge e was measured directly by Robert Millikan in 1909, Planck's constant h could also be determined from these measurements.

Electron energies and work functions are usually expressed in electronvolts (Section 18.4):

$$1 \text{ eV} = 1.602 \times 10^{-19} \text{ J}.$$

In terms of electronvolts, Planck's constant is

$$h = 4.136 \times 10^{-15} \text{ eV} \cdot \text{s}.$$

Table 28.1 lists a few typical work functions of elements. The values are approximate because they are sensitive to surface impurities. For example, a thin layer of cesium oxide can reduce the work function of a metal to about 1 eV.

TABLE 28.1 Work functions of elements

Element	Work function (eV)
Aluminum	4.3
Carbon	5.0
Copper	4.7
Gold	5.1
Nickel	5.1
Silicon	4.8
Silver	4.3
Sodium	2.7

Conceptual Analysis 28.1

Vary the frequency

Light falling on a metal surface causes electrons to be emitted from the metal by the photoelectric effect. In a particular experiment, the intensity of the incident light and the temperature of the metal are held constant. As we decrease the frequency of the incident light,

A. the work function of the metal increases.
B. the number of electrons emitted from the metal decreases steadily to zero.
C. the maximum speed of the emitted electrons decreases steadily until no electrons are emitted.
D. the stopping potential increases.

SOLUTION In the photoelectric effect, some of an incident photon's energy is used to remove an electron from the metal; the remainder becomes the electron's kinetic energy. The energy to remove the electron from the metal, called the work function ϕ, is constant for a given material. The smaller the frequency of light, the smaller the energy of each photon $(E = hf)$. Therefore, as we decrease the frequency of the incident light, less and less of each photon's energy is available to become kinetic energy of the emitted electrons, and the speed of the emitted electrons decreases steadily to zero. The correct answer is C.

EXAMPLE 28.1 ### Conductivity enhanced by photons

Silicon films become better electrical conductors when illuminated by photons with energies of 1.14 eV or greater. This behavior is called **photoconductivity.** What is the corresponding photon wavelength? In what portion of the electromagnetic spectrum does it lie?

SOLUTION

SET UP The energy E of a photon is related to its frequency f by $E = hf$. The wavelength λ is related to the frequency by the general wave relation $\lambda = c/f$. Combining these relations and solving for λ, we get $\lambda = hc/E$.

SOLVE It's easiest to perform our calculations using units of $(\text{eV} \cdot \text{s})$ for h. That is, $h = 4.136 \times 10^{-15} \text{ eV} \cdot \text{s}$. Using the preceding expression for λ, we get

$$\lambda = \frac{hc}{E} = \frac{(4.136 \times 10^{-15} \text{ eV} \cdot \text{s})(3.00 \times 10^8 \text{ m/s})}{1.14 \text{ eV}}$$
$$= 1.09 \times 10^{-6} \text{ m} = 1090 \text{ nm}.$$

The wavelengths of the visible spectrum are about 400 to 700 nm, so the wavelength we have found is in the near-infrared region of the spectrum.

REFLECT The frequency of a photon is inversely proportional to its wavelength, so the minimum energy of 1.14 eV corresponds to the maximum wavelength that causes photoconduction for silicon, in this case about 1090 nm. Thus, for silicon, all light with a wavelength less than 1090 nm, including all light in the visible spectrum, contributes to photoconductivity.

Practice Problem: If we need a material that is photoconductive for any wavelength in the visible spectrum (i.e., 400 to 700 nm), to what range of photon energies must it respond? *Answer:* 1.77 to 3.10 eV.

EXAMPLE 28.2   Reverse potential needed to stop photoelectric current

In a photoelectric-effect experiment with light of a certain frequency, a reverse potential difference of 1.25 V is required to reduce the current to zero. Using this value, together with the mass and charge of the electron, find the maximum kinetic energy and the maximum speed of the photoelectrons emitted.

SOLUTION

SET UP If the maximum electron kinetic energy doesn't exceed 1.25 eV, *all* the electrons will be stopped by a reverse potential difference of 1.25 V or greater. For consistency of units, we need to convert this energy to joules and then use the definition of kinetic energy to find the maximum speed.

SOLVE The maximum kinetic energy is

$$K_{max} = eV_0 = (1.60 \times 10^{-19} \text{ C})(1.25 \text{ V}) = 2.00 \times 10^{-19} \text{ J}.$$

From the definition of kinetic energy, $K_{max} = \frac{1}{2}mv_{max}^2$, we get

$$v_{max} = \sqrt{\frac{2K_{max}}{m}} = \sqrt{\frac{2(2.00 \times 19^{-19} \text{ J})}{9.11 \times 10^{-31} \text{ kg}}}$$
$$= 6.63 \times 10^5 \text{ m/s}.$$

Alternatively, from Equation 28.1,

$$\frac{1}{2}mv_{max}^2 = eV_0,$$

$$v_{max} = \sqrt{\frac{2eV_0}{m}} = \sqrt{\frac{2(1.60 \times 10^{-19} \text{ C})(1.25 \text{ V})}{9.11 \times 10^{-31} \text{ kg}}}$$
$$= 6.63 \times 10^5 \text{ m/s}.$$

REFLECT This speed is much *smaller* than *c*, the speed of light, meaning that we are justified in using the nonrelativistic kinetic-energy expression. An equivalent statement is that the kinetic energy $eV_0 = 1.25$ eV is much smaller than the electron's rest energy $mc^2 = 0.511$ MeV.

Practice Problem: What reverse potential is required to reduce the current to zero if the maximum electron speed is 9.38×10^5 m/s? *Answer:* 2.50 V.

EXAMPLE 28.3   How to measure Planck's constant

For a certain cathode material used in a photoelectric-effect experiment, a stopping potential of 3.0 V was required for light of wavelength 300 nm, 2.0 V for 400 nm, and 1.0 V for 600 nm. Determine the work function for this material and the value of Planck's constant, as obtained from these data.

SOLUTION

SET UP From the preceding discussion, a graph of V_0 (on the vertical axis) as a function of f (on the horizontal axis) should be a straight line, as given by Equation 28.4:

$$V_0 = \frac{h}{e}f - \frac{\phi}{e}.$$

Our graph is shown in Figure 28.4. We see that the slope of the line is h/e and the intercept on the vertical axis (corresponding to $f = 0$) is at $V_0 = -\phi/e$. (But note that the portion of the graph below the horizontal axis corresponds to negative values of V_0, for which f is less than the threshold value and no photoelectrons are actually emitted.)

SOLVE The frequencies, obtained from $f = c/\lambda$ and $c = 3.00 \times 10^8$ m/s, are 1.0, 0.75, and 0.5×10^{15} Hz. From the slope of the graph, we find that

$$\frac{h}{e} = \frac{1.0 \text{ V}}{0.25 \times 10^{15} \text{ s}^{-1}} = 4.0 \times 10^{-15} \text{ J} \cdot \text{s/C},$$
$$h = (4.0 \times 10^{-15} \text{ J} \cdot \text{s/C})(1.60 \times 10^{-19} \text{ C})$$
$$= 6.4 \times 10^{-34} \text{ J} \cdot \text{s}.$$

The intercept on the vertical axis occurs at

$$-\frac{\phi}{e} = -1.0 \text{ V}, \qquad \phi = 1.0 \text{ eV} = 1.60 \times 10^{-19} \text{ J}.$$

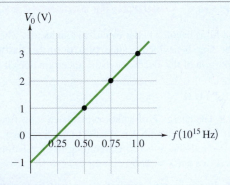

▲ **FIGURE 28.4** Our graph for this problem.

The work function ϕ is 1.0 eV. The graph shows that the minimum frequency for emission of electrons is 0.25×10^{15} Hz. The corresponding *maximum* wavelength for which electrons are emitted is

$$\lambda = \frac{c}{f} = \frac{3.0 \times 10^8 \text{ m/s}}{0.25 \times 10^{15} \text{ s}^{-1}} = 1.2 \times 10^{-6} \text{ m} = 1200 \text{ nm}.$$

REFLECT This experimental value of *h* differs by about 3% from the correct value.

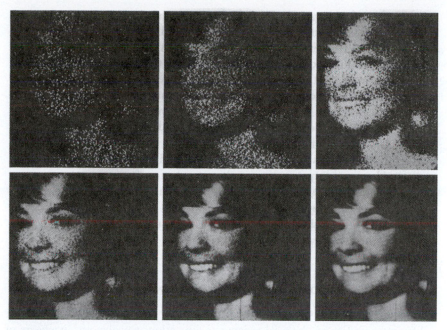

▲ **FIGURE 28.5** Comparison of photographs made with few and many photons.

Figure 28.5 shows a direct illustration of the particle aspects of light. When relatively few photons strike the photographic film, the impacts of individual photons are clearly visible. Only when the number of photons is much larger does a distinct picture emerge.

Photon frequency and wavelength

The concept of photons is applicable to *all* regions of the electromagnetic spectrum, including radio waves, x rays, and gamma rays. Photons always travel with the speed of light, c. A photon of any frequency f and wavelength $\lambda = c/f$ has energy E given by

$$E = hf = \frac{hc}{\lambda}. \qquad (28.5)$$

Furthermore, according to relativity theory, every particle that has energy must also have momentum, even if it has no rest mass. Photons have zero rest mass. According to Equation 27.25, a photon with energy E has momentum with magnitude p given by $E = pc$. Thus, the wavelength λ of a photon, its momentum p, and its frequency f are related simply by

$$p = \frac{E}{c} = \frac{hf}{c} = \frac{h}{\lambda}. \qquad (28.6)$$

EXAMPLE 28.4 Photons from a radio station

Radio station WQED in Pittsburgh broadcasts at 89.3 MHz with a radiated power of 43.0 kW. How many photons does it emit each second?

Continued

SOLUTION

SET UP First we need to find the energy of one photon. From that result, we can find the number of photons needed per second for a total power output of 43.0 kW. We also need the conversion factor $1 \text{ W} = 1 \text{ J/s}$. (Remember that the watt is a unit of power, not energy; power is energy per unit time. Remember also that $1 \text{ Hz} = 1 \text{ s}^{-1}$.)

SOLVE The station sends out 43.0×10^3 joules each second. The energy of each photon emitted is

$$E = hf = (6.626 \times 10^{-34} \text{ J} \cdot \text{s})(89.3 \times 10^6 \text{ Hz})$$
$$= 5.92 \times 10^{-26} \text{ J}.$$

The number of photons per second is therefore

$$\frac{43.0 \times 10^3 \text{ J/s}}{5.92 \times 10^{-26} \text{ J/photon}} = 7.26 \times 10^{29} \text{ photons/s}.$$

REFLECT With this huge number of photons leaving the station each second, the discreteness of the tiny individual bundles of energy isn't noticed; the radiated energy appears to be a continuous flow.

Practice Problem: During a diagnostic x ray, a total energy of 2.5×10^{-3} J is absorbed by about 5.0 kg of tissue. If the x-ray photons have an energy of 50 keV and the tissue has the same density as water, how many photons are absorbed by 1.0 cm³ of tissue? *Answer:* 6.2×10^7 photons.

28.2 Line Spectra and Energy Levels

Mastering PHYSICS

PhET: Neon Lights and Other Discharge Lamps
ActivPhysics 18.2: Spectroscopy

The existence of line spectra has been known for more than 200 years. A prism or a diffraction grating can be used to separate the various wavelengths in a beam of light into a *spectrum*. If the light source is a very hot solid or liquid (such as the filament in a lightbulb), the spectrum is *continuous;* light of all wavelengths is present (Figure 28.6a). But if the source is a gas carrying an electrical discharge (as in a neon sign), or if it is a volatile salt heated in a flame (as when table salt is thrown into a campfire), only a few colors appear, in the form of isolated sharp parallel lines (Figure 28.6b). (Each "line" is an image of the spectrograph slit, deviated through an angle that depends on the wavelength of the light forming the image. A spectrum of this sort is called a *line spectrum;* each line corresponds to a definite wavelength and frequency.

Early in the 19th century, it was discovered that each chemical element has a definite, unchanging set of wavelengths in its line spectrum, and the identification of elements by their spectra became a useful analytical technique. The characteristic spectrum for each element was assumed to be related to the internal structure

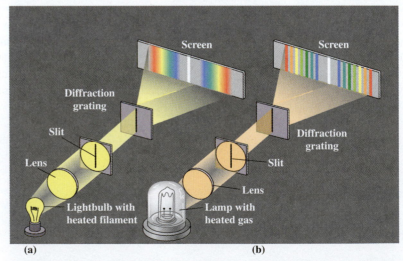

▲ **FIGURE 28.6** (a) Continuous spectrum from a very hot light source (the lightbulb filament). (b) Line spectrum emitted by a lamp containing a heated gas.

of its atoms, but attempts to understand this relation on the basis of classical mechanics and electrodynamics were not successful, even as recently as 1900. Two key pieces of the puzzle were missing: the photon concept and a new and revolutionary picture of the structure of the atom.

The key idea was finally found in 1913 by Niels Bohr, with an insight that, from a historical vantage point more than 90 years later, seems almost obvious; yet in its time, it represented a bold and brilliant stroke. Here's Bohr's reasoning: Every atom has some internal structure and internal motion and therefore some internal energy. But each atom has a set of possible energy levels. An atom can have an amount of internal energy corresponding to any one of these levels, but it cannot have an energy intermediate between two levels. If the atoms of a particular element can emit and absorb photons with only certain particular energies (corresponding to the line spectrum of that element), then *the atoms themselves must be able to possess only certain particular quantities of energy*. Bohr's hypothesis is shown schematically in Figure 28.7. Bohr assumed that while an atom is in one of these "permitted" energy states, it does not radiate. However, an atom can make a transition from one energy level to a lower level by emitting a photon with energy equal to the energy difference between the initial and final levels.

For the simplest atom, hydrogen, Bohr pictured these levels in terms of the electron revolving in various circular orbits around the proton. Only certain orbit radii were permitted. We'll return to this picture in the next section.

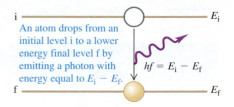

▲ **FIGURE 28.7** Energy levels and photon emission.

Bohr's hypothesis

If E_i is the initial energy of an atom before a transition from one energy level to another, E_f is the atom's (smaller) final energy after the transition, and the energy of the emitted photon is hf, then

$$hf = E_i - E_f. \qquad (28.7)$$

For example, a photon of orange light with wavelength $\lambda = 600$ nm has a frequency f given by

$$f = \frac{c}{\lambda} = \frac{3.00 \times 10^8 \text{ m/s}}{600 \times 10^{-9} \text{ m}} = 5.00 \times 10^{14} \text{ s}^{-1}.$$

The corresponding photon energy is

$$E = hf = (6.63 \times 10^{-34} \text{ J} \cdot \text{s})(5.00 \times 10^{14} \text{ s}^{-1})$$
$$= 3.31 \times 10^{-19} \text{ J} = 2.07 \text{ eV}.$$

This photon must be emitted in a transition between two states of the atom (each with a definite energy level) that differ in energy by 2.07 eV.

The same principle applies when a photon is *absorbed* by an atom. In this case, the atom's final energy is greater than its initial energy, and instead of Equation 28.7, we have

$$hf = E_f - E_i.$$

Hydrogen Spectrum

Let's see how Bohr's hypothesis fits in with what was known about spectra in 1913. The spectrum of hydrogen, the least massive atom, had been studied intensively. Under proper conditions, atomic hydrogen emits a series of four lines in the visible region of the spectrum, shown in Figure 28.8. The line with longest wavelength, or lowest frequency, in the red, is called H_α; the next line, in the blue-green, is H_β, and so on.

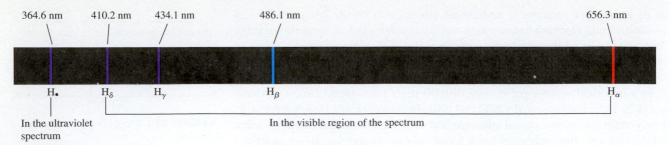

364.6 nm   410.2 nm   434.1 nm   486.1 nm   656.3 nm

$H_\bullet$ H_δ H_γ H_β H_α

In the ultraviolet spectrum

In the visible region of the spectrum

▲ **FIGURE 28.8** The Balmer series of spectral lines for atomic hydrogen.

In 1885, Johann Balmer (1825–1898) found (by trial and error) a formula that gives the wavelengths of these lines, which are now called the Balmer series.

Balmer's formula for the hydrogen spectrum
Balmer's formula is

$$\frac{1}{\lambda} = R\left(\frac{1}{2^2} - \frac{1}{n^2}\right), \tag{28.8}$$

where λ is the wavelength, R is a constant called the **Rydberg constant,** and n may have the integer values $3, 4, 5, \cdots$. If λ is in meters, the numerical value of R (determined from measurements of wavelengths) is

$$R = 1.097 \times 10^7 \text{ m}^{-1}.$$

Substituting $n = 3$ in Equation 28.8, we obtain the wavelength of the H_α line:

$$\frac{1}{\lambda} = \left(1.097 \times 10^7 \text{ m}^{-1}\right)\left(\frac{1}{2^2} - \frac{1}{3^2}\right), \quad \lambda = 656.3 \text{ nm}.$$

For $n = 4$, we obtain the wavelength of the H_β line, and so on.

Balmer's formula has a direct relation to Bohr's hypothesis about energy levels. Using the relations $f = c/\lambda$ and $E = hf$, we can find the photon energies E corresponding to the wavelengths of the Balmer series. Multiplying Equation 28.8 by hc, we find that

$$\frac{hc}{\lambda} = hf = E = hcR\left(\frac{1}{2^2} - \frac{1}{n^2}\right) = \frac{hcR}{2^2} - \frac{hcR}{n^2}. \tag{28.9}$$

Comparing Equations 28.7 and 28.9, we see that the two agree if we identify $-hcR/n^2$ as the initial energy E_i of the atom and $-hcR/2^2$ as its final energy E_f in a transition in which a photon with energy $hf = E_i - E_f$ is emitted.

The Balmer series (as well as others that we'll mention shortly) therefore suggests that the hydrogen atom has a series of energy levels, which we'll denote as E_n, given by

$$E_n = -\frac{hcR}{n^2}, \quad n = 2, 3, 4, \cdots. \tag{28.10}$$

(These energies are negative because we have arbitrarily chosen the potential energy to be zero at very large values of n. As we'll see, the state with $n = \infty$ corresponds to the state where the electron is completely separated from the nucleus—a single proton—and is at rest.) Each wavelength in the Balmer series corresponds to a transition from a state having n equal to 3 or greater to the state where $n = 2$.

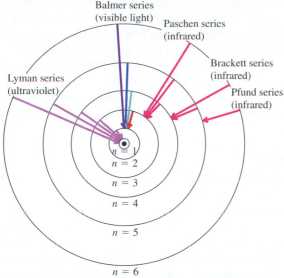

Balmer series
(visible light)

Paschen series
(infrared)

Brackett series
(infrared)

Pfund series
(infrared)

Lyman series
(ultraviolet)

$n = 1$
$n = 2$
$n = 3$
$n = 4$
$n = 5$
$n = 6$

"Permitted" orbits of an electron in the Bohr model of a hydrogen atom (not to scale). Arrows indicate the transitions responsible for some of the lines of various series.

(a)

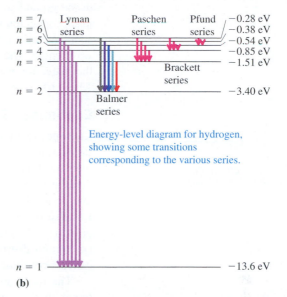

$n = 7$
$n = 6$
$n = 5$
$n = 4$
$n = 3$

Lyman series   Paschen series   Pfund series

-0.28 eV
-0.38 eV
-0.54 eV
-0.85 eV
-1.51 eV

$n = 2$

Brackett series

-3.40 eV

Balmer series

Energy-level diagram for hydrogen, showing some transitions corresponding to the various series.

$n = 1$ -13.6 eV

(b)

▲ **FIGURE 28.9** Bohr orbits and spectral series.

The numerical value of the product hcR is

$$hcR = (6.626 \times 10^{-34}\ \text{J} \cdot \text{s})(2.998 \times 10^8\ \text{m/s})(1.097 \times 10^7\ \text{m}^{-1})$$
$$= 2.179 \times 10^{-18}\ \text{J} = 13.6\ \text{eV}.$$

Thus, the magnitudes of the energy levels given by Equation 28.10 are approximately -13.6 eV, -3.40 eV, -1.51 eV,$\cdots$.

Other series of spectrum lines for hydrogen have since been discovered experimentally. They are named for their discoverers; their wavelengths are represented by formulas similar to the Balmer formula. One series is in the ultraviolet region, the others in the infrared. *All* the spectral series of hydrogen can be understood on the basis of transitions from one energy level (and corresponding electron orbit) to another, with the energy levels given by Equation 28.10. For the Lyman series, the final state is always $n = 1$; for the Paschen series, it is $n = 3$; and so on. Taken together, these spectral series give strong support to Bohr's picture of energy levels in atoms. The relation of the various spectral series to the energy levels and electron orbits is shown in Figure 28.9.

Energy Levels

Only atoms or ions with a single electron can be represented by a simple formula of the Balmer type. But it is *always* possible to analyze the more complicated spectra of other elements in terms of transitions among various energy levels and to deduce the numerical values of these levels from the measured spectrum wavelengths. Every atom has a lowest energy level (or energy state), representing the minimum energy the atom can have. This is called the **ground state,** and all states with energy greater than that of the ground state are called **excited states.** A photon corresponding to a particular spectrum line is emitted when an atom makes a transition from an excited state to a lower excited state or to the ground state.

In some cases, an atom can have two or more states with different electron configurations, but with the same energy. Thus, we'll sometimes need to distinguish between energy *states* and energy *levels;* one level can sometimes correspond to several states. We'll return to this distinction later; for now, we'll use the terms *state* and *level* interchangeably.

▲ **Application** **The physics of fireworks.** Fireworks were invented by the ancient Chinese, long before we understood the physics of how the various colors are produced by atomic emission. As you have learned, when atoms absorb energy, their electrons can be temporarily boosted to higher energy levels within the atom. When these electrons drop back to lower energy levels, they emit photons of characteristic wavelengths that correspond to energy differences between pairs of energy levels. When these wavelengths are in the visible region of the electromagnetic spectrum, we see different colors depending on which chemical element or compound has been excited. In practice, fireworks makers often use compounds of sodium to produce bright yellow, calcium for orange, strontium for red, barium for green, and copper for blue.

The sodium atom has two closely spaced levels called *resonance levels* at about 2.1 eV above the ground state. The characteristic yellow light with wavelengths 589.0 and 589.6 nm is emitted during transitions from one of these levels to the ground state. Conversely, a sodium atom initially in the ground state can *absorb* a photon with wavelength 589.0 or 589.6 nm. After a short time, the atom spontaneously returns to the ground state by emitting a photon with the same wavelength. The average time spent in the excited state is called the **lifetime** of the state; for the resonance levels of the sodium atom, the lifetime is about 1.6×10^{-8} s = 16 ns.

Continuous Spectra

Line spectra are produced by matter in the gaseous state, in which the atoms are far apart and interactions between them are negligible. If the atoms are identical, their spectra are also identical. But in condensed states of matter (liquid or solid), there are strong interactions between atoms. These interactions cause shifts in energy levels, and levels are shifted by different amounts for different atoms. Because of the very large numbers of atoms, practically any photon energy is possible. Therefore, hot condensed matter always emits a spectrum with a *continuous* distribution of wavelengths, not a line spectrum. Such a spectrum is called a **continuous emission spectrum.**

The total rate of radiation of energy from a hot liquid or solid is proportional to the fourth power of the absolute temperature T, as we learned in Section 14.7. The radiation consists of a continuous distribution of wavelengths; it is called **blackbody** radiation because of its relation to the radiation from a material that is an ideal absorber of radiation. Blackbody radiation from a hot material is most intense in the vicinity of a certain wavelength that is *inversely* proportional to the absolute temperature. As the material's temperature increases, the intensity peak shifts to shorter wavelengths, and the total intensity increases. When a body that is glowing dull red is heated further, it gets brighter and more orange or yellow. Historically, Planck's analysis of blackbody radiation initially led him to the photon concept five years before Einstein's analysis of the photoelectric effect.

PhET: Blackbody Spectrum
PhET: The Greenhouse Effect

PROBLEM-SOLVING STRATEGY 28.1 **Photons and energy levels**

SET UP

1. Remember that with photons, as with any other periodic wave, the wavelength λ and frequency f are related by $f = c/\lambda$. The energy E of a photon can be expressed as hf or hc/λ, whichever is more convenient for the problem at hand. Be careful with units: If E is in joules, h must be in joule-seconds, λ in meters, and f in s^{-1}, or hertz. The magnitudes are in such unfamiliar ranges that common sense may not help if your calculation is wrong by a factor of 10^{10}, so be careful when you add and subtract exponents with powers of 10.

SOLVE

2. It's often convenient to measure energy in electronvolts. The conversion 1 eV = 1.602×10^{-19} J is often useful. When energies are in eV, you may want to express h in electronvolt-seconds; in those units, $h = 4.136 \times 10^{-15}$ eV · s. We invite you to verify this value.

3. Keep in mind that an electron moving through a potential difference of 1 V gains or loses an amount of energy equal to 1 eV. You will use the electronvolt a lot in this chapter and the next two, so it's important that you get familiar with it now.

EXAMPLE 28.5 **A mythical atom**

A hypothetical atom has three energy levels: the ground-state level and levels 1.00 eV and 3.00 eV above the ground state. **(a)** Find the frequencies and wavelengths of the spectrum lines for this atom. **(b)** What wavelengths can be *absorbed* by the atom if it is initially in the ground state?

SOLUTION

SET UP First we draw an energy-level diagram (Figure 28.10a). **(a)** Draw arrows to show the possible transitions from higher to lower energy levels, and label the energy for each transition leading from a higher to a lower energy level. These are the photon energies; from these, we can find the frequencies and wavelengths. **(b)** Draw arrows on the diagram showing the possible transitions from lower to higher energy levels, starting from the lowest energy state. These are the energies of photons that can be absorbed, and again, we can find the associated frequencies and wavelengths.

SOLVE Part (a): The possible photon energies E, corresponding to the transitions shown, are 1.00 eV, 2.00 eV, and 3.00 eV. Each photon frequency f is given by $f = E/h$. For 1.00 eV, we get

$$f = \frac{E}{h} = \frac{1.00 \text{ eV}}{4.136 \times 10^{-15} \text{ eV} \cdot \text{s}} = 2.42 \times 10^{14} \text{ Hz}.$$

For 2.00 eV and 3.00 eV, $f = 4.84 \times 10^{14}$ Hz and 7.25×10^{14} Hz, respectively. We find the wavelengths from $\lambda = c/f$. For 1.00 eV,

$$\lambda = \frac{c}{f} = \frac{3.00 \times 10^8 \text{ m/s}}{2.42 \times 10^{14} \text{ Hz}} = 1.24 \times 10^{-6} \text{ m} = 1240 \text{ nm}.$$

For 2.00 eV and 3.00 eV, the wavelengths are 620 nm and 414 nm, respectively.

Part (b): For an atom that is initially in the ground state, only a 1.00 eV or 3.00 eV photon can be absorbed; a 2.00 eV photon cannot, because there is no energy level 2.00 eV above the

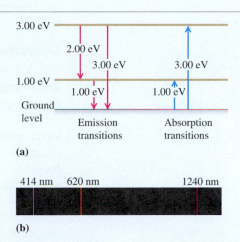

(a)

(b)

▲ FIGURE 28.10

ground state. From the preceding calculations, the corresponding wavelengths are 1240 nm and 414 nm, respectively; these two lines appear in the absorption spectrum for this atom.

REFLECT The lines in the emitted spectrum are 1240 nm (in the near infrared), 620 nm (red), and 414 nm (violet). Those in the absorption spectrum are 1240 nm and 414 nm. If the atom is initially in the state that is 1.00 eV above the ground state, then it can absorb a photon with an energy of 2.00 eV.

Practice Problem: How would your answers differ from those given above if the middle level were 2.00 eV above the ground state instead of 1.00 eV? *Answer:* (a) Same as before; (b) 620 nm, 1240 nm.

The Bohr hypothesis was successful in relating the wavelengths in line spectra to energy levels of atoms, but it provided no basis for predicting what the energy levels should be for any particular kind of atom. Bohr did provide a partial solution for this problem for the simplest atom, hydrogen; we'll discuss this solution in the next section. Then we'll introduce some general principles of quantum mechanics that are needed for a more general understanding of the structure and energy levels of atoms.

28.3 The Nuclear Atom and the Bohr Model

What does the inside of an atom look like? In one sense, this is a silly question: We know that atoms are much smaller than wavelengths of visible light, so there is no hope of actually *seeing* an atom. But we can still describe the interior of an atom by describing how the mass and electric charge are distributed through its volume.

Here's where things stood in 1910: J. J. Thomson had discovered the electron and measured its charge-to-mass ratio (e/m) in 1897; by 1910, Millikan had completed his first measurements of the electron charge e. These and other experiments showed that most of the mass of an atom had to be associated with the

MasteringPHYSICS

ActivPhysics 18.1: The Bohr Model
ActivPhysics 19.1: Particle Scattering

positive charge, not with the electrons. It was also known that the overall size of atoms is on the order of 10^{-10} m and that all atoms except hydrogen contain more than one electron. What was *not* known was how the mass and charge were distributed within the atom. Thomson had proposed a model of the atom that included a sphere of positive charge on the order of 10^{-10} m in diameter, with the electrons embedded in it like chocolate chips in a more or less spherical cookie.

Rutherford Scattering

The first experiments designed to probe the inner structure of the atom were the **Rutherford scattering** experiments, carried out in 1910–1911 by Sir Ernest Rutherford and two of his students, Hans Geiger and Ernest Marsden, at Cambridge University in England. Rutherford's experiment consisted of projecting alpha particles (emitted with speeds on the order of 10^7 m/s from naturally radioactive elements) at the atoms under study. The deflections of these particles provided information about the internal structure and charge distribution of the target atoms. (Alpha particles are now known to be identical to the nuclei of helium atoms: two protons and two neutrons bound together.)

Rutherford's experimental setup is shown schematically in Figure 28.11. A radioactive material at the left emits alpha particles. Thick lead screens stop all particles except those in a narrow beam defined by small holes. The beam then passes through a target consisting of a thin gold foil. (Gold was used because it can be beaten into extremely thin sheets.) After passing through the foil, the beam strikes a screen coated with zinc sulfide, similar in principle to the screen of an older model TV picture tube. A momentary flash or **scintillation** can be seen on the screen whenever it is struck by an alpha particle (just as spots on the TV screen glow when the electron beam strikes them). Rutherford's students were assigned the tedious task of counting the numbers of particles deflected through various angles. This unpleasant duty led Geiger to the invention of what is now called the Geiger counter, a commonly used radiation detector.

The fact that an alpha particle can pass through a thin metal foil shows that alpha particles must be able actually to penetrate into the interiors of atoms. The mass of an alpha particle is about 7300 times that of an electron. Momentum considerations show that the alpha particle can be scattered only a small amount by its interaction with the much lighter electrons. It's like driving a car through a hailstorm; the hailstones don't deflect the car much. Only interactions with the positive charge, which is tied to most of the mass of the atom, can deflect the alpha particle appreciably, perhaps comparable to colliding with an oncoming truck.

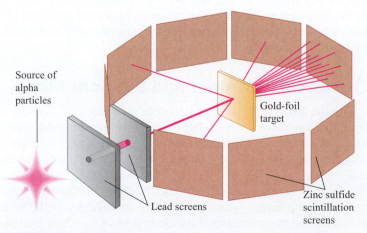

Source of alpha particles

Gold-foil target

Lead screens

Zinc sulfide scintillation screens

▲ **FIGURE 28.11** The scattering of alpha particles by a thin metal foil.

In the Thomson model, the positive charge is distributed through the whole atom. Rutherford calculated that the maximum deflection angle an alpha particle can have in this situation is only a few degrees, as Figure 28.12a suggests. The experimental results, however, were very different from this and were totally unexpected. Some alpha particles were scattered by nearly 180°—that is, almost straight backward! Rutherford wrote later:

> It was quite the most incredible event that ever happened to me in my life. It was almost as incredible as if you had fired a 15-inch shell at a piece of tissue paper and it came back and hit you.

Back to the drawing board! Suppose the positive charge, instead of being distributed through a sphere with atomic dimensions (on the order of 10^{-10} m), is all concentrated in a much *smaller* space. Rutherford called this concentration of positive charge the **nucleus** of the atom. Then the atom would act like a point charge down to much smaller distances. The maximum repulsive force on the alpha particle would be much larger, and large-angle scattering would be possible, as in Figure 28.12b. Rutherford again computed the numbers of particles expected to be scattered through various angles. Within the precision of his experiments, the computed and measured results agreed, down to distances on the order of 10^{-14} m.

Rutherford's experiments therefore established that the atom *does* have a nucleus—a very small, very dense structure no larger than 10^{-14} m in diameter. The nucleus occupies only about 10^{-12} of the total volume of the atom, but it contains all the positive charge and at least 99.95% of the total mass of the atom!

Figure 28.13 shows a computer simulation of the paths of 5.0 MeV alpha particles scattered by gold nuclei with a radius of 7.0×10^{-15} m (the actual value) and by nuclei with a hypothetical radius ten times this great. In the second case, there is no large-angle scattering.

The Bohr Model

At the same time (1913) that Bohr established the relationship between spectral wavelengths and energy levels, he also proposed a mechanical model of the hydrogen atom. Using this model, now known as the **Bohr model,** he was able to *calculate* the energy levels and obtain agreement with values determined from spectra.

In the light of Rutherford's discovery of the nucleus, the question arose as to what kept the negatively charged electrons at relatively large distances $(\sim 10^{-10}\,\text{m})$ from the extremely small $(\sim 10^{-15}\,\text{m})$, positively charged nucleus despite the electrostatic attraction that tended to pull the electrons into the

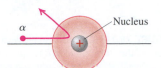

Thomson's model of the atom: An alpha particle is scattered through only a small angle.

(a)

Nucleus

Rutherford's model of the atom: An alpha particle can be scattered through a large angle by the compact, positively charged nucleus.

(b)

▲ **FIGURE 28.12** Rutherford scattering for the Thomson version of the atom vs. the nuclear atom. (Not drawn to scale.)

MasteringPHYSICS

PhET: Models of the Hydrogen Atom

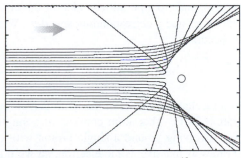
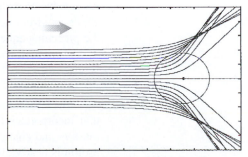

(a) A gold nucleus with radius 7.0×10^{-15} m gives large-angle scattering.

(b) A nucleus with 10 times the radius of the nucleus in (a) shows no large-scale scattering.

▲ **FIGURE 28.13** Computer simulation of scattering of 5.0-MeV alpha particles from a gold nucleus. Each curve shows a possible alpha-particle trajectory.

▲ FIGURE 28.14 A paradox in classical models of the atom: An electron, spiraling in toward the nucleus, emits a continuous spectrum of electromagnetic radiation.

nucleus. Rutherford suggested that perhaps the electrons *revolve* in orbits about the nucleus, just as the planets revolve around the sun, with the electrical attraction of the nucleus providing the necessary centripetal force.

But according to classical electromagnetic theory, any accelerating electric charge, either oscillating or revolving, radiates electromagnetic waves, just as a radio or TV transmitting antenna does. The total energy of a revolving electron should therefore decrease continuously, its orbit should become smaller and smaller, and it should rapidly spiral into the nucleus (Figure 28.14). Furthermore, according to classical theory, the *frequency* of the electromagnetic waves emitted should equal the frequency of revolution of the electrons in their orbits. As the electrons radiated energy, their angular velocities would change continuously, and they would emit a *continuous* spectrum (a mixture of all frequencies), not the *line* spectrum actually observed.

To solve this problem, Bohr made a revolutionary proposal. In effect, he postulated that an electron in an atom can revolve in certain **stable orbits,** with a definite energy associated with each orbit, without emitting radiation, contrary to the predictions of classical electromagnetic theory. According to Bohr, an atom emits or absorbs radiation only when it makes a transition from one of these stable orbits to another, at the same time emitting (or absorbing) a photon with appropriate energy and frequency given by Equation 28.7.

To determine the radii of the "permitted" orbits, Bohr introduced what we have to regard in hindsight as a brilliant intuitive guess. He postulated that the angular momentum L of the orbiting electron is **quantized;** that is, it can have only certain specific values. Specifically, Bohr proposed that the angular momentum must be an integer multiple of $h/2\pi$. (Note that the units of Planck's constant h, usually written as $J \cdot s$, are the same as the units of angular momentum, usually written as $kg \cdot m^2/s$.) Bohr postulated that in the hydrogen atom, only those orbits are permitted for which the angular momentum is an integral multiple of $h/2\pi$.

Recall (Equation 10.11) that the angular momentum L of a particle with mass m, moving with speed v in a circle of radius r, is $L = mvr$. So Bohr's assumption may be stated as follows:

Quantization of angular momentum in the Bohr model
In Bohr's hypothesis, the allowed values of angular momentum L for an electron in a circular orbit are given by

$$L = mvr = n\frac{h}{2\pi}, \qquad n = 1, 2, 3, \cdots.$$

Each value of n also corresponds to a permitted value of the orbit radius, which we denote from now on by r_n, and a corresponding speed v_n. With this notation, the preceding equation becomes

$$mv_n r_n = n\frac{h}{2\pi}. \qquad (28.11)$$

Now let's consider a mechanical model of the hydrogen atom (Figure 28.15) that incorporates Equation 28.11. This atom consists of a single electron with mass m and charge $-e$, revolving around a single proton with charge $+e$. The proton is nearly 2000 times as massive as the electron, so we'll assume that the proton doesn't move. We learned in Chapter 6 that when a particle with mass m moves with speed v_n in a circular orbit with radius r_n, its acceleration has magnitude v_n^2/r_n. According to Newton's second law, a force with magnitude

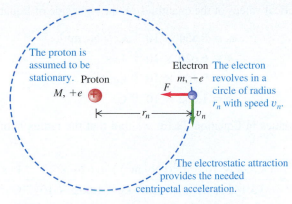

The proton is assumed to be stationary. Proton $M, +e$

Electron $m, -e$

The electron revolves in a circle of radius r_n with speed v_n.

F

r_n

v_n

The electrostatic attraction provides the needed centripetal acceleration.

▲ **FIGURE 28.15** Bohr model of the hydrogen atom.

$F = mv_n^2/r_n$ is needed to cause this acceleration. The force F is provided by the electrical attraction between the two charges, each with magnitude e, according to Coulomb's law, so

$$F = k\frac{e^2}{r_n^2} = \frac{1}{4\pi\epsilon_0}\frac{e^2}{r_n^2},$$

where $k = 1/4\pi\epsilon_0$ is the constant in Coulomb's law (Section 17.4).

The $F = ma$ equation is

$$\frac{1}{4\pi\epsilon_0}\frac{e^2}{r_n^2} = m\frac{v_n^2}{r_n}. \tag{28.12}$$

We can combine Equations 28.11 and 28.12 and rearrange terms to obtain separate expressions for r_n and v_n. When we do this, we find the following relationships:

Radius and speed of Bohr orbits

In the Bohr model, an electron in a state with quantum number n has orbit radius r_n and speed v_n given respectively by

$$r_n = \epsilon_0\frac{n^2h^2}{\pi me^2}, \tag{28.13}$$

$$v_n = \frac{1}{\epsilon_0}\frac{e^2}{2nh}. \tag{28.14}$$

Equation 28.13 shows that the orbit radius r_n is proportional to n^2; the smallest orbit r_1 corresponds to $n = 1$. This minimum radius r_1 is called the **Bohr radius**:

$$r_1 = \epsilon_0\frac{h^2}{\pi me^2}. \tag{28.15}$$

With this notation, we can rewrite Equation 28.13 as

$$r_n = n^2r_1. \tag{28.16}$$

The permitted, nonradiating orbits have radii $r_1, 4r_1, 9r_1$, and so on. The value of n for each orbit is called the **principal quantum number** for that orbit.

The numerical values of the quantities on the right side of Equation 28.15 are

$$\epsilon_0 = 8.854 \times 10^{-12} \, \text{C}^2/(\text{N} \cdot \text{m}^2),$$
$$h = 6.626 \times 10^{-34} \, \text{J} \cdot \text{s},$$
$$m = 9.109 \times 10^{-31} \, \text{kg},$$
$$e = 1.602 \times 10^{-19} \, \text{C}.$$

Using these values in Equation 28.15, we find that the radius r_1 of the smallest Bohr orbit is

$$r_1 = \frac{(8.854 \times 10^{-12} \, \text{C}^2/(\text{N} \cdot \text{m}^2))(6.626 \times 10^{-34} \, \text{J} \cdot \text{s})^2}{(3.142)(9.109 \times 10^{-31} \, \text{kg})(1.602 \times 10^{-19} \, \text{C})^2}$$
$$= 0.5293 \times 10^{-10} \, \text{m}.$$

This result gives us an estimate of the size of the hydrogen atom that is consistent with atomic dimensions estimated by various other methods.

We can also use the preceding result with Equation 28.14 to find the orbital speed of the electron for any value of n. We leave this calculation as an exercise. The result is that, for the $n = 1$ state, $v_1 = 2.19 \times 10^6$ m/s. This speed is less than 1% of the speed of light, so we don't need to be concerned about relativistic effects.

Energy Levels

We can use Equations 28.13 and 28.14 to find the kinetic and potential energies, K_n and U_n, for an electron in the orbit with quantum number n. For K_n, we use Equation 28.14:

$$K_n = \frac{1}{2}mv_n^2 = \frac{1}{\epsilon_0^2}\frac{me^4}{8n^2h^2}.$$

We use Equation 28.13 to express the potential energy U_n in terms of n:

$$U_n = -\frac{1}{4\pi\epsilon_0}\frac{e^2}{r_n} = -\frac{1}{\epsilon_0^2}\frac{me^4}{4n^2h^2}.$$

The total energy E_n for any n is the sum of the kinetic and potential energies:

$$E_n = K_n + U_n = -\frac{1}{\epsilon_0^2}\frac{me^4}{8n^2h^2}. \tag{28.17}$$

This expression has a negative sign because we have taken the reference level of potential energy to be zero when the electron is at rest at an infinite distance from the nucleus. We are interested only in energy *differences* between levels.

The preceding discussion shows that the possible energy levels and electron orbits for the atom are labeled by values of the quantum number n. For each value of n, there are corresponding values of the orbit radius r_n, speed v_n, angular momentum $L_n = nh/2\pi$, and total energy E_n. The energy of the atom is least when $n = 1$ and E_n has its largest negative value. This is the *ground state* of the atom—the state with the smallest orbit, with radius r_1. For $n = 2, 3, \cdots$, the absolute value of E_n is smaller and the energy is progressively larger (less negative). The orbit radius increases as n^2, as shown by Equation 28.13.

Comparing the expression for E_n in Equation 28.17 with Equation 28.10 ($E_n = -hcR/n^2$, the energy-level equation deduced from spectrum analysis), we see that they agree only if the coefficients are equal:

Relation of the Rydberg constant to fundamental constants:

$$hcR = \frac{1}{\epsilon_0^2}\frac{me^4}{8h^2}, \quad \text{or} \quad R = \frac{me^4}{8\epsilon_0^2 h^3 c}. \tag{28.18}$$

This equation shows us how to *calculate* the value of the Rydberg constant R from the fundamental physical constants m, c, e, h, and ϵ_0, all of which can be determined quite independently of the Bohr theory.

When we substitute the numerical values of these quantities into Equation 28.18, we obtain the value $R = 1.097 \times 10^7\ \text{m}^{-1}$. This value is within 0.01% of the value determined from wavelength measurements, a major triumph for Bohr's theory! We invite you to substitute the numerical values into Equation 28.18 and compute the value of R to confirm these statements.

We can also calculate the energy required to remove the electron completely from a hydrogen atom. This energy is called the **ionization energy.** Ionization corresponds to a transition from the ground state $(n = 1)$ to an infinitely large orbit radius $(n = \infty)$. The predicted energy is 13.6 eV. The ionization energy can also be measured directly, and the two values agree within 0.1%.

Quantitative Analysis 28.2

Transitions and frequency

According to the Bohr model of the hydrogen atom, as we look at higher and higher values of n, the wavelength of the photon emitted due to transitions between adjacent electron orbits

A. gets progressively smaller.
B. gets progressively larger.
C. remains constant.

SOLUTION In the Bohr model, the magnitude of the energy of a hydrogen atom in state n is proportional to $1/n^2$ (Equation 28.17). The energy of the photon emitted during a transition between two adjacent states is simply the difference of the energies $hf = E_i - E_f$ of the two states. As n gets larger, the magnitude of E_n and the difference $(E_i - E_f)$ get smaller. Therefore, the energy and frequency of the emitted photon are smaller for larger values of n, and, from $\lambda = c/f$, the wavelength increases. The correct answer is B.

EXAMPLE 28.6   The energies depend on the quantum number

Find the kinetic, potential, and total energies of the hydrogen atom in the $n = 2$ state, and find the wavelength of the photon emitted in the transition $n = 2 \rightarrow n = 1$.

SOLUTION

SET UP We'll need to use the expressions for kinetic and potential energy that we derived earlier, leading to Equation 28.17. We can simplify the calculations a lot by noting that the product of constants that appears in this equation is equal to hcR, where R is the Rydberg constant we encountered in Equation 28.18. The numerical value is 13.6 eV:

$$\frac{me^4}{8\epsilon_0^2 h^2} = hcR = 13.6\ \text{eV}.$$

SOLVE Using this equation, we can express the kinetic, potential, and total energies for the state with quantum number n respectively as

$$K_n = \frac{13.6\ \text{eV}}{n^2}, \quad U_n = \frac{-27.2\ \text{eV}}{n^2}, \quad E_n = \frac{-13.6\ \text{eV}}{n^2}.$$

For the $n = 2$ state, we find that $K_2 = 3.40\ \text{eV}$, $U_2 = -6.80\ \text{eV}$, and $E_2 = -3.40\ \text{eV}$.

The energy of the emitted photon is

$$E_i - E_f = -3.40\ \text{eV} - (-13.6\ \text{eV}) = 10.2\ \text{eV}$$
$$= 1.63 \times 10^{-18}\ \text{J}.$$

This is equal to hc/λ, so we obtain

$$\lambda = \frac{hc}{E_i - E_f} = \frac{(6.626 \times 10^{-34}\ \text{J}\cdot\text{s})(3.00 \times 10^8\ \text{m/s})}{1.63 \times 10^{-18}\ \text{J}}$$
$$= 122 \times 10^{-9}\ \text{m} = 122\ \text{nm}.$$

REFLECT The wavelength we just found is that of the Lyman alpha line, the longest-wavelength line in the Lyman series of ultraviolet lines in the hydrogen spectrum.

Practice Problem: Find the wavelength for the transition $n = 3 \rightarrow n = 1$ for singly ionized helium, which has one electron and a nuclear charge of $2e$. (Note that the value of the Rydberg constant is four times as great as for hydrogen because it is proportional to the square of the product of the nuclear charge and the electron charge. (See the next paragraph.) *Answer:* 25.6 nm.

The Bohr model can be extended to other one-electron atoms, such as the singly ionized helium atom, the doubly ionized lithium atom, and so on. If the nuclear charge is Ze (where Z is the atomic number) instead of just e, the effect in the preceding analysis is to replace e^2 everywhere by Ze^2. In particular, the orbit radii r_n given by Equation 28.13 become smaller by a factor of Z, and the energy levels E_n given by Equation 28.17 are multiplied by Z^2. We invite you to verify these statements.

Limitations of the Bohr Model

Although the Bohr model predicted the energy levels of the hydrogen atom correctly, it raised as many questions as it answered. It combined elements of classical physics with new postulates that were inconsistent with classical ideas. There was no clear justification for restricting the angular momentum to multiples of $h/2\pi$, except that it led to the right answer. The model provided no insight into what happens *during* a transition from one orbit to another. The stability of certain orbits was achieved at the expense of discarding the only picture available in Bohr's time of the electromagnetic mechanism by which the atom radiated energy. The angular velocities of the electron motion were not the same as the angular frequencies of the emitted radiation, a result that is contrary to classical electrodynamics. Attempts to extend the model to atoms with two or more electrons were not successful. In Chapter 29, we'll find that even more radical departures from classical concepts were needed before the understanding of atomic structure could progress further.

28.4 The Laser

A **laser** is a light source that produces a beam of highly coherent and very nearly monochromatic light as a result of "cooperative" emission from many atoms. The name *laser* is an acronym for "light amplification by stimulated emission of radiation." We can understand the principles of laser operation on the basis of photons and atomic energy levels.

If an atom has an excited state with energy level E above the ground state, then an atom in the ground state can absorb a photon whose frequency f is given by the Planck equation $E = hf$. This process is shown schematically in Figure 28.16a, which shows a gas in a transparent container. An atom A absorbs a photon, reaching an excited state, labeled A^* in the figure. Some time later, the excited atom returns to the ground state by emitting a photon with the same frequency and energy as the one originally absorbed. This process is called **spontaneous emission;** the direction of the emitted photon is random (Figure 28.16b), and its vibrations aren't synchronized with those of the incoming photon that started the process.

In **stimulated emission** (Figure 28.16c), an incident photon encounters an atom that's already in an excited state with the same energy above the ground state as the photon's energy. By a kind of resonance effect, the incoming photon causes the atom to emit another photon with the same frequency and direction as the incoming photon and with its vibrations synchronized with those of the incoming photon. One photon goes in, and two come out, with their vibrations synchronized—hence the term *light amplification.* Under proper conditions, the laser uses excitation and stimulated emission to produce a beam consisting of a large number of such synchronized photons. Such a beam of synchronized photons is said to be *coherent.*

Here's how one kind of laser works: Suppose we have a large number of identical atoms in a gas or vapor in a container with transparent walls, as in Figure 28.16a. At moderate temperatures, if no radiation is incident on the

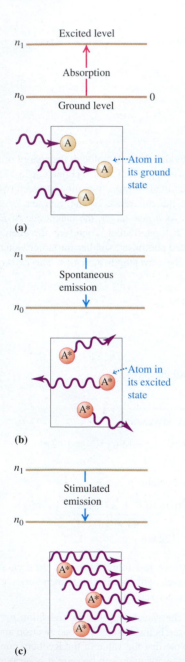

(a)

(b)

(c)

▲ **FIGURE 28.16** Three interaction processes for an atom and electromagnetic waves.

container, most of the atoms are in the ground state, and only a few are in excited states. If there is a state with energy level E above the ground state, the ratio of the number n_E of atoms in this state (the *population* of the state) to the number of atoms n_0 in the ground state is small.

Suppose we send through the container a beam of radiation with frequency f corresponding to the energy difference E. Some of the atoms absorb photons of energy E and are raised to the excited state, and the population ratio n_E/n_0 increases. Because n_0 is originally so much larger than n_E, an enormously intense beam of light would be required to increase n_E to a value comparable to n_0.

But now suppose that we can create a situation in which n_E is substantially increased compared with its normal equilibrium value; this condition is called a **population inversion.** If n_E can be increased enough, the rate of energy radiation by stimulated emission can actually exceed the rate of absorption. The system then acts as a source of radiation with photon energy E. Furthermore, because the photons are the result of stimulated emission, they all have the same frequency, phase, polarization, and direction, and they are all synchronized. The resulting emitted radiation is therefore much more *coherent* than light from ordinary sources, in which the emissions of individual atoms are *not* coordinated. This coherent emission is exactly what happens in a laser.

Lasers can be made to operate over a broad range of the electromagnetic spectrum, from microwaves to x rays. In recent years, lasers have found a wide variety of practical applications. The high intensity of a laser beam makes it a convenient drill. For example, a laser beam can drill a very small hole in a diamond for use as a die in making very small diameter wire. Because the photons in a laser beam are strongly correlated in their directions, the light output from a laser can be focused to a very narrow beam, and it can travel long distances without appreciable spreading. Surveyors often use lasers, especially in situations in which great precision is required, such as a long tunnel drilled from both ends.

Lasers are widely used in medical science. A laser can produce a narrow beam with extremely high intensity, high enough to vaporize anything in its path. This property is used in the treatment of a detached retina. A short burst of radiation damages a small area of the retina, and the resulting scar tissue "welds" the retina back to the choroid from which it has become detached. Laser beams have many surgical applications; blood vessels cut by the beam tend to seal themselves off, making it easier to control bleeding. Lasers are also used for selective destruction of tissue, as in the removal of tumors.

▲ **Application As red as rubies.** The beautiful red color of the ruby has been enjoyed for centuries, and in 1960 this gemstone was used in the first laser. A ruby laser uses a high-quality cylindrical ruby crystal surrounded by a high-intensity lamp that provides the energy to excite electrons of chromium atoms in the ruby to higher energy levels. As these excited electrons return to their ground state, they emit photons of red light at 694 nm, and photons from one atom stimulate emission of photons from other atoms, rapidly amplifying the light intensity. The result is a coherent beam of synchronized photons that can travel very long distances with little spreading, even to the moon and back. After Apollo astronauts placed mirrors on the moon in 1969, a ruby laser beam shot from Texas was used to accurately measure the earth-to-moon distance to within 10 cm, a precision of nine significant figures.

MasteringPHYSICS

PhET: Lasers
ActivPhysics 18.3: The Laser

28.5 X-Ray Production and Scattering

The production and scattering of **x rays** provides additional examples of the quantum view of electromagnetic radiation. **X rays** are produced when rapidly moving electrons that have been accelerated through a potential difference on the order of 10^3 to 10^6 V strike a metal target. X rays were first produced in 1895 by Wilhelm Röntgen (1845–1923), using an apparatus similar in principle to that shown in Figure 28.17. Electrons are "boiled off" from the heated cathode by thermionic emission and are accelerated toward the anode (the target) by a large potential difference V. The bulb is evacuated (to a residual gas pressure of 10^{-7} atm or less) so that the electrons can travel from cathode to anode without colliding with air molecules. When V is a few thousand volts or more, a penetrating radiation is emitted from the anode surface.

Electrons are emitted thermionically from the heated cathode and are accelerated toward the anode; when they strike it, x rays are produced.

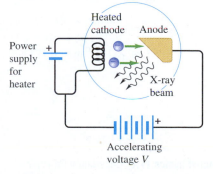

▲ **FIGURE 28.17** An apparatus used to produce x rays, similar to Röntgen's 1895 apparatus.

X-Ray Photons

Because of their electromagnetic origin, it is clear that x rays are electromagnetic waves. Like light, they are governed by quantum relations in their interaction with matter. Thus, we can talk about x-ray photons or quanta, and the energy of an x-ray photon is related to its frequency and wavelength in the same way as for photons of light: $E = hf = hc/\lambda$. Typical x-ray wavelengths are 0.001 to 1 nm $(10^{-12}$ to 10^{-9} m$)$. X-ray wavelengths can be measured quite precisely by crystal diffraction techniques, which we've discussed previously.

X-ray emission is the inverse of the photoelectric effect. In photoelectric emission, the energy of a photon is transformed into kinetic energy of an electron; in x-ray production, the kinetic energy of an electron is transformed into energy of a photon. The energy relation is exactly the same in both cases. In x-ray production, we can neglect the work function of the target and the initial kinetic energies of the electrons because they are ordinarily only a few electronvolts, very small in comparison to the energies of the accelerated electrons when they strike the target.

Each element has a set of atomic energy levels associated with x-ray photons and therefore also has a characteristic x-ray spectrum. These energy levels, called **x-ray energy levels,** are rather different in character from those associated with visible spectra. They are associated with vacancies in the inner electron configurations of complex atoms. The energy levels can be hundreds or thousands of electronvolts above the ground state, rather than a few electronvolts, as is typical with optical spectra.

EXAMPLE 28.7 Wavelength of an x-ray photon

Electrons are accelerated by a potential difference of 10.0 kV. If an electron produces a photon on impact with the target, what is the minimum wavelength of the resulting x rays?

SOLUTION

SET UP We simply note that the *maximum* photon energy $hf_{max} = hc/\lambda_{min}$ and *minimum* wavelength occur when *all* the initial kinetic energy eV of the electron is used to produce a single photon:

$$eV = hf = \frac{hc}{\lambda}.$$

All that's left is to solve for λ and substitute the numbers.

SOLVE Following the program just outlined, we get

$$\lambda = \frac{hc}{eV} = \frac{(6.626 \times 10^{-34}\,\text{J} \cdot \text{s})(3.00 \times 10^8\,\text{m/s})}{(1.602 \times 10^{-19}\,\text{C})(1.00 \times 10^4\,\text{V})}$$
$$= 1.24 \times 10^{-10}\,\text{m} = 0.124\,\text{nm}.$$

REFLECT We can measure the x-ray wavelength by crystal diffraction and confirm this prediction directly. The wavelength is *smaller* than that for typical visible-light photons by a factor on the order of 5000, and the photon energy is greater by the same factor.

Practice Problem: What accelerating voltage is needed to produce x rays with a wavelength of 0.050 nm? *Answer:* 24.8 kV.

ActivPhysics 17.6: Uncertainty Principle

Compton Scattering

A phenomenon called **Compton scattering,** first observed in 1924 by A. H. Compton, provides additional direct confirmation of the quantum nature of x rays. When x rays strike matter, some of the radiation is scattered, just as visible light falling on a rough surface undergoes diffuse reflection. Compton discovered that some of the scattered radiation has a smaller frequency and longer wavelength than the incident radiation and that the increase in wavelength, $\Delta\lambda$, depends on the angle through which the radiation is scattered.

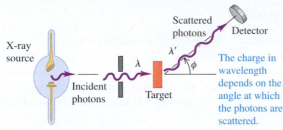

▲ FIGURE 28.18 A Compton-effect experiment.

If the scattered radiation emerges at an angle ϕ with respect to the incident direction, as shown in Figure 28.18, and if λ and λ' are the wavelengths of the incident and scattered radiation, respectively, then the wavelength increase $\Delta\lambda$ is given by a simple expression:

Wavelength increase in Compton scattering

$$\Delta\lambda = \lambda' - \lambda = \frac{h}{mc}(1 - \cos\phi), \qquad (28.19)$$

where m is the electron mass.

Compton scattering cannot be understood on the basis of classical electromagnetic theory, which predicts that the scattered wave has the *same* wavelength as the incident wave. In contrast, the quantum theory provides a beautifully clear explanation. We imagine the scattering process as a collision of two *particles:* the incident photon and an electron initially at rest, as in Figure 28.19a. The photon gives up some of its energy and momentum to the electron, which recoils as a result of the impact. The x-ray photon carries both energy and momentum; the *total* energy of the photon and the electron is conserved during the collision, and the same is true of the total momentum. The final scattered photon has less energy, smaller frequency, and longer wavelength than the initial one (Figure 28.19b).

Equation 28.19 can be derived from the principles of conservation of energy and momentum, treating the electron and the photons as particles, using the relativistic energy–momentum relations, and using the wavelength–momentum relation $p = h/\lambda$ for the incident and scattered photons. The quantity h/mc that appears in this equation has units of length, as it must for dimensional consistency. Its numerical value is

$$\frac{h}{mc} = \frac{6.626 \times 10^{-34} \text{ J} \cdot \text{s}}{(9.109 \times 10^{-31} \text{ kg})(2.998 \times 10^8 \text{ m/s})}$$
$$= 2.426 \times 10^{-12} \text{ m} = 2.426 \text{ pm}.$$

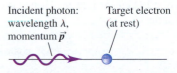

(a) Before collision: The target electron is at rest.

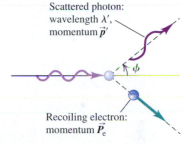

(b) After collision: The angle between the directions of the scattered photon and the incident photon is ϕ.

(c) Vector diagram showing the conservation of momentum in the collision: $\vec{p} = \vec{p}' + \vec{P}_\text{e}$.

▲ FIGURE 28.19 Schematic diagram of Compton scattering.

Conceptual Analysis 28.3

Photon strikes electron

A photon of light scatters off of an electron which is initially at rest. After the collision

A. The wavelength and momentum of the photon are less than before the collision.

B. The wavelength of the photon and the momentum of the electron are greater than before the collision.

C. The energy of the photon is less than before the collision, while its momentum is unchanged.

SOLUTION The scattering process can be thought of as a collision of two particles—the photon and the electron. Therefore, the energy and momentum lost by the photon are gained by the electron. When the photon loses energy, its wavelength increases, according to Equation 28.5 ($E = hc/\lambda$), and the electron gains momentum. The correct answer is B.

EXAMPLE 28.8 **Wavelength shift in Compton scattering**

For the x-ray photons in Example 28.7 ($\lambda = 0.124$ nm), at what angle do the Compton-scattered x rays have a wavelength that is 1.0% longer than that of the incident x rays?

SOLUTION

SET UP The scattered x-ray photons have less energy, lower frequency, and longer wavelength than the incident photons. Conservation of energy and momentum led us to the relation

$$\Delta\lambda = \lambda' - \lambda = \frac{h}{mc}(1 - \cos\phi)$$

for the wavelengths λ and λ' of the incident and scattered photons.

SOLVE In the preceding equation, we want $\Delta\lambda = \lambda' - \lambda$ to be 1.0% of 0.124 nm, or 1.24 pm. Using the value $h/mc = 2.426$ pm, we find that

$$\Delta\lambda = \frac{h}{mc}(1 - \cos\phi),$$
$$1.24 \text{ pm} = (2.426 \text{ pm})(1 - \cos\phi),$$
$$\phi = 60.7°.$$

REFLECT We note that if the scattering angle is less than this value, the interaction is weaker and the wavelength shift is also less.

Practice Problem: What is the scattering angle if the scattered wavelength is 0.050% longer than the incident wavelength? *Answer:* 13.0°.

Applications of X Rays

Because many materials that are opaque to ordinary light are transparent to x rays, this radiation can be used to visualize the interiors of materials such as broken bones or defects in structural steel. In the past 20 to 30 years, several highly sophisticated x-ray imaging techniques have been developed. One widely used system is *computerized axial tomography;* the corresponding instrument is called a CAT scanner. The x-ray source produces a thin, fan-shaped beam that is detected by an array of several hundred detectors in a line. The entire apparatus is rotated around the subject, and the changing photon-counting rates of the detectors are recorded digitally. A computer processes this information and reconstructs a picture of density over an entire cross section of the subject. Density differences as small as 1% can be detected with CAT scans, and tumors and other anomalies much too small to be seen with older x-ray techniques can be detected.

X rays cause damage to living tissues. As x-ray photons are absorbed in tissues, they break molecular bonds and create highly reactive free radicals (such as neutral H and OH), which in turn can disturb the molecular structure of proteins and especially genetic material. Young and rapidly growing cells are particularly susceptible; thus, x rays are useful for the selective destruction of cancer cells. However, a cell may be damaged by the radiation, but survive, continue dividing, and produce generations of defective cells; hence, paradoxically, x rays can *cause* cancer as well as treat it. The medical use of x rays requires careful assessment of the balance between risks and benefits of radiation exposure in each individual case.

28.6 The Wave Nature of Particles

Some aspects of emission and absorption of light can be understood on the basis of photons and atomic energy levels. But a complete theory should also offer some means of *predicting* the values of these energy levels for any particular atom. The Bohr model of the hydrogen atom was a start, but it could not be generalized to many-electron atoms. For more general problems, even more sweeping revisions of 19th-century ideas were needed. The wave–particle duality of electromagnetic radiation had to be extended to include *particles* as well as radiation. As a consequence, a particle can no longer be described as a single point moving in space. Instead, it is an inherently spread-out entity. As the particle moves, the spread-out character has some of the properties of a *wave*. The new theory, called *quantum mechanics,* is the key to understanding the structure of atoms and molecules, including their spectra, chemical behavior, and many other properties. It has the happy effect of restoring unity to our description of both particles and radiation, and wave concepts are a central theme.

De Broglie Waves

A major advance in the understanding of atomic structure began in 1924 (about 10 years after Bohr published his analysis of the hydrogen atom) with a proposition made by a young French physicist, Louis de Broglie. His reasoning, freely paraphrased, went like this:

Nature loves symmetry. Light is dualistic in nature; in some situations, it behaves like a wave, in others like a particle. If nature is symmetric, this duality should also hold for matter. Electrons and protons, which we usually think of as *particles,* may in some situations behave like *waves.*

If an electron acts like a wave, it should have a wavelength. Using the nonrelativistic definitions of momentum and energy, de Broglie postulated that a free electron with mass m, moving with speed v should have a wavelength λ related to the magnitude of its momentum, $p = mv$, in exactly the same way as for a photon, as expressed by Equation 28.6, $\lambda = h/p$.

De Broglie wavelength of an electron

$$\lambda = \frac{h}{p} = \frac{h}{mv}, \qquad (28.20)$$

where h is Planck's constant.

De Broglie asserted that the relation $\lambda = h/p$ should hold for *both* electrons and photons, and, indeed, for all particles. (But we shouldn't expect to see the wavelengths of *macroscopic* objects: A pitched baseball might have a momentum with magnitude $6\ kg \cdot m/s$; the corresponding wavelength would be about 10^{-34} m, much too small to be seen with even the most powerful microscope.)

De Broglie's proposal was radical, but it was clear that a radical idea was needed. The dual nature of electromagnetic radiation had led to adoption of the photon concept, also a radical idea. A similar revolution was needed in the mechanics of particles. An essential concept of quantum mechanics is that a particle is no longer described as located at a single point, but instead is an inherently spread-out entity described in terms of a function that has various values at various points in space. The spatial distribution describing a *free* electron has a recurring pattern characteristic of a *wave* that propagates through space. Electrons within atoms can be visualized as diffuse clouds surrounding the nucleus.

Conceptual Analysis 28.4

de Broglie wavelength

If the de Broglie wavelength of a particle is too small, its wavelike properties are difficult to observe. For which particle would the wavelike properties be easier to observe?

A. A low-mass, slow particle
B. A high-mass particle moving at nearly the speed of light.
C. A low-mass particle moving at nearly the speed of light.

SOLUTION The larger a particle's de Broglie wavelength, the easier it is to observe its wavelike properties. Since $\lambda = h/mv$ (Equation 28.20), slow, low-mass particles have the largest wavelengths and thus have the most easily observed wavelike properties. The correct answer is A.

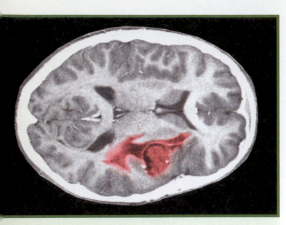

▲ BIO Application Boron neutron capture therapy. We have learned that particles exhibit wave-like behavior and that the wavelength and energy of a particle depend upon its velocity and mass. An experimental form of brain cancer therapy uses beams of slow-moving neutrons called thermal neutrons that are the proper wavelength to have just the right amount of energy to be absorbed by boron atoms, which are then converted into an unstable state. These unstable boron atoms then split into high-energy lithium nuclei and alpha particles that are deadly to nearby tissue. But because they can only travel a few micrometers, less than the width of a human cell, they will kill only the cell that had contained the boron atom. Medical researchers have developed boron-containing compounds that are selectively taken up by tumor cells, ensuring that only tumor cells will be destroyed by the beam of thermal neutrons.

PROBLEM-SOLVING STRATEGY 28.2   Atomic physics   (MP)

1. In atomic physics, the orders of magnitude of physical quantities are so unfamiliar that common sense often isn't much help in judging the reasonableness of a result. It helps to remind yourself of some typical magnitudes of various quantities:

   Size of an atom: 10^{-10} m
   Mass of an atom: 10^{-26} kg
   Mass of an electron: 10^{-30} kg
   Energy of an atomic energy level: 1 to 10 eV, or 10^{-18} J (but some interaction energies in many-electron atoms are much larger).
   Speed of an electron in the Bohr atom: 10^6 m/s
   Electron charge: 10^{-19} C
   kT at room temperature: 1/40 eV

   You may want to add items to this list. These values will also help you in Chapter 30 when we have to deal with magnitudes characteristic of *nuclear* rather than atomic structure, often different by factors of 10^4 to 10^6. In working out problems, be careful to handle powers of ten properly. A gross error might not be obvious.

2. As in preceding sections, energies may be expressed either in joules or in electronvolts. Be sure to use consistent units. Lengths, such as wavelengths, are always in meters if you use the other quantities consistently in SI units, such as $h = 6.626 \times 10^{-34}$ J · s. If you want nanometers or something else, don't forget to convert. In some problems, it's useful to express h in eV: $h = 4.136 \times 10^{-15}$ eV · s.

3. Kinetic energy can be expressed either as $K = \frac{1}{2}mv^2$ or (because $p = mv$) as $K = p^2/2m$. The latter form is often useful in calculations involving the de Broglie wavelength.

EXAMPLE 28.9   Energy of a thermal neutron

Neutrons have a wavelength, obeying the same relation as for electrons and protons: Equation 28.6 $(\lambda = h/p)$. Find the speed and kinetic energy of a neutron $(m = 1.675 \times 10^{-27}$ kg$)$ that has a de Broglie wavelength $\lambda = 0.200$ nm, typical of atomic spacing in crystals. Compare the energy with the average kinetic energy of a gas molecule at room temperature $(T = 20°C)$.

SOLUTION

SET UP We know the wavelength λ, the mass m, and Planck's constant h, so we can use the de Broglie relationship in the form $\lambda = h/mv$ to find the neutron's speed. From that result, we can find its kinetic energy.

SOLVE From the de Broglie relationship,

$$v = \frac{h}{\lambda m} = \frac{6.626 \times 10^{-34} \text{ J} \cdot \text{s}}{(0.200 \times 10^{-9} \text{ m})(1.675 \times 10^{-27} \text{ kg})}$$
$$= 1.98 \times 10^3 \text{ m/s};$$
$$K = \frac{1}{2}mv^2 = \frac{1}{2}(1.675 \times 10^{-27} \text{ kg})(1.98 \times 10^3 \text{ m/s})^2$$
$$= 3.28 \times 10^{-21} \text{ J} = 0.0205 \text{ eV}.$$

The average translational kinetic energy of a molecule of an ideal gas at absolute temperature $T = 293$ K (equal to 20°C or 68°F) is given by Equation 15.8, where k is Boltzmann's constant:

$$K = \frac{3}{2}kT = \frac{3}{2}(1.38 \times 10^{-23} \text{ J/K})(293 \text{ K})$$
$$= 6.07 \times 10^{-21} \text{ J} = 0.0379 \text{ eV}.$$

REFLECT The two energies are comparable in magnitude. In fact, a neutron with kinetic energy in this range is called a *thermal neutron*. Diffraction of thermal neutrons can be used to study crystal and molecular structure in the same way as x-ray diffraction is used. Neutron diffraction has proved especially useful in the study of large organic molecules.

Practice Problem: Find the kinetic energy of an electron with a de Broglie wavelength of 0.100 nm. *Answer:* 150 eV.

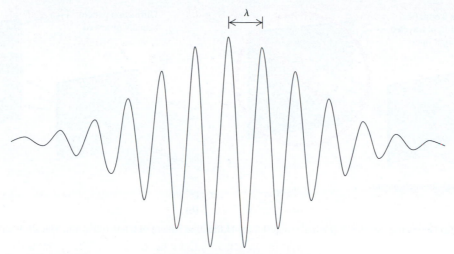

▲ **FIGURE 28.20** A wave that has both wave and particle aspects.

Davisson–Germer Experiment

De Broglie's wave hypothesis, unorthodox though it seemed at the time, received almost immediate direct experimental confirmation. We described in Section 26.7 how the atoms in a crystal can function as a three-dimensional diffraction grating for x rays. An x-ray beam is strongly reflected when it strikes a crystal at an angle that gives constructive interference in the various scattered waves. The existence of these strong reflections is evidence for the wave nature of x rays, and it provides a means for measuring the wavelengths of x rays.

In 1927, C. J. Davisson and L. H. Germer, working at Bell Telephone Laboratories, discovered that a beam of electrons was reflected in almost the same way that x rays would be reflected from the same crystal; that is, they were being diffracted. Davisson and Germer could determine the speeds of the electrons from the accelerating voltage, so they could compute the de Broglie wavelength $\lambda = h/mv$ from Equation 28.20. They found that the angles at which strong reflection took place were the same as those at which x rays with the same wavelength would be reflected. This phenomenon is called **electron diffraction,** and its discovery gave strong support to de Broglie's hypothesis.

Additional experiments were carried out in many laboratories. In Germany, Estermann and Stern demonstrated diffraction of alpha particles. More recently, diffraction experiments have been performed with low-energy neutrons and large molecules. Thus, the wave nature of particles, so strange in 1924, became firmly established in the ensuing years.

Figure 28.20 suggests how a possible wave function for a free electron might look. The wave is more or less localized in space; if we don't look too closely, it looks like a point particle. Yet it clearly has wave properties, including a definite wavelength. In short, it has both wave and particle properties.

28.7 Wave–Particle Duality

The discovery of the dual wave–particle nature of matter has forced us to reevaluate the kinematic language we use to describe the behavior of a particle. In classical Newtonian mechanics, we think of a particle as a point. At any instant in time, it has a definite position and a definite velocity. But in general, such a specific description isn't possible. When we look on a small enough scale, there are fundamental limitations on the precision with which we can describe the position and velocity of a particle. Many aspects of a particle's behavior can be stated only in terms of probabilities.

PhET: Davisson–Germer: Electron Diffraction

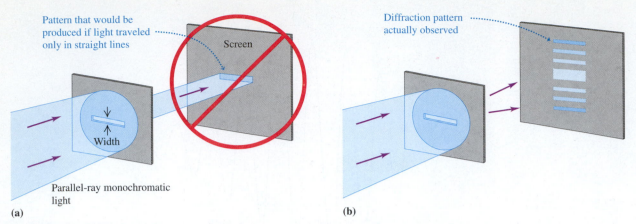

▲ **FIGURE 28.21** (a) Geometric "shadow" of a horizontal slit. (b) Diffraction pattern of a horizontal slit. The slit width has been greatly exaggerated.

Single-Slit Diffraction

To get some insight into the nature of the problem, let's review an optical single-slit diffraction experiment. Figure 28.21 shows a light beam incident on a narrow horizontal slit. If light traveled only in straight lines, the pattern on the screen would be as shown in Figure 28.21a. The actual pattern (Figure 28.21b), however, is a diffraction pattern, showing the wave nature of light. Detailed analysis shows that the width of the central maximum in the pattern is inversely proportional to the height of the aperture. Most (about 85%) of the total energy is concentrated in the central maximum of the pattern.

Now we perform the same experiment again, but using a beam of electrons instead of a beam of monochromatic light (Figure 28.22). We have to do the experiment in vacuum (10^{-7} atm or less) so that the electrons don't bump into air molecules. We can produce the electron beam with a setup similar in principle to the electron gun in a cathode-ray tube. This produces a narrow beam of electrons that all have the same direction and speed and therefore also the same de Broglie wavelength.

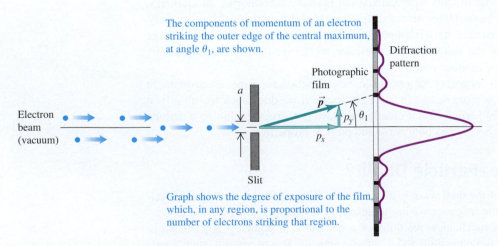

▲ **FIGURE 28.22** An electron diffraction experiment.

The result of this experiment, recorded on photographic film or with the use of more sophisticated detectors, is a diffraction pattern (Figure 28.22) identical to the one shown in Figure 28.21b. Most (85%) of the electrons strike the film in the region of the central maximum, but a few strike farther from the center, in the subsidiary maxima on both sides.

The electrons don't all follow the same path, even though they all have the same initial state of motion. In fact, we can't predict the trajectory of an individual electron from knowledge of its initial state. The best we can do is to say that most of the electrons go to a certain region, fewer go to other regions, and so on. That is, we can describe only the *probability* that an individual electron will strike each of various areas on the film. This fundamental indeterminacy has no counterpart in Newtonian mechanics, in which the motion of a particle or a system can always be predicted if we know the initial position and motion with great enough precision.

There are fundamental uncertainties in both the position and the momentum of an individual particle, and these two uncertainties are related inseparably and in a reciprocal way. The uncertainty Δy in the y position of a particle is determined by the slit width a, and the uncertainty Δp_y in the y component of momentum, p_y, is directly related to the width of the central maximum in the diffraction pattern, as shown in Figure 28.22. If we reduce the uncertainty in y by decreasing the slit width, the diffraction pattern becomes broader, corresponding to an *increase* in the uncertainty of p_y.

More detailed analysis shows that the *product* of the two uncertainties must be at least as great as h:

$$\Delta p_y a \geq h. \tag{28.21}$$

The slit width a represents the uncertainty in the *position* of an electron as it passes through the slit, and the width of the diffraction pattern determines the uncertainty Δp_y. We can reduce Δp_y only by increasing the slit width, which increases the position uncertainty. Conversely, when we make the slit narrower, the diffraction pattern broadens, and the corresponding uncertainty in momentum increases.

The Uncertainty Principle

In more general discussions of uncertainty relations, the uncertainty of a quantity is usually described in terms of the statistical concept of *standard deviation*, a measure of the spread or dispersion of a set of numbers around their average value. If a coordinate x has an uncertainty Δx defined in this way, and if the corresponding momentum component p_x has an uncertainty Δp_x, then the two uncertainties are found to be related in general by an inequality named for its discoverer, Werner Heisenberg:

PhET: Fourier: Making Waves
PhET: Quantum Wave Interference
ActivPhysics 17.6: Uncertainty Principle

Heisenberg uncertainty principle for position and momentum

$$\Delta x \, \Delta p_x \geq \frac{h}{2\pi}. \tag{28.22}$$

(This relationship differs from Equation 28.21 by a numerical factor of 2π because of the way "standard deviation" is defined.)

Equation 28.22 is one form of the Heisenberg uncertainty principle. It states that, in general, neither the momentum nor the position of a particle can be measured simultaneously with arbitrarily great precision, as classical physics would predict. Instead, the uncertainties in the two quantities play complementary roles, as we have described. There are corresponding uncertainty relations for the y and z coordinates and momentum components.

Uncertainty in Energy

There is also an uncertainty principle involving *energy*. It turns out that the energy of a system has an inherent uncertainty.

Heisenberg uncertainty principle: energy and time

The uncertainty ΔE in the energy of a system in a particular state depends on the *time interval* Δt during which the system remains in that state. The relation is

$$\Delta E \, \Delta t \geq \frac{h}{2\pi}. \tag{28.23}$$

A system that remains in a certain state for a long time (large Δt) can have a well-defined energy (small ΔE), but if it remains in that state for only a short time (small Δt), the uncertainty in energy must be correspondingly greater (large ΔE).

EXAMPLE 28.10 **Electron in a box**

An electron is confined inside a cubical box 1.0×10^{-10} m on a side. **(a)** Estimate the minimum uncertainty in the x component of the electron's momentum. **(b)** If the electron has momentum with magnitude equal to the uncertainty found in part (a), what is its kinetic energy? Express the result in joules and in electronvolts.

SOLUTION

SET UP Let's concentrate on x components. The uncertainty Δx is the size of the box: $\Delta x = 1.0 \times 10^{-10}$ m. (a) We get the uncertainty Δp_x in momentum from the uncertainty principle, Equation 28.22. (b) Knowing Δp_x, we can estimate the uncertainty in the kinetic energy K from the relation $K = p_x^2/2m$.

SOLVE **Part (a):** From Equation 28.22, the minimum uncertainty in p_x is

$$\Delta p_x = \frac{h}{2\pi \, \Delta x} = \frac{6.63 \times 10^{-34} \text{ J} \cdot \text{s}}{2\pi (1.0 \times 10^{-10} \text{ m})} = 1.1 \times 10^{-24} \text{ kg} \cdot \text{m/s}.$$

Part (b): Assuming that the energy is in the nonrelativistic range, an electron with this magnitude of momentum has kinetic energy

$$K = \frac{p_x^2}{2m} = \frac{(1.1 \times 10^{-24} \text{ kg} \cdot \text{m/s})^2}{2(9.11 \times 10^{-31} \text{ kg})}$$
$$= 6.1 \times 10^{-19} \text{ J} = 3.8 \text{ eV}.$$

REFLECT The box is roughly the same size as an atom, so it isn't surprising that the energy is of the same order of magnitude as typical electron energies in atoms. If our result had differed from typical atomic energy levels by a factor of 10^6, we might be suspicious.

Practice Problem: Suppose the box is the size of a nucleus (on the order of 10^{-15} m) and the particle is a proton. Make a rough estimate of the uncertainty in energy of the particle. *Answer:* 21 MeV.

EXAMPLE 28.11 **Energy uncertainty in the sodium atom**

A sodium atom in one of the "resonance levels" remains in that state for an average time of 1.6×10^{-8} s before it makes a transition back to the ground state by emitting a photon with wavelength 589 nm and energy 2.109 eV. Find the uncertainty in energy of the resonance level and the wavelength spread of the corresponding spectrum line.

Continued

SOLUTION

SET UP We need to use the uncertainty principle for energy given by Equation 28.23; we are given that $\Delta t = 1.6 \times 10^{-8}$ s, so we can find ΔE. Then the fractional uncertainty $\Delta\lambda/\lambda$ in wavelength should be proportional to the fractional uncertainty $\Delta E/E$ in energy.

SOLVE From Equation 28.23,

$$\Delta E = \frac{h}{2\pi\,\Delta t} = \frac{6.626 \times 10^{-34}\,\text{J}\cdot\text{s}}{(2\pi)(1.6 \times 10^{-8}\,\text{s})}$$
$$= 6.6 \times 10^{-27}\,\text{J} = 4.1 \times 10^{-8}\,\text{eV}.$$

The uncertainty in the resonance-level energy and in the corresponding photon energy amounts to about 2 parts in 10^8. (The atom remains an indefinitely long time in the ground state, so there is *no* uncertainty there.) Assuming that the corresponding spread in wavelength, or "width," of the spectrum line is also approximately two parts in 10^8, we find that

$$\Delta\lambda = (2 \times 10^{-8})(589\,\text{nm}) = 0.000012\,\text{nm}.$$

REFLECT The irreducible uncertainty $\Delta\lambda$ is called the *natural line width* of this particular spectrum line. Although small, it is within the limits of resolution of present-day spectrometers. Ordinarily, the natural line width is much smaller than line broadening from other causes, such as collisions among atoms.

28.8 The Electron Microscope

The **electron microscope** offers us an important and interesting example of the interplay between wave and particle properties of electrons. An electron beam can be used to form an image of an object in exactly the same way as a light beam. A ray of light is bent by reflection or refraction, and an electron trajectory is bent by an electric or magnetic field. Rays of light diverging from a point on an object can be brought to convergence by a converging lens or concave mirror, and electrons diverging from a small region can be brought to convergence by an electrostatic or magnetic lens. Figures 28.23a and 28.23b show the behavior of a simple type of electrostatic lens, and Figure 28.23c shows the analogous optical system. In each case, the image can be made larger than the object, so both devices can act as magnifiers.

The analogy between light rays and electrons goes deeper. The *ray* model of geometrical optics (ray optics) is an approximate representation of the more general *wave* model. Geometrical optics is valid whenever interference and diffraction effects can be neglected. Similarly, the model of an electron as a point

▲ **BIO Application** **Imaging the invisible.** Electrons accelerated by a potential of 200 kV have de Broglie wavelengths several orders of magnitude smaller than photons of visible light. The electron microscope creates such energetic electrons and then focuses them with magnetic and electrostatic lenses onto specimens. Biological materials sliced into thin sections can be examined in the electron microscope, producing images with far better resolution than is possible in the light microscope. Individual macromolecules can also be imaged by these means. The photograph shows the beautifully arranged alternating thin (actin) and thick (myosin) filaments of muscle. The diameter of the thin filaments is about 7 nm, and the myosin filaments are 10 to 11 nm thick. The bracket on the left (X) indicates the region of overlap between the actin and myosin molecules; the right bracket (Y) shows the region containing only myosin. The major uncertainty in using this technique is the degree to which the sample is altered by necessarily elaborate preparation for examination in a high-vacuum system.

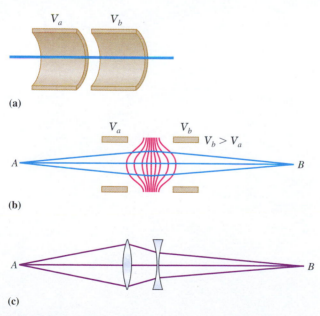

▲ **FIGURE 28.23** Electrostatic lens for an electron microscope.

particle following a line trajectory is an approximate description of the actual behavior of the electron; this model is useful when we can neglect effects associated with the wave nature of electrons.

How is an electron microscope superior to an optical microscope? The *resolution* of an optical microscope is limited by diffraction effects, as we discussed in Section 26.8. Using wavelengths in the visible spectrum, around 500 nm, an optical microscope can't resolve objects smaller than a few hundred nanometers, no matter how carefully its lenses are made. The resolution of an electron microscope is similarly limited by the wavelengths of the electrons, but these may be many thousands of times *smaller* than wavelengths of visible light. As a result, the useful magnification of an electron microscope can be thousands of times as great as that of an optical microscope.

Note that the ability of the electron microscope to form an image *does not* depend on the wave properties of electrons. We can compute the trajectories of electrons by treating them as classical charged particles under the action of electric- and magnetic-field forces. Only when we talk about *resolution* do the wave properties become important.

EXAMPLE 28.12 Electron wavelengths

An electron beam is formed by a setup similar to the electron gun in an older-model TV set, accelerating electrons through a potential difference (voltage) of several thousand volts. What accelerating voltage is needed to produce electrons with wavelength 10 pm = 0.010 nm (roughly 50,000 times smaller than typical visible-light wavelengths)?

SOLUTION

SET UP The wavelength λ is determined by the de Broglie relation, $\lambda = h/p = h/mv$. The momentum $p = mv$ is related to the kinetic energy $K = p^2/2m$, which in turn is determined by the accelerating voltage V: $K = eV$.

SOLVE When we put the preceding pieces together, we get

$$K = eV = \frac{p^2}{2m} = \frac{(h/\lambda)^2}{2m}.$$

Solving for V and inserting the appropriate numbers, we find that

$$V = \frac{h^2}{2me\lambda^2}$$
$$= \frac{(6.626 \times 10^{-34} \text{ J} \cdot \text{s})^2}{2(9.109 \times 10^{-31} \text{ kg})(1.602 \times 10^{-19} \text{ C})(10 \times 10^{-12} \text{ m})^2}$$
$$= 1.5 \times 10^4 \text{ V} = 15,000 \text{ V}.$$

REFLECT The voltage we just obtained is approximately equal to the accelerating voltage for the electron beams formerly used in older TV picture tubes and computer monitors. This example shows, incidentally, that the sharpness of a TV picture is *not* limited by electron diffraction effects. Also, the kinetic energy of an electron is 15,000 eV, a small fraction of the electron's *rest* energy of about 511,000 eV; this shows that relativistic effects need not be considered in solving this problem.

SUMMARY

The Photoelectric Effect

(**Section 28.1**) The energy E of an individual photon is proportional to its frequency f, $E = hf$ (Equation 28.2), where h is Planck's constant. The wavelength of a photon is given by $\lambda = c/f$, and the momentum p is inversely proportional to the wavelength:

$$p = \frac{E}{c} = \frac{hf}{c} = \frac{h}{\lambda} \qquad (28.6)$$

In the **photoelectric effect,** a photon striking a conductor is absorbed by an electron. If the photon energy is greater than the work function ϕ of the material, the electron can escape from the surface.

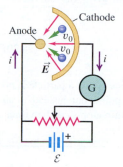

The photoelectric effect

Line Spectra and Energy Levels

(**Section 28.2**) If an atom makes a transition from an initial energy E_i to a lower final energy E_f, the energy of the emitted photon is hf, where $hf = E_i - E_f$ (Equation 28.7). The photon energies of the hydrogen atom's spectral lines are all differences between its energy levels. The energy levels are given by

$$E_n = -\frac{hcR}{n^2} = -\frac{13.6 \text{ eV}}{n^2}, \qquad n = 1, 2, 3, \cdots. \quad (28.10)$$

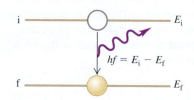

The Nuclear Atom and the Bohr Model

(**Section 28.3**) In Bohr's model of the hydrogen atom, the allowed values of angular momentum L for an electron in a circular orbit are integer multiples of the constant $h/2\pi$:

$$L = mv_n r_n = n\frac{h}{2\pi}, \qquad n = 1, 2, 3, \cdots. \quad (28.11)$$

The integer n is called the **principal quantum number;** it also determines the electron's radius and orbital speed for each level:

$$r_n = \epsilon_0 \frac{n^2 h^2}{\pi m e^2}, \qquad v_n = \frac{1}{\epsilon_0}\frac{e^2}{2nh} \quad (28.13, 28.14)$$

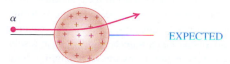

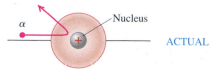

The Laser

(**Section 28.4**) A laser requires a population inversion in which the population of atoms in higher energy states is enhanced. Through stimulated emission, many photons are radiated with the same frequency, phase, and polarization—they are all synchronized.

X-Ray Production and Scattering

(**Section 28.5**) When rapidly moving electrons that have been accelerated through a potential difference of the order of 10^3 to 10^6 V strike a metal target, x rays are produced. In Compton scattering, x rays strike matter, some of the radiation is scattered with a larger wavelength than the incident radiation, and the increase in wavelength $\Delta\lambda$ depends on the angle through which the radiation is scattered.

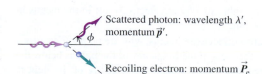

Continued

The Wave Nature of Particles

(Section 28.6) Just as photons have a wavelength $\lambda = h/p$ (Equation 28.6), using $p = mv$ in this equation gives the de Broglie wavelength λ for a particle of mass m:

$$\lambda = \frac{h}{p} = \frac{h}{mv}. \qquad (28.20)$$

An essential concept of quantum mechanics is that a particle is no longer described as located at a single point, but instead is an inherently spread-out entity. Electrons within atoms can be visualized as diffuse clouds surrounding the nucleus.

Wave–Particle Duality

(Section 28.7) Neither the momentum nor the position of a particle can be measured simultaneously with arbitrarily great precision, as classical physics would predict. Instead, the uncertainties in the two quantities play complementary roles: The less the uncertainty about a particle's position, the more uncertain is its momentum, and vice versa. This fundamental property of nature is described by Heisenberg's uncertainty principle:

$$\Delta x \, \Delta p_x \geq \frac{h}{2\pi}. \qquad (28.22)$$

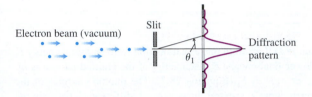

The Electron Microscope

(Section 28.8) An electron beam can be used to form an image of an object in exactly the same way as a light beam. While reflection and refraction are used to bend light rays, electric and magnetic fields are used to bend an electron's trajectory. Just as the resolution of an optical microscope is limited by the wavelength of light, the resolution of an electron microscope is limited by the wavelength of the electrons. Since the wavelengths of the electrons are much smaller than those for light used in an optical microscope, the resolution of an electron microscope is typically thousands of times better.

 For instructor-assigned homework, go to www.masteringphysics.com

Conceptual Questions

1. In the photoelectric effect, electrons are knocked out of metals by photons of light. So why are electrons left in the metals in an ordinary room?

2. In photoelectric-effect experiments, why are no photoelectrons ejected from the metal if the frequency of the light is below the threshold frequency, even though the intensity of the light is very strong?

3. In photoelectric-effect experiments, why does the stopping potential increase with the frequency of the light, but *not* with the intensity of the light?

4. In what ways do photons resemble familiar particles such as electrons? In what ways do they differ? For example, do they have mass? Electric charge? Can they be accelerated? What mechanical properties do they have?

5. Is the frequency of a light beam the same thing as the number of photons per second that it carries? If not, what is the difference between the light frequency and the number of photons per second? Could you change the number of photons per second in a beam of light without changing the frequency of the light? How?

6. Would you expect quantum effects to be generally more important at the low-frequency end of the electromagnetic spectrum (radio waves) or at the high-frequency end (x rays and gamma rays)? Why?

7. Except for some special-purpose films, most black-and-white photographic film is less sensitive at the far-red end of the

visible spectrum than at the blue end and has almost no sensitivity to infrared, even at high intensities. How can these properties be understood on the basis of photons?

8. Human skin is relatively insensitive to visible light, but ultraviolet radiation can be quite destructive to skin cells. Why do you think this is so?

9. Figure 28.3 shows that in a photoelectric-effect experiment the photocurrent i for large values of V_{AC} has the same value no matter what the light frequency f (provided that f is higher than the threshold frequency f_0). Explain why.

10. Particles such as electrons are sometimes whimsically called "wavicles." Explain why this is actually an appropriate name for them.

11. The phosphorescent materials that coat the inside of a fluorescent tube convert ultraviolet radiation (from the mercury-vapor discharge inside the tube) to visible light. Could one also make a phosphor that converts visible light to ultraviolet? Why or why not?

12. As an object is heated to a very high temperature and becomes self-luminous, the apparent color of the emitted radiation shifts from red to yellow and finally to blue with the increasing temperature. What causes the color shift?

13. If a proton and an electron have the same speed, which has the longer de Broglie wavelength? Explain.

14. Why go through the expense of building an electron microscope for studying very small objects such as organic molecules? Why not just use extremely short electromagnetic waves, which are much cheaper to generate?

15. Do the planets of the solar system obey a distance law $(r_n = n^2 r_1)$ like the electrons of the Bohr atom? Should they? Why (or why not)? (Consult Appendix E for the appropriate distances.)

16. Which has more total energy, a hydrogen atom with an electron in a high shell (large n) or in a low shell (small n)? Which is moving faster, the high-shell electron or the low-shell electron? Is there a contradiction here? Explain.

Multiple-Choice Problems

1. Light falling on a metal surface causes electrons to be emitted from the metal by the photoelectric effect. As we increase the *intensity* of this light, but keep its wavelength the same (there may be more than one correct answer),
   A. The number of electrons emitted from the metal increases.
   B. The number of electrons emitted from the metal does not change.
   C. The maximum speed of the emitted electrons does not change.
   D. The maximum speed of the emitted electrons increases.

2. Light falling on a metal surface causes electrons to be emitted from the metal by the photoelectric effect. As we increase the *frequency* of this light, but do not vary anything else (there may be more than one correct answer),
   A. The number of electrons emitted from the metal increases.
   B. The maximum speed of the emitted electrons increases.
   C. The maximum speed of the emitted electrons does not change.
   D. The work function of the metal increases.

3. According to the Bohr model of the hydrogen atom, as we look at higher and higher values of n, the distance between adjacent electron orbits
   A. gets progressively smaller and smaller.
   B. gets progressively larger and larger.
   C. remains constant.

4. According to the Bohr model of the hydrogen atom, as we look at higher and higher values of n, the difference in energy between adjacent electron orbits
   A. gets progressively smaller and smaller.
   B. gets progressively larger and larger.
   C. remains constant.

5. A photon of wavelength λ has energy E. If its wavelength were doubled, its energy would be
   A. $4E$.　　B. $2E$.　　C. $\frac{1}{2}E$.　　D. $\frac{1}{4}E$.

6. Consider a hypothetical single-electron Bohr atom for which an electron in the $n = 1$ shell has a total energy of -81.0 eV while one in the $n = 3$ shell has a total energy of -9.00 eV. Which statements about this atom are correct? (There may be more than one correct choice.)
   A. The energy needed to ionize the atom is 90.0 eV.
   B. It takes 90.0 eV to move an electron from the $n = 1$ to the $n = 3$ shell.
   C. If an electron makes a transition from the $n = 3$ to the $n = 1$ shell, it will give up 72.0 eV of energy.
   D. If an electron makes a transition from the $n = 3$ to the $n = 1$ shell, it must absorb 72.0 eV of energy.

7. Which of the following energies would be a physically *reasonable* photoelectric work function for a typical metal?
   A. 5.0 J.　　B. 5.0 GeV.　　C. 5.0 MeV.　　D. 5.0 eV.

8. If the Bohr radius of the $n = 3$ state of a hydrogen atom is R, then the radius of the ground state is
   A. $9R$.　　B. $3R$.　　C. $R/3$.　　D. $R/9$.

9. If the energy of the $n = 3$ state of a Bohr-model hydrogen atom is E, the energy of the ground state is
   A. $9E$.　　B. $3E$.　　C. $E/3$.　　D. $E/9$.

10. Electrons are accelerated from rest through a potential difference V. As V is increased, the de Broglie wavelength of these electrons
   A. increases.　　B. decreases.　　C. does not change.

11. Proton A has a de Broglie wavelength λ. If proton B has twice the speed (which is much less than the speed of light) of proton A, the de Broglie wavelength of proton B is
   A. 2λ.　　B. $\lambda\sqrt{2}$.　　C. $\lambda/\sqrt{2}$.　　D. $\lambda/2$.

12. Electron A has a de Broglie wavelength λ. If electron B has twice the kinetic energy (but a speed much less than that of light) of electron A, the de Broglie wavelength of electron B is
   A. 2λ.　　B. $\lambda\sqrt{2}$.　　C. $\lambda/\sqrt{2}$.　　D. $\lambda/2$.

13. When a photon of light scatters off of a free stationary electron, the wavelength of the photon
   A. decreases.　　B. increases.　　C. remains the same.
   D. could either increase or decrease, depending on the initial energy of the photon.

14. Which of the following is NOT a characteristic of laser light:
   A. All the photons in laser light are in phase with each other.
   B. Laser light contains a very broad spectrum of wavelengths.
   C. The photons produced by a laser all travel in the same direction.
   D. The photons in laser light all have the same frequency.

15. Some lasers emit light in pulses that are only 10^{-12} s in duration. Compared to a laser that emits a steady, continuous beam of light, the Heisenberg uncertainty principle leads us to expect the photons from a pulsed laser to have
    A. higher energy.
    B. lower energy.
    C. greater momentum.
    D. a broader range of frequencies.
    E. a narrower range of frequencies.

16. Light of frequency f falls on a metal surface and ejects electrons of maximum kinetic energy K by the photoelectric effect. If the frequency of this light is doubled, the maximum kinetic energy of the emitted electrons will be
    A. $K/2$.   B. K.
    C. $2K$.   D. greater than $2K$.

Problems

28.1 Photoelectric Effect

1. ● **Response of the eye.** The human eye is most sensitive to
BIO green light of wavelength 505 nm. Experiments have found that when people are kept in a dark room until their eyes adapt to the darkness, a *single* photon of green light will trigger receptor cells in the rods of the retina. (a) What is the frequency of this photon? (b) How much energy (in joules and eV) does it deliver to the receptor cells? (c) To appreciate what a small amount of energy this is, calculate how fast a typical bacterium of mass 9.5×10^{-12} g would move if it had that much energy.

2. ● An excited nucleus emits a gamma-ray photon with an energy of 2.45 MeV. (a) What is the photon's frequency? (b) What is the photon's wavelength? (c) How does this wavelength compare with a typical nuclear diameter of 10^{-14} m?

3. ● A laser used to weld detached retinas emits light with a
BIO wavelength of 652 nm in pulses that are 20.0 ms in duration. The average power expended during each pulse is 0.600 W. (a) How much energy is in each pulse, in joules? In electron volts? (b) What is the energy of one photon in joules? In electron volts? (c) How many photons are in each pulse?

4. ● A radio station broadcasts at a frequency of 92.0 MHz with a power output of 50.0 kW. (a) What is the energy of each emitted photon, in joules and electron volts? (b) How many photons are emitted per second?

5. ● The predominant wavelength emitted by an ultraviolet lamp is 248 nm. If the total power emitted at this wavelength is 12.0 W, how many photons are emitted per second?

6. ● A photon has momentum of magnitude 8.24×10^{-28} kg·m/s. (a) What is the energy of this photon? Give your answer in joules and in electron volts. (b) What is the wavelength of this photon? In what region of the electromagnetic spectrum does it lie?

7. ● In the photoelectric effect, what is the relationship between the threshold frequency f_0 and the work function ϕ?

8. ● A clean nickel surface is exposed to light of wavelength 235 nm. What is the maximum speed of the photoelectrons emitted from this surface? Use Table 28.1.

9. ● The photoelectric threshold wavelength of a tungsten surface is 272 nm. (a) What are the threshold frequency and work function (in eV) of this tungsten? (b) Calculate the maximum kinetic energy (in eV) of the electrons ejected from this tungsten surface by ultraviolet radiation of frequency 1.45×10^{15} Hz.

10. ● What would the minimum work function for a metal have to be for visible light (having wavelengths between 400 nm and 700 nm) to eject photoelectrons?

11. ●● When ultraviolet light with a wavelength of 400.0 nm falls on a certain metal surface, the maximum kinetic energy of the emitted photoelectrons is measured to be 1.10 eV. What is the maximum kinetic energy of the photoelectrons when light of wavelength 300.0 nm falls on the same surface?

12. ●● When ultraviolet light with a wavelength of 254 nm falls upon a clean metal surface, the stopping potential necessary to terminate the emission of photoelectrons is 0.181 V. (a) What is the photoelectric threshold wavelength for this metal? (b) What is the work function for the metal?

13. ●● The photoelectric work function of potassium is 2.3 eV. If light having a wavelength of 250 nm falls on potassium, find (a) the stopping potential in volts; (b) the kinetic energy, in electron volts, of the most energetic electrons ejected; (c) the speeds of these electrons.

14. ●● In a photoelectric effect experiment it is found that no current flows unless the incident light has a wavelength shorter than 289 nm. (a) What is the work function of the metal surface? (b) What stopping potential will be needed to halt the current if light of 225 nm falls on the surface?

15. ●● Light with a wavelength range of 145–295 nm shines on a silicon surface in a photoelectric effect apparatus, and a reversing potential of 3.50 V is applied to the resulting photoelectrons. (a) What is the longest wavelength of the light that will eject electrons from the silicon surface? (b) With what maximum kinetic energy will electrons reach the anode?

28.2 Line Spectra and Energy Levels

16. ● (a) How much energy is needed to ionize a hydrogen atom that is in the $n = 4$ state? (b) What would be the wavelength of a photon emitted by a hydrogen atom in a transition from the $n = 4$ state to the $n = 2$ state?

17. ● Use Balmer's formula to calculate (a) the wavelength, (b) the frequency, and (c) the photon energy for the H_γ line of the Balmer series for hydrogen.

18. ● Find the longest and shortest wavelengths in the Lyman and Paschen series for hydrogen. In what region of the electromagnetic spectrum does each series lie?

19. ● (a) Calculate the longest and shortest wavelengths for light in the Balmer, Lyman, and Brackett series. (b) Use your results from part (a) to decide in which part of the electromagnetic spectrum each of these series lies.

20. ●● The energy-level scheme for the hypothetical one-electron element searsium is shown in Fig. 28.24. The potential energy is taken to be zero for an electron at an infinite distance from the nucleus. (a) How much energy (in electron volts) does it take to ionize an electron from the ground level? (b) An 18 eV

$n = 4$ ———————————— -2 eV
$n = 3$ ———————————— -5 eV

$n = 2$ ———————————— -10 eV

$n = 1$ ———⬤——————— -20 eV

▲ **FIGURE 28.24** Problem 20.

photon is absorbed by a searsium atom in its ground level. As the atom returns to its ground level, what possible energies can the emitted photons have? Assume that there can be transitions between all pairs of levels. (c) What will happen if a photon with an energy of 8 eV strikes a searsium atom in its ground level? Why? (d) Photons emitted in the searsium transitions $n = 3 \to n = 2$ and $n = 3 \to n = 1$ will eject photoelectrons from an unknown metal, but the photon emitted from the transition $n = 4 \to n = 3$ will not. What are the limits (maximum and minimum possible values) of the work function of the metal?

21. •• In a set of experiments on a hypothetical one-electron atom, you measure the wavelengths of the photons emitted from transitions ending in the ground state $(n = 1)$, as shown in the energy-level diagram in Fig. 28.25. You also observe that it takes 17.50 eV to ionize this atom. (a) What is the energy of the atom in each of the levels $(n = 1, n = 2,$ etc.) shown in the figure? (b) If an electron made a transition from the $n = 4$ to the $n = 2$ level, what wavelength of light would it emit?

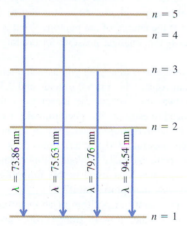

▲ **FIGURE 28.25** Problem 21.

28.3 The Nuclear Atom and the Bohr Model

22. • For a hydrogen atom in the ground state, determine, in electron volts, (a) the kinetic energy of the electron, (b) the potential energy, (c) the total energy, (d) the minimum energy required to remove the electron completely from the atom. (e) What wavelength does a photon with the energy calculated in part (d) have? In what region of the electromagnetic spectrum does it lie?

23. • Use the Bohr model for the following calculations: (a) What is the speed of the electron in a hydrogen atom in the $n = 1, 2,$ and 3 levels? (b) Calculate the radii of each of these levels. (c) Find the total energy (in eV) of the atom in each of these levels.

24. •• An electron in an excited state of hydrogen makes a transition from the $n = 5$ level to the $n = 2$ level. (a) Does the atom emit or absorb a photon during this process? How do you know? (b) Calculate the wavelength of the photon involved in the transition.

25. •• A hydrogen atom initially in the ground state absorbs a photon, which excites it to the $n = 4$ state. Determine the wavelength and frequency of the photon.

26. •• Light of wavelength 59 nm ionizes a hydrogen atom that was originally in its ground state. What is the kinetic energy of the ejected electron?

27. •• A triply ionized beryllium ion, Be^{3+} (a beryllium atom with three electrons removed), behaves very much like a hydrogen atom, except that the nuclear charge is four times as great. (a) What is the ground-level energy of Be^{3+}? How does this compare with the ground-level energy of the hydrogen atom? (b) What is the ionization energy of Be^{3+}? How does this compare with the ionization energy of the hydrogen atom? (c) For the hydrogen atom, the wavelength of the photon emitted in the transition $n = 2$ to $n = 1$ is 122 nm. (See Example 28.6.) What is the wavelength of the photon emitted when a Be^{3+} ion undergoes this transition? (d) For a given value of n, how does the radius of an orbit in Be^{3+} compare with that for hydrogen?

28.4 The Laser

28. • (a) Use the information for neon shown in Fig. 28.26 to compute the energy difference for the $5s$-to-$3p$ transition in neon. Express your result in electron volts and in joules. (b) Calculate the wavelength of a photon having this energy, and compare your result with the observed wavelength of the laser light. (c) What is the wavelength of the light from the $3p$-to-$3s$ transition in neon?

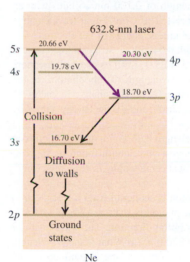

◀ **FIGURE 28.26** Problem 28.

29. • The diode laser keychain you use to entertain your cat has a wavelength of 645 nm. If the laser emits 4.50×10^{17} photons during a 30.0 s feline play session, what is its average power output?

30. • **Laser surgery.** Using a mixture of CO_2, N_2, and sometimes **BIO** He, CO_2 lasers emit a wavelength of 10.6 μm. At power outputs of 0.100 kW, such lasers are used for surgery. How many photons per second does a CO_2 laser deliver to the tissue during its use in an operation?

31. • **PRK surgery.** Photorefractive keratectomy (PRK) is a **BIO** laser-based surgery process that corrects near- and farsightedness by removing part of the lens of the eye to change its curvature and hence focal length. This procedure can remove layers 0.25 μm thick in pulses lasting 12.0 ns with a laser beam of wavelength 193 nm. Low-intensity beams can be used because each individual photon has enough energy to break the covalent bonds of the tissue. (a) In what part of the electromagnetic spectrum does this light lie? (b) What is the energy of a single photon? (c) If a 1.50 mW beam is used, how many photons are delivered to the lens in each pulse?

32. • **Removing birthmarks.** Pulsed dye lasers emit light of
BIO wavelength 585 nm in 0.45 ms pulses to remove skin blemishes such as birthmarks. The beam is usually focused onto a circular spot 5.0 mm in diameter. Suppose that the output of one such laser is 20.0 W. (a) What is the energy of each photon, in eV? (b) How many photons per square millimeter are delivered to the blemish during each pulse?

28.5 X-Ray Production and Scattering

33. • (a) What is the minimum potential difference between the filament and the target of an x-ray tube if the tube is to accelerate electrons to produce x rays with a wavelength of 0.150 nm? (b) What is the shortest wavelength produced in an x-ray tube operated at 30.0 kV? (c) Would the answers to parts (a) and (b) be different if the tube accelerated protons instead of electrons? Why or why not?

34. • The cathode-ray tubes that generated the picture in early color televisions were sources of x rays. If the acceleration voltage in a television tube is 15.0 kV, what are the shortest-wavelength x rays produced by the television? (Modern televisions contain shielding to stop these x rays.)

35. • An x ray with a wavelength of 0.100 nm collides with an electron that is initially at rest. The x ray's final wavelength is 0.110 nm. What is the final kinetic energy of the electron?

36. • If a photon of wavelength 0.04250 nm strikes a free electron and is scattered at an angle of 35.0° from its original direction, find (a) the change in the wavelength of this photon, (b) the wavelength of the scattered light, (c) the change in energy of the photon (is it a loss or a gain?), and (d) the energy gained by the electron.

37. • X rays with initial wavelength 0.0665 nm undergo Compton scattering. What is the longest wavelength found in the scattered x rays? At which scattering angle is this wavelength observed?

38. •• An incident x-ray photon is scattered from a free electron that is initially at rest. The photon is scattered straight back at an angle of 180° from its initial direction. The wavelength of the scattered photon is 0.0830 nm. (a) What is the wavelength of the incident photon? (b) What is the magnitude of the momentum of the electron after the collision? (c) What is the kinetic energy of the electron after the collision?

39. •• Protons are accelerated from rest by a potential difference of 4.00 kV and strike a metal target. If a proton produces one photon on impact, what is the minimum wavelength of the resulting x rays? How does your answer compare to the minimum wavelength if 4.00 keV electrons are used instead? Why do x-ray tubes use electrons rather than protons to produce x rays?

28.6 The Wave Nature of Particles

40. • (a) An electron moves with a speed of 4.70×10^6 m/s. What is its de Broglie wavelength? (b) A proton moves with the same speed. Determine its de Broglie wavelength.

41. • How fast would an electron have to move so that its de Broglie wavelength would be 1.00 mm?

42. • (a) Approximately what range of photon energies (in eV) corresponds to the visible spectrum? (b) Approximately what range of *wavelengths* and *kinetic energies* would electrons in this energy range have?

43. •• In the Bohr model of the hydrogen atom, what is the de Broglie wavelength for the electron when it is in (a) the $n = 1$ level and (b) the $n = 4$ level? In each case, compare the de Broglie wavelength to the circumference $2\pi r_n$ of the orbit.

44. • (a) What is the de Broglie wavelength of an electron accelerated through 800 V? (b) What is the de Broglie wavelength of a proton accelerated through the same potential difference?

45. •• Find the wavelengths of a photon and an electron that have the same energy of 25 eV. (The energy of the electron is its kinetic energy.)

28.7 Wave–Particle Duality

46. • (a) The uncertainty in the x component of the position of a proton is 2.0×10^{-12} m. What is the minimum uncertainty in the x component of the velocity of the proton? (b) The uncertainty in the x component of the velocity of an electron is 0.250 m/s. What is the minimum uncertainty in the x coordinate of the electron?

47. •• A certain atom has an energy level 3.50 eV above the ground state. When excited to this state, it remains 4.0 μs, on the average, before emitting a photon and returning to the ground state. (a) What is the energy of the photon? What is its wavelength? (b) What is the smallest possible uncertainty in energy of the photon?

48. •• A pesky 1.5 mg mosquito is annoying you as you attempt to study physics in your room, which is 5.0 m wide and 2.5 m high. You decide to swat the bothersome insect as it flies toward you, but you need to estimate its speed to make a successful hit. (a) What is the maximum uncertainty in the horizontal position of the mosquito? (b) What limit does the Heisenberg uncertainty principle place on your ability to know the horizontal velocity of this mosquito? Is this limitation a serious impediment to your attempt to swat it?

49. •• Suppose that the uncertainty in position of an electron is equal to the radius of the $n = 1$ Bohr orbit, about 0.5×10^{-10} m. Calculate the minimum uncertainty in the corresponding momentum component, and compare this with the magnitude of the momentum of the electron in the $n = 1$ Bohr orbit.

28.8 The Electron Microscope

50. •• (a) What accelerating potential is needed to produce electrons of wavelength 5.00 nm? (b) What would be the energy of photons having the same wavelength as these electrons? (c) What would be the wavelength of photons having the same energy as the electrons in part (a)?

51. •• (a) In an electron microscope, what accelerating voltage is needed to produce electrons with wavelength 0.0600 nm? (b) If protons are used instead of electrons, what accelerating voltage is needed to produce protons with wavelength 0.0600 nm? (*Hint:* In each case the initial kinetic energy is negligible.)

52. •• **Structure of a virus.** To investigate the structure of
BIO extremely small objects, such as viruses, the wavelength of the probing wave should be about one-tenth the size of the object for sharp images. But as the wavelength gets shorter, the energy of a photon of light gets greater and could damage or destroy the object being studied. One alternative is to use electron matter waves instead of light. Viruses vary considerably in size, but 50 nm is not unusual. Suppose you want to study such

a virus, using a wave of wavelength 5.00 nm. (a) If you use light of this wavelength, what would be the energy (in eV) of a single photon? (b) If you use an electron of this wavelength, what would be its kinetic energy (in eV)? Is it now clear why matter waves (such as in the electron microscope) are often preferable to electromagnetic waves for studying microscopic objects?

General Problems

53. •• **Exposing photographic film.** The light-sensitive compound on most photographic films is silver bromide (AgBr). A film is "exposed" when the light energy absorbed dissociates this molecule into its atoms. (The actual process is more complex, but the quantitative result does not differ greatly.) The energy of dissociation of AgBr is 1.00×10^5 J/mol. For a photon that is just able to dissociate a molecule of silver bromide, find (a) the photon's energy in electron volts, (b) the wavelength of the photon, and (c) the frequency of the photon. (d) Light from a firefly can expose photographic film, but the radiation from an FM station broadcasting 50,000 W at 100 MHz cannot. Explain why this is so, basing your answer on the energy of the photons involved.

54. •• A 2.50 W beam of light of wavelength 124 nm falls on a metal surface. You observe that the maximum kinetic energy of the ejected electrons is 4.16 eV. Assume that each photon in the beam ejects an electron. (a) What is the work function (in electron volts) of this metal? (b) How many photoelectrons are ejected each second from this metal? (c) If the power of the light beam, but not its wavelength, were reduced by half, what would be the answer to part (b)? (d) If the wavelength of the beam, but not its power, were reduced by half, what would be the answer to part (b)?

55. •• A sample of hydrogen atoms is irradiated with light with a wavelength of 85.5 nm, and electrons are observed leaving the gas. If each hydrogen atom were initially in its ground level, what would be the maximum kinetic energy, in electron volts, of these photoelectrons?

56. •• An unknown element has a spectrum for absorption from its ground level with lines at 2.0, 5.0, and 9.0 eV. Its ionization energy is 10.0 eV. (a) Draw an energy-level diagram for this element. (b) If a 9.0 eV photon is absorbed, what energies can the subsequently emitted photons have?

57. •• (a) What is the least amount of energy, in electron volts, that must be given to a hydrogen atom which is initially in its ground level so that it can emit the H_α line in the Balmer series? (b) How many different possibilities of spectral-line emissions are there for this atom when the electron starts in the $n = 3$ level and eventually ends up in the ground level? Calculate the wavelength of the emitted photon in each case.

58. •• A specimen of the microorganism *Gastropus hyptopus* **BIO** measures 0.0020 cm in length and can swim at a speed of 2.9 times its body length per second. The tiny animal has a mass of roughly 8.0×10^{-12} kg. (a) Calculate the de Broglie wavelength of this organism when it is swimming at top speed. (b) Calculate the kinetic energy of the organism (in eV) when it is swimming at top speed.

59. •• A photon with a wavelength of 0.1800 nm is Compton scattered through an angle of 180°. (a) What is the wavelength of the scattered photon? (b) How much energy is given to the electron? (c) What is the recoil speed of the electron? Is it necessary to use the relativistic kinetic–energy relationship?

60. •• (a) Calculate the maximum increase in photon wavelength that can occur during Compton scattering. (b) What is the energy (in electron volts) of the smallest-energy x-ray photon for which Compton scattering could result in doubling the original wavelength?

61. •• An incident x-ray photon of wavelength 0.0900 nm is scattered in the backward direction from a free electron that is initially at rest. (a) What is the magnitude of the momentum of the scattered photon? (b) What is the kinetic energy of the electron after the photon is scattered?

62. •• A photon with wavelength of 0.1100 nm collides with a free electron that is initially at rest. After the collision, the photon's wavelength is 0.1132 nm. (a) What is the kinetic energy of the electron after the collision? What is its speed? (b) If the electron is suddenly stopped (for example, in a solid target), all of its kinetic energy is used to create a photon. What is the wavelength of this photon?

63. • From the kinetic-molecular theory of an ideal gas (Chapter 15), we know that the average kinetic energy of an atom is $\frac{3}{2}kT$. What is the wavelength of a photon that has this energy for a temperature of 27°C?

64. •• **Doorway diffraction.** If your wavelength were 1.0 m, you would undergo considerable diffraction in moving through a doorway. (a) What must your speed be for you to have this wavelength? (Assume that your mass is 60.0 kg.) (b) At the speed calculated in part (a), how many years would it take you to move 0.80 m (one step)? Will you notice diffraction effects as you walk through doorways?

65. • What is the de Broglie wavelength of a red blood cell with a **BIO** mass of 1.00×10^{-11} g that is moving with a speed of 0.400 cm/s? Do we need to be concerned with the wave nature of the blood cells when we describe the flow of blood in the body?

66. •• **Removing vascular lesions.** A pulsed dye laser emits light **BIO** of wavelength 585 nm in 450 μs pulses. Because this wavelength is strongly absorbed by the hemoglobin in the blood, the method is especially effective for removing various types of blemishes due to blood, such as port-wine-colored birthmarks. To get a reasonable estimate of the power required for such laser surgery, we can model the blood as having the same specific heat and heat of vaporization as water $(4190 \text{ J/kg} \cdot \text{K}, 2.256 \times 10^6 \text{ J/kg})$. Suppose that each pulse must remove 2.0 μg of blood by evaporating it, starting at 33°C. (a) How much energy must each pulse deliver to the blemish? (b) What must be the power output of this laser? (c) How many photons does each pulse deliver to the blemish?

67. •• (a) What is the energy of a photon that has wavelength 0.10 μm? (b) Through approximately what potential difference must electrons be accelerated so that they will exhibit wave nature in passing through a pinhole 0.10 μm in diameter? What is the speed of these electrons? (c) If protons rather than electrons were used, through what potential difference would protons have to be accelerated so they would exhibit wave nature in passing through this pinhole? What would be the speed of these protons?

68. •• In a parallel universe, the value of Planck's constant is 0.0663 J · s. Assume that the physical laws and all other physical constants are the same as in our universe. In this other universe, two physics students are playing catch with a baseball. They are 50 m apart, and one throws a 0.10 kg ball with a

speed of 5.0 m/s. (a) What is the uncertainty in the ball's horizontal momentum in a direction perpendicular to that in which it is being thrown if the student throwing the ball knows that it is located within a cube with volume 1000 cm³ at the time she throws it? (b) By what horizontal distance could the ball miss the second student?

69. •• The neutral π^0 meson is an unstable particle produced in high-energy particle collisions. Its mass is about 264 times that of the electron, and it exists for an average lifetime of 8.4×10^{-17} s before decaying into two gamma-ray photons. Assuming that the mass and energy of the particle are related by the Einstein relation $E = mc^2$, find the uncertainty in the mass of the particle and express it as a fraction of the particle's mass.

70. •• A beam of electrons is accelerated from rest and then passes through a pair of identical thin slits that are 1.25 nm apart. You observe that the first double-slit interference dark fringe occurs at ± 18.0° from the original direction of the beam when viewed on a distant screen. (a) Are these electrons relativistic? How do you know? (b) Through what potential difference were the electrons accelerated?

Passage Problems

BIO Radiation therapy of tumors. Malignant tumors are often treated with targeted x-ray radiation therapy. To generate these medical x rays, a linear accelerator directs a high-energy beam of electrons toward a metal target (typically tungsten). As the electrons pass near the heavy metal nuclei, they are deflected and accelerated, emitting high-energy photons in a process known as Bremsstrahlung (from the German for "braking radiation"). The resultant x rays are collimated into a beam that is focused on the tumor. The tissue absorbs the energy predominately via Compton interactions, so it is important to know how many Compton interactions occur and how many ionizations a single Compton electron produces. A linear accelerator used in radiation therapy produces x ray photons with an average energy of about 2 MeV, each of which impart 1 MeV to the Compton electrons. A typical tumor has 10^8 cells/cm³ and a full treatment may involve a dose of 70 Gy in 35 fractional exposures on different days. The gray (Gy) is a measure of the absorbed energy dose of radiation per unit mass of tissue, expressed in the units of J/kg.

71. How much energy is imparted to a cell during one day's treatment? Assume that the specific gravity of the tumor is 1 and that $1J = 6 \times 10^{18}$eV.
    A. 12 MeV/cell
    B. 120 MeV/cell
    C. 120 MeV/cell
    D. 120×10^3 MeV/cell

72. Suppose the answer to problem 71 is 12 MeV/cell. How many Compton interactions will occur per cell in a single day's treatment?
    A. 120×10^6
    B. 120×10^4
    C. 120×10^2
    D. 12

73. Each Compton electron causes a series of ionization in the tissue as it interacts with molecules (mainly water), and each ionization takes about 40 eV. How many ionizations occur in a single cell as a result of a day's treatment?
    A. 3
    B. 3×10^2
    C. 3×10^4
    D. 3×10^6

29 Atoms, Molecules, and Solids

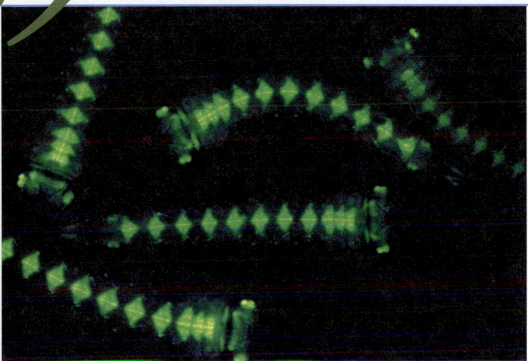

The basic concepts of quantum mechanics that we studied in Chapter 28 provide the key to understanding many aspects of the structure of atoms, molecules, and solid materials. We'll use the concept of a wave function as the basic language for describing electron configurations and motions. An additional property of electrons, called electron spin, plays a central role in describing these configurations. And we need one more crucial principle: the exclusion principle, which states that two electrons may not occupy the same quantum-mechanical state.

Using these three ideas, we can apply the principles of quantum mechanics to the analysis of a variety of atomic systems, such as the structure and chemical behavior of multielectron atoms (including the periodic table of the elements), molecular bonds (the binding of two or more atoms in a stable structure), and the large-scale binding of many atoms into solid structures. Many properties of solid materials can be understood at least qualitatively on the basis of electron configurations. We'll discuss semiconductors, a particular class of solid materials, in some detail because of their inherent interest and their great practical importance in present-day technology. Finally, we'll look briefly at the phenomenon of superconductivity: the complete disappearance of electrical resistance at low temperatures.

These mosquito larvae glow green because they are labeled with green fluorescent protein, a protein isolated from jellyfish that fluoresces (glows) green when illuminated with blue or UV light. In this chapter, you'll learn how such phenomena work.

29.1 Electrons in Atoms

We've learned that the classical description of the motion of a particle, in terms of three coordinates and three velocity components, can't be applied to motion on an atomic scale. In some situations, electrons behave like waves rather than particles, and their wavelength can be measured by diffraction experiments, as we discussed in Section 28.6. So we need to develop a description of electron motion that incorporates the idea of a spatially spread-out entity. In developing this description, we can take a cue from the language of classical wave motion.

Wave Functions

In Chapter 12, we described the motion of transverse waves on a string by specifying the position of each point in the string at each instant of time. To do this, we used a *wave function,* a concept we introduced in Section 12.4. If a point on the string, at some distance x from the origin, is displaced transversely a distance y from its equilibrium position at some time t, then there is a function $y = f(x, t)$, or $y(x, t)$, that represents the displacement of that point at *any* x at *any* time t. Once we know the wave function that describes a particular wave motion, we can find the position, velocity, and acceleration of any point on the string at any time, and so on.

In some situations, electrons behave like waves, so it's natural to adapt the language of waves to describe the motion of electrons. We'll use a **wave function** as the central element of our generalized language for describing the possible dynamic states (analogous to position and motion) of electrons (and, indeed, particles in general). The symbol usually used for such a wave function is Ψ. For a single electron, Ψ is a function of the space coordinates (x, y, z) and time. For a system with many electrons, it is a function of all the space coordinates of all the particles, and time. Just as the wave function $y(x, t)$ for mechanical waves on a string provides a complete description of the motion, the wave function $\Psi(x, y, z, t)$ for a particle contains all the information that can be known about the dynamic state of that particle.

Two questions arise. First, how do we relate the wave function to observable behavior of the particle? Second, how do we determine what Ψ is for any particular dynamic state of the system? With reference to the first question, the wave function describes the distribution of the particle in space. We can think of it as a cloudlike entity, more dense in some regions and less dense in others. The wave function is related to the *probability* of finding the particle in each of various regions: The particle is most likely to be found in regions where Ψ is large. We've already used this interpretation in our discussion of electron diffraction experiments. If the particle has a charge, the wave function can be used to find the charge density at any point in space.

In addition, from Ψ we can calculate the average position of the particle, its average velocity, and dynamic quantities such as momentum, energy, and angular momentum. The mathematical techniques required are far beyond the scope of this discussion, but they are well established and well supported by experimental results. Figure 29.1 shows cross sections of wave functions for three possible

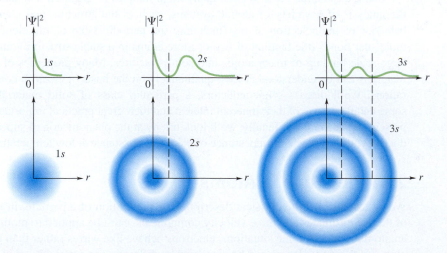

▲ **FIGURE 29.1** Three-dimensional probability distribution functions $|\Psi|^2$ for the spherically symmetric $1s$, $2s$, and $3s$ wave functions.

spherically symmetric states of the hydrogen atom. Of course, there is no possibility of actually seeing this distribution, since the atom is much smaller than wavelengths of visible light.

The answer to the second question is that the wave function Ψ must be one of a set of solutions of a certain differential equation. The **Schrödinger equation,** as it is called, was developed by Erwin Schrödinger (1887–1961) in 1926. In principle, we can set up a Schrödinger equation for any given physical situation. The wave functions that are solutions of this equation represent various possible physical states of the system, the analog of the orbits in the Bohr model.

Furthermore, it turns out that, for many systems, it is not even *possible* to find acceptable solutions of the Schrödinger equation unless some physical quantity, such as the energy of the system, has certain special values. Thus, the solutions of the Schrödinger equation for any particular system also yield a set of allowed energy levels. This discovery is of the utmost importance. Before the development of the Schrödinger equation, there was no way to predict energy levels from any fundamental theory, except for the very limited success of the Bohr model for hydrogen.

Soon after the Schrödinger equation was developed, it was applied to the problem of the hydrogen atom. The predicted energy levels E_n for this, the simplest of all atoms, turned out to be identical to those from the Bohr model, Equation 28.17. Therefore, these results also agreed with experimental values obtained from spectrum analysis.

In addition, the solutions have quantized values of angular momentum. As we learned in Section 10.7, angular momentum is fundamentally a vector quantity. In the Bohr model, the angular momentum is a vector perpendicular to the plane of the electron's orbit. The quantization of angular momentum was put into the Bohr model as an *ad hoc* assumption with no fundamental justification. With the Schrödinger equation, it appears naturally as a condition for the existence of acceptable solutions. Specifically, it is found that, for an electron in a hydrogen atom in a state with energy E_n and quantum number n, acceptable solutions of the Schrödinger equation exist only when the magnitude L of the vector angular momentum $\vec{L}$ is given by the following expression:

Permitted values of angular momentum

The possible values of angular momentum of the electron in a hydrogen atom are

$$L = \sqrt{l(l + 1)}\,\frac{h}{2\pi} \qquad (l = 0, 1, 2, \cdots, n - 1). \qquad (29.1)$$

The *component* of $\vec{L}$ in a given direction—say, the z component L_z, can have only the set of values

$$L_z = m_l \frac{h}{2\pi} \qquad (|m_l| = 0, 1, 2, \cdots, l). \qquad (29.2)$$

That is, m_l is a positive or negative integer or zero, with magnitude no greater than l.

Thus, instead of a single quantum number n as in the Bohr model, the possible wave functions for the hydrogen atom (corresponding to the various solutions of the Schrödinger equation) are labeled according to the values of *three* integers (quantum numbers) n, l, and m_l. These are called the **principal quantum number** (n), the **angular momentum quantum number** (l), and the **magnetic quantum number** (m_l). (There is also a fourth quantum number associated with

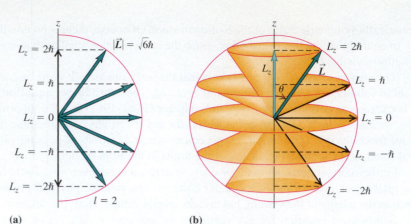

▲ **FIGURE 29.2** Magnitude and components of angular momentum, L and L_z.

electron spin; we'll discuss it later in this section.) For each energy level E_n, there are several distinct states having the same energy (and the same value of n), but different values of l and m_l. The only exception is the ground state, for which $n = 1$ and the only possibility for the other two quantum numbers is $(l = 0, m_l = 0)$. For any value of n, l can be no greater than $n - 1$. The term *magnetic quantum number* refers to the slight shifts in energy levels (and thus in spectral-line wavelengths) that occur when the atom is placed in a magnetic field. These shifts are different for states having the same n and l, but different values of m_l.

The quantity $h/2\pi$ appears so often in quantum mechanics that we give it a special symbol, $\hbar$, pronounced "h-bar." That is,

$$\hbar = \frac{h}{2\pi} = 1.054 \times 10^{-34}\,\text{J}\cdot\text{s}.$$

In terms of $\hbar$, Equations 29.1 and 29.2 respectively become

$$L = \sqrt{l(l + 1)}\,\hbar \qquad (l = 0, 1, 2, \cdots, n - 1) \qquad (29.3)$$

and

$$L_z = m_l \hbar \qquad (m_l = 0, \pm 1, \pm 2, \cdots, \pm l). \qquad (29.4)$$

Note that the component L_z can never be quite as large as L. For example, when $l = 2$ and $m_l = 2$, we find that

$$L = \sqrt{2(2 + 1)}\,\hbar = 2.45\hbar, \qquad L_z = 2\hbar.$$

Figure 29.2 shows the situation. For $l = 2$ and $m_l = 2$, the angle between the vector $\vec{L}$ and the z axis is given by

$$\cos\theta = \frac{L_z}{L} = \frac{2}{2.45}, \qquad \theta = 35°.$$

We can't know the precise direction of $\vec{L}$; it can have any direction on the cone in the figure at an angle of $\theta = 35°$ with the z axis. This result also follows from the uncertainty principle, which makes it impossible to predict the *direction* of the angular momentum vector with complete certainty.

Figure 29.2 shows the angular momentum relations for the particular states with $l = 2$. In each state, L and L_z have definite values, but the component of $\vec{L}$ perpendicular to the z axis does not; we can say only that this component is confined to a cone of directions, as shown in the figure.

EXAMPLE 29.1 **Angular momentum states**

Consider a hydrogen atom in a state with $n = 4$. Find expressions for the largest magnitude L of angular momentum, the largest positive value of L_z, and the corresponding values of the quantum numbers l and m_l. For the corresponding quantum state, find the smallest angle that the angular momentum vector can make with the $+z$ axis.

SOLUTION

SET UP If $n = 4$, the largest possible value of l is 3, and the maximum positive value of m_l is 3. The possible values of L and L_z are given by Equations 29.3 and 29.4.

SOLVE From Equation 29.3, when $l = 3$, $L = \sqrt{l(l + 1)}\hbar = \sqrt{3(3 + 1)}\hbar = 3.46\hbar$. From Equation 29.4, $L_z = m_l\hbar = 3\hbar$. The geometry is similar to Figure 29.1; the minimum value of θ is given by

$$\cos\theta_{min} = \frac{L_z}{L} = \frac{m_l\hbar}{\sqrt{l(l + 1)}\hbar} = \frac{3}{\sqrt{3(3 + 1)}}, \qquad \theta_{min} = 30.0°.$$

REFLECT For the same value of l, but other values of the quantum number l_z, the angle is greater. For $m_l = -3$, it is 150°.

Practice Problem: In this problem, L_x and L_y don't have definite values, but the quantity $L_x^2 + L_y^2$ does. What is it? *Answer:* $[l(l + 1) - m_l^2]\hbar^2$.

The new quantum mechanics we have just described is much more complex, both conceptually and mathematically, than Newtonian mechanics. It deals with probabilities rather than certainties, and it predicts discrete rather than continuous behavior. However, quantum mechanics enables us to understand physical phenomena and to analyze physical problems for which classical mechanics is completely powerless.

Electron Spin

All atoms show shifts or splitting of their energy levels and spectral wavelengths, when placed in a magnetic field. This isn't surprising; electric charges moving in a magnetic field are acted upon by forces, as we discussed in Chapter 20. But some atoms show unexpected shifts even when *no* magnetic field is present. It was suggested in 1925 (initially in the context of the Bohr model) that perhaps the electron should be pictured as a spinning sphere of charge rather than an orbiting point charge, a concept called **electron spin.** If so, the electron has not only angular momentum associated with its orbital motion, but also additional angular momentum associated with rotation on its axis. The resulting additional magnetic-field interactions might cause the observed energy-level anomalies.

Here's an analogy: The earth travels in a nearly circular orbit around the sun, and at the same time it rotates on its axis. Each motion has its associated angular momentum, which we call the orbital and spin angular momentum, respectively. The total angular momentum of the system is the vector sum of the two.

Precise spectroscopic analysis, as well as a variety of other experimental evidence, has shown conclusively that the electron *does* have angular momentum that doesn't depend on its orbital motion, but is intrinsic to the particle itself. Like orbital angular momentum, spin angular momentum (usually denoted by $\vec{S}$) is found to be quantized. Suppose we have an apparatus that measures the component of $\vec{S}$ in a particular direction, such as the z axis. Denoting the z component of $\vec{S}$ by S_z, we find that the only possible values are

$$S_z = \pm\tfrac{1}{2}\hbar.$$

This relation is similar to Equation 29.4 for L_z, the z component of orbital angular momentum, but the magnitude of S_z is one-half of $\hbar$ instead of an integral

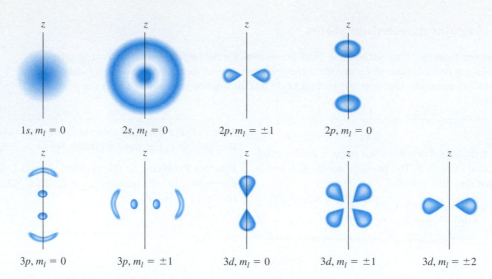

$1s, m_l = 0$ $2s, m_l = 0$ $2p, m_l = \pm1$ $2p, m_l = 0$

$3p, m_l = 0$ $3p, m_l = \pm1$ $3d, m_l = 0$ $3d, m_l = \pm1$ $3d, m_l = \pm2$

▲ **FIGURE 29.3** Cross sections of three-dimensional probability distributions for a few quantum states of the hydrogen atom.

multiple. Thus, an electron is often called a "spin-$\frac{1}{2}$ particle." The magnitude $S = |\vec{S}|$ of the spin angular momentum (in analogy to Equation 29.3) is

$$S = \sqrt{\tfrac{1}{2}\left(\tfrac{1}{2} + 1\right)}\hbar = \frac{\sqrt{3}}{2}\hbar.$$

In quantum mechanics, in which the Bohr orbits are superseded by wave functions, we can't really picture an electron as a solid object that spins. But if we visualize the electron wave functions as "clouds" surrounding the nucleus, then we can imagine many tiny arrows distributed throughout the cloud, all pointing in the same direction, either all $+z$ or all $-z$. But don't take this picture too seriously!

In any event, the concept of electron spin is well established by a variety of experimental evidence. To label completely the state of the electron in a hydrogen atom, we now need a *fourth* quantum number s, which we'll call the *spin quantum number,* to specify the electron spin orientation. If s can take the value $\frac{1}{2}$ or $-\frac{1}{2}$, then the z component of spin angular momentum is given by

$$S_z = s\hbar \quad (s = \pm\tfrac{1}{2}). \tag{29.5}$$

Thus, the possible states of the electron in the hydrogen atom are labeled with a set of four quantum numbers (n, l, m_l, s). The principal quantum number n determines the energy, and the other three quantum numbers specify the angular momentum and its component in a specified direction, usually taken to be the z axis.

Figure 29.3 shows cross-sectional representations of several hydrogen-atom wave functions.

Many-Electron Atoms

The hydrogen atom, with one electron and a nucleus consisting of a single proton, is the simplest of all atoms. All other electrically neutral atoms contain more than one electron. The analysis of atoms with more than one electron increases in complexity very rapidly. Each electron interacts not only with the positively charged nucleus, but also with all the other electrons. In principle, the motion of the electrons is still governed by the Schrödinger equation, but the mathematical problem of finding appropriate solutions of that equation is so complex that it has not been solved exactly even for the helium atom, which has a mere two electrons.

Various approximation schemes can be used to apply the Schrödinger equation to multielectron atoms. The most drastic is to ignore the interactions of the electrons with each other and assume that each electron moves under the influence only of the electric field of the nucleus, which is considered to be a point charge. A less drastic and more useful model is to think of all the electrons together as making up a charge cloud that is, on the average, spherically symmetric. In this model, each individual electron moves under the action of the total electric field due to the nucleus and this averaged-out electron cloud. It turns out that when an electron moves in an electric field produced by a spherically symmetric charge distribution, the angular momentum states given by the quantum numbers l, m_l, and s are exactly the same as those of the hydrogen atom.

This model is called the **central-field approximation;** it provides a useful starting point for the understanding of the structure of complex atoms. The energy levels now depend on both n and l; usually, for a given value of n, the levels increase in energy with increasing l. But individual levels can still be labeled by the set of four quantum numbers (n, l, m_l, s), with the same restrictions as before:

$$0 \le l \le n - 1, \qquad |m_l| \le l, \qquad \text{and} \qquad s = \pm\tfrac{1}{2}. \qquad (29.6)$$

The Exclusion Principle

To implement the central-field approximation, we need one additional principle: the **exclusion principle.** To understand the essential role this principle plays, we need to consider the lowest energy state, or **ground state,** of a multielectron atom. In the central-field model, which ignores interactions between electrons, there is a lowest energy state (corresponding roughly to the $n = 1$ state for the single electron in the hydrogen atom). We might expect that, in the ground state of a complex atom, *all* the electrons should be in this lowest state. If so, then the behavior of atoms with increasing numbers of electrons should show gradual changes in physical and chemical properties as the number of electrons in the atoms increases.

A variety of evidence shows conclusively that this is *not* what happens at all. For example, the elements fluorine, neon, and sodium respectively have 9, 10, and 11 electrons per atom. Fluorine is a halogen; it tends strongly to form compounds in which each atom acquires an extra electron. Sodium, an alkali metal, forms compounds in which it *loses* an electron, and neon is an inert gas (or noble gas) that ordinarily forms no compounds at all. These observations, and many others, show that, in the ground state of a complex atom, the electrons *cannot* all be in the lowest energy states.

The key to resolving this puzzle was discovered by the Swiss physicist Wolfgang Pauli in 1925, and it is accordingly named for its discoverer:

The Pauli exclusion principle

No two electrons in an atom can occupy the same quantum-mechanical state. Alternatively, no two electrons in an atom can have the same values of all four of their quantum numbers.

Different quantum states correspond to different spatial distributions, including different distances from the nucleus, and to different electron-spin orientations. Roughly speaking, in a complex atom there isn't enough room for all the electrons in the states nearest the nucleus. Some electrons are forced by the Pauli principle into states farther away, with higher energies.

TABLE 29.1 **Quantum states of electrons in the first four shells**

n	l	m_l	Spectroscopic notation	Number of states	Shell
1	0	0	$1s$	2	K
2	0	0	$2s$	2	
2	1	$-1, 0, 1$	$2p$	6	} 8 L
3	0	0	$3s$	2	
3	1	$-1, 0, 1$	$3p$	6	} 18 M
3	2	$-2, -1, 0, 1, 2$	$3d$	10	
4	0	0	$4s$	2	
4	1	$-1, 0, 1$	$4p$	6	
4	2	$-2, -1, 0, 1, 2$	$4d$	10	} 32 N
4	3	$-3, -2, -1, 0, 1, 2, 3$	$4f$	14	

We can now make a list of all the possible sets of quantum numbers, and thus of the possible states of electrons, in an atom. Such a list is given in Table 29.1, which also indicates two alternative notations. It is customary to designate the value of l by a letter, according to this scheme:

$$l = 0: s \text{ state}$$
$$l = 1: p \text{ state}$$
$$l = 2: d \text{ state}$$
$$l = 3: f \text{ state}$$
$$l = 4: g \text{ state}$$

This peculiar choice of letters originated in the early days of spectroscopy and has no fundamental significance. A state for which $n = 2$ and $l = 1$ is called a $2p$ state, and so on, as shown in Table 29.1, which also shows the relationship between values of n and the x-ray levels $(K, L, M, \cdots)$ that we'll describe in the next section. The $n = 1$ levels are designated as K, $n = 2$ levels as L, and so on.

EXAMPLE 29.2 **Quantum states for the hydrogen atom**

How many different $4f$ states does the hydrogen atom have? Make a list.

SOLUTION

SET UP AND SOLVE The "4" in $4f$ means $n = 4$, and the f means $l = 3$. The maximum magnitude of m_l is 3, so there are seven possible values of m_l: $m_l = -3, -2, -1, 0, 1, 2, 3$. For each of these, there are two possible values of the spin quantum number s: $\pm \frac{1}{2}$. The total number of states is 14.

REFLECT For any value of l, the number of possible values of m_l is $2l + 1$.

Practice Problem: How many $4d$ states does the hydrogen atom have? How many $4g$ states? *Answer:* 10; 0.

With some exceptions, the average distance of an electron from the nucleus increases with n. Thus, each value of n corresponds roughly to a region of space around the nucleus in the form of a spherical **shell.** Hence we speak of the K shell as the region occupied by the electrons in the $n = 1$ states (closest to the nucleus), the L shell as the region of the $n = 2$ states, and so on. States with the same n but different l, form *subshells,* such as the $3p$ subshell.

Each quantum state corresponds to a certain distribution of the electron cloud in space. Therefore, the exclusion principle says, in effect, "No more than two electrons (with opposite values of the spin quantum number s) can occupy the same region of space." The wave functions that describe electron distributions don't have sharp, definite boundaries, but the exclusion principle limits the degree of overlap of electron wave functions that is permitted. The maximum numbers of electrons in each shell and subshell are shown in Table 29.1.

29.2 Atomic Structure

We're now ready to use the exclusion principle, along with the electron energy states described in Section 29.1, to derive the most important features of the structure and chemical behavior of multielectron atoms, including the periodic table of the elements. The number of electrons in an atom in its normal (electrically neutral) state is called the **atomic number,** denoted by Z. The nucleus contains Z protons and some number of neutrons. The neutron has *no* charge; the proton and electron charges have the same magnitude, but opposite sign, so in the normal atom the total electric charge is zero. Because the electrons are attracted to the nucleus, the quantum states corresponding to regions nearest the nucleus have the lowest energies.

We can imagine constructing an atom by starting with a bare nucleus with Z protons and adding Z electrons, one by one. To obtain the ground state, we fill the lowest-energy states (those closest to the nucleus, with the smallest values of n and l) first, and we use successively higher states until all the electrons are in place. In filling up these states, we must give careful attention to the limitations imposed by the exclusion principle, as shown in Table 29.1. The chemical properties of an atom are determined principally by interactions involving the outermost electrons, so we particularly want to learn how those electrons are arranged.

Let's look at the ground-state electron configurations for the first few atoms (in order of increasing Z). For hydrogen, the ground state is $1s$; the single electron is in the state $n = 1$, $l = 0$, $m_l = 0$, and $s = \pm\frac{1}{2}$. In the helium atom $(Z = 2)$, both electrons are in $1s$ states, with opposite spin components; we denote this configuration as $1s^2$. For helium, the K shell $(n = 1)$ is completely filled, and all others are empty. Helium is an inert gas; it has no tendency to gain or lose an electron, and ordinarily it forms no compounds.

Lithium $(Z = 3)$ has three electrons; in the ground state, two are in $1s$ states (filling the $1s$ level), and one is in a $2s$ state. We denote this configuration as $1s^2 2s$. (The exclusion principle forbids all three electrons being in $1s$ states.) On the average, the $2s$ electron is considerably farther from the nucleus than the $1s$ electrons, as shown schematically in Figure 29.4. Thus, the net charge influencing the $2s$ electron is approximately $+e$, rather than $+3e$, as it would be without the two $1s$ electrons present. The $2s$ electron is loosely bound; only 5.4 eV is required to remove it, compared with 13.6 eV needed to ionize the hydrogen atom. In chemical behavior, lithium is an *alkali metal*. It forms ionic compounds in which each atom *loses* an electron, corresponding to a valence of $+1$.

Next comes beryllium $(Z = 4)$; its ground-state configuration is $1s^2 2s^2$, with a filled K shell and two electrons in the L shell. Beryllium is the first of the *alkaline-earth* elements, forming ionic compounds in which the valence of the atoms is $+2$.

Table 29.2 on the next page shows the ground-state electron configurations of the first 30 elements. The L shell can hold a total of eight electrons. At $Z = 10$, both the K and L shells are filled, and there are no electrons in the M shell. We expect this to be a particularly stable configuration with little tendency either to gain or to lose electrons. This element is neon, a noble gas with no known compounds. The next element after neon is sodium $(Z = 11)$, with filled K and L shells and one electron in the M shell. Its "filled-shell-plus-one-electron" configuration resembles that of lithium, and indeed, both are alkali metals (a group that also includes potassium, rubidium, cesium, and francium). The element before neon is fluorine, with $Z = 9$. It has a vacancy in the L shell and has an affinity for

On average, $2s$ electron is considerably farther from the nucleus than the $1s$ electrons. Therefore, it experiences a net nuclear charge of approximately $+e$ (rather than $+3e$).

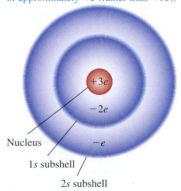

▲ **FIGURE 29.4** Schematic representation of the charge distribution in a lithium atom.

an extra electron to fill that shell. Fluorine forms ionic compounds in which it has a valence of -1. This behavior is characteristic of the halogens (fluorine, chlorine, bromine, iodine, and astatine), all of which have similar configurations.

▲ **Application It's a gas.** An interesting and sometimes colorful group of elements are the noble gases, located in the extreme right column of the periodic table. As with any group of elements in a column of the periodic table, they share similar chemical characteristics. The noble gases, including helium, neon, argon, and krypton, are unusual because all have completely filled electron shells or subshells. They have almost no tendency to form molecular bonds with other atoms and therefore exist as monatomic gases. When noble gases are confined in a thin glass tube and subjected to high voltage, electrons in the atoms are excited to higher energy levels. As the electrons return to the ground state, the atoms emit photons of visible light. Each element produces a characteristic color and, in various combinations, they can produce the wide variety of eye-catching colors that we see in "neon" lights today.

TABLE 29.2 Ground-state electron configurations of atoms

Element	Symbol	Atomic number (Z)	Electron configuration
Hydrogen	H	1	$1s$
Helium	He	2	$1s^2$
Lithium	Li	3	$1s^2 2s$
Beryllium	Be	4	$1s^2 2s^2$
Boron	B	5	$1s^2 2s^2 2p$
Carbon	C	6	$1s^2 2s^2 2p^2$
Nitrogen	N	7	$1s^2 2s^2 2p^3$
Oxygen	O	8	$1s^2 2s^2 2p^4$
Fluorine	F	9	$1s^2 2s^2 2p^5$
Neon	Ne	10	$1s^2 2s^2 2p^6$
Sodium	Na	11	$1s^2 2s^2 2p^6 3s$
Magnesium	Mg	12	$1s^2 2s^2 2p^6 3s^2$
Aluminum	Al	13	$1s^2 2s^2 2p^6 3s^2 3p$
Silicon	Si	14	$1s^2 2s^2 2p^6 3s^2 3p^2$
Phosphorus	P	15	$1s^2 2s^2 2p^6 3s^2 3p^3$
Sulfur	S	16	$1s^2 2s^2 2p^6 3s^2 3p^4$
Chlorine	Cl	17	$1s^2 2s^2 2p^6 3s^2 3p^5$
Argon	Ar	18	$1s^2 2s^2 2p^6 3s^2 3p^6$
Potassium	K	19	$1s^2 2s^2 2p^6 3s^2 3p^6 4s$
Calcium	Ca	20	$1s^2 2s^2 2p^6 3s^2 3p^6 4s^2$
Scandium	Sc	21	$1s^2 2s^2 2p^6 3s^2 3p^6 3d 4s^2$
Titanium	Ti	22	$1s^2 2s^2 2p^6 3s^2 3p^6 3d^2 4s^2$
Vanadium	V	23	$1s^2 2s^2 2p^6 3s^2 3p^6 3d^3 4s^2$
Chromium	Cr	24	$1s^2 2s^2 2p^6 3s^2 3p^6 3d^5 4s$
Manganese	Mn	25	$1s^2 2s^2 2p^6 3s^2 3p^6 3d^5 4s^2$
Iron	Fe	26	$1s^2 2s^2 2p^6 3s^2 3p^6 3d^6 4s^2$
Cobalt	Co	27	$1s^2 2s^2 2p^6 3s^2 3p^6 3d^7 4s^2$
Nickel	Ni	28	$1s^2 2s^2 2p^6 3s^2 3p^6 3d^8 4s^2$
Copper	Cu	29	$1s^2 2s^2 2p^6 3s^2 3p^6 3d^{10} 4s$
Zinc	Zn	30	$1s^2 2s^2 2p^6 3s^2 3p^6 3d^{10} 4s^2$

Conceptual Analysis 29.1

Spectroscopic notation

The atom having the electron configuration $1s^2$, $2s^2$, $2p^6$ has (there may be more than one correct choice)

A. electrons in states $n = 1, 2$.
B. electrons in states $n = 2, 6$.
C. 6 orbital electrons.
D. 10 orbital electrons.

SOLUTION In spectroscopic notation, the numbers preceding s, p, d, f, etc., indicate the quantum number n. Thus, $1s^2$ corresponds to electrons in state $n = 1$, and $2s^2$ and $2p^6$ correspond to $n = 2$ (choice A). Furthermore, the *exponent* in the notation $1s^2$ indicates that there are 2 orbital electrons in an s state, $2s^2$ indicates another 2 electrons in an s state, and $2p^6$ indicates 6 more electrons in $2l$ states. Therefore, there are $2 + 2 + 6 = 10$ orbital electrons, and choice D is also correct.

The Periodic Table of the Elements

Proceeding down the list, we can understand many of the regularities in chemical behavior displayed by the **periodic table of the elements** (Appendix D) on the basis of electron configurations. A slight complication occurs with the M and N

shells because the $3d$ and $4s$ subshells ($n = 3$, $l = 2$ and $n = 4$, $l = 0$, respectively) overlap in energy. Argon ($Z = 18$) has all the $1s$, $2s$, $2p$, $3s$, and $3p$ states filled, but in potassium ($Z = 19$) the additional electron goes into a $4s$ level rather than a $3d$ level (because the $4s$ level has slightly lower energy).

The next several elements have one or two electrons in the $4s$ states and increasing numbers in the $3d$ states. These elements are all metals with rather similar chemical and physical properties; they form the first *transition series,* starting with scandium ($Z = 21$) and ending with zinc ($Z = 30$), for which all the $3d$ and $4s$ levels are filled. Something similar happens with $Z = 57$ through $Z = 71$, which have two electrons in the $6s$ levels but only partially filled $4f$ and $5d$ levels. These are the rare-earth elements, with similar physical and chemical properties. Yet another such series, called the *actinide series,* starts with $Z = 91$.

The similarity of the elements in each group (vertical column) of the periodic table reflects a corresponding similarity in outer electron configuration. All the noble gases (helium, neon, argon, krypton, xenon, and radon) have filled-shell or filled-subshell configurations. All the alkali metals (lithium, sodium, potassium, rubidium, cesium, and francium) have "filled-shell-or-subshell-plus-one" configurations. All the alkaline-earth metals (beryllium, magnesium, calcium, strontium, barium, and radium) have "filled-shell-or-subshell-plus-two" configurations, and all the halogens (fluorine, chlorine, bromine, iodine, and astatine) have "filled-shell-or-subshell-minus-one" configurations.

PROBLEM-SOLVING STRATEGY 29.1 **Atomic structure**

1. Be sure you know how to count the energy states for electrons in the central-field approximation. There are four quantum numbers: n, l, m_l, and s; n is always positive, l can be zero or positive, m_l can be zero, positive, or negative, and $s = \pm\frac{1}{2}$. Be sure you know how to count the number of states in each shell and subshell; study Tables 29.1 and 29.2 carefully.

2. As in Chapter 28, familiarizing yourself with some numerical magnitudes is useful. Here are two examples to work out: The magnitude of the electrical potential energy of a proton and an electron (or two electrons) 0.10 nm apart (typical of atomic dimensions) is 14.4 eV, or 2.31×10^{-18} J. Think of other examples like this, and work them out to help you know what kinds of magnitudes to expect in atomic physics.

3. As in Chapter 28, you'll need to use both electronvolts and joules. The conversion $1\ \text{eV} = 1.602 \times 10^{-19}$ J and Planck's constant in eV, $h = 4.136 \times 10^{-15}$ eV $\cdot$ s, are useful quantities. Nanometers are convenient for atomic and molecular dimensions, but don't forget to convert to meters in calculations.

EXAMPLE 29.3 **Periodic table of the elements**

The element fluorine has atomic number $Z = 9$ and ground-state electron configuration $1s^2 2s^2 2p^5$. What element of next-larger Z has chemical properties that are similar to those of fluorine? Explain the reasoning behind your answer by giving the ground-state electron configuration for that element.

SOLUTION

SET UP Each s subshell ($l = 0$ levels) can accommodate two electrons, and each p subshell ($l = 1$ levels) can hold six elec-

trons. Subshells fill in the order $1s$, $2s$, $2p$, $3s$, $3p$, $4s$, etc. Fluorine is one electron short of having a filled $2p$ subshell.

Continued

SOLVE The next element with similar properties is one electron short of having a filled $3p$ subshell. The $1s$, $2s$, $2p$, $3s$, and $3p$ levels together can accommodate $2 + 2 + 6 + 2 + 6 = 18$ electrons. With one $3p$ vacancy, the number of electrons in the atom in question is 17. The element we seek is chlorine ($Z = 17$).

REFLECT Fluorine and chlorine are in adjacent rows of the same column of the periodic table, and both are halogens.

Practice Problem: The element lithium has atomic number $Z = 3$ and ground-state electron configuration $1s^2 2s$. What element of next-larger Z has chemical properties that are similar to those of lithium? *Answer:* Sodium ($Z = 11$).

The periodic table of the elements was first formulated by the Russian chemist Dmitri Ivanovich Mendeleev in about 1870 as a purely *empirical* structure based on observations of the similarities in behavior of groups of chemical elements. The understanding of this behavior on the basis of atomic structure came nearly 50 years later.

X-Ray Energy Levels

The outer electrons of an atom are responsible for optical spectra. Their excited states are usually only a few electronvolts above the energy of the ground state. In transitions from excited states to the ground state, they usually emit photons in or near the visible region, with photon energies of about 2 to 3 eV. There are also **x-ray energy levels,** corresponding to vacancies in the *inner* shells of a complex atom. We mentioned these levels briefly in Section 28.5. In an x-ray tube, an electron may strike the target with enough energy to knock an electron out of an inner shell of the target atom. These inner electrons are much closer to the nucleus than the electrons in the outer shells; they are therefore much more tightly bound, and hundreds or thousands of electronvolts may be required to remove them.

Suppose an electron is knocked out of the K shell. The atom is then left with a vacancy; we'll call this state a K *x-ray energy level.* This vacancy can subsequently be filled by an electron falling in from one of the outer shells, such as the L, M, N, $\cdots$ shell. The transition is accompanied by a *decrease* in the energy of the atom (because *less* energy would be needed to remove an electron from the L, M, N, $\cdots$, shell), and an x-ray photon is emitted with energy equal to this decrease. Each state (for any specific element) has a definite energy, so the emitted x rays have definite wavelengths, and the spectrum is a *line spectrum.*

If the outermost electrons are in the N shell, the x-ray spectrum has three lines, resulting from transitions in which the vacancy in the K shell is filled by an L, M, or N electron. This series of lines is called a K series. Figure 29.5 shows the K series for tungsten, molybdenum, and copper.

There are other series of x-ray lines, called the L, M, and N series, produced by the ejection of an electron from the L, M, or N shell rather than the K shell. Electrons in these outer shells are farther away from the nucleus and are not held as tightly as those in the K shell. Their removal thus requires less energy, and the x-ray photons emitted when these vacancies are filled have lower energy than those in the K series have.

The three lines in each series are called the K_α, K_β, and K_γ lines. The K_α line is produced by the transition of an L electron to the vacancy in the K shell, the K_β line by an M electron, and the K_γ line by an N electron.

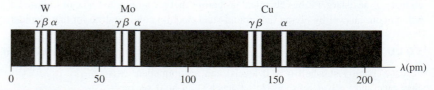

▲ **FIGURE 29.5** K series for tungsten (W), molybdenum (Mo), and copper (Cu).

X-ray line emission

Consider x-ray line emission from an iron (Fe) atom. Which of the following statements is correct?

A. The energy of x-ray photons is less than the energy of visible-light photons.

B. X rays are emitted when electrons make a transition from higher energy shells to the K shell.

C. A vacancy or hole is left behind in the K shell after the x ray is emitted.

SOLUTION For line emission of x rays, an electron must first be knocked out from one of the lowest-energy shells, such as the K shell. Then an electron from a higher energy shell drops down to fill this vacancy, emitting an x ray in the process. The correct answer is B.

EXAMPLE 29.4 Counting electrons in shells

(a) For silicon (Si, $Z = 14$) and germanium (Ge, $Z = 32$), make lists of the numbers of electrons in each subshell $(1s, 2s, 2p, \cdots)$. Use the allowed values of the quantum numbers along with the exclusion principle; do not refer to Table 29.2. Explain why these two elements have similar chemical behavior.

SOLUTION

SET UP Each s subshell ($l = 0$ levels) can accommodate two electrons $(l = 0,\ s = \pm\tfrac{1}{2})$. Each p subshell ($l = 1$ levels) can hold six) $(l = 1,\ m = -1, 0, +1,\ s = \pm\tfrac{1}{2})$, and so on.

SOLVE (a) For silicon, the $(1s)$, $(2s)$, and $(2p)$ subshells together contain $2 + 2 + 6 = 10$ electrons. The remaining 4 of the total of 14 must be in the $(3s)$ and $(3p)$ subshells, with two electrons in each. For germanium, the $(1s)$, $(2s)$, $(2p)$, $(3s)$, $(3p)$, and $(3d)$ subshells together contain $2 + 2 + 6 + 2 +$ $6 + 10 = 28$ electrons. The remaining 4 electrons go in the $(4s)$ and $(4p)$ subshells, with 2 electrons in each. Both elements have filled shells plus 4 extra electrons in outer, unfilled shells.

REFLECT The element carbon $(Z = 6)$ also has two s and two p electrons in its outermost shell. Its chemical behavior is similar to that of silicon and germanium.

Practice Problem: Carry out a similar analysis for the elements having $Z = 12$ and $Z = 20$, and predict the similarity in chemical behavior of these two elements.

EXAMPLE 29.5 Potential energy of two particles

Verify the statement in the Problem-Solving Strategy that the magnitude of the potential energy of an electron and a proton separated by 0.100 nm is 14.4 eV.

SOLUTION

SET UP The potential energy of two point charges q_1 and q_2 separated by a distance r is

$$U = \frac{1}{4\pi\epsilon_0} \frac{q_1 q_2}{r},$$

a relationship we derived in Section 18.1 (Equation 18.8).

SOLVE The magnitude of the charge of each particle is

$$q_1 = q_2 = 1.60 \times 10^{-19}\ \text{C}.$$

When the distance r between them is

$$r = 0.100\ \text{nm} = 1.00 \times 10^{-10}\ \text{m},$$

the potential energy is

$$U = \frac{1}{4\pi\epsilon_0} \frac{|q_1 q_2|}{r} = (8.99 \times 10^9\ \text{N} \cdot \text{m}^2/\text{C}^2) \frac{(1.60 \times 10^{-19}\ \text{C})^2}{1.00 \times 10^{-10}\ \text{m}}$$
$$= 2.30 \times 10^{-18}\ \text{J} = 14.4\ \text{eV}.$$

REFLECT This result is the same order of magnitude as the ionization energy of hydrogen (the energy required to remove the electron completely—i.e., 13.6 eV).

Practice Problem: If one of the two electrons in a helium atom has been removed, what is the potential energy of the remaining electron at a distance of 0.100 nm from the nucleus? Express your result both in joules and in eV. *Answer:* $-28.8\ \text{eV} = -4.60 \times 10^{-18}\ \text{J}$.

29.3 Diatomic Molecules

The study of electron configurations in atoms provides valuable insight into the nature of chemical bonds—the interactions that hold atoms together to form stable structures such as molecules and solids. There are several types of chemical bonds, including ionic, covalent, van der Waals, and hydrogen bonds.

PhET: Double Wells and Covalent Bonds
PhET: The Greenhouse Effect

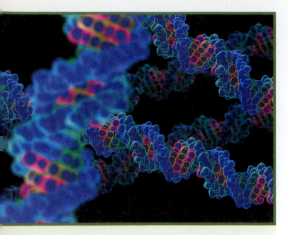

▲ **BIO** Application **Bonding for life.** The structure of the DNA double helix is based on several of the types of chemical bonds we have discussed. The atoms in each of the two strands are held together by covalent bonds resulting from the sharing of electron pairs between adjacent atoms. Each strand also contains numerous negatively charged phosphate groups that are involved in ionic bonding via electrostatic attractions to positive ions that help stabilize the structure. Weaker hydrogen bonds hold the two DNA strands together, and can be easily broken to allow the DNA strands to be copied during cellular reproduction. And, in processes involving DNA–protein interactions, the DNA typically fits into a precisely shaped pocket of the protein; it is held there in part by close-range interactions involving van der Waals bonds.

Ionic Bond

The **ionic bond,** also called the electrovalent or heteropolar bond, is an interaction between two ionized atoms.. The most familiar example is sodium chloride (NaCl), in which the sodium atom gives its one $3s$ electron to the chlorine atom, filling the vacancy in the $3p$ subshell of chlorine.

Let's look at the energy balance in this transaction. Removing the $3s$ electron from the sodium atom requires 5.1 eV of energy; this is called the **ionization energy,** or ionization potential, of sodium. Chlorine has an **electron affinity** of 3.6 eV. That is, the neutral chlorine atom can attract an extra electron into the vacancy in its $2p$ subshell, where the electron is attracted to the nucleus, with an attractive potential energy of magnitude 3.6 eV. Thus, creating the separated Na^+ and Cl^- ions requires a net expenditure of only 5.1 eV − 3.6 eV = 1.5 eV. When the two mutually attracting ions come together, the magnitude of their negative potential energy is determined by the closeness to which they can approach each other. This in turn is limited by the exclusion principle, which forbids extensive overlap of the electron clouds of the two ions.

The minimum potential energy for NaCl turns out to be −5.7 eV at a separation (between centers) of 0.24 nm. At distances less than this, the interaction becomes repulsive. The net energy given up by the system in creating the ions and letting them come together to the equilibrium separation of 0.24 nm is −5.7 eV + 1.5 eV = −4.2 eV, and this is the binding energy of the molecule. That is, 4.2 eV of energy is needed to dissociate the molecule into separate neutral atoms.

Ionic bonds are interactions between charge distributions that are nearly spherically symmetric. Their electrical interaction is similar to that of two point charges, so they are not highly directional. They can involve more than one electron per atom. The alkaline-earth elements form ionic compounds in which an atom loses two electrons; an example is $Mg^{++}(Cl^-)_2$. Loss of more than two electrons is relatively rare; instead, a different kind of bond comes into operation.

Covalent Bond

The **covalent,** or homopolar, bond is characterized by a more nearly symmetric participation of the two atoms, in contrast to the asymmetry of the electron-transfer process of the ionic bond. The simplest example of a covalent bond is that in the hydrogen molecule, a structure containing two protons and two electrons. This bond is shown schematically in Figure 29.6. As the separate atoms come together, the electron wave functions are distorted from the configurations of isolated atoms and become more concentrated in the region between the two protons. The net attraction of the electrons for each proton more than balances the repulsion of the protons and that of the two electrons. The energy of the covalent bond in the hydrogen molecule H_2 is −4.48 eV.

The exclusion principle permits two electrons to occupy the same region of space only when they have opposite spins. When the spins are parallel, the state that would be most favorable from energy considerations (i.e., both electrons in the region between atoms) is forbidden by the exclusion principle. Thus, opposite spins are an essential requirement for an electron-pair bond, and no more than two electrons can participate in such a bond.

However, an atom with several electrons in its outermost shell can form several electron-pair bonds. The bonding of carbon and hydrogen atoms, of central importance in organic chemistry, is an example. In the methane molecule (CH_4), the carbon atom is at the center of a regular tetrahedron, with a hydrogen atom at

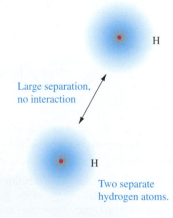

Large separation, no interaction

H

Two separate hydrogen atoms.

The covalent bond; the charge clouds for the two electrons with opposite spin s are concentrated in the region between the nuclei (protons).

H_2

▲ **FIGURE 29.6** Formation of covalent bond in hydrogen.

each corner. The carbon atom has four electrons in its L shell $(n = 2)$, and one of these electrons forms a covalent bond with each of the four hydrogen atoms, as shown in Figure 29.7. Similar patterns occur in more complex organic molecules.

Ionic and covalent bonds represent two extremes in the nature of **molecular bonds,** but there is no sharp division between the two types. Often, there is a partial transfer of one or more electrons (corresponding to a greater or smaller distortion of the electron wave functions) from one atom to another. As a result, many molecules having dissimilar atoms have electric dipole moments (a preponderance of positive charge at one end and of negative charge at the other). Such molecules are called **polar molecules.** Water molecules have exceptionally large electric dipole moments that are responsible for the exceptionally large dielectric constant of liquid water.

Weak Bonds

Ionic and covalent bonds, with typical bond energies of 1 to 5 eV, are considered strong bonds. There are also two types of much *weaker* bonds with typical energies of 0.5 eV or less. One of these, the **van der Waals bond,** is an interaction between the electric dipole moments of two atoms or molecules. The bonding of water molecules in the liquid and solid states results partly from dipole–dipole interactions. The interaction potential energy drops off very quickly with distance r between molecules, usually in proportion to $1/r^6$.

Even when an atom or molecule has no permanent dipole moment, fluctuating charge distributions can lead to fluctuating dipole moments that, in turn, can induce dipole moments in neighboring structures. The resulting dipole–dipole interaction can be attractive and can lead to weak bonding of atoms or molecules. The low-temperature liquefaction and solidification of the inert gases and of such molecules as H_2, O_2, and N_2 is due to induced-dipole van der Waals interactions. Not much thermal agitation energy is needed to break these weak bonds, so such substances usually exist in the liquid and solid states only at very low temperatures.

Another type of weak bond, the **hydrogen bond,** is analogous to the covalent bond, in which an electron pair binds two positively charged structures. In the hydrogen bond, a proton (H^+ ion) gets between two atoms, polarizing them and attracting them by means of the induced dipoles. This bond is unique to hydrogen-containing compounds, because only hydrogen has a singly ionized state with no remaining electron cloud. (The hydrogen ion is a bare proton, much smaller than any other singly ionized atom.) The energy required to break a hydrogen bond is small, usually less than 0.5 eV. Hydrogen bonding plays an essential role in many organic molecules. For example, it provides the cross-linking of polymer chains such as polyethylene and the cross-link bonding between the two strands of the famous double-helix DNA molecule.

All these types of bonds play roles in the structure of solids as well as of molecules. Indeed, a solid is in many respects a giant molecule. Still another type of bonding, the *metallic bond,* comes into play in the structure of metallic solids. We'll return to this subject in the next section.

Molecular Spectra

All molecules have quantized energy levels associated with the internal motion of their electrons. In addition, an entire molecule can rotate. The simplest example, a diatomic molecule, can be thought of as a rigid dumbbell (Figure 29.8a) that can rotate about an axis through its center of mass. The angular momentum

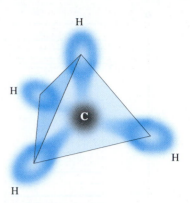

▲ **FIGURE 29.7** Schematic diagram of the methane molecule. The carbon atom is at the center of a regular tetrahedron and forms four covalent bonds with the hydrogen atoms at the corners. Each covalent bond includes two electrons with opposite spins, forming a charge cloud that is concentrated between the carbon atom and a hydrogen atom.

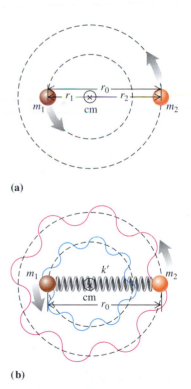

▲ **FIGURE 29.8** A model of (a) a rotating diatomic molecule and (b) vibration and rotation. Vibrational energies are typically on the order of 0.1 eV, much smaller than those of atomic spectra, but usually much larger than energies of the rotational levels.

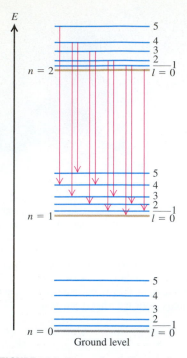

▲ FIGURE 29.9 Energy-level diagram for vibrational and rotational energy levels of a diatomic molecule. For each vibrational level (n), there is a series of more closely spaced rotational levels (l). Several transitions corresponding to a single band in a band spectrum are shown.

associated with this rotation is quantized, just as the angular momentum of an electron in an atom is quantized. Thus, the kinetic energy associated with the rotational motion of a molecule is also quantized: The molecule has a series of rotational energy levels. Because a molecule is always much more massive than an individual electron, these levels are much more closely spaced (typically on the order of 10^{-3} eV) than are the usual atomic energy levels (typically a few eV). The corresponding photon energies associated with transitions among these levels are in the far-infrared region of the spectrum.

In addition, a molecule is never completely rigid; the bonds can stretch and flex. Hence, a more realistic model of a diatomic molecule (Figure 29.8b) consists of two masses connected by a spring rather than a rigid rod. Then, in addition to rotating, the molecule can vibrate, with the atoms moving along the line joining them. The additional energy (both kinetic and potential) associated with vibrational motion is quantized. A typical scheme of vibrational and rotational levels is shown in Figure 29.9. Transitions among these levels lead to complex infrared spectra with *bands* of spectral lines (Figure 29.10). All molecules can also have excited states of their *electrons,* in addition to the rotational and vibrational states we have described. In general, these lie at much higher energies than the rotational and vibrational states.

Infrared spectroscopy has proved to be an extremely valuable analytical tool. It provides information about the strength and rigidity of molecular bonds and the structure of complex molecules. Also, because every molecule (like every atom) has its unique characteristic spectrum, infrared spectroscopy can be used to identify unknown compounds.

EXAMPLE 29.6 Wavelengths for atomic and molecular transitions

The separation between adjacent energy levels is typically a few eV for atomic energy levels, on the order of 0.1 eV for vibrational levels, and on the order of 10^{-3} eV for rotational levels. Find the wavelength of the photon emitted during a transition in which the energy of the molecule decreases by 5.00 eV, by 0.500 eV, and by 5.00×10^{-3} eV. In each case, in what region of the electromagnetic spectrum does the photon lie?

SOLUTION

SET UP The transition energy for the molecule equals the energy of the photon. The photon wavelength λ is related to its energy E by Equation 28.2:

$$E = hf = \frac{hc}{\lambda}, \quad \text{or} \quad \lambda = \frac{hc}{E}.$$

The energies are given in electronvolts, so we use the value of Planck's constant h in electronvolt seconds: $h = 4.136 \times 10^{-15}$ eV · s.

SOLVE For a transition energy of 5.0 eV,

$$\lambda = \frac{hc}{E} = \frac{(4.136 \times 10^{-15}\ \text{eV·s})(3.00 \times 10^{8}\ \text{m/s})}{5.00\ \text{eV}}$$
$$= 2.48 \times 10^{-7}\ \text{m} = 248\ \text{nm}.$$

This wavelength is in the ultraviolet region of the spectrum. (The visible spectrum is from about 400 to 700 nm.)

For a photon energy of 0.500 eV (smaller by a factor of 10), the wavelength is 10 times as great, 2480 nm = 2.48 μm, in the infrared region. For $E = 5.00 \times 10^{-3}$ eV, $\lambda = 0.248$ mm, in the microwave region.

REFLECT These photon energies (and corresponding wavelengths) range over a factor of a thousand, yet they represent only a small segment of the entire electromagnetic spectrum. Compared with photons of visible light, x-ray energies are typically higher by a factor of 10^{4} to 10^{6}, gamma-ray energies are higher still, and radio and TV photon energies are usually smaller by a factor on the order of 10^{6} to 10^{8}.

Practice Problem: What is the smallest energy of a transition that produces a photon in the visible region of the spectrum? *Answer:* 1.77 eV.

▲ **FIGURE 29.10** A typical molecular band spectrum.

29.4 Structure and Properties of Solids

The term *condensed matter* includes both solids and liquids. In both states, the interactions between atoms or molecules are strong enough to give the material a definite volume that changes relatively little with applied stress. The distances between centers of adjacent atoms or molecules in condensed matter are of the same order of magnitude as the sizes of the atoms or molecules themselves, typically 0.1 to 0.5 nm.

A crystalline solid is characterized by **long-range order,** a recurring pattern of atomic positions that extends over many atoms. This pattern is called the **crystal structure** or the **lattice structure** of the solid. Four examples of crystal lattices are shown in Figure 29.11. Most liquids have only **short-range order** (correlations between neighboring atoms or molecules) but not long-range order. There are also amorphous (noncrystalline) solids, such as glass. Conversely, some liquids show long-range order, such as the organic compounds used in liquid crystal digital display devices.

Ionic and Covalent Crystals

In some cases, the forces responsible for the regular arrangement of atoms in a crystal are the same as those involved in molecular bonds. Ionic and covalent molecular bonds are found in ionic and covalent crystals, respectively. The most familiar **ionic crystals** are the alkali halides, such as ordinary salt (NaCl). The positive sodium ions and the negative chlorine ions occupy alternate positions in a cubic crystal lattice, as shown in Figure 29.12. The forces are the familiar Coulomb's-law forces between charged particles. These forces have no preferred direction, and the arrangement in which the material crystallizes is determined by the relative size of the two ions.

An example of a **covalent crystal** is the diamond structure, found in the diamond form of carbon and also in silicon, germanium, and tin (Figure 29.13). All these elements are in Group IV of the periodic table, with four electrons in the outermost shell. In the diamond structure, each atom is situated at the center of a regular tetrahedron, with four nearest-neighbor atoms at the corners. The centrally located atom forms a covalent bond with each of these nearest-neighbor atoms. These bonds are strongly directional because of the asymmetrical electron distributions, and the result is a tetrahedral structure.

A third type of crystal, less directly related to the chemical bond than are ionic or covalent crystals, is the **metallic crystal.** In this structure, the outermost electrons are not localized at individual atomic lattice sites, but are detached from their parent atoms and are free to move through the crystal. The corresponding charge clouds (and their associated wave functions) extend over many atoms. Thus, we can picture a metallic crystal roughly as an array of positive ions (atoms from which one or more electrons have been removed) immersed in a sea of electrons whose attraction for the positive ions holds the crystal together (Figure 29.14). This "sea" has many of the properties of a gas, and indeed we speak of the electron-gas model of metallic solids.

In a metallic crystal, the atoms would form shared-electron bonds if they had enough valence electrons, but they don't. Instead, electrons are shared among

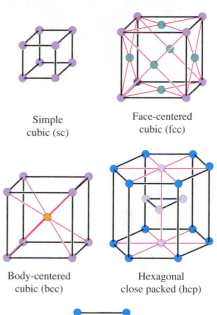

Simple cubic (sc)

Face-centered cubic (fcc)

Body-centered cubic (bcc)

Hexagonal close packed (hcp)

Top view, hexagonal close packed

▲ **FIGURE 29.11** Portions of some common types of crystal lattices.

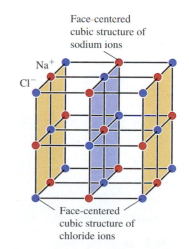

Face-centered cubic structure of sodium ions

Na⁺

Cl⁻

Face-centered cubic structure of chloride ions

▲ **FIGURE 29.12** Representation of part of the sodium chloride crystal structure. The distances between ions are exaggerated.

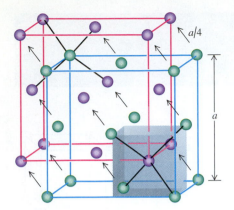

▲ FIGURE 29.13 The diamond structure, shown as two interpenetrating face-centered cubic structures, with distances between atoms exaggerated.

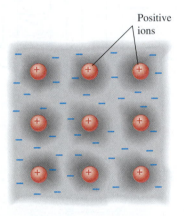

▲ FIGURE 29.14 In a metallic solid, one or more electrons are detached from each atom and are free to wander around the crystal, forming an "electron gas." The wave functions for these electrons extend over many atoms. The positive ions vibrate around fixed locations in the crystal.

The irregularity is seen most easily by viewing the figure from various directions at a grazing angle with the page.

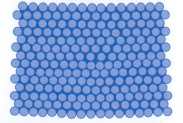

▲ FIGURE 29.15 An edge dislocation in two dimensions. In three dimensions, an edge dislocation would look like an extra plane of atoms slipped partway into the crystal.

many atoms. This bonding is not strongly directional. The shape of the crystal structure is determined primarily by considerations of **close packing**—that is, the maximum number of atoms that can fit into a given volume. The two most common metallic crystal structures—the face-centered cubic structure and the hexagonal close-packed structure—are shown in Figure 29.11. In each of these structures, each atom has 12 nearest neighbors.

In a perfect crystal, the repeating crystal structure extends uninterrupted throughout the entire material. Real crystals show a variety of departures from this idealized structure. Materials are often polycrystalline, composed of many small perfect single crystals bonded together at grain boundaries. Within a single crystal, interstitial atoms may occur in places where they do not belong, and there may be vacancies—lattice sites that should be occupied by an atom but aren't. An imperfection of particular interest in semiconductors, which we'll discuss in Section 29.6, is the *impurity atom*, a foreign atom (e.g., arsenic in a silicon crystal) substituting for the usual occupant of a lattice site.

A more complex kind of imperfection is the *dislocation*, shown schematically in Figure 29.15, in which one plane of atoms slips in relation to another. The mechanical properties of metallic crystals are influenced strongly by the presence of dislocations. The ductility and malleability of some metals depend on the presence of dislocations that move through the crystal during plastic deformations.

Resistivity

We can understand many macroscopic properties of solids on the basis of the microscopic structure of materials. Electrical properties are of particular interest because of their dominant role in present-day technology. The electrical **resistivity** of a crystalline solid material is determined by the amount of freedom the electrons have to move within the crystal lattice. In a metallic crystal, the valence electrons are not bound to individual lattice sites, but are free to move through the crystal. Metals are usually good conductors.

In a covalent crystal, the valence electrons are tied up in the bonds responsible for the crystal structure and are therefore not free to move. There are no mobile charges available for conduction, and such materials are usually insulators. Similarly, an ionic crystal such as NaCl has no charges that are free to move, and solid NaCl is an insulator. However, when salt is melted, the ions are no longer locked to their individual lattice sites, but are free to move, and molten NaCl is a good conductor. There are, of course, no perfect conductors (except for superconductors at low temperatures) or perfect insulators, but the resistivities of good insulators are greater than those of good conductors by an enormous factor, on the order of at least 10^{15}.

In addition, the resistivities of all materials depend on temperature; in general, the large resistivity of an insulator decreases with temperature, but that of a good conductor usually increases at increased temperatures. Two competing effects are responsible for this difference. In metals, the number of electrons available for conduction is nearly independent of temperature, and the resistivity is determined by the frequency of collisions between electrons and the positively charged ion cores in the lattice. As the temperature increases, the amplitude of vibration of the ion cores increases, and the electrons collide more frequently with the ion cores. This effect causes the resistivities of most metals to increase with temperature.

In insulators, the small amount of conduction that does take place is due to electrons that have gained enough energy from thermal motion of the lattice to break away from their "home" atoms and wander through the lattice. The number of electrons able to gain this much energy is very strongly dependent on temperature; a twofold increase in the number of mobile electrons for a 10 C° rise in temperature is typical. Partially offsetting this energy gain is the increased frequency of collisions at higher temperatures, as with metals, but the increased number of carriers is a far larger effect. Resistivities of insulators invariably decrease rapidly (i.e., insulators become better conductors) as the temperature increases.

Electrical resistivity (or conductivity) is closely related to thermal conductivity, which involves the transport of microscopic mechanical energy rather than electric charge. Wave motion associated with vibrations of the crystal lattice is one mechanism for energy transfer. In metals, the mobile electrons also carry kinetic energy from one region to another. This effect turns out to be much larger than that of the lattice vibrations. As a result, metals are usually much better thermal conductors than are nonmetals, which have few free electrons. Good electrical conductors are nearly always also good thermal conductors.

29.5 Energy Bands

The concept of **energy bands** in solids offers us additional insight into several properties of solids. To introduce the idea, suppose we have a large number N of identical atoms, far enough apart that their interactions are negligible. Every atom has the same energy-level diagram. We can draw an energy-level diagram for the entire system. It looks just like the diagram for a single atom, but the exclusion principle, applied to the entire system, permits each state to be occupied by N electrons instead of just one.

Now we begin to push the atoms closer together. Because of the electrical interactions and the exclusion principle, the wave functions—especially those of the outer, or *valence,* electrons—begin to distort. The energy levels also shift somewhat; some move upward and some downward, depending on the environment of each individual atom. Thus, each valence electron state for the system, formerly a sharp energy level that could accommodate N electrons, splits into a band containing N closely spaced levels, as shown in Figure 29.16. Ordinarily, N is very large, on the order of Avogadro's number (10^{23}), so we can think of the levels as forming a continuous distribution of energies within a band. Between adjacent energy bands are gaps, or forbidden regions, where there are *no* allowed energy levels. The width of a gap is called a *band gap,* denoted by E_g. The inner electrons in an atom are affected much less by nearby atoms than the valence electrons are, and their energy levels remain relatively sharp.

What does all this have to do with electrical conductivity? In insulators and semiconductors, the valence electrons fill completely the highest occupied band, called the **valence band.** The next-higher band, called the **conduction band,** is completely empty. The **energy gap** E_g separating the two may be on the order of 1 to 5 eV. This situation is shown in Figure 29.17a. The electrons in the valence

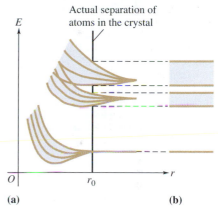

FIGURE 29.16 Origin of energy bands in a solid. (a) As the distance r between atoms decreases, the energy levels spread into bands. The vertical line at r_0 shows the actual atomic spacing in the crystal. (b) Symbolic representation of energy bands.

In an insulator at absolute zero, there are no electrons in the conduction band.

A semiconductor has the same band structure as an insulator but a smaller gap between the valence and conduction bands.

A conductor has a partially filled conduction band.

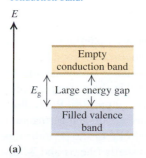

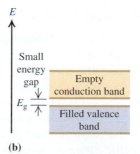

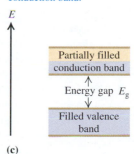

FIGURE 29.17 Three types of energy-band structures.

▲ **Application Packing them in.** This beautiful diamond is made of pure carbon and is the hardest natural substance known. The pile of graphite, a soft mineral used in powdered form as a dry lubricant, is also pure carbon. How is this possible? It's all in the packing—how the individual carbon atoms are bonded together in space. In a diamond, each atom is covalently bonded to four other atoms in a regular tetrahedral arrangement, giving diamond its extreme toughness. But in graphite, each carbon atom is tightly bound only to three others, forming a series of planar hexagonal sheets. Although the bonds within the sheets are stronger than those in diamond, the sheets themselves are held together only weakly. Therefore, individual sheets can easily slide past each other, accounting for the slippery feel of pure graphite and its use in pencils and lubricants.

At absolute zero, a completely filled valence band is separated by a narrow energy gap E_g of 1 eV or less from a completely empty conduction band. At ordinary temperatures, a number of electrons are excited to the conduction band.

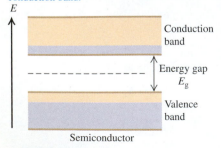

▲ **FIGURE 29.18** Band structure of a semiconductor.

band are not free to move in response to an applied electric field; to move, an electron would have to go to a different quantum-mechanical state with slightly different energy, but all the nearby states are already occupied. The only way an electron can move is to jump into the conduction band. This would require an additional energy of a few electronvolts (at least as great as the band-gap energy E_g), and that much energy is ordinarily not available. (The average energy available from thermal motion is typically on the order of kT, which, at room temperature, is about 1/40 eV.) The situation is like a completely filled parking lot: None of the cars can move because there is no place to go. If a car could jump over the others, it could move.

However, at any temperature above absolute zero, the atoms of the crystal have some vibrational motion, and there is some probability that an electron can gain enough energy (at least as great as E_g) from thermal motion to jump to the conduction band. Once in the conduction band, an electron is free to move in response to an applied electric field because there are plenty of nearby empty states available. There are always a few electrons in the conduction band, so no material is a perfect insulator. Furthermore, as the temperature increases, the population in the conduction band increases very rapidly. A doubling of the number of conduction electrons for a temperature rise of 10 C° is typical.

With metals, the situation is different, because the valence band is only partly filled. The metal sodium is an example. For an isolated sodium atom in its ground state, the valence electron is in a 3s state. There are 3p excited states at about 2.1 eV above the 3s ground state. But in the crystal lattice of *solid* sodium, the 3s and 3p *bands* spread out enough that they actually overlap, forming a single band that is only one-quarter filled. The situation is similar to the one shown in Figure 29.17c. Electrons in states near the top of the filled portion of the band have many adjacent unoccupied states available, and they can easily gain or lose small amounts of energy in response to an applied electric field. Therefore, these electrons are mobile and can contribute to electrical and thermal conductivity. Metallic crystals always have partly filled bands. In the *ionic* NaCl crystal, by contrast, there is no overlapping of bands; the valence band is completely filled, and solid sodium chloride is an insulator.

The band picture also adds insight into the phenomenon of dielectric breakdown, in which materials that are normally insulators become conductors when they are subjected to a large enough electric field. If the electric field in a material is so large that there is a potential difference of a few volts over a distance comparable to atomic sizes (i.e., a field on the order of 10^{10} V/m), then the field can do enough work on a valence electron to boost it over the forbidden region and into the conduction band. Real insulators usually have dielectric strengths much less than this because of structural imperfections that provide some energy states in the forbidden region.

The concept of energy bands is useful in understanding the properties of semiconductors, which we'll study in the next section.

29.6 Semiconductors

As the name implies, a **semiconductor** is a material with an electrical resistivity that is intermediate between the resistivities of good conductors and those of good insulators. The tremendous importance of semiconductors in present-day electronics stems in part from the fact that their electrical properties are highly sensitive to very small concentrations of impurities. We'll discuss the basic concepts, using the semiconductor elements silicon and germanium as examples.

Silicon and germanium are in Group IV of the periodic table. Each has four electrons in the outermost electron subshells (the 3s and 3p levels for Si, the 4s and 4p levels for Ge). Both crystallize in the diamond structure (Section 29.4)

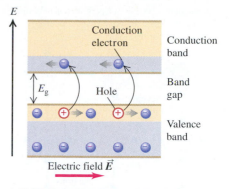

with covalent bonding. Each atom lies at the center of a regular tetrahedron, forming a covalent bond with each of four nearest neighbors at the corners of the tetrahedron. All the valence electrons are involved in the bonding. The band structure (Section 29.5) has a valence band (which, at a temperature of absolute zero, would be completely filled), separated by an energy gap E_g from a nearly empty conduction band (Figure 29.17b).

At low temperatures, these materials are insulators. Electrons in the valence band have no nearby levels available into which they can move in response to an applied electric field. However, the energy gap E_g between the valence and conduction bands is unusually small: 1.14 eV for silicon and only 0.67 eV for germanium, compared with 5 eV or more for many insulators. Thus, even at room temperature, a substantial number of electrons are dissociated from their parent atoms and jump the gap into the conduction band (Figure 29.18). This number increases rapidly with increasing temperature.

When an electron is removed from a covalent bond, it leaves a vacancy where there would ordinarily be an electron. An electron from a neighboring atom can easily drop into this vacancy, leaving the neighbor with the vacancy. In this way, the vacancy, usually called a **hole,** can travel through the crystal and serve as an additional current carrier. A hole behaves like a positively charged particle, even though the moving charges are electrons. It's like describing the motion of a bubble in a liquid. In a pure semiconductor, holes and electrons are always present in equal numbers. When an electric field is applied, they move in opposite directions (Figure 29.19). This conductivity is called **intrinsic conductivity,** to distinguish it from conductivity due to impurities, which we'll discuss later.

The parking-lot analogy we mentioned earlier helps to clarify the mechanisms of conduction in a semiconductor. A covalent crystal with no bonds broken is like a filled floor of a parking garage. No cars (electrons) can move because there is nowhere for them to go (i.e., no unoccupied energy states with nearly the same energy as the occupied states). But if one car is moved to the empty floor above, it can move freely, and the empty space it leaves also permits other cars to move on the nearly filled floor. The motion of the vacant space corresponds to a hole in the normally filled valence band.

Impurities

Suppose we mix into melted germanium ($Z = 32$) a small amount of arsenic ($Z = 33$), the next element after germanium in the periodic table. (The deliberate addition of impurity elements is often called *doping.*) Arsenic is in Group V of the periodic table (Appendix D); it has five valence electrons. When one of these electrons is removed, the remaining electron structure is essentially that of germanium. The only difference is that it is scaled down in size by the insignificant factor 32/33, because the arsenic nucleus has a charge of $+33e$ rather than $+32e$. An arsenic atom can comfortably take the place of a germanium atom in the lattice (Figure 29.20a). Four of its five valence electrons form the necessary covalent bonds with the nearest neighbors. The fifth valence electron is very loosely bound, with a binding energy of only about 0.01 eV. In the band picture, this valence electron corresponds to an isolated energy level lying in the gap, 0.01 eV below the bottom of the conduction band (Figure 29.20b). This level is called a *donor level,* and the impurity atom that is responsible for it is called a *donor impurity.*

All Group V elements, including nitrogen, phosphorus, arsenic, antimony, and bismuth, can serve as donor impurities. Even at ordinary temperatures, substantial numbers of electrons in donor levels can gain enough energy to climb into the conduction band, where they are free to wander through the lattice. The corresponding positive charge is associated with the nuclear charge of the arsenic atom ($+33e$ instead of $+32e$). It is *not* free to move; in contrast to the situation

▲ FIGURE 29.19 Motion of electrons in the conduction band, and of holes in the valence band of a semiconductor under the action of an applied electric field $\vec{E}$.

MasteringPHYSICS

PhET: Semiconductors
PhET: Conductivity

A donor (*n*-type) impurity atom has a fifth valence electron that does not participate in the covalent bonding and is very loosely bound.

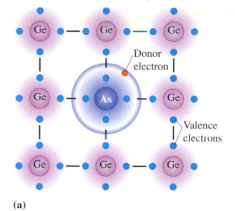

(a)

Energy-band diagram for an *n*-type semiconductor at a low temperature. One donor electron has been excited from the donor levels into the conduction band.

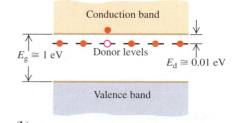

(b)

▲ FIGURE 29.20 *n*-type impurity conductivity.

An acceptor *p*-type impurity atom has only three valence electrons, so it can borrow an electron from a neighboring atom.

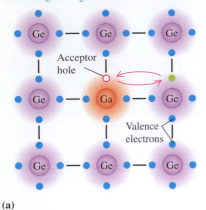

(a)

Energy-band diagram for a *p*-type semiconductor at a low temperature. One acceptor level has accepted an electron from the valence band, leaving a hole behind.

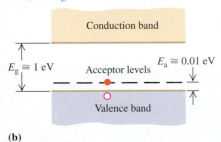

(b)

▲ **FIGURE 29.21** *p*-type semiconductor.

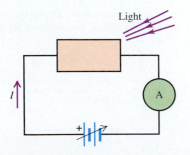

▲ **FIGURE 29.22** A semiconductor photocell in a circuit.

with electrons and holes in pure germanium, this positive charge does not participate in conduction.

At ordinary temperatures, only a very small fraction of the valence electrons in pure germanium are able to escape their sites and participate in conduction. A concentration of arsenic atoms as small as one part in 10^{10} can increase the conductivity so drastically that conduction due to impurities becomes by far the dominant mechanism. In this case, the conductivity is due almost entirely to negative-charge (electron) motion. We call the material an ***n*-type semiconductor,** with *n*-type impurities.

Adding atoms of an element in Group III, with only three valence electrons, has an analogous effect. An example is gallium ($Z = 31$); placed in the germanium lattice, the gallium atom would like to form four covalent bonds, but it has only three outer electrons. It can, however, "steal" an electron from a neighboring germanium atom to complete the required four covalent bonds. The resulting atom has the same electron configuration as Ge, but is larger by a factor of $32/31$ because the nuclear charge is $+31e$ instead of $+32e$. This "theft" leaves the neighboring atom with a hole, or missing electron, which can then move through the lattice just as in intrinsic conductivity.

The "stolen" electron is bound to the gallium atom in a level called an *acceptor level*, about 0.01 eV above the top of the valence band (Figure 29.21). The gallium atom thus completes the needed four covalent bonds, but it has a net charge of $-e$ (because there are 32 electrons and a nuclear charge of $+31e$). In this case, the corresponding negative charge is associated with the deficiency of positive charge of the gallium nucleus ($+31e$ instead of $+32e$), so that negative charge is not free to move. A semiconductor with Group III impurities is called a ***p*-type semiconductor**—a material with *p*-type impurities. The two types of impurities, *n* and *p*, are also called *donors* and *acceptors,* respectively. The deliberate addition of these impurity elements is called *doping.*

29.7 Semiconductor Devices

The first transistor was invented in 1947. Since then, semiconductor devices have revolutionized the electronics industry, with applications in communications, computer systems, control systems, and many other areas. Semiconductor devices play an indispensable role in contemporary electronics. Semiconductor-based large-scale integrated circuits can incorporate the equivalent of many thousands of transistors, capacitors, resistors, and diodes on a silicon chip less than 1 cm square. Such chips form the heart of every pocket calculator, personal computer, and mainframe computer.

One simple semiconductor device is the *photocell.* In Figure 29.22, a thin slab of pure silicon or germanium (with small intrinsic conductivity) is connected to the terminals of a battery. The resulting current is small because there are very few mobile charges. But now we irradiate the slab with electromagnetic waves whose photons have at least as much energy as the band gap E_g between the valence and conduction bands. (This radiation corresponds to photon energies in the visible or near-infrared region of the electromagnetic spectrum.) Now an electron in the valence band can absorb a photon and jump to the conduction band, where it and the hole it left behind contribute to the conductivity of the slab. Therefore, the conductivity and the circuit current increase with the intensity of the radiation.

Detectors for charged particles operate on the same principle. When a high-energy charged particle passes through the semiconductor material, it collides inelastically with valence electrons, exciting them from the valence to the conduction

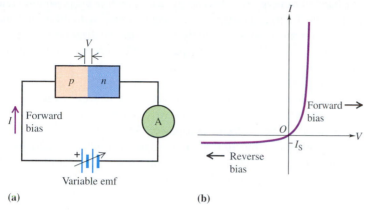

▲ FIGURE 29.23 (a) Schematic diagram of a *p–n* junction in a circuit. (b) Graph of the current–voltage relationship.

band and creating pairs of holes and conduction electrons. The conductivity increases momentarily, causing a pulse of current in an external circuit. Semiconductor detectors are widely used in nuclear and high-energy physics research.

The *p-n* Junction

In many semiconductor devices, the essential principle is that the conductivity of the material is controlled by impurity concentrations, which can be varied within wide limits from one region of a device to another. An example is the **p–n junction**—the boundary layer between two regions of a semiconductor, one with *p*-type impurities and the other with *n*-type impurities. This boundary region is called a *junction;* such a region can be fabricated by the deposition of *n*-type material on a clean surface of some *p*-type material, or in various other ways.

When a *p–n* junction is connected in an external circuit, as in Figure 29.23a, and the potential difference *V* across the junction is varied, the current varies as shown in Figure 29.23b. The device conducts much more readily in the direction from *p* to *n* than the reverse, and the current *I* is not proportional to the potential difference *V*. This behavior is in sharp contrast to the symmetric and linear behavior of a resistor that obeys Ohm's law. Such a one-way device is called a **diode.**

We can understand the behavior of a *p–n* junction diode qualitatively on the basis of the conductivity mechanisms in the two regions. In Figure 29.23, we connect the *p* region to the positive terminal of the battery, so it is at higher potential than the *n* region. The resulting electric field is in the direction from *p* to *n*. This is called the *forward direction,* and the positive potential difference is called **forward bias.** Holes in the *p* region flow into the *n* region, and electrons in the *n* region move into the *p* region; this flow constitutes a forward current.

When the polarity is reversed, the condition is called **reverse bias.** In this case, the field tends to push electrons from *p* to *n* and holes from *n* to *p*. But there are very few mobile electrons in the *p* region, only those associated with intrinsic conductivity and some that diffuse over from the *n* region. Similarly, there are very few holes in the *n* region. As a result, the current in the reverse direction is much smaller than that obtained with the same potential in the forward direction.

A **light-emitting diode** (LED) is, as the name implies, a *p–n* junction that emits light. When the junction is forward biased, many holes are injected across the junction into the *n* region, and electrons are injected into the *p* region. When an electron falls into a hole, the pair can emit a photon with energy approximately

▲ Application Is there a room available? The electronic behavior of semiconductors involves not only the movement of electrons but also the movement of electron vacancies known as "holes" where electrons are absent. If a Group IV element such as germanium contains a very tiny impurity of a group V element such as arsenic, the arsenic fits into the tetrahedral lattice of the germanium crystal, but has an extra electron. This extra electron is free to move when an external voltage is applied, creating an *n*-type semiconductor. If, on the other hand, the germanium contains a tiny impurity of a Group III element such as gallium, there is a deficiency of one electron, creating a hole. Under an applied voltage, a hole in this *p*-type semiconductor behaves like a mobile positive charge, moving as an electron would, but in the opposite direction.

When $V_e = 0$, the current is very small. When a potential V_e is applied between emitter and base, holes travel from the emitter to the base. When V_e is sufficiently large, most of the holes continue into the collector.

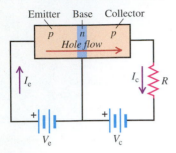

▲ **FIGURE 29.24** Schematic diagram of a p–n–p transistor and circuit.

When $V_b = 0$, I_c is very small, and most of the voltage V_c appears across the base–collector junction. As V_b increases, the base–collector potential decreases, and more holes can diffuse into the collector; thus, I_c increases. Ordinarily, I_c is much larger than I_b.

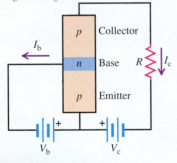

▲ **FIGURE 29.25** A common-emitter circuit.

equal to the band-gap energy E_g. This energy (and therefore the photon wavelength and the color of the light) can be varied by using materials with different band-gap energies. Light-emitting diodes are widely used for digital displays in clocks, electronic equipment, automobile instrument panels, and many other applications.

The reverse process is called the *photovoltaic effect*. Here the material absorbs photons, and electron–hole pairs are created. Pairs are created close enough to a p–n junction so that they can migrate to it. The electrons are swept to the n side and the holes to the p side. When we connect this device to an external circuit, it becomes a source of emf and power. Such a device is often called a solar cell, although it can function with any light with photon energies greater than the band-gap energy. The production of low-cost photovoltaic cells for large-scale solar energy conversion is an active field of research.

Transistors

One type of **transistor** includes two p–n junctions in a "sandwich" configuration, which may be either p–n–p or n–p–n. A p–n–p transistor is shown in Figure 29.24. The outer regions are called the **emitter** and **collector**, and the center region is called the **base**. When there is no current in the left loop of the circuit, there is only a small current through the resistor R, because the voltage across the base–collector junction is in the reverse direction. But when a forward-bias voltage is applied between emitter and base, as shown in the figure, the holes traveling from emitter to base can travel through the base to the second junction, where they come under the influence of the collector-to-base potential difference and flow on through the resistor.

In this way, the current in the collector circuit is controlled by the current in the emitter circuit. Furthermore, V_c may be considerably larger than V_e, so the power dissipated in R may be much larger than the power supplied to the emitter circuit by the battery. Thus, the device functions as a **power amplifier.** If the potential drop across R is greater than V_e, it may also be a voltage amplifier.

In this configuration, the base is the common element between the "input" and "output" sides of the circuit. Another widely used arrangement is the common-emitter circuit, shown in Figure 29.25. In this circuit, the current in the collector side of the circuit is much larger than that in the base side, and the result is current amplification.

An important variation is the field-effect transistor (Figure 29.26). A slab of p-type silicon is made with two n-type regions on the top, called the *source* and the *drain;* a metallic conductor is fastened to each. Separated from the slab by an insulating layer of silicon dioxide (SiO_2) is a third electrode called the *gate.* When there is no charge on the gate and a potential difference of either polarity is

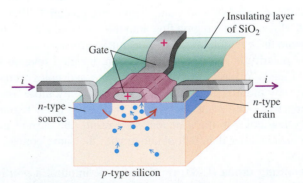

▲ **FIGURE 29.26** A field-effect transistor. The current from the source to the drain is controlled by the potential difference between the source and the drain and by the charge on the gate; no current flows through the gate.

applied between the source and the drain, there is very little current, because one of the *p–n* junctions is reverse biased.

Now we place a positive charge on the gate. There aren't many free electrons in the *p*-type material, but there are some, and they are attracted toward the positively charged gate. The concentration of electrons near the gate (and between the two junctions) is greatly enhanced, and the electron current between source and drain increases accordingly. The conductivity of the region between source and drain depends critically on the electron concentration. Thus, the current is extremely sensitive to the gate charge and potential, and the device functions as an amplifier. Note that there is very little current into or out of the gate. The device just described is called a metal-oxide-semiconductor field-effect transistor (MOSFET).

A further refinement in semiconductor technology is the **integrated circuit,** commonly referred to as a **chip.** By successively depositing layers of material and etching patterns to define current paths, we can combine the functions of several MOSFET transistors, capacitors, and resistors on a single square of semiconductor material that may be only a few millimeters on a side. An elaboration of this idea leads to large-scale integrated circuits and very-large-scale integration (VLSI). Starting with a silicon chip base, various layers are built up, including evaporated metal layers for conducting paths and silicon dioxide layers for insulators and for dielectric layers in capacitors. Appropriate patterns are etched into each layer by the use of photosensitive etch-resistant materials, onto which optically reduced patterns are projected.

A circuit containing the functional equivalent of many thousands of transistors, diodes, resistors, and capacitors can be built up on a single chip. These metal-oxide-semiconductor (MOS) chips are the heart of pocket calculators and nearly all present-day computers, large and small. An example is shown in Figure 29.27.

▲ **FIGURE 29.27** An integrated circuit chip the size of your thumb can contain millions of transistors.

29.8 Superconductivity

The electrical resistivity of all conducting materials varies with temperature. Some materials, including several metallic alloys and oxides, show a phenomenon called **superconductivity,** or the complete disappearance of electrical resistance at low temperatures. As the temperature decreases, the resistivity at first decreases smoothly. But then, at a certain temperature T_c, called the **critical temperature,** a phase transition occurs and the resistivity suddenly drops to exactly zero, or at most to a value so small that it cannot be measured. When a current is established in a superconducting ring, it continues indefinitely with no need for an external energy source.

Superconductivity was discovered in 1911 by the Dutch physicist H. Kamerlingh Onnes. He had just discovered how to liquefy helium, which has a boiling temperature of 4.2 K at atmospheric pressure (lower at reduced pressure). Measurements of the resistivity of mercury at very low temperatures showed that when very pure solid mercury is cooled to 4.16 K, it undergoes a phase transition in which its resistivity suddenly drops to zero.

There are two types of superconducting materials. A magnetic field can never exist inside a type-I superconductor. When we place such a material in a magnetic field, eddy currents are induced that exactly cancel the applied field everywhere inside the material (except for a surface layer a hundred or so atoms thick). The critical temperature always *decreases* when the material is placed in a magnetic field. When the field is sufficiently strong, the superconducting phase transition is eliminated.

Each superconducting material has a critical temperature T_c above which it is no longer a superconductor. Until 1986, the highest T_c attained was about 23 K, with a niobium–germanium alloy. This meant that superconductivy occurred only when the material was cooled by means of (expensive and scarce) liquid helium or (explosive) liquid hydrogen. But in January 1986, Karl Muller and Johannes

Bednorz discovered an alloy of barium, lanthanum, and copper with a T_c of nearly 40 K. By 1987, a ceramic material consisting of a complex oxide of yttrium, copper, and barium had been found that has $T_c = 93$ K. This temperature is significant, because it is above the 77 K boiling temperature of (inexpensive and safe) liquid nitrogen. The current (2005) record for T_c for these ceramic high-temperature superconductors is about 160 K. Materials that are superconductors at room temperatures may well become a reality in the future.

Conceptual Analysis 29.3 The critical temperature

A superconducting material is cooled below its critical temperature. Which statement is correct?

A. The critical temperature is the temperature at which the resistivity of a material remains constant with temperature.
B. The critical temperature is always close to absolute zero.
C. The critical temperature is the temperature at which the resistivity of a material becomes zero.

SOLUTION The electrical resistivity of a superconductor makes an abrupt drop to zero at temperatures below the critical temperature. This transition to superconductivity doesn't need to be near absolute zero; indeed, some materials have a critical temperature near 160 K. The correct answer is C.

▲ **Application We've come a long way, baby.** The invention of the transistor in 1947 sparked a revolution in many areas of electronics, including the ways we are able to enjoy our music. Transmitted signals in the earliest radios were amplified using bulky vacuum tubes, making these radios large and relatively immobile. The development of n-type and p-type semiconductors and their combined use to make a transistor allowed amplification of radio signals in a pocket-sized set, aptly named the "transistor radio." But the one-inch-long transistors in these portable units were enormous by today's standards. Now, using photoetching and metal vapor deposition on a microscopic scale, thousands of individual transistors and circuit elements can be produced on a single 1-mm-square integrated circuit. These "chips" are used in computers, cell phones, and nearly all modern electronic devices that we now take for granted.

Superconductivity was not well understood on a theoretical basis until 1957. In that year, Bardeen, Cooper, and Schrieffer published the theory, now called the BCS theory, that was to earn them the Nobel Prize in 1972. The key to the BCS theory is an interaction between *pairs* of conduction electrons, called *Cooper pairs,* caused by an interaction with the positive ion cores of the crystal structure. A free electron exerts attractive forces on nearby positive ion cores, pulling them slightly closer together and distorting the structure. The resulting slight concentration of positive charge then exerts an attractive force on another electron with momentum opposite that of the first electron. At ordinary temperatures this electron-pair interaction is small in comparison to energies of thermal motion, but at very low temperatures it becomes significant.

Thus bound together, the pairs of electrons cannot *individually* gain or lose small amounts of energy, as they could ordinarily do in a partly filled conduction band. The result is an energy gap in the allowed states of the pairs, and at low temperatures there is not enough energy for a pair to jump this gap. Therefore, the electrons can move freely through the lattice without any exchange of energy through collisions.

When a type-II superconductor is placed in a magnetic field, the bulk of the material is superconducting, but there are thin filaments of material in the normal state, aligned parallel to the field. Currents circulate around the boundaries of these filaments, and there is magnetic flux inside the filaments. Type-II superconductors are used for electromagnets because much larger magnetic fields can usually be present without destroying the superconducting state than is possible with ordinary superconductors.

Superconducting electromagnets are widely used in research laboratories. Once a current is established in the coil of such a magnet, no additional power input is required because there is no resistive energy loss. The coils can also be made more compact because there is no need to provide channels for cooling fluids. Thus, superconducting magnets can attain large fields much more easily and economically than conventional magnets can; fields on the order of 10 T are fairly routine. These considerations also make superconductors attractive for long-distance electric-power transmission, an active area of development.

One of the most glamorous applications of superconductors is in the field of magnetic levitation. Imagine a superconducting ring mounted on a railroad car that runs on a conducting guideway. The current induced in the guideway leads to a repulsive interaction with the rail, and levitation is possible. A magnetically levitated train is now in regular service on a 30 km guideway to and from the Pudong International Airport, outside Shanghai, China. The usual maximum speed is 267 mi/h.

The search for room-temperature superconductors continues. The implications of such materials, if they can be found, are breathtaking. They would have important applications for long-distance power transmission, magnetic levitation, computer design, and many other areas. The high-temperature superconductors discovered thus far are mechanically weak and brittle, like many ceramics, and are often chemically unstable. Fabricating conductors from them will pose difficult technological problems.

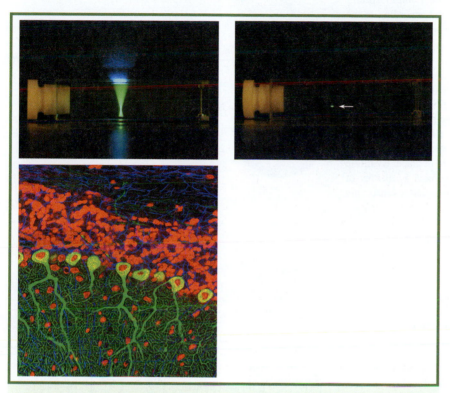

▲ **BIO** Application **Looking into living tissue via quantum physics.** Molecules have electrons with quantized excited states; in some cases, the energy gap between states matches the energies of photons from the near-UV and the visible regions of the spectrum. An electron excited by a photon rapidly loses some energy through vibrational relaxation (see Fig. 29.9) and then returns to the ground state with the emission of a photon that has somewhat less energy (longer wavelength) than the exciting photon. This process of fluorescence has important applications in biology and medicine. Fluorescent molecules can be introduced into cells by a number of means, including engineering the genes so that the cells themselves produce the molecules. These fluorophores then give information about many processes and structures in the living cells.

One problem with fluorescence microscopy is the absorption of energy by molecules that are not in the plane of focus of the microscope objective. The fluorescent emission from the out-of-focus region degrades the image, and the absorbed energy can be damaging to cells. One solution to this problem is an application of the two-photon effect, a quantum mechanical phenomenon described in the 1930s. If two photons each of approximately half the energy required to excite an electron are absorbed within a very short time, the electron moves to the excited state. A fluorophore that is efficiently excited by absorbing single photons of 450 nm can be excited by absorbing two photons at about 900 nm. However, the photon density must be very high; sufficient photon density is achieved only in the focal plane of the objective. For the remainder of the optical path, the 900 nm photons are not absorbed, so no out-of-focus light is generated and no energy is imparted to the cell.

The upper figure shows two side views (perpendicular to the optical axis of the microscope) of the fluorescent emissions from beams of focused exciting light passing through a medium containing a fluorophore. The right image is of single photon fluorescence and the left image is of two-photon fluorescence; the difference in out-of-focus fluorescence is obvious. The lower photograph is of two-photon imaging of brain tissue showing Purkinje cells in yellow.

SUMMARY

Electrons in Atoms

(Section 29.1) The Bohr model fails to fully describe atoms because it combines elements of classical physics with some principles of quantum mechanics. To explain the observed properties of an atom, electrons must be described with a **wave function** that is determined by solving the Schrödinger equation. As a result, some properties of the atom are quantized, including orbital and spin angular momentum. The **quantum numbers** (n, l, m_l, s) specify the quantum-mechanical states of the atom, and according to the **Pauli exclusion principle,** no two electrons in an atom can occupy the same state.

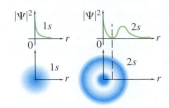

Atomic Structure

(Section 29.2) We can imagine constructing a neutral atom by starting with a bare nucleus with Z protons and adding Z electrons, one by one, respecting the Pauli exclusion principle. To obtain the ground state, we fill the lowest-energy states (those closest to the nucleus, with the smallest values of n and l) first, and we use successively higher states until all the electrons are in place. By filling the atomic states we begin to learn the chemical properties of atoms, which are determined principally by interactions involving the *outermost* electrons.

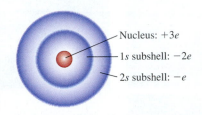

Nucleus: $+3e$
$1s$ subshell: $-2e$
$2s$ subshell: $-e$

Diatomic Molecules

(Section 29.3) An **ionic bond** is a bond between two ionized atoms—one atom gives at least one electron to fill a vacancy in a shell of the other. In a **covalent bond,** the electron cloud tends to concentrate between the atoms—the positive nucleus of each atom is attracted to the somewhat centralized electron cloud. There are also weaker bonds such as the **van der Waals** and **hydrogen bonds.**

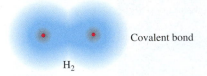

Covalent bond

H_2

Structure and Properties of Solids

(Section 29.4) A crystalline solid is characterized by long-range order, a recurring pattern of atomic positions that extends over many atoms. Liquids have short-range order. We can understand many macroscopic properties of solids, including mechanical, thermal, electrical, magnetic, and optical properties, by considering their relation to the microscopic structure of the material.

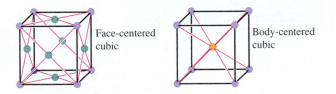

Face-centered cubic

Body-centered cubic

Semiconductors

(Sections 29.5–29.7) A **semiconductor** is a material with an electrical resistivity that is intermediate between those of good conductors and those of good insulators. A **hole** is a vacancy in a bond where there would normally be an electron—the hole acts like a positive charge. In **n-type** semiconductors, the conductivity is due to the motion of electrons; in **p-type,** holes act as the moving charges. A **transistor** can be made by layering two p-n junctions—the resulting devices can act as power, current, or voltage amplifiers.

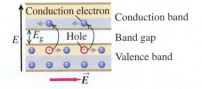

Conduction electron
Conduction band
E E_g Hole Band gap
Valence band
$\vec{E}$

Superconductivity

(Section 29.8) As the temperature decreases in a superconductor, the resistivity at first decreases smoothly. But then at the **critical temperature,** a phase transition occurs and the resistivity suddenly drops to zero. One recently discovered superconductor has a critical temperature of 160 K. When we place a superconductor in a magnetic field, eddy currents are induced that exactly cancel the applied field everywhere inside the material—a magnetic field can never exist inside a type-I superconducting material.

Conceptual Questions

1. In the ground state of the helium atom, the electrons must have opposite spins. Why?
2. What does it mean for a quantity to be *quantized?* Give several examples of quantized quantities.
3. The exclusion principle is sometimes described as the quantum version of the principle that no two objects can be in the same place at the same time. Is this description appropriate?
4. In describing the atom, we often refer to the "electron cloud." Does this mean that the electron is spread out in space in the form of a cloud? Just what is that cloud?
5. In what ratio would you expect Mg and Cl to combine? Why?
6. A student asserted that any filled shell (i.e., all the levels for a given n occupied by electrons) must have zero total angular momentum and hence must be spherically symmetric. Do you believe this? What about a filled *subshell* (all values of m_l for given values of n and l)?
7. The central-field approximation is more accurate for alkali metals than for transition metals such as iron, nickel, or copper. Why do you think this is? (Figure 29.3 may be helpful in answering this question.)
8. The nucleus of a gold atom contains 79 protons. How would you expect the energy required to remove a $1s$ electron from a gold atom to compare with the energy required to remove a $1s$ electron from a hydrogen atom? In what region of the electromagnetic spectrum would a photon of the appropriate energy lie?
9. Elements can be identified by their visible spectra. Could analogous techniques be used to identify compounds from their molecular spectra? In what region of the electromagnetic spectrum would the appropriate radiation lie?
10. The ionization energies of the alkali metals (i.e., the energies required to remove an outer electron from the metal) are in the range from 4 to 6 eV, while those of the inert gases are in the range from 15 to 25 eV. Why the difference?
11. The energy required to remove the $3s$ electron from a sodium atom in its ground state is about 5 eV. Would you expect the energy required to remove an additional electron to be about the same, more, or less? Why?
12. Individual atoms have discrete energy levels, but certain solids (which are made up of only individual atoms) show energy bands and gaps. What causes the solids to behave so differently from the atoms of which they are composed?
13. Increasing the temperature of an ordinary conductor normally makes it a poorer conductor, whereas increasing the temperature of an ordinary nonconductor normally makes it a *better* conductor. Why this difference in behavior? Explain at the atomic level.
14. Why does a diode conduct current in one direction better than in the opposite direction? Why do ordinary resistors *not* behave this way?
15. Why is it advantageous to add impurities to semiconductors? That is, why are they doped?
16. What is the difference between an n-type and a p-type semiconductor?

Multiple-Choice Problems

1. Which statement about an electron in the $5f$ state is correct? (There may be more than one correct choice.)
   A. The magnitude of its orbital angular momentum is $2\sqrt{3}\hbar$.
   B. The largest orbital angular momentum it could have in any given direction is $5\hbar$.
   C. Its magnetic quantum number could have the value ± 3, ± 2, ± 1, or 0.
   D. Its spin angular momentum is $2\sqrt{3}\hbar$.
2. The atom having the electron configuration $1s^2\,2s^2\,2p^6\,3s^2\,3p^5$ has (there may be more than one correct choice)
   A. 17 orbital electrons.
   B. 11 orbital electrons.
   C. electrons with $l = 0, 1, 2$.
   D. electrons with $m_l = 0$ and ± 1.
3. Which statement about x rays is correct? (There may be more than one correct choice.)
   A. Their wavelengths are greater than those of visible light.
   B. They are emitted when electrons in complex atoms (those having a large atomic number) make transitions from higher shells to the K shell.
   C. They are emitted when electrons in hydrogen make transitions from higher shells to the K shell.
   D. The energy of x-ray photons is greater than the energy of visible-light photons.
4. Table 29.2 shows that for the ground state of the potassium atom the outermost electron is in a $4s$ state. This tells you that, compared to the $3d$ level for this atom, the $4s$ level
   A. has a lower energy.
   B. has a higher energy.
   C. has approximately the same energy.
5. Which statement about semiconductors is correct?
   A. Their resistivity is normally somewhat greater than that of most metals.
   B. A "hole" in a semiconductor is due to the removal of a proton.
   C. Electrical conduction in an n-type semiconductor is due to the transfer of neutrons.
   D. Electrical conduction in a p-type semiconductor is due to the transfer of electrons.
6. An electron in the N shell can have which of the following quantum numbers? (There may be more than one correct choice.)
   A. $s = -\frac{1}{2}$.   B. $l = 1$.   C. $m_l = 4$.
   D. $n = -4$.   E. $m_l = -3$.
7. Bonds that involve the essentially complete transfer of electrons from one atom to another are
   A. van der Waals bonds.
   B. hydrogen bonds.
   C. ionic bonds.
   D. covalent bonds.
8. Consider an atom having electron configuration $1s^2\,2s^2\,2p^4$. The atom with the next higher Z and having similar chemical properties would have Z equal to
   A. 9.   B. 15.   C. 16.   D. 34.
9. The number of $5f$ states for hydrogen is
   A. 22.   B. 18.   C. 14.   D. 7.
10. If an electron has the quantum numbers $n = 4$, $l = 2$, $m_l = -1$, $s = \frac{1}{2}$, it is in the state
    A. $2p$.
    B. $2g$.
    C. $4p$.
    D. $4d$.
    E. $4f$.

Problems

29.1 Electrons in Atoms

1. • The orbital angular momentum of an electron has a magnitude of 4.716×10^{-34} kg $\cdot$ m^2/s. What is the angular-momentum quantum number l for this electron?

2. • Consider states with $l = 3$. (a) In units of $\hbar$, what is the largest possible value of L_z? (b) In units of $\hbar$, what is the value of L? Which is larger, L or the maximum possible L_z? (c) Assume a model in which $\vec{L}$ is described as a classical vector. For each allowed value of L_z, what angle does the vector $\vec{L}$ make with the $+z$ axis?

3. •• An electron is in the hydrogen atom with $n = 3$. (a) Find the possible values of L and L_z for this electron, in units of $\hbar$. (b) For each value of L, find all the possible angles between L and the z axis.

4. •• An electron is in the hydrogen atom with $n = 5$. (a) Find the possible values of L and L_z for this electron, in units of $\hbar$. (b) For each value of L, find all the possible angles between L and the z axis. (d) What are the maximum and minimum values of the magnitude of the angle between L and the z axis?

5. •• Consider an electron in the N shell. (a) What is the smallest orbital angular momentum it could have? (b) What is the largest orbital angular momentum it could have? Express your answers in terms of $\hbar$ and in SI units. (c) What is the largest orbital angular momentum this electron could have in any chosen direction? Express your answers in terms of $\hbar$ and in SI units. (d) What is the largest spin angular momentum this electron could have in any chosen direction? Express your answers in terms of $\hbar$ and in SI units. (e) For the electron in part (c), what is the ratio of its spin angular momentum in the z direction to its orbital angular momentum in the z direction?

6. • (a) How many different $3d$ states does hydrogen have? Make a list showing all of them. (b) How many different $3f$ states does it have?

7. •• (a) How many different $5g$ states does hydrogen have? (b) Which of the states in part (a) has the largest angle between $\vec{L}$ and the z axis and what is that angle? (c) Which of the states in part (a) has the smallest angle between $\vec{L}$ and the z axis, and what is that angle?

8. •• In a particular state of the hydrogen atom, the angle between the angular momentum vector $\vec{L}$ and the z axis is $\theta = 26.6°$. (See Figure 29.2.) If this is the smallest angle for this particular value of the angular momentum quantum number l, what is l?

29.2 Atomic Structure

9. • Make a list of the four quantum numbers n, l, m_l, and s for each of the 12 electrons in the ground state of the magnesium atom.

10. • (a) List the different possible combinations of quantum numbers l and m_l for the $n = 5$ shell. (b) How many electrons can be placed in the $n = 5$ shell?

11. • For bromine $(Z = 35)$, make a list of the number of electrons in each subshell $(1s, 2s, 2p,$ etc.$)$.

12. • (a) Write out the electron configuration $(1s^2 2s^2,$ etc.$)$ for Li and Na. (b) How many electrons does each of these atoms have in its outer shell?

13. •• (a) Write out the ground-state electron configuration $(1s^2, 2s^2,$ etc.$)$ for the carbon atom. (b) What element of next-larger Z has chemical properties similar to those of carbon? (See Example 29.3.) Give the ground-state electron configuration for this element.

14. •• (a) Write out the ground-state electron configuration $(1s^2 2s^2,$ etc.$)$ for the beryllium atom. (b) What element of next-larger Z has chemical properties similar to those of beryllium? (See Example 29.3.) Give the ground-state electron configuration of this element. (c) Use the procedure of part (b) to predict what element of next-larger Z than in (b) will have chemical properties similar to those of the element you found in part (b), and give its ground-state electron configuration.

15. •• Write out the electron configuration $(1s^2 2s^2,$ etc.$)$ for Ne, Ar, and Kr. (b) How many electrons does each of these atoms have in its outer shell? (c) Predict the chemical behavior of these three atoms. Explain your reasoning.

16. •• Calculate, in units of $\hbar$, the magnitude of the maximum orbital angular momentum for an electron in a hydrogen atom for states with a principal quantum number of 2, 20, and 200. Compare each with the value of $n\hbar$ postulated in the Bohr model. What trend do you see?

17. • (a) What is the orbital angular momentum of any s-subshell electron? (b) If we try to model the atom classically as a scaled-down version of a solar system, with the electrons orbiting the nucleus the way the planets orbit the sun, what does the result in part (a) tell us would be the speed of an s-subshell electron? Is this result physically possible? What would happen to an electron with that speed?

18. •• The energies for an electron in the K, L, and M shells of the tungsten atom are $-69,500$ eV, $-12,000$ eV, and -2200 eV, respectively. Calculate the wavelengths of the K_α and K_β x rays of tungsten.

29.3 Diatomic Molecules

19. • If the energy of the H$_2$ covalent bond is -4.48 eV, what wavelength of light is needed to break that molecule apart? In what part of the electromagnetic spectrum does this light lie?

20. • (a) A molecule decreases its vibrational energy by 0.250 eV by giving up a photon of light. What wavelength of light does it give up during this process, and in what part of the electromagnetic spectrum does that wavelength of light lie? (b) An atom decreases its energy by 8.50 eV by giving up a photon of light. What wavelength of light does it give up during this process, and in what part of the electromagnetic spectrum does that wavelength of light lie? (c) A molecule decreases its rotational energy by 3.20×10^{-3} eV by giving up a photon of light. What wavelength of light does it give up during this process, and in what part of the electromagnetic spectrum does that wavelength of light lie?

21. •• **An ionic bond.** (a) Calculate the electric potential energy for a K$^+$ ion and a Br$^-$ ion separated by a distance of 0.29 nm, the equilibrium separation in the KBr molecule. Treat the ions as point charges. (b) The ionization energy of the potassium atom is 4.3 eV. Atomic bromine has an electron affinity of 3.5 eV. Use these data and the results of part (a) to estimate the binding energy of the KBr molecule. Do you expect the actual binding energy to be higher or lower than your estimate? Explain your reasoning.

29.4 Structure and Properties of Solids

22. •• The spacing of adjacent atoms in a NaCl crystal is 0.282 nm, and the masses of the atoms are 3.82×10^{-26} kg (Na) and

5.89 × 10^{-26} kg (Cl). Use this information to calculate the density of sodium chloride.

23. •• Potassium bromide (KBr) has a density of 2.75 × 10^3 kg/m^3 and the same crystal structure as NaCl. The mass of potassium is 6.49 × 10^{-26} kg, and that of bromine is 1.33 × 10^{-25} kg. (a) Calculate the average spacing between adjacent atoms in a KBr crystal. (b) Compare the spacing for KBr with the spacing for NaCl. (See previous problem.) Is the relation between these two values qualitatively what you would expect? Explain your reasoning.

29.5 Energy Bands
29.6 Semiconductors
29.7 Semiconductor Devices

24. • Look at the graph for a diode in Figure 29.23(b) in the text. Why does this graph go below the horizontal axis to the left of the origin? Would it do this if the diode were replaced by an ordinary resistor?

25. •• The gap between valence and conduction bands in diamond is 5.47 eV. (a) What is the maximum wavelength of a photon that can excite an electron from the top of the valence band into the conduction band? In what region of the electromagnetic spectrum does this photon lie? (b) Explain why pure diamond is transparent and colorless. (*Hint:* Will photons of visible light that strike a diamond be absorbed or transmitted?) (c) Most gem diamonds have a yellow color. Explain how impurities in the diamond can cause this color.

26. •• A variable dc power supply having reversible polarity is connected to a diode having a *p–n* junction as shown in Figure 29.28. Starting with the power supply's polarity as shown in the figure, its potential is gradually decreased to zero and then gradually increased

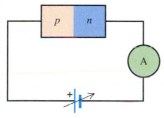

▲ **FIGURE 29.28** Problem 26.

in the reverse direction. (a) Sketch a graph of the reading in the ammeter as a function of the potential difference V across the power supply. Make sign differences clear. (b) Suppose now that the terminals of the diode are reversed and the same procedure is followed. Sketch a graph of the reading in the ammeter as a function of the potential difference V across the power supply. Make sign differences clear. (c) Suppose now that the diode is replaced by an ordinary resistor and the same procedure is followed. Sketch a graph of the reading in the ammeter as a function of the potential difference V across the power supply. Make sign differences clear. (d) Explain the reasons for the differences between the graphs in (i) parts (a) and (b), (ii) parts (a) and (c).

27. •• The gap between valence and conduction bands in silicon is 1.12 eV. A nickel nucleus in an excited state emits a gamma-ray photon with wavelength 9.31 × 10^{-4} nm. How many electrons can be excited from the top of the valence band to the bottom of the conduction band by the absorption of this gamma ray?

29.8 Superconductivity

28. • Sketch a qualitative (no numbers) graph of the resistance as a function of temperature for (a) an ordinary conductor, such as Cu, including temperatures approaching 0 K; (b) a superconductor; include temperatures above and below the critical temperature, and let the temperature approach 0 K.

General Problems

29. •• For magnesium, the first ionization potential is 7.6 eV; the second (the additional energy required to remove a second electron) is almost twice this, 15 eV, and the third ionization potential is much larger, about 80 eV. Why do these numbers keep increasing?

30. •• **Failure of the classical model.** (a) What is the spin angular momentum of an electron along the z direction, in SI units? (b) In the rest of this problem, we shall try to understand the behavior of an electron by modeling it as a classical ball of matter. We shall also use the nonrelativistic formulas, although a thorough treatment would require use of the more complicated relativistic ones. If we model this electron classically as a solid uniform sphere of diameter 1.0 × 10^{-15} m, what would be its angular velocity in rad/s? (c) What would be the speed of the surface of this classical sphere? Is there anything suspicious about this result? (d) In part (c), the result cannot be correct because the surface of the electron cannot move faster than the speed of light. Suppose we try to "correct" this model by assuming that the surface is moving at the speed of light— the upper limit of its speed. What would have to be the diameter of an electron in that case? Given that atomic nuclei are about 10 × 10^{-15} m in diameter, does this result seem plausible for the size of an electron? (e) Would the bizarre results in parts (c) and (d) be affected appreciably if we instead modeled the electron as a hollow spherical shell instead of a solid? Why? Notice that we encounter results that contradict observation when we try to think of the electron simply as a spinning sphere. We simply cannot understand the behavior of an electron (and other subatomic particles) by using our ordinary classical models.

31. •• An electron has spin angular momentum and orbital angular momentum. For the 3d electron in scandium, what percent of its total orbital angular momentum is its spin angular momentum in the z direction?

32. •• The dissociation energy of the hydrogen molecule (i.e., the energy required to separate the two atoms) is 4.48 eV. In the gas phase (treated as an ideal gas), at what temperature is the average translational kinetic energy of a molecule equal to this energy?

33. •• The maximum wavelength of light that a certain silicon photocell can detect is 1.11 μm. (a) What is the energy gap (in electron volts) between the valence and conduction bands for this photocell? (b) Explain why pure silicon is opaque. (*Hint:* Will visible light that strikes silicon be transmitted or absorbed?)

34. •• Use the electron configurations of He, Ne, and Ar to explain why these atoms normally do not combine chemically with other atoms.

35. •• Use the electron configurations of H and O to explain why these atoms combine chemically in a two-to-one ratio to form water.

36. •• Use the electron configurations of Si and O to explain why these atoms combine chemically in a one-to-two ratio to form sand.

37. •• Consider an electron in hydrogen having total energy −0.5440 eV. (a) What are the possible values of its orbital angular momentum (in terms of $\hbar$)? (b) What wavelength of light would it take to excite this electron to the next higher shell? Is this photon visible to humans?

38. •• The energy of the van der Waals bond, which is responsible for a number of the characteristics of water, is about 0.50 eV. (a) At what temperature would the average translational kinetic energy of water molecules be equal to this energy? (b) At that temperature, would water be liquid or gas? Under ordinary everyday conditions, do van der Waals forces play a role in the behavior of water?

39. •• (a) What is the lowest possible energy (in electron volts) of an electron in hydrogen if its orbital angular momentum is $\sqrt{12}\hbar$? (b) What are the largest and smallest values of the z component of the orbital angular momentum (in terms of $\hbar$) for the electron in part (a)? (c) What are the largest and smallest values of the spin angular momentum (in terms of $\hbar$) for the electron in part (a)? (d) What are the largest and smallest values of the orbital angular momentum (in terms of $\hbar$) for an electron in the M shell of hydrogen?

40. •• An electron in hydrogen is in the $5f$ state. (a) Find the largest possible value of the z component of its angular momentum. (b) Show that for the electron in part (a), the corresponding x and y components of its angular momentum satisfy the equation $\sqrt{L_x^2 + L_y^2} = \hbar\sqrt{3}$.

30 Nuclear and High-Energy Physics

During the 20th century, nuclear physics has had enormous effects on humankind, many beneficial, some catastrophic. Many people have strong opinions about the uses of nuclear physics. Ideally, these opinions should be based on understanding rather than on prejudice or emotion, and we hope that this chapter will help you reach that understanding.

Every atom contains at its center an extremely dense, positively charged *nucleus*, much smaller than the overall size of the atom, but containing most of its mass. In this chapter, we'll look first at several important properties of nuclei and of the interactions that hold them together. The stability or instability of a particular nucleus is determined by the competition between the attractive nuclear forces among the protons and neutrons and the repulsive electrical interactions among the protons. Unstable nuclei decay by a variety of processes, transforming themselves spontaneously into other structures. Structure-altering nuclear reactions can also be induced by collision of a nucleus with a particle or another nucleus. Two classes of reactions of special interest are fission and fusion. Research into the nature of, and interactions among, fundamental particles has required the construction of large experimental facilities, such as particle accelerators, as scientists probe more deeply into this most fundamental aspect of the nature of our physical universe.

This bog mummy was found in a peat deposit in Denmark. Radiocarbon dating, performed on a sample of his liver, indicates that he died in the late third century BCE. His body was preserved by burial in a peat bog, which also turned his dark hair red.

30.1 Properties of Nuclei

Nuclei are composed of protons and neutrons (collectively referred to as *nucleons*). These two particles have nearly equal masses (within about 0.1%), about 1.67×10^{-27} kg. They aren't fundamental particles, but are composed of more

1003

▲ Application Where's the nucleus? If a hydrogen atom were the size of this baseball stadium, how large do you think the nucleus would be? The answer: Only the size of a pencil eraser in the center. Therefore, textbook drawings of atoms that show the nucleus are far out of scale. If the nucleus were the size of a baseball, the electron cloud would be about 20 km across. In fact, because so much of the atom is empty space, it has been estimated that, without this space, all the solid matter in the human body would fit on the head of a pin!

basic entities called *quarks*. We'll begin our discussion by describing the behavior of nuclei in terms of protons and neutrons; later we'll return to their relation to quarks.

The total number of protons in a nucleus is the **atomic number** Z. In an electrically neutral atom, this is equal to the number of electrons. Hence, Z determines the chemical properties of each element. The number of neutrons, denoted by N, is called the **neutron number.** The total number of protons and neutrons is called the **nucleon number,** denoted by A:

$$A = Z + N. \tag{30.1}$$

The mass of a nucleus is approximately proportional to the total number of nucleons, so A is also (and more commonly) called the **mass number.**

The most obvious feature of the atomic nucleus is its size: The nucleus is 20,000 to 200,000 times smaller than the atom itself. We can picture a nucleus as roughly spherical in shape. The radius depends on the mass, which in turn depends primarily on the total number A of nucleons. The radii of most nuclei are represented fairly well by an empirical equation:

Radii of nuclei

The radii of nuclei are given approximately by the formula

$$R = R_0 A^{1/3}, \tag{30.2}$$

where R_0 is an empirical constant:

$$R_0 = 1.2 \times 10^{-15} \text{ m} = 1.2 \text{ fm}.$$

The volume V of a sphere is equal to $4\pi R^3/3$, so Equation 30.2 shows that the *volume* of a nucleus is proportional to its mass number A and thus to its total mass. It follows that the *density* (mass per unit volume, proportional to A/R^3) is the same for all nuclei. That is, **all nuclei have approximately the same density.** This fact is of crucial importance in understanding nuclear structure.

EXAMPLE 30.1   Nuclear density

The most common variety of iron nucleus has $A = 56$. Find the radius, approximate mass, and approximate density of the iron nucleus.

SOLUTION

SET UP The mass number A and the radius R are related by Equation 30.2. The proton and neutron masses are approximately equal, about 1.67×10^{-27} kg each. The total mass M of the nucleus is A times the mass m of a single proton or neutron.

SOLVE From Equation 30.2, the radius R of the nucleus is given by

$$R = R_0 A^{1/3} = (1.2 \times 10^{-15} \text{ m})(56^{1/3}) = 4.6 \times 10^{-15} \text{ m}.$$

The total mass M (neglecting the small difference in mass between protons and neutrons) is

$$M = (56)(1.67 \times 10^{-27} \text{ kg}) = 9.4 \times 10^{-26} \text{ kg}.$$

The volume V is

$$V = \tfrac{4}{3}\pi R^3 = \tfrac{4}{3}\pi (4.6 \times 10^{-15} \text{ m})^3 = 4.1 \times 10^{-43} \text{ m}^3,$$

and the density ρ is

$$\rho = \frac{M}{V} = \frac{9.4 \times 10^{-26} \text{ kg}}{4.1 \times 10^{-43} \text{ m}^3} = 2.3 \times 10^{17} \text{ kg/m}^3.$$

REFLECT The radius of the iron nucleus is about 4.6×10^{-15} m, on the order of 10^{-5} of the overall radius of an atom of iron. The density of the element iron (in the solid state) is about 8000 kg/m^3 (8 g/cm^3), so we see that the nucleus is on the order of 10^{13} times as dense as the bulk material. Densities of this magnitude are also found in white dwarf stars, which are similar to gigantic nuclei. A 1 cm cube of material with this density would have a mass of 2.3×10^{11} kg, or 230 million metric tons!

Practice Problem: Find the approximate mass, radius, and density of the uranium nucleus with $A = 238$, and compare the results with the values for iron. Assume that the total mass is 238 times the mass of a single neutron, neglecting binding-energy corrections. *Answer:* 4.0×10^{-25} kg, 7.4×10^{-15} m, $2.3 \times 10^{17} \text{ kg/m}^3$.

Nuclear Masses

A single nuclear species having specific values of both Z and N is called a **nuclide.** Table 30.1 lists values of A, Z, and N for several nuclides. An atom's electron structure, which determines its chemical properties, is determined by the charge Ze of the nucleus. The table shows some nuclei with the same Z but different N. These are nuclei of the same element, but they have different masses. Nuclei of a given element with different mass numbers are called **isotopes** of the element. An example that is abundant in nature is chlorine $(Cl, Z = 17)$. About 76% of chlorine nuclei have $N = 18$; the other 24% have $N = 20$. Because of the differing nuclear masses, different isotopes of an element have slightly different physical properties, such as melting and boiling temperatures and diffusion rates. The two common isotopes of uranium, with $A = 235$ and $A = 238$, are separated industrially by means of centrifuges or by taking advantage of the different diffusion rates of gaseous uranium hexafluoride (UF_6) containing the two isotopes.

Table 30.1 also shows the usual notation for individual nuclides: the symbol of the element, with a pre-subscript equal to the atomic number Z and a pre-superscript equal to the mass number A. For example, $^{13}_6C$ denotes the isotope of carbon with $Z = 6$, $A = 13$, and $N = 7$. The general format for an element El is $^A_Z El$. The isotopes of chlorine, with $A = 35$ and $A = 37$, are written as $^{35}_{17}Cl$ and $^{37}_{17}Cl$, respectively. This notation is redundant, because the name of the element determines the atomic number Z, but it's a useful aid to memory. The pre-subscript (the value of Z) is sometimes omitted, as in ^{35}Cl.

The proton and neutron masses are, respectively,

$$m_p = 1.67262171(29) \times 10^{-27} \text{ kg,}$$
$$m_n = 1.67492728(29) \times 10^{-27} \text{ kg.}$$

(The two digits in parentheses indicate the uncertainty in the last two digits of each value.) These values are nearly equal, so it is not surprising that many nuclear masses are approximately integer multiples of the proton or neutron mass. For precise measurements, it's useful to define a new mass unit equal to

TABLE 30.1 Compositions of some common nuclei

Nucleus	Mass number (total number of nuclear particles), A	Atomic number (number of protons), Z	Neutron number, $N = Z - A$
^1_1H	1	1	0
^2_1D	2	1	1
^4_2He	4	2	2
^6_3Li	6	3	3
^7_3Li	7	3	4
^9_4Be	9	4	5
$^{10}_5B$	10	5	5
$^{11}_5B$	11	5	6
$^{12}_6C$	12	6	6
$^{13}_6C$	13	6	7
$^{14}_7N$	14	7	7
$^{16}_8O$	16	8	8
$^{23}_{11}Na$	23	11	12
$^{65}_{29}Cu$	65	29	36
$^{200}_{80}Hg$	200	80	120
$^{235}_{92}U$	235	92	143
$^{238}_{92}U$	238	92	146

one-twelfth of the mass of the neutral carbon atom with mass number $A = 12$. This mass is called one **unified atomic mass unit** (1 u):

$$1\text{ u} = 1.66053886(28) \times 10^{-27}\text{ kg}.$$

In unified atomic mass units, the masses of the proton, neutron, and electron are

$$m_\text{p} = 1.00727646688(13)\text{ u},$$
$$m_\text{n} = 1.00866491560(55)\text{ u},$$
$$m_\text{e} = 0.00054857990945(24)\text{ u},$$
$$m_\text{H} = m_\text{p} + m_\text{e} = 1.0078250\text{ u}.$$

These values are more precise than the u-to-kg conversion factor because atomic masses can be compared with each other with greater precision than they can be compared with the standard kilogram. Also note that 1 mole (see Section 15.1) of particles with a mass of 1 u each would have a total mass of exactly 1 gram.

We can find the energy equivalent of 1 u from the relation $E = mc^2$:

$$E = (1.66054 \times 10^{-27}\text{ kg})(2.99792 \times 10^8\text{ m/s})^2$$
$$= 1.49242 \times 10^{-10}\text{ J} = 931.494\text{ MeV}.$$

The masses of some common atoms, including their electrons, are shown in Table 30.2. Such tables always give masses of *neutral* atoms (including Z electrons), rather than masses of bare nuclei (atoms with all their electrons removed), because it is much more difficult to measure masses of bare nuclei with high precision. To obtain the mass of a bare nucleus, subtract Z times the electron mass from the atomic mass.

The total mass of a nucleus is always *less* than the total mass of its constituent parts because of the mass equivalent ($E = mc^2$) of the internal kinetic energy and of the (negative) potential energy associated with the attractive forces that hold the nucleus together. This mass difference is called the **mass defect,** denoted by ΔM. The magnitude of the total potential energy is called the **binding energy,** denoted by E_B. Thus, $E_\text{B} = (\Delta M)c^2$.

PhET: Simplified MRI
ActivPhysics 19.2: Nuclear Binding Energy

TABLE 30.2 Atomic masses of light elements

Element	Atomic number, Z	Neutron number, N	Atomic mass, u	Mass number, A
Hydrogen, H	1	0	1.007825	1
Deuterium, H	1	1	2.014101	2
Helium, He	2	1	3.016029	3
Helium, He	2	2	4.002603	4
Lithium, Li	3	3	6.015123	6
Lithium, Li	3	4	7.016005	7
Beryllium, Be	4	5	9.012182	9
Boron, B	5	5	10.012937	10
Boron, B	5	6	11.009305	11
Carbon, C	6	6	12.000000	12
Carbon, C	6	7	13.003355	13
Nitrogen, N	7	7	14.003074	14
Nitrogen, N	7	8	15.000109	15
Oxygen, O	8	8	15.994915	16
Oxygen, O	8	9	16.999132	17
Oxygen, O	8	10	17.999161	18

Source: Atomic Mass Evaluation 2003, Nuclear Physics A 729 (2003).

Mass defect

The mass defect ΔM for a nucleus with mass M containing Z protons and N neutrons is defined as

$$\Delta M = Zm_p + Nm_n - M. \qquad (30.3)$$

(The term *mass defect* is somewhat misleading. This quantity doesn't represent a defect or discrepancy in our calculations; it is simply the mass difference defined in Equation 30.3.)

The simplest nucleus, that of hydrogen, is a single proton. Next comes the nucleus of ^2_1H, the isotope of hydrogen with mass number 2, usually called *deuterium*. Its nucleus consists of a proton and a neutron bound together to form a particle called the *deuteron*. To find the binding energy of the deuteron, we calculate the mass defect ΔM, using the values in Table 30.2:

$$\Delta M = m_H + m_n - m_D$$
$$= 1.007825 \text{ u} + 1.008665 \text{ u} - 2.014101 \text{ u} = 0.00239 \text{ u}.$$

(Note that m_H and m_D each include one electron, so in the calculations of ΔM, these atomic masses can be used in place of nuclear masses; the electron masses cancel out.) The energy equivalent is

$$E_B = (0.00239 \text{ u})(931.5 \text{ MeV/u}) = 2.23 \text{ MeV}.$$

This is the amount of energy required to pull the deuteron apart into a proton and a neutron; we call it the *binding energy* of the deuteron. This value is unusually small; binding energies for most nuclei are about 8 MeV per nucleon.

PROBLEM-SOLVING STRATEGY 30.1   Nuclear structure

1. As in Chapters 28 and 29, familiarity with numerical magnitudes is helpful. The scale of things in nuclear structures is very different from that of atomic structures. The size of a nucleus is on the order of 10^{-15} m; the potential energy of interaction between two protons at this distance is 2.31×10^{-13} J, or 1.44 MeV. Typical nuclear interaction energies are on the order of a few million electronvolts (MeV), rather than a few electronvolts (eV) as with atoms. Protons and neutrons are about 1840 times as massive as electrons. The binding energy per nucleon is roughly 1% of the rest energy of a nucleon; compare this percentage with that of the ionization energy of the hydrogen atom, which is only 0.003% of the electron rest energy. By contrast, angular momentum is of the same order of magnitude in both atoms and nuclei, because it is determined by the value of Planck's constant. Compare the results of your calculations with these ranges of magnitude.

2. When you're doing energy calculations that involve the mass defect, binding energies, and so on, note that mass tables nearly always list the masses of *neutral* atoms, with their full complements of electrons. If you need the mass of a bare nucleus, you have to subtract the masses of these electrons. The binding energies of the electrons are much smaller than nuclear binding energies, and we won't worry about them. Calculations of binding energies often involve subtracting two nearly equal quantities. To get enough precision in the difference, you often have to carry five or six significant figures, if that many are available. If not, you may have to be content with an approximate result.

3. Nuclear masses are usually measured in atomic mass units (u). A useful conversion factor is the energy equivalence 1 u = 931.5 MeV.

EXAMPLE 30.2 **Mass defect for carbon**

Find the mass defect, the total binding energy, and the binding energy per nucleon for the common isotope of carbon, $^{12}_{6}C$.

SOLUTION

SET UP The neutral carbon atom consists of six protons, six neutrons, and six electrons. The total mass of six protons and six electrons is equal to the mass of six hydrogen atoms. Therefore, the total mass of the separate parts is $6m_H + 6m_n$, so the mass defect ΔM is given by $\Delta M = 6m_H + 6m_n - m_C$. We find the masses from Table 30.2 and the preceding discussion.

SOLVE We substitute numerical values in this expression for ΔM, obtaining

$$\Delta M = 6(1.007825\ u) + 6(1.0086649\ u) - 12.000000\ u$$
$$= 0.09894\ u.$$

The energy equivalent of this mass is

$$(0.09894\ u)(931.5\ MeV/u) = 92.16\ MeV.$$

Thus, the total binding energy for the 12 nucleons is 92.16 MeV, and the binding energy per nucleon is $92.16\ MeV/12 = 7.68$ MeV per nucleon.

Alternate Solution: We could first find the mass of the *bare* carbon nucleus by subtracting the mass of six electrons from the mass of the neutral atom (including six electrons), given in Table 30.2 as 12.00000 u. Then we find the total mass of six protons and six neutrons. The mass defect is that total minus the mass of the bare nucleus. This method is more complicated than the first solution because the electron masses have to be subtracted explicitly. We don't recommend this alternate solution method.

REFLECT To pull the carbon nucleus completely apart into 12 separate nucleons would require a minimum of 92.16 MeV. The binding energy *per nucleon* is one-twelfth of this, or 7.68 MeV per nucleon. Nearly all stable nuclei, from the lightest to the most massive, have binding energies in the range from 6 to 9 MeV per nucleon, for reasons we'll discuss in the next section.

Practice Problem: Find the mass defect, the total binding energy, and the binding energy per nucleon for the common isotope of helium, ^{4}He. *Answer:* 0.03038 u, 28.30 MeV, 7.075 MeV.

In our discussion of masses, mass defects, and binding energies of nuclei, we have always assumed (without actually saying so) that the nuclei are in their lowest energy state, or ground state. But just as with atoms and molecules, nuclei have internal motion and sets of allowed energy levels, including a ground state (state of lowest energy) and excited states. Because of the great strength of nuclear interactions, excitation energies of nuclei are typically on the order of 1 MeV, compared with a few eV for atomic energy levels. In ordinary physical and chemical transformations, the nucleus always remains in its ground state. When a nucleus is placed in an excited state, either by bombardment with high-energy particles or by a radioactive transformation, it can decay to the ground state by the emission of one or more photons, called in this case gamma rays or gamma-ray photons, with typical energies of a few tenths MeV to a few MeV.

In Section 29.1, we discussed the orbital and spin angular momentum of electrons in atoms. The proton and neutron are also spin-$1/2$ particles; each has a spin angular momentum that can have a component $\pm\frac{1}{2}\hbar$ in any given axis direction. In addition, each particle can have orbital angular momentum due to motion within the nucleus, with possible components in a given axis direction that are integer multiples of $\hbar$. The total angular momentum of a nucleus is usually called **nuclear spin,** although in general it is associated with both orbital and spin angular momentum of the protons and neutrons.

30.2 Nuclear Stability

Nearly all stable nuclei, from the lightest to the most massive, have binding energies in the range of 6 to 9 MeV per nucleon (Figure 30.1). The forces that hold protons and neutrons together in the nucleus, despite the electrical repulsion of the protons, are an example of the strong interactions we mentioned in Section 5.5.

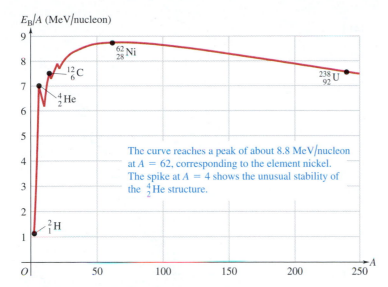

E_B/A (MeV/nucleon)

The curve reaches a peak of about 8.8 MeV/nucleon at $A = 62$, corresponding to the element nickel. The spike at $A = 4$ shows the unusual stability of the $^4_2$He structure.

▲ **FIGURE 30.1** Approximate binding energy per nucleon as a function of mass number A (the total number of nucleons) for stable nuclides.

In the present context, this kind of interaction is called the **nuclear force.** Here are its most important characteristics:

1. The nuclear force does not depend on charge; both neutrons and protons are bound, and the binding is the same for both.
2. The nuclear force has a short range, on the order of nuclear dimensions, 10^{-15} m. (Otherwise, the nucleus would pull in additional protons and neutrons.) But within its range, the nuclear force is much stronger than electrical forces; otherwise, the nucleus could never be stable because of the mutual repulsion of the protons.
3. The nearly constant density of nuclear matter and the nearly constant binding energy per nucleon show that a particular nucleon cannot interact simultaneously with *all* the other nucleons in a nucleus, but only with those few in its immediate vicinity. This property is different from its counterpart in electrical forces: Every proton in the nucleus repels every other one. The limited number of interactions of which nucleons are capable is called *saturation;* it is analogous in some respects to covalent bonding in molecules. (For example, a carbon atom can form at most four covalent bonds with other atoms.)
4. The nuclear force favors the binding of *pairs* of protons or neutrons with opposite spins and of pairs of pairs (i.e., a pair of protons and a pair of neutrons), with each pair having a total spin of zero. Thus, the alpha particle (two protons and two neutrons) is an exceptionally stable nucleus.

These qualitative features of the nuclear force are very helpful in understanding why some nuclear structures are stable and others are unstable. Of about 2500 different nuclides now known, only about 300 are stable. The others are **radioactive**—unstable structures formed during the early history of the universe or from the decay of other unstable nuclei. They decay to form other nuclides by emitting particles and electromagnetic radiation. The time scale of these decay processes ranges from a small fraction of a microsecond to billions of years.

The stable nuclides are plotted on the graph in Figure 30.2, where the neutron number N and proton number (or atomic number) Z for each nuclide are shown. (This chart is called a *Segrè chart* after its inventor, Emilio Segrè.) Each light blue line perpendicular to the line $N = Z$ represents a specific value of the mass number $A = Z + N$. Most lines of constant A pass through only one or two stable nuclides. This means that there are usually only one or two stable nuclides

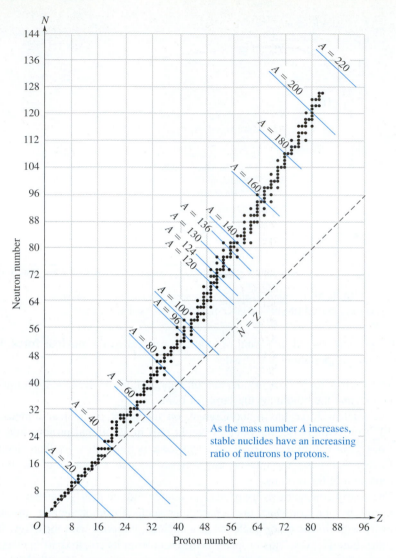

▲ **FIGURE 30.2** Segrè chart showing neutron number and proton number for stable nuclides.

with a given mass number. The lines at $A = 20$, $A = 40$, $A = 60$, and $A = 80$ are examples. In four cases, these lines pass through *three* stable nuclides, namely, at $A = 96$, 124, 130, and 136. Only four stable nuclides have both odd Z and odd N:

$$^2_1\text{H}, \qquad ^6_3\text{Li}, \qquad ^{10}_5\text{B}, \qquad ^{14}_7\text{N};$$

these are called odd–odd nuclides. Also, there is no stable nuclide with $A = 5$ or $A = 8$. The ^4He nucleus (alpha particle), with a pair of protons and a pair of neutrons, does not readily accept a fifth particle into its structure, and a collection of eight nucleons splits rapidly into two ^4He nuclei.

The points representing stable nuclides define a rather narrow region of stability. For small mass numbers, the numbers of protons and neutrons are approximately equal: $N = Z$. The ratio N/Z increases gradually with A, up to about 1.6 at large mass numbers. The graph shows that no nuclide with A greater than 209 or Z greater than 83 is stable. Note also that there is *no* stable nuclide with $Z = 43$ (technetium) or 61 (promethium).

The stability or instability of a nucleus is determined primarily by the competition between the attractive nuclear force and the repulsive electrical force. As we've mentioned, the nuclear force favors pairs of nucleons and pairs of pairs. If there were no electrical interactions, the most stable nuclei would be those with equal numbers of neutrons and protons, $N = Z$. The electrical repulsion shifts the balance to favor a greater numbers of neutrons, but a nucleus with too many neutrons is unstable because not enough of them are paired with protons. A nucleus with too many protons has too much repulsive electrical interaction, compared with the attractive nuclear interaction, to be stable.

As A increases, the positive electric energy per nucleon grows faster than the negative nuclear energy per nucleon, because of the saturation effects with the (attractive) nuclear force that aren't present with the (repulsive) electrical force. At sufficiently large A, a point is reached at which stability is impossible. Thus, the competition between electric and nuclear forces accounts for the fact that the neutron–proton ratio in stable nuclei increases with Z and also for the fact that a nucleus cannot be stable if either A or Z is too large.

 Conceptual Analysis 30.1

The force between nucleons

Which, if any, of the following statements concerning the nuclear force are false?

A. The nuclear force is a very short-range force.
B. The nuclear force is attractive.
C. The nuclear force acts on both protons and neutrons.

SOLUTION All of the preceding statements about the nuclear force are correct.

30.3 Radioactivity

As mentioned earlier, of the 2500 known nuclides, only about 300 are stable. The remaining ones decay by emitting particles and transforming themselves into other nuclides, which also may be unstable. The spontaneous disintegration of nuclides that don't meet the stability requirements we discussed in the preceding section is called **radioactivity.** A nucleus is unstable if it is too big $(Z > 83$ or $A > 209)$ or if its neutron–proton ratio is wrong.

Mastering**PHYSICS**

PhET: Alpha Decay
ActivPhysics 19.4: Radioactivity

Alpha Decay

The two most common decay modes are the emission of alpha (α) particles and the emission of beta (β) particles. When a nucleus disintegrates and emits an alpha particle, it is said to undergo **alpha decay.** An **alpha particle** is identical to the ^{4}He nucleus: two protons and two neutrons bound together, with zero total spin. Alpha emission occurs principally with nuclei that are too large to be stable. When a nucleus emits an α particle, its N and Z values each decrease by two, and A decreases by four. Figure 30.2 shows that the resulting nucleus is closer to stable territory on the Segrè chart than the original nucleus is.

A familiar example of an alpha emitter is radium, ^{226}Ra (Figure 30.3). The speed of the emitted α particle can be determined from the curvature of its trajectory when it passes through a magnetic field. This speed turns out to be about 1.5×10^7 m/s. Although large, it is only 5% of the speed of light, so we can use the nonrelativistic kinetic-energy expression. The α particle mass is 6.64×10^{-27} kg, so the kinetic energy K is

$$K = \tfrac{1}{2}Mv^2 = \tfrac{1}{2}\left(6.64 \times 10^{-27}\,\text{kg}\right)\left(1.5 \times 10^7\,\text{m/s}\right)^2$$
$$= 7.5 \times 10^{-13}\,\text{J} = 4.7 \times 10^6\,\text{eV} = 4.7\,\text{MeV}.$$

Alpha particles are always emitted with a definite kinetic energy, determined by conservation of momentum and energy. They can travel several centimeters in air or a few tenths or hundredths of a millimeter through solids before they are brought to rest by collisions.

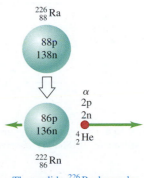

The nuclide $^{226}_{88}$Ra decays by alpha emission to $^{222}_{86}$Rn.

▲ **FIGURE 30.3** Radium decay.

▲ **BIO** Application **Radiation protection.**
Depending on their charge and kinetic
energy, different types of radiation penetrate matter to different degrees. Alpha particles are electrically charged, and because
of their mass they are ejected from radioactive nuclei at relatively low speeds. Therefore, it takes only a little matter to stop
them—a sheet of paper, ordinary clothing,
or the outer layer of your skin. These clothes
are ample protection against an external
alpha emitter. However, if an alpha emitter
(such as radon gas) decays within the body,
it can do serious damage to nearby cells.
Beta particles (electrons) leave the nucleus
at high speeds and thus have more penetrating power than alpha particles. They can
be stopped by heavy clothing and penetrate
only a few millimeters through body tissues. ^{131}I, a beta emitter, is used to treat
thyroid cancer. The isotope is taken up by
the cancerous tissue and kills it. Gamma
rays—being uncharged, energetic photons—
require heavy shielding, such as a lead
shield or concrete wall.

Beta Decay

Early in the 20th century, it was established that the β particle is an *electron.* You
might well ask how a nucleus can emit an electron if there aren't any electrons in
the nucleus. The emission of a β particle involves the transformation of a neutron
in the nucleus into a proton and an electron. We'll discuss such transformations in
more detail later in this chapter. Even a free neutron decays into a proton and an
electron, with an average lifetime of about 15 minutes. Today, this emitted particle is called a **beta-minus (β^-) particle,** to distinguish it from another particle,
the positron (β^+), that is identical, except that it has a positive rather than negative charge. We'll talk more about the whole menagerie of fundamental particles
in Section 30.8.

Beta-minus decay usually occurs with nuclei for which the neutron-to-proton
ratio N/Z is too great for stability. In β^- decay, N decreases by one, Z increases
by one, and A doesn't change. Note that both α and β^- emission, by changing the
Z value of a nucleus, have the effect of changing one element into another—the
dream of medieval alchemists. When a nucleus has a neutron-to-proton ratio that
is too *small* for stability, it may emit a positron (β^+), increasing N by one and
decreasing Z by one.

Beta particles can be identified and their speeds measured by measuring the
curvature of their paths in a magnetic field. The speeds range up to 0.9995 of the
speed of light, in the extreme-relativistic range. If the only two particles involved
were the β^- particle and the recoiling nucleus, conservation of momentum and
energy would require each to have a definite energy. But in fact, β^- particles are
found to have a continuous spectrum of energies. Thus, in 1930, the Swiss theoretical physicist Wolfgang Pauli proposed that a third particle, electrically neutral
and unseen in the original experiments, must be involved.

Enrico Fermi christened this particle the **neutrino.** It wasn't observed directly
until 1953, although by then its existence had been firmly established by indirect
evidence. Today, this particle is called the **antineutrino,** symbolized $\bar{\nu}_e$. The basic
process of β^- decay can be represented as

$$n \rightarrow p + \beta^- + \bar{\nu}_e. \tag{30.4}$$

After α or β^- emission, the remaining nucleus is sometimes left in an excited
state. It can then decay to its ground state by emitting a photon, often called a
gamma-ray photon or simply a **gamma** (γ). Typical γ energies are 10 keV to
5 MeV. For example, α particles emitted from radium have two possible kinetic
energies, either 4.784 MeV or 4.602 MeV, corresponding to a *total* released
energy of 4.871 MeV or 4.685 MeV. When an α with the smaller energy is emitted, the resulting nucleus (which corresponds to the element *radon*), is left in an
excited state. It then decays to its ground state by emitting a γ photon with energy

$$4.871 \text{ MeV} - 4.685 \text{ MeV} = 0.186 \text{ MeV.}$$

A photon with that amount of energy is in fact observed during this decay. These
processes are shown in Figure 30.4.

Decay Series

When a radioactive nucleus decays, the resulting nucleus may also be unstable.
In this case, there is a series of successive decays until a stable configuration is
reached. The most abundant radioactive nucleus found on earth is uranium ^{238}U,
which undergoes a series of 14 decays, including eight α emissions and six β^-
emissions, terminating at the stable isotope of lead, ^{206}Pb.

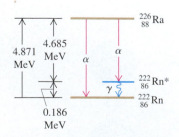

▲ **FIGURE 30.4** Energy-level diagram of γ
photon with energy 0.186 MeV observed during decay.

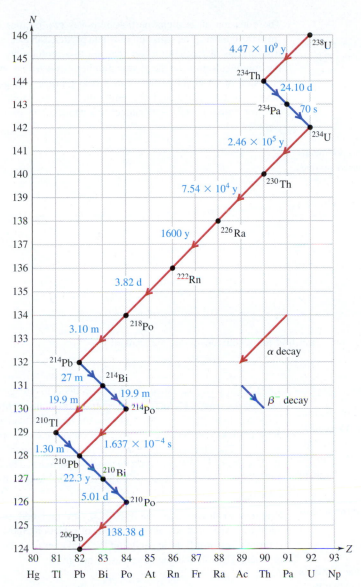

▲ **FIGURE 30.5** Segrè chart showing the uranium ^{238}U decay series, terminating with the stable nuclide ^{206}Pb.

Radioactive decay series can be represented on a Segrè chart, as in Figure 30.5. In alpha emission, N and Z each decrease by two; in beta emission, N decreases by one and Z increases by one. The half-lives (discussed later in this section) of the individual decays are given in years (y), days (d), hours (h), minutes (m), or seconds (s). Figure 30.5 shows the uranium decay series, which begins with the common uranium isotope ^{238}U and ends with an isotope of lead, ^{206}Pb. We note that the series includes unstable isotopes of several elements that also have *stable* isotopes, including thallium (Tl), lead (Pb), and bismuth (Bi). The unstable isotopes of these elements that occur in the ^{238}U series all have too many neutrons to be stable, as we discussed in Section 30.1.

Another decay series starts with the uncommon isotope ^{235}U and ends with ^{207}Pb; a third starts with thorium (^{232}Th) and ends with ^{208}Pb. A fourth series, produced in nuclear reactors, starts with neptunium (^{237}Np) and ends with ^{209}Bi.

Quantitative Analysis 30.2   A change in charge

An element of atomic number 84 decays radioactively to an element of atomic number 83. Which of the following emissions achieves this result?

A. alpha particle only.
B. beta particle only.
C. alpha particle plus beta particle.

SOLUTION   A change in atomic number from 84 to 83 means that the number of protons in the nucleus has decreased by one.

The emission of an alpha particle removes two protons from the nucleus. The emission of a beta particle (an electron) means that a neutron has changed to a proton plus the emitted electron, so the atomic number increases by one. The emission of an alpha particle and a beta particle will give the desired result. (So would the emission of a positron; during positron emission, a proton changes to a neutron plus the emitted positron.)

EXAMPLE 30.3   Radium decay

Suppose you are given the following atomic masses:

$$^{226}_{88}\text{Ra}: \quad 226.025410 \text{ u},$$
$$^{222}_{86}\text{Rn}: \quad 222.017578 \text{ u},$$
$$^{4}_{2}\text{He}: \quad 4.002603 \text{ u}.$$

Show that α emission is energetically possible for radium, and find the kinetic energy of the emitted α particle.

SOLUTION

SET UP First note that the masses given are those of the neutral atoms, including 88, 86, and 2 electrons, respectively, so we have the same numbers of electrons in the initial and final states. This is essential. Next, we have to compare the mass of ^{226}Ra with the sum of the masses of ^{222}Rn and ^{4}He. If the former is greater, then α decay is possible.

SOLVE The difference between the mass of ^{226}Ra and the sum of the masses of ^{222}Rn and ^{4}He is

$$226.025410 \text{ u} - (222.017578 \text{ u} + 4.002603 \text{ u}) = 0.005229 \text{ u}.$$

The fact that this quantity is positive shows that the decay is energetically possible. The energy equivalent of 0.005229 u is

$$E = (0.005229 \text{ u})(931.5 \text{ MeV/u}) = 4.871 \text{ MeV}.$$

REFLECT The total mass of the system has decreased by 0.005229 u, and the corresponding increase in kinetic energy is

4.871 MeV. Thus, we expect the decay products to emerge with a total kinetic energy of 4.871 MeV. Momentum is also conserved: If the parent nucleus is initially at rest, the daughter nucleus and the α particle have momenta with equal magnitude and opposite direction. The kinetic energy is $K = p^2/2m$, so the kinetic energy divides inversely as the masses of the two particles. The α gets $222/(222 + 4)$ of the total, or 4.78 MeV, equal to the observed α energy.

Practice Problem: The nuclide ^{60}Co, an odd–odd unstable nucleus, is used in medical applications of radiation. Show that it is unstable against β decay, and find the total energy of the decay products. The following atomic masses are given:

$$^{60}_{27}\text{Co}: \quad 59.933817 \text{ u},$$
$$^{60}_{28}\text{Ni}: \quad 59.930786 \text{ u}.$$

Answer: 2.82 MeV.

Decay Rates

In any sample of a radioactive element, the number of radioactive nuclei decreases continuously as the nuclei decay. This is a statistical process; there is no way to predict when any individual nucleus will decay. The rates of decay of various nuclei cover a very wide range, from microseconds to billions of years.

Let N be the number of radioactive nuclei in a sample at time t, and let ΔN be the (negative) change in that number during a short time interval Δt. (*Note:* We've previously used N to denote the number of neutrons in a nucleus. The present N is something completely different, so be careful.) The rate of change of N is $\Delta N / \Delta t$. The larger the number of nuclei in the sample, the larger is the number that decay during any interval, so the rate of change of N is proportional to N. That is, it is equal to a constant λ multiplied by N:

Rate of decay of radioactive nuclei

$$\frac{\Delta N}{\Delta t} = -\lambda N. \tag{30.5}$$

The constant λ is called the **decay constant.** It has different values for different nuclides.

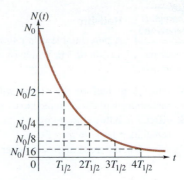

▲ **FIGURE 30.6** Graph of exponential decay.

A large value of λ corresponds to rapid decay, a small value of λ to slower decay. The situation is analogous to a discharging capacitor, which we studied in Section 19.8. Figure 30.6 shows the number N of nuclei as a function of time for the decay of polonium, ^{214}Po.

It's possible to derive an expression for the number N of nuclei remaining after any time t if the number at time $t = 0$ is N_0. The derivation requires calculus, and the result is an exponential function:

Number of radioactive nuclei remaining after time t
If there are N_0 nuclei with decay constant λ at time $t = 0$, the number N remaining after time t is

$$N = N_0 e^{-\lambda t}. \tag{30.6}$$

The **half-life** $T_{1/2}$ of a radioactive substance is the time required for the number of radioactive nuclei to decrease to half the original number N_0. Then half of those remaining nuclei decay during a second interval $T_{1/2}$, and so on. The numbers remaining after successive intervals of $T_{1/2}$ are $N_0/2, N_0/4, N_0/8, \ldots$.

The half-life $T_{1/2}$ and the decay constant λ are related by the following equation:

Decay constant and half-life

$$T_{1/2} = \frac{\ln 2}{\lambda} = \frac{0.693}{\lambda}. \tag{30.7}$$

In particle physics, the life of an unstable nucleus or particle is usually described by the lifetime, not the half-life. The **lifetime** (or mean lifetime) T_{mean} is the average time for a nucleus or particle to decay:

$$T_{\text{mean}} = \frac{1}{\lambda} = \frac{T_{1/2}}{\ln 2} = \frac{T_{1/2}}{0.693}. \tag{30.8}$$

The **activity** of a specimen is the number of decays per unit time. A common unit is the **curie,** abbreviated Ci, defined to be 3.70×10^{10} decays per second:

$$1 \text{ Ci} = 3.70 \times 10^{10} \text{ decays/s.}$$

This is approximately equal to the activity of 1 g of radium. The number of decays is proportional to the number of radioactive nuclei, so the activity decreases with time. Thus, Figure 30.6 is also a graph of the *activity* of polonium, ^{210}Po, which has a half-life of 140 days.

The SI unit of activity is the **becquerel,** abbreviated Bq. One becquerel is one decay per second:

$$1 \text{ Bq} = 1 \text{ decay/s}, \qquad 1 \text{ Ci} = 3.70 \times 10^{10} \text{ Bq.}$$

Conceptual Analysis 30.3

Half-life

A pure sample of a radioactive substance in a container has a half-life of 1 h. Which of the following statements is or are true?

A. After 1 h, half of the atoms will have decayed, and the remainder will decay in the next 1 h.
B. After 1 h, half of the atoms will have decayed, and half of the remainder decay in the next 1 h.
C. After 1 h, the mass in the container is half its original value.

SOLUTION During a half-life, half of the sample of the element is converted, usually to a different element. Half of the remainder of the sample is converted during the next half-life. Thus, B is correct. Because the atoms do not disappear, the mass of the sample remains the same, except for the tiny loss representing the conversion of mass to energy.

Radiocarbon Dating

An interesting application of radioactivity is the dating of archeological and geological specimens by measuring the concentration of radioactive isotopes. The most familiar example is *radiocarbon dating*. The unstable isotope ^{14}C is produced by nuclear reactions in the atmosphere caused by cosmic-ray bombardment; thus, there is a small proportion of ^{14}C in the carbon dioxide in the atmosphere. Plants that obtain their carbon from this source contain the same proportion of ^{14}C as the atmosphere. When a plant dies, it stops taking in carbon, and the ^{14}C it has already absorbed decays, with a half-life of 5730 years. By measuring the proportion of ^{14}C in the remains, we can determine how long ago the organism died.

One difficulty with radiocarbon dating is that the concentration of ^{14}C in the atmosphere has varied by a few percent during the last 100 years because of the burning of fossil fuels and atmospheric testing of nuclear weapons. Corrections can be made on the basis of other data, such as measurements on annual-growth rings of trees. Similar techniques are used with other isotopes for dating geological specimens. Some rocks, for example, contain the unstable potassium isotope ^{40}K, which decays to the stable nuclide ^{40}Ar, with a half-life of 1.3×10^9 y. The age of the rock can be determined by comparing the concentrations of ^{40}K and ^{40}Ar.

EXAMPLE 30.4 **Radiocarbon dating**

The activity of atmospheric carbon due to the presence of ^{14}C is about 0.255 Bq per gram of carbon. What fraction of carbon atoms are ^{14}C? If the activity of an archeological specimen is 0.0637 Bq per gram, what is its approximate age?

SOLUTION

SET UP We know that the half-life of ^{14}C is $T_{1/2} = 5730$ y. We convert years to seconds and then find the decay constant λ from Equation 30.7, $T_{1/2} = 0.693/\lambda$. The number of decays per second in the atmosphere is 0.255, and we can find the *total* number of ^{14}C atoms from Equation 30.5:

$$\frac{\Delta N}{\Delta t} = -\lambda N.$$

Finally, we compare this N with the total number of all carbon atoms, using the fact that one gram of the common isotope ^{12}C is $\frac{1}{12}$ mol.

SOLVE First, the half-life is

$$T_{1/2} = (5730 \text{ y})(3.156 \times 10^7 \text{ s/y}) = 1.81 \times 10^{11} \text{ s}.$$

The decay constant is $\lambda = 0.693/T_{1/2} = 3.38 \times 10^{-12} \text{ s}^{-1}$.

Next, with $\Delta t = 1$ s, Equation 30.5 gives

$$N = -\frac{1}{\lambda}\frac{\Delta N}{\Delta t} = -\left(\frac{1}{3.83 \times 10^{-12} \text{ s}^{-1}}\right)\left(\frac{-0.255}{1 \text{ s}}\right)$$
$$= 6.66 \times 10^{10}.$$

The total number of C atoms in 1 g $(=1/12 \text{ mol})$ is $(1/12 \text{ mol})(6.023 \times 10^{23} \text{ atoms/mol})$, or 5.02×10^{22}. The ratio of ^{14}C atoms to all carbon atoms is

$$\frac{6.66 \times 10^{10}}{5.02 \times 10^{22}} = 1.33 \times 10^{-12}.$$

The activity of the archeological specimen, per unit mass, is one-fourth the activity of ^{14}C in the present atmosphere. Thus, the ^{14}C in the specimen has been decaying for a time equal to twice the half-life, or about 11,400 y.

Continued

REFLECT Only about one carbon atom in a trillion in the earth's atmosphere is ^{14}C, yet such small concentrations can be used to date archeological specimens with surprising precision.

Practice Problem: An archeological specimen with carbon mass 0.25 g has an activity of 8.0×10^{-3} Bq due to ^{14}C decays. Find the approximate age of the specimen. *Answer:* 17,000 y.

30.4 Radiation and the Life Sciences

The interaction of radiation with living organisms is a vitally important topic. In this discussion, we define **radiation** to include radioactivity (alpha, beta, gamma, and neutrons) and electromagnetic radiation such as x rays and gamma rays. As these particles pass through matter, they lose energy, breaking molecular bonds and creating ions—hence the term *ionizing radiation.* Charged particles interact through direct electrical forces on the electrons in the material. X rays and gamma rays interact by the photoelectric effect, in which an electron absorbs a photon and is freed from its molecular bond, or by Compton scattering (Section 28.5). Neutrons cause ionization indirectly by colliding with nuclei or by being absorbed by nuclei, with subsequent alpha or beta decay of the resulting unstable nucleus.

The interactions of radiation with living tissue are extremely complex. It is well known that excessive exposure to radiation, including sunlight, x rays, and all the nuclear radiations, can destroy tissues. In mild cases this results in a burn, as with common sunburn. Greater exposure can cause severe illness or death by a variety of mechanisms, including massive destruction of tissue cells, the alteration of genetic material, and destruction of the components of bone marrow that produce red blood cells.

Radiation Doses

The quantitative description of the effect of radiation on living tissue is called **radiation dosimetry.** The **absorbed dose** of radiation is defined as the energy delivered to the tissue, per unit mass of tissue. The SI unit of absorbed dose, the joule per kilogram, is called the **gray** (Gy); that is, 1 Gy = 1 J/kg. Another unit, in more common use at present, is the **rad,** defined as 0.01 J/kg = 1 cGy:

▲ **BIO** Application **The oldest American** This young woman's skull, found in Brazil, is the oldest human skull known from the Americas. Radiocarbon dating puts its age at about 11,500 years old—more than 2000 years older than the celebrated Kennewick Man. Some scientists argue that her skull measurements and teeth resemble those of Africans or South Sea islanders, raising the possibility that the Americas were peopled from the sea as well as by migration from northeast Asia across the Bering land bridge during the last Ice Age.

Definition of unit of absorbed radiation dose

$$1 \text{ Gy} = 1 \text{ J/kg}, \qquad 1 \text{ rad} = 0.01 \text{ J/kg} = 0.01 \text{ Gy} = 1 \text{ cGy}. \quad (30.9)$$

By itself, the absorbed dose isn't an adequate measure of biological effect, because equal energies of different kinds of radiation cause different degrees of biological effect. This variation is described by a numerical factor called the **relative biological effectiveness (RBE)** of each specific radiation. X rays with 200 keV energy are defined to have an RBE of unity, and the effects of other kinds of radiation can be compared experimentally. Table 30.3 shows approximate values of RBE for several types of radiation. All these values depend somewhat on the energy of the radiation and on the kind of tissue in which the radiation is absorbed.

The biological effect of radiation is measured by the product of the absorbed dose and the RBE of the radiation; this product is called the *biologically equivalent dose,* or simply, equivalent dose. The SI unit of equivalent dose for humans is the **Sievert** (Sv):

TABLE 30.3 Relative biological effectiveness (RBE) for several types of radiation

Radiation	RBE
X rays and gamma rays	1
Electrons	1
Protons	5
Alpha particles	20
Heavy ions	20
Slow neutrons	5–20
	(energy dependent)

Units of equivalent dose

$$\text{equivalent dose (Sv)} = \text{RBE} \times \text{absorbed dose (Gy)}. \qquad (30.10)$$

A more common unit, corresponding to the rad, for equivalent dose for humans is the **rem:**

$$\text{equivalent dose (rem)} = \text{RBE} \times \text{absorbed dose (rad)}. \qquad (30.11)$$

Thus, 1 rem = 0.01 Sv.

Conceptual Analysis 30.4

Absorbing radiation

Suppose a person who has one hand in a wide beam of x rays receives an equivalent dose of 90 millirem in 30 sec. If the person puts both hands in this beam, side by side, for 30 sec, then the equivalent dose the person will receive is

A. 180 millirem, because twice the number of ions will be created.
B. 90 millirem, because the number of ions created per gram is unchanged.
C. 45 millirem, because the radiation is spread over twice as much of the body.

SOLUTION The millirem is related to the number of molecules ionized by radiation in one gram of tissue. So whether your body receives an equivalent dose of 90 millirem to the tip of your finger or 90 millirem to the whole body, the dose is the same. The correct answer is B. However, the number of ions created in the body and the possible destructive effects, and the potential for cancer, are vastly different for the two cases. Any time you receive ionizing radiation (such as that from diagnostic x rays), the area exposed should be kept as small as possible.

EXAMPLE 30.5 **Radiation exposure during an x ray**

During a diagnostic x ray, a broken leg with a mass of 5 kg receives an equivalent dose of 50 mrem. Determine the total energy absorbed and the number of x-ray photons absorbed if the x-ray energy is 50 keV.

SOLUTION

SET UP We note that for x rays, RBE = 1, so the absorbed dose is equal to the equivalent dose. We express the absorbed dose in J/kg, find the energy of one x-ray photon, and then find the number of x-ray photons needed to supply the specified total energy.

SOLVE For x rays, RBE = 1, so the absorbed dose is

$$50 \text{ mrad} = 0.050 \text{ rad} = (0.050)(0.01 \text{ J/kg})$$
$$= 5.0 \times 10^{-4} \text{ J/kg}.$$

The total energy absorbed is

$$(5.0 \times 10^{-4} \text{ J/kg})(5 \text{ kg}) = 2.5 \times 10^{-3} \text{ J}$$
$$= 1.56 \times 10^{16} \text{ eV}.$$

The energy of one x-ray photon is 50 keV = 5.0×10^4 eV.

The number of x-ray photons is

$$\frac{1.56 \times 10^{16} \text{ eV}}{5.0 \times 10^4 \text{ eV}} = 3.1 \times 10^{11} \text{ photons}.$$

REFLECT If the ionizing radiation had been a beam of alpha particles, for which RBE = 20, the absorbed dose needed for an equivalent dose of 50 mrem would be 2.5 mrad, corresponding to a total absorbed energy of 1.25×10^{-4} J.

Practice Problem: In a diagnostic x-ray procedure, 5.00×10^{10} photons are absorbed by tissue with a mass of 0.600 kg. The x-ray wavelength is 0.0200 nm. Find the total energy absorbed by the tissue and the equivalent dose in rem. *Answer:* 4.97×10^{-4} J, 0.0828 rem.

Radiation Hazards

Ionizing radiation is a double-edged sword; it poses serious health hazards, yet it also provides many benefits to humanity, including the diagnosis and treatment of disease and a wide variety of analytical applications. Here are a few numbers for perspective: An ordinary chest x ray delivers about 0.20 to 0.40 mSv to about 5 kg of tissue. (To convert from Sv to rem, simply multiply by 100.) Radiation exposure from cosmic rays and natural radioactivity in soil, building materials, and so on is on the order of 1.0 mSv (0.10 rem) per year at sea level and twice that at an elevation of 1500 m (about 5000 ft). It's been estimated that the average total radiation exposure for the entire U.S. population is roughly 360 mrem or 3.6 mSv per year, with about 80% of this from natural sources and 20% from human-made sources.

A whole-body dose of up to about 0.2 Sv causes no immediately detectable effect. A short-term whole-body dose of 5 Sv or more usually causes death within a few days or weeks. A localized dose of 100 Sv causes complete destruction of the exposed tissue.

The long-term hazards of radiation exposure in causing various cancers and genetic defects have been widely publicized, and the question of whether there is any "safe" level of radiation exposure has been hotly debated. U.S. government regulations are based on a predetermined maximum yearly exposure, from all except natural sources, of 2 to 5 mSv. Workers with occupational exposure to radiation are permitted 50 mSv per year. Recent studies suggest that these limits are too high and that even extremely small exposures carry hazards, but it is very difficult to gather reliable statistics on the effects of low dosages. It has become clear that any use of x rays for medical diagnosis should be preceded by a very careful consideration of the relation of risk to possible benefit.

Another sharply debated question is that of radiation hazards from nuclear power plants. The radiation level from these plants is *not* negligible. However, to make a meaningful evaluation of hazards, we must compare these levels with the alternatives, such as levels from coal-powered plants. The health hazards of coal smoke are serious and well documented, and the natural radioactivity in the smoke from a coal-fired power plant is believed to be roughly 100 times as great as that from a properly operating nuclear plant with equal capacity.

The comparison is complicated by the possibility of a catastrophic nuclear accident and the very serious problem of disposing of radioactive waste from nuclear plants. It is clearly impossible to eliminate *all* hazards to health. Our goal should be rather to try to take a rational approach to the problem of *minimizing* the total hazard from all sources. Figure 30.7 shows one estimate of the various sources of radiation exposure for the U.S. population.

Radon Hazards

A serious health hazard in some regions is the accumulation in houses of radon, ^{222}Rn, an inert, colorless, odorless radioactive gas. Looking at the ^{238}U decay chain (Figure 30.5), we see that radon (^{222}Rn) is produced by the decay of ^{226}Ra, which occurs in minute quantities in the rocks and soil on which houses are built. In turn, ^{222}Rn decays to polonium, ^{218}Po, with a half-life of 3.82 days. It's a dynamic equilibrium situation in which the rate of production equals the rate of decay. The hazard from ^{222}Rn is greater than for the other elements in the ^{238}U decay series because ^{222}Rn is a *gas*. During its short half-life, it can migrate from the soil into your house, especially via basements. If a ^{222}Rn nucleus decays in your lungs, it emits a damaging α particle. And the radioactive daughter nucleus, ^{218}Po, which is *not* chemically inert, is likely to stay there until *it* decays, emits another damaging α particle, and so on through the ^{238}U decay series.

So how great a hazard *is* radon? The average activity in the air inside American homes due to ^{222}Ra is about 1.5 pCi per liter of air, although in some local areas it is as high as 3500 pCi/L. (Recall that 1 pCi = 10^{-12} Ci and 1 Ci = 3.7×10^{10} Bq.) It has been estimated that lifetime exposure to a concentration of 1.5 pCi per liter of air reduces average life expectancy by about 40 days. For comparison, life-long smoking of one pack of cigarettes per day reduces life expectancy by about 6 years, and lifelong exposure to the average emission from all the nuclear power plants in the world reduces life expectancy by something in the range from 0.01 day to 5 days. These figures include catastrophes such as the 1986 nuclear reactor disaster at Chornobyl in Ukraine, for which the *local* effect on life expectancy is much greater and more difficult to determine.

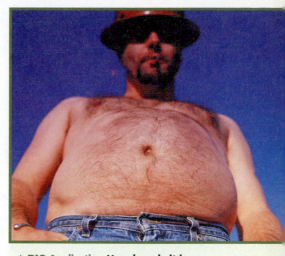

▲ **BIO** Application **How long is it in here?** When studying human exposure to radioactive isotopes, it is important to consider not only an element's physical half-life but also its biological half-life, a measure of how long an element is retained in the human body. As you know, the physical half-life is the time it takes for half of the atoms in a radioactive isotope to decay. The biological half-life is the time it takes for half of an element to be excreted from the body. Although the physical half-lives of radioactive ^3H (12 years) and ^{14}C (5700 years) are relatively long, both are cleared from the body quickly, having biological half-lives of only 12 days. On the other hand, radioactive strontium (^{90}Sr), with a physical half-life of 29 years, is a more serious problem because it is a chemical cousin to calcium and is deposited in bones, giving it a biological half-life of 50 years.

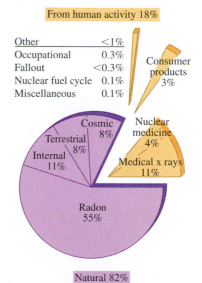

From human activity 18%

Other	<1%
Occupational	0.3%
Fallout	<0.3%
Nuclear fuel cycle	0.1%
Miscellaneous	0.1%

Consumer products 3%

Cosmic 8%

Terrestrial 8%

Internal 11%

Nuclear medicine 4%

Medical x rays 11%

Radon 55%

Natural 82%

▲ **FIGURE 30.7** Contributions of various sources to the total average radiation exposure in the U.S. population, expressed as percentages of the total.

EXAMPLE 30.6 Radioactivity in the air

What is the average activity of a liter of air, in Bq? In decays per hour?

SOLUTION

SET UP We use the average value just cited, 1.5 pCi/L, along with the conversion factor $1\ \text{Ci} = 3.7 \times 10^{10}\ \text{Bq}$.

SOLVE $1.5\text{pCi/L} = (1.5 \times 10^{-12}\ \text{Ci/L})\left(\dfrac{3.70 \times 10^{10}\ \text{Bq}}{1\ \text{Ci}}\right)$

$$= 0.056\ \text{Bq}.$$

Since $1\ \text{Bq} = 1\ \text{decay/s}$, the number of decays in one hour is

$$(0.056\ \text{decays/s})(3600\ \text{s/h}) = 200\ \text{decays/h}.$$

REFLECT A level of activity of air of 1.5 pCi/L represents, on average, the decay of one Rn nucleus every 18 s in each liter of air. If the basement of a medium-size house has a volume of 10^6 L, the total activity of the air in this basement is about 1.5×10^{-6} Ci, or about 5×10^4 disintegrations per second.

Practice Problem: If your lungs contain an average of 5.0 L of air, how many decays occur each minute? *Answer:* 17.

▲ **BIO Application Not science fiction.** This woman is wearing a gamma knife radiosurgery helmet—a device that allows gamma rays to kill a tumor deep in the brain without harming the intervening tissue. Gamma beams aimed through the 101 holes in the helmet converge on the precise site of the tumor. Each beam alone is too weak to do significant harm to the tissue it passes through, but where the beams add up they are lethal. Radiation is used in many other medical techniques, both diagnostic and therapeutic, including CT, PET, and MRI scans, plus others mentioned in the text.

Benefits of Radiation

Radiation is widely used in medicine for intentional selective destruction of tissue such as tumors. The hazards are not negligible, but if the disease would be fatal without treatment, any hazard may be preferable. Artificially produced isotopes are often used as sources of radiation treatment. Such artificial sources have several advantages over naturally radioactive isotopes. They can have shorter half-lives and correspondingly greater activity. Isotopes can be chosen that do not emit alpha particles, which are usually not wanted, and the electrons they emit are easily stopped by thin metal sheets without blocking the desired gamma radiation. Photon and electron beams from linear accelerators have also been used as sources of radiation.

Here are several examples of the expanding field called *nuclear medicine.* Radioactive isotopes have the same electron configurations and chemical behavior as stable isotopes of the same element. But the location and concentration of radioactive isotopes can be detected easily by measurements of the radiation they emit. A familiar example is the use of an unstable isotope of iodine for thyroid studies. A minute quantity of ^{131}I is fed or injected into the patient, and the speed with which it becomes concentrated in the thyroid provides a measure of thyroid function. The half-life is about 8 days, so there are no long-lasting radiation hazards. Nearly all the iodine taken in is either eliminated or stored in the thyroid, and the body does not discriminate between the unstable isotope ^{131}I and the stable isotope ^{127}I.

With the use of more sophisticated scanning detectors, one can also obtain a "picture" of the thyroid, which shows enlargement and other abnormalities. This procedure, a type of autoradiography, is comparable to photographing the glowing filament of an incandescent lightbulb, using the light emitted by the filament itself. If cancerous thyroid nodules are detected, they can be destroyed by much larger doses of ^{131}I.

Similar techniques are used to visualize coronary arteries. A thin tube or catheter is threaded through a vein into the heart, and a radioactive material is injected. Narrowed or blocked arteries can actually be photographed by use of a scanning detector; such a picture is called an *angiogram* or *arteriogram.* A useful isotope for such purposes is technetium (^{99}Tc), formed in the beta decay of molybdenum-99 (^{99}Mo). Technetium-99 is formed in an excited state, from which it decays to the ground state by γ emission with a half-life of about 6 hours, unusually long for γ emission. The chemistry of technetium is such that it can readily be attached to organic molecules that are taken up by various organs of the body. A small quantity of such technetium-bearing molecules is injected into a patient, and a scanning detector or gamma camera is used to produce an image that reveals which parts of the body take up these γ-emitting molecules. This technique, in which ^{99}Tc acts as a radioactive *tracer*, plays an important role in locating cancers, embolisms, and other pathologies (Figure 30.8).

Tracer techniques have many other applications. Tritium (^{3}H), a radioactive isotope of hydrogen, is used to tag molecules in complex organic reactions. In the world of machinery, radioactive iron can be used to study piston ring wear; radioactive tags on pesticide molecules can be used to trace their passage through food chains. Laundry detergent manufacturers test the effectiveness of their products by using radioactive dirt.

Many direct effects of radiation are also useful, such as strengthening polymers by cross-linking, sterilizing surgical tools, dispersing unwanted static electricity in the air, and ionizing the air in smoke detectors. Gamma rays are used to sterilize and preserve some food products.

30.5 Nuclear Reactions

In Section 30.3, we studied the decay of unstable nuclei by the spontaneous emission of an α or β particle, sometimes followed by γ emission. Nothing was done to initiate this emission, and nothing could be done to control it. Now let's consider some processes in which nuclear particles are rearranged as a result of the bombardment of a nucleus by a particle, rather than through a spontaneous natural process.

Rutherford suggested in 1919 that a massive particle with sufficient kinetic energy might be able to penetrate a nucleus. The result would be either a new nucleus with greater atomic number and mass number or a decay of the original nucleus. Rutherford bombarded nitrogen with alpha particles from naturally radioactive sources and obtained an oxygen nucleus and a proton, according to the equation

$$^{4}_{2}\text{He} + ^{14}_{7}\text{N} \rightarrow ^{17}_{8}\text{O} + ^{1}_{1}\text{H}. \tag{30.12}$$

Such a process is called a **nuclear reaction.**

Nuclear reactions obey the classical conservation principles for electric charge, momentum, angular momentum, and energy (including kinetic and rest energies). Another conservation law, not anticipated by classical physics, is conservation of the total number of nucleons. The numbers of protons and neutrons need not be conserved separately. (We've already seen that in β decay a neutron changes into a proton.) We'll look at the basis for the principle of conservation of nucleon number in Section 30.8.

When two nuclei interact, charge conservation requires the sum of the initial atomic numbers (Z) to be equal to the sum of the final atomic numbers. From conservation of the total number of nucleons, the sum of the initial mass numbers (A) is also equal to the sum of the final mass numbers. But the initial rest mass is, in general, *not* equal to the final rest mass, reflecting the fact that the collisions may be inelastic.

Reaction Energy

The difference between the rest masses before and after the reaction corresponds to the **reaction energy,** according to the mass–energy relation $E = mc^2$. If initial particles A and B interact to produce final particles C and D, the reaction energy Q is defined as

$$Q = (M_A + M_B - M_C - M_D)c^2. \tag{30.13}$$

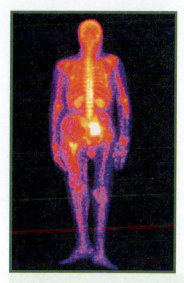

▲ **BIO Application A high "tech" helper.** The most commonly used radioisotope in medicine is technetium-99 (^{99}Tc), which is quite useful in diagnostic procedures for a number of reasons. It decays to produce a low-energy beta particle, which is not very damaging to tissue, plus a non-destructive gamma ray that is energetic enough to leave the body to provide an image, such as the one shown in the bone scan above. ^{99}Tc has a half-life of only 6 hours, which is long enough to monitor metabolic processes in the body but short enough to disappear quickly afterward. Due to its versatility in chemical bonding, it can be incorporated into a wide variety of chemical compounds that can be targeted to specific tissues or organs such as red blood cells, heart muscle, or the spleen. One promising derivative of ^{99}Tc binds selectively to tumor tissues, allowing imaging of cancers that might otherwise be difficult to detect.

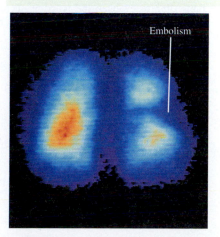

▲ **FIGURE 30.8** Lung scan with radioactive tracer ^{99}Tc. The orange glow in the lung on the left indicates strong γ-ray emission by the ^{99}Tc, which shows that the chemical was able to pass into this lung through the bloodstream. The lung on the right shows weaker emission, indicating an embolism that is restricting blood flow.

To balance the electrons, we use the neutral atomic masses in Equation 30.13 ($_1^1\text{H}$ for a proton, $_2^4\text{He}$, and so on). When Q is positive, the total mass decreases and the total kinetic energy increases. Such a reaction is called an **exoergic reaction** or (borrowing a term from chemistry) an **exothermic reaction.** When Q is negative, the mass increases and the kinetic energy decreases, and the reaction is said to be **endoergic** or **endothermic.** Note that an endoergic reaction cannot occur at all unless the initial kinetic energy is at least as great as $|Q|$; this energy is called the **threshold energy** for the reaction. (1 u is equivalent to 931.5 MeV.)

EXAMPLE 30.7 An exoergic reaction

When lithium is bombarded by a proton, two alpha particles are produced. Find the reaction energy Q.

SOLUTION

SET UP The reaction can be represented as

$$_1^1\text{H} + _3^7\text{Li} \rightarrow _2^4\text{He} + _2^4\text{He}.$$

We need to use the definition of reaction energy, Equation 30.13. The masses we need are given in Table 30.2

SOLVE From Table 30.2, we find the initial and final masses:

$A\!:_1^1\text{H}$	1.007825 u	$C\!:_2^4\text{He}$	4.002603 u
$B\!:_3^7\text{Li}$	7.016005 u	$D\!:_2^4\text{He}$	4.002603 u
	8.023830 u		8.005206 u

Four electron masses are included on each side. We see that

$$M_A + M_B - M_C - M_D = 0.018624 \text{ u}.$$

The mass decreases by 0.018624 u; from Equation 30.13, the reaction energy is

$$Q = (0.018624 \text{ u})(931.5 \text{ MeV/u}) = 17.35 \text{ MeV}.$$

REFLECT The total mass decreases, so the total kinetic energy increases. The final total kinetic energy of the two separating α particles is 17.35 MeV greater than the initial total kinetic energy of the proton and the lithium nucleus.

Practice Problem: Find the Q value for the reaction $_0^1\text{n} + _5^{10}\text{B} \rightarrow _3^7\text{Li} + _2^4\text{He}$. Is this reaction exoergic or endoergic? *Answer:* 2.79 MeV; exoergic.

EXAMPLE 30.8 An endoergic reaction

Find the reaction energy for Rutherford's experiment, described by Equation 30.12.

SOLUTION

SET UP As in Example 30.7, we use the rest masses found in Table 30.2 to evaluate the reaction energy Q, given by Equation 30.12. The initial and total masses each include nine electron masses.

SOLVE The mass calculation, in tabular form, is

$A\!:_2^4\text{He}$	4.002603 u	$C\!:_8^{17}\text{O}$	16.999132 u
$B\!:_7^{14}\text{N}$	14.003074 u	$D\!:_1^1\text{H}$	1.007825 u
	18.005677 u		18.006957 u

We see that the total rest mass increases by 0.001280 u, and the corresponding reaction energy is

$$Q = (-0.001280 \text{ u})(931.5 \text{ MeV/u}) = -1.192 \text{ MeV}.$$

REFLECT This amount of energy is absorbed in the reaction. In a head-on collision with zero total momentum, the minimum total initial kinetic energy for this reaction to occur is 1.192 MeV. Ordinarily, though, this reaction would be produced by bombarding stationary ^{14}N nuclei with α particles. In this case, the α energy must be *greater* than 1.192 MeV. The α can't give up *all* of its kinetic energy, because then the final total kinetic energy would be zero and momentum would not be conserved. It turns out that, to conserve momentum, the initial α energy must be at least 1.533 MeV.

Practice Problem: Consider the reaction $_3^6\text{Li} + _2^4\text{He} \rightarrow _4^9\text{Be} + _1^1\text{H}$, produced by bombarding a solid lithium target with α particles. Show that this reaction is endoergic, and find the amount by which the total initial kinetic energy exceeds the total final value. *Answer:* 2.125 MeV.

For a charged particle such as a proton or an α particle to penetrate the nucleus of another atom, it usually must have enough initial kinetic energy to overcome the potential-energy barrier caused by the repulsive electrostatic forces. For example, in the reaction of Example 30.7, suppose we treat the proton and the ^7Li nucleus as spherically symmetric charges with radius given by

$R = R_0 A^{1/3}$ (Equation 30.2) and with a distance of about 3.5×10^{-15} m between their centers. Then the repulsive potential energy U of the proton (charge $+e$) and the lithium nucleus (charge $+3e$) at this distance is

$$U = \frac{1}{4\pi\epsilon_0} \frac{(e)(3e)}{r} = \frac{(9.0 \times 10^9 \, \text{N} \cdot \text{m}^2 \cdot \text{C}^{-2})(3)(1.6 \times 10^{-19} \, \text{C})^2}{3.5 \times 10^{-15} \, \text{m}}$$
$$= 1.98 \times 10^{-13} \, \text{J} = 1.24 \times 10^6 \, \text{eV} = 1.24 \, \text{MeV}.$$

Even though energy is liberated in this reaction, the proton must have a minimum, or *threshold,* energy of about 1.2 MeV for the reaction to occur.

The absorption of neutrons by nuclei forms an important class of nuclear reactions. Heavy nuclei bombarded by neutrons in a nuclear reactor can undergo a series of neutron absorptions alternating with beta decays, in which the mass number A increases by as much as 25. Some of the transuranic elements (elements having Z larger than 92, which don't occur in nature) are produced in this way. Many transuranic elements, having Z as high as 116, have been identified.

The analytical technique of neutron activation analysis uses similar reactions. When stable nuclei are bombarded by neutrons, some nuclei absorb neutrons and then undergo β^- decay. The energies of the β^- and γ emissions depend on the unstable parent nuclide and provide a means of identifying it. The presence of quantities of elements far too small for conventional chemical analysis can be detected in this way.

30.6 Nuclear Fission

Nuclear fission is a decay process in which an unstable nucleus splits into two fragments of comparable mass instead of emitting an α or β particle. Fission was discovered in 1939, when Otto Hahn and Fritz Strassman bombarded uranium $(Z = 92)$ with neutrons. The resulting radiation did not coincide with that of any known radioactive nuclide. Meticulous chemical analysis led to the astonishing conclusion that they had found radioactive isotopes of barium $(Z = 56)$ and krypton $(Z = 36)$. They concluded, correctly, that the uranium nuclei were splitting into two massive fragments, which they called **fission fragments.** A few free neutrons usually appeared along with the fission fragments. The energy released during fission, almost 200 MeV per nucleus, emerged as kinetic energy of the fission fragments.

Both the common (99.3%) isotope ^{238}U and the uncommon (0.7%) isotope ^{235}U (as well as several other nuclides) can be split by neutron bombardment, ^{235}U by slow neutrons, but ^{238}U only by neutrons with at least 1.2 MeV of energy. Fission resulting from neutron absorption is called **induced fission.** Some nuclei can also undergo **spontaneous fission,** which occurs without initial neutron absorption. More than 100 different nuclides representing more than 20 different elements have been found among the fission products. Figure 30.9 shows the distribution of mass numbers for fission fragments from the nuclide ^{235}U.

Fission reactions are always exoergic, because the fission fragments have greater binding energy per nucleon than the original nucleus. The total kinetic energy of the fission fragments is enormous, about 200 MeV (compared with typical α and β energies of a few MeV).

Fission fragments always have too many neutrons to be stable. The N/Z value for stable nuclides is about 1.3 at $A = 100$ and 1.4 at $A = 150$. The fragments have approximately the same N/Z as ^{235}U, about 1.55. They respond to this surplus of neutrons by emitting two or three free neutrons and by undergoing a series of β^- decays (each of which increases Z by one and decreases N by one), until a stable value of N/Z is reached.

PhET: Nuclear Fission

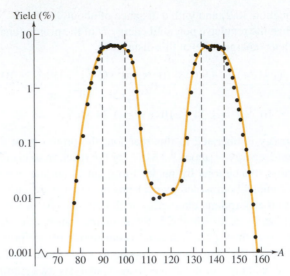

▲ **FIGURE 30.9** Mass distribution of fission fragments.

Chain Reactions

Fission of a uranium nucleus, triggered by neutron bombardment, releases other neutrons that can trigger more fissions; thus, a **chain reaction** is possible (Figure 30.10). The reaction can be made to proceed slowly and in a controlled manner, as in a nuclear reactor, or explosively, as in a bomb. The energy released in a chain reaction is enormous, far greater than in any chemical reaction. For example, when uranium is oxidized, or "burned," to uranium dioxide (UO_2), the heat of combustion is about 4500 J/g. Expressed as energy per atom, this is about 11 eV per uranium atom. By contrast, fission liberates about 200 MeV per atom, 20 million times as much energy.

Nuclear Reactors

A **nuclear reactor** is a system in which a controlled nuclear chain reaction is used to liberate energy. In a nuclear power plant, this energy is used to generate steam, which operates a turbine and turns an electrical generator. On the average, each fission of a ^{235}U nucleus produces about 2.5 free neutrons, so 40% of the neutrons are needed to sustain a chain reaction. The probability of neutron absorption by a nucleus is much greater for low-energy (less than 1 eV) neutrons than for the higher-energy (1 MeV) neutrons liberated during fission. The emitted neutrons are slowed down by collisions with nuclei in the surrounding material, called the *moderator*, so that they can cause further fissions. In nuclear power plants the moderator is often water, occasionally graphite.

The rate of the reaction is controlled by inserting or withdrawing control rods made of elements (often cadmium) whose nuclei absorb neutrons without undergoing any additional reaction. The isotope ^{238}U can also absorb neutrons, leading to ^{239}U, but not with high enough probability for it to sustain a chain reaction by itself. Thus, uranium used in reactors is "enriched" by increasing the proportion of ^{235}U from the natural value of 0.7%, typically to 3% or so, by isotope-separation processing.

In a nuclear power plant, the fission energy appears as kinetic energy of the fission fragments, and its immediate result is to heat the fuel elements and the surrounding water. This heat generates steam to drive turbines, which in turn drive the electrical generators (Figure 30.11). The steam generator is a heat exchanger that takes heat from this highly radioactive water and generates non-radioactive steam to run the turbines.

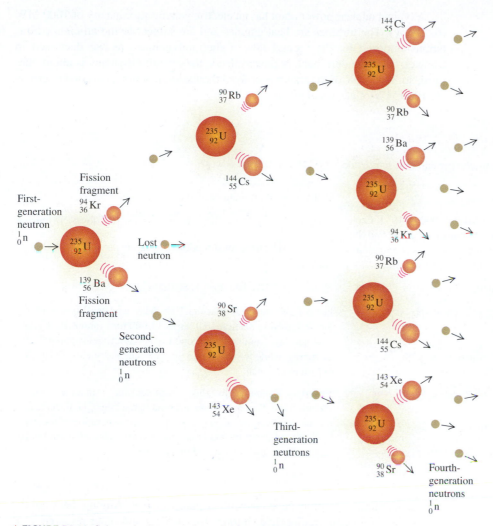

▲ **FIGURE 30.10** Schematic diagram of a nuclear fission chain reaction.

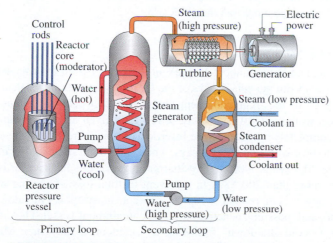

▲ **FIGURE 30.11** Schematic diagram of a nuclear power plant.

A typical nuclear power plant has an electric generating capacity of 1000 MW (or 10^9 W). The turbines are heat engines and are subject to the efficiency limitations imposed by the second law of thermodynamics, as we discussed in Chapter 16. In modern nuclear power plants, the overall efficiency is about one-third, so 3000 MW of thermal power from the fission reaction is needed to generate 1000 MW of electrical power.

EXAMPLE 30.9 Fuel consumption in a reactor

How much uranium has to undergo fission per unit time to provide 3000 MW of thermal power?

SOLUTION

SET UP Each second, we need 3000 MJ, or 3000×10^6 J. Each fission provides 200 MeV. From these numbers, we can find the number of fissions per second. Then, using the mass of a uranium atom, we can find the total mass of U needed per unit time.

SOLVE Each fission provides an amount of energy

$$200 \text{ MeV} = (200 \text{ MeV})(1.6 \times 10^{-13} \text{ J/MeV})$$
$$= 3.2 \times 10^{-11} \text{ J}.$$

The number of fissions needed per second is

$$\frac{3000 \times 10^6 \text{ J}}{3.2 \times 10^{-11} \text{ J}} = 9.4 \times 10^{19}.$$

Each uranium atom has a mass of about

$$(235)(1.67 \times 10^{-27} \text{ kg}) = 3.9 \times 10^{-25} \text{ kg},$$

so the mass of uranium needed per second is

$$(9.4 \times 10^{19})(3.9 \times 10^{-25} \text{ kg}) = 3.7 \times 10^{-5} \text{ kg} = 37 \text{ mg}.$$

REFLECT The total consumption of uranium per day (86,400 seconds) is

$$(3.7 \times 10^{-5} \text{ kg/s})(86,400 \text{ s/d}) = 3.2 \text{ kg/d}.$$

For comparison, note that the 1000 MW coal-fired power plant we described in Section 16.9 burns 14,000 tons (about 10^7 kg) of coal per day. Combustion of one carbon atom yields about 2 eV of energy, while fission of one uranium nucleus yields 200 MeV, 10^8 times as much.

Practice Problem: In this example, the calculated mass includes only the mass of the fissionable nuclide $^{235}_{92}$U. If the reactor fuel has been enriched to 3% by isotope separation of the natural mix of uranium nuclides, what total mass of uranium is required per day? *Answer:* About 100 kg/d.

Nuclear fission reactors have many other practical uses. Among these are the production of artificial radioactive isotopes for medical and other research, producing high-intensity neutron beams for research in nuclear structure, and producing fissionable transuranic elements such as plutonium ^{239}Pu from the common uranium isotope ^{238}U. The last is the function of *breeder reactors,* which can actually produce more fuel than they use.

Hazards of Nuclear Reactors

We mentioned earlier that about 15 MeV of the energy from the fission of a ^{235}U nucleus comes from the beta decays of the fission fragments rather than from the kinetic energy of the fragments themselves. This fact poses a serious problem with respect to the control and safety of reactors. Even after the chain reaction has been completely stopped by the insertion of control rods into the core, heat continues to be evolved by the β decays, which cannot be stopped. For the reactor in Example 30.9 (with 3000 MW of thermal power), this heat power is initially very large, more than 200 MW. In the event of total loss of cooling water, this amount of power is more than enough to cause a catastrophic "meltdown" of the reactor core and, possibly, penetration of the containment vessel. The difficulty in achieving a "cold shutdown" following an accident at the Three Mile Island nuclear power plant in Pennsylvania in March 1979 was a result of the continued evolution of heat due to β decays.

The Chornobyl catastrophe of April 26, 1986 resulted from a combination of an inherently unstable design and several human errors committed during a test of the emergency core cooling system. Too many control rods were withdrawn to compensate for a decrease in power caused by a buildup of neutron absorbers such as ^{135}Xe. The power level rose from 1% of normal to 100 times normal in 4 seconds; a steam explosion ruptured pipes in the core cooling system and blew the heavy concrete cover off the reactor vessel. The graphite moderator caught fire and burned for several days. The total activity of the radioactive material released into the atmosphere has been estimated as about 10^8 Ci.

30.7 Nuclear Fusion

In a **nuclear fusion** reaction, two or more small light nuclei combine to form a larger nucleus. Fusion reactions release energy for the same reason as fission reactions: The binding energy per nucleon after the reaction is greater than before. Referring to Figure 30.1, we see that the binding energy per nucleon increases with A up to about $A = 60$, so the fusion of nearly any two light nuclei to make a nucleus with A less than 60 is likely to be an exoergic reaction.

ActivPhysics 19.3: Fusion

Here are three examples of energy-liberating fusion reactions:

$$^1_1\text{H} + ^1_1\text{H} \rightarrow ^2_1\text{H} + \beta^+ + \nu_e,$$
$$^1_1\text{H} + ^2_1\text{H} \rightarrow ^3_2\text{He} + \gamma,$$
$$^3_2\text{He} + ^3_2\text{He} \rightarrow ^4_2\text{He} + ^1_1\text{H} + ^1_1\text{H}.$$

In the first reaction, two protons combine to form a deuteron, a β^+ or positron (a positively charged electron that we'll talk about in Section 30.8), and an electron neutrino. In the second, a proton and a deuteron (^2_1H, a proton and a neutron bound together) combine to form the light isotope of helium, ^3_2He. In the third, two ^3_2He nuclei unite to form ordinary helium (^4_2He) and two protons.

The positrons produced during the first step of this sequence collide with electrons; mutual annihilation takes place, and their energy is converted into γ radiation. The net effect of the sequence is therefore the combination of four hydrogen nuclei into a helium nucleus and gamma radiation. The net energy release, which can be calculated from the mass balance, turns out to be 26.7 MeV.

These fusion reactions, collectively known as the proton–proton chain, take place in the interior of the sun and other stars. Each gram of the sun's mass contains about 4.5×10^{23} protons. If all of these protons were fused into helium, the energy released would be about 130,000 kWh. If the sun were to continue to radiate at its present rate, it would take about 75 billion years to exhaust its supply of protons.

For two nuclei to fuse, they must come together to within the range of the nuclear force, typically on the order of 2×10^{-15} m. To do this, they must overcome the electrical repulsion of their positive charges. For two protons a distance of 2×10^{-15} m apart, the corresponding potential energy is on the order of 1.1×10^{-13} J, or 0.7 MeV; this amount represents the initial kinetic energy the fusing nuclei must have.

Such energies are available at extremely high temperatures. According to Section 15.3 (Equation 15.8), the average translational kinetic energy of a gas molecule at temperature T is $\frac{3}{2}kT$, where k is Boltzmann's constant. For this energy to be equal to 1.1×10^{-13} J, the temperature must be on the order of 5×10^9 K. Not all the nuclei have to have this energy, but the temperature must be on the order of millions of kelvins if any appreciable fraction of the nuclei are to have enough kinetic energy to surmount the electrical repulsion and achieve fusion. Fusion chain reactions with such high temperatures occur in the interiors of stars. Because of these extreme temperature requirements, such reactions are called **thermonuclear reactions.** An uncontrolled thermonuclear reaction occurs in a hydrogen bomb, with enormous destructive power.

Intensive efforts are underway in many laboratories to achieve controlled fusion reactions, which potentially represent an enormous new energy resource. In one kind of experiment, a plasma is heated to an extremely high temperature by an electrical discharge, while being contained by appropriately shaped magnetic fields. In another experiment, pellets of the material to be fused are heated by a high-intensity laser beam. As yet, no one has succeeded in producing fusion reactions under controlled conditions to yield a net surplus of usable energy.

Methods of achieving fusion that don't require high temperatures are also being studied; these are called **cold fusion.** A few researchers have claimed to have achieved cold fusion in an electrolytic process, but their results have not been confirmed by other investigators.

30.8 Fundamental Particles

It can be argued that the study of fundamental particles began in about 400 B.C., when the Greek philosophers Democritus and Leucippus suggested that matter is made of indivisible particles that they called *atoms.* This idea lay dormant until about 1804, when John Dalton (1766–1844), often called the father of modern chemistry, discovered that many chemical phenomena could be explained on the basis of atoms of each element as the fundamental, indivisible building blocks of matter. But toward the end of the 19th century, it became clear that atoms are *not* indivisible. The electron, the photon, and the proton were all discovered within a span of 20 years at the beginning of the 20th century. By 1925, quantum mechanics was in full flower as the key to understanding atomic structure, although many details remained to be worked out. At that time, it appeared that the proton and the electron were the basic building blocks of all matter.

The Neutron

The discovery of the neutron in 1930 was an important milestone. In that year, two German physicists, Bothe and Becker, observed that when beryllium, boron, or lithium was bombarded by high-energy (several MeV) α particles from the radioactive element polonium, the bombarded material emitted a radiation that had much greater penetrating ability than the original α particles. Experiments by James Chadwick the following year showed that this emanation consisted of uncharged (electrically neutral) particles with mass approximately equal to that of the proton. Chadwick christened these particles *neutrons.* A typical reaction, using a beryllium target, is

$$\ ^4_2\text{He} + \ ^9_4\text{Be} \rightarrow \ ^{12}_6\text{C} + \ ^1_0\text{n}, \tag{30.14}$$

where ^1_0n denotes a neutron.

Because neutrons have no charge, they produce no ionization when they pass through gases, and they are not deflected by electric or magnetic fields. They interact only with nuclei; they can be slowed down during elastic scattering with a nucleus, and they can penetrate the nucleus. Slow neutrons can be detected by means of another nuclear reaction—the ejection of an α particle from a boron nucleus, according to the reaction

$$\ ^1_0\text{n} + \ ^{10}_5\text{B} \rightarrow \ ^7_3\text{Li} + \ ^4_2\text{He}. \tag{30.15}$$

Because of its electric charge, the ejected α particle is easy to detect with a Geiger counter or other particle detector

The discovery of the neutron cleared up a mystery about the composition of the nucleus. Before 1930, it had been thought that the total mass of a nucleus was due only to its protons, but no one understood why the charge-to-mass ratio wasn't the same for all nuclei. It soon became clear that all nuclei (except hydrogen) contain both protons and neutrons. In fact, the proton, the neutron, and the electron are the basic building blocks of atoms. One might think that that would be the end of the story. On the contrary, it is barely the beginning!

The Positron

The positive electron, or **positron** (denoted by β^+ or e^+), was first observed in 1932 by Carl D. Anderson during an investigation of cosmic rays. Anderson used a cloud chamber, a common experimental tool in the early days of particle physics. In the cloud chamber, supercooled vapor condenses around a line of ions created by the passage of a charged particle; the result is a visible track whose density depends on the particle's speed. When the cloud chamber is placed in a magnetic field, the paths of charged particles are curved; measurements of the curvature of their trajectories can be used to determine the charge and momentum, and thus the mass, of the particles.

In one historic cloud-chamber photograph (Figure 30.12), the density and curvature of a certain track suggested a mass equal to that of the electron. But the track curved the wrong way, showing that the particle had positive charge. Anderson concluded correctly that the track had been made by a positive electron, or **positron.** The mass of the positron is equal to that of an ordinary (negative) electron; its charge is equal in magnitude, but opposite in sign, to the electron charge. Pairs of particles related to each other in this way are said to be **antiparticles** of each other.

Positrons do not form a part of ordinary matter. They are produced in high-energy collisions of charged particles or gamma rays with matter in a process called **pair production,** in which an ordinary electron (e^- or β^-) and a positron (e^+ or β^+) are produced simultaneously (Figure 30.13). Electric charge is conserved in this process, but enough energy E must be available to account for the energy equivalent of the rest masses m of the two particles. The minimum energy for pair production is

$$E = 2mc^2 = 2(9.11 \times 10^{-31}\,\text{kg})(3.00 \times 10^8\,\text{m/s})^2$$
$$= 1.64 \times 10^{-13}\,\text{J} = 1.02\,\text{MeV}.$$

The inverse process, e^+e^- annihilation, occurs when a positron and an electron collide. Both particles disappear, and two or three gamma-ray photons appear, with total energy $2mc^2$. Decay into a single photon is impossible because such a process cannot conserve both energy and momentum.

Positrons also occur in the decay of some unstable nuclei. Recall that nuclei with too many neutrons for stability often emit a β^- particle (an electron), decreasing N by one and increasing Z by one. A nucleus with too few neutrons for stability may respond by converting a proton to a neutron and emitting a positron (β^+), thereby increasing N by one and decreasing Z by one. Such nuclides don't occur in nature, but they can be produced artificially by neutron bombardment of stable nuclides in nuclear reactors. An example is the unstable odd–odd nuclide ^{22}Na, which has one less neutron than the stable ^{23}Na. The nuclide ^{22}Na decays with a half-life of 2.6 y by emitting a positron, leaving the stable even–even nuclide ^{22}Ne, with the same mass number $A = 22$.

Mesons as Force Mediators

In classical physics, we describe the interaction of charged particles in terms of Coulomb's-law forces. In quantum mechanics, we can describe this interaction in terms of the emission and absorption of photons. Two electrons repel each

▲ **FIGURE 30.12** The cloud-chamber track made by the first positron ever identified. The lead plate in this photograph is 6 mm thick.

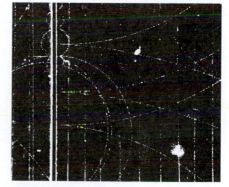

(a)

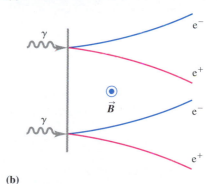

(b)

▲ **FIGURE 30.13** (a) Photograph of bubble-chamber tracks of electron–positron pairs that are produced when 300 MeV photons strike a lead sheet. A magnetic field directed out of the photograph made the electrons and positrons curve in opposite directions. (b) Diagram showing the pair-production process for two of the photons.

Two skaters exert repulsive forces on each other by tossing a ball back and forth.

(a)

Two skaters exert attractive forces on each other when one tries to grab the ball out of the other's hands.

(b)

▲ **FIGURE 30.14** Exchange of a ball as an example of a force mediator.

ActivPhysics 19.5: Particle Physics

other as one emits a photon and the other absorbs it, just as two skaters can push each other apart by tossing a ball back and forth between them (Figure 30.14a). If the charges are opposite and the force is attractive, we imagine the skaters grabbing the ball away from each other (Figure 30.14b). We say that the electromagnetic interaction between two charged particles is *mediated,* or transmitted, by photons.

In 1935, the nature of the nuclear force, which binds protons and neutrons together despite the electrical repulsions of the protons, was a complete mystery. In that year, the Japanese physicist Hideki Yukawa (1907–1981) proposed a hypothetical particle he called a **meson,** with mass intermediate between those of the electron and the proton. Yukawa suggested that nucleons could interact by emitting and absorbing mesons in a process analogous to the electrical interaction between charged particles by the exchange of photons. At the time, there was not the slightest shred of evidence that mesons actually existed. But it was clear that a radical idea was needed to understand nuclear forces, and the scientific world was receptive to Yukawa's proposal.

There were some blind alleys encountered in developing Yukawa's idea. In 1937, two new particles were discovered. These particles are now called **muons.** The μ^- has a charge equal to that of the electron, and its antiparticle, the μ^+, has a positive charge of equal magnitude. The two particles have equal mass, about 207 times the electron mass (that is, $207m_e$). The energy equivalent is about 106 MeV, so the rest mass can also be expressed as $106\ \text{MeV}/c^2$. Muons, like electrons, have spin $\frac{1}{2}$. They are unstable; each decays, usually into an electron with the same sign, plus two neutrinos, with a lifetime of about 2.2×10^{-6} s. In short, muons act very much like massive electrons.

It soon became clear that these particles had no strong interactions with nuclei and therefore could not be Yukawa's mesons. But in 1947 yet another family of unstable particles was discovered. These were christened π mesons or **pions;** they have charges $+e$, $-e$, and zero. The charged pions (π^+ and π^-) have masses of about 273 times the electron mass ($273m_e$), with rest energy 140 MeV. Each usually decays into a muon with the same sign, plus a neutrino, with a lifetime of about 2.6×10^{-8} s. The neutral pion (π^0) has a smaller mass, about 264 electron masses ($264m_e$), with a corresponding rest energy of 135 MeV, and decays into two gamma-ray photons with an extremely short lifetime of about 8.4×10^{-17} s. The pions *do* interact strongly with nuclei, and they *are* the particles predicted by Yukawa as mediators of the nuclear force.

The discovery of pions in 1947 resolved some questions about the nuclear force. But more surprises loomed on the horizon, and an entire new branch of physics, now called *high-energy physics,* began to unfold. In the next section, we'll sketch briefly some aspects of this new field.

30.9 High-Energy Physics

From the security of 1931, when it was thought that there were three permanent, unchanging fundamental particles, we enter the partly mapped territory of present-day particle physics, in the midst of a veritable deluge of new particles. The entities that we're accustomed to calling particles are not permanent; they can be created and destroyed during interactions with other particles. Each particle has an associated antiparticle; for a few, the particle and its antiparticle are identical. Many particles are unstable, decaying spontaneously into other particles. Some particles serve as mediators, or transmitters, of the various interactions.

The Four Forces

Here's where things stood in 1950: Four categories of interactions (often called *forces*) had been identified at that time. They are, in order of decreasing strength,

1. the strong interaction,
2. the electromagnetic interaction,
3. the weak interaction, and
4. the gravitational interaction.

The electromagnetic and gravitational forces are familiar from classical physics. Both have a $1/r^2$ dependence on distance, but the gravitational force is very much weaker than the electromagnetic force. For example, the gravitational attraction of two protons is smaller than their electrical repulsion by a factor of about 10^{36}. The gravitational force is of primary importance in the structure of stars and the large-scale behavior of the universe, but it is not believed to play a significant role in particle interactions at currently attainable energies.

The other two forces are less familiar. The strong interaction is responsible for the nuclear force and also for the production of pions and several other particles in high-energy collisions. Within its range, the strong interaction is roughly 100 times as strong as the electromagnetic interaction, but it drops off with distance more quickly than $1/r^2$. The fourth interaction, called the weak interaction, is responsible for beta decay, such as the conversion of a neutron into a proton, an electron, and an antineutrino. It is also responsible for the decay of many unstable particles, such as pions into muons, muons into electrons, Σ particles into protons, and so on. Like the strong interaction, the weak interaction is a short-range interaction, but it is weaker by a factor of about 10^9.

More Particles

We've outlined the most important properties of muons and pions. The next major event was the discovery of the **antiproton** (denoted by $\bar{\text{p}}$). The existence of this particle had been suspected ever since the discovery of the positron in 1932. The $\bar{\text{p}}$ was finally found in 1955, when proton–antiproton (p–$\bar{\text{p}}$) pairs were created with the use of a beam of 6 GeV protons. The antineutron was found soon afterward. Especially after 1960, as higher-energy accelerators and more sophisticated detectors were developed, a veritable blizzard of new unstable particles were identified. To describe and classify them, we need a small blizzard of new terms.

It became clear that it's useful to classify particles in terms of their interactions. The two principal categories are **hadrons,** which have strong interactions, and **leptons,** which do not. Hadrons include baryons and mesons; baryons are nucleons and more massive particles that resemble nucleons. We'll also distinguish between **bosons,** which always have zero or integer spins, and **fermions,** which have half-integer spins. Fermions obey the exclusion principle; bosons do not.

Leptons

Leptons include the electrons ($\text{e}^{\pm}$), the muons ($\mu^{\pm}$), the tau particles ($\tau^{\pm}$), and three kinds of neutrinos. The taus, discovered in 1975, have spin $\frac{1}{2}$ and mass 1784 MeV/c^2. In all, there are six leptons and six antileptons. All leptons have spin $\frac{1}{2}$. Taus and muons are unstable; a tau can decay into a muon plus two neutrinos, and a muon decays into an electron plus two neutrinos. Originally, the neutrinos were believed to have zero rest mass, but recent experimental evidence shows that they have small nonzero masses. Note that the τ particles are more massive than nucleons; they are classified as leptons rather than baryons because they have no strong interactions. Leptons obey a *conservation principle*. For the three kinds of

TABLE 30.4 **The leptons**

Particle name	Symbol	Anti-particle	Mass (MeV/c^2)	L_e	L_μ	L_τ	Lifetime (s)	Principal decay modes
Electron	e^-	e^+	0.511	+1	0	0	Stable	
Electron neutrino	ν_e	$\bar\nu_e$	$<3 \times 10^{-6}$	+1	0	0	Stable	
Muon	μ^-	μ^+	105.7	0	+1	0	2.20×10^{-6}	$e^-\bar\nu_e\nu_\mu$
Muon neutrino	ν_μ	$\bar\nu_\mu$	<0.19	0	+1	0	Stable	
Tau	τ^-	τ^+	1777	0	0	+1	2.9×10^{-13}	$\mu^-\bar\nu_\mu\nu_\tau, e^-\bar\nu_e\nu_\tau$
Tau neutrino	ν_τ	$\bar\nu_\tau$	<18.2	0	0	+1	Stable	

leptons there are three lepton numbers L_e, L_μ, and L_τ. In all interactions, each lepton number is separately conserved. Table 30.4 lists all the known leptons.

Hadrons

Hadrons, the strongly interacting particles, are a much more complex family than leptons. There are two subclasses: mesons and baryons. **Mesons** have spin 0 or 1, and baryons have spin $\frac{1}{2}$ or $\frac{3}{2}$. Therefore, all mesons are bosons, and all baryons are fermions. Mesons include the pions, already mentioned, and several other particles, including K mesons (or *kaons*), η mesons, and other particles that we'll mention later. **Baryons** include the nucleons and several particles called *hyperons,* including the Λ, Σ, Ξ, and Ω. These resemble the nucleons but are more massive. All the hyperons are unstable, decaying by various processes to other hyperons or to nucleons. All the mesons are unstable because all can decay to less massive particles, in accordance with all the conservation laws governing such decays. Each hadron has an associated antiparticle, denoted with an overbar, like the antiproton $\bar p$. Table 30.5 is a sampling of the many known hadrons (including mesons and baryons). We'll discuss the terms *strangeness, isospin,* and *quark content* later.

TABLE 30.5 **Some known hadrons and their properties**

Symbol	Mass, MeV/c^2	Charge	Spin	Isospin	Strangeness	Mean lifetime, s	Typical decay modes	Quark content
Mesons								
π^0	135.0	0	0	1	0	8.4×10^{-17}	$\gamma\gamma$	$u\bar u, d\bar d$
π^+	139.6	+1	0	1	0	2.60×10^{-8}	$\mu^+\nu_\mu$	$u\bar d$
π^-	139.6	−1	0	1	0	2.60×10^{-8}	$\mu^-\bar\nu_\mu$	$\bar u d$
K^+	493.7	+1	0	1/2	+1	1.24×10^{-8}	$\mu^+\nu_\mu$	$u\bar s$
K^-	493.7	−1	0	1/2	−1	1.24×10^{-8}	$\mu^-\bar\nu_\mu$	$\bar u s$
η^0	547.3	0	0	0	0	$\sim 10^{-18}$	$\gamma\gamma$	$u\bar u, d\bar d, s\bar s$
Baryons								
p	938.3	+1	1/2	1/2	0	stable	—	uud
n	939.6	0	1/2	1/2	0	886	$pe^-\bar\nu_e$	udd
Λ^0	1115	0	1/2	0	−1	2.63×10^{-10}	$p\pi^-$ or $n\pi^0$	uds
Σ^+	1189	+1	1/2	1	−1	8.02×10^{-11}	$p\pi^0$ or $n\pi^+$	uus
Δ^{++}	1232	+2	3/2	3/2	0	$\sim 10^{-23}$	$p\pi^+$	uuu
Ξ^-	1321	−1	1/2	1/2	−2	1.64×10^{-10}	$\Lambda^0\pi^-$	dss
Ω^-	1672	−1	3/2	0	−3	8.2×10^{-11}	$\Lambda^0 K^-$	sss
Λ_c^+	2285	1	1/2	0	0	2.0×10^{-13}	$pK^-\pi^+$	udc

Baryons obey the principle of **conservation of baryon number.** We assign a baryon number $B = 1$ to each baryon and $B = -1$ to each antibaryon. **In all interactions, the total baryon number is conserved.** This principle is the reason that in all nuclear reactions the total mass number A must be conserved. Protons and neutrons are baryons, and A is the total number of nucleons in a nucleus.

EXAMPLE 30.10 **Conservation of baryon number**

Which of the following reactions obey the law of conservation of baryon number?

A. $n + p \rightarrow n + p + p + \bar{p}$;
B. $n + p \rightarrow n + p + \bar{p}$

SOLUTION

SET UP AND SOLVE In each case, the initial baryon number is $1 + 1 = 2$. (A) The final baryon number for this reaction is $1 + 1 + 1 + (-1) = 2$, so baryon number is conserved. (B) The final baryon number for this reaction is $1 + 1 + (-1) = 1$.

REFLECT Baryon number is *not* conserved, and reaction B does not occur in nature.

Strangeness

The K mesons and the Λ and Σ hyperons appeared on the scene during the late 1950s. Because of their unusual behavior, they were called *strange particles.* They were produced in high-energy collisions such as $\pi^+ + p$, and a K and a hyperon were always produced together. The frequency of occurrence of the production process suggested that it was a strong-interaction process, but the relatively long lifetimes of these particles suggested that their decay was a weak-interaction processes. Even stranger, the K^0 appeared to have two lifetimes, one (about 9×10^{-11} s) characteristic of strong-interaction decays, the other nearly 600 times longer. So were the K mesons hadrons or not?

This question led physicists to introduce a new quantity called **strangeness.** The hyperons Λ^0 and $\Sigma^{\pm,0}$ were assigned a strangeness value $S = -1$, and the associated K^0 and K^+ mesons were assigned a value of $S = 1$ (as shown in Table 30.5). The corresponding antiparticles have opposite strangeness, $S = +1$ for $\overline{\Lambda}^0$ and $\overline{\Sigma}^{\pm,0}$ and $S = -1$ for $\overline{K}^0$ and K^-. Strangeness is *conserved* in production processes such as

$$p + \pi^- \rightarrow \Sigma^- + K^+ \qquad \text{and} \qquad p + \pi^- \rightarrow \Lambda^0 + K^0.$$

The process $p + \pi^- \rightarrow p + K^-$ is forbidden by conservation of strangeness, and it doesn't occur in nature.

When strange particles decay individually, strangeness is usually not conserved. Typical processes include

$$\Sigma^+ \rightarrow n + \pi^+,$$
$$\Lambda^0 \rightarrow p + \pi^-,$$
$$K^- \rightarrow \pi^+ + \pi^- + \pi^-.$$

In each of these processes, the initial strangeness is 1 or -1, and the final strangeness is zero. All observations of these particles are consistent with the conclusion that strangeness is conserved in strong interactions, but that it can change by zero or one unit in weak interactions.

Conservation Laws

The classical conservation laws of energy, momentum, angular momentum, and electric charge are believed to be obeyed in all interactions. These laws are called

absolute conservation laws. Baryon number is also conserved in all interactions. The decay of strange particles provides our first example of a **conditional conservation law,** one that is obeyed in some interactions but not in others. Strangeness is conserved in strong and electromagnetic interactions, but not in all weak interactions.

Conceptual Analysis 30.5

Particle creation

One reason a photon could not create an odd number of electrons plus positrons is that such a process would

A. not conserve charge.
B. not conserve energy.
C. require photon energies that are not attainable.
D. result in the creation of mass.

SOLUTION One of the observed rules of nature is that during any type of particle interaction or change, the net charge (the number of positive charges minus the number of negative charges) does not change, a rule called conservation of charge. If a photon, a zero-charge particle, converts into one positron and one electron, the net charge still adds to zero. An odd number of particles must include either an extra electron or an extra positron, violating conservation of charge.

Quarks

The leptons form a fairly neat package: three mass particles and three neutrinos, each with its antiparticle, together with a conservation law relating their numbers. Physicists now believe that leptons are genuinely fundamental particles. The hadron family, by contrast, is a mess. Table 30.5 shows only a small sample of more than 250 hadrons that have been discovered since 1960. It has become clear that these particles *do not* represent the most fundamental level of the structure of matter; instead, there is at least one additional level of structure.

Our present understanding of the nature of this level is built on a proposal made initially in 1964 by Murray Gell-Mann and his collaborators. In this proposal, hadrons are *not* fundamental particles, but are composite structures whose constituents are spin-$\frac{1}{2}$ fermions called **quarks.** Each baryon is composed of three quarks (qqq), and each meson is a quark–antiquark pair $(q\bar{q})$. No other combinations seem to be necessary. This scheme requires that quarks have electric charges with magnitudes one-third and two-thirds of the electron charge e, which was previously thought to be the smallest unit of charge. Quarks also have fractional values of the baryon number B. Two quarks can combine with their spins parallel to form a particle with spin 1 or with their spins antiparallel to form a particle with spin 0. Similarly, three quarks can combine to form a particle with spin $\frac{1}{2}$ or $\frac{3}{2}$.

The first quark theory included three types (flavors) of quarks, labeled u (up), d (down), and s (strange), as shown in Table 30.6. The corresponding antiquarks $\bar{u}$, $\bar{d}$, and $\bar{s}$ have opposite values of Q, B, and S. Protons, neutrons, π and K mesons, and several hyperons can be constructed from these three quarks. (We describe the charge Q of a particle as a multiple of the magnitude e of the electron charge.) For example, the proton quark content is uud; a proton has $Q/e = +1$, baryon number $B = +1$, and strangeness $S = 0$. From Table 30.6, the u quark has $Q/e = \frac{2}{3}$ and $B = \frac{1}{3}$, and the d quark has $Q/e = -\frac{1}{3}$ and $B = \frac{1}{3}$. So the values of Q/e add to 1, and the values of the baryon number B also add to 1, as we would expect.

TABLE 30.6 Properties of the three original quarks

Particle	Q/e	Spin	Baryon number	Strangeness	Charm	Bottomness	Topness
u	$\frac{2}{3}$	$\frac{1}{2}$	$\frac{1}{3}$	0	0	0	0
d	$-\frac{1}{3}$	$\frac{1}{2}$	$\frac{1}{3}$	0	0	0	0
s	$-\frac{1}{3}$	$\frac{1}{2}$	$\frac{1}{3}$	-1	0	0	0

The neutron is **udd**, with total $Q = 0$ and $B = 1$. The π^+ meson is $u\bar{d}$, with $Q/e = 1$ and $B = 0$, and the K^+ meson is $u\bar{s}$. The antiproton is $\bar{p} = \bar{u}\bar{u}\bar{d}$, the negative pion is $\pi^- = \bar{u}d$, and so on. Table 30.5 lists the quark content of some of the hadrons (mesons and baryons).

In the standard model, the attractive interactions that hold quarks together are assumed to be mediated by massless spin-1 bosons called **gluons,** just as photons mediate the electromagnetic interaction and pions mediate the nucleon–nucleon interaction in the old Yukawa theory. Quarks, having spin $\frac{1}{2}$, are fermions and so are subject to the exclusion principle. They come in three colors, and the exclusion principle is assumed to apply separately to each color. A baryon always contains one red, one green, and one blue quark, so the baryon itself has no color. Each gluon has a color and an anticolor, and color is conserved during the emission and absorption of a gluon by a quark. The color of an individual quark changes continually because of gluon exchange.

The theory of strong interactions is known as **quantum chromodynamics** (QCD). Individual free quarks have not been observed. In most QCD theories, there are phenomena associated with the creation of quark–antiquark pairs that make it impossible to observe a single free, isolated quark. Nevertheless, an impressive body of experimental evidence supports the correctness of the quark structure of hadrons and the belief that quantum chromodynamics is the key to understanding the strong interactions.

More Quarks

Although the three quarks we've described seem to fit the particles known in 1970, they are not by any means the end of the story. Additional particles discovered since 1970, as well as theoretical considerations based on symmetry properties, point to the existence of three additional quarks: **c**, **b**, and **t**. Experimental confirmation of this conjecture came in 1995 with the discovery of the **t** quark. Table 30.7 lists some properties of the six quarks.

The Standard Model

The particles and interactions we've described, including six quarks (from which all the hadrons are made), six leptons, and the particles that mediate those interactions, form a reasonably comprehensive picture of the fundamental building blocks of nature; this scheme has come to be called the **standard model.** The strong interaction among quarks is mediated by gluons, and the electromagnetic interaction among charged particles is mediated by photons, both spin-1 bosons. The weak interaction is mediated by exchange of the weak bosons $(W^\pm$ and $Z^0)$, spin-1 particles with enormous masses, 80 GeV/c^2 and 91 GeV/c^2. The gravitational interaction is thought to be mediated by a massless spin-2 boson called the graviton, which has not yet been observed experimentally.

It has long been a dream of particle theorists to be able to combine all four interactions of nature into a single unified theory. In 1967, Steven Weinberg and

TABLE 30.7 Properties of quarks

Particle	Q/e	Spin	Baryon number	Strangeness	Charm	Bottomness	Topness
u	$\frac{2}{3}$	$\frac{1}{2}$	$\frac{1}{3}$	0	0	0	0
d	$-\frac{1}{3}$	$\frac{1}{2}$	$\frac{1}{3}$	0	0	0	0
s	$-\frac{1}{3}$	$\frac{1}{2}$	$\frac{1}{3}$	-1	0	0	0
c	$\frac{2}{3}$	$\frac{1}{2}$	$\frac{1}{3}$	0	$+1$	0	0
b	$-\frac{1}{3}$	$\frac{1}{2}$	$\frac{1}{3}$	0	0	$+1$	0
t	$\frac{2}{3}$	$\frac{1}{2}$	$\frac{1}{3}$	0	0	0	$+1$

Abdus Salam proposed a theory that treats the weak and electromagnetic forces as two aspects of a single interaction at sufficiently high energies. This **electroweak theory** was successfully verified in 1983 with the discovery of the weak-force intermediary particles, the Z^0 and $W^{\pm}$ bosons by two experimental groups working at the $p\overline{p}$ collider at the CERN Laboratory in Geneva, Switzerland. Weinberg, Salam, and Sheldon Glashow, who also contributed to the theory, received the 1979 Nobel Prize in Physics for their work.

Grand Unified Theories

Can the theory of strong interaction and the electroweak theory be unified to give a comprehensive theory of strong, weak, and electromagnetic interactions? Such schemes, called **grand unified theories** (GUTs), are still speculative in nature. One interesting feature of some grand unified theories is that they predict the decay of the proton (violating the conservation of baryon number), with an estimated lifetime on the order of 10^{30} years. Experiments are under way that, theoretically, should have detected the decay of the proton if its lifetime is 10^{30} years or less. Such decays have not been observed, but experimental work continues.

In the standard model, the neutrinos have zero mass, although experiments designed to determine neutrino masses are extremely difficult to perform. In most GUTs, neutrinos must have nonzero masses. If they do have mass, transitions called neutrino oscillations can occur in which one type of neutrino ($\nu_{\rm e}$, ν_{μ}, or ν_{τ}) changes into another type. Recent (1998) experiments have indeed confirmed the existence of neutrino oscillations, showing that neutrinos do have nonzero masses. The present upper limits on neutrino masses are shown in Table 30.4. This discovery has cleared up a long-standing mystery about neutrinos coming from the sun. The observed flux of solar electron neutrinos is only one-third of the predicted value. It now appears that the sun is indeed producing electron neutrinos at the predicted rate, but that two-thirds of them are transformed into muon or tau neutrinos as they pass through and interact with the material of the sun.

The ultimate dream of theorists is to unify all four fundamental interactions, including gravitation as well as the strong and electroweak interactions included in GUTs. Such a unified theory is whimsically called a **theory of everything** (TOE). Such theories range from the speculative to the fantastic. One popular ingredient is a space-time continuum with more than four dimensions, containing structures called *strings*. Another element is supersymmetry, which gives every boson a fermion "superpartner." Such concepts lead to the prediction of whole new families of particles, including sleptons, photinos, squarks bound together by gluinos, and even winos and wimps, none of which has been observed experimentally. Theorists are still very far away from a satisfactory TOE.

30.10 Cosmology

In this final section of our book, we'll explore briefly the connections between the early history of the universe and the interactions among fundamental particles. It is surprising and remarkable that at the dawn of the 21st century physicists have found such close ties between the smallest things we know about (the range of the weak interaction, on the order of 10^{-18} m) and the largest (the universe itself, on the order of at least 10^{26} m).

The Expanding Universe

Until early in the 20th century, it was usually assumed that the universe was *static;* stars might move relative to each other, but without any general expansion or contraction. But if everything is initially sitting still in the universe, why doesn't gravity just pull it all together into one big glob? Newton himself recognized the seriousness of this troubling question.

About 1930, astronomers began to find evidence that the universe is *not* static. The motions of distant galaxies relative to earth can be measured by observing the shifts in the wavelengths of their spectra due to the Doppler effect. (This is the same effect for *light* that we studied with *sound* in Section 12.12.) The shifts are called **red shifts,** because they are always toward longer wavelengths, showing that distant galaxies appear to be receding from us and from each other. The astronomer Edwin Hubble measured red shifts from many distant galaxies and came to the astonishing conclusion that the speed of recession, v, of a galaxy is approximately proportional to its distance r from us. This relation is now called **Hubble's law.**

Another aspect of Hubble's observations was that distant galaxies appeared to be receding in all directions. There is no particular reason to think that our galaxy is at the center of the universe; if we were moving along with some other galaxy, everything else would still seem to be receding from us. Thus, the universe looks the same from all locations, and we believe that the laws of physics are the same everywhere.

The Big Bang

An appealing hypothesis suggested by Hubble's law is that at some time in the past, all the matter in the universe was concentrated in a very small space and was blown apart in an immense explosion labeled the **Big Bang,** giving all observable matter more or less the velocities we observe today. By correlating distances with speeds of recession, it has been established that the Big Bang occurred about 14 billion years ago. In this calculation, we've neglected any slowing down due to gravitational attraction. Whether or not this omission is justified depends on the average density of matter; we'll return to this point later.

Critical Density

We've mentioned that the law of gravitation isn't consistent with a static universe. But what is its role in an expanding universe? Gravitational attractions should slow the initial expansion, but by how much? If they are strong enough, the universe should expand more and more slowly, eventually stop, and then begin to contract, perhaps all the way down to what's been called a "Big Crunch." But if the gravitational forces are much weaker, they slow the expansion only a little, and the universe continues to expand forever.

The situation is analogous to the problem of the escape velocity of a projectile launched from earth. A similar analysis can be carried out for the universe. Whether or not the universe continues to expand indefinitely depends on the average density of matter. If matter is relatively dense, there is a lot of gravitational attraction to slow, and eventually stop, the expansion and make the universe contract again. If not, the expansion continues indefinitely. The critical density, denoted by ρ_c, turns out to be

$$\rho_c = 5.8 \times 10^{-27} \text{ kg/m}^3.$$

If the average density of the universe is less than this, the universe continues to expand indefinitely; if it is greater, it eventually stops expanding and begins to contract, possibly leading to the Big Crunch and then another Big Bang. Note that ρ_c is a very small number by terrestrial standards; the mass of a hydrogen atom is 1.67×10^{-27} kg, so ρ_c corresponds to about three hydrogen atoms per cubic meter.

Recent research shows that the average density of all matter in the universe is about 27% of the critical density ρ_c, but that the average density of *luminous* matter (i.e., matter that emits some sort of radiation) is only 4% of ρ_c. In other words, most of the matter in the universe does not emit electromagnetic radiation of any kind. At present, the nature of this **dark matter** remains an outstanding mystery.

We mentioned some candidates for dark matter at the end of Section 30.9, including WIMPs (weakly interactive massive particles) and other subatomic particles far more massive than any that can be produced in accelerator experiments. But whatever the true nature of dark matter, it is by far the dominant form of matter in the universe.

Because the average density of matter in the universe is less than the critical density, we might expect that the universe will continue to expand indefinitely. Gravitational attractions should slow the expansion down, but not enough to stop it. One way to test this prediction is to look at redshifts of extremely distant objects. The surprising results from such measurements show that distant galaxies actually have *smaller* redshifts than values predicted by Hubble's law. The implication is that the expansion has been speeding up rather than slowing down.

Why is the expansion speeding up? The explanation generally accepted by astronomers and physicists is that space is permeated with a kind of energy that has no gravitational effect and emits no electromagnetic radiation, but rather acts as a kind of "antigravity" that produces a universal repulsion. This invisible, immaterial energy is called **dark energy.** The nature of dark energy is poorly understood, but it is the subject of very active research. Present estimates indicate that the energy *density* of dark energy is about three times greater than that of matter, a result which suggests that the total energy density of the universe is greater than the critical density ρ_c and that the universe will continue to expand forever.

The Beginning of Time

The evolution of the universe has been characterized by the continuous growth of a scale factor R, which we can think of, very roughly, as characterizing the size of the universe. In Section 30.9, we mentioned the unification of the electromagnetic and weak interactions at energies of several hundred GeV, constituting the electroweak interaction. Most theorists believe that the strong and electroweak interactions become unified at energies on the order of 10^{14} GeV (as in the GUT models) and that at energies on the order of 10^{19} GeV all four of the fundamental forces (strong, electromagnetic, weak, and gravitational) become unified. Average particle energies were in the latter range at a time on the order of 10^{-43} s after the Big Bang. If we mentally go backward in time, we have to stop at $t = 10^{-43}$ s because we have no adequate theory that unifies all four interactions.

The Standard Cosmological Model

The brief chronology of the Big Bang that follows is called the *standard model.* The figure on pages 1040–1041 presents a graphical description of this history, with the characteristic sizes, particle energies, and temperatures at various times. Referring to this chart frequently will help you understand the discussion that follows. In this model, the temperature at time $t = 10^{-43}$ s was about 10^{32} K, and the average energy per particle was on the order of 10^{19} GeV. In the unified theories, this is about the energy below which gravity begins to behave as a separate interaction. The time $t = 10^{-43}$ s therefore marks the end of any proposed TOE and the beginning of the GUT period.

During the GUT period, from $t = 10^{-43}$ s to $t = 10^{-35}$ s, the strong and electroweak forces were still unified, and the universe consisted of a soup of quarks and leptons transforming into each other so freely that there was no distinction between the two families of particles. One important characteristic of GUTs is that at sufficiently high energies, baryon number is not conserved. Thus, by the end of the GUT period, the numbers of quarks and antiquarks may have been unequal. This point has important implications; we'll return to it shortly.

By $t = 10^{-35}$ s, the temperature had decreased to about 10^{27} K and the average energy to about 10^{14} GeV. At this energy, the strong force separated from the electroweak force, and baryon number and lepton number began to be separately conserved. Some models postulate an enormous and very rapid expansion at this time, a factor on the order of 10^{50} in 10^{-32} s. As the expansion continued and the average particle energy continued to decrease, quarks began to bind together to form nucleons and antinucleons. After the average energy fell below the threshold for nucleon–antinucleon pair production, many of the nucleons annihilated nearly all the less numerous antinucleons. Later still, the average energy dropped below the threshold for electron–positron pair production; most of the remaining positrons were annihilated, leaving the universe with many more protons and electrons than the antiparticles of each.

By time $t = 225$ s, the average energy dropped below the binding energy of the deuteron, about 2.23 MeV, and the age of *nucleosynthesis* had begun. At first, only the simplest bound states—^{2}H, ^{3}He, and ^{4}He—appeared. Further nucleosynthesis began very much later, at about $t = 2 \times 10^{13}$ s (approximately 700,000 y). By this time, the average energy was a few eV, and electrically neutral H and He atoms could form. With the electrical repulsions of the nuclei canceled out, gravitational attraction could pull the neutral atoms together to form galaxies and, eventually, stars. Thermonuclear reactions in stars are believed to have produced all of the more massive nuclei and, ultimately, the nuclides and chemical elements we know today.

Matter and Antimatter

One of the most remarkable features of our universe is the asymmetry between matter and antimatter. One might think that the universe should have equal numbers of protons and antiprotons, and of electrons and positrons, but this doesn't appear to be the case. There is no evidence for the existence of substantial amounts of antimatter (matter composed of antiprotons, antineutrons, and positrons) anywhere in the universe. Theories of the early universe must confront this imbalance.

We've mentioned that most GUTs allow for the violation of conservation of baryon number at energies at which the strong and electroweak interactions have converged. If particle–antiparticle symmetry is also violated, we have a mechanism for making more quarks than antiquarks, more leptons than antileptons, and, eventually, more nucleons than antinucleons. But any asymmetry created in this way during the GUT era might be wiped out by the electroweak interaction after the end of the GUT era. The problem of the matter–antimatter asymmetry is still very much an open one.

We hope that this qualitative discussion has conveyed at least a hint as to the close connections between particle theory and cosmology. There are many unanswered questions. We don't know what happened during the first 10^{-43} s after the Big Bang because we have no quantum theory of gravity; there are many versions of GUTs from which to choose. We don't know what dark matter and dark energy are. We are still very far from having a suitable theory that unifies all four interactions in nature. But this search for understanding of the physical world we live in continues to be one of the most exciting adventures of the human mind.

1040

AGE OF QUARKS AND GLUONS (GUT Period)

Dense concentration of matter and antimatter; gravity a separate force; more quarks than antiquarks. Inflationary period (10^{-35} s): rapid expansion, strong force separates from electroweak force.

AGE OF NUCLEONS AND ANTINUCLEONS

Quarks bind together to form nucleons and antinucleons; energy too low for nucleon-antinucleon pair production at 10^{-2} s

AGE OF NUCLEOSYNTHESIS

Stable deuterons; matter 74% H, 25% He, 1% heavier nuclei

AGE OF LEPTONS

Leptons distinct from quarks; $W^{\pm}$ and Z^0 bosons mediate weak force (10^{-12} s)

BIG BANG 10^{-43} s 10^{-32} s 10^{-6} s   225 s   10

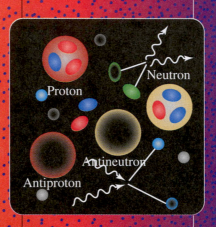

Neutrino

Quarks

Antiquarks

Antineutrino

Proton

Neutron

Antineutron

Antiproton

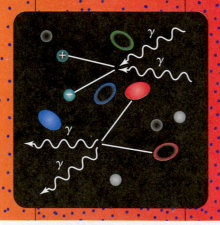

γ

γ

$+$

$-$

γ

γ

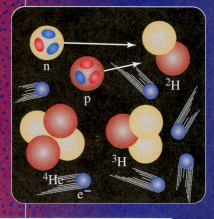

n

p

^{2}H

^{3}H

^{4}He

e$^-$

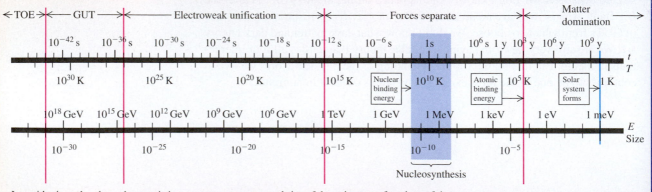

$\leftarrow$ TOE $\rightarrow$ | $\leftarrow$ GUT $\rightarrow$ | Electroweak unification | Forces separate | Matter domination

10^{-42} s 10^{-36} s 10^{-30} s 10^{-24} s 10^{-18} s 10^{-12} s 10^{-6} s   1s 10^6 s 1 y 10^3 y 10^6 y 10^9 y t

T

10^{30} K 10^{25} K 10^{20} K 10^{15} K   Nuclear binding energy $\rightarrow$ 10^{10} K   Atomic binding energy $\rightarrow$ 10^5 K   Solar system forms $\rightarrow$ 1 K

10^{18} GeV 10^{15} GeV 10^{12} GeV 10^9 GeV 10^6 GeV 1 TeV 1 GeV 1 MeV 1 keV 1 eV 1 meV

E

Size

10^{-30} 10^{-25} 10^{-20} 10^{-15} 10^{-10} 10^{-5}

Nucleosynthesis

Logarithmic scales show characteristic temperature, energy, and size of the universe as functions of time.

AGE OF IONS
Expanding, cooling
gas of ionized
H and He

AGE OF ATOMS
Neutral atoms form, pulled
together by gravity; universe becomes
transparent to most light

**AGE OF STARS
AND GALAXIES**
Thermonclear fusion
begins in stars, forming
heavier nuclei

10^{13} s

10^{15} s

NOW

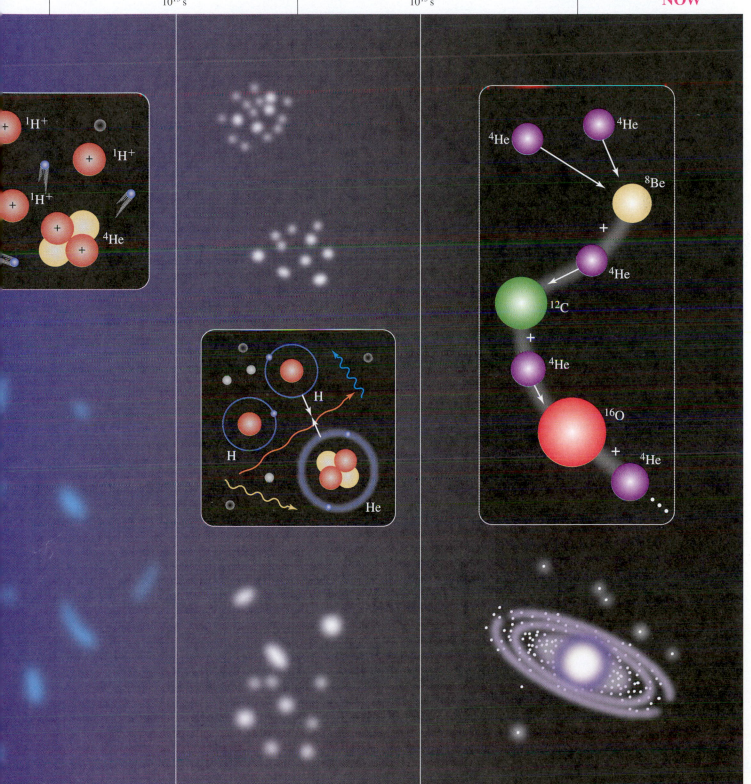

SUMMARY

Properties of Nuclei

(Section 30.1) An atomic nucleus is roughly spherical in shape, with a radius that increases with the number of nucleons (protons and neutrons) A in the nucleus: $R = R_0 A^{1/3}$ (Equation 30.1). The total mass of a nucleus is always *less* than the total mass of its constituent parts because of the mass equivalent $(E = mc^2)$ of the binding energy that hold the nucleus together.

Nuclear Stability

(Section 30.2) The most important reason that some nuclei are stable and others are not is the competition between the attractive nuclear force and the repulsive electrical force. The nuclear force favors *pairs* of nucleons with opposite spin, and *pairs of pairs*. Electrical repulsion tends to favor greater numbers of neutrons, but a nucleus with too many neutrons is unstable because not enough of them are paired with protons.

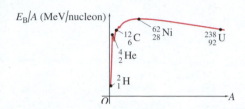

Radioactivity, Radiation, and the Life Sciences

(Sections 30.3 and 30.4) The two most common decay modes are the emission of alpha (α) and beta (β) particles. The **alpha particle** has two protons and two neutrons bound together, with zero total spin. Alpha emission occurs principally with nuclei that are too large to be stable. The **beta particle** (or more specifically, the beta-minus β^- particle) is an electron. The antineutrino $\bar{\nu}_e$, is also involved in the process of β^- decay represented as $n \rightarrow p + \beta^- + \bar{\nu}_e$ (Equation 30.4).

Radiation exposure can cause cancers and genetic defects, yet it can be effective in medicine for intentional selective destruction of tissue such as tumors. Radiation can also help diagnose a problem: By injecting low doses of radioactive isotopes into the body, "images" can be made of potentially problematic regions where the radioactive material builds up.

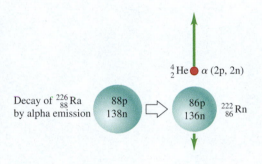

Nuclear Reactions, Fission, and Fusion

(Sections 30.5–30.7) A **nuclear reaction** is the result of bombardment of a nucleus by a particle, rather than a spontaneous natural process as in radioactive decay. For a charged particle such as a proton or α particle to penetrate the nucleus of another atom, it must usually have enough initial kinetic energy to overcome the potential-energy barrier caused by the repulsive electrostatic forces.

Nuclear fission is a decay process in which an unstable nucleus splits into two fragments. The total mass of the resulting fragments is less than the original atom, and this released mass-energy appears mostly as kinetic energy of the fragments. In a **nuclear fusion** reaction, two or more small light nuclei come together or *fuse*, to form a larger nucleus. Fusion reactions release energy for the same reason as fission reactions; the binding energy per nucleon after the reaction is greater than before.

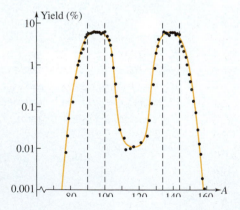

Continued

Fundamental Particles and High-Energy Physics

(Sections 30.8 and 30.9) For each type of massive particle (electron, proton, neutron, etc.) there exists an antiparticle. When a particle and antiparticle collide, they can annihilate. Protons and neutrons, as well as other hadrons, are made of quarks of fractional charge. The proton consists of three quarks, **uud**, while the neutron is made of **udd**. Mesons are also made of quarks, combining a quark and antiquark, as in the π^+ meson, $u\bar{d}$. The world we encounter every day has numerous substances with varying properties. While the properties are many, the fundamental building blocks are few—**u** quarks, **d** quarks, electrons, and photons.

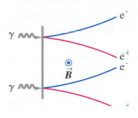

Cosmology

(Section 30.10) The universe has been expanding since the Big Bang, about 14 billion years ago. As expressed in **Hubble's law,** the more distant the galaxy, the greater its speed of recession. The energy content of the universe falls into three categories—regular luminous matter, dark matter, and dark energy. Only a few percent of the universe is regular matter made of protons and neutrons (forming stars and galaxies). About 25% of energy in the universe is **dark matter,** and we still don't know what it's made of. More recently, there's evidence that about 75% of the energy in the universe is **dark energy**—we don't know how this energy comes about, but we know it isn't mass. Taken together, the total energy appears to exceed critical density; if so, we will expand forever.

 For instructor-assigned homework, go to www.masteringphysics.com

Conceptual Questions

1. Since different isotopes of the same element have the same chemical behavior, how can they be separated from each other?

2. In beta decay, a neutron becomes a proton, an electron, and an antineutrino. This type of decay also occurs with free neutrons, with a half-life of about 15 min. Could a free *proton* undergo a similar decay to become a neutron, a positron (positive electron), and a neutrino? Why or why not?

3. In the ^{238}U decay chain shown in Figure 30.5, some nuclides in the chain are found much more abundantly in nature than others, despite the fact that every ^{238}U nucleus goes through every step in the chain before finally becoming ^{206}Pb. Why are the abundances of the intermediate nuclides not all the same?

4. Why aren't the masses of all nuclei integer multiples of the mass of a single nucleon?

5. True or false? During one half-life, the *mass* of a radioisotope is reduced by half. Explain.

6. Why is the decay of an unstable nucleus not affected by the *chemical* situation of the atom, such as the nature of the molecule in which it is bound?

7. Changing the temperature of atoms affects their chemical reaction rate, but has no effect on their rate of radioactive decay. Why is one rate affected, but not the other?

8. The only two stable nuclides with more protons than neutrons are $^1_1$H and $^3_2$He. Why is $Z > N$ so uncommon?

9. Why do high-Z nuclei require more neutrons than protons to be stable, as indicated in the Segrè diagram in Figure 30.2?

10. The binding energy curve in Figure 30.1 slopes *downward* for heavy nuclei. Why does this happen? Why does the binding energy per nucleon *not* keep increasing?

11. Nuclear power plants use nuclear fission reactions to generate steam to run steam-turbine generators. How does the nuclear reaction produce *heat*?

12. **Electron capture.** There are cases in which a nucleus having too few neutrons for stability can capture one of the electrons in the K shell of the atom. This process is known as *electron capture* (or, sometimes, *K*-capture). What is the effect of electron capture on N, A, and Z? The atom ^{103}Pd decays this way; identify the daughter nucleus that is produced.

13. **Positron emission.** One form of beta decay (known as *positron emission*) is due to the decay of a proton to a positron (a positive electron, β^+), a neutron, and a neutrino. What is the effect of positron emission on N, A, and Z? The atom ^{40}K decays this way; identify the daughter nucleus that is produced. (Note that a *free* proton cannot decay this way, only a proton in the nucleus.)

14. Is it possible that some parts of the universe contain antimatter whose atoms have nuclei made of antiprotons and antineutrons, surrounded by positrons? How might we detect this condition without actually going there? What problems could arise if we actually *did* go there?

15. Why are so many health hazards associated with fission fragments that are produced during the fission of heavy nuclei?

16. According to the standard model of the fundamental particles, what are the similarities between baryons and leptons? What are the most important differences?

Multiple-Choice Problems

1. Which of the following statements about the atomic nucleus are correct? (There may be more than one correct choice.)
   A. It is held together by the strong force that affects protons and neutrons, but not electrons.
   B. It has a typical radius on the order of 10^{-10} m.
   C. Its volume is nearly the same for all atoms.
   D. The density of heavy nuclei is considerably greater than the density of light nuclei.

2. A nucleus containing 60 nucleons has a radius R. The radius of a nucleus containing 180 nucleons would be closest to
   A. $3R$.
   B. $3^{2/3}R = 2.08R$.
   C. $3^{1/3}R = 1.44R$.
   D. $\sqrt{3}R = 1.73R$.

3. The hypothetical atom $^{37}_{17}X$ contains (there may be more than one correct choice)
   A. 17 orbital electrons.
   B. 37 protons.
   C. 17 protons.
   D. 20 neutrons.
   E. 54 nucleons.

4. Which of the following statements about a typical atomic nucleus are correct? (There may be more than one correct choice.)
   A. The number of neutrons is equal to the number of protons.
   B. It contains protons, neutrons, and neutrinos.
   C. It usually contains more neutrons than protons.
   D. It has a typical radius on the order of 10^{-15} m.

5. Which nuclide X would properly complete the following reaction:
$$^1_0n + \,^{235}_{92}U \rightarrow \,^{88}_{38}Sr + X + 12^1_0n$$
   A. $^{146}_{54}Xe$
   B. $^{148}_{52}Te$
   C. $^{136}_{54}Xe$
   D. $^{136}_{42}Mo$

6. One problem in radiocarbon dating of biological samples, especially very old ones, is that they can easily be contaminated with modern biological material during the measurement process. What effect would this contamination have on the measurements?
   A. Contaminated samples would be measured to be older than they really are.
   B. The half-life of the radioactive carbon would appear to be shorter.
   C. Contaminated samples would be measured to be younger than they really are.
   D. The decay rate of the radioactive carbon would appear to be increased.

7. In a nuclear accident, a lab worker receives a dose of D rads of radiation of x rays having an RBE of 1. If instead he had been exposed to the same amount of energy from alpha particles having an RBE of 20, his exposure would have been (there may be more than one correct choice)
   A. D rads.
   B. $20D$ rads.
   C. D rem.
   D. $20D$ rem.

8. Atom A has twice as long a half-life as atom B. If both of them start out with an amount N_0, then in the time that A has been reduced to $N_0/2$, B has been reduced to
   A. $N_0/2$.
   B. $N_0/4$.
   C. $N_0/8$.
   D. $N_0/16$.

9. If the binding energy per nucleon for hypothetical element $^{13}_7X$ is 5 MeV/nucleon, the total binding energy of its nucleus is
   A. 5 MeV.
   B. 30 MeV.
   C. 35 MeV.
   D. 65 MeV.
   E. 100 MeV.

10. A certain atomic nucleus containing 50 nucleons has a density ρ. If another nucleus contains 150 nucleons, its density will be closest to
    A. 3ρ.
    B. $3^{2/3}\rho = 2.08\rho$.
    C. $3^{1/3}\rho = 1.44\rho$.
    D. $\sqrt{3}\rho = 1.73\rho$.
    E. ρ.

11. A beaker contains a pure sample of a radioactive substance having a half-life of 1 h. Which statement about this substance is correct? (There may be more than one correct choice.)
    A. After 1 h, half of the atoms will have decayed, and the remainder will decay in the next 1 h.
    B. After 1 h, half of the atoms will have decayed, and half of the remainder will decay in the next 1 h.
    C. After 1 h, the mass in the beaker will be only half its original value.

12. After a particle was emitted from the nucleus of an atom, it was found that the atomic number of the atom had *increased*. The emitted particle could have been (there may be more than one correct choice)
    A. a β^- particle.
    B. an alpha particle.
    C. a positron (β^+).
    D. a gamma ray.

13. In a nuclear reactor accident, one worker receives a dose of 20 rad of x rays, while a second worker receives a dose of 10 rad of alpha particles. How do the two workers' biologically equivalent doses compare?
    A. The first worker receives twice the equivalent dose of the second worker.
    B. The first worker receives 40 times the equivalent dose of the second worker.
    C. The first worker receives 1/200 the equivalent dose of the second worker.
    D. The first worker receives 1/10 the equivalent dose of the second worker.

14. When copper-64, $^{64}_{29}Cu$, undergoes β^- decay, the daughter nucleus contains
    A. 34 protons and 30 neutrons.
    B. 30 protons and 34 neutrons.
    C. 28 protons and 36 neutrons.
    D. 27 protons and 33 neutrons.

15. The radiation from a radioactive sample of a single isotope decreases to one-eighth of the original intensity I_0 in 30 years. What would the intensity be after 10 *more* years?
    A. $I_0/16$.
    B. $I_0/32$.
    C. $I_0/64$.

Problems

30.1 Properties of Nuclei

1. • How many protons and how many neutrons are there in a nucleus of (a) neon, $^{21}_{10}$Ne (b) zinc, $^{65}_{30}$Zn (c) silver, $^{108}_{47}$Ag.

2. • Calculate the approximate (a) radius, (b) volume, and (c) density (in kg/m^3) of a nucleus of gold, $^{197}_{79}$Au. (You can ignore the binding energy of the nucleus for this calculation.)

3. •• **Density of the nucleus.** (a) Using the empirical formula for the radius of a nucleus, show that the volume of a nucleus is directly proportional to its nucleon number A. (b) Give a reasonable argument concluding that the mass m of a nucleus of nucleon number A is approximately $m = m_p A$, where m_p is the mass of a proton. (c) Use the results of parts (a) and (b) to show that all nuclei should have about the same *density*. Then calculate this density in kg/m^3, and compare it with the density of lead (which is $11.4 \, g/cm^3$) and a neutron star (about $10^{17} \, kg/m^3$).

4. • For the common isotope of nitrogen, ^{14}N, calculate (a) the mass defect, (b) the binding energy, and (c) the binding energy per nucleon.

5. • Calculate the binding energy (in MeV) and the binding energy per nucleon of (a) the deuterium nucleus, $^2_1$H, and (b) the helium nucleus, $^4_2$He. (c) How do the results of parts (a) and (b) compare?

6. •• (a) Calculate the total binding energy (in MeV) of the nuclei of ^{56}Fe (of atomic mass 55.934937 u) and of ^{207}Pb (of atomic mass 206.975897 u). (b) Calculate the binding energy per nucleon for each of these atoms. (c) How much energy would be needed to totally take apart each of these nuclei? (d) For which of these atoms are the individual nucleons more tightly bound? Explain your reasoning.

7. • What is the maximum wavelength of a γ ray that could break a deuteron into a proton and a neutron? (This process is called photodisintegration.)

8. •• A photon with a wavelength of 3.50×10^{-13} m strikes a deuteron, splitting it into a proton and a neutron. (a) Calculate the kinetic energy released in this interaction. (b) Assuming the two particles share the energy equally, and taking their masses to be 1.00 u, calculate their speeds after the photodisintegration.

30.2 Nuclear Stability

30.3 Radioactivity

9. • The isotope ^{218}Po decays via α decay. The measured atomic mass of ^{218}Po is 218.008973 u, and the daughter nucleus atomic mass is 213.999805 u. (See also Table 30.2.) (a) Identify the daughter nucleus by name, nucleon number, atomic number, and neutron number. (b) Calculate the kinetic energy (in MeV) of the α particle if we can ignore the recoil of the daughter nucleus.

10. • Tritium ($^3_1$H) is an unstable isotope of hydrogen; its mass, including one electron, is 3.016049 u. (a) Show that tritium must be unstable with respect to beta decay because the decay products ($^3_2$He plus an emitted electron) have less total mass than the tritium. (b) Determine the total kinetic energy (in MeV) of the decay products, taking care to account for the electron masses correctly.

11. •• **Thorium series.** The following decays make up the thorium decay series (the X's are unknowns for you to identify):

$$^{232}\text{Th} \xrightarrow{\alpha} X_1, \qquad ^{228}\text{Ra} \xrightarrow{\beta^-} {}^{228}\text{Ac}, \qquad X_2 \xrightarrow{\beta^-} {}^{228}\text{Th},$$

$$^{228}\text{Th} \xrightarrow{X_3} {}^{224}\text{Ra}, \qquad ^{224}\text{Ra} \xrightarrow{\alpha} {}^{220}\text{Rn}, \qquad ^{220}\text{Rn} \xrightarrow{\alpha} X_4,$$

$$X_5 \xrightarrow{\alpha} {}^{212}\text{Pb}, \text{ and } ^{212}\text{Pb} \xrightarrow{X_6} {}^{212}\text{Bi}. \text{ The } {}^{212}\text{Bi then decays}$$

by an α decay and a β^- decay, which can occur in either order (α followed by β or β followed by α). (a) Identify each of the six unknowns (X_1, X_2, etc.) by nucleon number, atomic number, neutron number, and name. (b) Write out the decays of ^{212}Bi and indicate the end product of this series. (For some guidance, see the discussion under "Decay Series" in Section 30.3.) (c) Draw a Segrè chart for the thorium series, similar to the one shown in Figure 30.5.

12. •• Suppose that 8.50 g of a nuclide of mass number 105 decays at a rate of 6.24×10^{11} Bq. What is its half-life? (*Hint:* Use the fact that $\Delta N/\Delta t = -\lambda N$. You are given $\Delta N/\Delta t$ and can figure out N knowing the mass number and mass of your sample.)

13. • A radioisotope has a half-life of 5.00 min and an initial decay rate of 6.00×10^3 Bq. (a) What is the decay constant? (b) What will be the decay rate at the end of (i) 5.00 min, (ii) 10.0 min, (iii) 25.0 min?

14. •• A radioactive isotope has a half-life of 76.0 min. A sample is prepared that has an initial activity of 16.0×10^{10} Bq. (a) How many radioactive nuclei are initially present in the sample? (b) How many are present after 76.0 min? What is the activity at this time? (c) Repeat part (b) for a time of 152 min after the sample is first prepared.

15. •• Calcium-47 is a β^- emitter with a half-life of 4.5 days. If a bone sample contains 2.24 g of this isotope, at what rate will it decay?

16. •• A 12.0-g sample of carbon from living matter decays at the
BIO rate of 180.0 decays/min due to the radioactive ^{14}C in it. What will be the decay rate of this sample in (a) 1000 years and (b) 50,000 years? (*Hint:* The decay rate is proportional to the number of radioactive carbon atoms remaining; you can therefore replace N and N_0 in Eq. 30.6 with decay rates once you have a value for λ.)

17. •• **Radioactive tracers.** Radioactive isotopes are often intro-
BIO duced into the body through the bloodstream. Their spread through the body can then be monitored by detecting the appearance of radiation in different organs. ^{131}I, a β^- emitter with a half-life of 8.0 d, is one such tracer. Suppose a scientist introduces a sample with an activity of 375 Bq and watches it spread to the organs. (a) Assuming that the sample all went to the thyroid gland, what will be the decay rate in that gland 24 d (about $2\frac{1}{2}$ weeks) later? (b) If the decay rate in the thyroid 24 d later is actually measured to be 17.0 Bq, what percent of the tracer went to that gland? (c) What isotope remains after the I-131 decays?

18. •• The common isotope of uranium, ^{238}U, has a half-life of 4.47×10^9 years, decaying to ^{234}Th by alpha emission. (a) What is the decay constant? (b) What mass of uranium is required for an activity of 1.00 curie? (c) How many alpha particles are emitted per second by 10.0 g of uranium?

19. •• A sample of the radioactive nuclide ^{199}Pt is prepared that has an initial activity of 7.56×10^{11} Bq. (a) 92.4 min after the sample is prepared, the activity has fallen to 9.45×10^{10} Bq. What is the half-life of this nuclide? (b) How many radioactive nuclei were initially present in the sample?

20. •• A sample of charcoal from an archaeological site contains 55.0 g of carbon and decays at a rate of 0.877 Bq. How old is it?

21. •• **We are stardust.** In 1952, spectral lines of the element technetium-99 (^{99}Tc) were discovered in a red-giant star. Red giants are very old stars, often around 10 billion years old, and near the end of their lives. Technetium has *no* stable isotopes, and the half-life of ^{99}Tc is 200,000 years. (a) For how many half-lives has the ^{99}Tc been in the red-giant star if its age is 10 billion years? (b) What fraction of the original ^{99}Tc would be left at the end of that time? This discovery was extremely important because it provided convincing evidence for the theory (now essentially known to be true) that most of the atoms heavier than hydrogen and helium were made inside of stars by thermonuclear fusion and other nuclear processes. If the Tc had been part of the star since it was born, the amount remaining after 10 billion years would have been so minute that it would not have been detectable. This knowledge is what led the late astronomer Carl Sagan to proclaim that "we are stardust."

22. •• Radioactive isotopes used in cancer therapy have a "shelf-life," like pharmaceuticals used in chemotherapy. Just after it has been manufactured in a nuclear reactor, the activity of a sample of ^{60}Co is 5000 Ci. When its activity falls below 3500 Ci, it is considered too weak a source to use in treatment. You work in the radiology department of a large hospital. One of these ^{60}Co sources in your inventory was manufactured on October 6, 2008. It is now April 6, 2011. Is the source still usable? The half-life of ^{60}Co is 5.271 years.

30.4 Radiation and the Life Sciences

23. • **Radiation overdose.** If a person's entire body is exposed to
BIO 5.0 J/kg of x rays, death usually follows within a few days. (Consult Table 30.3.) (a) Express this lethal radiation dose in Gy, rad, Sv, and rem. (b) How much total energy does a 70.0 kg person absorb from such a dose? (c) If the 5.0 J/kg came from a beam of protons instead of x rays, what would be the answers to parts (a) and (b)?

24. • A radiation specialist prescribes a dose of 125 rem for a
BIO patient, using an apparatus that emits alpha particles. (Consult Table 30.3.) (a) How many rads does this dose provide to the patient? (b) How much energy does a 5.0 g sample of irradiated tissue receive? (c) Suppose your hospital has only an electron source available. How many rads should you administer to this patient to achieve the same 125 rem dose?

25. •• A nuclear chemist receives an accidental radiation dose
BIO of 5.0 Gy from slow neutrons (RBE = 4.0). What does she receive in rad, rem, and J/kg?

26. •• (a) If a chest x ray delivers 0.25 mSv to 5.0 kg of tissue, how
BIO many *total* joules of energy does this tissue receive? (b) Natural radiation and cosmic rays deliver about 0.10 mSv per year at sea level. Assuming an RBE of 1, how many rem and rads is this dose, and how many joules of energy does a 75 kg person

receive in a year? (c) How many chest x rays like the one in part (a) would it take to deliver the same *total* amount of energy to a 75 kg person as she receives from natural radiation in a year at sea level, as described in part (b)?

27. •• **To scan or not to scan?** It has become popular for some
BIO people to have yearly whole-body scans (CT scans, formerly called CAT scans), using x rays, just to see if they detect anything suspicious. A number of medical people have recently questioned the advisability of such scans, due in part to the radiation they impart. Typically, one such scan gives a dose of 12 mSv, applied to the *whole body.* By contrast, a chest x ray typically administers 0.20 mSv to only 5.0 kg of tissue. How many chest x rays would deliver the same *total* amount of energy to the body of a 75 kg person as one whole-body scan?

28. •• In an industrial accident a 65-kg person receives a lethal
BIO whole-body equivalent dose of 5.4 Sv from x rays. (a) What is the equivalent dose in rem? (b) What is the absorbed dose in rad? (c) What is the total energy absorbed by the person's body? How does this amount of energy compare to the amount of energy required to raise the temperature of 65 kg of water 0.010°C?

29. •• **Food irradiation.** Food is often irradiated with the use of
BIO either x rays or electron beams, to help prevent spoilage. A low dose of 5–75 kilorads (krad) helps to reduce and kill inactive parasites, a medium dose of 100–400 krad kills microorganisms and pathogens such as salmonella, and a high dose of 2300–5700 krad sterilizes food so that it can be stored without refrigeration. (a) A dose of 175 krad kills spoilage microorganisms in fish. If x rays are used, what would be the dose in Gy, Sv, and rem, and how much energy would a 150 g portion of fish absorb? (See Table 30.3.) (b) Repeat part (a) if electrons of RBE 1.50 are used instead of x rays.

30. •• In a diagnostic x-ray procedure, 5.00×10^{10} photons are
BIO absorbed by tissue with a mass of 0.600 kg. The x-ray wavelength is 0.0200 nm. (a) What is the total energy absorbed by the tissue? (b) What is the equivalent dose in rem?

31. •• A person ingests an amount of a radioactive source with a
BIO very long lifetime and activity 0.72 μCi. The radioactive material lodges in the lungs, where all of the 4.0-MeV α particles emitted are absorbed within a 0.50-kg mass of tissue. Calculate the absorbed dose and the equivalent dose for one year.

32. •• **Irradiating ourselves!** The radiocarbon in our bodies is
BIO one of the naturally occurring sources of radiation. Let's see how large a dose we receive. ^{14}C decays via β^- emission, and 18% of our body's mass is carbon. (a) Write out the decay scheme of carbon-14 and show the end product. (A neutrino is also produced.) (b) Neglecting the effects of the neutrino, how much kinetic energy (in MeV) is released per decay? The atomic mass of C-14 is 14.003242 u. (See Table 30.2.) (c) How many grams of carbon are there in a 75 kg person? How many decays per second does this carbon produce? (d) Assuming that all the energy released in these decays is absorbed by the body, how many MeV/s and J/s does the C-14 release in this person's body? (e) Consult Table 30.3 and use the largest appropriate RBE for the particles involved. What radiation dose does the person give himself in a year, in Gy, rad, Sv, and rem?

30.5 Nuclear Reactions

33. • Consider the nuclear reaction $^4_2\text{He} + ^7_3\text{Li} \rightarrow ^{10}_5\text{B} + ^1_0\text{n}$. Is energy absorbed or liberated? How much energy?

34. •• Consider the nuclear reaction

$$^2_1\text{H} + ^{14}_7\text{N} \rightarrow \text{X} + ^{10}_5\text{B}$$

where X is a nuclide. (a) What are Z and A for the nuclide X? (b) Calculate the reaction energy Q (in MeV).

35. •• How much energy (in J and MeV) would a proton need for its surface to reach the surface of the nucleus of the ^{17}O atom? Assume that oxygen and the proton obey the empirical radius formula of Section 30.1.

30.6 Nuclear Fission

36. • Assuming that 200 MeV is released per fission, how many fissions per second take place in a 100 MW reactor?

37. •• The United States uses 1.0×10^{19} J of electrical energy per year. If all this energy came from the fission of ^{235}U, which releases 200 MeV per fission event, (a) how many kilograms of ^{235}U would be used per year; (b) how many kilograms of uranium would have to be mined per year to provide that much ^{235}U? (Recall that only 0.70% of naturally occurring uranium is ^{235}U.)

38. •• At the beginning of Section 30.6, a fission process is illustrated in which ^{235}U is struck by a neutron and undergoes fission to produce ^{144}Ba, ^{89}Kr, and three neutrons. The measured masses of these isotopes are 235.043930 u (^{235}U), 143.922953 u (^{144}Ba), 88.917630 u (^{89}Kr), and 1.0086649 u (neutron). (a) Calculate the energy (in MeV) released by each fission reaction. (b) Calculate the energy released per gram of ^{235}U, in MeV/g.

30.7 Nuclear Fusion

39. • Calculate the energy released in the fusion reaction $^2_1\text{H} + ^3_1\text{He} \rightarrow ^4_2\text{H} + ^1_0\text{n}$. The atomic mass of ^3_1H (tritium) is 3.016049 u.

40. • Consider the fusion reaction $^2_1\text{H} + ^2_1\text{H} \rightarrow ^3_2\text{He} + ^1_0\text{n}$. (a) Compute the energy liberated in this reaction in MeV and in joules. (b) Compute the energy *per mole* of deuterium, remembering that the gas is diatomic, and compare it with the heat of combustion of diatomic molecular hydrogen, which is about 2.9×10^5 J/mol.

41. •• **Comparison of energy released per gram of fuel.** (a) When gasoline is burned, it releases 1.3×10^8 J per gallon (3.788 L) of energy. Given that the density of gasoline is 737 kg/m^3, express the quantity of energy released in J/g of fuel. (b) During fission, when a neutron is absorbed by a ^{235}U nucleus, about 200 MeV of energy is released for each nucleus that undergoes fission. Express this quantity in J/g of fuel. (c) In the proton–proton chain that takes place in stars like our sun, the overall fusion reaction can be summarized as six protons fusing to form one ^4He nucleus with two leftover protons and the liberation of 26.7 MeV of energy. The fuel is the six protons. Express the energy produced here in units of J/g of fuel. Notice the huge difference between the two forms of nuclear energy, on the one hand, and the chemical energy from gasoline, on the other. (d) Our sun produces energy at a measured rate of 3.92×10^{26} W. If its mass of 1.99×10^{30} kg were all gasoline, how long could it last before consuming all its fuel? (*Historical note:* Before the discovery of nuclear fusion

and the vast amounts of energy it releases, scientists were confused. They knew that the earth was at least many millions of years old, but could not explain how the sun could survive that long if its energy came from chemical burning.)

42. • Show that the net result of the proton–proton fusion chain that occurs inside our sun can be summarized as

$$6\,\text{p}^+ \rightarrow ^4_2\text{He} + 2\,\text{p}^+ + 2\,\beta^+ + 2\,\gamma + 2\,\nu_e.$$

30.8 Fundamental Particles

30.9 High-Energy Physics

43. •• **Pair annihilation.** Consider the case where an electron e$^-$ and a positron e$^+$ annihilate each other and produce photons. Assume that these two particles collide head-on with equal, but small, speeds. (a) Show that it is not possible for only *one* photon to be produced. (*Hint:* Consider the conservation law that must be true in any collision.) (b) Show that if only two photons are produced, they must travel in opposite directions and have equal energy. (c) Calculate the wavelength of each of the photons in part (b). In what part of the electromagnetic spectrum do they lie?

44. •• **Radiation therapy with π^- mesons.** Beams of π^-
BIO mesons are used in radiation therapy for certain cancers. The energy comes from the complete decay of the π^- to *stable* particles. (a) Write out the complete decay of a π^- meson to stable particles. What are these particles? (*Hint:* See the end of Section 30.8.) (b) How much energy is released from the complete decay of a single π^- meson to stable particles? (You can ignore the very small masses of the neutrinos.) (c) How many π^- mesons need to decay to give a dose of 50.0 Gy to 10.0 g of tissue? (d) What would be the equivalent dose in part (c) in Sv and in rem? Consult Table 30.3 and use the largest appropriate RBE for the particles involved in this decay.

45. • If a Σ^+ at rest decays into a proton and a π^0, what is the total kinetic energy of the decay products?

46. • A positive pion at rest decays into a positive muon and a neutrino. (a) Approximately how much energy is released in the decay? (Assume the neutrino has zero rest mass. Use the muon and pion masses given in terms of the electron mass near the end of Section 30.8.) (b) Why can't a positive muon decay into a positive pion?

47. •• A proton and an antiproton annihilate, producing two photons. Find the energy, frequency, and wavelength of each photon emitted (a) if the initial kinetic energies of the proton and antiproton are negligible and (b) if each particle has an initial kinetic energy of 830 MeV.

48. • Which of the following reactions obey the conservation of baryon number? (a) $\text{p} + \text{p} \rightarrow \text{p} + \text{e}^+$, (b) $\text{p} + \text{n} \rightarrow 2\,\text{e}^+ + \text{e}^-$, (c) $\text{p} \rightarrow \text{n} + \text{e}^- + \bar{\nu}_e$, (d) $\text{p} + \bar{\text{p}} \rightarrow 2\,\gamma$.

49. •• **Comparing the strengths of the four forces.** Both the strong and the weak interactions are short-range forces having a range of around 1 fm. If you have two protons separated by this distance, they will be influenced by all four of the fundamental forces (or interactions). (a) Calculate the strengths of the electrical and gravitational interactions that will influence each of these protons. (b) Use the information given under "The Four Forces" at the beginning of Section 30.9 to estimate the approximate strengths of the strong interaction and the

weak interaction on each of these protons. (c) Arrange the four forces in order of strength, starting with the strongest. (d) Express each force as a multiple of the weakest of the four.

50. •• Determine the electric charge, baryon number, strangeness quantum number, and charm quantum number for the following quark combinations: (a) **uus**, (b) $c\bar{s}$, (c) $\overline{ddu}$, and (d) $\bar{c}b$.

30.10 Cosmology

51. •• The critical density of the universe is 5.8×10^{-27} kg/m³. (a) Assuming that the universe is all hydrogen, express the critical density in the number of H atoms per cubic meter. (b) If the density of the universe is equal to the critical density, how many atoms, on the average, would you expect to find in a room of dimensions 4 m × 7 m × 3 m? (c) Compare your answer in part (b) with the number of atoms you would find in this room under normal conditions on the earth.

General Problems

52. •• The results of activity measurements on a radioactive sample are given in the table. (a) Estimate the half-life of the sample. (b) Find the sample's decay constant. (c) How many radioactive nuclei were present in the sample at $t = 0$? (d) How many were present after 7.0 h?

Time (h)	Decays/s
0	20,000
0.5	15,900
1.0	12,600
1.5	9,980
2.0	7,940
2.5	6,300
3.0	4,970
4.0	3,150
5.0	1,980
6.0	1,250
7.0	790

53. • The starship *Enterprise,* of television and movie fame, is powered by the controlled combination of matter and antimatter. If the entire 400 kg antimatter fuel supply of the *Enterprise* combines with matter, how much energy is released?

54. •• A 70.0 kg person experiences a whole-body exposure to
BIO alpha radiation with energy of 1.50 MeV. A total of 5.00×10^{12} alpha particles is absorbed. (a) What is the absorbed dose in rad? (b) What is the equivalent dose in rem? (c) If the source is 0.0100 g of ^{226}Ra (half-life 1600 years) somewhere in the body, what is the activity of the source? (d) If all the alpha particles produced are absorbed, what time is required for this dose to be delivered?

55. •• A ^{60}Co source with activity 15.0 Ci is imbedded in a tumor
BIO that has a mass of 0.500 kg. The Co source emits gamma-ray photons with average energy of 1.25 MeV. Half the photons are absorbed in the tumor, and half escape. (a) What energy is delivered to the tumor per second? (b) What absorbed dose (in rad) is delivered per second? (c) What equivalent dose (in rem) is delivered per second if the RBE for these gamma rays is 0.70? (d) What exposure time is required for an equivalent dose of 200 rem?

56. •• The nucleus $^{15}_{8}$O has a half-life of 2.0 min. $^{19}_{8}$O has a half-life of about 0.5 min. (a) If, at some time, a sample contains equal amounts of $^{15}_{8}$O and $^{19}_{8}$O, what is the ratio of $^{15}_{8}$O to $^{19}_{8}$O after 2.0 min? (b) After 10.0 min?

57. •• The unstable isotope ^{40}K is used to date rock samples. Its half-life is 1.28×10^{8} years. (a) How many decays occur per second in a sample containing 6.00×10^{-6} g of ^{40}K? (b) What is the activity of the sample in curies?

58. •• **Radiation treatment of prostate cancer.** In many cases,
BIO prostate cancer is treated by implanting 60 to 100 small seeds of radioactive material into the tumor. The energy released from the decays kills the tumor. One isotope that is used (there are others) is palladium (^{103}Pd), with a half-life of 17 days. If a typical grain contains 0.250 g of ^{103}Pd, (a) what is its initial activity rate in Bq, and (b) what is the rate 68 days later?

59. •• An unstable isotope of cobalt, ^{60}Co, has one more neutron in its nucleus than the stable ^{59}Co and is a beta emitter with a half-life of 5.3 years. This isotope is widely used in medicine. A certain radiation source in a hospital contains 0.0400 g of ^{60}Co. (a) What is the decay constant for that isotope? (b) How many atoms are in the source? (c) How many decays occur per second? (d) What is the activity of the source, in curies?

60. •• **An oceanographic tracer.** Nuclear weapons tests in the 1950s and 1960s released significant amounts of radioactive tritium ($^{3}_{1}$H, half-life 12.3 years) into the atmosphere. The tritium atoms were quickly bound into water molecules and rained out of the air, most of them ending up in the ocean. For any of this tritium-tagged water that sinks below the surface, the amount of time during which it has been isolated from the surface can be calculated by measuring the ratio of the decay product, $^{3}_{2}$He, to the remaining tritium in the water. For example, if the ratio of $^{3}_{2}$He to $^{3}_{1}$H in a sample of water is 1:1, the water has been below the surface for one half-life, or approximately 12 years. This method has provided oceanographers with a convenient way to trace the movements of subsurface currents in parts of the ocean. Suppose that in a particular sample of water, the ratio of $^{3}_{2}$He to $^{3}_{1}$H is 4.3 to 1.0. How many years ago did this water sink below the surface?

61. •• A bone fragment found in a cave believed to have been
BIO inhabited by early humans contains 0.21 times as much ^{14}C as an equal amount of carbon in the atmosphere when the organism containing the bone died. (See Example 30.4 in Section 30.3.) Find the approximate age of the fragment.

62. • **Radioactive fallout.** One of the problems of in-air testing
BIO of nuclear weapons (or, even worse, the *use* of such weapons!) is the danger of radioactive fallout. One of the most problematic nuclides in such fallout is strontium-90 (^{90}Sr), which breaks down by β^{-} decay with a half-life of 28 years. It is chemically similar to calcium and therefore can be incorporated into bones and teeth, where, due to its rather long half-life, it remains for years as an internal source of radiation. (a) What is the daughter nucleus of the ^{90}Sr decay? (b) What percent of the original level of ^{90}Sr is left after 56 years? (c) How long would you have to wait for the original level to be reduced to 6.25% of its original value?

63. • Consider the nuclear reaction $^{2}_{1}$H + $^{14}_{7}$N → $^{6}_{3}$Li + $^{10}_{5}$B. Is energy absorbed or liberated? How much?

64. •• The atomic mass of $^{56}_{26}$Fe is 55.934939 u, and the atomic mass of $^{56}_{27}$Co is 55.939847 u. (a) Which of these nuclei will decay into the other? (b) What type of decay will occur? (c) How much kinetic energy will the products of the decay have?

65. • A K^{+} meson at rest decays into two π mesons. (a) What are the allowed combinations of π^{0}, π^{+}, and π^{-} as decay products? (b) Find the total kinetic energy of the π mesons.

66. •• The measured energy width of the ϕ meson is 4.0 MeV, and its mass is 1020 MeV/c^2. Using the uncertainty principle (in the form $\Delta E\, \Delta t \geq h/2\pi$), estimate the lifetime of the ϕ meson.

67. •• Given that each particle contains only combinations of $\boldsymbol{u}$, $\boldsymbol{d}$, $\boldsymbol{s}$, $\overline{\boldsymbol{u}}$, $\overline{\boldsymbol{d}}$, and $\overline{\boldsymbol{s}}$, deduce the quark content of (a) a particle with charge $+e$, baryon number 0, and strangeness $+1$; (b) a particle with charge $+e$, baryon number -1, and strangeness $+1$; (c) a particle with charge 0, baryon number $+1$, and strangeness -2.

Passage Problems

BIO **Looking under the hood of PET.** Positron Emission Tomography (PET), a kind of imaging, involves injecting a patient with artificially produced atoms that have nuclei containing an excess of neutrons. As these neutrons decay into protons, they emit positrons at fairly slow, non-relativistic speeds. When a positron encounters an electron, they annihilate each other and emit two x ray photons in opposite directions. The patient is enclosed in a circular array of photodetectors, with the tissue to be imaged centered in the detector array. If two photons strike detectors simultaneously (within 10 ns), we can conclude that they are produced by an e^+e^- annihilation event somewhere along a line connecting the two photodetectors. By observing many such simultaneous events, we can create a map of the distribution of the positron-emitting atoms in the tissue. The index of refraction of biological tissue for x rays is 1.

68. What is the energy of each of the photons resulting from an annihilation event?

A. $\dfrac{1}{2}\, m_e v^2$, where v is the speed of the positron.

B. $m_e v^2$

C. $\dfrac{1}{2}\, m_e c^2$

D. $m_e c^2$

69. What is the wavelength of each photon produced in an annihilation event?

A. $\dfrac{m_e v^2}{hc}$

B. $\dfrac{hc}{m_e v^2}$

C. $\dfrac{h}{m_e c}$

D. $\dfrac{m_e c^2}{h}$

70. Suppose that an annihilation event occurs on the line 3 cm from the center of the line connecting the two photodetectors that receive the resultant x rays. Often, a section of the whole brain is being imaged, so the x rays do not come from a point source. Will those photons be counted as having arrived simultaneously?

A. No, because the time difference will be 100 ms.

B. No, because the time difference will be 200 ms.

C. Yes, because the time difference will be 0.1 ns.

D. Yes because the time difference will be 0.2 ns.

A

The International System of Units

The Système International d'Unités, abbreviated SI, is the system developed by the General Conference on Weights and Measures and adopted by nearly all the industrial nations of the world. The following material is adapted from B. N. Taylor, ed., National Institute of Standards and Technology Spec. Pub. 811 (U.S. Govt. Printing Office, Washington, DC, 1995). See also **http://physics.nist.gov/cuu**

Quantity	Name of unit	Symbol	Equivalent units
SI base units			
length	meter	m	
mass	kilogram	kg	
time	second	s	
electric current	ampere	A	
thermodynamic temperature	kelvin	K	
amount of substance	mole	mol	
luminous intensity	candela	cd	
SI derived units			
area	square meter	m^2	
volume	cubic meter	m^3	
frequency	hertz	Hz	s^{-1}
mass density (density)	kilogram per cubic meter	kg/m^3	
speed, velocity	meter per second	m/s	
angular velocity	radian per second	rad/s	
acceleration	meter per second squared	m/s^2	
angular acceleration	radian per second squared	rad/s^2	
force	newton	N	$kg \cdot m/s^2$
pressure (mechanical stress)	pascal	Pa	N/m^2
kinematic viscosity	square meter per second	m^2/s	
dynamic viscosity	newton-second per square meter	$N \cdot s/m^2$	
work, energy, quantity of heat	joule	J	$N \cdot m$
power	watt	W	J/s
quantity of electricity	coulomb	C	$A \cdot s$
potential difference, electromotive force	volt	V	J/C, W/A
electric field strength	volt per meter	V/m	N/C
electric resistance	ohm	Ω	V/A
capacitance	farad	F	$A \cdot s/V$
magnetic flux	weber	Wb	$V \cdot s$
inductance	henry	H	$V \cdot s/A$
magnetic flux density	tesla	T	Wb/m^2
magnetic field strength	ampere per meter	A/m	
magnetomotive force	ampere	A	
luminous flux	lumen	lm	$cd \cdot sr$
luminance	candela per square meter	cd/m^2	
illuminance	lux	lx	lm/m^2
wave number	1 per meter	m^{-1}	
entropy	joule per kelvin	J/K	
specific heat capacity	joule per kilogram-kelvin	$J/(kg \cdot K)$	
thermal conductivity	watt per meter-kelvin	$W/(m \cdot K)$	

Quantity	Name of unit	Symbol	Equivalent units
radiant intensity	watt per steradian	W/sr	
activity (of a radioactive source)	becquerel	Bq	s^{-1}
radiation dose	gray	Gy	J/kg
radiation dose equivalent	sievert	Sv	J/kg
	SI supplementary units		
plane angle	radian	rad	
solid angle	steradian	sr	

Definitions of SI Units

meter (m) The *meter* is the length equal to the distance traveled by light, in vacuum, in a time of $1/299{,}792{,}458$ second.

kilogram (kg) The *kilogram* is the unit of mass; it is equal to the mass of the international prototype of the kilogram. (The international prototype of the kilogram is a particular cylinder of platinum-iridium alloy that is preserved in a vault at Sèvres, France, by the International Bureau of Weights and Measures.)

second (s) The *second* is the duration of 9,192,631,770 periods of the radiation corresponding to the transition between the two hyperfine levels of the ground state of the cesium-133 atom.

ampere (A) The *ampere* is that constant current that, if maintained in two straight parallel conductors of infinite length, of negligible circular cross section, and placed 1 meter apart in vacuum, would produce between these conductors a force equal to 2×10^{-7} newton per meter of length.

kelvin (K) The *kelvin,* unit of thermodynamic temperature, is the fraction 1/273.16 of the thermodynamic temperature of the triple point of water.

ohm (V) The *ohm* is the electric resistance between two points of a conductor when a constant difference of potential of 1 volt, applied between these two points, produces in this conductor a current of 1 ampere, this conductor not being the source of any electromotive force.

coulomb (C) The *coulomb* is the quantity of electricity transported in 1 second by a current of 1 ampere.

candela (cd) The *candela* is the luminous intensity, in a given direction, of a source that emits monochromatic radiation of frequency 540×10^{12} hertz and that has a radiant intensity in that direction of 1/683 watt per steradian.

mole (mol) The *mole* is the amount of substance of a system that contains as many elementary entities as there are carbon atoms in 0.012 kg of carbon 12. The elementary entities must be specified and may be atoms, molecules, ions, electrons, other particles, or specified groups of such particles.

newton (N) The *newton* is that force that gives to a mass of 1 kilogram an acceleration of 1 meter per second per second.

joule (J) The *joule* is the work done when the point of application of a constant force of 1 newton is displaced a distance of 1 meter in the direction of the force.

watt (W) The *watt* is the power that gives rise to the production of energy at the rate of 1 joule per second.

volt (V) The *volt* is the difference of electric potential between two points of a conducting wire carrying a constant current of 1 ampere, when the power dissipated between these points is equal to 1 watt.

weber (Wb) The *weber* is the magnetic flux that, linking a circuit of one turn, produces in it an electromotive force of 1 volt as it is reduced to zero at a uniform rate in 1 second.

lumen (lm) The *lumen* is the luminous flux emitted in a solid angle of 1 steradian by a uniform point source having an intensity of 1 candela.

farad (F) The *farad* is the capacitance of a capacitor between the plates of which there appears a difference of potential of 1 volt when it is charged by a quantity of electricity equal to 1 coulomb.

henry (H) The *henry* is the inductance of a closed circuit in which an electromotive force of 1 volt is produced when the electric current in the circuit varies uniformly at a rate of 1 ampere per second.

radian (rad) The *radian* is the plane angle between two radii of a circle that cut off on the circumference an arc equal in length to the radius.

steradian (sr) The *steradian* is the solid angle that, having its vertex in the center of a sphere, cuts off an area of the surface of the sphere equal to that of a square with sides of length equal to the radius of the sphere.

SI Prefixes The names of multiples and submultiples of SI units may be formed by application of the prefixes listed in Appendix F.

The Greek Alphabet

Name	Capital	Lowercase	Name	Capital	Lowercase
Alpha	A	α	Nu	N	ν
Beta	B	β	Xi	Ξ	ξ
Gamma	Γ	γ	Omicron	O	o
Delta	Δ	δ	Pi	Π	π
Epsilon	E	ϵ	Rho	P	ρ
Zeta	Z	ζ	Sigma	Σ	σ
Eta	H	η	Tau	T	τ
Theta	Θ	θ	Upsilon	Υ	υ
Iota	I	ι	Phi	Φ	ϕ
Kappa	K	κ	Chi	X	χ
Lambda	Λ	λ	Psi	Ψ	ψ
Mu	M	μ	Omega	Ω	ω

Periodic Table of the Elements

Group	1	2	3	4	5	6	7	8	9	10	11	12	13	14	15	16	17	18
Period																		
1	1 **H** 1.008																	2 **He** 4.003
2	3 **Li** 6.941	4 **Be** 9.012											5 **B** 10.811	6 **C** 12.011	7 **N** 14.007	8 **O** 15.999	9 **F** 18.998	10 **Ne** 20.180
3	11 **Na** 22.990	12 **Mg** 24.305											13 **Al** 26.982	14 **Si** 28.086	15 **P** 30.974	16 **S** 32.065	17 **Cl** 35.453	18 **Ar** 39.948
4	19 **K** 39.098	20 **Ca** 40.078	21 **Sc** 44.956	22 **Ti** 47.867	23 **V** 50.942	24 **Cr** 51.996	25 **Mn** 54.938	26 **Fe** 55.845	27 **Co** 58.933	28 **Ni** 58.693	29 **Cu** 63.546	30 **Zn** 65.409	31 **Ga** 69.723	32 **Ge** 72.64	33 **As** 74.922	34 **Se** 78.96	35 **Br** 79.904	36 **Kr** 83.798
5	37 **Rb** 85.468	38 **Sr** 87.62	39 **Y** 88.906	40 **Zr** 91.224	41 **Nb** 92.906	42 **Mo** 95.94	43 **Tc** (98)	44 **Ru** 101.07	45 **Rh** 102.906	46 **Pd** 106.42	47 **Ag** 107.868	48 **Cd** 112.411	49 **In** 114.818	50 **Sn** 118.710	51 **Sb** 121.760	52 **Te** 127.60	53 **I** 126.904	54 **Xe** 131.293
6	55 **Cs** 132.905	56 **Ba** 137.327	71 **Lu** 174.967	72 **Hf** 178.49	73 **Ta** 180.948	74 **W** 183.84	75 **Re** 186.207	76 **Os** 190.23	77 **Ir** 192.217	78 **Pt** 195.078	79 **Au** 196.967	80 **Hg** 200.59	81 **Tl** 204.383	82 **Pb** 207.2	83 **Bi** 208.980	84 **Po** (209)	85 **At** (210)	86 **Rn** (222)
7	87 **Fr** (223)	88 **Ra** (226)	103 **Lr** (262)	104 **Rf** (261)	105 **Db** (262)	106 **Sg** (266)	107 **Bh** (264)	108 **Hs** (269)	109 **Mt** (268)	110 **Ds** (271)	111 **Rg** (272)	112 **Uub** (285)	113 **Uut** (284)	114 **Uuq** (289)	115 **Uup** (288)	116 **Uuh** (292)	117 **Uus** (294)	118 **Uuo**

Lanthanoids

57 **La** 138.905	58 **Ce** 140.116	59 **Pr** 140.908	60 **Nd** 144.24	61 **Pm** (145)	62 **Sm** 150.36	63 **Eu** 151.964	64 **Gd** 157.25	65 **Tb** 158.925	66 **Dy** 162.500	67 **Ho** 164.930	68 **Er** 167.259	69 **Tm** 168.934	70 **Yb** 173.04

Actinoids

89 **Ac** (227)	90 **TH** (232)	91 **Pa** (231)	92 **U** (238)	93 **Np** (237)	94 **Pu** (244)	95 **Am** (243)	96 **Cm** (247)	97 **Bk** (247)	98 **Cf** (251)	99 **Es** (252)	100 **Fm** (257)	101 **Md** (258)	102 **No** (259)

For each element the average atomic mass of the mixture of isotopes occurring in nature is shown. For elements having no stable isotope, the approximate atomic mass of the longest-lived isotope is shown in parentheses. For elements that have been predicted but not yet confirmed, no atomic mass is given. All atomic masses are expressed in atomic mass units $(1 \text{ u} = 1.660538728(83) \times 10^{-27} \text{ kg})$, equivalent to grams per mole (g/mol).

Unit Conversion Factors

LENGTH

1 m = 100 cm = 1000 mm = 10^6 μm = 10^9 nm
1 km = 1000 m = 0.6214 mi
1 m = 3.281 ft = 39.37 in.
1 cm = 0.3937 in.
1 in. = 2.540 cm
1 ft = 30.48 cm
1 yd = 91.44 cm
1 mi = 5280 ft = 1.609 km
1 Å = 10^{-10} m = 10^{-8} cm = 10^{-1} nm
1 nautical mile = 6080 ft
1 light year = 9.461 $\times$ 10^{15} m

AREA

1 cm^2 = 0.155 in.^{2}
1 m^2 = 10^4 cm^2 = 10.76 ft^2
1 in.^{2} = 6.452 cm^2
1 ft^2 = 144 in.^{2} = 0.0929 m^2

VOLUME

1 liter = 1000 cm^3 = 10^{-3} m^3
       = 0.03531 ft^3 = 61.02 in.^{3}
1 ft^3 = 0.02832 m^3 = 28.32 liters = 7.477 gallons
1 gallon = 3.788 liters

TIME

1 min = 60 s
1 h = 3600 s
1 d = 86,400 s
1 y = 365.24 d = 3.156 $\times$ 10^7 s

ANGLE

1 rad = 57.30° = 180°/π
1° = 0.01745 rad = π/180 rad
1 revolution = 360° = 2π rad
1 rev/min (rpm) = 0.1047 rad/s

SPEED

1 m/s = 3.281 ft/s
1 ft/s = 0.3048 m/s
1 mi/min = 60 mi/h = 88 ft/s
1 km/h = 0.2778 m/s = 0.6214 mi/h
1 mi/h = 1.466 ft/s = 0.4470 m/s = 1.609 km/h
1 furlong/fortnight = 1.662 $\times$ 10^{-4} m/s

ACCELERATION

1 m/s^2 = 100 cm/s^2 = 3.281 ft/s^2
1 cm/s^2 = 0.01 m/s^2 = 0.03281 ft/s^2
1 ft/s^2 = 0.3048 m/s^2 = 30.48 cm/s^2
1 mi/h $\cdot$ s = 1.467 ft/s^2

MASS

1 kg = 10^3 g = 0.0685 slug
1 g = 6.85 $\times$ 10^{-5} slug
1 slug = 14.59 kg
1 u = 1.661 $\times$ 10^{-27} kg
1 kg has a weight of 2.205 lb when g = 9.80 m/s^2

FORCE

1 N = 10^5 dyn = 0.2248 lb
1 lb = 4.448 N = 4.448 $\times$ 10^5 dyn

PRESSURE

1 Pa = 1 N/m^2 = 1.450 $\times$ 10^{-4} lb/in.^{2} = 0.209 lb/ft^2
1 bar = 10^5 Pa
1 lb/in.^{2} = 6895 Pa
1 lb/ft^2 = 47.88 Pa
1 atm = 1.013 $\times$ 10^5 Pa = 1.013 bar
      = 14.7 lb/in.^{2} = 2117 lb/ft^2
1 mm Hg = 1 torr = 133.3 Pa

ENERGY

1 J = 10^7 ergs = 0.239 cal
1 cal = 4.186 J (based on 15° calorie)
1 ft $\cdot$ lb = 1.356 J
1 Btu = 1055 J = 252 cal = 778 ft $\cdot$ lb
1 eV = 1.602 $\times$ 10^{-19} J
1 kWh = 3.60 $\times$ 10^6 J

MASS–ENERGY EQUIVALENCE

1 kg $\leftrightarrow$ 8.988 $\times$ 10^{16} J
1 u $\leftrightarrow$ 931.5 MeV
1 eV $\leftrightarrow$ 1.074-10^{-9} u

POWER

1 W = 1 J/s
1 hp = 746 W = 550 ft $\cdot$ lb/s
1 Btu/h = 0.293 W

Numerical Constants

Fundamental Physical Constants*

Name	Symbol	Value
Speed of light in vacuum	c	2.99792458×10^8 m/s
Magnitude of charge of electron	e	$1.60217653(14) \times 10^{-19}$ C
Gravitational constant	G	$6.6742(10) \times 10^{-11}$ N $\cdot$ m^2/kg^2
Planck's constant	h	$6.6260693(11) \times 10^{-34}$ J $\cdot$ s
Boltzmann constant	k	$1.3806505(24) \times 10^{-23}$ J/K
Avogadro's number	N_A	$6.0221415(10) \times 10^{23}$ molecules/mol
Gas constant	R	$8.314472(15)$ J/(mol $\cdot$ K)
Mass of electron	m_e	$9.1093826(16) \times 10^{-31}$ kg
Mass of proton	m_p	$1.67262171(29) \times 10^{-27}$ kg
Mass of neutron	m_n	$1.67492728(29) \times 10^{-27}$ kg
Permeability of vacuum	μ_0	$4\pi \times 10^{-7}$ Wb/T $\cdot$ m/A
Permittivity of vacuum	$\epsilon_0 = 1/\mu_0 c^2$	$8.854187817 \cdots \times 10^{-12}$ C^2/(N $\cdot$ m^2)
	$1/4\pi\epsilon_0$	$8.987551787 \cdots \times 10^9$ N $\cdot$ m^2/C^2

Other Useful Constants*

Name	Symbol	Value
Mechanical equivalent of heat		4.186 J/cal ($15°$ calorie)
Standard atmospheric pressure	1 atm	1.01325×10^5 Pa
Absolute zero	0 K	$-273.15°$C
Electron volt	1 eV	$1.60217653(14) \times 10^{-19}$ J
Unified atomic mass unit	1 u	$1.66053886(28) \times 10^{-27}$ kg
Electron rest energy	$m_e c^2$	$0.510998918(44)$ MeV
Volume of ideal gas (0°C and 1 atm)		$22.413996(39)$ liter/mol
Acceleration due to gravity (standard)	g	9.80 m/s^2

*Source: National Institute of Standards and Technology (**http://physics.nist.gov/cuu**). Numbers in parentheses show the uncertainty in the final digits of the main number; for example, the number 1.6454(21) means 1.6454 ± 0.0021. Values shown without uncertainties are exact.

Astronomical Data[†]

Body	Mass (kg)	Radius (m)	Orbit radius (m)	Orbit period
Sun	1.99×10^{30}	6.96×10^{8}	—	—
Moon	7.35×10^{22}	1.74×10^{6}	3.84×10^{8}	27.3 d
Mercury	3.30×10^{23}	2.44×10^{6}	5.79×10^{10}	88.0 d
Venus	4.87×10^{24}	6.05×10^{6}	1.08×10^{11}	224.7 d
Earth	5.97×10^{24}	6.38×10^{6}	1.50×10^{11}	365.3 d
Mars	6.42×10^{23}	3.40×10^{6}	2.28×10^{11}	687.0 d
Jupiter	1.90×10^{27}	6.91×10^{7}	7.78×10^{11}	11.86 y
Saturn	5.68×10^{26}	6.03×10^{7}	1.43×10^{12}	29.45 y
Uranus	8.68×10^{25}	2.56×10^{7}	2.87×10^{12}	84.02 y
Neptune	1.02×10^{26}	2.48×10^{7}	4.50×10^{12}	164.8 y
Pluto	1.31×10^{22}	1.15×10^{6}	5.91×10^{12}	247.9 y

[†]Source: NASA Jet Propulsion LaboratorySolar System Dynamics Group (**http://ssd.jpl.nasa.gov**), and P. Kenneth Seidelmann, ed., ***Explanatory Supplement to the Astronomical Almanac*** (University Science Books, Mill Valley, CA, 1992), pp. 704–706. For each body, "radius" is its radius at its equator and "orbit radius" is its average distance from the sun (for the planets) or from the earth (for the moon).

Prefixes for Powers of 10

Power of ten	Prefix	Abbreviation	Pronunciation
10^{-24}	yocto-	y	*yoc*-toe
10^{-21}	zepto-	z	*zep*-toe
10^{-18}	atto-	a	*at*-toe
10^{-15}	femto-	f	*fem*-toe
10^{-12}	pico-	p	*pee*-koe
10^{-9}	nano-	n	*nan*-oe
10^{-6}	micro-	μ	*my*-crow
10^{-3}	milli-	m	*mil*-i
10^{-2}	centi-	c	*cen*-ti
10^{3}	kilo-	k	*kil*-oe
10^{6}	mega-	M	*meg*-a
10^{9}	giga-	G	*jig*-a or *gig*-a
10^{12}	tera-	T	*ter*-a
10^{15}	peta-	P	*pet*-a
10^{18}	exa-	E	*ex*-a
10^{21}	zetta-	Z	*zet*-a
10^{24}	yotta-	Y	*yot*-a

Examples:

1 femtometer = 1 fm = 10^{-15} m

1 picosecond = 1 ps = 10^{-12} s

1 nanocoulomb = 1 nC = 10^{-9} C

1 microkelvin = 1 μK = 10^{-6} K

1 millivolt = 1 mV = 10^{-3} V

1 kilopascal = 1 kPa = 10^{3} Pa

1 megawatt = 1 MW = 10^{6} W

1 gigahertz = 1 GHz = 10^{9} Hz

Answers to Selected Odd-Numbered Problems

Chapter 0

Problems

1. $9x^8y^4$
3. $4x^4y^{-6}$
5. 4.75×10^5
7. 1.23×10^{-4}
9. $x = 4$
11. $x = \pm 2\sqrt{3}$
13. $x = 2, 3$
15. $t = 1.8$ s, and $t = -2.25$
17. $x = 2, y = 3$
19. 200 N
21. 1.2×10^{-6} m
23. 186 N
25. (a) $\log\left(\dfrac{x^4y}{(x+y)^3}\right)$ (b) $\log(xz)$
27. 10^{-6}
29. (a) 0.75 m, 0.045 m^2 (b) 0.55 m^2, 0.039 m^3
    (c) 0.6 m^2, 0.02 m^3 (d) 0.58 m^2, 0.034 m^3
31. 22 cm

Chapter 17

Conceptual Questions

1. The electrified object induces a charge separation in the neutral object. For example, a positively charged rod displaces negative charge in the bit of paper toward the rod. Since the negative charge in the paper is then closer to the rod than the positive charge in the paper and since the electrical force is inversely proportional to the distance squared, the attractive force between positive and negative charges is larger in magnitude than the repulsive force between positive and positive charges and the net force is attractive.

3. *Similarities:* Both forces are proportional to the square of the distance between the objects. Both forces obey Newton's third law. *Differences:* The gravitational force is always attractive. The electrical force can be either attractive or repulsive, depending on the signs of the charges. The electrical force is stronger than the gravitational force. For example, the electrical force between two 1 Coulomb charges separated by 1 meter is immense whereas the gravitational force between two 1 kilogram masses separated by 1 meter is very small.

5. Positive and negative are just names given to the sign of charge that protons and electrons have. There is no inherent significance.

7. (a) The charged rod pulls charge of the opposite sign onto the metal ball. This leaves the gold leaf and metal tube with a net charge of the same sign as the rod. Since the gold leaf and the tube have charge of the same sign, they repel. (b) When the rod is removed the charge on the ball spreads back over the tube and leaf and they become neutral again. There is no net charge and no electrical force so the leaf hangs vertically. (c) When the charged rod touches the ball it transfers charge to it. This net charge spreads over the ball, tube and leaf. When the rod is removed this net charge stays on the leaf and tube and they continue to repel. The gold leaf continues to hang at an angle away from the tube.

11. The dipole consists of equal amounts of positive and negative charge. In a uniform electric field the force on the positive charge is equal in magnitude and opposite in direction to the force on the negative charge, and the net force is zero. If the field is not uniform and is stronger at one of the charges in the dipole, the force on that charge is greater in magnitude than the force on the other charge and the net force is not zero.

13. The lightning is attracted to the sharp point of the rod. Copper is an excellent conductor and the lightning current travels through it rather than through the wood of the barn, since wood is a poor conductor of electricity.

15. (a) In a solid material the molecules are held to each other by electrical forces. When one surface slides over another, molecules of the surface are pulled off. The resistance to this happening, due to the electrical bonding, gives rise to the friction force. (b) The steel atoms are held together by strong electrical forces. In order to penetrate the surface, the surface atoms must be displaced and the electrical bonding forces oppose this. (c) Chemical bonding arises from electrical forces within the molecules, due to charge displacement within the molecules.

Multiple-Choice Problems

1. D
3. C
5. E
7. C
9. B
11. D
13. B
15. A

Problems

5. (a) 1.56×10^{13} electrons
   (b) 1.56×10^{10} electrons
7. (a) 4.27×10^{24} protons; 6.83×10^5 C
   (b) 4.27×10^{24} electrons
9. 1.07×10^{-14} m; 1.07×10^{-14} m
11. (a) 230 N; yes; 52 lbs
    (b) 2.30×10^{-8} N; no
13. (a) 3.4×10^{14}
    (b) $+55 \mu$C
15. (a) 7.25×10^{24} electrons
    (b) 5.27×10^{15} electrons
    (c) 7.27×10^{-10}
17. 3.7 km
19. 0.750 nC
21. 2.59×10^{-6} N; $-y$ direction
23. (a) 1680 N; from the $+5.00 \mu$C charge toward the -5.00μC charge
    (b) 22.3 N $\cdot$ m, clockwise
25. 1.26×10^{-8} N; attractive
27. 4.27×10^{-10} N, to the left
29. 2.21×10^4 m/s^2
31. (a) 2.50 N/C, upward
    (b) 4.00×10^{-19} N, upward
33. (a) 1.62×10^3 N/C
    (b) 4.27×10^6 m/s
35. 7.23×10^{-9} C
37. (a) 5.8×10^{19} N/C (b) 5.8×10^9 N/C
39. (a) 9.0×10^{-12} C (b) 32 N/C; away from axon (c) 280 m
41. (a) 1050 N/C; $-x$ direction (b) 312 N/C; $+x$ direction (c) 845 N/C; $+x$ direction
43. (a) between the charges, 0.24 m from the $+0.500$ nC charge
    (b) 0.40 m from ±0.500 nC charge, 1.60 m from ∓8.00 nC charge
45. (a) zero (b) vector from $-q$ to $+q$ parallel to $\vec{E}$; vector from $-q$ to $+q$ antiparallel to $\vec{E}$ (c) stable: vector from $-q$ to $+q$

is parallel to $\vec{E}$; unstable: vector from $-q$ to $+q$ is antiparallel to $\vec{E}$

47. (a) $E/4$ (b) $E/9$

49. (a) top, positive; middle, negative; bottom, positive; (b) on horizontal line through middle charge

55. S_1: 0; S_2: 7.9×10^5 N $\cdot$ m^2/C; S_3: 0; S_4: 1.1×10^5 N $\cdot$ m^2/C; S_5: 6.8×10^5 N $\cdot$ m^2/C

57. (a) zero
    (b) 3.75×10^7 N/C; radially inward
    (c) 1.11×10^7 N/C; radially inward

59. (a) 1.80×10^{10}
    (b) 414 N/C

61. (a) zero

63. (a) yes, $-12\ \mu C$
    (b) $+12\ \mu C$

65. (a) 3.2 nC (b) to the right
    (c) $x = -1.76$ m

67. (a) $F_x = 8.64 \times 10^{-5}$ N; $F_y = -5.52 \times 10^{-5}$ N
    (b) 1.03×10^{-4} N; 32.6° below $+x$ axis

69. (a) 1.35×10^{-3} N; away from vacant corner (b) 1.29×10^{-3} N; toward center of square

71. 3.41×10^4 N/C, to the left

73. 1.7×10^6 electrons

75. $q_1 = 6.14 \times 10^{-6}$ N/C; $q_2 = -6.14 \times 10^{-6}$ N/C

77. 2.2×10^6 m/s

Passage Problems

79. C

Chapter 18

Conceptual Questions

1. The electrical force at a point is tangent to the electric field lines at that point. Therefore, there is no component of electrical force in the direction perpendicular to the field line. No work is done on a charge when it moves perpendicular to a field line so the potential doesn't change in this direction and this direction is along an equipotential surface.

3. Zero electric field means zero force which in turn means no electrical work done on a test charge when it moves from point A to point B within the region. Therefore, any two points A and B in the region are at the same potential. In this region the potential is constant but not necessarily zero.

5. Current flows between points of different potential. (a) When they are in the car and touch two points on the car body no current flows through them because the two points are at the same potential. (b) The body of the car and the ground have a large potential difference so when an occupant touches both the car and the ground a dangerously large current flows through him.

7. Since the capacitor is disconnected from the battery before the plates are pulled apart, the charge on the plates stays constant while the plates are moved apart. (a) The electric field between the plates depends only on the charge per unit area on the plates. This doesn't change and the electric field stays constant. (b) The charge is trapped on the plates and doesn't change. (c) Since the distance between the plates increases while the electric field stays the same, more work is done by the electric force on a test charge when it moves from one plate to the other and the potential difference increases. (d) The plates attract each other so work must be done to pull them apart and this increases the stored energy. We can also see this from the equation $U = QV/2$. V increases while Q stays the same, so U increases.

11. As the temperature of the liquid increases the random motion of the molecules increases and this decreases the alignment of the molecular dipoles along the electric field direction. This reduces the induced electric field within the dielectric and therefore reduces the effect of the dielectric on the net field between the plates. The dielectric constant is the ratio of the net field without the dielectric to the field with the dielectric. Therefore, the dielectric constant becomes closer to 1.0 as the temperature increases.

13. The stored energy for a capacitor is given by $U = \frac{1}{2}CV^2$. In parallel the potential difference V across each capacitor equals the battery voltage whereas in series the potentials add to give the battery voltage. Therefore, the voltage for each capacitor is greater in parallel and the stored energy is greater when they are connected in parallel.

15. (a) No. At a point midway between two point charges of equal magnitude and opposite sign the potential (relative to infinity) is zero but the electric field is not zero. The electric field is directed toward the negative charge. (b) No. At a point midway between two equal point charges (equal in magnitude and sign) the electric field is zero and the potential (relative to infinity) is not zero.

Multiple-Choice Problems

1. A, C
3. A
5. C
7. B, D
9. B
11. D
13. A
15. C

Problems

1. (a) 0 (b) $+7.50 \times 10^{-4}$ J
   (c) -2.06×10^{-3} J

3. 0.373 m

5. (a) -2.88×10^{-17} J
   (b) -5.04×10^{-17} J

7. -1.42×10^{-18} J

9. 0.078 J

11. (a) 8000 N/C (b) 1.92×10^{-5} N

13. (a) 1.25×10^4 V/m
    (b) 2.20×10^{15} m/s^2

15. 1.5×10^3 km

17. 25 V/m

19. (a) 2.5 mm (b) 7.5 mm

21. 7.42 m/s; faster

23. (a) 0.199 J (b) (i) 26.6 m/s (ii) 36.7 m/s (iii) 37.6 m/s

25. (b) 5.8×10^6 m/s

27. 4.2×10^6 V

29. (b) yes (c) flat sheets parallel to the plates

31. (a) 27.0 kV (b) 0.98 cm; 2.40 m
    (c) the increasing spacing shows that the field is weaker at greater distances from the charged sphere

35. (a) electron: 5.93×10^5 m/s; proton: 1.38×10^4 m/s
    (b) electron: 1.87×10^7 m/s; proton: 4.38×10^5 m/s
    (c) electron: 0.0256 keV; proton: 47.0 keV

37. (a) 3.29 pF (b) 13.2 kV
    (c) 4.02×10^6 V/m

39. (a) 400 V (b) 3.39×10^{-2} m^2
    (c) 6.67×10^5 V/m
    (d) 5.90×10^{-6} C/m^2

41. (a) 120 μC (b) 60 μC (c) 480 μC

43. (a) 0.447 μF (b) 60.0 V

45. 2.8 mm

47. (a) 20 pF (b) 8.6 pF

49. (a) $Q_1 = 80.0 \times 10^{-6}$ C; $Q_3 = 120.0 \times 10^{-6}$ C (b) 37.4 V

51. $V_2 = 50$ V; $V_3 = 70$ V

53. (a) 3.00 μF: 30.8 μC; 5.00 μF: 51.3 μC; 6.00 μF: 82.1 μC (b) 3.00 μF: 10.3 V; 5.00 μF: 10.3 V; 6.00 μF: 13.7 V

55. (a) 3.47 μF (b) 174 μC (c) 174 μC on each

57. (a) 90 μC (b) 20 μC (c) 5.4×10^{-4} J (parallel); 1.2×10^{-4} J (series)

59. (a) 7.5 μC; 5.6 μJ
    (b) 3.2×10^{-3} C; 630 V

61. (a) 4.19 J (b) 16.8 J

63. (a) 2.4 μC (b) 2.4 μC (c) 43.2 μJ
    (d) 150 nF: 19.2 μJ; 120 nF: 24.6 μJ
    (e) 150 nF: 16 V; 120 nF: 20 V

65. (a) 0.0160 C (b) 533 V for each
    (c) 4.27 J (d) 2.13 J

67. (a) 2770 V (b) 5540 V (c) 3.53×10^{-3} J

69. (a) 1.18 μF per cm^2 (b) 1.13×10^6 V/m

71. (a) 3.6 mJ; 13.5 mJ
    (b) increased by 9.9 mJ

73. (a) 6.3×10^{-6} C (b) 6.3×10^{-6} C
    (c) no effect

75. (a) 0.415 m (b) 2.30×10^{-10} C
    (c) away from the point charge

77. (a) 5.67×10^7 V/m
    (b) 0.28 V; outside (c) 7×10^{-15} J
    (d) 1.0×10^7 V/m; 0.052 V

79. (a) $\dfrac{mv^2}{r} = \dfrac{ke^2}{r^2}$ and $v = \sqrt{\dfrac{ke^2}{mr}}$
    (b) $K = \dfrac{1}{2}mv^2 = \dfrac{1}{2}\dfrac{ke^2}{r} = -\dfrac{1}{2}U$

(c) $E = K + U = \frac{1}{2}U = -\frac{1}{2}\frac{ke^2}{r} =$

$-\frac{1}{2}\frac{k(1.60 \times 10^{-19}\,\text{C})^2}{5.29 \times 10^{-11}\,\text{m}} = -2.17 \times$

$10^{-18}\,\text{J} = -13.6\,\text{eV}$

81. (a) $4.43 \times 10^{-11}\,\text{F}$ (b) $5.31 \times 10^{-9}\,\text{C}$
    (c) $1.50 \times 10^4\,\text{V/m}$ (d) $3.19 \times 10^{-7}\,\text{J}$
    (e) (a) $2.21 \times 10^{-11}\,\text{F}$;
    (b) $5.31 \times 10^{-9}\,\text{C}$; (c) $1.50 \times 10^4\,\text{V/m}$;
    (d) $6.37 \times 10^{-7}\,\text{J}$
83. (a) $3.54 \times 10^{-11}\,\text{F}$ (b) $1.06 \times 10^{-8}\,\text{C}$
    (c) $1.59 \times 10^{-6}\,\text{J}$
85. (a) $2.3\,\mu\text{F}$ (b) $Q_1 = 9.7 \times 10^{-4}\,\text{C}$;
    $Q_2 = 6.4 \times 10^{-4}\,\text{C}$

Passage Problems

89. B
91. D

Chapter 19

Conceptual Questions

1. Yes. The open circuit voltage V_{open} is the emf $\mathcal{E}$ of the source. When the source is short-circuited, $\mathcal{E} - I_{\text{short}}r = 0$, where r is the internal resistance of the source.
$I_{\text{short}} = \frac{\mathcal{E}}{r} \cdot \frac{V_{\text{open}}}{I_{\text{short}}} = \frac{\mathcal{E}}{(\mathcal{E}/r)} = r.$

5. (a) False. Consider a circuit containing a single resistor R_1. Adding a second resistor R_2 in parallel with R_1 lowers the net resistance of the circuit, because the second resistor adds an alternative current path. (b) False. A simple example is two resistors in parallel. If one is removed, the net resistance doubles.

7. A capacitor stores energy in the electric field that is created between its plates when the capacitor is charged. A resistor cannot store energy. A resistor dissipates electrical energy by converting it to thermal energy and the thermal energy cannot be converted back to electrical energy in the resistor at some later time.

9. No. The terminal voltage V_{ab} is equal to $\mathcal{E} - Ir$. The terminal voltage can be noticeably less than the emf when a large current passes through the battery.

11. In series, the emf of the combination is twice the emf of a single battery and the bulb is much brighter than if a single battery were used. In parallel the emf applied to the bulb is the emf of a single battery. In parallel the current through each battery is half the current through the bulb. The batteries last longer, but the brightness of the bulb is the same as if only a single battery is used.

15. Energy is charge times potential and it is the energy dissipated in your body that is harmful. When you receive a net charge by scuffing your shoes across the carpet, the net charge you receive is very small. A large current can flow when you discharge, but only for a very short time. The total charge that flows is very small

and the energy generated is also very small. A power line is a continuous source of current and the current can flow long enough for a fatal amount of energy to be transferred to your body.

Multiple-Choice Problems

1. C
3. D
5. A
7. C
9. C
11. A
13. C
15. E

Problems

1. (a) 9.38×10^{18} electrons/s (b) 450 C
   (c) 5.00 s
3. $9.0\,\mu\text{A}$
5. (a) 7.0×10^{22} (b) 0.035 mm/s
7. 121 m
9. $1.75 \times 10^{-6}\,\Omega \cdot \text{m}$
11. (a) $2.00 \times 10^{-4}\,\Omega$ (b) 2.05 cm
13. $9R$
15. (a) $1.06 \times 10^{-5}\,\Omega \cdot \text{m}$
    (b) $1.05 \times 10^{-3}\,(°\text{C})^{-1}$
17. (a) yes, graph of V versus I is a straight line (b) R is the slope (c) $150\,\Omega$
19. $666°\text{C}$
21. 5.0 V; yes
23. 27.4 V
25. (a) $I/2$ (b) $4I$ (c) $2I$
27. 3.08 V; $0.067\,\Omega$; $1.8\,\Omega$
29. (a) 2.50 V; $2.14\,\Omega$ (b) 1.91 V
31. (a) $0.700\,\Omega$ (b) $5.30\,\Omega$
33. (a) 21.8 A (b) $0.688\,\Omega$
35. (a) 1.70 W (b) $24.5\,\Omega$
37. (a) $240\,\Omega$ (b) less; $144\,\Omega$
39. 40 W; 0.40 J
41. 0.450 A
43. (a) $26.7\,\Omega$ (b) 4.50 A (c) 453 W (d) larger
45. (a) $1000\,\Omega$ (b) 100 V (c) 10 W
47. (a) $134\,\Omega$ (b) $15\,\Omega$
49. (a) $27.7\,\Omega$ (b) 4.33 A (c) $40\,\Omega$: 3.00 A;
    $90\,\Omega$: 1.33 A
51. (a) $8.80\,\Omega$ (b) 3.18 A in each (c) 3.18 A
    (d) $1.60\,\Omega$: 5.09 V; $2.40\,\Omega$: 7.64 V;
    $4.80\,\Omega$: 15.3 V (e) $1.60\,\Omega$: 16.2 W;
    $2.40\,\Omega$: 24.3 W; $4.80\,\Omega$: 48.5 W
    (f) greatest resistance; $P = I^2R$ and I is the same for each
53. $3.00\,\Omega$; $1.00\,\Omega$ and $3.00\,\Omega$: 12.0 A;
    $7.00\,\Omega$ and $5.00\,\Omega$: 4.00 A
55. Solder a $76.5\,\text{k}\Omega$ resistor in parallel with the $69.8\,\text{k}\Omega$ resistor.
57. (a) 0.333 A (b) 0.250 A (c) 0.583 A
59. (a) $20.0\,\Omega$ (b) A_2: 4.00 A; A_3: 12.0 A;
    A_4: 14.0 A; A_5: 8.00 A
61. 0.714 A
63. $4.00 \times 10^{-3}\,\text{F}$
65. (a) 0 s, 0 C; 5 s, $2.7 \times 10^{-4}\,\text{C}$; 10 s, $4.42 \times 10^{-4}\,\text{C}$; 20 s, $6.21 \times 10^{-4}\,\text{C}$; 100 s, $7.44 \times 10^{-4}\,\text{C}$ (b) 0 s, $6.70 \times 10^{-5}\,\text{A}$; 5 s, $4.27 \times 10^{-5}\,\text{A}$; 10 s, $2.72 \times 10^{-5}\,\text{A}$; 20 s; $1.11 \times 10^{-5}\,\text{A}$; 100 s, $8.20 \times 10^{-9}\,\text{A}$

67. (a) $8.49 \times 10^{-7}\,\text{F}$ (b) 2.89 s
69. (a) $Q_{10} = 500\,\text{pC}$; $Q_{20} = Q_{30} = Q_{40} = 461\,\text{pC}$
    (b) $V_{10} = 50.0\,\text{V}$; $V_{20} = 23.1\,\text{V}$;
    $V_{30} = 15.4\,\text{V}$; $V_{40} = 11.5\,\text{V}$
    (c) 2.50 A (d) 384 ps
73. (a) $18.7\,\Omega$ (b) $7.5\,\Omega$
75. (a) $237°\text{C}$ (b) (i) 162 W (ii) 148 W
77. (a) 0.529 W (b) 9530 J (c) $6.8\,\Omega$
79. (a) $80\,\text{C}°$ (b) no; person would vaporize
81. 48.0 W
83. (a) toaster: 15.0 A; frypan: 11.7 A; lamp: 0.625 A (b) yes
85. 243 W
87. (a) 77% (b) 0.0077%
89. (a) 0.22 V (b) point a
91. (a) $5.00 \times 10^5\,\Omega$ (b) $1.20 \times 10^{-5}\,\text{F}$

Passage Problems

93. B
95. B

Chapter 20

Conceptual Questions

1. No. There could be magnetic field along the direction of motion. Such a field would produce no force on the electrons.

3. The magnetic force is always perpendicular to the velocity and hence does no work and doesn't change the speed of the particle. But the force perpendicular to the velocity changes the direction of the velocity and alters the path of the particle. Another example is a rock swinging on the end of a string. The tension in the string does no work but is necessary in order to keep the rock moving in an arc of a circle.

5. A stream of charged particles produces a magnetic field, just as current flow in a wire does. The field of the particles adds vectorially to the earth's field and alters the net field.

9. The energy comes from the source that is producing the current in the wires. The magnetic forces do work because the current-carrying charges (electrons) are confined to the wires. Forces perpendicular to the motion of the electrons produce bulk motion of the wires. There is no contradiction, it is just a different context.

11. You could use either a current carrying wire or a bar magnet that does have its poles marked to first determine which ends of the compass needle is a north pole and which is a south pole. Then the north pole of the large bar magnet attracts the south pole of the compass needle.

13. No, this is not plausible. In such a reversal the direction of the earth's angular momentum would reverse. A huge external torque would have to be applied to the earth to change the direction of its angular momentum. Conservation of angular momentum prohibits a spin reversal.

15. No. The force on a moving charge is perpendicular to the magnetic field and hence to the tangent to the field line at a

point. An example is a charge moving in a circular path whose plane is perpendicular to the magnetic field. In this case the charge moves perpendicular to the field lines.

Multiple-Choice Problems

1. B, C
3. C
5. D
7. C
9. B
11. C
13. A
15. A

Problems

1. (a) positive (b) 0.0505 N
3. (a) 3.54×10^{-16} N, into page
   (b) maximum: $F = 4.32 \times 10^{-16}$ N, $\vec{v}$ perpendicular to $\vec{B}$; minimum: $F = 0$, $\vec{v}$ parallel or antiparallel to $\vec{B}$
   (c) 3.45×10^{-16} N, out of page
5. (a) 5.60×10^{-3} N, $+z$ direction
   (b) 1.12 m/s^2
7. 6.45×10^{10} m/s^2, north
9. 1.79×10^{-1} m/s^2, in the $-z$ direction
11. (a) 1.02 T (b) out of page
13. 1.47×10^6 m/s
15. (a) 1.33×10^7 m/s; 1.33×10^7 m/s
    (b) 27.7 cm
17. $R/3$
19. (a) 4.51×10^{-3} T (b) 2.46×10^{-6} T
21. 8.38×10^{-4} T
23. (a) 4920 m/s (b) 9.95×10^{-26} kg
25. 2.0 cm
27. 2.5 N
29. (a) 2.1×10^{-4} N, upward
    (b) 2.1×10^{-4} N, west
    (c) zero (d) no
31. (a) *da:* zero; *ab:* 1.2 N, into page; *bc:* zero; *cd:* 1.2 N, out of page
    (b) zero
33. 0.724 N, 63.4° below the direction that is horizontal and to the right
35. (a) 4.71×10^{-3} N · m (b) 0.025 A · m^2
37. (a) zero, zero
    (b) $F = 0$, $\tau = 0.0809$ N · m
39. (a) 1.25 N · m (b) the north side of the coil will tend to rise
41. 4.0×10^{-5} T; same order as the earth's field
43. 25 μA
45. 4.3×10^{-5} T
47. (a) 4.0 μT (b) 16.0 cm (c) 16.0 μT
49. (a) 1.44×10^{-7} T (b) 4.80×10^{-8} T
51. (a) zero (b) 6.67 μT; toward top of page
53. 0.0161 N, attractive
55. 2×10^{-3} N; repulsive
57. (a) 8.8 A
59. 24
61. 2.77 A
63. $\dfrac{\mu_0 I}{4R}$; into page
65. 41.8 A

67. (a) 2.51×10^{-5} T (b) 5.03×10^{-4} T; no
69. 1.1 A
71. (a) 1.57×10^{-7} T; out of page
    (b) 8.00×10^{-7} T; out of page
73. 1.16×10^{-7} T
75. (a) 1.257×10^{-6} T · m/A (b) 1.3504 T
77. 7.93×10^{-10} N, south
79. 817 V
81. 7.81 mm
83. (a) 2.90×10^{-5} T, east (b) yes
85. (a) along line passing through intersection of wires and with slope -1.00
    (b) along line passing through intersection of wires and with slope $+0.333$
    (c) along line passing through intersection of wires and with slope $+1.00$
87. 5.88 m/s^2
89. (b) same for both (c) the ion in the large circle
91. $\frac{1}{2} q\omega r^2 B$

Passage Problems

93. B
95. D

Chapter 21

Conceptual Questions

1. Let A be the loop that is connected to the current source. View the two loops from above and assume the current in A is clockwise. Inside A its magnetic field is directed away from you and outside A and at points in the plane of the loops its field is directed toward you. Therefore, the flux through the second loop is toward you and is increasing. Inside the second loop the field of the induced current is away from you, to oppose the increase in flux in the opposite direction. To produce a field in this direction, the induced current in the second loop is clockwise. Therefore, when the current in the first loop is increasing, the currents in the two loops are in the same direction. On the other hand, when the current is decreasing the flux through the second loop is decreasing and is toward you, so the field of the induced current is toward you inside the second loop. This means the induced current is counterclockwise. Therefore, when the current in the first loop is decreasing the currents in the two loops are in opposite directions.

3. The field lines for the magnetic field of the wire are sets of concentric circles whose planes are perpendicular to the wire. The magnetic field of the wire is parallel to the plane of the ring and produces no flux through the ring. There is no current induced in the ring.

5. To store useful amounts of energy, large currents and large values of inductance L are needed. Large L for a solenoid means a large number of turns. This requires a long length of wire and this means the inductor will have resistance that isn't

small. Large I^2R heating will dissipate electrical energy. To avoid this, superconducting solenoids would have to be used.

7. Transformer operation depends on induced emf. For dc, the current is constant and the current in the primary doesn't cause a changing flux in the secondary. There is no induced emf and no current and voltage in the secondary.

9. The induced emf doesn't depend on the size of the magnetic flux. It instead depends on the rate of change of the flux.

13. Let the magnetic field of the magnet be directed away from you, perpendicular to the plane of the ring. When the ring is pushed into the field, the induced current is counterclockwise in the figure and the force on the ring due to the induced current is to the left. To move the ring into the field an external force directed into the field region must be applied. When the ring is pulled from the field, the induced current is clockwise in the figure and the force on the ring due to the induced current is to the right. To move the ring out of the field region an external force directed away from the field region must be applied. These results agree with Lenz's law.

15. The voltage across the resistor is proportional to the current through it, the voltage across the inductor depends on the rate of change of the current, and the voltage across the capacitor depends on the charge on its plates. (a) In the *R-C* circuit, just after the switch is closed the charge on the capacitor is zero and the full battery voltage appears across the resistor. The current has its maximum value. In the *R-L* circuit, just after the switch is closed the current is still zero but is increasing at its maximum rate. The full battery voltage is across the inductor. (b) In the *R-C* circuit, the full battery voltage is across the capacitor and the current is zero. In the *R-L* circuit, the current is no longer changing, the full battery voltage is across the resistor and

the current is a maximum. The capacitor has no effect initially but causes the current to go to zero as time progresses. The inductor initially limits the rate at which the current can increase but after a long time it has no effect on the circuit.

Multiple-Choice Problems

1. A
3. B
5. B
7. C
9. C
11. C
13. D
15. A

Problems

1. (a) 3.05×10^{-3} Wb
   (b) 1.83×10^{-3} Wb (c) 0
3. -7.8×10^{-4} Wb
5. 10.6 T/s
7. 58.4 V
9. 3.0 mA
11. (a) 3.70 rad/s (b) 0
13. (a) counterclockwise (b) clockwise
    (c) induced current is zero
15. counterclockwise; clockwise
17. A: counterclockwise; C: clockwise
19. clockwise
21. (a) 2.10 V (b) east (c) zero
23. 1.25 mV; higher
25. (a) 3.00 V (b) from b to a (c) 0.800 N
27. 4.00×10^{-5} H
29. (a) 6.82 mH (b) 3.27×10^{-4} Wb
    (c) 2.45 mV
31. (a) 1.96 H (b) 7.11×10^{-3} Wb
33. 0.250 H
35. (a) 1940 (b) 800 A/s
37. 238 turns
39. (a) use step-down transformer
    (b) 6.67 A (c) 36.0 Ω
41. (a) 108 (b) 110 W (c) 0.918 A
43. (a) 6.74 mJ (b) 14 A; no
45. (a) 0.161 T (b) 1.03×10^4 J/m^3
    (c) 0.129 J (d) 4.02×10^{-5} H
47. $I\sqrt{3}$
49. (a) 2.40 A/s (b) 6.00 V (c) 0.475 A
    (d) 0.750 A
51. (a) $V_1 = 0$, $V_2 = 25.0$ V, $A = 0$
    (b) $V_1 = 25.0$ V, $V_2 = 0$ V, $A = 1.67$ A
    (c) none would change
53. (a) 1.39 μs (b) 2.46 μs
55. (a) 0.640 mJ (b) 0.584 A
57. (a) 1860 Ω (b) 0.963 H
61. 222 μF; 9.31 μH

Passage Problems

65. B

Chapter 22

Conceptual Questions

1. (a) The individual voltages achieve their maximum values at different times.
   (b) Yes, Kirchhoff's loop rule requires

that at any instant of time the instantaneous voltages obey this equation.
3. The inductor doesn't consume electrical energy. It stores energy but later releases that energy back to the circuit. A resistor dissipates electrical energy by converting it to unwanted thermal energy.
5. At high frequencies the current is changing rapidly and the induced emf across the inductor is very large. This large induced emf opposes the current and current doesn't flow, just as if there were a break (an infinite resistance) in the circuit.
7. When the power factor is small, a large current is needed to supply a given amount of power. This results in large i^2R rates of electrical energy losses in the transmission lines.
9. The voltages across the capacitor and inductor are 180° out of phase so the voltage across the combination is zero at all times. The current is limited only by the resistance in the circuit, and the source voltage amplitude is equal to IR. If $X_C > R$, then the voltage amplitude across the capacitor exceeds the voltage amplitude across the resistor and hence exceeds the source voltage amplitude.
11. An advantage of ac is that transformers allow ac voltage to be stepped up or down. Power can be transmitted at high voltage and low current to minimize I^2R losses in the transmission lines, and the voltage can be stepped down to safer values for distribution to consumers. A disadvantage of ac is that the power delivered depends on the rms voltage and peak voltages are larger by $\sqrt{2}$.
13. ac voltage is generated by application of Faraday's law, by rotating coils in a magnetic field.
15. There is I^2R heating in the wires of the primary and secondary windings. There is generation of thermal energy through hysteresis in the transformer core. And eddy currents induced in the core leads to I^2R heating.

Multiple-Choice Problems

1. D
3. A
5. A
7. B
9. D
11. C
13. C
15. C

Problems

1. 1.06 A
3. (a) $\dfrac{1}{\sqrt{LC}}$ (b) 7560 rad/s; 37.8 Ω
5. 1.63×10^6 Hz
7. (a) 0.250 mA (b) 25.0 mA
   (c) 2.50 A
9. (a) 10.6 mA (b) 8.48 V

11. 50.0 V
15. (a) 40.0 W (b) 0.167 A (c) 720 Ω
17. (a) 12.5 W (b) 12.5 W (c) 0
19. (a) 1.0×10^4 rad/s (b) 6.28×10^{-4} s
21. (a) 1.00 (b) 75.0 W (c) 75.0 W
23. (a) 115 Ω (b) 146 Ω (c) 146 Ω
25. (a) 250 rad/s (b) 400 Ω
    (c) $V_C = V_L = 30.0$ V; $V_R = 240$ V
27. (a) 3.24×10^3 rad/s (b) 125 Ω
29. (a) 0.400 A (b) 0.800 A
31. 0.124 H
33. 3.59×10^7 rad/s
35. (a) 20.5 V (b) 21.2 V
37. (a) 7.32 W (b) 7.32 W
39. (a) 102 Ω (b) 0.882 W (c) 270 V
41. (a) 568 Ω (b) 0.866 A (c) 226 V
43. 9.64 mA (b) 241 V (c) 1.16 V
    (d) 1.39×10^{-5} J

Passage Problems

45. C

Chapter 23

Conceptual Questions

3. The momentum change of the light, and hence the momentum transferred to the surface, is greater when the light reflects. For absorption, the final momentum is zero. For reflection, the final momentum is equal in magnitude and opposite in direction to the initial momentum, so the change in momentum is twice as great as for absorption.
5. The hot air has a slightly smaller index of refraction than room temperature air. As light passes into and out of the hot air its rays are bent by the change in refractive index.
7. The refractive index of air is only slightly greater than unity and the distance the light travels through the atmosphere is a very, very small fraction of the distance from the sun to the earth, so the delay introduced by the earth's atmosphere is much, much less than 8 minutes.
9. Atmospheric refraction causes the sun to still be seen after the sun's disk has passed below the horizon. The same effect occurs at sun rise, and allows the sun to be seen before it rises above the horizon. This effect does increase the time between when the sun is seen to rise until it is seen to set.
11. As described in Figure 23.38, light reflected at an angle θ_p is polarized parallel to the reflecting surface. And, as described in Figure 23.46, skylight from overhead is polarized. Either of these sources of polarized light could be used to determine the direction of the axis of a polarizer, by observing what orientation of the polarizer gave maximum transmitted intensity.
13. Any wave exhibits both reflection and refraction. An echo is one example of reflection of sound. When we stand in a

boat we hear sounds that are generated underwater because sound waves refract from the water into the air.

15. The incident and reflected rays travel in the same material. The reflection occurs because of the abrupt change in refractive index but in reflection the light stays in the same material. The refracted ray leaves one material and enters the other, and there is a change in wavelength.

Multiple-Choice Problems

1. D
3. A, B, E
5. C
7. A
9. C
11. B
13. B
15. D

Problems

1. 8.33 min
3. (a) 1.28 s
   (b) 8.16×10^{13} km
5. (a) (i) 6.0×10^4 Hz (ii) 6.0×10^{13} Hz
   (iii) 6.0×10^{16} Hz (b) (i) 4.62×10^{-14} m $= 4.62 \times 10^{-5}$ nm
   (ii) 508 m $= 5.08 \times 10^{11}$ nm
7. (a) 375 V/m; 1.25 μT
   (b) 9.52×10^{14} Hz; 316 nm; 1.05×10^{-15} s; not visible (c) 3.00×10^8 m/s
9. (a) 4.92×10^{14} Hz
   to 7.50×10^{14} Hz
   (b) 2.70×10^{15} rad/s
   to 4.71×10^{15} rad/s
   (c) 8.98×10^6 rad/m to 1.57×10^7 rad/m
11. 3.0×10^{18} Hz; 3.3×10^{-19} s; 6.3×10^{10} rad/m
13. (a) 1.74×10^5 V/m, parallel to the $-z$-axis (b) $E = -(1.74 \times 10^5 \text{ V/m}) \cdot \sin[(3.83 \times 10^{15} \text{ rad/s})t - (1.28 \times 10^7 \text{ rad/m})x]$; $B = (5.80 \times 10^{-4} \text{ T}) \sin[(3.83 \times 10^{15} \text{ rad/s})t - (1.28 \times 10^7 \text{ rad/m})x]$
15. (a) $-x$ direction (b) 6.59×10^{11}
    (c) $E = (2.48 \text{ V/m}) \sin[(4.14 \times 10^{12} \text{ rad/s})t + (1.38 \times 10^4 \text{ rad/m})x]$
17. (a) 330 W/m^2 (b) 500 V/m; 1.7 μT
19. (a) 80 J (b) 1.0×10^{21} W/m^2
    (c) 8.7×10^{11} V/m; 2.9×10^3 T
21. (a) 2.00×10^{-8} Pa
    (b) 4.00×10^{-8} Pa
23. $\sqrt{2}$
25. (a) 8.47×10^8 Hz
    (b) 1.80×10^{-10} T
    (c) 3.87×10^{-6} W/m^2
27. (b) 45°
31. 1.38
33. 1.65
35. (a) 389 nm (b) 584 nm
39. (a) 47.5° (b) 66.0°

43. 23.4°
45. 24.4°
47. 401 m^2
49. (a) 48.9° (b) 28.7°
51. 9.43°
55. (a) violet: 1.67; red: 1.62 (b) violet: 240 nm; red: 432 nm (c) 1.03; red (d) 1.1°
57. (a) $0.285I_0$ (b) linearly polarized
59. (a) A: $I_0/2$; B: $0.125I_0$; C: $0.0938I_0$
    (b) zero
61. (a) 63.4° (b) 71.6°
63. $\alpha = \arccos\left(\dfrac{\cos\theta}{\sqrt{2}}\right) = \cos^{-1}\left(\dfrac{\cos\theta}{\sqrt{2}}\right)$
65. (a) 61.8°
67. (a) 7.81×10^9 Hz
    (b) 4.50×10^{-9} T
    (c) 2.42×10^{-3} W/m^2
    (d) 1.93×10^{-12} N
69. (a) 0.375 mJ (b) 4.08×10^{-3} Pa
    (c) 604 nm; 3.70×10^{14} Hz
    (d) 3.03×10^4 V/m; 1.01×10^{-4} T
71. 173 km
73. (a) reflective (b) 6.42 km^2 (c) both gravitational force and radiation pressure due to sun are inversely proportional to the square of the distance from the sun
75. 51.7°
77. $n_1 = 1.10$; $n_2 = 1.14$
79. 1.23
81. 39.2°
83. 42.5°
85. (a) red: 1.73; blue: 2.75 (b) blue
87. (b) angle of incidence for first slab and index of refraction of first and last slabs
89. (a) 48.6° (b) 48.6°
91. (a) $n = 1.11$ (b) (i) 9.78 ns (ii) 4.09 ns; total $= 8.98$ ns

Passage Problems

93. C

Chapter 24

Conceptual Questions

1. The image formation for a spherical mirror is based on the law of reflection. This law is the same whether the light is propagating in air or water. So, no the focal length of the spherical mirror does not change. The image formation for a lens depends on the law of refraction. The bending of light by the lens depends on the change in refractive index as the light enters and leaves the lens. When a lens is immersed in water its focal length changes.
3. Real images can be projected onto a screen but virtual images cannot.
5. The image formed by the first mirror serves as the object for the second mirror. Each successive image is farther behind the mirror surface than the previous one. Each successive image arises from one additional reflection. Some intensity is lost at each reflection, so the images are progressively dimmer.

7. It is desired to have an enlarged, upright image. The image formed by a convex mirror is always smaller than the object. The image formed by a concave mirror is enlarged and upright when the object distance is smaller than the focal length. The radius of curvature of the mirror is related to the focal length f by $f = R/2$. So, the radius of curvature of the mirror is chosen so that person's face is inside the focal point.
9. The spoon has a small radius of curvature and hence a small focal length. When she looks at the concave side of the spoon, her face is outside the focal point and the image is real and inverted. When she looks at the convex side of the spoon the image is virtual and upright.
11. (a) For a concave mirror the image of a distant object is formed at the focal point in front of the mirror. (b) For a convex mirror the image of a distant object is formed at the focal point behind the mirror.
13. If light passes through a lens in the opposite direction, the focal length of the lens is unchanged. This can be seen as follows: $\dfrac{1}{f} = (n-1)\left(\dfrac{1}{R_1} - \dfrac{1}{R_2}\right)$. When the lens is turned around, $R_1 \rightarrow -R_2$ and $R_2 \rightarrow -R_1$, so $\dfrac{1}{f}$ is unchanged.
15. (a) A convex mirror always forms a virtual, erect image, for all object distances, so this mirror must be concave. (b) The erect image is virtual and the upside down image is real.

Multiple-Choice Problems

1. C
3. C
5. D
7. A
9. B
11. E
13. B, D
15. B

Problems

1. 39.2 cm to the right of the mirror; 4.85 cm
3. (a) 3
5. (a) 7.5 cm in front of mirror; 4.00 mm tall
   (b) 10.0 cm in front of mirror; 8.00 mm tall
   (c) 5.00 cm behind mirror; 16.0 mm tall
   (d) 5.00 cm in front of mirror; 0.040 mm tall
7. (a) 0.213 mm (b) 1.75 m in front of mirror
9. (a) 0.0103 m (b) height of image is much less than height of car so car appears to be farther away than its actual distance
11. 18.0 cm from the vertex; 50.0 cm tall, erect, virtual
13. (a) 4.00 (b) 48.0 cm behind mirror; virtual
15. (b) 6.60 cm to right of mirror; 0.240 cm tall; erect; virtual

17. (a) 10.0 cm to left of the mirror; 2.20 mm tall (b) 4.29 cm to right of mirror; 0.944 mm tall
19. (a) 8.00 cm to right of vertex (b) 13.7cm to right of vertex (c) 5.33 cm to left of vertex
21. 14.8 cm to right of vertex; 0.578 mm tall; inverted
23. (a) at center of bowl; +1.33 (b) no
25. 2.67 cm
27. 6.53 m
29. object: 4.85 cm from lens; image: 15.75 cm to left of lens; virtual
31. (a) 0.600 m (b) inverted (c) +0.450 m; converging
33. (a) −4.80 cm; diverging (b) 2.44 mm; virtual
35. $n = 1.55$
37. (a) +11.0 cm; +11.0 cm (b) +17.0 cm; +17.0 cm (c) +133 cm; −133 cm (d) −18.4 cm; −18.4 cm (e) −11.0 cm; −11.0 cm
39. (a) ±7.0 mm (b) 8.2 mm on other side of lens; 4.4 mm; real; inverted
41. (a) 107 cm to right of lens; 17.8 mm tall; real; inverted (b) same as in part (a)
43. 71.2 cm to right of lens, $m = -2.97$
45. (a) 200 cm to right of first lens; 4.80 cm tall (b) 150 cm to right of second lens; 7.20 cm tall
47. (a) 53.0 cm to right (b) real (c) 2.50 mm; inverted
49. $s = 18.0$ cm: (a) 63.0 cm to right of lens (b) −3.50 (c) real (d) inverted
    $s = 7.00$ cm: (a) 14.0 cm to left of lens (b) +2.00 (c) virtual (d) erect
51. (a) 26.3 cm from lens with height 12.4 mm; image is erect; same side
53. (a) converging (b) 8.0 cm (c) 24.0 cm to right of lens (d) 24.0 cm to right of lens
55. (a) converging (b) 18.0 cm (c) 22.5 cm to left of lens (d) 22.5 cm to left of lens
57. −6.62 cm
59. 7.20 m; 4.43 m
61. (a) image is 58 cm from lens, on same side as object; (b) $m = 1.7$
63. 220 cm
65. light directly through lens:
    (b) (i) 51.3 cm to right of lens (ii) real (iii) inverted
    light reflecting off mirror:
    (b) (i) 51.3 cm to right of lens (ii) real (iii) erect

Passage Problems

69. D

Chapter 25

Conceptual Questions

1. The filled wine glass acts as a thick, converging lens. The light is bent when it passes from air to glass, from glass to wine, from wine to glass and from glass to air. The refracting properties of the glass are changed when it is empty. If the glass is thick, an image can still be formed but with a different image distance. Water has a different refractive index from wine. When the glass is filled with water an image can be formed, but with a different image distance.
3. The eye is similar to a camera since a lens forms an image on a screen, either the retina or the film. For a camera, the lens has fixed focal length and the distance from the lens to the film is changed in order to focus the image on the film. For the eye, the lens to retina distance is fixed and the focal length can be changed.
5. Compared to a glass lens in air, the refraction air → glass is replaced by water → air and the refraction glass → air is replaced by air → water. In each case the light is bent in the opposite direction relative to the normal to the lens surface. A shape that is a diverging lens for glass in air is a converging lens for air surrounded by water. The air pocket must be thinner in the center than at its edges in order to serve as a converging lens.
7. To start a fire you need a converging lens, that forms a real image at the focal point of the lens. A converging lens corrects farsightedness and a diverging lens corrects nearsightedness. Your eyeglasses work if you are farsighted.
9. Yes, the fishbowl acts as a converging lens and can form a bright image at its focus.
11. The angular magnification M is proportional to $1/f$, so the lens with a smaller focal length gives the larger M. For a lens with one flat side, the lensmaker's equation gives $f = \left(\dfrac{1}{n-1}\right)R$, where R is the radius of curvature of the curved side. Small R gives small f, so the lens with one highly curved side gives the greater angular magnification.
13. No. The angular magnification depends on the focal lengths of the lenses and is independent of the diameter of the lenses. The larger diameter lenses will gather more light and will therefore produce brighter images.
15. No. To correct the severe nearsightedness, a diverging lens with power of large magnitude is required. For an object at the normal near point, this lens will produce an image closer to the eye than the near point. The lens that corrects far vision impairs close vision. Bifocals can be used to correct far vision without impairing, or even correcting, near vision, but they essentially have two lenses.

Multiple-Choice Problems

1. C
3. C
5. C
7. D
9. A, D
11. D
13. D
15. D

Problems

1. (a) 75 mm (b) $\dfrac{1}{62.5}$ s
3. (a) $f/22.1$ (b) 660 inches (c) $f/3.1$ to $f/13$
5. 10.2 m
7. (a) 7.63 cm (b) 3.33 cm; inverted; real (c) 4.2×10^3 pixels
9. (a) 2.83 m (b) 5.09 cm
11. (a) 2.76 m behind lens (b) 0.552 m by 0.828 m
13. (a) 15.3 cm (b) 1.41 m by 2.12 m
15. (a) 462 mm (b) 50 cm
17. 2.00 cm to 2.50 cm
19. (a) 150 diopters; 115 diopters (b) 6.67mm to 8.70 mm (c) 8.70 mm (d) 6.85 mm
21. +49.4 cm; +2.02 diopters
23. 4.17 diopters
25. −3.66 diopters
27. (a) +2.33 diopters (b) −1.67 diopters
29. (a) 2.17 (b) 11.5 cm
31. (a) 6.06 cm (b) 4.13 mm
33. (a) 60.0 mm (b) 100
35. (a) 620× (b) $f_e = 2.50$ cm; $f_o = 0.226$ cm (c) $s = 0.230$ cm
37. (a) 640 (b) 43
39. 27, 4.8
41. (a) eyepiece: 980 mm; objective: 1710 mm; 2.69 m (b) 1.74 (c) 0.87°
43. 0.504 m
45. (a) red: 23.9 cm; violet: 22.1 cm (b) 118 cm; violet: 83.9 cm
47. (a) 140 cm (b) red: 16 cm beyond screen; violet: 18 cm in front of screen
49. 2.60 cm
51. (a) 50 μm (b) 0.70 min
55. (a) 19.8 mm (b) 0.106 cm
57. (a) 1.58 cm (b) 0.183 cm (c) no
59. (a) 30.9 cm (b) 29.2 cm
61. (a) 1.38 (b) 2.29 m

Chapter 26

Conceptual Questions

1. Yes. Use two coherent sources of sound, such as two speakers connected to the same amplifier. The distance from the speakers should be on the order of the wavelength of the sound. Observe the sound intensity at distances from the speakers that are much larger than the separation between the speakers. All that is required is that the

waves obey the principle of superposition and exhibit interference; this is the case for both longitudinal and transverse waves. To actually observe interference effects reflections of the sound waves must be suppressed.

3. The water alters the wavelength of the light. $y_m = R\dfrac{m\lambda}{d}$, so the shorter wavelength in water means the bright fringes are closer together.

7. No. The Bragg condition is $2d \sin\theta = m\lambda$. For visible light $\lambda \gg d$ and the first maximum is not observed; for $m = 1$ the equation says $\sin\theta > 1$, which doesn't occur.

9. Each thickness of oil produces constructive interference in the reflected light for a particular wavelength (color). Contours of constant thickness of oil are closed lines so the fringes of each color are closed lines.

11. A large-diameter objective lens has more light gathering power and produces brighter images. The magnification depends on the focal length of the lens, which in turn depends on the radius of curvature of the lens surface and not on its diameter.

13. The number of bright spots is related to the maximum m in $d \sin\theta = m\lambda$. The largest value $\sin 90° =$ can have is 1, so the maximum m is less than d/λ. The number of bright spots depends on the wavelength of the light and the separation of the two slits.

15. The diffraction limit on resolution is given by $\sin\theta = 1.22\dfrac{\lambda}{D}$, where θ is the smallest angular separation of objects whose images can be resolved. When λ is larger, as it is for radio waves as compared to visible light, a larger aperture diameter D is needed to achieve the same resolution.

Multiple-Choice Problems

1. B
3. B
5. A
7. B
9. C
11. A
13. D
15. C

Problems

1. (a) 150 cm, 116 cm, 82 cm, 48 cm, 14 cm
   (b) 133 cm, 99 cm, 65 cm, 31 cm
3. (a) 240 m (b) 120 m
5. (a) destructive (b) 68.0 cm (c) 34.0 cm
7. 590 nm
9. (a) 0.0389 mm (b) 2.60 cm
11. 0.833 mm
13. (a) 39 (b) $\pm 73.4°$
15. (a) 514 nm; green (b) 603 nm; orange

17. 80.5 nm
19. (a) 96.4 nm (b) 192 nm, 289 nm, 386 nm
21. 27.1 fringes/cm
23. 95.8 nm
25. $\pm 9.71°$, $\pm 19.7°$, $\pm 30.4°$, $\pm 42.4°$, $\pm 57.5°$
27. 0.800 mm
29. (a) 10.9 mm (b) 5.4 mm
31. $\pm 16.0°$, $\pm 33.4°$, $\pm 55.6°$
33. 634 nm
35. 13.9°, 28.7°, 46.1°
37. (a) 472 nm (b) 54.7 cm
39. (a) 23.3°, 52.3° (b) 58.8°
41. 20.2°
43. 0.559 nm
45. 429 m
47. 1.88 m
49. 0.084 mm; no
51. 81 cm
53. 1.73
55. (a) 103 nm (b) no; no (c) destructive: 600 nm; constructive: none
57. 30.2 μm
59. (a) 594 nm, 424 nm (b) 495 nm
61. 1.82 mm
63. 198 m
65. 22.3 yr

Passage Problems

67. C

Chapter 27

Conceptual Questions

1. Relativistic effects, such as time dilation and length contraction, would be part of our everyday experience.

3. No. If two events occurring at different points in a particular frame appear to be simultaneous in that frame, they need not be simultaneous in other frames. But if they occur at the same space point in one frame and are simultaneous in that frame, then they must be simultaneous in all frames. If they are both at the same space point and same time point in a frame then there is nothing to distinguish one from the other with respect to time and space.

7. They are massless in the sense of the equation $E^2 = (mc^2)^2 + (pc)^2$. For a photon there is no mc^2 term and energy and momentum are related by $E = pc$.

9. No. In the relativistic expressions for energy and momentum, both these quantities approach infinity as the speed of the particle approaches the speed of light.

11. c

13. A force in the direction of $\vec{v}$ increases the speed of an object. For a force in this direction, $a = F/\gamma^3 m$. As $v \to c$, $\gamma \to \infty$ and the acceleration produced by the force approaches zero. The closer the speed approaches c, the more difficult it is to increase the speed still further. Or, since $K \to \infty$ as $v \to c$, an infinite

amount of work would be required to bring an object to the speed of light.

15. There are many experimental verifications of the theory of relativity. For example, experiments with atomic clocks and decay of unstable elementary particles have verified time dilation and length contraction. The Doppler effect for electromagnetic waves is based on relativity and is commonly used in radar guns. And the expression $E = mc^2$ is used to accurately predict energy release in the decay of unstable nuclei and elementary particles.

Multiple-Choice Problems

1. A, B
3. A, D
5. C
7. D
9. D
11. C
13. A, C
15. A

Problems

1. From the passenger's point of view, the signal from earth was sent before the one from the asteroid.
3. 0.301 ms
5. (a) 2.555555556×10^5 h $= 29.2$ y; (b) 0.8 s
7. (a) 12.0 ms (b) 0.998c
9. 1.12 h; clock in spacecraft
11. 2.86×10^8 m/s
13. (a) 1.05 m (b) 3.80 m; no (c) 2.00 m
15. (a) 9.17 km (b) 0.65 km; 7.1% (c) 1.32×10^{-5} s; 3.90 km; 7.1%
17. 0.837c; away from
19. 0.385c
21. (a) 0.116c (b) 0.229c
23. (a) 4.21×10^7 m/s (b) greater
25. yes, for nonrelativistic momentum
27. (a) 1.14×10^{-15} J; 1.16×10^{-15} J; 1.02 (b) 3.08×10^{-14} J; 8.26×10^{-14} J; 2.68
29. 3.01×10^{-10} J $= 1.88 \times 10^9$ eV
31. (a) 4.5×10^{-10} J (b) 1.94×10^{-18} kg·m/s (c) 0.968c
33. (a) 1.11×10^3 kg (b) 52.1 cm
35. (a) 2.06 MeV (b) 3.30×10^{-13} J $= 2.06$ MeV
37. 2.10×10^8 m/s
39. 4.53×10^2 m
41. 0.83c
43. (a) $4c/5$ (b) c (c) (i) 145 MeV (ii) 625 MeV (d) (i) 117 MeV (ii) 469 MeV
45. 1.49×10^{-11} kg
49. (a) 7.20×10^{13} J (b) 1.80×10^{19} W (c) 7.35×10^9 kg
51. (a) 8.34 y (b) 3.02 ly
53. m $= 1.67 \times 10^{-27}$ kg, proton

Passage Problems

55. C
57. B

Chapter 28

Conceptual Questions

1. The threshold wavelength is generally shorter than the wavelength for visible light, even more so if the surface is not clean and not free of coatings.

3. Increasing the frequency of the light increases the energy of each photon in the light and thereby increases the energy given to each photoelectron. Increasing the intensity means more photons but doesn't change their energy. Increasing the intensity of the light therefore means more photoelectrons are produced but doesn't change the maximum energy each can have.

5. The frequency of a light beam is totally different from the number of photons per second. The frequency is related to the wavelength and to the energy of each photon. The number of photons per second depends on the intensity of the light. To change the intensity of the light without changing the frequency you could for example pass a monochromatic laser beam though an absorbing medium. The emerging light has lower intensity but the same frequency.

7. Photons of different wavelength have different energy, with longer wavelength photons having less energy. Infrared photons individually have too little energy to produce the chemical change that exposes the film.

11. No. An ultraviolet photon has more energy than a visible-light photon. The phosphor reduces the energy of the photons by taking energy away. There is no way to increase the energy of the photons with a phosphor. Converting visible light to ultraviolet would violate conservation of energy.

15. If we apply Bohr's angular momentum quantization to a planet, the quantum numbers n are huge and orbits for successive n are infinitesimally close in radius. No discrete nature of the orbit radii of the planets is observable.

Multiple-Choice Problems

1. A, C
3. B
5. C
7. D
9. A
11. D
13. B
15. D

Problems

1. (a) 5.94×10^{14} Hz
   (b) 3.94×10^{-19} J $= 2.46$ eV
   (c) 9.1 mm/s
3. (a) 0.0120 J $= 7.49 \times 10^{16}$ eV
   (b) 3.04×10^{-19} J $= 1.90$ eV
   (c) 3.94×10^{16} photons

5. 1.50×10^{19} photons/s
7. $\phi = hf_0$
9. (a) 1.10×10^{15} Hz; 4.55 eV
   (b) 1.44 eV
11. 2.13 eV
13. (a) 2.7 V (b) 2.7 eV (c) 9.7×10^5 m/s
15. (a) 259 nm (b) 0.27 eV
17. (a) 433 nm (b) 6.93×10^{14} Hz
    (c) 2.87 eV
19. (a) Balmer: 656 nm, 365 nm; Lyman: 122 nm, 91.2 nm; Brackett: 4051 nm, 1459 nm (b) Balmer: mostly visible; Lyman: ultraviolet; Brackett: infrared
21. (a) -17.50 eV; -4.38 eV; -1.95 eV; -1.10 eV; -0.71 eV (b) 378 nm
23. (a) 2.16×10^6 m/s; 1.09×10^6 m/s; 7.29×10^5 m/s (b) 0.529×10^{-10} m; 2.12×10^{-10} m; 4.76×10^{-10} m
    (c) -13.6 eV; -3.40 eV; -1.51 eV
25. 97.2 nm; 3.09×10^{15} Hz
27. (a) -218 eV; 16 times hydrogen value
    (b) 218 eV; 16 times hydrogen value
    (c) 7.63 nm
    (d) 1/4 times hydrogen value
29. 4.62 mW
31. (a) ultraviolet (b) 6.42 eV
    (c) 1.75×10^7 photons
33. (a) 8.29 kV (b) 0.0414 nm (c) no
35. 1.13 keV
37. 0.0714 nm, 180°
39. 3.11×10^{-10} m, the same
41. 0.727 m/s
43. (a) 3.32×10^{-10} m (b) 1.33×10^{-9} m
45. photon: 49.6 nm; electron: 0.245 nm
47. (a) 3.50 eV; 354 nm
    (b) 1.65×10^{-10} eV
49. 2.0×10^{-24} kg · m/s; comparable
51. (a) 419 V (b) 0.229 V
53. (a) 1.04 eV (b) 1.20×10^{-6} m
    (c) 2.51×10^{14} Hz (d) FM radio photons individually have too little energy
55. 14.5 eV
57. (a) 12.1 eV (b) 3; 658 nm, 103 nm, 122 nm
59. (a) 0.1849 nm (b) 183 eV
    (c) 8.02×10^6 m/s; no
61. (a) 6.99×10^{-24} kg · m/s (b) 705 eV
63. 32.0 μm
65. 1.66×10^{-17} m; no
67. (a) 12 eV (b) 1.5×10^{-4} V; 7.3×10^3 m/s
    (c) 8.2×10^{-8} V; 4.0 m/s
69. 1.4×10^{-35} kg; 5.8×10^{-8}

Passage Problems

71. B
73. D

Chapter 29

Conceptual Questions

1. The n, l and m_l quantum numbers are all zero for both electrons, so the Pauli exclusion principle requires that their spin quantum number s be different. The spin quantum number determines the z component of the spin angular momentum and for different s this component has opposite sign.

3. Somewhat. The Exclusion Principle refers to quantum numbers not position and the quantum description is in terms of position probabilities and not orbits of precise radii. But the quantum numbers of an electron determine the spatial distribution of the position probability.

5. Mg has two electrons outside a filled shell and Cl is one electron short of a filled shell, so we expect them to combine in a one to two ratio and form $MgCl_2$.

9. Yes, and in fact this is routinely done. The wavelengths of light emitted or absorbed in molecular transitions between vibrational or rotational levels are characteristic of each molecule. The radiation is in the infrared and micro-wave regions.

11. More. Due to screening by the other electrons, the $3s$ electron experiences an effective nuclear charge of about $+e$. There is less screening for the outer electron in Mg^+ and this electron is more tightly bound.

13. In a p-n junction diode a forward bias causes the plentiful holes in the p region to flow into the n region and the plentiful free electrons in the n region to flow into the p region. Both these flows correspond to conventional current in the p to n direction. But a reverse bias pushes the charges to flow in the opposite directions, and the p region has few electrons to flow into the n region and the n has few holes to flow into the p region. Current easily flows in the forward direction but very little current flows with reverse bias. The conduction by a resistor has no directionality and current flows with equal ease (and difficulty) in either direction.

15. An n-type semiconductor has impurities that add free electrons that can conduct current. A p-type semiconductor has impurities that create holes (missing electrons) that are free to move through the material and thereby conduct electricity.

Multiple-Choice Problems

1. A, C
3. B, D
5. A, D
7. C
9. C

Problems

1. $l = 4$
3. (a) $L = 0$, $L_z = 0$. $L = \sqrt{2}\hbar$, $L_z = 0, \pm\hbar$. $L = \sqrt{6}\hbar$, $L_z = 0, \pm\hbar, \pm 2\hbar$
   (b) $L = 0$: θ not defined. $L = \sqrt{2}\hbar$: 45.0°, 90.0°, 135.0°.
   $L = \sqrt{6}\hbar$: 35.3°, 65.9°, 90.0°, 114.1°, 144.7°

5. (a) zero

   (b) $2\sqrt{3}\hbar = 3.65 \times 10^{-34}$ kg · m²/s

   (c) $3\hbar = 3.16 \times 10^{-34}$ kg · m²/s

   (d) $\hbar/2 = 5.27 \times 10^{-35}$ kg · m²/s

   (e) 1/6

7. (a) 18 (b) $m_l = -4$; 153.4° (c) $m_l = 4$; 26.6°

9. $n = 1, l = 0, m_l = 0, s = \frac{1}{2}$;

   $n = 1, l = 0, m_l = 0, s = -\frac{1}{2}$;

   $n = 2, l = 0, m_l = 0, s = \frac{1}{2}$;

   $n = 2, l = 0, m_l = 0, s = -\frac{1}{2}$;

   $n = 2, l = 1, m_l = 1, s = \frac{1}{2}$;

   $n = 2, l = 1, m_l = 1, s = -\frac{1}{2}$;

   $n = 2, l = 1, m_l = 0, s = \frac{1}{2}$;

   $n = 2, l = 1, m_l = 0, s = -\frac{1}{2}$;

   $n = 2, l = 1, m_l = -1, s = \frac{1}{2}$;

   $n = 2, l = 1, m_l = -1, s = -\frac{1}{2}$;

   $n = 3, l = 0, m_l = 0, s = \frac{1}{2}$;

   $n = 3, l = 0, m_l = 0, s = -\frac{1}{2}$

11. $1s^2 2s^2 2p^6 3s^2 3p^6 3d^{10} 4s^2 4p^5$

13. (a) $1s^2 2s^2 2p^2$

    (b) silicon; $1s^2 2s^2 2p^6 3s^2 3p^2$

15. (a) Ne: $1s^2 2s^2 2p^6$;

    Ar: $1s^2 2s^2 2p^6 3s^2 3p^6$;

    Kr: $1s^2 2s^2 2p^6 3s^2 3p^6 3d^{10} 4s^2 4p^6$

    (b) six (c) chemically inert

17. (a) zero (b) zero; no; would fall into nucleus

19. 277 nm; ultraviolet

21. (a) 5.0 eV (b) −4.2 eV

23. (a) 0.330 nm (b) larger for KBr

25. (a) 227 nm; ultraviolet

27. 1.20×10^6 electrons

31. ±20.4%

33. (a) 1.12 eV

35. H has one electron in an unfilled shell and O is two electrons short of a filled shell

37. (a) 0, $\sqrt{2}\,\hbar$, $\sqrt{6}\,\hbar$, $\sqrt{12}\,\hbar$, $\sqrt{20}\,\hbar$,

    (b) 7470 nm, infrared, not visible

39. (a) −0.8500 eV (b) largest: $3\hbar$, smallest: $-3\hbar$ (c) $S = \sqrt{3/4}\,\hbar$ for all electrons

    (d) largest: $\sqrt{6}\,\hbar$, smallest: 0

Chapter 30

Conceptual Questions

1. They have slightly different masses and this mass difference can be used to separate them. Techniques include diffusion and ultracentrifuging of gaseous compounds.

3. The abundance is greater for the intermediate nuclides that have a longer half-life.

5. False. The number of radioactive nuclei is reduced by half, but they have decayed to daughter nuclei that retain most of the mass.

7. Chemical reactions are between atoms and molecules, and the average speeds with which they collide and the frequency of the collisions depend on the temperature. Radioactivity involves forces within the nucleus and the nucleus is unaffected by the temperature of the atoms. Also, typical kinetic energies of atoms near room temperature are similar to energy changes in chemical reactions whereas energies in radioactive decay correspond to kinetic energies at extremely high temperatures.

9. Every proton exerts an electrical force on every other proton and the number of pairs of protons and hence the total electrical energy increases more than linearly as the number of protons increases. More neutrons are needed, to increase the average distance between protons and to add nuclear force without adding electrical repulsion.

11. The kinetic energy of the reaction products is converted to thermal energy when they are stopped in a material.

13. A is unchanged. Z decreases by one, so N increases by one. β^+ decay of $^{40}_{19}$K produces the nucleus $^{40}_{18}$Ar.

15. Many fission fragments have a long half-life so stay radioactive for many years after they are produced. Also, many of them are taken up into the biosphere. For example, radioactive strontium has a half-life of 30 years and is deposited in bones because it is a chemical cousin to calcium.

Multiple-Choice Problems

1. A

3. A, C, D

5. C

7. A, D

9. D

11. B

13. D

15. A

Problems

1. (a) 10, 11 (b) 30, 35 (c) 47, 61

3. (c) 2.3×10^{17} kg/m³

5. (a) 2.23 MeV; 1.11 MeV/nucleon

   (b) 28.3 MeV; 7.07 MeV/nucleon

   (c) binding energy per nucleon is much larger for $^4_2$He

7. 5.575×10^{-13} m

9. (a) lead, $A = 214$, $Z = 82$, $N = 132$

   (b) 6.12 MeV

11. (a) X_1: $A = 228$, $Z = 88$, $N = 140$, radon;

    X_2: $A = 228$, $Z = 89$, $N = 139$, actinium;

    X_3: $A = 4$, $Z = 2$, $N = 2$, α particle;

    X_4: $A = 216$, $Z = 84$, $N = 132$, polonium;

    X_5: $A = 216$, $Z = 84$, $N = 132$, polonium;

    X_6: $A = 0$, $Z = -1$, $N = 0$, electron (β^-)

    (b) $^{212}_{83}$Bi $\xrightarrow{\alpha}$ $^{208}_{81}$Tl $\xrightarrow{\beta^-}$ $^{208}_{82}$Pb or

    $^{212}_{83}$Bi $\xrightarrow{\beta^-}$ $^{212}_{84}$Po $\xrightarrow{\alpha}$ $^{208}_{82}$Pb; end product is $^{208}_{82}$Pb

13. (a) 2.31×10^{-3} s⁻¹

    (b) (i) 3.00×10^3 Bq (ii) 1.50×10^3 Bq

    (iii) 188 Bq

15. 5.30×10^{16} Bq

17. (a) 46.9 Bq (b) 36.2% (c) $^{131}_{54}$Xe

19. (a) 30.8 min (b) 2.02×10^{15}

21. (a) 5.0×10^4 (b) $\times 10^{-15,000}$

23. (a) 5.0 Gy = 500 rad; 5.0 Sv = 500 rem

    (b) 350 J (c) 5.0 Gy, 500 rad, 50 Sv, 5000 rem, 350 J

25. 500 rad, 2000 rem, 5.0 J/kg

27. 900

29. (a) 1.75 kGy, 1.75 kSv, 175 krem; 262 J

    (b) 1.75 kGy, 2.62 kSv, 262 krem; 262 J

31. absorbed dose: 108 rad, equivalent dose: 2160 rem

33. 2.80 MeV absorbed

35. 4.3×10^{-13} J = 2.7 MeV

37. (a) 1.23×10^5 kg (b) 1.76×10^7 kg

39. 17.6 MeV

41. (a) 4.7×10^4 J/g (b) 8.2×10^{10} J/g

    (c) 4.26×10^{11} J/g (d) 7600 yr

43. (c) 2.42 pm; gamma rays

45. 116 MeV

47. 938.3 MeV, 2.27×10^{23} Hz, 1.32×10^{-15} m (b) 1768 MeV, 42.8×10^{22} Hz, 7.02×10^{-16} m

49. (a) electrical: 200 N; gravitational: 2×10^{-34} N (b) strong: 2×10^4 N; weak: 2×10^{-5} N (c) strong, electrical, weak, gravitational

    (d) $F_{strong} \approx 1 \times 10^{38} F_g$; $F_e \approx 1 \times 10^{36} F_g$; $F_{weak} \approx 1 \times 10^{29} F_g$

51. (a) 3.5 atoms/m³ (b) 294 atoms

    (c) 2×10^{27} atoms

53. 7.19×10^{19} J

55. (a) 0.0556 J/s (b) 11.1 rad/s

    (c) 7.78 rem/s (d) 25.7 s

57. (a) 15.4 decays/s (b) 4.16×10^{-10} Ci

59. (a) 4.14×10^{-9} s⁻¹

    (b) 4.01×10^{20} atoms

    (c) 1.7×10^{12} decays/s (d) 45 Ci

61. 1.287×10^4 y

63. 10.1 MeV absorbed

65. (a) $\pi^0 + \pi^+$ (b) 219 MeV

67. (a) $u\bar{s}$ (b) $d\bar{d}\bar{s}$ (c) uss

Passage Problems

69. C

CREDITS

Chapter 0

Page **0–1**: Shutterstock.

Chapter 1

Page **1**: Scott Cunningham/Getty Images. Page **2** TL: Shutterstock. Page **2** TR: Space Telescope Science Institute. Page **2** B: Bruce Ayres/Getty Images. Page **3**: Gerard Lacz/Peter Arnold/Photolibrary. Page **4** T: National Institute of Standards and Technology. Page **4** B: Bureau International des Poids et Mesures. Page **6** TL: Anglo-Australian Observatory/David Malin. Page **6** TR: Janice Carr/CDC. Page **6** ML: NASA. Page **6** MM: Getty Images Inc.— PhotoDisc. Page **6** MR: National Institute of Standards and Technology. Page **6** BL: NASA. Page **7**: NASA. Page **10**: Roger Viollet/Getty Images. Page **14**: Tom Walker/Getty Images.

Chapter 2

Page **29**: iStockphoto. Page **32**: Shutterstock. Page **37**: AP Photo/Dave Parker. Page **39**: AP Photo/Alastair Grant. Page **42**: Getty Images— BC. Page **49**: Dr. Kevin Eggan. Page **51**: James Sugar/Stock Photo/Black Star. Page **52**: Richard Megna/Fundamental Photographs.

Chapter 3

Page **68**: Daisy Gilardini/The Image Bank/Getty Images. Page **69**: AP Photo/Toby Talbot. Page **72**: Cindy Lewis/Carphotos/Alamy. Page **73**: Mark Boulton/Alamy. Page **75**: Richard Megna/Fundamental Photographs. Page **77**: Ken Davies/Masterfile. Page **78** T: Richard Megna/Fundamental Photographs. Page **78** B: Joe Raedle/Getty Images. Page **79**: Pascal Ribollet. Page **82**: Elsa/Getty Images. Page **87**: National Executive Committee for Space-Based PNT. Page **88**: Shutterstock.

Chapter 4

Page **99**: Ulrich Doering/Alamy. Page **102**: Sami Sarkis/Getty Images—Photodisc. Page **103**: Loomis Dean/Time & Life Pictures/Getty Images. Page **107**: CERN/Geneva. Page **110**: NASA. Page **111**: Shutterstock. Page **113**: Agence Zoom/Getty Images. Page **115**: NASA. Page **116** T: Rensselaer County Historical Society. Page **116** (a): John McDonough. Page **116** (b): John McDonough/Getty Images. Page **116** (c): Mark M. Lawrence/Corbis.

Chapter 5

Page **128**: Eric Draper/Getty Images. Page **129**: European Southern Observatory. Page **134**: Richard Megna/Fundamental Photographs. Page **138**: iStock-photo. Page **139**: Dean Conger/Corbis. Page **140**: Dr. Paul Selvin. Page **148** (a): NASA. Page **148** (b): Helen Hansma. Page **148** (c): Shutterstock. Page **148** (d): Anglo-Australian Observatory/David Malin.

Chapter 6

Page **161**: Clive Mason/Getty Images. Page **162**: Cornell University Press. Page **163**: Jed Jacobsohn/Getty Images. Page **170**: David P. Hall/Masterfile. Page **173**: NASA. Page **174**: NASA. Page **176**: Corbis. Page **178**: NASA. Page **179** TL: NASA. Page **179** TR: NASA. Page **179** B: NASA. Page **186**: NASA. Page **187**: NASA.

Chapter 7

Page **188**: Audio-kinetic ball machine by George Rhoads. Page **190**: Karlene V. Schwartz. Page **191**: Martin Harvey/Photo Researchers, Inc. Page **192**: NASA. Page **193**: Johannes Simon/ AFP/Getty Images. Page **201**: Biology Media/Photo Researchers, Inc. Page **203**: Michael Yamashita/Corbis. Page **207** T: Adamsmith/Getty Images. Page **207** B: Bechara Kachar/NIH. Page **208**: Stephen Dalton/Photo Researchers, Inc. Page **216**: Thomas, D. D., D. Kast, and V. Korman. 2009. Site-Directed Spectroscopic Probes of Actomyosin Structural Dynamics. Annu Rev Biophys. 38:347–369. Page **218**: Shutterstock. Page **225**: Getty Images Inc.—PhotoDisc.

Chapter 8

Page **231**: Steve Dunwell/Getty Images. Page **232**: Archives of Ontario. Page **235**: Jean Louis Batt/Getty Images. Page **241**: Franck Seguin/Corbis. Page **244**: Mark Garlick/Photo Researchers, Inc. Page **248** T: Getty Images. Page **248** B: Stephen Dalton/Photo Researchers, Inc. Page **253**: John Slater/Corbis. Page **254** T: Richard Megna/Fundamental Photographs. Page **254** B: Shutterstock. Page **255**: Berit Myrekrok/Getty Images/Royalty Free. Page **256**: NASA.

Chapter 9

Page **267**: EFE/Javier Lizón/NewsCom. Page **268**: Mike Powell/Getty Images. Page **269**: Image Source/Getty Images/Royalty Free. Page **274**: SSPL/The Image Works. Page **277**: f8 Imaging/Hulton Archive/Getty Images. Page **282**: JPL/NASA.

Chapter 10

Page **294**: NASA. Page **297**: David Cumming/Corbis. Page **303**: Timothy Ryan. Page **304**: Chris Pelkie and Daniel Ripoll. Page **305**: Mark Dadswell/Getty Images. Page **306**: NASA. Page **308** T: NASA. Page **308** BL: Chris Trotman/Corbis. Page **308** BR: Duomo/Corbis. Page **313**: Corbis. Page **318**: Walter Sanders/Time & Life Pictures/Getty Images.

Chapter 11

Page **333**: Ahmad Masood/Reuters. Page **335**: Eric Cabanis/AFP/ Getty Images. Page **336**: Lowell Georgia/Corbis. Page **337**: Ken Robinson. Page **340**: Hans Pfletschinger/Peter Arnold/Photolibrary. Page **342**: Gabe Palmer/Corbis. Page **348**: Wald De Heer. Page **352**: Frank Herholdt/Getty Images. Page **353**: Javier Larrea/AGE Fotostock. Page **354**: Pixtal/Age Fotostock. Page **356**: AP Photo.

Chapter 12

Page **365**: Shutterstock. Page **366**: Shutterstock. Page **367**: Photolibrary. Page **368**: Super Stock/AGE Fotostock. Page **373**: Education Development Center, Inc. Page **375** T: Richard Megna/Fundamental Photographs. Page **375** B: Andrew Davidhazy. Page **384**: AP Photo. Page **385**: Roger Ressmeyer/Corbis. Page **386**: Susumu Nishinaga/Photo Researchers, Inc. Page **390**: Shutterstock. Page **391**: NOAA. Page **395**: Jim Reed/Corbis. Page **396**: Kretztechnik/ Photo Researchers, Inc. Page **403**: Dorling Kindersley Media Library.

Chapter 13

Page **407**: Shutterstock. Page **408**: Shutterstock. Page **411**: Sargent-Welch/VWR International. Page **412**: Richard T. Nowitz/Photo Researchers, Inc. Page **413**: Alan Becker/Getty Images. Page **418**: Dallas and John Heaton/Stock Connection. Page **420** (a): Mario Beauregard/Stock Connection. Page **420** (b): Adam Hart-Davis/ Photo Researchers, Inc. Page **420** (c): Shaun Lowe/iStockphoto. Page **421**: Adam Hart-Davis/Photo Researchers, Inc.

Page **422**: Creatas/Thinkstock. Page **423** T: Pearson Education. Page **423** B: Alix/Photo Researchers, Inc. Page **424**: Peter/Georgina Bowater/Creative Eye/Mira. Page **425**: Mark Wilson/Getty Images. Page **430**: Harold E. Edgerton/Palm Press, Inc. Page **431** L: Getty Images, Inc.—Photodisc./Royalty Free. Page **431** R: Colin Barker/Getty Images.

Chapter 14

Page **441**: Rick & Nora Bowers/Alamy. Page **444** T: NOAA. Page **444** B: Sargent-Welch/VWR International. Page **447**: Jeff Daly/Fundamental Photographs. Page **450**: Shutterstock. Page **452**: Hugh D. Young. Page **455** T: Ted Kinsman/Photo Researchers, Inc. Page **455** B: Richard Megna/Fundamental Photographs. Page **457**: Image Source/Photolibrary. Page **458**: USDA/ARS. Page **461** T: Nature Picture Library. Page **461** B: Russ Underwood/Lockheed Martin. Page **464**: Getty Images Inc.—Punchstock. Page **465**: Philip Rosenberg/Pacific Stock/Photolibrary. Page **466**: OSF/Photolibrary. Page **467**: David Mauriuz/Corbis.

Chapter 15

Page **477**: Shutterstock. Page **478** T: Richard Megna/Fundamental Photographs. Page **478** B: NASA. Page **479**: Colin Garratt/Corbis. Page **480**: NASA. Page **481**: Shutterstock. Page **482**: Shutterstock. Page **483**: Miguel Angelo Silva/iStockphoto. Page **485**: National Researcher Council of Canada. Page **486**: ThermoMicroscopes. Page **489**: NASA. Page **490**: Calvin Hamilton. Page **494**: USGS. Page **495**: U.S. Navy photo by Mass Communication Specialist Seaman Ryan Steinhour (Released).

Chapter 16

Page **516**: Shutterstock. Page **517**: Shutterstock. Page **518**: iStockphoto. Page **522**: Mike Kepka/Corbis. Page **523**: Lambert/Hulton Archive/Getty Images. Page **527**: Shutterstock. Page **530**: Sinclair Stammers/Photo Researchers, Inc. Page **535**: NASA. Page **536**: Samantha Brown/AFP/Getty Images. Page **537** (b): Amos Zezmer/Omni-Photo Communications, Inc. Page **537** (c): Shutterstock.

Chapter 17

Page **545**: Shutterstock. Page **546**: NOAA. Page **549**: Shutterstock. Page **550**: Richard Megna/Fundamental Photographs. Page **552**: Getty Images, Inc.—Photodisc./Royalty Free. Page **553**: Massachusetts Institute of Technology. Page **558**: Dave Watts/Nature Picture Library. Page **560**: Kenneth Robinson. Page **565**: Gary Retherford/Photo Researchers, Inc. Page **572** T: Peter Terren/Tesladownunder.com. Page **572** B: Tom Bean/Corbis. Page **581**: David Parker/Photo Researchers, Inc.

Chapter 18

Page **582**: Shutterstock. Page **583**: American Association for the Advancement of Science. Page **587**: Gandee Vasan/Getty Images. Page **589**: Shutterstock. Page **595**: Eric Schrader—Pearson Education. Page **596**: Sandia National Laboratories. Page **597**: David M. Phillips/Photo Researchers, Inc. Page **604**: Thinkstock. Page **606** L: Stanford Linear Accelerator/SPL/Photo Researchers, Inc. Page **606** R: C. Mooney/Getty Images.

Chapter 19

Page **618**: Shutterstock. Page **619**: Alfred Pasteka/SPL/Photo Researchers, Inc. Page **622**: Shutterstock. Page **625** (a): Getty Images—Digital Vision. Page **625** (b): The M. C. Escher Company BV. Page **625** R: George Grall/Getty Images. Page **629**: Kenneth Robinson. Page **630**: Alan Senior. Page **631**: Reuters/Corbis. Page **640**: Richard Megna/Fundamental Photographs. Page **643**: Hemera/AGE Fotostock.

Chapter 20

Page **658**: Alex Bartel/Photo Researchers, Inc. Page **659**: Charles D. Winters/Photo Researchers, Inc. Page **661**: Iain MacDonald.

Page **662**: Richard Megna/Fundamental Photographs. Page **664**: Simon Fraser/Photo Researchers, Inc. Page **667** T: Bettmann/Corbis. Page **667** B: Sargent-Welch/VWR International. Page **668** L: NASA. Page **668** R: JPL/NASA. Page **673**: Shutterstock. Page **677**: Eric Schrader, Pearson Education. Page **683**: Haim Bau.

Chapter 21

Page **698**: ITAR-TASS/Nikolai Kuznetsov/Newscom. Page **699**: Peter Anderson-Dorling Kindersley Media Library. Page **705**: American Association for the Advancement of Science. Page **710**: Tethers Unlimited, Inc. Page **712**: Splashpower Ltd. Page **716**: Martyn Goddard/Corbis. Page **717**: Shutterstock. Page **719**: Los Alamos National Laboratory. Page **725**: H. Armstrong Roberts/Corbis.

Chapter 22

Page **735**: Casey Fleser/CC-By-2.0—http://creativecommons.org/licenses/by/2.0/deed.en. Page **738**: The Image Bank/Getty Images. Page **748**: Richard Nuccitelli et al. Int. J. Cancer: 125, 438–445 (2009). Page **749**: Dynamic Graphics/Photis/Alamy.

Chapter 23

Page **761**: Shutterstock. Page **762**: Shutterstock. Page **763**: Oakridge National Laboratory. Page **765** (a): Image courtesy of NRAO/AUI and David Thilker (JHU), Robert Braun (Astron), WSRT. Page **765** (b): NASA. Page **765** (c): NASA. Page **765** (d): Max-Planck-Institut für extraterrestrische Physik (MPE)/NASA. Page **765** B: Andrew Davidhazy. Page **772** T: Jerry Lodriguss/Photo Researchers, Inc. Page **772** B: NASA. Page **773** T: Thinkstock. Page **773** B: Alxander Tsiaras. Page **780**: Susan Schwartzenberg. Page **781** T: Dr. Dina Mandoli. Page **781** B: Barry Blanchard. Page **783**: Randy O' Rourke. Page **785** T: Nadav Shashar. Page **785** B: Kristen Brochmann/Fundamental Photographs. Page **789** L: Sepp Seitz/Woodfin Camp & Associates, Inc. Page **789** R: Peter Aprahamian/Sharples Stress Engineers Ltd./Photo Researchers, Inc. Page **791**: K. Nomachi/Fundamental Photographs. Page **794**: Kristen Brochmann.

Chapter 24

Page **803**: Derrick Alderman/Alamy. Page **804**: Ed Kashi/Corbis. Page **806**: Martin Bough/Fundamental Photographs. Page **811**: Pearson Education. Page **816**: Thinkstock. Page **817**: Tom Fleming/Photo Researchers, Inc. Page **818**: Richard Megna/Fundamental Photographs. Page **821**: Technodiamant USA.

Chapter 25

Page **837**: Stefan Schuetz/Getty Images. Page **838**: Heidi Hofer. Page **839** (a–c): Marshall Henrichs. Page **840**: Fotolia. Page **841** T: Susumu Nishinaga/Photo Researchers, Inc. Page **841** B: Lennart Nilsson/Scanpix Sweden AB. Page **842**: Shutterstock. Page **843**: Photodisc Red/Getty Images. Page **848**: Jan Hinsch/Photo Researchers, Inc. Page **850**: Eastman Kodak. Page **852** (a): Large Binocular Telescope Corporation. Page **852** (b): Southern African Large Telescope.

Chapter 26

Page **862**: Fotolia. Page **863**: Shutterstock. Page **865**: Roger Freedman. Page **866**: Pearson Education. Page **869**: Shutterstock. Page **872** T: Bausch & Lomb Inc. Page **872** M: Bausch & Lomb Inc. Page **872** B: Graeme Harris/Getty Images. Page **873**: Dr. Peter Vukusic. Page **874** T: Pearson Education. Page **874** M: Pearson Education. Page **874** B: Pearson Education/PH College. Page **877** L: Photographed for Pearson Science by Elisabeth Pierson, Radboud University, Nijmegen, Netherlands. Page **877** R: Michael W. Davidson, National High Magnetic Field Laboratory, The Florida State University. Page **880** T: TEK Image/Photo Researchers, Inc. Page **880** B: Photodisc Green/Getty Images. Page **882**: Estate of Bertram Eugene Warren. Page **884** T: SPL/Photo Researchers, Inc. Page **884** B: SPL/Photo Researchers, Inc. Page **886** T: Springer-Verlag GmbH & Co. KG. Page **886** B: Pearson Education.

888 T: South African Astronomical Observatory. Page **888** B: European Southern Observatory. Page **890**: Paul Silverman/Fundamental Photographs.

Chapter 27

Page **899**: Bettmann/Corbis. Page **901**: A. Inden/Zefa/Corbis. Page **906**: Kimimasa/Corbis. Page **909**: iStockphoto. Page **914**: Henrick Sorensen/Getty Images—BC. Page **918**: David Parker/Photo Researchers, Inc. Page **920**: SuperStock, Inc.

Chapter 28

Page **932**: Peter Arnold, Inc./Photolibrary. Page **934**: Shutterstock. Page **937**: Corning Corporation. Page **941**: BrandX. Page **951**: Hank Morgan/Photo Researchers, Inc. Page **956**: Scott Camazine/Photo Researchers, Inc. Page **961**: V. Brockhaus/zefa/Corbis.

Chapter 29

Page **971**: Sinclair Stammers/Photo Researchers, Inc. Page **980**: Shutterstock. Page **984**: iStockphoto. Page **987**: Roger Freedman. Page **990**: Paul Silverman/Fundamental Photographs. Page **993**: Shutterstock. Page **995**: Getty Images—Photodisc-Royalty Free. Page **996**: Lebrecht Music & Arts Photo Library. Page **997**: Tom Deerinck.

Chapter 30

Page **1003**: C Gascoigne/Robert Harding. Page **1004**: Buddy Mays/Corbis. Page **1012**: Shutterstock. Page **1017**: Kenneth Garrett/National Geographic Image Collection. Page **1019**: David Stuart/Masterfile. Page **1020**: Stockbroker/Photolibrary. Page **1021** T: GJLP/Photo Researchers, Inc. Page **1021** B: Dept. of Nuclear Medicine, Charing Cross Hospital/Photo Researchers, Inc. Page **1029** T: Ernest Orlando Lawrence Berkeley National Laboratory. Page **1029** B: Carl D. Anderson/Ernest Orlando Lawrence Berkeley National Laboratory.

INDEX